D1477705

ELECTROCHEMISTRY

ELECTROCHEMISTRY

THEORETICAL PRINCIPLES AND PRACTICAL APPLICATIONS

by

GIULIO MILAZZO

Istituto Superiore di Sanità,
Rome (Italy)

with the co-operation of

G. BOMBARA P. GALLONE
I. EPELBOIN S. HJERTÉN
C. FURLANI M. LEDERER
A. LIBERTI

First English Edition,
translated from the Italian manuscript by

P. J. MILL, B.A., Ph.D.

ELSEVIER PUBLISHING COMPANY
AMSTERDAM / LONDON / NEW YORK
1963

SOLE DISTRIBUTORS FOR THE UNITED STATES AND CANADA
AMERICAN ELSEVIER PUBLISHING COMPANY
52 VANDERBILT AVENUE, NEW YORK 17, N.Y.

SOLE DISTRIBUTORS FOR GREAT BRITAIN
ELSEVIER PUBLISHING COMPANY LIMITED
12B, RIPPLESIDE COMMERCIAL ESTATE
RIPPLE ROAD, BARKING, ESSEX

LIBRARY OF CONGRESS CATALOG CARD NUMBER 62-13019

WITH 131 ILLUSTRATIONS AND 108 TABLES

PREFACE

Like the first Italian edition, this book seeks to set out the fundamental concepts of electrochemistry in the simplest and most readily accessible form; it is intended to be suitable even for non-specialists but does presuppose a certain minimum familiarity with chemical and physico-chemical concepts. It may thus serve as a teaching text but can also be used as a reference book, since it collects together and tabulates many numerical data which are often dispersed in books and original papers which are not readily consulted.

It is not possible to include in a short text all the theory and applications of electrochemistry, for it would then become an enormous treatise. This publication, as a specialized text, first develops that part of the theory which is too specialized to be included in a general textbook of physical chemistry. It then studies the application to electrochemical processes of the general concepts of chemical change, acquired in the first place from physical chemistry and developed by further theoretical studies carried out especially for this purpose.

An attempt has been made in the text to describe not only the results obtained, but also the difficulties and uncertainties of both theoretical and experimental nature which still exist.

It has not been thought convenient to deal with certain topics which are sometimes found in books on electrochemistry (pH, indicators, hydrolysis, buffer solutions, electrothermic reactions, electric furnaces, etc.) in that these do not form a part of electrochemistry but rather of the theory of equilibria (ionic equilibria), of the theory of reactions at high temperatures (electrothermic reactions) and of technology (electric furnaces) etc.

It has also not been thought opportune to describe in detail *all* the electrochemical processes of industrial interest. Above all, the impossibility of consulting the vast number of original papers and books published in recent years, would make useless any attempt to describe the various industrial processes in the actual state which they have reached in modern technology. Secondly, plant technology is undergoing a continuous evolution and many processes differ not in matters of principle but simply in constructional details of the plant with the aim of not infringing pre-existing patents. Again, it is very difficult to extract from the patent literature the true essence of a process, since often the most important points are 'top secret' so that one runs the risk of describing a patented process which has in fact never been used. It has thus been thought preferable to deal only with the conceptual part of some of the more important or typical industrial electrochemical processes, limiting the description of the plants to the parts of interest for an understanding of the process.

I should like to thank my colleagues G. Bombara (ENI Research Laboratories, Milan), I. Epelboin (Faculty of Sciences, Paris), C. Furlani (University of Rome), P. Gallone (De Nora Inc., Milan), S. Hjertén (University of Uppsala), M. Lederer (Consiglio Nazionale delle Ricerche, Roma) and A. Liberti (University of Naples) for their contributions.

I should particularly like to thank also Professor D. Marotta for the facilities which he afforded me during the preparation of the manuscript.

This edition shows no fundamental novelties in the choice or subdivision of the material with respect to the preceding editions. It has naturally been brought up to date to take account of the developments and of the new knowledge acquired in the last decade of research. I hope that with this edition I may make a modest contribution to a wider knowledge of electrochemistry, comforted in this by the favourable acceptance obtained by the first Italian edition and the second edition in German.

Rome, October 1962 G. Milazzo

SYMBOLS AND ABBREVIATIONS

a	activity; acceleration
A	affinity
A	ampere
Ah	ampere-hour
c	concentration (in general); molarity (moles. litre^{-1})
C	capacitance
°C	temperature in degrees centigrade
C	coulomb
d	exact differential
d	density
D	coefficient of diffusion
e	base for natural logarithms
e^-	electron
E	energy (in general); internal energy; energy of activation
f	activity coefficient
f_Λ	coefficient of conductance
F	free energy (Helmholtz); force
F	faraday (96,500 C)
G	free enthalpy (Gibbs free energy)
h	partial molar enthalpy
H	enthalpy
I	ionic strength; current intensity
I_+	anodic current intensity
I_-	cathodic current intensity
I_0	exchange current intensity
$\vec{I}$	progressive current intensity
$\overleftarrow{I}$	regressive current intensity
J	current density
J	joule
k	constants (in general); Boltzmann constant
K	equilibrium constant; colloidal equivalent
$\vec{k}$, $\overleftarrow{k}$	kinetic constants
°K	temperature in degrees Kelvin
l	length, depth or thickness of a layer
L	solubility product
ln	natural logarithm
log	decimal logarithm
m	mass; molality (moles.kg solvent^{-1})
M	molecular weight
n	number of moles; number of elementary charges

N Avogadro's number; preceded by a number, normality (equivalents litre^{-1})
p pressure
q quantity of heat
Q quantity of electricity
r crystalline ionic radius
R resistance; gas constant
R_{cur} current efficiency
R_{en} energy efficiency
s partial molar entropy
S entropy; surface area; solubility
t centigrade temperature; time; transference numbers
t_+ cationic transference number
t_- anionic transference number
T absolute temperature
u electrical mobility
u_+ cationic electrical mobility
u_- anionic electrical mobility
U electric tension
v volume
V electric tension at terminals
V volt
x molar fraction
w work; migration velocity
W watt
Wh watt-hour
z charge on an ion in general
z_+ charge on a cation
z_- charge on an anion
Z impedance
α degree of dissociation
α, β transfer coefficients
δ infinitesimal increment
∂ partial differential
ε dielectric constant
η viscosity; overtension
ζ electrokinetic potential
$\varkappa$ specific conductance
λ equivalent ionic conductance; Donnan partition coefficient
Λ equivalent conductance of an electrolyte
μ chemical potential
$\tilde{\mu}$ electrochemical potential
ν stoichiometric reaction coefficient
ξ advancement of a reaction
π membrane electric tension

Π	product
ϱ	specific resistance
Σ	sum
φ	internal electric potential of a phase (galvani potential); dilution
χ	conductance; surface electric potential of a phase
ψ	potential (in general); external electric potential of a phase (volta potential)
τ	transition time
Ω	ohm

Symbols for other quantities which occur less frequently in this book are described directly in the text.

CONTENTS

Chapter IV. Electrolysis and electrochemical kinetics in aqueous solutions

by C.FURLANI and G.MILAZZO

Chapter V. Analytical applications

by A.LIBERTI and G.MILAZZO

Chapter VI. The electrochemistry of colloids and electrokinetic phenomena

Chapter VII. General considerations about electrochemical plants

Chapter VIII. Electrometallurgy in aqueous solutions

Chapter IX. Nonmetallurgical electrolytic processes

by P.GALLONE and G.MILAZZO

A. Electrolysis of alkali halides

B. Other nonmetallurgical processes

Chapter X. Electrolysis in molten electrolytes

Chapter XI. Practical primary cells and storage batteries (accumulators)

by G.BOMBARA and G.MILAZZO

Chapter XII. The electrochemistry of gases

Chapter X. Electrolysis in molten electrolytes

Chapter XI. Practical primary cells and storage batteries (accumulators)

by G. BOMBARA and G. MILAZZO

Chapter XII. The electrochemistry of gases

CHAPTER I

INTRODUCTION

1. Outline

Electrochemical phenomena have been known for more than a century and a half, particularly through the studies of Faraday. However, electrochemistry began to emerge as an independent branch of physical chemistry at the end of the last century with a series of fundamental studies carried out by some of the leading chemists and physicists of that time; these included Helmholtz, Arrhenius, Ostwald, Nernst, Lodge, Whetham, Kohlrausch, Caspar, Lewis and others. The electrochemical industry began to develop at the same time, principally through the invention of the dynamo. Today electrochemistry may be defined as *that branch of physical chemistry which studies the relationship between chemical transformations and energy in those reactions which involve electrical energy external to the system.* Thus, in particular it concerns electrolysis, galvanic cells, some reactions in the gaseous phase and many of the phenomena of colloids etc.

The relationships between chemical changes and external electrical work are varied and complex.

(1) The passage of an electrical current through an electrolyte gives rise to a migration of material and to chemical reactions. If, for example, a current is passed through a solution of an alkaline chloride, without any particular experimental arrangement, it will be observed that there is an evolution of gaseous chlorine at the positive pole, that the concentration of the alkaline chloride diminishes, and that according to the particular conditions, there is a formation of alkaline hydroxide, hypochlorite and finally even chlorate. Thus, in this way there has been a transformation of material at the expense of external electrical energy.

(2) Some spontaneous chemical reactions when carried out in a suitable manner, produce external electrical energy at the expense of the free energy of the system. A classical example of this is the Daniell cell, composed of a copper electrode immersed in a solution of copper sulphate and of a zinc electrode immersed in a solution of zinc sulphate. If the two solutions are brought into contact without being allowed to mix (*e.g.* by

means of a porous diaphragm) and if the two poles are connected with an external metallic wire, it will be found that a current passes through this wire whilst at the same time a certain amount of metallic copper is deposited upon the copper electrode, the concentration of the copper sulphate diminishes, a certain amount of zinc dissolves and the concentration of the zinc sulphate increases.

(3) The passage of a so-called dark discharge through a stream of air or oxygen produces ozone. The formation of ozone from oxygen is endothermic and thus occurs during the dark discharge at the expense of a part of the electrical energy dissipated in this discharge. This is also a forced reaction and thus in this sense analogous to that given in the first example.

(4) Many colloidal phenomena are of an electrochemical nature. If a suitable amount of an electrolyte is added to a colloidal solution it will coagulate. If a colloidal solution is placed in an electrical field generated by two conductors, to which an electric tension is applied, immersed in the solution, the colloidal particles will all migrate in the same direction; either towards the end at the highest positive potential or towards that at the lowest negative potential.

It would be possible to cite in this way an endless number of examples to illustrate the multiplicity of aspects and relationships of a predominantly electrochemical nature.

There exist also particular fields of electrochemistry which besides being somewhat too specialized to be included in a textbook of this type, are still not completely clear. Such, for example, are the studies on electrolysis in glow discharges in which one of the two electrodes is the solution-gas interface, or the studies on electrolysis by means of sparks etc. Such topics will not be dealt with.

2. Outline of Thermodynamics

All electrochemical reactions, like all other chemical reactions, always involve a transformation of the internal energy of the system into work and finally into heat, or *vice versa*, depending upon the laws of thermodynamics. There is no general agreement in the chemical and physico-chemical literature on conventions for the signs which indicate if the stated quantity is lost by the system to the environment or gained from it. It will therefore be opportune to establish from the start the criteria which have been chosen for the positive or negative signs necessary to characterize the direction of the transformation. The so-called *egoistic* criteria, referred to the system, are followed. Energy and work *given up by the system to the environment* are thus characterized by a *negative sign;* the same quantity, however *given up by the environment to the system* is characterized by a *positive sign.*

(I) *Internal Energy*

As far as thermodynamics is concerned the various forms in which energy may be transferred from one body to another, and in particular from a thermodynamic system to the environment or *vice versa,* may be reduced to two – heat and work. When a system passes from an initial to a final state its energy content does not usually remain constant. The *internal energy* of the system (E) is defined as *the sum of all the forms of energy possessed by the system apart from that derived from the position of the system in space or in reference to other bodies,* which latter does not directly concern thermodynamics. The function E is a function of the state of the system, *i.e.* it depends exclusively on the variables which define the thermodynamic state of the system and not on its preceding history. Its absolute value is not known, but changes in it may be known either by measurement or calculation. These changes depend exclusively on the initial and final states of the system and are independent of the way by which the system passes from the initial to the final state. This property is expressed by the equation

$$\Delta E = E_{final} - E_{initial} \tag{1}$$

By means of the first principle of thermodynamics, which establishes the equivalence of all forms of energy and work[1] and the impossibility of creating or destroying energy, it is possible to re-express equation (1) in the form

$$\Delta E = \Sigma (w + q) \tag{2}$$

[1] Naturally the limiting case of the transformation of matter into energy according to Einstein's equation is not considered since this does not concern normal physico-chemical processes.

in which the algebraic signs for work (w) and heat (q) follow the accepted convention. For an infinitesimal process

$$dE = \delta w + \delta q \quad [1] \tag{3}$$

In particular, for chemical reactions occurring at constant volume, the change in internal energy is equal to their heat of reaction at constant volume.

(II) *Entropy*

The Carnot cycle[2] will not be repeated here in the interest of brevity, but from this it is possible to deduce the efficiency of any ideal engine[3] converting heat into work and functioning between the absolute temperatures T_2 and T_1; these are respectively the temperatures of the warmer and of the cooler reservoirs. This, therefore, gives the relationship between the quantity of heat effectively converted into work and the quantity of heat taken from the warmer reservoir; the quantity of heat effectively converted into work is equal to the algebraic sum of the quantities of heat q_2 and q_1 exchanged respectively with the reservoir at temperature T_2 and the reservoir at temperature T_1.

This relationship is

$$\left(\frac{q_2 + q_1}{q_2}\right)_{rev} = \left(\frac{T_2 - T_1}{T_2}\right) \tag{4}$$

This may be re-written as

$$\left(\frac{q_2}{T_2}\right)_{rev} = \left(\frac{-q_1}{T_1}\right)_{rev}$$

For a reversible cycle (see below) this gives

$$\sum \left(\frac{q}{T}\right)_{rev} = 0 \tag{5}$$

For every reversible cycle the summation q/T is zero. Any temperature T_1 and T_2 and any initial quantity of heat q_2, *i.e.* the point of the cycle at which the inversion occurs, may be chosen, and the routes followed

[1] The symbol d indicates an exact differential whilst the symbol δ signifies that the corresponding expression is not in general an exact differential.

[2] The word 'cycle' stands for any series of transformations which leads to a final state identical to the initial state.

[3] That is, one which functions without loss (by friction, irreversibility etc.).

from the initial point to the inversion point and back again may be different. Consequently, q/T is independent of the route followed for the transformation provided that it is reversible. This fraction is equal to a change of a certain state function (S) of the system. It follows further from (5) that for a non-cyclic transformation, *i.e.* from a certain initial state to a certain final state, the expression $\int_{initial}^{final}(\delta q/T)_{rev}$ is not zero[1].

$$\int_{initial}^{final}(\delta q/T)_{rev} = \Delta S \neq 0 \tag{6}$$

All this holds true for any transformation, even a non-isothermal one, provided that it is *reversible*, *i.e.* that it is possible to reverse it and to travel in the exactly opposite sense through all the previous states and thus to return to the initial state of the system with no variation in the physicochemical environments connected with the system finally persisting. If it were possible to have such an ideal transformation in practice, it would be extremely slow and would take place through a series of equilibrium states for which the values of the variables involved in the transformation would in any given state differ by an infinitesimal amount from the values which they would assume in the immediately successive states. In such a case, it is permissible to divide the total transformation into a certain number of partial transformations whose changes in temperature between the initial and final states are so small that they may be considered as constant. If these partial isothermal transformations are infinitesimally small, the quantity of heat exchanged becomes δq for every transformation, and the summation, which gives the total change in the function S, is the integral

$$\int_{initial}^{final} \frac{\delta q}{T} = \Delta S$$

Since S is a state function its differential must be an exact differential and one may write

$$\int_{initial}^{final} \frac{\delta q}{T} = \int_{initial}^{final} \mathrm{d}S = S_{final} - S_{initial} = \Delta S$$

Clausius called the function S *entropy;* it is measured in calories degrees^{-1} (quantity of substance)$^{-1}$.

[1] With the exception of adiabatic transformations in which $q = 0$.

It follows from the sign convention adopted that the total entropy change in a reversible transformation ($\Delta S_{system} + \Delta S_{environment}$) is always zero and the change of entropy of a system is equal to, but of opposite sign to that of the environment.

For an irreversible cycle, however, with its lower efficiency, (4) becomes

$$\left(\frac{q_2 + q_1}{q_2}\right)_{irr} < \left(\frac{T_2 - T_1}{T_2}\right)$$

i.e.

$$\sum \left(\frac{q}{T}\right)_{irr} < 0 \qquad (7)$$

For (7) to hold true it is only necessary for one of the individual processes which form the cycle to be irreversible. Considering for the sake of simplicity, a non-isothermal cycle between the extremes I and II, composed of only two transformations (I → II and II → I), of which only I → II is irreversible, gives

$$\int_{\mathrm{I}}^{\mathrm{II}} \left(\frac{\delta q}{T}\right)_{irr} + \int_{\mathrm{II}}^{\mathrm{I}} \left(\frac{\delta q}{T}\right)_{rev} < 0 \qquad (8)$$

But by definition, (6) is valid for the reversible transformation II → I

$$\Delta S_{\mathrm{II} \to \mathrm{I}} = S_{\mathrm{I}} - S_{\mathrm{II}} = \int_{\mathrm{II}}^{\mathrm{I}} \left(\frac{\delta q}{T}\right)_{rev} \qquad (9)$$

and combining (8) and (9) gives

$$\int_{\mathrm{I}}^{\mathrm{II}} \left(\frac{\delta q}{T}\right)_{irr} + S_{\mathrm{I}} - S_{\mathrm{II}} < 0$$

i.e.

$$\int_{\mathrm{I}}^{\mathrm{II}} \left(\frac{\delta q}{T}\right)_{irr} < S_{\mathrm{II}} - S_{\mathrm{I}}$$

The entropy change of the system is thus greater than $\int_{\mathrm{I}}^{\mathrm{II}} (\delta q/T)_{irr}$, which is the entropy change of the environment, even if the heat is exchanged irreversibly. Thus the total entropy change of an irreversible cycle is always positive

$$(S_{\mathrm{II}} - S_{\mathrm{I}}) - \int_{\mathrm{I}}^{\mathrm{II}} \left(\frac{\delta q}{T}\right)_{irr} > 0$$

According to Clausius it is possible to formulate the two fundamental principles of thermodynamics as follows.

(1) *The total amount of energy in the universe is constant.*

(2) *The total entropy in the universe is increasing.*

In other words, entropy is the arrow which points the direction in which a spontaneous natural process will go; all such processes are irreversible.

It follows from this that in the special case of reversible isothermal processes, the quantity of heat exchanged can be expressed by the equation

$$q = T\Delta S \tag{10}$$

which for an infinitesimal transformation becomes

$$\delta q = T\mathrm{d}S \tag{11}$$

(III) *Enthalpy*

For many processes which take place at constant pressure it is more convenient to use another state function which takes account of both the internal energy and of the external volume work involved during the transformation. This function is called *enthalpy* and it is defined by the equation

$$H = E + pv \tag{12}$$

(where p = pressure and v = volume). Since internal energy, pressure and volume are all state functions it follows that enthalpy is a state function independent of the route followed to reach the state. Hence its differential $\mathrm{d}H$ is also exact. Its physical significance is simple. Its variation ΔH represents *the sum of the changes in the internal energy of a system and of the volume work exchanged with the environment during a transformation.*

$$\Delta H = H_{final} - H_{initial} = \Delta E + \Delta(pv) \tag{13}$$

At constant pressure (13) becomes

$$\Delta H = \Delta E + p\Delta v$$

and for an infinitesimal transformation

$$\mathrm{d}H = \mathrm{d}E + p\mathrm{d}v \tag{14}$$

Although this function is used almost exclusively for processes at constant pressure, it is obvious from (13) that a change in enthalpy exists also for processes taking place at constant volume. It is given by the equation

$$\Delta H = \Delta E + v\Delta p$$

This equation is less frequently used.

As for internal energy the absolute value of the enthalpy is not known but its changes may be known by measurement or calculation. In particular, for a chemical reaction at constant pressure the change in enthalpy is equal to the heat of reaction at constant pressure.

(IV) *Free Energy and Free Enthalpy*[1]

The two functions *free energy* (F) and *free enthalpy* (G) are defined by the equations

$$F = E - TS \tag{15}$$

$$G = E - TS + pv \tag{16}$$

$$= H - TS \tag{16a}$$

$$= F + pv \tag{16b}$$

These two new functions, which are relationships between state functions, are themselves state functions and thus their differentials are exact. The change in the free energy for an infinitesimal transformation is given by

$$\mathrm{d}F = \mathrm{d}E - T\mathrm{d}S - S\mathrm{d}T \tag{17}$$

and combining this with (3) and (11) gives, for an isothermal process where $\mathrm{d}T = 0$,

$$\mathrm{d}F = \delta w_{rev} \tag{18}$$

Thus, in an isothermal and reversible process *the change in the free energy*

[1] The free energy and the free enthalpy are often called the Helmholtz free energy and the Gibbs free energy, respectively.

represents the total work exchange during the transformation. For a transformation which is not infinitesimal

$$\Delta F = F_{final} - F_{initial} = w_{rev}$$

The value of ΔF is independent of the route followed during the transformation, since F is a state function and thus it is the same for an irreversible transformation. In this case (18) is not valid and must be replaced by

$$\mathrm{d}F > \delta w \tag{19}$$

The external work δw_{rev} in turn arises as the sum of the mechanical work of expansion or compression (volume work $\delta w_{vol} = -p\mathrm{d}v$[1]) and of all the other forms of work (*e.g.* electrical work, work of solution etc.) which are usable in other ways (δw_{use})

$$\delta w_{rev} = \delta w_{use} - p\mathrm{d}v \tag{20}$$

Combining this with (18) gives

$$\delta w_{use} = \mathrm{d}F + p\mathrm{d}v \tag{21}$$

But (21) is the change in free enthalpy in a process at constant pressure where $\mathrm{d}p = 0$. In fact (16b) gives

$$\mathrm{d}G = \mathrm{d}F + p\mathrm{d}v \tag{22}$$

On the other hand

$$\mathrm{d}G = \mathrm{d}E - T\mathrm{d}S - S\mathrm{d}T + p\mathrm{d}v + v\mathrm{d}p$$

and for a process at constant pressure where $\mathrm{d}p = 0$, combining with (17) gives (22) once more.

Comparing (21) with (22) it becomes obvious that the changes in G in a process at constant pressure *are equal to the useful work exchanged during the transformation.* The change of free enthalpy may also be expressed as a function of the enthalpy. Thus, from (16a)

$$\mathrm{d}G = \mathrm{d}H - T\mathrm{d}S - S\mathrm{d}T \tag{23}$$

[1] In an expansion against the external pressure p the volume increases and thus $\mathrm{d}v$ is positive, but in view of the sign convention one must write

$$\delta w_{vol} = -p\mathrm{d}v$$

since the work is done by the system against the environment.

Subtracting (17) from this and combining with (14) gives

$$dG = dF + pdv$$

which is equal to (22).

Since F and G are state functions the mode of calculation of their changes for finite transformations is obvious and will not be discussed.

Differentiating the free enthalpy function in the form given in (16) and combining with equations (3), (11) and (19) gives

$$dG = - SdT + vdp + \delta w_{use} \tag{24}$$

In a simple system which does not undergo changes of composition the only possible form of work is the mechanical one of compression or expansion. In this case $\delta w_{use} = 0$ and (24) becomes

$$dG = - SdT + vdp$$

At constant pressure when $dp = 0$

$$dG = - SdT$$

hence

$$\left(\frac{\partial G}{\partial T}\right)_p = - S \tag{25}$$

whilst at constant temperature ($dT = 0$)

$$dG = vdp$$

and hence

$$\left(\frac{\partial G}{\partial p}\right)_T = v \tag{26}$$

(V) *Chemical Potential*

The thermodynamic relationships described are strictly applicable in the forms given, only to closed systems, *i.e.* to those in which the composition does not change. They may, however, be applied to open systems also, *i.e.* to those in which the composition of the system is variable. A special example of an open system is one in which a chemical reaction occurs, since this system may be assimilated to open systems in a general sense. In fact, in such a system it is possible to pass from the initial to the final state, either by means of a chemical reaction, or by the elimination of the reacting substances and the addition of the products of the reaction.

Since the functions discussed are all extensive, *i.e.* dependent on the quantity of the substance, it is obviously necessary, in order to deal with open systems in general, to take account of this new variable expressed as the number of moles n_i. Thus, choosing the most convenient independent variables, the various functions must be re-written as

$$\begin{aligned} E &= f_1 (S, v, n_1, n_2, \ldots, n_i) \\ S &= f_2 (E, v, n_1, n_2, \ldots, n_i) \\ H &= f_3 (S, p, n_1, n_2, \ldots, n_i) \\ F &= f_4 (T, v, n_1, n_2, \ldots, n_i) \\ G &= f_5 (T, p, n_1, n_2, \ldots, n_i) \end{aligned} \tag{27}$$

The function which will be mostly used is the free enthalpy and the discussion will be limited to this.

For a chemical reaction carried out reversibly the change in free enthalpy during an infinitesimal transformation may be written in the form

$$dG = \left(\frac{\partial G}{\partial T}\right)_{p, n_1, n_2 \ldots n_i} dT + \left(\frac{\partial G}{\partial p}\right)_{T, n_1, n_2 \ldots n_i} dp + \left(\frac{\partial G}{\partial n_1}\right)_{T, p, n_2 \ldots n_i} dn_1 + \ldots + \left(\frac{\partial G}{\partial n_i}\right)_{T, p, n_1 \ldots n_{(j \neq i)}} dn_i \tag{28}$$

which combined with (25) and (26) leads to

$$dG = -SdT + vdp + \left(\frac{\partial G}{\partial n_1}\right)_{T, p, n_2 \ldots n_i} dn_1 + \left(\frac{\partial G}{\partial n_2}\right)_{T, p, n_1, n_3 \ldots n_i} dn_2 + \ldots + \left(\frac{\partial G}{\partial n_i}\right)_{T, p, n_1 \ldots n_{(j \neq i)}} dn_i \tag{29}$$

Combining (24) with (29) gives for a change at constant temperature and pressure

$$dG = \partial w_{use} = \left(\frac{\partial G}{\partial n_1}\right)_{T, p, n_2 \ldots n_i} dn_1 + \left(\frac{\partial G}{\partial n_2}\right)_{T, p, n_1, n_3 \ldots n_i} dn_2 + \ldots + \left(\frac{\partial G}{\partial n_i}\right)_{T, p, n_1 \ldots n_{(j \neq i)}} dn_i \tag{30}$$

The coefficients $\left(\frac{\partial G}{\partial n_i}\right)_{T, p, n_j \neq i}$ etc. are called the *chemical potentials* of the components 1,2 etc. and are represented by the letter μ.

In other words the chemical potential of a component i represents the change in free enthalpy of the system, *i.e.* the useful work, when one mole of the component is produced or lost reversibly at constant temperature and pressure in a system so large that concentration remains virtually unchanged. For a chemical reaction at constant temperature and pressure.

$$p\text{A} + q\text{B} = r\text{C} + s\text{D}$$

one may write

$$\Delta G = \sum \left(\frac{\partial G}{\partial n_i}\right)_{T,\,p,\,n_j \neq i} \mathrm{d}n_i$$

$$= r\mu_{\text{C}} + s\mu_{\text{D}} - p\mu_{\text{A}} - q\mu_{\text{B}}$$

For a spontaneous reaction $\Delta G = \Sigma \nu_i \mu_i$, which must be less than zero. For convenience the *affinity*, A is defined as equal in absolute magnitude but of opposite sign to the change in free enthalpy

$$A = -\Delta G$$

so that the affinity of a spontaneous chemical reaction is positive.

An analogous expression can be obtained starting from the free energy

$$\mu_i = \left(\frac{\partial F}{\partial n_i}\right)_{T,\,p,\,n_j \neq i}$$

and this is normally used for processes carried out at constant temperature and volume.

In order to obtain the change in free enthalpy with changes in the composition at constant temperature of a system containing only the component 1, it is necessary to differentiate equation (16a) with respect to n

$$\mu_1 = \left(\frac{\partial G}{\partial n_1}\right)_{T,p} = \left(\frac{\partial H}{\partial n_1}\right)_{T,p} - \left(\frac{T\partial S}{\partial n_1}\right)_{T,p} = h_1 - Ts_1 \tag{31}$$

where h_1 and s_1 are the molar enthalpy and molar entropy respectively. For a homogeneous system of more than one component, the proportions of the mixture must be borne in mind. The dependence on the proportions may be easily demonstrated for a mixture of gases, where the change in the free enthalpy induced in a system consisting of n_1 moles of component 1 and n_2 moles of component 2 at a total pressure p, by the addition of one mole of component 1 at a constant temperature, is by definition

$$\left(\frac{\partial G}{\partial n_1}\right)_{T,\,p,\,n_2} = \mu_1 \tag{32}$$

The chemical potential of component 1 at pressure p is $\mu_{1_{init}}$. To introduce one mole isothermally into the system consisting of n_1 moles of component 1 and n_2 moles of component 2 at a total pressure p, it is first necessary to expand isothermically and reversibly component 1 from the pressure p to the partial pressure p_1 and then to introduce it at constant volume *e.g.* through a semipermeable membrane, into the system. It is assumed that the system is so large as to undergo no appreciable change in pressure through the introduction or elimination of one mole of one of the components. The total work exchanged is thus equal to that of the initial expansion given by the expression

$$w = RT \ln \frac{p_1}{p} \tag{33}$$

Since the partial and total pressures of a mixture of ideal gases are directly proportional to the numbers of partial and total moles, (33) becomes

$$w = RT \ln \frac{n_1}{\Sigma n_i} = RT \ln x_1 \tag{34}$$

where Σn_i indicates the total number of moles and x_1 the mole fraction or thermodynamic concentration $n_1/\Sigma n_i$. Adding the value of the work, given by (34), to the initial value $\mu_{1_{init}}$ gives

$$\mu_1 = \mu_{1_{init}} + RT \ln x_1 \tag{35}$$

When $x = 1$, (35) becomes

$$\mu_1 = \boldsymbol{\mu}_1 \tag{36}$$

Combining (36) with (31) gives finally

$$\boldsymbol{\mu}_1 = h_1 - Ts_1$$

and thence

$$\mu_1 = h_1 - Ts_1 + RT \ln x_1 \tag{37}$$

For an ideal solution the calculation is analogous, except that the osmotic pressure replaces the gaseous pressure. If as is usually the case, the system is not ideal then, according to Lewis the concentration must be replaced by the *activity* (*cf.* Chap. III, 11), which is related to the concentration by

$$a = f_a x$$

in which f_a is the activity coefficient (*cf.* Chap. III, 11). Thus (37) becomes

$$\mu_i = h_i - Ts_i + RT \ln a_i \tag{38}$$

$$= \boldsymbol{\mu}_i + RT \ln a_i \tag{38a}$$

Changing the unit of measurement for the concentration (mole fraction; mole/kg solvent; mole/litre solvent, etc.) naturally changes the numerical value of the standard chemical potential (see below) but its significance remains unchanged.

For normal calculations it is convenient to refer to a state with fixed conditions of temperature, pressure and concentration, from which to compute the chemical potentials for other states. The chemical potential in this case is called the *standard chemical potential.* The conditions selected for pure bodies are – one atmosphere pressure, 25° C and a physical state which is stable at the given temperature and pressure. For mixed phases, and especially for solutions, the concentration must also be considered and should rigorously be expressed in mole fractions. However, it is normally expressed in activity per kg (molality), or per litre (molarity) of the solvent[1] and for approximate calculations simply in mole per kg or per litre of solution[1].

Equations for the transformation from one unit of concentration to another are collected in Table I, 1 for mixed phases of two components.

TABLE I, 1

THE RELATIONSHIPS BETWEEN MOLE FRACTIONS, MOLALITY AND MOLARITY

Unitage	x_2	m_2	c_2
$x_2 =$	x_2	$\dfrac{M_1 m_2}{1000 + M_1 m_2}$	$\dfrac{M_1 c_2}{1000\,d - c_2 (M_2 - M_1)}$
$m_2 =$	$\dfrac{1000\,x_2}{M_1 (1 + x_2)}$	m_2	$\dfrac{1000\,c_2}{1000\,d - c_2 (M_2 - M_1)}$
$c_2 =$	$\dfrac{1000\,x_2 d}{M_1 + x_2 (M_2 - M_1)}$	$\dfrac{1000\,m_2 d}{1000 + m_2 M_2}$	c_2

The subscripts 1 and 2 refer to component 1 (solvent) and component 2 (solute), x = mole fraction, m = molality, c = molarity, M = molecular weight, d = density of the solution.

[1] The numerical value for the activity differs from that defined above in that the thermodynamic activity of equations (38) and (38a) is virtually equal to the mole fraction neglecting the activity coefficient, and is thus unitary only for pure substances, whilst with solutions it is unitary when it corresponds to 1 mole per kg or per litre of solvent. Thus, obviously the standard state of the solution is not a practicable possibility since such a solution would have to have unitary activity but the properties of an ideally dilute solution. The standard state of a solution thus has only a theoretical value.

For the most usual case of solutions in water, the last two methods of expressing the concentration differ least with very dilute solutions. Equation (38a) thus assumes the general forms

$$\mu = \mu_0 + RT \ln a \tag{38b}$$

$$\mu = \mu_0' + RT \ln c \tag{38c}$$

The molal or molar activity may however be expressed also as a function of the concentration c through the equation

$$a = f_a c$$

in which naturally the numerical value of the activity coefficient differs from that defined above. Equation (38b) can consequently assume the form

$$\mu = \mu_0 + RT \ln (f_a c) \tag{38d}$$

All this is naturally valid also for polyphasic systems in which an analogous expression exists for each phase.

The other functions considered have a similar dependence upon concentration but this will not be discussed in detail here. Reference may be made to a general textbook of thermodynamics for such discussions.

(VI) *The Gibbs–Helmholtz Equation*

The general relationship of the thermodynamic functions, free energy and free enthalpy, to the temperature is described by the Gibbs–Helmholtz equation. It will be illustrated for the particular case of a chemical reaction at constant pressure. In this case the variables $n_1, n_2, \ldots n_i$ are no longer independent since the numbers of moles of the reactants and of the products of the reaction are connected by the stoichiometric coefficient of the reaction equation. Referring the calculation to the quantity of substance indicated by the chemical reaction (1 advancement or stoichiometric unit)[1], the total quantity of the substance becomes fixed and thus it is possible to use (16a) as a starting point. Making $\mathrm{d}p$ equal to 0 and

[1] It will be remembered that one advancement or stoichiometric unit is defined as follows. Starting from a given initial state, the reaction is said to have proceeded by one advancement or stoichiometric unit if a number of moles has been produced of each member of the second half of the stoichiometric equation, equal to the stoichiometric coefficient of that member.

referring the reaction to a finite quantity (1 stoichiometric unit) gives, with T constant,

$$\Delta G = \Delta H - T\Delta S$$

whence

$$\frac{\Delta G - \Delta H}{T} = -\Delta S \tag{39}$$

Although such a relationship was rigorously derived for a closed system, *i.e.* one in which there is no change of composition, it may also be applied in this case provided that it is limited to the quantity of substance indicated. The following reasoning will make this clear. Equation (12) may be differentiated to

$$\mathrm{d}H = \mathrm{d}E + p\mathrm{d}v + v\mathrm{d}p$$

Substituting for $\mathrm{d}E$ the value given in equation (3) and combining with equations (11) and (20) gives the following transformation

$$\mathrm{d}E = \mathrm{d}H - p\mathrm{d}v - v\mathrm{d}p = \delta w + \delta q = \delta w + T\mathrm{d}S = \delta w_{use} - p\mathrm{d}v + T\mathrm{d}S$$

$$\mathrm{d}S = \frac{\mathrm{d}H - v\mathrm{d}p - \delta w_{use}}{T}$$

At constant pressure ($\mathrm{d}p = 0$), $\delta w_{use} = \mathrm{d}G$ from equations (21) and (22), which gives

$$\mathrm{d}S = \frac{\mathrm{d}H - \mathrm{d}G}{T}$$

and this for a finite change at constant temperature becomes

$$\Delta S = \frac{\Delta H - \Delta G}{T}$$

which is equal to equation (39).

If the reaction is first carried out at constant pressure at temperature T and then again at temperature $(T + \mathrm{d}T)$ the dependence of the free enthalpy of the reaction ΔG upon temperature may be determined in the following way. To pass from the initial system I at temperature T to the final system II at temperature $(T + \mathrm{d}T)$ it is possible to follow two equivalent routes.

(a) The reaction is carried out at temperature T with a change in free enthalpy ΔG_T and then the system is heated from temperature T to temperature $(T + \mathrm{d}T)$ with a change in free enthalpy which from (25) is given by the equation

$$\mathrm{d}G_{\mathrm{II}} = - S_{\mathrm{II}}\mathrm{d}T$$

(b) The initial system is heated by $\mathrm{d}T$ with a change of free enthalpy

$$\mathrm{d}G_{\mathrm{I}} = -S_{\mathrm{I}}\mathrm{d}T$$

and then the reaction is carried out at temperature $(T+\mathrm{d}T)$ with a change of free enthalpy $\Delta G_{(T+\mathrm{d}T)}$, which is given by the equation.

$$\Delta G_{(T+\mathrm{d}T)} = \Delta G_T + \left(\frac{\partial \Delta G}{\partial T}\right)_p \mathrm{d}T$$

Obviously the following relationship holds

$$\Delta G_T - S_{\mathrm{II}}\mathrm{d}T = \Delta G_T + \left(\frac{\partial \Delta G}{\partial T}\right)_p \mathrm{d}T - S_{\mathrm{I}}\mathrm{d}T$$

hence

$$(S_{\mathrm{I}} - S_{\mathrm{II}})\mathrm{d}T = -\Delta S\mathrm{d}T = \left(\frac{\partial \Delta G}{\partial T}\right)_p \mathrm{d}T \qquad (40)$$

and finally replacing the value ΔS in (40) with that shown in (39) gives

$$\left\{\frac{\partial(\Delta G)}{\partial T}\right\}_p = \frac{\Delta G - \Delta H}{T} \qquad (41)$$

Equation (41) is an expression of the Gibbs–Helmholtz equation.

3. Electrical Units[1]

Before beginning the systematic study of electrochemical phenomena it will be convenient to recall some of the concepts and magnitudes of electrical units used in electrochemistry. These are: *the unit of quantity of electricity; the unit of potential difference; the unit of current intensity; the unit of electrical resistance; the unit of energy.*

Units of measurement are divided into *absolute* and *practical* units. The former are either defined by convention (primary units) or are derived from these according to physical laws (derived units). However, these are often not suitable for general use because they are too big or too small in comparison with the amounts which have to be measured, and they are therefore replaced by the so-called *practical* units. These practical units

[1] At the present time a tendency is developing — which is supported by the International Union of Pure and Applied Chemistry — to replace the C.G.S. system (centimetre, gram, second) by the M.K.S.A. system (metre, kilogram, second, ampere). Since this change has not yet been officially adopted, the traditional system of units will be followed for its advantage in teaching even though the M.K.S.A. system shows considerable practical advantages which recommend its adoption.

are either multiples or fractions of the absolute units. It is thus necessary to refer always to the absolute units and to derive the practical units from these.

(I) *Unit of Charge*

This is defined in the absolute C.G.S.$_{es}$[1] system by Coulomb's Law; if two electrical charges of the same sign Q_1 and Q_2 which are virtually points, are placed l cm apart in vacuum, then the force of repulsion between them is given by

$$F = \frac{Q_1 Q_2}{l^2}$$

Thus, an electrical charge is unitary when it exerts a force of one dyne on another electrical charge of the same sign and magnitude placed 1 cm from it in a vacuum. This quantity of electricity is very small in comparison with the amounts involved in applied electrical technology. The amount of electricity which passes every second, through the incandescent filament of a lamp of a few candle power is of the order of thousands of millions of C.G.S.$_{es}$ units. The Coulomb (C) is used as the practical unit: $1\ \mathrm{C} = 2.997_7 \cdot 10^9$ C.G.S.$_{es}$ units[2]. The ampere-hour (Ah) is often used and is equivalent to 3,600 C.

(II) *Unit of Potential Difference*

The establishment of a force of attraction or repulsion between two electrical charges shows that throughout the space around each of them there is a field of force. The intensity of the field generated by an electrical charge Q at any point x, l cm from the electrical charge producing the field, is defined as the force acting upon a unit electrical charge at the point x. By Coulomb's Law this is

$$F = \pm \frac{Q}{l^2} \quad [3]$$

If any electrical charge is placed at a point x instead of the unit electrical

[1] The centimetre, gram, second, electrostatic system.

[2] The value $2.997_7 \cong 3$ is derived from theoretical considerations on the electrostatic and electromagnetic systems of measurement (*cf.* below) and will not be discussed further here.

[3] By convention the force is given a negative sign when it exerts an attraction and a positive sign when it exerts a repulsion.

charge, the force acting upon it is given by the product of the field intensity and the electrical charge. On moving the unit charge, work is obtained or performed, depending on whether the movement is produced by the field of force or against it. For an infinitesimal movement dl the work, given by the product of the force and the movement, is

$$\mp \, \delta w = \pm \, F\mathrm{d}l = \pm \frac{Q}{l^2} \, \mathrm{d}l$$

The sign is negative or positive depending on whether the work is done by the field of force or against it. For a movement of a unit positive charge from the point x, at l_0 cm, to infinity, the resultant work is

$$\mp \, w = \pm \int_{x = l_0}^{\infty} \frac{Q}{l^2} \, \mathrm{d}l$$

$$= \pm \frac{Q}{l_0} = \mp \, \psi$$

The function

$$\psi = \frac{Q}{l}$$

is called the *potential* of the field at the point x. It is positive if during the movement of a positive electrical charge the work is performed by the field of force and it is negative when the work is performed against this field.

For any electrical charge Q_1 the work done or obtained during its movement from the point x to infinity, in a field whose potential at the starting point is ψ, is

$$w = \pm \, \psi Q_1$$

In other words, ignoring the sign, the work is given by the product of the potential and the electrical charge. This allows the measurement of the potential at any point x in the field, from the work done or obtained during the movement of any electrical charge Q_1 from the point x to infinity[1]

$$\psi = \frac{w}{Q_1}$$

[1] This is true in a vacuum. In any medium whose dielectric constant (*cf.* Chap. II, 8) is ε the potential ψ depends upon the dielectric constant, according to the equation

$$\psi = \frac{1}{\varepsilon} \frac{w}{Q_1}$$

The potential of the earth's surface is taken as the zero for the potential scale.

The potential difference (p.d.) between two points a and b of the field, $(\psi_b - \psi_a)$, in the C.G.S.$_{es}$ system, is thus equal to the work obtained, or performed, in moving the unit charge from the point a to the point b. It is unitary when this work is 1 erg. The practical unit of p.d., called the Volt (V), is given by the potential difference between two points in an electrical field when the work performed, by or against the field of force, in moving a charge of 1 C from one point to another is 1 Joule (J). Introducing the corresponding absolute units (1 C $= 2.997_7 \cdot 10^9$ C.G.S.$_{es}$ units; 1 J $= 10^7$ ergs) gives

$$1\ \text{V} = \frac{10^7}{2.997_7 \cdot 10^9} = \frac{1}{2.997_7 \cdot 10^2}\ \text{C.G.S.}_{es}\ \text{units},$$

i.e. 1 absolute unit of p.d. $= 299.7_7$ V.

The difference $(\psi_a - \psi_b)$ is called the *electric tension.*

(III) *Unit of Current Intensity*

The current intensity I is defined as the quantity of electricity which passes through a section of the conductor normal to the direction of the current flow in unit time. The current intensity is unitary in the absolute electrostatic system when the quantity of electricity passing through this section of the conductor in one second is unitary. Since this unit is too small, it is replaced in ordinary use by the practical unit: the Ampere (A). The current is one ampere when 1 C passes through the section of the conductor in one second. Thus, 1 A $= 2.997_7 \cdot 10^9$ C.G.S.$_{es}$ units.

(IV) *Unit of Resistance*

If any constant electric tension is applied to the ends a and b of a conductor which has no branches or divisions, nor any other inner electric tension (*cf.* Chapter III), then the current within the conductor will be defined by Ohm's Law

$$\frac{\psi_a - \psi_b}{I} = R$$

The constant R is called the *electric resistance.* It is unitary when $(\psi_a - \psi_b) = 1$ and $I = 1$. The absolute unit is too large for ordinary use. The practical unit is obtained by making $(\psi_a - \psi_b) = 1$ V and $I = 1$ A, when R becomes 1 Ohm (Ω). Replacing the Volt and the Ampere by their values expressed in absolute units gives

$$1\ \Omega = \frac{1}{2.997_7 \cdot 10^2} \cdot \frac{1}{2.997_7 \cdot 10^9} = \frac{1}{2.997_7{}^2 \cdot 10^{11}}\ \text{C.G.S.}_{es}\ \text{units}$$

(V) *Unit of Energy*

The electrical energy is characterized as the product of the electric tension and the quantity of electricity. The practical unit, the Volt–Coulomb, is equal to the product of 1 V · 1 C = 1 Joule (J). In absolute units

$$\frac{1}{299.7_7 \cdot 10^2} \cdot 2.997_7 \cdot 10^9 = 10^7\ \text{ergs} = 1\ \text{J}.$$

Since 1 C = 1 A · 1 sec, 1 J = 1 V · 1 A · 1 sec = 1 Watt-second (Watt = = W = the practical unit of electrical power = 1 V · 1 A). Other units in common use are the Watt-hour (Wh) = 3,600 Watt-seconds and the kilowatt-hour (kWh) = 3,600,000 Watt-seconds.

In addition to an electrostatic basis, the units of measurement may be derived on an electromagnetic basis, giving the C.G.S.$_{em}$ system[1]. Here, that current intensity is considered as unity which when circulating in a coil of 1 cm radius, exerts a force of 2π dynes on a unit magnetic pole placed at the centre of the coil. All of the other electrical measurements are derived from this definition, the definition of the unit magnetic pole and the laws of electrodynamics and electromagnetism.

The magnitudes of the same quantities are not the same in the two systems. This early led to a replacement of the electrostatic system of measurement by the electromagnetic one for most of the electrical quantities, in view of the greater facility of making electromagnetic determinations.

Each of the systems of measurement – electrostatic, electromagnetic and practical – offer advantages and disadvantages. The Giorgi system adopted by the International Electrotechnical Commission in 1935 eliminates many of these disadvantages. In this system the primary units are the metre (M), the kilogram (K), the second (S) and the absolute Ohm (Ω). The latter here becomes a primary unit equal to the resistance of a column of mercury of constant cross-section area of 1 mm^2, 106.246 $\pm$ 0.002 cm long, held at the temperature of melting ice (*cf.* below). This system is also

[1] The electromagnetic centimetre, gram, second system.

called the M.K.S.Ω system from the initials of its units. The absolute ampere is derived from the absolute Ohm by the relationship that the amount of heat q, expressed in Joules, developed in a resistance is proportional to the square of the current intensity I, expressed in amperes, to the resistance R expressed in Ohms and to the time T expressed in seconds

$$q = I^2RT$$

If $q = 1$ J, $R = 1\ \Omega$ and $T = 1$ sec, then $I = 1$ A.

The units for the other quantities are derived by the laws of electrostatics, electrodynamics and electromagnetism. Many of the units of the Giorgi system (M.K.S.Ω) are equivalent to the practical units.

For many of these units it is possible to establish standards, as for the metre, the kilogram, the ohm and the ampere – which as an independent primary unit is defined as that intensity of current which will deposit $0.00111806 \pm 0.00000005$ grams of silver per second at the cathode of a normal silver voltameter (*cf.* Chapter V, 7).

It was decided at a conference held in London in 1908 that it was then convenient to take as the fundamental legal units, conventionally defined values with the numbers somewhat rounded off. For distinction these were called *international units*. One international Ω is the resistance of a column of mercury of constant cross-section area of one mm^2, of 106.300 cm length, having a mass of 14.4521 g, at the temperature of melting ice. One international A is that current intensity which will deposit 0.001118000 g of silver per second at the cathode of a normal silver voltameter. At a conference held in Washington, in 1910, a standard was also defined for the electric tension. This is the tension of a international Weston cell (*cf.* Chapter III, 2) which was taken conventionally as 1.01830 international V at 20°. The other international units may be derived from these using the relationships which link the units of measurement.

In 1954 the Xth General Weights and Measures Conference adopted as a complete system of primary units those of length (1 metre = the Sèvres standard), time (1 second = 1/86,400 of the mean solar day of 1 January 1900), mass (1 kilogram = the Sèvres standard), current intensity (1 ampere), temperature interval (1 degree) and luminous intensity (1 candle). The other two fundamental units involved in electrochemistry are defined as follows. The ampere is the unit of current intensity (replacing the ohm as a fundamental unit in the Giorgi system); it is that current intensity which when maintained in two parallel, straight, infinitely long wires of negligible circular cross-section, placed one metre apart in a vacuum, produces a force of $2 \cdot 10^{-7}$ newtons[1] between them, equivalent in the C.G.S. system to $2 \cdot 10^{-2}$ dynes. The temperature interval is 1 celsius or 1 kelvin which is the temperature interval equal to that of the thermodynamic scale, resulting from the division into 273.16 parts of the temperature interval between absolute zero and the triple point of water. This temperature is expressed in centigrade or Kelvin degrees depending on whether the triple point of water is given the value 0.00 or 273.16.

The so-called absolute electrical units are simply multiples or fractions of the units of the $C.G.S._{em}$ system such that the resulting units are practically the same as the corresponding international units.

No perfect identity exists between the electrostatic and electromagnetic units of the C.G.S. system, the practical Giorgi units and the absolute units, both because of the difference of the dimensions of the various systems and because of the rounding off of the numerical values of the conventionally established international units with respect to the numerical values of the standards.

Table I,2 collects together the units of measurement which concern electrochemistry, expressed in the various systems[2] according to data given by the Bureau of Standards of the USA. It may be mentioned as a point of interest that there is a marked tendency to introduce a new primary measurement relating to the quantity of the substance; its unitage has not yet been definitely selected between mole and kilomole.

TABLE I,2

UNITS OF ELECTRICAL MEASUREMENT

Measurement	*Unit*		*System*			
			Absolute MKSA	*International*	*C.G.S.$_{es}$*	*C.G.S.$_{em}$*
Quantity of electricity	Coulomb	C	1.00000	1.00007	$2.99789 \cdot 10^{9}$	$1.00000 . 10^{-1}$
Electric tension	Volt	V	1.00000	0.99966	$3.33554 \cdot 10^{-3}$	$1.00000 \cdot 10^{8}$
Current intensity	Ampere	A	1.00000	1.00007	$2.99789 \cdot 10^{9}$	$1.00000 \cdot 10^{-1}$
Resistance	Ohm	Ω	1.00000	0.99951	$1.11263 \cdot 10^{-12}$	$1.00000 \cdot 10^{9}$
Energy	Joule	J	1.00000	0.99984	$0.99996 \cdot 10^{7}$	$1.00000 \cdot 10^{7}$

[1] One newton is that force which when applied to the centre of gravity of a body of one kilogram mass induces an acceleration of 1 m sec^{-2}; it is equal to 10^5 dynes.

[2] Any good textbook of physics may be consulted for a fuller discussion of the concepts of units of measurement.

CHAPTER II

ELECTROLYTES AND ELECTROLYTIC CONDUCTION

1. Electrolysis and Electrolytes

If two plates of a suitable metal, such as platinum, are immersed in a solution of an alkali chloride and connected with the terminals of a source of electricity of a sufficiently high electric tension, then it will be found that a current passes through the solution. At the same time a whole series of chemical reactions will be provoked by the current. This experiment may be repeated, with the same result, using any of the substances which are classified chemically as salts, acids or bases. If, for example, the current is passed through hydrochloric acid solution it will be found that there is an evolution of gaseous chlorine at the positive pole and of gaseous hydrogen at the negative. Repeating the experiment with a solution of copper sulphate and copper plates it will be found that metallic copper is deposited at the negative pole and that copper is dissolved from the positive pole.

In these two cases it is immediately possible to observe at what point the chemical reaction provoked by the passage of the current occurs. It is found in fact that these reactions take place exclusively at the surface between the solution and the metal. These are called *primary reactions* since they are provoked directly by the passage of the current. In more complex cases, such as those involving a solution of an alkali chloride, other chemical reactions can be observed in the bulk of the solution which are provoked indirectly by the passage of the current. These take place in fact, even in the absence of a current flow, when the substances formed by the primary reaction come into contact. These are called *secondary reactions.* The primary reactions are those occurring between the particles originally present in solution and the electrical charges taken from, or given up to, the external circuit; whilst the secondary reactions involve initially substances formed by the primary reactions. These substances may react amongst themselves, with particles originally present in the solution, with the metal plates, with the solvent or with the electrical charges of the external circuit[1] etc.

[1] Strictly speaking in this case the reaction would be another concomitant primary reaction.

It has already been mentioned that, under the action of an electric tension existing between two electrodes immersed in a solution of hydrochloric acid, the components of the molecules of the acid migrate separately, one part going towards the negative pole and the other to the positive. At these metallic terminals they are then evolved in the form of elementary molecules of gaseous hydrogen and chlorine. It must be emphasized that

(1) the chlorine migrates *completely* towards the positive pole where it is then released after the reaction, in the elementary state, and the hydrogen migrates *completely* towards the negative pole, where an analogous reaction occurs with the release of elementary hydrogen;

(2) the passage of the current occurs *immediately* the electric tension is applied and there is not even the smallest time interval between the application of the external electric tension and the passage of the current.

These two facts show that in the hydrochloric acid solution exist neither the HCl molecule nor the atoms of H or Cl, but only electrically charged particles which are respectively positive and negative, *i.e.* the H^+ and Cl^- ions. The first observation leads to the conclusion that the particles are electrically charged because otherwise the electrical field created by the applied electric tension would be unable to make them migrate to either pole; in fact all the hydrogen particles are positively charged and all the chlorine particles negatively charged. The absence of even the slightest delay between the application of the electric tension and the passage of the current, mentioned in the second observation, shows moreover that the ions exist in the solution before the application of the electric tension.

That the ions already exist in the solution before the passage of the current may readily be shown also by the fact that under suitable conditions (*e.g.* with two plates of copper immersed in an aqueous solution of a copper salt) there is no lower limit to the size of the electric tension required to pass a current through a salt solution. This implies that no work has to be performed by the current to split the molecule of the salt into ions, and that the ions are already present independently of the passage of the current.

When the ions reach the surface between the solution and the terminal metal of the external circuit, they lose their charge by reacting with the electrical charge of the metallic terminals themselves. In the present example, where the solution contains hydrochloric acid, atoms of hydrogen and chlorine are formed which reacting amongst themselves give the corresponding gaseous molecules. Thus the passage of the current is accompanied by a migration of material or rather, the electrical charges are applied to material particles which carry them towards one or other pole depending on their sign. Faraday proposed that those conductors in which the passage of a current was accompanied by a simultaneous migration of material should be called *electrolytes*[1] and that the complex of phenomena induced by the passage of a current under the action of an electric tension applied externally to conductors of this type, and the associated chemical reactions, should be called *electrolysis.* The ions which are the material carriers of the electrical charges are divided into *cations* (positively charged and migrating towards the *cathode*, which is here the negatively charged electrode) and *anions* (negatively charged and migrating towards the *anode*, which is here the positively charged electrode)[2].

The metallic ends of the external circuit which are immersed in the electrolytes have the generic name of *electrodes.* They are always conductors of the first class (see below).

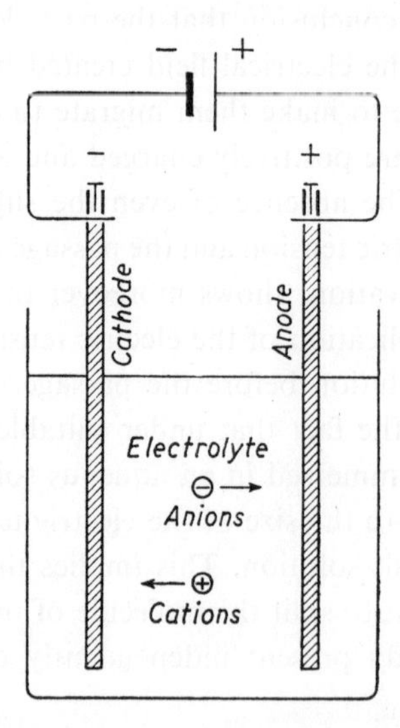

Fig. II, 1. Electrochemical symbols and terms

[1] See below for a definition of electrolysis and conductors of the second class.
[2] Care must be taken with the terms *anode* and *cathode*, not to confuse their electrical charges when they are employed for *electrolytic cells* and *galvanic cells* for the production of electrical energy (see Chapter III). In *electrolytic* cells the *anode* is *positively*, and the *cathode negatively* charged, whereas in *galvanic* cells the reverse is true: the *cathode* is *positively* charged and the *anode negatively.*

Fig. II, 1 illustrates the terms and symbols employed in electrochemistry. The phenomenon of electrolysis is not limited to aqueous solutions at about room temperatures; it also occurs with solvents, other than water, which have certain definite properties, particularly a high dielectric constant (*cf.* 8). It occurs too in melted and solid electrolytes and at various temperatures. For example, a little lead chloride may be melted in a graphite crucible and a graphite rod immersed in the molten mass without touching the crucible. Then on connecting its base with the negative terminal and the rod with the positive terminal of an accumulator, a current will be found to pass, with a simultaneous evolution of chlorine gas at the graphite rod and a deposition of metallic lead on the wall of the crucible.

Thus the phenomenon of conduction in electrolytes is fundamentally different from that of conduction in metals. In these, conduction is electronic, which implies that it is actually the elementary electric charges, or electrons, which move and give rise to the electric current; it is not possible to show any simultaneous migration of material[1]. However, in electrolytes conduction is ionic since the electric charges are transported by ions, *i.e.* by material particles whose migration under the action of the electric tension gives rise to the movement of electric charges – the electric current. Here then is found a migration of the material simultaneously with the passage of the electric current.

Conductors which have electronic conduction are also called *first class conductors*. They include metals, certain forms of carbon (*e.g.* graphite) as well as some oxides and sulphides in the solid state. Conductors in which the flow of the electric current is accompanied by a migration of material are called *second class conductors* or electrolytes. In such conductors the flow of the current is always accompanied by chemical reactions at the surface between the first and second class conductors at which a change of the mechanism of conduction occurs. Second class conductors are always chemical compounds and never elements since they must be able to yield oppositely charged ions which obviously could not happen with elements[2]. Second class conductors are also known amongst substances in the solid state *e.g.* silver halides.

[1] With the exception of certain alloys which will not be considered.

[2] Unless otherwise stated (*e.g.* Chapter XII) this treatise does not consider conduction during a discharge through gases at low pressures, where the conduction is also partly ionic. In such cases the gaseous ions are formed either from free electric charges and elementary molecules independently of their chemical nature, or from molecules or atoms which have lost one or more electrons.

Ions are particles that can be formally derived from atoms, or groups of atoms, with one or more elementary charges in excess[1] with respect to the electrically-neutral state. Hydrogen, the metals and those groups (*e.g.* NH_4^+) which form bases with the hydroxyl ion, give rise to cations, whilst the halogens and those atomic groups which form salts with metals and acids with hydrogen, give rise to anions.

Every cation is formed of an atom or group of atoms with a number of elementary negative charges (electrons) deficient when referred to the electrically-neutral state. This number corresponds to the electrochemical valency. For anions the definition is analogous, with the difference that the charge is given by a number of negative charges *i.e.* of electrons, in excess of the electrically-neutral state. If an element or group of atoms shows various valency levels then it also gives rise to ions with differing numbers of charges (*e.g.* the ferrocyanide and ferricyanide anions, the ferrous and ferric cations, etc.). Ions are symbolized by the same chemical symbol as the element or atomic group followed by a number and a $-$ or $+$ sign placed above the line to indicate the number of positive or negative charges in excess, *e.g.* Na^+, Ba^{2+}, PO_4^{3-}, etc. In accordance with the usage by which the symbol of an element represents at the same time the weight of a gram-atom, the symbol of an ion represents the weight of a gram-ion and the corresponding electric charge.

It is thus implicitly admitted that electricity is also of an atomistic and corpuscular nature which in fact corresponds to reality. A whole series of physical methods exist (cathode rays, radioactivity, Edison effect, photoelectric effects, etc.) which have allowed the elementary negative charges (e) to be isolated. These charges are called *electrons*. In 1932 Anderson isolated the elementary positive charges which are called positrons. Their electrical charges in absolute values are the same: $1.591 \cdot 10^{-19}$ C or $4.80273 \cdot 10^{-10}$ C.G.S.$_{es}$ units; their stationary mass is 1/1835 of that of a hydrogen atom. One equivalent of electricity is thus given by the product of one elementary charge and Avogadro's number ($N = 6.02312 \cdot 10^{23}$). This product, which is thus the quantity of positive or negative electricity linked to a gram-equivalent of any ion, is called one Faraday (**F**). This has also been measured experimentally and its value

[1] This description is really over-simplified in that the distribution of the electrons in the outer orbitals of ions is often different from that in the corresponding neutral atoms (or groups of atoms). This explains the profound chemical differences between ions and atoms.

is known much more exactly than are the values of e and N. The most probable value known today is 96491.4 C_{abs}; in practice the value used is 96.500.

The following definitions have been adopted by CITCE Nomenclature Commission and recommended by the International Union of Physical and Applied Chemistry.

(a) A conductor is of the second class when the moving charges are anions and cations.

(b) An electrolyte is a second class conductor in the form of a solution and, sometimes, of a pure liquid or solid.

Although these definitions may not respect historical precedence, they are immediately clear on the basis of what has been said so far.

2. Transference Numbers

The quantity of electricity carried per second by any ionic species in solution, depends upon the velocity with which the ions migrate under the influence of the electric tension and upon the specific charge carried. The total charge carried by one equivalent (1 Na^+, $\frac{1}{2}$ Ba^{2+}, $\frac{1}{3}$ PO_4^{3-}) is always constant and is the same for the two ions of opposite charge provided that the solution is electro-neutral. Thus, the quantity of electricity carried per second by each of the two ions can only be the same if the rates of effective migration of the two ions are equal. When an ion is subjected to an electric tension, a force acts upon it to move it, overcoming the friction which it encounters in its movement through the medium. For a material point of mass m subjected in vacuum to a force F

$$ma = F \tag{1}$$

where a represents the acceleration. When the material point moves within the bulk of a fluid it must overcome the frictional resistance, and (1) becomes

$$ma = F - kw \tag{2}$$

where kw represents a frictional resistance proportional to the velocity w with a coefficient k. The frictional resistance R encountered during the movement through a liquid is defined by Stokes' Law.

$$R = 6\pi\eta rw = kw \tag{3}$$

where η = the coefficient of viscosity and r = the radius of the sphere. Substituting (3) into (2) gives

$$ma = F - 6\pi\eta rw \tag{4}$$

If F is kept constant and the velocity w is gradually increased from zero, a certain value of w is reached such that

$$6\pi\eta rw = F$$

so that

$$F - 6\pi\eta rw = 0 \tag{5}$$

Thus from (4)

$$ma = 0$$

and since the mass cannot be zero, the acceleration must be so. Hence, the velocity w must be constant and (5) gives

$$w = \frac{F}{6\pi\eta r}$$

and the velocity under steady conditions becomes constant and proportional to the applied force. This reasoning may be applied to the motion of ions considered, to a first approximation, as spherical, in a fluid in which they are subjected to an electric tension[1]. The force acting in this case is proportional to the intensity of the field $\mathrm{d}U/\mathrm{d}l$. If two points a and b are $\mathrm{d}l$ apart and have an electric tension $\mathrm{d}U$ between them, then the work $\mathrm{d}w$[2] corresponding to the movement of an ion with a charge Q, is given by[3]

$$\mathrm{d}w = Q\mathrm{d}U = F\mathrm{d}l \text{ and hence } F = Q\frac{\mathrm{d}U}{\mathrm{d}l}$$

The constant velocity of an ion, in cm sec^{-1}, under the action of a field of 1 V per cm is called the *electrical mobility*, indicated by the symbol u_+ for cations and u_- for anions.

The relationship between electrical mobility and the quantity of current transported can be obtained as follows. Fig. II, 2 represents an electrolytic

[1] Stokes' Law is insufficient for a rigorous treatment in that the solvent cannot be considered as continuous with respect to ions in solution (*cf.* H. MUKHERJEE, *Indian J. Phys.*, 23 (1949) 503; 24 (1950) 137).

[2] Note that the symbol w may represent either velocity or work.

[3] This equation omits the positive or negative signs of the electrical charge, of the electric tension and of the direction of movement with or against the field of force. In each specific case, the + or − sign must be introduced depending upon the convention adopted.

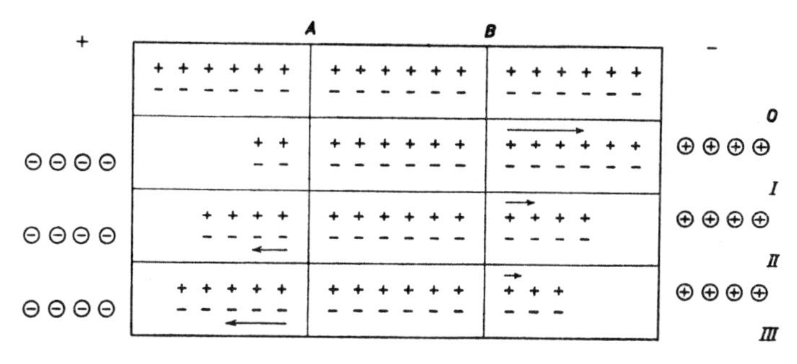

Fig. II, 2. Schematic diagrams for transference numbers

cell ideally divided into three regions by diaphragms *A* and *B*. Before the start of the experiment the concentrations are the same in the three compartments, as shown in section 0. If a certain amount of electricity is passed then there will be a discharge of cations at the cathode and of anions at the anode in amounts which are strictly equivalent both to each other and to the number of equivalents of electricity passed through the cell. The quantity of electricity must naturally not be too great since if it were sufficient to induce total electrolysis of the electrolyte present it would not be possible to make any further measurement. The required condition for the experiments is that the amount of electricity passed through the cell is such as to cause no change of concentration in the central region.

The following three cases, illustrated by sections I, II and III, may be considered.

(I) The current is transported exclusively by the cations.

(II) The current is transported by both the cations and the anions migrating with the same velocity.

(III) The current is transported by the cations and the anions migrating with different velocities.

In Fig. II, 2 each + sign corresponds to 1 equivalent of cations and each — sign to 1 equivalent of anions. The signs enclosed in circles represent the respective equivalents after they have lost their charge. The arrows indicate by their length the number of equivalents of each ionic species which pass through the diaphragms in unit time. In order to avoid useless complications, it is supposed that the electrolyte is such that the electrochemical reactions consist exclusively in the discharge of the ions and that there are no secondary reactions of any type, and also that the electrolyte is totally dissociated.

During the electrolytic process with the discharge of ions at the electrodes there is a disappearance of electrolyte and a consequent fall in its concentration[1].

In the first case only the cations transport the current and migrate under the influence of the electric field. If it is supposed that there are initially six equivalents in each of the three compartments and that four equivalents of electricity are passed through the cell, then four equivalents each of cations and of anions will be discharged. However, whilst the anions stay still, the cations migrate towards the cathode and so disappear from the anodic region. The cations discharged at the cathode are replaced by other cations migrating towards the cathode. On the other hand, the anions discharged at the anode are not replaced. Thus, only the anodic region loses electrolyte. The total balance of the effects of the migration and the discharge on the concentration are shown in scheme I.

SCHEME I

+	*A*	*B* —
Anodic region	*Central region*	*Cathodic region*
4 equivalents of cations migrate towards the central region	4 equivalents of cations come from the anodic region	
	4 equivalents of cations migrate towards the cathodic region	4 equivalents of cations come from the central region
4 equivalents of anions are discharged		4 equivalents of cations are discharged
Anodic loss		*Cathodic loss*
4 equivalents of electrolyte	Concentration unaltered	—

[1] In these theoretical considerations no account is taken of accessory phenomena which occur in contact with the electrodes and in their immediate neighbourhood, such as diffusion, convective currents in the liquid, turbulent mixing caused by the evolution of gas at the electrode, etc. The considerations developed here apply to the mean concentrations of the regions, unaffected by these phenomena.

In the second case both the anions and the cations transport the current, migrating in opposite directions, with the same velocity. For both the ionic species there is a partial replacement of the ions which have been discharged. Thus at the cathode there is a fall in concentration due to the discharge of cations and the simultaneous migration of anions. At the anode there is a simultaneous migration of cations and discharge of anions. Since equal rates of migration have been assumed, the number of cations which migrate from the anodic region is equal to the number of anions which migrate from the cathodic region and thus the fall in concentrations in the anodic and cathodic regions are equal. The relative balance is shown in scheme II.

SCHEME II

+	*A*	*B* −
Anodic region	*Central region*	*Cathodic region*
2 equivalents of cations migrate towards the central region	2 equivalents of cations come from the anodic region	
	2 equivalents of cations migrate towards the cathodic region	2 equivalents of cations come from the central region
	2 equivalents of anions come from the cathodic region	2 equivalents of anions migrate towards the central region
2 equivalents of anions come from the central region	2 equivalents of anions migrate towards the anodic region	
4 equivalents of anions are discharged		4 equivalents of cations are discharged
Anodic loss		*Cathodic loss*
2 equivalents of electrolyte	Concentration unaltered	2 equivalents of electrolyte

In the third case the current is transported by both the ionic species, but with different velocities, *e.g.* the anions might migrate with a velocity three times that of the cations. For each cation which migrates towards the cathode three anions migrate towards the anode. After the discharge of four equivalents of electricity, the resulting concentrations are those shown in scheme III. The fall in concentration in the cathodic region would be three times as great as that in the anodic region. The relative balance of migration and discharge is illustrated in scheme III.

SCHEME III

+ Anodic region	*A* Central region	*B* Cathodic region −
1 equivalent of cations migrates towards the central region	1 equivalent of cations comes from the anodic region	
	1 equivalent of cations migrates towards the cathodic region	1 equivalent of cations comes from the central region
	3 equivalents of anions come from the cathodic region	3 equivalents of anions migrate towards the central region
3 equivalents of anions come from the central region	3 equivalents of anions migrate towards the anodic region	
4 equivalents of anions are discharged		4 equivalents of cations are discharged
Anodic loss		*Cathodic loss*
1 equivalent of electrolyte	Concentration unaltered	3 equivalents of electrolyte

If p_c indicates the fall in concentration in the cathodic compartment and p_a the fall in the anodic compartment, then in the three cases described

$$\frac{p_a}{p_c} = \frac{u_+}{u_-} = \frac{4}{0}; \frac{2}{2} \text{ and } \frac{1}{3}$$

respectively, *i.e.* the anodic loss is to the cathodic loss as the electrical mobility of the cation is to that of the anion. In general for the discharge of one equivalent of ions the loss in the anodic region is not 1 but $1 - t$ ($t < 1$) because t equivalents of anions have migrated simultaneously into the anodic region; at the same time $1 - t$ equivalents of cations have migrated towards the cathode. The loss in the cathodic region is thus t equivalents. Replacing the fraction p_a/p_c by $(1 - t)/t$ gives

$$\frac{u_+}{u_-} = \frac{1-t}{t} = \frac{1}{t} - 1$$

$$\frac{u_+}{u_-} + 1 = \frac{1}{t}$$

$$\frac{u_+ + u_-}{u_-} = \frac{1}{t}$$

from which

$$t = \frac{u_-}{u_+ + u_-}$$

Similarly from

$$\frac{p_a}{p_c} = \frac{1-t}{t}$$

is obtained

$$t = \frac{p_c}{p_a + p_c} = \frac{p_c}{p_{tot}}$$

For equal charges the quantity of electricity transported in unit time by each ionic species is as the respective electrical mobilities, so that t_- indicates the fraction of the current transported by the anion, whilst $1 - t_- = t_+$ indicates the fraction of the current transported by the cations. This is

$$t_+ = \frac{u_+}{u_+ + u_-} = \frac{p_a}{p_{tot}}$$

The fractions t_- and t_+ are respectively the *transference numbers* of the anion and of the cation (Hittorf).

The apparatus of Jahn is very suitable for the measurement of transference numbers using the change in the concentrations in the anodic and cathodic regions. The apparatus is shown in Fig. II, 3 where K represents the cathode which consists of a glass bulb containing a little mercury which is covered by a layer of a concentrated solution of copper nitrate. This avoids any evolution of gas at the cathode with consequent mixing of the solvent. The electrolytic process consists exclusively in the deposition of copper. The electrical connection is made with a thick copper wire dipping into the mercury and insulated by a glass tube. The anode A is

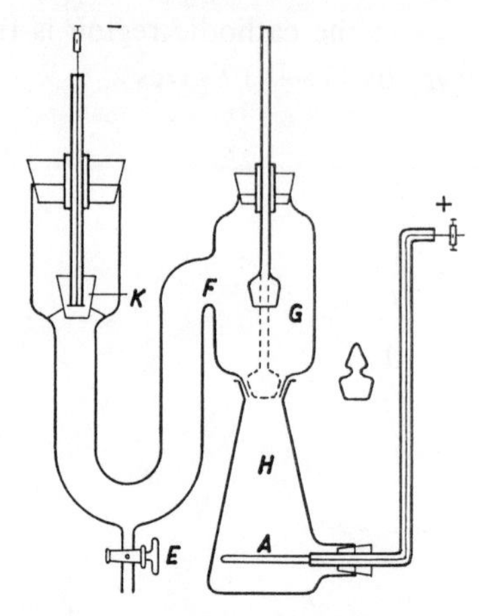

Fig. II,3. Jahn's apparatus for transference numbers

attached with a rubber bung and consists of a thick copper, silver or cadmium wire so that an evolution of the gas is avoided at the anode also. Here the anodic electrolytic process consists in the dissolution of metal. The upper part of the apparatus, which carries the U-shaped side arm, is attached to the lower part by the ground glass joint *H*. The lower aperture is closed at the end of the experiment, by lowering the ground glass stopper *G*. By opening the stop-cock *E* the liquid is drained from the cathodic region up to the point of attachment of the side arm *F*, whilst the liquid of the central region is contained between the attachment of the side arm *F* and the stopper *G* placed in position *H*. After the stopper *G* is closed the liquid of the anodic region is isolated in the lower part of the apparatus. For an experiment, the apparatus is connected in series with a source of direct current (usually an accumulator), an ammeter, a resistance to regulate the current intensity and a coulometer (see page 283

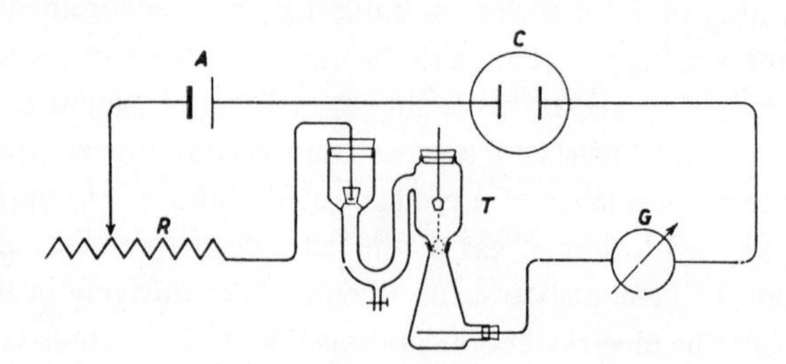

Fig. II,4. Electrical wiring for measurements of transference numbers

et seq.) for the measurement of the quantity of electricity. This is illustrated in Fig. II, 4, where *A* is the accumulator, *R* is the resistance, *T* the apparatus for the measurement of the transference number, *G* the ammeter and *C* the coulometer.

Naturally, in the measurement of the quantity of electricity which has passed through the solution, allowance must be made for that transported by the solvent itself[1].

Another, more precise, method for the determination of transference numbers is based on the measurement of the movement of a boundary between two electrolyte solutions on the passage of a current. This method has been brought to a high level of precision, particularly by MacInnes and his colleagues. It is based on the principle illustrated in Fig. II, 5, which represents the section of a tube containing solutions of electrolytes AK′ and AK″ with for example a common anion A, which are in contact along the boundary $a-b$. By suitably selecting both the cation K″ in

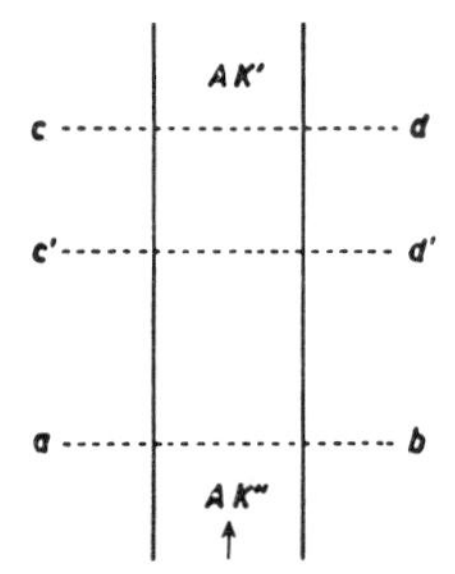

Fig. II, 5. Moving boundary

relation to cation K′ (whose transference number is to be determined) and the concentration of the two electrolytes, when an electric current is passed through the tube, in the direction indicated by the arrow, the cations will migrate in the same direction whilst the anions will migrate in the opposite direction. Thus, the boundary will migrate in the same direction as the current since the cation K′ will be replaced by the cation K″; in other words the solution of the electrolyte AK″ replaces the solution of the electrolyte AK′.

If on passing one **F** the boundary moves from position $a-b$ to $c-d$ then all of the volume v of solution AK′, expressed in cm^3, contained

[1] A discussion of the experimental difficulties in the measurement of transference numbers by the Hittorf method is given by A. L. LEVY, *J. Chem. Educ.*, 29 (1952) 384.

between the sections $a - b$ and $c - d$ will have been replaced by the solution AK″. If c is the concentration expressed in equivalents per cm^3, cv equivalents of cation will have migrated under the action of the electric field. But by definition, the transference number is equal to the fraction of the current carried by the ionic species; so that if cv equivalents are moved by the passage of one **F**, the product cv gives the actual transference number of the cation K′, *i.e.*

$$t_{K'} = cv \tag{6}$$

Usually in practice an amount of electricity Q is passed, which is much smaller than 1 **F** so that the boundary migrates from position $a - b$ to the nearer position $c' - d'$, moving through the volume v'. Since the volumes moved are directly proportional to the quantity of electricity passed

$$v' : v = Q : \mathbf{F} \tag{7}$$

and eliminating v from equations (6) and (7) gives

$$t_{K'} = (cv'\mathbf{F})/(It)$$

where the quantity of electricity Q has been replaced by the product of the current intensity I and the time t[1].

A similar argument may be applied to a pair of electrolytes A′K and A″K, with a common cation, for the determination of the transference number of the anions.

It is implicitly assumed that:

(a) there are no significant diffusion or mixing phenomena between the two solutions in contact,

(b) the movement of the boundary is independent of the nature and concentration of the following ion (otherwise called the *indicator ion*),

(c) there are no volume changes to affect the position of the boundary. By working under suitable conditions these assumptions may be actually fulfilled or else it may be possible to calculate accurately the necessary corrections. The following conditions are particularly important.

(i) The indicator ion must always follow the ion being examined.

(ii) The indicator ion must have a lower electrical mobility than the ion being examined.

[1] The current intensity must be constant for such a substitution to be valid. Note that the symbol t represents both the transference number and the time.

(iii) The solution with the higher density must be under that with the lower density.

(iv) The concentration of the indicator ion must fulfil the relationship

$$c : t = c_{ind} : t_{ind}$$

(where c and c_{ind} are the concentrations of the ion being examined and of the indicator ion respectively, and t and t_{ind} are their respective transference numbers) so that the two ions acquire the same effective rate of migration (see page 65 *et seq.*)[1].

The apparatus developed for this purpose by MacInnes and his colleagues is shown in Fig. II, 6, where E and E' are the two electrodes and BC and $B'C'$ are two pairs of accurately ground discs which may be rotated about each other around a common axis, so that a sharply defined boundary can be produced. D is the graduated tube used for measuring the movement of the boundary. This method is quicker and possibly more exact than that of Hittorf but requires careful operation.

Table II, 1 shows how, with careful work, the two methods will yield the same results.

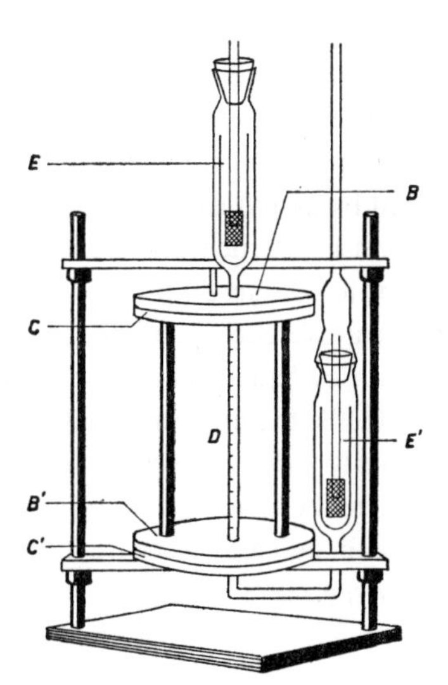

Fig. II, 6. MacInnes apparatus for transference numbers

[1] A detailed discussion of this method of measuring transference numbers is given in D. A. MacInnes, *Principles of Electrochemistry*, Reinhold, New York, 1939, p. 68 *et seq.*

TABLE II, 1

TRANSFERENCE NUMBERS OF THE K^+ AND Li^+ IONS MEASURED BY THE METHODS OF HITTORF AND MACINNES

Electrolyte	*Method*	*Transference number at the concentration*			
		0.02 *m*	0.05 *m*	0.2 *m*	0.5 *m*
KCl	Hittorf	0.4893	0.4894	–	0.4896
KCl	MacInnes	0.4901	0.4899	–	0.4888
LiCl	Hittorf	0.327	0.323	0.312	–
LiCl	MacInnes	0.326	0.321	0.311	–

Other methods for the determination of transference numbers are based on measurements of conductivity (see page 65 *et seq.*) and of the electric tensions of special types of galvanic cells (see page 142 *et seq.*). A method has recently been developed based on polarographic measurements[1].

When transference numbers are measured by analytical determinations of the concentration changes in the anodic and cathodic regions it must be borne in mind that the motions of the ions themselves may also cause unwanted changes in concentration. Allowance must be made for these which are due to the movement of the solvent either because it is transported by more or less solvated ions (*cf.* 6 of this chapter) or because of the electrokinetic effect (*cf.* Chapter VI). The transference number which is obtained by analytical determinations is called the *apparent transference number* whilst the true transference number is that defined by the equation from the rate of migration. The correction which has to be made to the apparent transference number to give the true one depends on the concentration. For aqueous solutions of a concentration of the order of magnitude of 0.1 N the difference between the true and apparent transference numbers only rarely exceeds a few units in the third decimal place. The correction to be made to the apparent transference number to give the true one can be calculated by determining the solvation of the ions (see page 69 *et seq.*) and thence the amount of solvent transported.

The transference number indicates the relationship between the electrical mobility of a particular ion and the sum of the electrical mobilities of all the ionic species present in the solution. This relationship, when

[1] S. Khalafalla and Z. Hanafy, *J. chim. phys.*, 53 (1956) 986.

suitably developed, is valid for mixtures of electrolytes also; in this case the equivalent concentrations of the various electrolytes must be borne in mind since the quantity of electricity Q_{ci} which each cation can transport is proportional to the concentration and the electrical mobility of that cation

$$Q_{ci} = K(c_i u_{i+})$$

If the total amount of electricity transported is Q, given by the sum of the individual quantities Q_i for each of the anions in one direction and each of the cations in the other, then

$$\begin{aligned} Q &= K(c_1 u_{1+} + c_2 u_{2+} \ldots + c_1 u_{1-} + c_2 u_{2-} \ldots) \\ &= K\Sigma c_i u_i \end{aligned}$$

but the transference numbers of each of the ions indicate that fraction of the total electricity transported *i.e.*

$$t_{1+} = \frac{Q_{c_1}}{Q} = \frac{Kc_1 u_{1+}}{K\Sigma c_i u_{i+}} = \frac{c_1 u_{1+}}{c_1 u_{1+} + c_2 u_{2+} \ldots + c_1 u_{1-} + c_2 u_{2-} \ldots}$$

$$t_{2+} = \frac{Q_{c_2}}{Q} = \frac{Kc_2 u_{2+}}{K\Sigma c_i u_{i+}} = \frac{c_2 u_{2+}}{c_1 u_{1+} + c_2 u_{2+} \ldots + c_1 u_{1-} + c_2 u_{2-} \ldots}$$

$$t_{1-} = \frac{Q_{a_1}}{Q} = \frac{Kc_1 u_{1-}}{K\Sigma c_i u_{i-}} = \frac{c_1 u_{1-}}{c_1 u_{1+} + c_2 u_{2+} \ldots + c_1 u_{1-} + c_2 u_{2-} \ldots}$$

and so on.

For a binary electrolyte the transference number also indicates directly the relative velocities of the two ions. The transference number does not have a characteristic value for each ionic species since its magnitude depends upon the other ionic species present. It is independent of the current intensity carrying out the electrolysis but it depends upon the temperature and the concentration of the electrolyte.

The dependence upon the temperature is such that as it increases the transference number tends to the value 0.5, *i.e.* at higher temperatures the current tends to be equally divided between the two ionic species.

The dependence upon the concentration ceases when the electrolyte is very dilute because only there do the quantities u_+ and u_- become truly constant (*cf.* 8 of this chapter). For non-ideally dilute solutions, the dependence of the transference number on the concentration c may be expressed to a sufficient approximation by

$$t = t_0 - A\sqrt{c}$$

TABLE II,2

TRANSFERENCE NUMBERS OF ELECTROLYTES IN AQUEOUS SOLUTION*

Substance	*Concentration*		°C	t_-	t_+
HF	0.031	*N*	25	0.150	—
HCl	0.01	*N*	25	—	0.8251
HBr	0.1	*N*	25	—	0.792
HI	0.2 — 0.06	*N*	25	0.174	—
HIO_3	0.010	*N*	25	0.09835	—
HNO_3	0		25	—	0.8303
H_2SO_4	0.01	*N*	25	0.185	—
CH_3COOH	1 — 0.1	*N*	25	0.108	—
PiH**	1 — 0.1	%	—	—	0.910
LiOH	0.20	*N*	18	0.848	—
NaOH	0.04	*N*	25	0.799	—
KOH	0		25	—	0.274
LiCl	0.01	*N*	25	—	0.3289
LiI	0.1	*N*	—	0.682	—
Li_2SO_4	0.05	*m*	25	—	0.39
NaCl	0.01	*N*	25	—	0.3918
NaBr	0.05	*N*	25	0.609	—
NaI	0.05	*N*	25	0.619	—
Na_2SO_4	0.01	*N*	25	—	0.3848
$NaNO_3$	0.1	*N*	25	0.5903	—
NaN_3	0.3	*N*	20	0.523	—
Na_2CO_3	0.05	*N*	23	0.590	—
$NaCH_3COO$	0.01	*N*	25	—	0.5537
KCl	0.01	*N*	25	—	0.4902
KBr	0.01	*N*	25	—	0.4833
KI	0.01	*N*	25	—	0.4884
$KClO_3$	0.02	*N*	18	0.466	—
$KClO_4$	0.1	*N*	18	0.477	—
$KBrO_3$	0.02	*N*	18	0.433	—
$KMnO_4$	0		23	0.457	—
K_2SO_4	0.01	*N*	25	—	0.4829
KNO_3	0.01	*N*	25	—	0.5084
K_2CO_3	0.04	*N*	22	0.435	—
$K_3[Fe(CN)_6]$	0		25	0.574	—
$K_4[Fe(CN)_6]$	0		25	0.601	—
KCH_3COO	0.01	*N*	25	—	0.6948
RbCl	0.02	*N*	18	0.503	—
RbBr	0.02	*N*	18	0.505	—

Substance	*Concentration*		°C	t_-	t_+
RbI	0.02	*N*	18	0.502	—
CsCl	0.02	*N*	18	0.496	—
CsBr	0.02	*N*	18	0.503	—
CsI	0.02	*N*	18	0.503	—
NH_4Cl	0.01	*N*	25	—	0.490
NH_4Br	0.02	*N*	18	0.517	—
NH_4I	0.02	*N*	18	0.511	—
NH_4NO_3	0.1	*N*	25	0.4870	—
NH_4Pi**	0.05 – 0.03	*N*	—	0.292	—
$MgCl_2$	0.052	*m*	25	—	0.375
$MgBr_2$	0.02	*N*	18	0.615	—
MgI_2	0.02	*N*	18	0.612	—
$MgSO_4$	0.02	*N*	25	—	0.36
$CaCl_2$	0.01	*N*	25	—	0.4277
$CaBr_2$	0.02	*N*	18	0.591	—
CaI_2	0.02	*N*	18	0.584	—
$CaSO_4$	0.0045	*N*	18	0.559	—
$Ca(NO_3)_2$	0.005	*N*	18	0.550	—
$SrCl_2$	0.01	*m*	25	—	0.424
$SrBr_2$	0.02	*N*	18	0.590	—
SrI_2	0.02	*N*	18	0.584	—
$BaCl_2$	0.001	*N*	25	—	0.4444
$BaBr_2$	0.0025	*m*	18	0.564	—
BaI_2	0.02	*N*	18	0.574	—
$Ba(NO_3)_2$	0.05	*N*	18	0.544	—
$CuCl_2$	0.05	*N*	23	0.595	—
$CuBr_2$	0.106	*m*	25	0.555	—
$CuSO_4$	0.053	*N*	16 – 19	—	0.375
$AgClO_3$	0.02	*N*	25	0.505	—
$AgClO_4$	0.02	*N*	25	0.514	—
Ag_2SO_4	0.05	*N*	17	0.554	—
$AgNO_3$	0.01	*N*	25	—	0.4648
$AgCH_3COO$	0.01	*N*	25	0.376	—
$Zn(ClO_4)_2$	0.1	*m*	25	—	0.409
$ZnCl_2$	0.5	*m*	25	—	0.331
$ZnBr_2$	0.02	*m*	25	—	0.389
ZnI_2	0.05	*m*	25	—	0.382
$ZnSO_4$	0.005	*m*	--	—	0.384

(continued) →

(Table II, 2 continued)

Substance	*Concentration*		°C	t_-	t_+
$CdCl_2$	0.02	N	25	—	0.486
$CdBr_2$	0.02	N	25	—	0.434
CdI_2	0.017 – 0.007	N	18	0.556	—
$CdSO_4$	0.008	N	18	0.613	—
$Hg_2(NO_3)_2$	0.05	N	20	—	0.480
$NiSO_4$	0.1	N	40	—	0.366
$LaCl_3$	0.01	N	25	—	0.4625
$TlCl$	0.01	N	22	0.516	—
$TlClO_4$	0.05	m	25	—	0.531
Tl_2SO_4	0.03	N	25	0.521	—
$TlNO_3$	0.101	m	25	0.481	—
$Pb(NO_3)_2$	0.1 – 0.03	N	25	0.513	—
$MnCl_2$	0.05	N	18	0.613	—
$CoCl_2$	0		20	—	0.489
$(UO_2)(NO_3)_2$	0.0024	N	25	0.81	—

0 indicates a concentration tending towards zero

N indicates that the concentration is expressed in equivalents per litre

m indicates that the concentration is expressed in moles per kg

* Other values obtained under various experimental conditions may be found in: *Critical Tables*, Vol. VI, McGraw Hill, New York, 1929, pp.309 *et seq*. See also: R. Parsons, *Handbook of Electrochemical Constants*, Butterworth, London, 1959; and for many salts of rare earths: F. H. Spedding *et al.*, *J. Am. Chem. Soc.*, 74 (1952) 2778, 4751; 76 (1954) 879, 882, 884.

** Pi indicates the picrate ion.

where t is the transference number at the given concentration c, t_0 is the transference number at infinite dilution and A is an empirical constant which may have a negative or a positive value. It is evident from the condition $t_+ + t_- = 1$, that if the constant A is positive for an ionic species in a given electrolyte then it must be negative for the other ionic species of the same electrolyte.

Table II, 2 reports the majority of the transference numbers so far measured in aqueous solutions and the experimental conditions employed. The transference number of an ionic species may also depend on the presence of non-electrolytes, if these can form complex ions with the ionic species being considered, with consequent changes in the rate of migration. Transference numbers in general do not differ greatly from

0.5 for most ions; H^+ and OH^- however, are exceptions, having very high values. Sometimes measurements of transference numbers give abnormal or even negative values and in such cases a marked dependence of the value on the concentration is also noted. This dependence cannot be expressed by the equation relating the transference number to the square root of the concentration, or by any similar equation. It is very probable that in such cases there is a formation of a new chemical entity such as incompletely dissociated, complex or polyatomic ions. In, for example, a concentrated solution of zinc iodide there is probably a formation of the complex ion ZnI_3^- in addition to the presence of the Zn^{2+} and I^- ions. Thus, part of the zinc migrates towards the anode falsifying the analytical determinations of the concentration. At concentrations above 3.5 *m* the transference number of zinc becomes negative[1].

Similar formations of more or less dissociable and more or less dissociated complex ions may be encountered in mixtures of electrolytes. In a solution containing platinum chloride and an excess of hydrochloric acid, the platinum migrates exclusively towards the anode. This implies the formation of a complex ion between the Pt^{4+} ion and the Cl^- ion and that this anion is virtually undissociated. The compound H_2PtCl_6 is in fact known and this could dissociate according to two schemes

$$\text{I. } H_2PtCl_6 \rightleftharpoons 2\,H^+ + Pt^{4+} + 6\,Cl^-$$

$$\text{II. } H_2PtCl_6 \rightleftharpoons 2\,H^+ + PtCl_6^{2-}$$

The transference number measurements favour the second scheme of dissociation.

In pure molten electrolytes the exact determination of transference numbers is very difficult. It is not possible to talk of changes in the concentration at the anodic and cathodic regions with a pure electrolyte. Determinations of this type could however be made with solutions of electrolytes in other molten electrolytes; but in such cases the solvent, being also an electrolyte, will play a greater or smaller part in the conduction process[2]. Another method is based on the transport of electrolyte from one compartment to another of a special cell[3]. Besides these theoretical

[1] R. H. Stokes and B. J. Levien, *J. Am. Chem. Soc.*, 68 (1946) 1852.

[2] A summary of the literature on transference numbers in molten electrolytes up to 1938 is given by Baimokov and Samusenko, *Trans. Leningrad Ind. Inst.*, (1938) 3.

[3] For the experimental technique see F. R. Duke and R. W. Laity, *J. Am. Chem. Soc.*, 76 (1954) 4096; *J. Phys. Chem.*, 59 (1955) 549; H. Bloom and N. J. Doull, *J. Phys. Chem.*, 60 (1956) 620.

TABLE II, 3

TRANSFERENCE NUMBERS OF SOLID ELECTROLYTES

Substance	*Melting point* °C	*Temperature* °C	t_-	t_+
NaF	992	500	—	1.00
NaF	992	625	0.139	0.861
NaCl*	801	400 – 425	—	1.00
KCl	776	435	—	0.956
KCl	776	600	0.116	0.884
KBr	730	605	—	0.5
KI	680	610	—	0.9
CuCl**	422	18 – 366	—	1.00
βCuBr	—	395 – 445	—	1.00
γCuBr**	—	27 – 390	—	1.00
αCuI	605	450 – 500	—	1.00
βCuI	—	400 – 440	—	1.00
γCuI**	—	250 – 400	—	1.00
Cu_2O	—	1000	1.00	—
αCu_2S	1130	220	—	1.00
AgCl	455	200 – 350	—	1.00
AgBr	434	200 – 300	—	1.00
αAgI	552	150 – 400	—	1.00
βAgI	—	20	—	1.00
BaF_2	1289	500	1.00	—
$BaCl_2$	962	400 – 700	1.00	—
$BaBr_2$	847	350 – 450	1.00	—
PbF_2	824	200	1.00	—
$PbCl_2$	501	90 – 484	1.00	—
$PbBr_2$	373	250 – 365	1.00	—
PbI_2	402	255	—	0.39
PbI_2	402	290	—	0.67
$PbBr_2 - PbF_2$	—	255	Br^- 0.133 F^- 0.867	— —
Ag_2HgI_4	—	60	—	Ag^+ 0.94 Hg^{2+} 0.06

* According to W. Jost and H. Schweizer (*Z. physik. Chem. B*, 20 (1933) 118) the value of t_+ for NaCl between 557° and 710° varies from 1.00 to 0.12.

** At temperatures below 300° for CuCl, 360° for CuBr and 390° for CuI, electronic conductance also occurs, becoming total at 18°. However t_+ always remains constant at 1.00 for that part of the current which is conducted ionically.

difficulties there are marked experimental problems which are much greater than with solutions. Irregular convection currents are particularly troublesome and are caused by the high temperatures at which the experiments must be carried out and the difficulties of obtaining a uniform temperature distribution throughout the bulk of the mass. Certain special precautions have permitted the identification of some ionic species in molten electrolytes. It was observed, for example, that in the electrolysis of pure lead chloride the lead migrates towards the cathode, whilst with lead chloride dissolved in potassium chloride the lead migrates towards the anode: an obvious demonstration of the formation of a complex anion involving the lead. It has been shown in other cases that in certain molten electrolytes only one ionic species transports the electricity. In a molten mixture of sodium and aluminium chlorides in approximately equal molar proportions, only the Na^+ ion migrates so that its transference number is 1.00.

In solid crystalline electrolytes in general only one of the ionic species present transports the current, so that its transference number is 1 and there is unipolar conduction. However, cases are known of crystals showing bipolar conduction. With electrolytes in the solid state conduction is not always exclusively ionic, quite often the current is carried by both ions and electrons, so that the conductor is behaving partly as a second class and partly as a first class conductor. In some cases of oxides, sulphides and nitrides there is in fact a virtual absence of ionic conduction. Table II, 3 collects together most of the transference numbers so far measured with solid electrolytes.

3. The Conductance of Electrolytes and its Measurement

The electrical conductance of any conductor is defined as the reciprocal of its resistance. Thus, every measurement of conductance reduces simply to a measurement of resistance. The electrical resistance R of any conductor is directly proportional to its length l and inversely proportional to the cross-sectional area S normal to the direction of the current. It also depends on a characteristic constant ϱ for every material which is called the specific resistance, or resistivity, such that

$$R = \frac{l}{S} \varrho$$

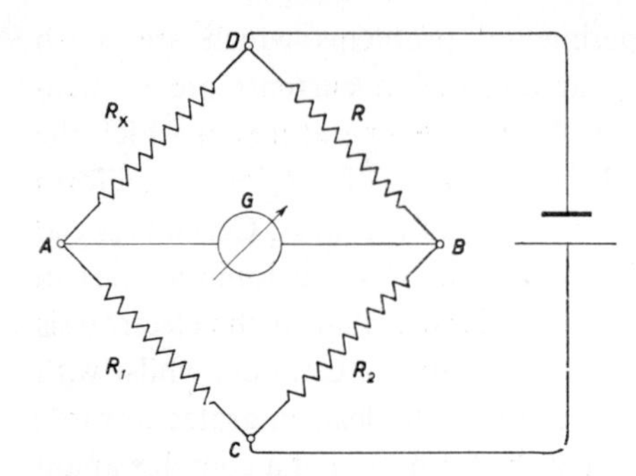

Fig. II,7. Wheatstone's bridge

Where $l = 1$ cm and $S = 1$ cm^2, $\varrho = R$, *i.e.* ϱ is the resistance of a conductor 1 cm long with a cross-sectional area in a plane normal to the direction of the current of 1 cm^2. The reciprocal of the resistance is the conductance

$$\chi = \frac{1}{\varrho} \frac{S}{l}$$

The constant $1/\varrho = \varkappa$ is called the *specific conductance* or *conductivity* and indicates the conductance of a 1 cm length of material having a cross-sectional area of 1 cm^2. The conductance is directly proportional to the cross-sectional area and inversely proportional to the length. If the resistance is expressed in Ω the conductance will be expressed[1] in Ω^{-1}. It can be measured on the Wheatstone bridge principle or by measurements of electric tension.

The Wheatstone bridge consists of a standard resistance R, (Fig. II, 7), two variable resistances R_1 and R_2 and the unknown resistance R_x. By applying an electric tension across points C and D a current is produced in the two limbs formed by the resistances R_1R_x and R_2R. A zero indicator G inserted between points A and B serves to show when no electric tension exists between them. In this case the bridge is balanced and R_x may be determined. From Ohm's Law the electric tension between points C and A is R_1I_1 (where I_1 represents the current intensity in limb R_1R_x), and that between C and B is R_2I_2 (where I_2 represents the current intensity in limb R_2R). When, therefore, A and B are at the same potential

$$R_1I_1 = R_2I_2 \qquad (1)$$

[1] 1 $\Omega^{-1} = 1/\Omega$ = reciprocal ohms. In some texts the conductance is expressed in mhos (the inverse of ohm).

Similarly the electric tension between the points A and D and between the points B and D is given by the relationship of $R_x I_1$ to RI_2, and these must be equal when the points A and B are at the same potential

$$R_x I_1 = RI_2 \tag{2}$$

Dividing (2) by (1) gives

$$\frac{R_x}{R_1} = \frac{R}{R_2}$$

i.e.

$$R_x = R\frac{R_1}{R_2}$$

Then, if the values of the resistance R and of the ratio R_1/R_2 are known, the value of R_x is easily calculated. However, whilst it is easy to balance the bridge with respect to the resistance using a direct current this is not suitable for measuring the resistance of an electrolyte since it induces polarization at the electrodes and electrolysis (*cf.* Chapter IV) which cause continuous variations of the resistance. It is thus necessary to make the measurement with alternating current of a sufficiently high and undistorted frequency to eliminate these effects. In this case the bridge must also be balanced with respect to capacitance and inductance so as to have a true absence of current in the zero indicator G which is inserted in the bridge. As this cannot at present be completely achieved, the zero indicator will show a minimum of current rather than a complete absence. Errors due to the heating of the solution of the electrolyte must also be avoided. The form of the Wheatstone bridge used for alternating currents is called the Kohlrausch bridge.

In practice the Kohlrausch bridge (Fig. II, 8) for the measurement of conductance consists of a taut, homogeneous, calibrated wire AB on which runs a movable contact C, a graduated resistance box R, a source of alternating current of audio frequency s, an earphone T which serves as the

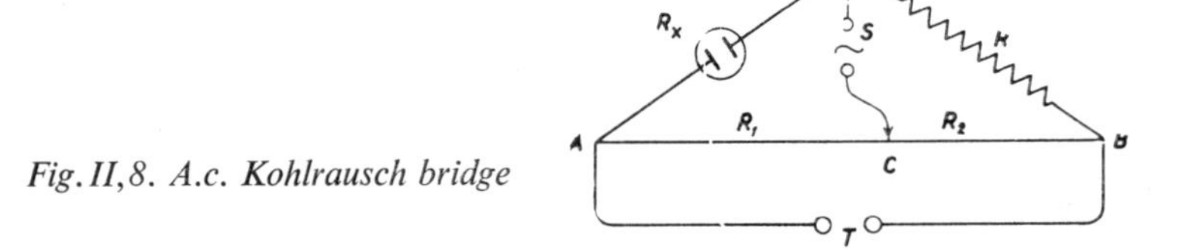

Fig. II, 8. A.c. Kohlrausch bridge

zero indicator and a cell R_x containing the electrolyte whose conductance is to be measured. An electronic valve oscillator is preferably used as the source of alternating current: having the advantage of producing an alternating current of the desired frequency and of virtually pure sinusoidal form. With high resistances it is difficult to locate exactly the position of the minimum and it is then advisable to insert an amplifier between the earphone and the bridge outlet to make the minimum more sharply defined[1].

Visual reading instruments have recently been introduced for the identification of the minimum and the earphone has also been replaced by an alternating current galvanometer. Again, a rectifier may be inserted between the bridge outlet and a galvanometer; this rectifier may be of the thermal (thermocouple), crystal (galena, copper oxide or selenium), electronic or mechanical types. The latter has a synchronous motor which reverses the polarity of a commutator at each inversion of the alternating current. Performing the measurements with direct reading instruments has the advantage of allowing the progress of the measurement to be followed continuously and very weak minima to be interpolated more exactly. The cathode ray oscillograph has also been used in place of the earphone for measurements for special purposes[2]. Another interesting method for the determination of the conductance of electrolytes is based on the measurement of the electric tension existing between two points in a solution of an electrolyte. The electric tension is detected with two auxilliary test electrodes fixed to the cell through which passes a continuous current. The tension is measured potentiometrically (*cf.* Chap. III, 2) and is compared, for example, with the electric tension generated in a standard resistance through which flows an equal current[3].

[1] A very careful discussion of the technique, the measuring instruments and the errors of this method is given by G. Jones *et al.*, *J. Am. Chem. Soc.*, 50 (1928) 1049; 51 (1929) 2407; 53 (1931) 411,1207; 57 (1935) 272,280 and by T. Shedlovsky, *J. Am. Chem. Soc.*, 52 (1930) 1793,1806. See also H. Gerischer, *Z. Electrochem.*, 58 (1954) 9.

[2] See for example G. Jones, K. J. Musels and W. Juda, *J. Am. Chem. Soc.*, 62 (1940) 2919; R. N. Haszeldine and A. A. Woolt, *Chem. Ind.*, (1950) 544.

[3] The following publications are amongst the most recent on this method of measuring conductance, which is also called the four electrode method (two carrying the current and two for the measurement): A. R. Gordon *et al.*, *J. Am. Chem. Soc.*, 75 (1953) 2855; R. F. Palmer and A. B. Scott, *J. Am. Chem. Soc.*, 72 (1950) 4821; D. J. G. Ives and H. Swaroopa, *Trans. Faraday Soc.*, 49 (1953) 788; H. H. Lim, *Austr. J. Chem.*, 9 (1956) 443; L. Elias and H. J. Schiff, *J. Phys. Chem.*, 60 (1956) 595; F. Spillner, *Chem. Ing. Techn.*, 27 (1957) 24.

The measurement of the conductivity of an electrolyte in solution is carried out in vessels provided with electrodes whose surface area S and distance apart l must be known. In general, however, electrolytes whose conductivity $\varkappa$ is well known are used to calibrate the cell. In the equation which gives the resistance

$$R = \frac{1}{\varkappa} \cdot \frac{l}{S}$$

the fraction l/S is called the cell constant C and can be determined when the resistance R of a particular electrolyte of known conductivity $\varkappa$, is measured in the cell:

$$C = R\varkappa$$

The conductivity of any other electrolyte is readily calculated from this

$$\varkappa = C/R$$

Table II,4 lists the conductivities of some of the electrolytic solutions which are most commonly used for the calibration of cells.

TABLE II,4

THE CONDUCTIVITY OF KCl IN AQUEOUS SOLUTIONS FOR THE CALIBRATION OF CELLS

*Concentration** g/1000 g *solution*	$\varkappa\ 10^4$**		*Cell constant range*
	20° C	25° C	
71.3828	1020.2_4	1117.3_3	5 – 200
7.43344	116.67_6	128.86_2	0.5 – 20
0.746558	12.757_2	14.114_5	0.05 – 2

* values corrected for the buoyancy of the air
** values corrected for the conductance of the water itself

Some types of cells are shown in Fig. II,9. Cells of type A are suitable for everyday measurements whilst those of type B are suitable for more delicate or precise measurements.

The conductance which is measured in this way is naturally the sum of the conductances of the solvent and the solute. With solutions whose total conductance is very small, account must be taken of the conductance of the solvent itself if a measurement of that of the dissolved electrolyte

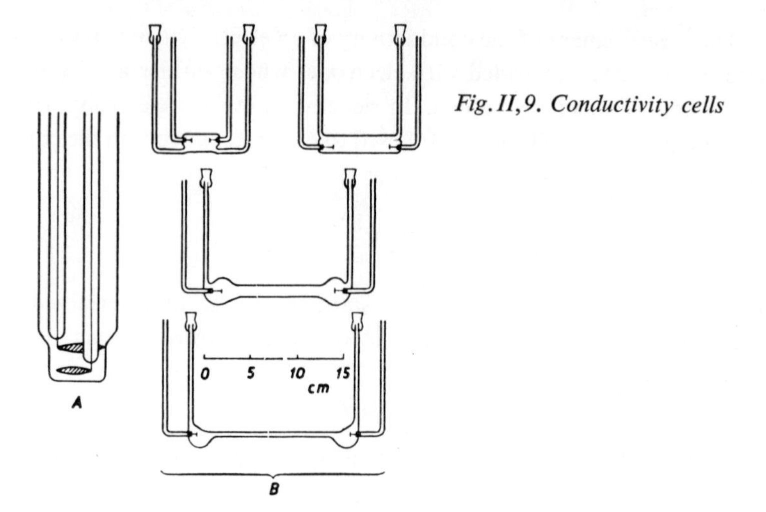

Fig. II,9. Conductivity cells

alone is required. For the most usual case of aqueous solutions the values of the conductivity $\varkappa$ and of the equivalent conductance Λ (see below) of water at 18° are $3.84 \cdot 10^{-8}$ and $6.928 \cdot 10^{-7}$ respectively.

4. Equivalent Conductance

When comparing the conductance of solutions of various electrolytes it is convenient to refer to one equivalent. If the concentration is referred to cm^3, the *equivalent conductance* is $\varkappa/c_{eq}$: the ratio of the specific conductance to the concentration expressed in equivalents per cm^3. In other words Λ is the conductance of one equivalent of substance under the given experimental conditions[1]. The reciprocal of the concentration c, *i.e.* the dilution φ, indicates in how many cm^3 one equivalent[2] of electrolyte is dissolved.

$$\Lambda = \frac{\varkappa}{c_{eq}} = \varkappa\varphi_{eq}$$

The conductance of a solution of an electrolyte in a given solvent depends not only on the electrolyte dissolved, but also on various physical factors

[1] Completely analogous considerations and definitions apply to the *molar conductance* μ which refers to 1 mole of electrolyte.

[2] Unless otherwise indicated, the concentrations and dilutions are expressed in equivalents in the following discussion.

such as the quantity of electrolyte, the concentration, the temperature and the viscosity.

The dependence on the concentration is complex. Fig. II, 10 shows the typical picture of this relationship in terms of a graph whose ordinates show the conductivity and abscissae the concentration (or dilution), with its origin at concentration zero. Conductivity at zero concentration is obviously zero and rises with increasing quantity of electrolyte, expressed as concentration, until it passes through a maximum and then decreases again. For some electrolytes saturation is reached before the maximum conductivity can be achieved. This behaviour can be expressed mathematically by an equation of the type

$$\varkappa = cf_{\Lambda}L \tag{1}$$

where c is the concentration; f_{Λ} is a factor less than 1, called the *coefficient of conductance* which rises, tending towards 1, as the concentration tends towards zero; L is a constant which is characteristic for the substance. Multiplying both sides of the equation by φ gives

$$\varkappa\varphi = \Lambda = c\varphi f_{\Lambda}L$$

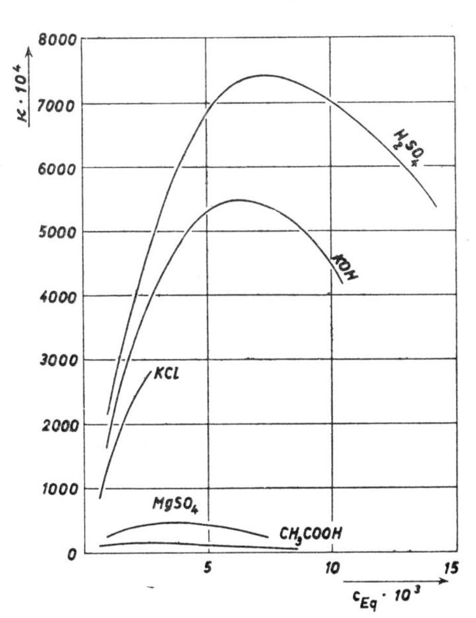

Fig. II, 10. Specific conductance as a function of the concentration

However, φ is the reciprocal of c and hence $c\varphi$ equals 1 and

$$\Lambda = f_\Lambda L \tag{2}$$

Thus, the conductance of one equivalent rises with increasing dilution to a limiting value which is reached when dissociation is complete. In fact, the conductance of an electrolyte at a given concentration depends on the number of ions present per cm^3: the ions originating from the dissociation of the electrolyte. The number of ions present, for a given concentration of electrolyte, depends on the degree of dissociation. The degree of dissociation α is defined as the ratio of the number of dissociated moles to the total number of moles before dissociation. It increases with diminishing concentration, tending towards 1, *i.e.* at infinite dilution the electrolytes are completely dissociated into their ions and at this dilution the equivalent conductance reaches its maximum.

Fig. II, 11 shows the dependence of the equivalent conductance of various substances on dilution.

The conductance of a certain substance in a particular solvent at a given temperature depends on the number of ions present and on their rate of

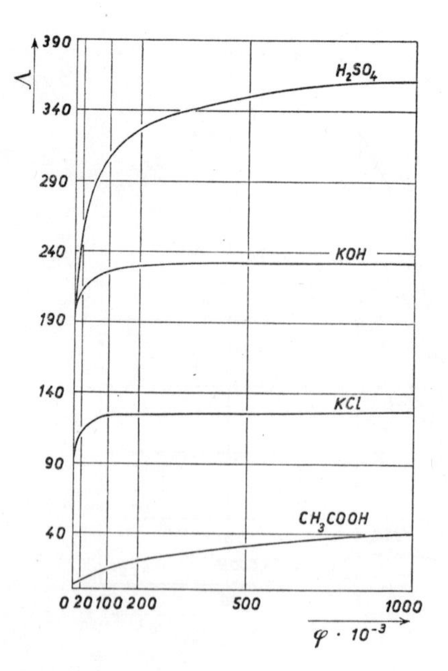

Fig. II, 11. Equivalent conductance as a function of the dilution

migration[1] which will for the moment be considered as constant. The number of ions present depends in turn on the quantity of electrolyte and the degree of dissociation. If then, the quantity of electrolyte is kept constant, the factor f_Λ must be a function of the degree of dissociation[2]. Comparison of the values for f_Λ and the degree of dissociation α as functions of the concentration, shows that they have the same behaviour. Sufficiently dilute solutions of weak electrolytes may be considered as ideal solutions and the thermodynamic equations may be applied in their limiting form to interpret various phenomena in solutions of electrolytes, *e.g.* the cryoscopic lowering, using the classical theory of dissociation. For solutions of such electrolytes which are not too concentrated, the values of the degree of dissociation and of the coefficient of conductance are virtually coincident until infinite dilution is reached, when they both become 1. The constant L thus becomes

$$\Lambda_0 = L$$

and is thus the equivalent conductance at infinite dilution or the limiting equivalent conductance. If Λ_v is the equivalent conductance at dilution φ, then (2) gives

$$\Lambda_v = f_\Lambda \Lambda_0 \qquad (3)$$

Since in sufficiently dilute solutions of weak electrolytes the numerical value of the coefficient of conductance is virtually the same as that of the degree of dissociation, (3) may be written in the form

$$\Lambda = \alpha \Lambda_0$$

Thus, the degree of dissociation of an electrolyte may also be determined by measurements of the conductance.

The physical significance of the coefficient of conductance will be clarified by the study of strong electrolytes (*cf.* 8 of this chapter). If the values of the equivalent conductance are plotted, not against the concentration itself, but against its square root (Fig. II, 12), it will be apparent that $\Lambda = f\sqrt{c}$ becomes a straight line, particularly for those substances

[1] In fact, the rate of migration is not constant but depends upon the concentration and the presence of other ions especially with the so-called strong electrolytes (*cf.* 8 of this chapter).

[2] The factor f_Λ is also a function of the electrical mobility which has however been considered as constant for the time being.

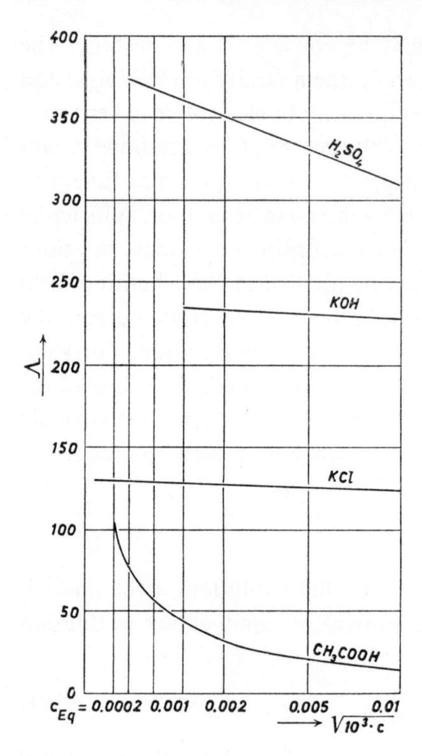

Fig. II, 12.
Equivalent conductance as a function of the square root of the concentration

which are called strong electrolytes. Thus, the equation for conductance may be written empirically as

$$\Lambda = a - b\sqrt{c}$$

where a and b are constant at constant temperature. If c tends to zero

$$\Lambda = \Lambda_0 = a$$

and

$$\Lambda_v = \Lambda_0 - b\sqrt{c}$$

This allows the limiting equivalent conductance to be estimated by extrapolation of the results obtained for the equivalent conductance at various concentrations. This equation is valid not only for aqueous solutions but also for those in other solvents. The constant b in particular, varies with the nature of the solvent since it depends upon its dielectric constant and viscosity.

These relationships were empirically determined by Kohlrausch; but they can now be derived theoretically by the theory of strong electrolytes (*cf.* 8 of this chapter) and the constants can be calculated from the properties and behaviour of strong electrolytes as described by Debye and Hückel. Onsager[1] derived a general expression which is strictly valid as a limiting equation for strong electrolytes and takes account of the asymmetry and electrophoretic effects (*cf.* 8 of this chapter). It can be written in the form

$$\Lambda_v = \Lambda_0 - \left(\frac{e^2 a}{6\varepsilon kT}\Lambda_0 w + \frac{ea(z_+ + z_-)}{300 \cdot 6\pi\eta}\mathbf{F}\right) \tag{4}$$

where e = the charge of an electron; a = the effective radius of the ionic atmosphere (*cf.* 8 of this chapter); k = the Boltzmann constant; ε = the dielectric constant; T = the absolute temperature; z_+, z_- = the electrochemical valency of the individual ionic species; η = the viscosity; and $\mathbf{F}$ = 1 Faraday. The value of w is defined by

$$w = \frac{2\,q|z_+||z_-|}{1 + \sqrt{q}}$$

where

$$q = \frac{|z_+||z_-|\Lambda_0}{(|z_+| + |z_-|)\,(|z_+|\lambda_{0-} + |z_-|\lambda_{0+})}$$

(λ_{0+} and λ_{0-} are the equivalent ionic conductances at infinite or limiting dilution; see below). Substituting into (4), with the numerical values of the constants and remembering that a is a function of the square root of the concentration gives

$$\Lambda_v = \Lambda_0 - \left(\frac{9.90 \cdot 10^5}{(\varepsilon T)^{\frac{3}{2}}}\Lambda_0 w + \frac{29.15(|z_+| + |z_-|)}{(\varepsilon T)^{\frac{1}{2}}\eta}\right)\sqrt{c(|z_+| + |z_-|)}$$

For the special case of uni-univalent electrolytes, assuming complete dissociation,

$$\Lambda_v = \Lambda_0 - \left(\frac{8.20 \cdot 10^5}{(\varepsilon T)^{\frac{3}{2}}}\Lambda_0 + \frac{82.4}{(\varepsilon T)^{\frac{1}{2}}\eta}\right)\sqrt{c}$$

i.e.

$$\Lambda_v = \Lambda_0 - (\alpha\Lambda_0 + \beta)\sqrt{c} \tag{5}$$

[1] It is not within the scope of this book to explain in detail the calculations by which Onsager obtained the equation which carries his name. The original literature may be consulted: *Physik. Z.*, 27 (1926) 388; 28 (1927) 277; *Trans. Faraday Soc.*, 23 (1927) 341.

With concentrations referred to 1 cm^3, the constants α and β have, at 25° C, the values shown in Table II, 5.

TABLE II, 5

THE VALUES OF CONSTANTS α AND β IN ONSAGER'S EQUATION

Solvents	α	β
Water	0.229	60.2
Methanol	0.923	156.1
Ethanol	1.33	89.7

Equation (5) is valid experimentally for solutions of less than about 10^{-3} *N*, which confirms its value as a limiting equation. For higher concentrations Shedlovsky showed that if (5) were expressed in the form

$$\Lambda_0 = \frac{\Lambda_v + \beta\sqrt{c}}{1 - \alpha\sqrt{c}}$$

the value of Λ_0 so obtained was not constant and varied with changes in the concentration above 10^{-3} *N*. If these values for Λ_0 (which Shedlovsky calls Λ_0') are plotted against *c*, then many uni-univalent electrolytes give straight lines which intersect the conductance axis at the true value of the limiting conductance. This is equivalent to putting the conductance equation into the form

$$\Lambda_0 = \Lambda_0' - Bc$$

where $\Lambda_0' = (\Lambda_v + \beta\sqrt{c})/(1 - \alpha\sqrt{c})$, and this may be expanded into

$$\Lambda_v = \Lambda_0 - (\alpha\Lambda_0 + \beta)\sqrt{c} + Bc(1 - \alpha\sqrt{c})$$

This equation with an empirical factor *B* describes well the conductance of many strong uni-univalent electrolytes at concentrations up to about 0.1 *N*. Other empirical equations have also been devised; in general they take the form of exact mathematical developments of Onsager's equation with additional corrective terms in *log c*, c^2 etc.[1].

The exact value of the limiting equivalent conductance of an electrolyte

[1] A review article on the literature containing the results obtained by Onsager, Shedlovsky etc., has been published by D. A. McInnes, *J. Franklin Inst.*, 225 (1938) 661.

may be calculated as follows. The value of the limiting equivalent conductance shows various regularities when dealing with electrolytes with common anions or cations. Table II, 6 gives values for the limiting conductance of various binary electrolytes.

TABLE II, 6

VALUES OF Λ_0 FOR SOME ELECTROLYTES IN AQUEOUS SOLUTION AT 25°

Anions	*Cations*			
	K^+	*Na^+*	*Li^+*	*Tl^+*
Cl^-	149.87	126.46	115.03	151.25
NO_3^-	144.96	121.55	110.12	146.34
F^-	128.92	105.51	—	130.3

If the difference is taken between the values of the limiting equivalent conductances for salts with common ions, the following values are obtained

KNO_3	144.96 −	KF	128.92 −	KCl	149.87 −
$NaNO_3$	121.55	NaF	105.51	NaCl	126.46
	23.41		23.41		23.41
LiCl	115.03 −	TlCl	151.25 −	NaCl	126.46 −
$LiNO_3$	110.12	$TlNO_3$	146.34	$NaNO_3$	121.55
	4.91		4.91		4.91

Thus, the differences between the values for the limiting equivalent conductances for salts of potassium and sodium with common anions are constant and so are those for chlorides and nitrates with the same cations. Analogous results are obtained for the differences between other pairs of salts with common cations or anions. This implies that the value of the limiting equivalent conductance for a binary electrolyte is the sum of two additive constants which are characteristic for each of the two ions.

This may also be readily demonstrated as follows. The current passes through the electrolyte because it is transported by ions migrating under the action of the applied electric tension. In calculating the total quantity

TABLE II,7

LIMITING EQUIVALENT IONIC CONDUCTANCES IN AQUEOUS SOLUTIONS *

Ions	18° C	25° C	*Ions*	18° C	25° C
		CATIONS			
H^+	315	349.8	$\frac{1}{2}$ Zn^{2+}	45.0	52.8
Li^+	32.55	38.69	$\frac{1}{2}$ Cd^{2+}	45.1	54
Na^+	42.6	50.11	$\frac{1}{2}$ Pb^{2+}	60.5	70
K^+	63.65	73.50	$\frac{1}{2}$ Mn^{2+}	44.5	53.5
Rb^+	66.3	77.8	$\frac{1}{2}$ Fe^{2+}	(44.5)	(53.5)
Cs^+	66.8	77.3	$\frac{1}{2}$ Co^{2+}	(45)	55
NH_4^+	63.6	73.4	$\frac{1}{2}$ Ni^{2+}	(45)	54
Ag^+	53.25	61.92	$\frac{1}{3}$ Al^{3+}	—	63
Tl^+	64.8	74.7	$\frac{1}{3}$ Fe^{3+}	—	68
$\frac{1}{2}$ Be^{2+}	—	45	$\frac{1}{3}$ Cr^{3+}	—	67
$\frac{1}{2}$ Mg^{2+}	44.6	53.06	$\frac{1}{3}$ La^{3+}	—	69.8
$\frac{1}{2}$ Ca^{2+}	50.4	59.50	$\frac{1}{3}$ Sc^{3+}	—	64.7
$\frac{1}{2}$ Sr^{2+}	50.6	59.46	$\frac{1}{3}$ Ce^{3+}	—	69.9
$\frac{1}{2}$ Ba^{2+}	54.35	63.64	$\frac{1}{3}$ Pr^{3+}	—	69.6
$\frac{1}{2}$ Ra^{2+}	56.5	(66.8) **	$\frac{1}{3}$ Nd^{3+}	—	69.4
$\frac{1}{2}$ Cu^{2+}	45.3	56.6	$\frac{1}{3}$ Sm^{3+}	—	68.6
$\frac{1}{2}$ Hg_2^{2+}	—	68.6	$\frac{1}{3}$ Er^{3+}	—	65.9
$\frac{1}{2}$ Hg^{2+}	—	63.6	$\frac{1}{3}$ Eu^{3+}	—	67.8
			$\frac{1}{3}$ Y^{3+}	—	65.2
			$\frac{1}{3}$ $Co(NH_3)_6^{3+}$	—	99.2
			$N(CH_3)_4^+$	—	44.92
			$NH(CH_3)_3^+$	—	42
			$N(C_2H_5)_4^+$	—	32.66
			$N(C_3H_7)_4^+$	—	23
		ANIONS			
OH^-	174	197.6	ClO_3^-	55.8	64.6
OD^-	—	119	ClO_4^-	59.1	67.4
F^-	47.6	55.4	BrO_3^-	49.0	55.7
Cl^-	66.3	76.34	IO_3^-	34.8	40.7
Br^-	68.2	78.14	IO_4^-	49	54.5
I^-	66.8	76.97	MnO_4^-	53	61
CN^-	—	82	ReO_4^-	46.5	54.97
CNO^-	54.8	64.6	HCO_3^-	—	44.5
CNS^-	57.4	66	$H_2PO_4^-$	28	36
NO_2^-	59	(72)	$H_2AsO_4^-$	—	34
NO_3^-	62.6	71.44	$H_2SbO_4^-$	—	31
ClO_2^-	—	52	HS^-	57	65

Ions	18° C	25° C	*Ions*	18° C	25° C
HSO_3^-	—	58	monochloracetate$^-$	—	39.8
HSO_4^-	—	52	dichloracetate$^-$	—	38
N_3^-	—	69.5	trichloracetate$^-$	—	35
$\frac{1}{2}$ CO_3^{2-}	60.5	69.3	cyanacetate$^-$	—	41.8
$\frac{1}{2}$ HPO_4^{2-}	—	57	*n*-propionate$^-$	—	35.8
$\frac{1}{2}$ SO_3^{2-}	—	72	*n*-butyrate$^-$	—	32.6
$\frac{1}{2}$ SO_4^{2-}	68.7	80	benzoate$^-$	—	32.3
$\frac{1}{2}$ $S_2O_3^{2-}$	—	87.4	*o*-chlorbenzoate$^-$	—	30.5
$\frac{1}{2}$ $S_2O_4^{2-}$	—	66.5	*o*-nitrobenzoate$^-$	—	31.7
$\frac{1}{2}$ CrO_4^{2-}	—	83	3,5-dinitrobenzoate$^-$	—	28.7
$\frac{1}{3}$ PO_4^{3-}	—	92.8	picrate$^-$	25.14	31.39
$\frac{1}{2}$ SeO_4^{2-}	65	75.7	$\frac{1}{2}$ oxalate^{2-}	—	24.1
$\frac{1}{2}$ CrO_4^{2-}	72	83	$\frac{1}{2}$ tartrate^{2-}	55	59.6
$\frac{1}{2}$ MoO_4^{2-}	—	74.5	$\frac{1}{2}$ *o*-phthalate^{2-}	—	52
$\frac{1}{2}$ WO_4^{2-}	(59)	69.4	$\frac{1}{3}$ citrate^{3-}	61.2	71.5
$\frac{1}{3}$ $Fe(CN)_6^{3-}$	—	99.1	ethylbenzene-		
$\frac{1}{4}$ $Fe(CN)_6^{4-}$	—	111	*p*-sulphonate$^-$	—	29.3
$(CN)_2N^-$	46.5	54.3	*n*-butylbenzene$^-$		
$(CN)_3C^-$	38.5	46.4	*p*-sulphonate$^-$	—	25.6
$(NO_2)_3C^-$	—	46	*n*-octylbenzene$^-$		
formate$^-$	48	54.6	*p*-sulphonate$^-$	—	23.1
acetate$^-$	35	40.9	salicylate$^-$	—	35

* The values given for 25° C are more accurate than those for 18° C.
** The values in parentheses are not quite certain.

of electricity which passes in unit time it must be borne in mind that the cations transport positive charges in one direction whilst the anions transport negative charges in the opposite direction, and hence their effects are additive. Considering an electrolyte in a cylindrical vessel of S cm² right cross sectional area[1], between two electrodes l cm apart to which an electric tension U is applied, the ions will move with a velocity

$$w_+ = u_+ \frac{U}{l} \tag{6}$$

$$w_- = u_- \frac{U}{l} \tag{7}$$

[1] A section normal to the axis of the cylinder.

TABLE II, 8

EQUIVALENT CONDUCTANCES OF ELECTROLYTES

Electrolyte	Concentrations					
	0.001 *N*		0.01 *N*		0.1 *N*	
	18° C	25° C	18° C	25° C	18° C	25° C
HCl	377	421.36	370	412.0	351	392.32
HIO_3	343.3	284.38	323.9	358.64	—	—
HNO_3	375	—	360	—	301	—
H_2SO_4	361	—	309	—	233	—
$HClO_4$	—	—	—	—	—	362
HCH_3COO	41	—	14.3	—	4.6	5.2
LiCl	96.5	112.40	92.1	107.32	82.4	100.11
$LiClO_4$	—	103.44	—	98.61	—	88.56
$LiIO_3$	65.3	—	61.2	—	51.5	—
$LiNO_3$	92.9	—	88.6	—	79.2	—
Li_2SO_4	96.4	—	86.85	—	68.2	—
NaOH	208	244.71	200	283.3	183	—
NaF	87.8	—	83.5	—	73.1	—
NaCl	106.5	123.74	101.95	118.51	92.0	111.06
NaI	—	124.25	—	119.24	—	112.79
$NaIO_3$	75.2	—	70.9	—	60.45	—
$NaClO_4$	—	114.87	—	109.59	—	98.43
$NaNO_3$	102.85	—	98.2	—	87.2	—
$NaCH_3COO$	75.2	88.5	70.2	83.76	61.1	72.80
Na_2SO_4	106.7	124.15	96.8	112.44	78.4	89.98
Na_2CO_3	112	—	96.2	—	94.1	—
Na_2HPO_4	58.4	—	54	—	44	—
KOH	234	—	228	—	213	—
KF	108.9	—	104.3	—	94.0	—
KCl	127.3	146.95	122.4	141.27	112.0	128.96
KBr	129.4	—	124.4	143.43	114.2	131.39
KI	128.2	—	123.4	142.18	114.0	113.11
$KClO_3$	116.9	—	111.6	—	99.2	—
KIO_3	96.0	—	91.2	—	79.7	—
$KClO_4$	119.0	137.87	114.2	131.46	—	115.20
KNO_3	123.7	141.84	118.2	132.82	104.8	120.40
KIO_4	—	124.94	—	118.51	—	98.12
KCNS	118.65	—	113.95	—	104.3	—
$KHCO_3$	—	115.34	—	110.08	—	—
K_2CO_3	133.0	—	115.5	—	94.1	—
KCH_3COO	98.3	112.79	94	108.16	83.8	—
$K_4[Fe(CN)_6]$	—	167.24	—	134.83	53	97.87

Electrolyte	*Concentrations*					
	0.001 *N*		0.01 *N*		0.1 *N*	
	18° C	25° C	18° C	25° C	18° C	25° C
$K_3[Fe(CN)_6]$	—	163.1	—	—	—	—
RbCl	130.3	—	125.3	—	113.9	—
CsCl	130.7	—	125.2	—	113.5	—
$CsNO_3$	127.6	—	121.3	—	—	—
NH_4OH	28	—	9.6	—	3.3	—
NH_4Cl	127.3	—	122.1	141.28	110.7	128.75
NH_4NO_3	124.25	—	118	—	106.6	—
$AgNO_3$	113.15	130.51	107.8	124.76	94.3	109.14
$CuSO_4$	98.5	115.26	71.7	83.12	43.8	50.58
$MgCl_2$	106.35	124.11	98.1	114.55	83.4	97.10
$MgSO_4$	99.8	—	76.2	—	49.7	—
Ag_2SO_4	—	135.7	—	119.9	—	—
$CaCl_2$	111.95	130.36	103.4	120.36	88.2	108.47
$Ca(NO_3)_2$	108.5	—	99.5	—	82.5	—
$CaSO_4$	104.3	—	77	—	—	—
$Ca(CH_3COO)_2$	79.6	—	71.9	—	54	—
$Ca_2[Fe(CN)_6]$	—	—	—	—	—	40.2
$SrCl_2$	114.5	130.33	105.4	115.54	90.2	102.19
$Sr(NO_3)_2$	108.3	—	99.0	—	80.9	—
$Sr(CH_3COO)_2$	80.1	—	72.8	—	56.7	—
$BaCl_2$	115.6	134.34	106.7	123.94	90.8	105.19
$Ba(OH)_2$	207.0	235.0	180.1	204.2	—	—
$Ba(NO_3)_2$	111.7	—	100.96	—	78.9	—
$BaBrO_3$	—	113.61	—	102.7	—	—
$Ba(CH_3COO)_2$	85	—	77.1	—	60.2	—
$ZnCl_2$	107	—	98	—	82	—
$ZnSO_4$	98.4	114.53	72.75	84.91	45.3	52.64
$CdCl_2$	—	—	83	—	50	—
$CdBr_2$	—	—	76.3	—	44.6	—
CdI_2	—	—	65.6	—	31.0	—
$Cd(NO_3)_2$	—	—	96	—	80.8	—
$CdSO_4$	97.7	—	70.3	—	42.2	—
$Pb(NO_3)_2$	116.1	—	103.5	—	77.3	—
TlF	113.3	—	105.4	—	92.6	—
TlCl	128.2	—	120.2	—	—	—
$TlNO_3$	124.7	—	118.4	—	101.2	—
Tl_2SO_4	—	147.8	—	130.0	—	96.0
$LaCl_3$	—	137.0	—	121.8	—	99.1
$La_2(SO_4)_3$	—	—	—	—	21.5	23.9
$CoSO_4$	—	112.7	—	82.78	—	51.12

In each second all the ions pass through the right section which are located in a cylindrical zone of area S and length w.

The number of ions present here is given by the product of the effective ionic concentration c_i[1] and the volume of this cylinder. For the cations this is $c_{+_i} S u_+ \frac{U}{l}$ and for the anions $c_{-_i} S u_- \frac{U}{l}$. If the ions have a valency z and the concentration is expressed in moles/cm³, (c_m) the quantity of electricity transported per second through the right section of the cylinder by each of the two ionic species becomes

$$I_{cat} = \mathbf{F} c_{+m} z_+ S u_+ \frac{U}{l} \tag{8}$$

and

$$I_{an} = \mathbf{F} c_{-m} z_- S u_- \frac{U}{l} \tag{8'}$$

The total current intensity I is the sum of these two quantities

$$I = \mathbf{F} \frac{U}{l} S(u_+ c_{+m} z_+ + u_- c_{-m} z_-)$$

Combining with Ohm's Law ($I = U/R$), gives

$$R = \frac{l}{S\mathbf{F}(u_+ c_{+m} z_+ + u_- c_{-m} z_-)}$$

Since, however,

$$R = \frac{l}{S} \varrho = \frac{l}{S} \frac{1}{\varkappa}$$

$$\varkappa = \mathbf{F}(u_+ c_{+m} z_+ + u_- c_{-m} z_-) \tag{9}$$

Expressing the concentration in equivalents rather than in moles per cm³ gives equation (10). The equivalent ionic concentration is equal to the equivalent concentration of the electrolyte multiplied by the degree of dissociation which in sufficiently dilute solutions is numerically equal to the coefficient of conductance; the concentration expressed in equivalents is the same for both cations and anions since the solution remains electrically neutral.

$$\varkappa = \mathbf{F} c_{eq} f_\Lambda (u_+ + u_-) \tag{10}$$

[1] The number of ions per cm³.

At infinitely low concentration, $f_\Lambda = 1$ and multiplying by the dilution φ gives

$$\varkappa\varphi = \Lambda_0 = \mathbf{F}c_{eq}\varphi(u_+ + u_-)$$

whence

$$\Lambda_0 = \mathbf{F}u_+ + \mathbf{F}u_- = \lambda_{0\,cat} + \lambda_{0\,an}$$

In other words the limiting conductance of an electrolyte is given by the sum of two additive constants: one for the cations ($\lambda_{0\,cat}$) and one for the anions ($\lambda_{0\,an}$); these are called the limiting equivalents ionic conductances[1] and are independent of each other (Kohlrausch's Law of the independent migration of ions).

Table II, 7 collects figures on the limiting equivalent ionic conductances of many inorganic and organic ions in aqueous solutions. It is interesting to note that for almost all ions the value lies around 50 with the exception of the H^+ and OH^- ions which have much higher values. Table II, 8 collects figures on the equivalent conductances of most common electrolytes under various conditions of concentration and temperature.

5. The Calculation and Measurement of the Rate of Migration

The equation

$$\Lambda_0 = \mathbf{F}(u_+ + u_-) \tag{1}$$

taken together with that derived from the study of transference numbers

$$t_+ = \frac{u_+}{u_+ + u_-} \tag{2}$$

forms a simultaneous equation system with two unknowns which when resolved allows the electrical mobilities u_+ and u_- to be deduced. Multiplying the numerator and denominator of the right-hand side of equation (2) by **F** gives

$$t_+ = \frac{\mathbf{F}u_+}{\mathbf{F}(u_+ + u_-)} \tag{3}$$

and replacing $\mathbf{F}(u_+ + u_-)$ with its value from (1) gives

$$t_+ = \frac{\mathbf{F}u_+}{\Lambda_0}$$

[1] This is often called the 'ionic mobility'.

and hence

$$u_+ = \frac{\Lambda_0}{\mathbf{F}} t_+$$

and similarly

$$u_- = \frac{\Lambda_0}{\mathbf{F}} t_-$$

Equation (3) gives at the same time the method of finding the transference number from conductance measurements

$$t_+ = \frac{\mathbf{F}u_+}{\mathbf{F}(u_+ + u_-)} = \frac{\lambda_{0\,cat}}{\Lambda_0}$$

Similarly

$$t_- = \frac{\mathbf{F}u_-}{\mathbf{F}(u_+ + u_-)} = \frac{\lambda_{0\,an}}{\Lambda_0}$$

The electrical mobilities u_+ and u_- may however also be measured directly by observing the movement of a boundary between two solutions of electrolytes placed in a vertical vessel, with electrodes at the top and bottom, to which an electric tension is applied. Under suitable experimental conditions it is possible to carry out the experiment so that the boundary remains sharp and easily recognised despite its movement.

Consider a very tall vertical cylinder of right cross-sectional area S cm^2 with electrodes at the top and bottom, to which an electric tension is applied. At a certain distance from the electrodes, the electrical field becomes homogeneous throughout all the rest of the cylinder. The lower part of the cylinder contains a binary uni-univalent electrolyte[1], which is assumed to be completely dissociated, of concentration c_1. Above this is placed a layer of another binary uni-univalent electrolyte with a common anion (*e.g.* potassium and sodium chlorides) at a concentration c_2 so that a clear boundary is formed between the solutions. When an electric tension is applied to the electrodes the cations migrate towards the cathode which is placed at the upper end of the cylinder. The cations of the upper layer have a higher electrical mobility than those of the lower. A convenient concentration of electrolytes must be chosen in order that the bound-

[1] When suitably generalized this argument is valid for any type of electrolyte; the example chosen for the calculation allows the essentials of the phenomenon to be illustrated more clearly and simply.

ary remains sharp and visible despite its constant movement towards the cathode. Under these conditions the two cations move at the same rate.

It is apparent that for two cations with different electrical mobilities to move at the same effective velocity, they must be subjected to different forces which compensate for their difference in electrical mobility. In other words, they are subjected to different electric tensions. This condition is readily achieved by a suitable choice of concentrations to make the conductances of the two layers different. Since the current intensity is constant at every point, the fall in potential must be different, and by Ohm's Law this is inversely proportional to the conductances of each of the two layers of electrolytes. The ratio of the concentrations of the two cations which is necessary for them to move with the same effective velocity may be determined as follows. Let t_{1+} be the transference number of the cation of the lower electrolyte and t_{2+} that of the upper electrolyte. From equation (8) on page 64 and the definition of transference number, and bearing in mind equation (6) on page 61, are obtained

$$t_{1+}I = S\mathbf{F}c_1w_1z_1 \tag{4}$$

and

$$t_{2+}I = S\mathbf{F}c_2w_2z_2 \tag{5}$$

where c_1 and c_2 are the concentrations of the two electrolytes, w_1 and w_2 are the effective velocities of the cations, and z_1 and z_2 are their respective valencies. But it has been postulated that $z_1 = z_2 = 1$, so that

$$I = \frac{\mathbf{F}c_1w_1S}{t_{1+}} = \frac{\mathbf{F}c_2w_2S}{t_{2+}}$$

Choosing concentrations such that $c_1/t_{1+} = c_2/t_{2+}$ gives $w_1 = w_2 = w$, *i.e.* the effective velocities of the two cations become equal. Knowing the conductivities and measuring the current intensity and the effective velocity of migration gives at once the values of u_{1+} and u_{2+}. In fact

$$w = u_+ \frac{U}{l}$$

In the very thin boundary layer of thickness dl, the expression U/l must be replaced by $\mathrm{d}U/\mathrm{d}l$. By Ohm's Law $\mathrm{d}U = I\mathrm{d}R$, and since $\mathrm{d}R = (1/\varkappa)(\mathrm{d}l/S)$,

$$\mathrm{d}U = (I\mathrm{d}l)/(\varkappa S)$$

whence

$$w = \frac{u_+ I}{\varkappa S}$$

i.e.

$$u_+ = \frac{\varkappa w S}{I}$$

The direct measurement of the electrical mobilities u_+ and u_- offers another experimental method of determining transference numbers.

For further details of the apparatus and techniques for this measurement see K. JELLINECK, *Lehrbuch der physikalischen Chemie*, Vol. III, p. 472 *et seq.*, F. Enke, Stuttgart, 1930. More recently J. CLÉRIN (*Ann. chim.*, [XI] 20 (1945) 244) used a new method based on the phenomena of electrolysis at the electrodes, which taken together are equivalent to an overall electrophoresis of the electrolyte. Under certain conditions it is possible to obtain a distinct migration front formed by one of the boundaries separating the electrolyte and the solvents. By measuring the time taken for such a front to pass from one electrical taster to another, and bearing in mind certain correction factors, it is possible to obtain the absolute velocities of the cations and the anions. Although, the results do not always agree with those obtained by classical methods, the technique is cited as it should be capable of further development.

TABLE II, 9

THE ELECTRICAL MOBILITIES OF CERTAIN IONS AT 18° C IN AQUEOUS SOLUTION
($cm \cdot sec^{-1} \cdot 10^{-5}$)

Cation	u_+	*Cation*	u_+	*Anion*	u_-	*Anion*	u_-
H^+	325	Zn^{2+}	48	OH^-	176	CrO_4^{2-}	74
Li^+	35	Cd^{2+}	49	F^-	48	$Cr_2O_7^{2-}$	47
Na^+	45	Al^{3+}	41	Cl^-	68	NO_3^-	64
K^+	67	Lu^{3+}	52	Br^-	70	PO_4^{3-}	49
Cs^+	71	Sm^{3+}	55	I^-	69	CO_3^{2-}	72
NH_4^+	68	Tl^+	68	ClO_3^-	57	Formate$^-$	48
Cu^{2+}	44	Pb_4^{2+}	70	ClO_4^-	67	Acetate$^-$	36
Ag^+	56	Th^{4+}	24	BrO_3^-	48	Chlor-acetate$^-$	28
Be^{2+}	29	Cr^{3+}	47	IO_3^-	35		
Mg^{2+}	48	Mn^{2+}	46	IO_4^-	49	Benzoate$^-$	27
Ca^{2+}	54	Fe^{2+}	47	MnO_4^-	55		
Sr^{2+}	54	Fe^{3+}	63	SO_4^{2-}	71		
Ba^{2+}	57	Co^{2+}	45	$S_2O_8^{2-}$	72		
Ra^{2+}	60	Ni^{2+}	46	SCN^-	59		

Table II,9 shows the electrical mobilities of some ions. Discrepancies occur between the values given by different authors which are due, in particular, to the marked experimental difficulties; but the figures do give some idea of the order of magnitude of electrical mobilities.

According to Kohlrausch the electrical mobilities u_+ and u_- should be characteristic for each ionic species and independent of the concentration and of the presence of other ionic species. In general, this has been confirmed, but only within certain limits; in practice the electrical mobility of an ion varies when different ions of opposite sign form the electrolyte with it. This can be seen in Table II, 10.

TABLE II, 10

THE ELECTRICAL MOBILITY OF THE K^+ ION IN SOLUTIONS OF VARIOUS SALTS at $c = 0.1\ N$

Salt	$t = 18°$ C	$t = 25°$ C
KCl	0.000563	0.000654
KBr	562	656
KI	564	652
$KClO_3$	549	631
$KBrO_3$	551	636
KNO_3	536	621
K_2SO_4	510	540

The values for the electrical mobility of the K^+ ion may be considered constant to a first approximation but more exact measurements show up the differences reported in Table II, 10, which can be interpreted by the theory of strong electrolytes (p. 79 *et seq.*).

6. The Dependence of the Conductance upon Experimental Conditions (Temperature, Viscosity, Pressure, Electrical Field and Frequency)

A rise in temperature always leads to an increase in the electrical mobility. The limiting equivalent conductance consequently shows the same behaviour which is characteristic of the conductors of the 2nd class. This occurs because the viscosity of the medium falls with increasing temperature so that the ions encounter less mechanical resistance to their movement through the medium under the action of the electrical field. The equivalent conductance of electrolytes at finite concentrations, however, may not show the same behaviour. This in fact depends upon the coefficients of conductance and the solvation which are in turn functions of the temperature. Their behaviour, especially with concentrated solutions, produces a maximum value for the equivalent conductance at high temperatures. This maximum shifts to increasingly high temperatures as the concentration tends towards zero, and finally disappears. Table II, 11 illustrates the formation of maxima of equivalent conductance, as a function of temperature, for certain electrolytes. The maximum value for the equivalent conductance is shown in bold type.

TABLE II, 11

THE DEPENDENCE OF THE EQUIVALENT CONDUCTANCE UPON TEMPERATURE

Electrolyte	*Concentration*	18° C	50° C	75° C	100° C	128° C	156° C	218° C	281° C	306° C
KCl	0.08 *N*	113.5	—	—	341.5	—	498	638	**723**	720
$AgNO_3$	0.08 *N*	96.5	—	—	294	—	432	552	**614**	604
$Ba(NO_3)_2$	0.08 *N*	81.6	—	—	257.5	—	372	**449**	430	—
$MgSO_4$	0.08 *N*	52	—	—	**136**	—	133	—	75.2	—
H_2SO_4	0.002 *N*	353.9	501.3	560.8	**571.0**	551	536	563*	—	637

* The tendency of the equivalent conductance of sulphuric acid towards a probable second maximum is related to the successive dissociation of two H^+ ions.

The limiting equivalent conductance thus depends essentially on the viscosity of the solvent, particularly for temperatures not far removed from room temperature. This has been shown by two series of experiments. The first was initiated by Kohlrausch and developed by Walden and his colleagues, and showed how with increasing temperature the values of the limiting equivalent conductance of each ion varied in inverse proportion to the variation of the viscosity of the solvent; thus, the value of $\Lambda_0\eta$ remained constant with changes in temperature. This is called Walden's Rule. Table II, 12 shows the constant value of the product $\lambda_0\eta$ for several ions in aqueous solution at various temperatures.

TABLE II, 12

THE CONSTANT VALUE OF $\lambda_0\eta$ AT VARIOUS TEMPERATURES IN AQUEOUS SOLUTIONS

ηH_2O		0° C	18° C	25° C	50° C	100° C	mean
		0.01792	0.01056	0.00894	0.0055	0.00284	
Li^+	λ	19.1	32.5	38.7	—	120	
	$\lambda\eta$	0.342	0.343	0.346	—	0.341	0.343
$N(C_2H_5)_4^+$	λ	16.2	28.1	33.3	53.4	103	
	$\lambda\eta$	0.290	0.296	0.298	0.294	0.293	0.294
Ba^{2+}	λ	33	54.4	63.7	104	200	
	$\lambda\eta$	0.591	0.574	0.569	0.572	0.568	0.574
OH^-	λ	105	174	200	—	(446)*	
	$\lambda\eta$	1.88	1.84	1.79	—	(1.27)	1.84
SO_4^{2-}	λ	41	68.6	(79.8)	132	256	
	$\lambda\eta$	0.73	0.724	(0.714)	0.726	0.727	0.724
CH_3COO^-	λ	20.4	35.0	40.87	66	129	
	$\lambda\eta$	0.367	0.369	0.365	0.363	0.366	0.366

* The values in parentheses are not quite certain.

The same rule is also valid for non-aqueous solvents; this is shown by Table II, 13 which collects together data on the conductance of various salts dissolved in benzonitrile. Similar tables could be constructed to show the constant value of $\lambda_0\eta$ for other solvents.

TABLE II, 13

THE CONSTANT VALUE OF $\lambda_0\eta$ AT VARIOUS TEMPERATURES IN BENZONITRILE

Electrolyte	0° C	10° C	20° C	25° C	30° C	40° C	50° C	60° C	70° C	*mean*
KI	0.62	0.63	0.63	—	0.64	0.64	0.63	0.62	0.60	0.626
NaI	0.59	0.59	0.59	—	0.59	0.59	0.59	0.57	0.55	0.584
LiI	0.57	0.57	0.57	—	0.57	0.57	0.57	0.55	—	0.567
LiBr	0.45	0.45	0.45	—	0.44	0.44	0.44	0.43	0.43	0.441
$AgNO_3$	0.64	0.65	0.65	—	0.65	0.65	0.64	0.62	—	0.643
$N(C_2H_5)_4I$	0.65	—	—	0.66	—	—	0.66	—	0.63	0.650

TABLE II, 14

THE CONSTANT VALUE OF $\Lambda_0\eta$ FOR SALTS IN VARIOUS SOLVENTS AT A CONSTANT TEMPERATURE OF 25° C

Electrolyte	$(CH_3)_4N$ Pi*	$(C_2H_5)_4N$ Pi	$(nC_3H_7)_4N$ Pi	$(C_5H_{11})_4N$ Pi	$(C_2H_5)_4N$ I	$(C_2H_5)_4N$ Cl	KI
Solvent	*Values of $\Lambda_0\eta$*						
Water	0.686	0.563	0.486	0.486	0.981	0.911	1.354
Methanol	0.627	0.593	—	—	0.678	0.683	0.626
Ethanol	0.585	0.564	—	—	0.586	0.636	0.559
Phenol (50° C)	0.590	0.562	0.487	0.487	0.631	0.661	0.539
Acetone	0.591	0.563	0.500	0.500	0.662	0.662	0.586
Methyl ethyl ketone	0.579	0.561	0.503	0.503	0.620	0.636	0.580
Acetonitrile	0.586	0.563	0.501	0.501	0.643	0.655	0.642
Ethylcyanoacetic ester	—	—	—	—	0.646	—	0.628
Benzonitrile	—	—	—	—	0.659	—	0.646
o-Toluolnitrile	—	—	—	—	0.650	—	0.645
Ethylene chloride	0.587	0.563	0.489	0.489	0.604	0.639	—
Nitromethane	0.604	0.586	0.515	0.515	0.698	0.698	0.765
Nitrobenzene	—	0.598	—	—	0.673	0.671	—
Pyridine	0.666	0.635	0.652	0.652	0.760	0.756	—

* Pi indicates the picrate ion.

The second series of experiments which were initiated and developed particularly by Walden and his colleagues also, concern the effect of the viscosity of various solvents at constant temperature on $\Lambda_0\eta$. Walden observed initially that with tetraethylammonium iodide dissolved in some 30 solvents, the value of the product of the limiting conductance in any particular solvent and the viscosity of that solvent was constant at constant temperature. In other words, at constant temperature the limiting conductance is inversely proportional to the viscosity of the solvent which is in fact a consequence of Stokes' Law. This relationship is illustrated by Table II, 14 which contains some more recent data[1].

It may be concluded from these two groups of experiments that for the motion of ions travelling within a solvent, the resistance appears as if it

[1] Further details on conductance in non-aqueous solutions are given particularly by P. Walden, *Elektrochemie nichtwässeriger Lösungen*, J. A. Barth, Leipzig, (1924) and recently by L. Fischer, G. Winkler and G. Jander, *Z. Elektrochem.*, 62 (1958) 1.

were due to the friction amongst the particles of solvent themselves rather than to the friction between the ions and the particles of solvent. This can be interpreted as due to the formation of one or more layers of solvent molecules completely enveloping the ions and bound about them by definite forces of an electrostatic nature. This phenomenon is called *solvation* in general, and *hydration* in the particular case of aqueous solutions. It is also a function of the concentration and the temperature and this partially explains why it is that non-ideally dilute solutions, where the solvation state may vary as a function of the temperature, do not show a simple monotonic increase of the equivalent conductance but rather the formation of a maximum.

In fact, either a rise in temperature or an increase in concentration can cause a diminution of the solvation, which may consequently lead to a diminution of the degree of dissociation, *i.e.* of the number of ions present in solution, and hence of the equivalent conductance at that volume. The diminution in the degree of dissociation, which is comprised in the value of the coefficient of conductance f_{Λ}, may be insufficient to compensate for the fall in viscosity with a rise in temperature alone. This would then lead to the formation of a maximum. To this must be added the fact that the dielectric constant of the solvent, and hence its ionizing power (p. 79 *et seq.*), is a function of both the concentration and the temperature and contributes to the formation of the maximum of conductance[1]. Moreover, it must be borne in mind that the limiting equivalent conductances of inorganic ions, apart from the H^+ and OH^- ions, always lie about 50, notwithstanding their varying sizes and charges. This confirms the existence of a hydration shell which is greater for the small ions and increasingly smaller for the larger ones. This phenomenon can also be interpreted by the nature of the forces which cause the solvation. Solvents for electrolytes are always substances with a high dielectric constant, *i.e.* having molecules with a marked dipole moment. It is apparent that solvation is brought about by the electrostatic attraction of the dipoles of the solvent molecules. The modern viewpoint considers the electrical charge of an ion to be distributed uniformly over its surface. However, in calculations of the forces of attraction and repulsion the charge may be considered as localized at the centre of the ion. The bigger an ion is, the greater is the distance between its centre and its surface which is the nearest point to

[1] See U. TESEI, *Gazz. chim. ital.*, 71 (1941) 351.

which a solvent molecule may approach. The two forces, of attraction (ion-dipole charge of opposite sign) and of repulsion (ion-dipole charge of the same sign) diminish, to a first approximation, as the square of the distance. The former is, however, always of a higher absolute value than the latter since the dipole spontaneously orientates itself so that the end with the charge of opposite sign to that of the ion, is closest to it. Thus, with ions of equal charge, the difference between these two forces, *i.e.* the resultant attractive force, diminishes and tends towards zero as the dimensions of the ions increase and hence the solvation tends to diminish in the same manner.

Various methods exist for the determination of the solvation of ions, both experimentally and by thermodynamic calculations. For example, Nernst's technique involves adding an inactive substance to the electrolyte during transference measurements and determining its change in concentration in the anodic and cathodic regions after the passage of the current. Remy separated the anodic compartment from the cathodic by a diaphragm and measured the quantity of water carried through this. Again, from measurements of the limiting ionic conductance the actual radius of the ions moving within the solvent may be calculated by Stokes' Law and then compared with the ionic radii determined by X-ray measurements on crystals. However, marked discrepancies still exist in determinations of the number of solvent molecules bound by each ion[1].

The concept of solvation satisfactorily explains many of the deviations from Walden's Rule that $\lambda_0 \eta$ is a constant.

This relationship is not completely true. Deviations are found especially with ions of small radius dissolved in water. Walden's Rule must be considered as a limiting relationship valid particularly for ions of large dimensions. The normal behaviour of ions of large dimensions is based on Stokes' Law (*cf.* 2 of this chapter) on the movement of a sphere in a fluid medium. However, as the dimensions of the moving body approach atomic size and thus become further removed from macroscopic dimensions, it becomes more probable that the ionic dimensions will not remain constant in passing from one solvent to another. In this way the basic assumption of Stokes' Law is lost. In fact, the ions of smaller size show the greater deviations and these ions are known to have a stronger tend-

[1] See J. O'M. Bockris, *Quart. Revs. (London)*, 3 (1949) 173; G. Journet and J. Vadon, *Bull. soc. chim. France*, (1955) 593; J. O'M. Bockris and B. E. Conway, *Modern Aspects of Electrochemistry*, Butterworth, London, 1954, Chap. II.

ency to solvation; this also depends on the nature of the solvent. Thus, the dimensions of the ions may actually change in passing from one solvent to another. The bigger the ion and the smaller its tendency to solvation, the more constant will be its dimensions and the value of $\lambda_0\eta$ will approach a constant[1].

It must further be emphasized that the most notable variations from Walden's Rule are to be observed when the solvent is water. This may be due to an inhomogeneity of the solvent, in the sense that it contains polymers of various size. These polymers have been demonstrated by X-ray measurements and the determination of the absorption spectrum of water under various experimental conditions. It is thus necessary to distinguish between a macroviscosity which acts according to Stokes' Law on the motion of spheres of dimensions large compared to those of of the polymers, and a microviscosity[2] which varies within certain limits from point to point within the solvent and acts on the motion of spheres of smaller dimensions than those of the polymers. It will readily be appreciated that small ions may move freely in the midst of large aggregates of water molecules since they are not subjected to the macroviscosity which is greater than the microviscosity. This explains the observed deviations from Walden's Rule and the abnormal behaviour of some aqueous solutions with respect to temperature. In some cases Walden's Rule may be modified to an expression of the type

$$\Lambda_0\eta^s = constant$$

where s is less than 1[3].

The conductance also depends upon the pressure, the intensity of the electrical field and the frequency. The effects of these are shown only when they reach very high values and reference may be made to specialized texts for a detailed treatment[4].

It may merely be mentioned qualitatively that with pressures of the order of 100 atmospheres there is a rise in specific conductance due to the

[1] See E. DARMOIS, *J. chim. phys.*, 43 (1946) 1.

[2] A. SPERNOL and K. WIRTZ, *Z. Naturforsch.*, 8a (1953) 522; A. GIERER and K. WIRTZ, *Z. Naturforsch.*, 8a (1953) 532.

[3] B. B. OWEN and G. W. WATERS, *J. Am. Chem. Soc.*, 60 (1938) 2377.

[4] EUCKEN–WOLFF, *Hand- und Jahrbuch der chemischen Physik*, Vol. VI, Akademische Verlagsgesellschaft, Leipzig (1933); WIEN–HARMS, *Handbuch der Experimentalphysik*, Vol. XXI, Akademische Verlagsgesellschaft, Leipzig (1932); H. FALKENHAGEN, *Elektrolyte*, S. Hirzel, Leipzig, 2*nd* Edition, 1953; H. S. HARNED and B. B. OWEN, *The Physical Chemistry of Electrolytic Solutions*, Reinhold, New York, 3*rd* Edition, 1958.

increase in concentration caused by the compression of the solution, the increase in the degree of dissociation (with weak electrolytes), a change in the coefficient of conductance caused by variations of the interionic forces (*cf.* 8 of this chapter), and a change in the limiting equivalent conductance, which is inversely proportional to the change in the viscosity of the solvent so long as Stokes' Law retains its validity.

When the field intensity rises from the value of the order of volts $\cdot$ cm^{-1}, which it has during normal measurements of conductance, to values of the order of 10^5 volts $\cdot$ cm^{-1}, the conductance increases when other conditions remain constant. This is called the Wien effect. The conductance similaıly increases when the frequency of the alternating current rises from values of the order of 10^3 sec^{-1}, which it has in normal measurements, to those of the order of 10^6 sec^{-1} or more: the Debye-Falkenhagen effect. The explanation of these two effects is given in 8 of this chapter (page 79).

7. Conductance in Molten Electrolytes

The conductance of molten electrolytes is measured in a similar way to that described for solutions. Here too the specific conductance $\varkappa$, the equivalent conductance Λ, and the molar conductance μ, are distinguished. The specific conductance of a molten electrolyte is defined in a similar mode to that of a solution. It varies with the temperature according to the equation

$$\varkappa = a + b \cdot 10^{-2}(t - t_1)$$

Table II, 15 shows the values of the specific conductances of some molten halides, together with the melting points and the molar volumes at this temperature.

The molar and equivalent conductances are given by $\varkappa/c_{mol}$ and $\varkappa/c_{eq}$ respectively, where c_{mol} represents the number of moles and c_{eq} the number of equivalents, contained in 1 cm^3. This is equivalent to multiplying the specific conductance by the molar volume or the equivalent volume respectively, *i.e.* by the volumes, expressed in cm^3, which contain one mole or one equivalent of electrolyte.

It is interesting to compare the conductances of the chlorides of the main groups of the periodic system at temperatures just above their melting points. The relevant values are shown in Table II, 16.

The numbers are the equivalent conductances. For some compounds

TABLE II, 15

CONDUCTIVITIES OF MOLTEN HALIDES

$$\varkappa = a + b \cdot 10^{-2}(t - t_1)$$

Electrolyte	*a*	*b*	t_1 (°C)	*m.p.* (°C)	*Molar volume*
LiF	20.3	100.0	905	870	—
LiCl	7.59	1.0	780	613	28.3
NaF	3.15	8.3	1000	992	—
NaCl	3.66	2.2	850	801	37.7
KF	4.14	4.5	860	856	—
KCl	2.19	2.1	800	776	48.8
KBr	1.66	2.0	760	730	—
KI	1.35	2.3	710	680	—
RbCl	1.49	2.1	733	715	53.7
CsCl	1.14	2.0	660	646	59.9
Cu_2Cl_2	3.27	2.45	430	422	26.9
Cu_2I_2	1.82	1.78	605	605	—
AgCl	4.44	1.84	600	457.5	29.6
AgBr	3.39	1.70	600	434	—
AgI	2.17	0.61	600	552	—
$BeCl_2$	0.0032	26	451	440	52.7
$MgCl_2$	1.05	1.7	729	708	56.6
$CaCl_2$	1.99	3.5	795	772	60
$SrCl_2$	1.98	2.9	900	873	58.7
$BaCl_2$	1.71	3.0	—*	962	66.3
$ZnCl_2$	0.051	1.5	460	513	53.8
$CdCl_2$	1.93	2.0	576	568	54.8
$CdBr_2$	1.06	2.0	571	567	—
CdI_2	0.19	2.1	389	388	—
Hg_2Cl_2	1.0	1.8	529	525	58.1
$HgCl_2$	0.00052	0.0005	294	276	—
$AlCl_3$	$0.56 \cdot 10^{-6}$	—	—	190**	101
$ScCl_3$	0.56	2.8	959	939	91
YCl_3	0.40	2.0	714	680	77.5
$LaCl_3$	1.14	3.3	868	860	77.8
$InCl_3$	0.42	9.0	594	586	103
TlCl	1.17	3.5	450	430	—
$ThCl_4$	0.67	1.8	814	765	—
$SnCl_2$	0.89	5.7	263	246	—
$PbCl_2$	1.48	4.6	508	501	—
$BiCl_3$	0.44	1.4	266	230	—
$MoCl_5$	$1.8 \cdot 10^{-6}$	—	—	194	—
WCl_6	$1.9 \cdot 10^{-6}$	—	—	275	—
WCl_5	0.67	2.3	250	248	—
UCl_4	0.34	2.8	570	—	—
$TeCl_4$	0.12	1.1	236	244	—

* Error in original. ** At 2.5 atm.

TABLE II, 16

THE EQUIVALENT CONDUCTANCES OF SOME MOLTEN CHLORIDES AT THE MELTING POINT

HCl $\sim 10^{-6}$					
LiCl 166	$BeCl_2$ 0.086	BCl_3 0	CCl_4 0		
NaCl 133.5	$MgCl_2$ 28.8	$AlCl_3$ $15 \cdot 10^{-6}$	$SiCl_4$ 0	PCl_5 0	
KCl 103.5	$CaCl_2$ 51.9	$ScCl_3$ 15	$TiCl_4$ 0	VCl_5 0	
RbCl 78.2	$SrCl_2$ 55.7	YCl_3 9.5	$ZrCl_4$ —	$NbCl_5$ $\varkappa = 2 \cdot 10^{-7}$	$MoCl_5$ $\varkappa = 1.8 \cdot 10^{-6}$
CsCl 66.7	$BaCl_2$ 64.6	$LaCl_3$ 29.0	$HfCl_4$ —	$TaCl_5$ $\varkappa = 3 \cdot 10^{-7}$	WCl_6 $\varkappa = 2 \cdot 10^{-6}$
			$ThCl_4$ 16		UCl_4 $\varkappa = 0.34$

whose equivalent volume is unknown, the specific conductances are shown. It can readily be seen that these chlorides may be divided into two groups: good conductors and insulators, and that these groups are separated by the stepwise line.

TABLE II, 17

THE EQUIVALENT CONDUCTANCES OF MOLTEN CHLORIDES OF ELEMENTS WITH VARIABLE VALENCY

Salt	Λ	*Salt*	Λ
Hg_2Cl_2	40	TlCl	46.5
$HgCl_2$	$2.5 \cdot 10^{-3}$	$TlCl_3$	$< 2.5 \cdot 10^{-3}$
InCl	130	$SnCl_2$	21.9
$InCl_2$	29	$SnCl_4$	0
$InCl_3$	17		
		$PbCl_2$	40.7
		$PbCl_4$	$< 2 \cdot 10^{-5}$

In general the values of the conductance decrease across each horizontal line with the increasing valency of the cation. This phenomenon is paralleled by the variations in conductance of the chlorides of metals with variable valency. The chloride with the greatest conductance is always that with the lowest valency, as is shown in Table II, 17.

It is interesting to note that those chlorides which are good conductors are only difficultly volatile, with high melting points, whilst the insulators are volatile at low temperatures and have low melting-points; some are in fact liquids at room temperature. According to Biltz, this is due to the conductors having an ionic crystal lattice whilst the insulators consist mainly of undissociated molecules. An ionic crystal lattice is maintained by the electrostatic forces between an ion and the surrounding ions of opposite charge. It is thus difficult to break the equilibrium of forces in such a system, *i.e.* to make it melt or sublime. In the molten state, however, it is highly probable that the electrolyte is in the main dissociated, since the ions already exist in the solid state. The conductance will thus naturally be high. Substances which show a molecular crystal lattice, or are liquid at room temperature, have this lattice maintained by forces which are much weaker than those maintaining an ionic one. Thus, even a relatively small increase of energy will overcome the forces and destroy the crystalline form. In passing into the liquid state the molecules remain, initially, largely undissociated as they are in the crystal, and the conductance is low.

Some chlorides exist, however, whose conductance is intermediate between that of good conductors and that of insulators. This implies that strong, medium and weak electrolytes exist even in the molten state (see page 92 *et seq.*).

8. The State of Electrolytes

The electrical conductance of electrolytes can be interpreted on the assumption that these are more or less dissociated into their ions. This was first suggested by Arrhenius to explain the observed facts and his hypothesis and its consequences were experimentally confirmed within the limits of accuracy available at that time, particularly with electrolytes of low conductance. However, it was shown to be invalid for a particular group of electrolytes, characterized by a very high conductance. Another theory had then to be developed to take account of these observations.

According to Arrhenius, an electrolyte in solution splits into its ions giving a true chemical equilibrium for the dissociation *e.g.*

$$CH_3COOH \rightleftharpoons CH_3COO^- + H^+$$

which shifts to the right with increasing dilution. The degree of dissociation can be determined by conductance measurements

$$\alpha = \frac{\Lambda_v}{\Lambda_0}$$

(p. 55) and since $\Lambda_0 = \mathbf{F}(u_+ + u_-)$

$$\alpha = \frac{\Lambda_v}{\mathbf{F}(u_+ + u_-)}$$

This relationship holds whenever the assumptions of a constant rate of migration of the ions and their independence of concentration and the presence of other ions, are true. The values of the degree of dissociation of various electrolytes at different concentrations, are distributed freely between 1 and very low values of the order of $10^{-2} - 10^{-3}$. Still lower values exist but are difficult to detect by conductance measurements and, as will appear, are of little interest to the present argument.

Those substances having a high degree of dissociation, of at least 0.5 even at high concentrations, are called *strong electrolytes;* whilst those whose conductance is very low even at high dilutions are called *weak electrolytes.*

The degree of dissociation is related to the temperature, depending on whether the dissociation process is endothermic or exothermic, according to the normal laws of chemical equilibria. It also depends on the nature of the solvent or, more exactly, on its dielectric constant. The dielectric constant ε of a medium is defined as the ratio of the forces F_0 and F with which two charges of opposite sign attract each other electrostatically when placed at the same distance apart in a vacuum, and in the medium, respectively.

$$\varepsilon = F_0/F \tag{1}$$

Thus, by definition, the dielectric constant of a vacuum is 1 and it is found that those of all other media are greater than 1. The dissociating power of a medium on which the degree of dissociation and hence the dissociation equilibrium depend, is also a function of the dielectric constant. This is made apparent by calculating the work necessary to dissociate the ions of a binary electrolyte[1] carrying the charges z_+ and z_-, placed at a distance l apart. This work is given by the product of the electrostatic force, holding the ions together, and the distance moved by one of the ions from its position in the molecule or crystal to infinity.

$$-w = \int_{x=l}^{\infty} F\mathrm{d}x \qquad (2)$$

Substituting the value of F given in (1) into (2) and introducing Coulomb's Law ($F_0 = (z_+z_-)/x^2$), gives

$$w = -\frac{1}{\varepsilon} z_+z_- \int_{x=l}^{\infty} \frac{1}{x^2}\mathrm{d}x = -\frac{1}{\varepsilon}\frac{z_+z_-}{l}$$

The final value is positive since the product z_+z_- is negative. With a high dielectric constant a very small electrostatic force exists between the ions. If the value of this force diminishes, the amount of work required to dissociate the molecule will fall and hence the degree of dissociation will rise, provided that other conditions remain constant. In fact, some parallelity is observed between the degree of dissociation and the dielectric constant. This is not, however, strict, in that many factors have been ignored *e.g.* screening effects, dipole moments etc., whose effects are not yet sufficiently well understood.

Table II, 18 shows this parallelity between the dielectric constant and the dissociating power[2] expressed as the percentage apparent dissociation measured with one electrolyte – $N(C_2H_5)_4I$ – throughout, at 25° C and 0.01 N concentration.

As mentioned before, however, the dielectric constant of the solvent is not the only factor which effects the dissociation. An equally important, if not decisive, factor is the solvation affinity of the ions, *i.e.* the change in free enthalpy with the interaction of the ions with the solvent molecules. Finally, there is the possibility of forming ionic pairs under the action of non-Coulombic forces.

Bearing in mind that the conductance is a function both of the number of ions present (which is affected by the degree of dissociation) and of their electrical mobilities (which are affected by the interionic forces, see

[1] A similar calculation may be applied to other types of electrolytes.
[2] See also A. Gemant, *J. Chem. Phys.*, 10 (1942) 723.

TABLE II, 18

THE DIELECTRIC CONSTANTS AND DISSOCIATING POWERS OF VARIOUS SOLVENTS

Solvent	*Dielectric constant*	*Temperature* °C	*Apparent percentage dissociation*
Formamide	109	20	93
Water	78.5	25	91
Succinonitrile	56.5	57.4	90
Citraconic anhydride	39.5	25	82
Nitromethane	38.5	30	78
Ethylene glycol	37.7	25	78
Acetonitrile	37.5	20	74
Nitrobenzene	34.8	25	71
Methanol	32.6	25	73
Benzonitrile	25.2	25	61
Epichlorhydrine	26	25	60
Ethanol	24.3	25	54
Acetone	20.7	25	50
Benzaldehyde	17.8	20	51
Acetyl bromide	16.2	20	47
Acetyl chloride	15.9	20	46

below), it is not difficult to account for the variations in the behaviour of weak and strong electrolytes. With weak electrolytes, where the degree of dissociation is low, the number of ions present is also low; their distance apart will be relatively large and thus interionic forces have little influence whilst the degree of dissociation is decisive. Hence to a first approximation, the behaviour of these electrolytes may be described as a function of the dissociation in accordance with Arrhenius' theory. Dissociation with strong electrolytes is virtually complete, the number of ions is large and the distance between them small. Thus, the effect of the interionic forces becomes dominant over that of the degree of dissociation, which shows little variation. The behaviour of such electrolytes may be described as a function of the interionic forces according to Debye and Hückel's theory of strong electrolytes.

(I) *Weak Electrolytes*

An equilibrium exists in the solution between the ions and the undissociated molecules, to which the Law of Mass Action must be applicable. In the case of acetic acid

$$CH_3COOH \rightleftharpoons CH_3COO^- + H^+$$

the concentrations of the H^+ and CH_3COO^- ions are equal and dependent on the degree of dissociation. They can, therefore, be represented by $\alpha \cdot c$ where c is the original concentration of the acetic acid. The concentration of the undissociated molecules will be $(1 - \alpha)c$. By the Law of Mass Action

$$K = \frac{\alpha^2 c^2}{(1 - \alpha)c} = \frac{\alpha^2 c}{1 - \alpha}$$

and replacing α by Λ_v/Λ_0 gives

$$K = \frac{(\Lambda_v^2/\Lambda_0^2)c}{1 - (\Lambda_v/\Lambda_0)} = \frac{\Lambda_v^2 c}{\Lambda_0(\Lambda_0 - \Lambda_v)}$$

The calculation is perfectly analogous for an electrolyte which yields p ions and gives

$$K = \frac{\Lambda_v^p \cdot c^{(p-1)}}{\Lambda_0^{(p-1)}(\Lambda_0 - \Lambda_v)}$$

This is Ostwald's Dilution Law which is valid for weak electrolytes, such as acetic acid, which have low equivalent conductances and degrees of dissociation even at high dilutions. The value of the equilibrium, or dissociation, constant determined conductometrically for this group of electrolytes remains virtually unchanged over the wide range of concentrations for which a sufficiently accurate determination can be made. Table II, 19 shows the actual constancy of the dissociation constant K especially when allowance is made for the change in viscosity which parallels the increase in concentration. The introduction of a viscosity correction factor improves the result to some extent, but whereas Arrhenius' theory leads to a linear relationship of the conductance to the concentration, at low concentrations, in fact the conductance is dependent upon the square root of the concentration.

Arrhenius' theory and the consequent Ostwald Dilution Law may be considered as limiting cases for infinitely dilute electrolytes. Here the

TABLE II,19

OSTWALD'S LAW $\left(K = \frac{\Lambda_v^2 c}{\Lambda_0(\Lambda_0 - \Lambda_v)}\right)$ FOR ACETIC ACID

Concentration	Λ_0	$100\frac{\Lambda_v}{\Lambda_0}$	$K \cdot 10^5$	$\frac{\eta_{sol}}{\eta_{H_2O}}$	$100 \cdot \frac{\Lambda_v}{\Lambda_0} \cdot \frac{\eta_{sol}}{\eta_{H_2O}}$	$K \cdot 10^{-5}$ *corr.*
0*	392	100	—	1.000	100	—
$4.94 \cdot 10^{-4}$	68.22	17.4	1.81	1.000	17.4	1.81
$9.88 \cdot 10^{-4}$	49.50	12.6	1.80	1.000	12.6	1.80
$1.98 \cdot 10^{-3}$	35.67	9.10	1.80	1.000	9.10	1.80
$3.95 \cdot 10^{-3}$	25.60	6.53	1.80	1.000	6.53	1.80
$7.91 \cdot 10^{-3}$	18.30	4.67	1.81	1.001	4.67	1.81
$1.58 \cdot 10^{-2}$	13.03	3.32	1.81	1.002	3.33	1.81
$3.16 \cdot 10^{-2}$	9.260	2.36	1.79	1.004	2.37	1.80
$6.32 \cdot 10^{-2}$	6.561	1.67	1.79	1.008	1.68	1.82
$1.265 \cdot 10^{-1}$	4.61	1.18	1.78	1.015	1.20	1.84
$2.529 \cdot 10^{-1}$	3.221	0.822	1.72	1.031	0.849	1.84
$5.06 \cdot 10^{-1}$	2.211	0.564	1.62	1.060	0.598	1.82
1.011	1.443	0.368	1.37	1.113	0.410	1.71

* Concentration tending towards zero.

electrical mobility may validly be considered as constant and independent of the concentration, which are the conditions necessary for calculating the degree of dissociation from conductance measurements. Here too the activities (*cf.* Chapter III) may be replaced by the ionic concentrations calculated from the degree of dissociation and are independent of the presence of other electrolytes. None of these conditions remain valid for effective ionic concentrations which are not infinitely small; Arrhenius' theory can only be considered as a first approximation for the description of the behaviour of solutions of electrolytes. The weaker the electrolyte, the more truly does it correspond to reality.

The degrees of dissociation of many weak acids and bases have been determined conductometrically allowing for these limitations. The values thus obtained correspond to the equilibrium constants defined by the laws of the classical theory of equilibria applied to ideal solutions. The discrepancies between the calculated and measured values become greater as the behaviour of the solutions differ from that of ideal solutions.

Using conductance measurements it is possible to follow the progressive dissociation of ternary electrolytes, particularly when they are weak, *e.g.* tartaric acid which dissociates in two stages,

$$C_4H_6O_6 \rightleftharpoons C_4H_5O_6^- + H^+$$
$$C_4H_5O_6^- \rightleftharpoons C_4H_4O_6^{2-} + H^+$$

Dissociation constants may be calculated for each of the two equilibria. The results of such studies and the methods employed to obtain more reliable values by means of successive approximations, which involve the theory of equilibria much more than electrochemistry, are fully described in physicochemical texts to which reference may be made[1].

Water alone will be briefly discussed here in view of its especial importance as a solvent in electrochemistry. Water is also a weak electrolyte, dissociating as

$$H_2O \rightleftharpoons H^+ + OH^-$$

The specific conductance of very pure water, determined as accurately as possible, is $\varkappa = 3.84 \cdot 10^{-8}$ at 18°. Since 1 cm^3 at 18° contains 0.05543 moles of H_2O, the equivalent conductance will be

$$\Lambda = \frac{\varkappa}{c} = \frac{3.84 \cdot 10^{-8}}{5.543 \cdot 10^{-2}} = 6.928 \cdot 10^{-7}$$

If water were completely dissociated, its equivalent conductance would be

$$\Lambda_0 = \lambda_{0_{H^+}} + \lambda_{0_{OH^-}} = 489$$

and the ratio of these two conductances gives the degree of dissociation

$$\frac{\Lambda_v}{\Lambda_0} = \frac{6.928 \cdot 10^{-7}}{489} = 1.417 \cdot 10^{-9} = \alpha$$

Hence the concentration of the H^+ ion is $55.43 \cdot 1.417 \cdot 10^{-9} = 0.785 \cdot 10^{-7}$ moles/litre at 18° and $1.049 \cdot 10^{-7}$ at 25°.

(II) *Strong Electrolytes*

The dissociation constants of strong electrolytes, calculated in this way, are not unaltered with changes in concentration, even at very high dilutions. This is apparent from the figures given in Table II,20 for potassium chloride.

[1] For example, S. GLASSTONE, *Textbook of Physical Chemistry*, Van Nostrand, New York, 1953, 2*nd* Edition.

TABLE II, 20

OSTWALD'S LAW $K = \frac{a^2c\ 1000}{1 - a}$ FOR KCl

$c \cdot 1000$	$a = \Lambda_v/\Lambda_0$	$K = \frac{a^2c\ 1000}{1 - a}$
0.001	0.980	0.048
0.01	0.943	0.156
0.1	0.864	0.549

This is a logical consequence since the solution of a strong electrolyte can in no way be considered as ideal. Hence, particularly with non-binary electrolytes, there can be no concordance between the values of the degree of dissociation calculated from conductance measurements and determined by other methods *e.g.* the lowering of the freezing point. Table II, 21 reports the degree of dissociation for lanthanum nitrate calculated from the lowering of the freezing point ΔT and from the conductance Λ_v.

Changes in the solubility product of difficultly soluble electrolytes are also found in the presence of electrolytes with no common ion. This has led to a quantitative elaboration of the influence of various other factors not considered in Arrhenius' theory. The results are collected in the theory of strong electrolytes or total dissociation (Debye and Hückel's theory). The detailed quantitative treatment[1] falls outside the scope of

TABLE II, 21

THE DEPENDENCE OF THE DEGREE OF DISSOCIATION OF $La(NO_3)_3$ ON CONCENTRATION

$c \cdot 1000$	*Degree of dissociation*	
	calculated from ΔT	*calculated from Λ_v*
0.001	0.946	0.920
0.01	0.865	0.788
0.1	0.715	0.635

[1] Which was first attempted by Milner in 1912.

this book but the assumptions and hypotheses on which were based the calculations – carried out principally by Debye and Hückel, by Onsager and by Bonino – will be mentioned. Various experimental facts strongly suggested that the degree of dissociation in solutions of strong electrolytes, determined according to the classical Arrhenius concepts from either conductance or cryoscopic measurements, does not correspond to the true degree of dissociation even when the values obtained by the two methods are virtually the same. In fact, it appears that the true degree of dissociation is very close to 1, *i.e.* that the dissociation is virtually complete. The following arguments support this.

(i) Strong electrolytes, almost without exception, have an ionic crystal lattice, *i.e.* no individual molecules are present, but only ions. The distribution of the ions is such that each is surrounded by others of opposite sign as indicated in Fig. II, 13. The crystalline form is maintained by coulombic electrostatic forces. A vacuum exists between each ion and its neighbours, *i.e.* a medium with a dielectric constant of 1. When the crystal dissolves, the vacuum between the ions is replaced by the solvent, *i.e.* a

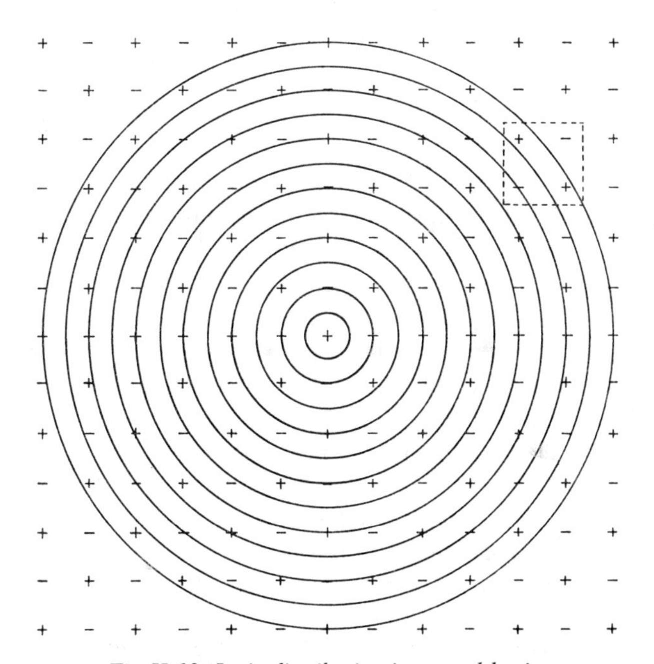

Fig. II, 13. Ionic distribution in crystal lattice

medium with a dielectric constant greater than 1. Thus the attractive force between ions of opposite charge is weakened. It is apparent that when the crystal dissolves it is very easy for the ions, which already exist in the crystal, to remain free in the solution without forming undissociated molecules.

(ii) Experiments based on the absorption of light by solutions of electrolytes have also led to the conclusion that the actual dissociation in solutions of strong electrolytes is virtually independent of the concentration and much higher than the values determined experimentally from either conductance or cryoscopic measurements.

The Bouguer–Lambert–Beer Law is valid for strong electrolytes, not only for low concentrations but often for very high ones. This law may be put in the form[1]

$$\log \frac{I_0}{I} = \varepsilon c l$$

where I_0 is the intensity of the incident light, I is the intensity of the transmitted light, ε is the molecular extinction coefficient (a characteristic constant for each wave length of incident light) of the absorbing substance, c is the concentration and l is the depth of the absorbing solution. This expresses the constancy of the absorbance ($\log I_0/I$) when the value of the product cl – *i.e.* the number of absorbing particles which the light traverses – remains constant. The individual values of the concentration and depth are not important so long as cl is constant.

A solution of a strong electrolyte contains ions derived from the dissociation of the electrolyte. If the concentration is varied but the value of cl is kept constant, there are no observable changes in the spectrum often until very high concentrations are reached. This implies that the total number of absorbing particles is unchanged; or, when both the ions and the undissociated molecules absorb light, that the numbers of both the ions and the undissociated molecules are unchanged. In other words, the degree of dissociation remains constant and independent of the concentration. Since the Bouguer–Lambert–Beer Law is a limiting relationship, especially valid for solutions tending towards zero concentration where the degree of dissociation tends towards 1, it may be concluded that strong electrolytes must in practice be completely dissociated in solutions

[1] Care should be taken not to confuse ε = the dielectric constant with ε = the molar extinction coefficient.

at all concentrations for which the Bouguer–Lambert–Beer Law remains valid. This is confirmed by the finding that with solutions of weak electrolytes, even small changes in concentration lead to marked changes in the absorption spectra even at very low concentrations.

Under controlled experimental conditions it is possible to determine the true degree of dissociation of an electrolyte from the absorption spectrum.

(iii) Measurements of the Raman spectra provide a third, highly significant, argument. Such studies have in certain cases disclosed undissociated molecules of electrolytes, *e.g.* with nitric or sulphuric acids, either pure or in the presence of traces of water. With ordinary solutions of strong electrolytes it is not possible to demonstrate the presence of undissociated electrolyte molecules [1].

The theory of strong electrolytes, according to which these electrolytes are always completely dissociated, has been based on these foundations. The following considerations underly this theory. The ions of opposite sign, present in the solution, attract each other electrostatically so that they do not exist in a state of perfect disorder; there is not an absence of reciprocal effects between ions and thus the characteristic nature of an ideal solution at infinite dilution is lacking. It can be assumed that each ion lies at the centre of an atmosphere, or cloud, of ions of opposite sign, which has statistically spherical symmetry (see Fig. II, 13). The radius of the sphere depends on the valency of the ions, the dilution, the temperature and the dielectric constant of the solvent. If an electric tension is applied the central ion will move towards one electrode whilst the ionic atmosphere will move in the opposite direction. Because of this the ionic atmosphere will tend to reform in front of the ion and to disappear behind it. The process of reforming the ionic atmosphere, however, requires a certain time – the relaxation time – so that a definite asymmetry of the ionic atmosphere is induced. It is less dense before and more dense behind the central ion and will thus produce a braking effect. Thus, the hypothesis of the constancy of the ionic mobilities u_+ and u_-, and their independence of the concentration and presence of other ions, is no longer valid.

The ionic atmosphere also exerts a braking effect by dragging along with it, in its migration, solvent molecules associated with its component

[1] This third argument is apparently against the theory of strong electrolytes, but it confirms the weakness of the Arrhenius' theory.

ions. This, so called, electrophoretic effect will oppose and restrict the movement of the central ion. Although there is no overall transfer of solvent from one point to another, due to this effect, it will nevertheless exert a braking action on each individual ion.

These effects are the reason for the diminution of the equivalent conductance of strong electrolytes with increasing concentration, despite their being completely dissociated; they also explain the dependence on all the ions present even if these are derived from other electrolytes. Deviations from the behaviour of ideal solutions with respect to osmotic pressure and its associated quantities (vapour pressure, lowering of the freezing point, etc.) may also be satisfactorily explained on the basis of these electrostatic forces.

This theory and its derived relationships are valid, however, only for very high dilutions. On increasing the concentration very marked deviations are found. The following summary of the assumptions underlying the Debye and Hückel theory will clarify the reasons for this.

(1) The electrolyte is completely dissociated into its ions.

(2) The ions are ideal points, which can not be deformed by polarization, and have a spherical field symmetry.

(3) Only coulombic forces exist (forces due to dipole moments, intermolecular forces etc., are not considered).

(4) The dielectric constant of the solution is independent of its concentration and equal to that of the solvent.

(5) Certain mathematical simplifications had to be introduced into the calculation.

The theory developed on these assumptions leads, for example, to a linear relationship between the conductance and the square root of the concentration (Onsager's equation). These assumptions tend to lose their validity as the concentration increases. It is necessary to consider the following points.

(1) Ions are not ideal points but have finite dimensions and the effects of the ionic size will be increasingly noted as the ions become closer together, *i.e.* as the concentration increases.

(2) Refractometric measurements show that the ions may be deformed by polarization.

(3) It is not permissible to consider only the coulombic electrostatic forces; other forces acting between the ions should also be included and their influence will increase at smaller distances.

(4) At high concentrations the dielectric constant of the solution is not equal to that of the solvent and will depend upon the concentration.

(5) The radius of the ionic atmosphere changes with the concentration and the solvation state of the ions may also vary.

(6) The number of molecules of solvent fixed by the solvation of each ion becomes greater with rising concentration, so increasing the true concentration still further.

(7) Certain simplifications of the mathematical calculation can no longer be permitted.

(8) The complete dissociation of an electrolyte into its ions can no longer be accepted (see argument (iii), p. 89).

Optical procedures are of considerable help here too; in fact, complete dissociation and the existence of only coulombic electrostatic forces can not explain the optical behaviour of solutions of strong electrolytes. Absorption spectrum measurements on solutions of strong electrolytes at concentrations where the Bouguer–Lambert–Beer Law is not valid indicate that there is probably an association of ions. This can be of two types. The first consists of 'associated ion pairs'[1] which differ from undissociated molecules in that the ions forming the pairs are solvated and located at the minimum possible distance apart at which repulsive forces begin to appear, whilst in undissociated molecules the ions are not solvated and are generally closer together. The second type consists of ionic swarms. Each case implies the formation either of partially dissociated molecules, *e.g.* $Co^{2+} + Cl^- = CoCl^+$, whose existence has been demonstrated experimentally through the observation of anomalous transference numbers (*cf.* page 29 *et seq.*), or of undissociated molecules in *chemical* equilibrium with their ions.

In non-aqueous solvents it is difficult to distinguish between associated ions and undissociated molecules. In each case, however, the difference exists between strong and weak electrolytes depending on the specific interactions of solvents and solute. The concepts and calculations of

[1] See for example C. A. Kraus, *J. Phys. Chem.*, 60 (1956) 129.

Debye and Hückel have been greatly developed and in some ways perfected by Bonino and his colleagues; reference may be made to their original papers[1].

The ratio $\Lambda_v/\Lambda_0 = f_\Lambda$ (the coefficient of conductance) expresses the ratio of the sum of the electrical mobilities of the cations and anions in the volume involved and in infinite dilution. It is thus a measure of the effect of interionic forces on the velocity of migration and hence on conductance.

The Wien and Debye–Falkenhagen effects (see 6 of this chapter) can also be satisfactorily interpreted by this theory. When the applied electrical field is so intense that the effective velocity of the ions reaches values of the order of decimetres or metres per second, each ion passes through the thickness of an ionic atmosphere (of the order of 10^{-8} cm) many times within the relaxation time (of the order of 10^{-9} sec). Thus, the migrating ion virtually loses its ionic atmosphere and is no longer subject to the braking action which this causes. The conductance will then rise. With weak electrolytes, where the effect of the interionic forces is small however, the high electrical field intensity brings about an increase in their dissociation and hence an increase in the concentration of ions. Once more the conductance will rise.

Similarly, when the frequency reaches a vibration period equal to, or less than, the relaxation time the braking effect due to the formation of an asymmetry in the ionic atmosphere will be diminished or destroyed. Here too an increase in conductance will occur.

(III) *Molten Electrolytes*

According to Lorenz molten electrolytes are also completely dissociated, for the following reasons.

(i) Firstly, molten electrolytes which are good conductors, give in the solid state an ionic crystal lattice as noted in 7 of this chapter, in which the ions are not bound into molecules as such. On increasing the thermal energy content of the crystal, by raising the temperature, the kinetic energy of each particle reaches, at the temperature of the melting

[1] G. B. Bonino *et al.*, *Mem. reale accad. Italia*, 4 (1933) 415, 445, 465; *Gazz. chim. ital.*, 78 (1948) 63; *Rend. accad. naz. Lincei* [VIII], 3 (1948) 442, 520.

point, a sufficiently high level to overcome the electrostatic forces which maintain the structure and thus the crystal melts. There is no reason to think that the ions will reunite at the melting point to form undissociated molecules. The molten weak electrolytes fall into a different category and in the solid state show layer lattice or even molecular lattice structures. Here then, it is highly probable that at the moment of fusion the substance still contains undissociated molecules in equilibrium with the ions.

(ii) When a small amount of a strong electrolyte is dissolved in another molten strong electrolyte, the laws of ideal dilute solutions are followed until relatively high concentrations, assuming a virtually complete dissociation. No conclusions on the degree of dissociation of weak electrolytes can be obtained from experiments of this type, however, since the dissociation may be partially compensated for by associations with solvent ions to form complex ions. For example, in a molten mixture of lead and potassium chlorides, complex ions are formed containing lead and these migrate towards the anode; it is very difficult to determine the position of the equilibrium between the simple ions, complex ions and any undissociated molecules.

(iii) It can also be deduced that dissociation is practically complete, from measurements of viscosity and conductance. It has been seen (*cf.* 2 of this chapter) that according to Stokes' Law, a sphere moving with constant velocity within a fluid must exert a force equal to the friction

$$F = 6\pi\eta r w$$

(η = the viscosity coefficient, r = the radius of the sphere and w = the velocity). If the sphere is a cation in a particular medium and if F is the force, equal in any medium, which acts on it when the potential gradient is 1 V cm^{-1}, then where r' = the radius, u_+' = the electrical mobility and η' = the coefficient of viscosity in aqueous solution and r'', u_+'' and η'' are the same quantities in the molten mixture

$$F = 6\pi\eta' r' u_+' = 6\pi\eta'' r'' u_+''$$

Thus

$$\frac{u_+'}{u_+''} = \frac{\eta'' r''}{\eta' r'}$$

and assuming, to a first approximation, that the ionic radius is the same in both the aqueous solution and in the molten mixture

$$\frac{u_+'}{u_+''} = \frac{\eta''}{\eta'}$$

Similarly, for anions

$$\frac{u_-'}{u_-''} = \frac{\eta''}{\eta'}$$

whence

$$u_+' = u_+'' \frac{\eta''}{\eta'}$$

and

$$u_-' = u_-'' \frac{\eta''}{\eta'}$$

Now, since $\Lambda_0 = \mathbf{F}(u_+ + u_-)$

$$\frac{\Lambda_0'}{\Lambda_0''} = \frac{u_+' + u_-'}{u_+'' + u_-''} = \frac{\eta''}{\eta'} \cdot \frac{u_+'' + u_-''}{u_+'' + u_-''} = \frac{\eta''}{\eta'}$$

i.e. Walden's rule.

In other words, it is possible, in this way, to calculate the equivalent conductance in the molten mass from the equivalent conductance at infinite dilution in aqueous solution, assuming complete dissociation, from the equation

$$\Lambda_0'' = \Lambda_0' \frac{\eta'}{\eta''}$$

TABLE II,22

COMPARISON OF THE MEASURED AND CALCULATED EQUIVALENT CONDUCTANCES FOR MOLTEN ELECTROLYTES

Substance	°C	Λ *measured*	η'/η''	Λ *calculated*
NaCl	850	39.10	0.361	39.31
NaCl	896	50.30	0.471	51.19
NaCl	924	59.43	0.577	62.71
$NaNO_3$	308	106.28	0.879	92.48
$NaNO_3$	368	110.63	1.044	109.83
$NaNO_3$	418	112.83	1.087	114.35

The equivalent conductance measured at various temperatures for sodium chloride and nitrate coincides reasonably well with the values obtained by calculation, which is evidence in favour of a strong dissociation for these two salts, as shown in Table II, 22.

The following objections may, however, be raised to this argument. Firstly, it is not permissible to consider the radius of an ion as the same in a solution, where the ion is solvated, and in a molten mass, where it can not be solvated. Secondly, if the validity of Stokes' Law is assumed then the value of $\Lambda_0\eta$ should be constant even in the molten state and this is not true for many inorganic electrolytes. The inconstancy of $\Lambda_0\eta$ with respect to temperature may be attributed in part to interionic forces of an electrostatic nature and in part to incomplete dissociation. At the present time, however, it is not possible to distinguish these two effects quantitatively.

CHAPTER III

GALVANIC CELLS

1. Introduction

Many chemical processes when suitably carried out, can give rise to a production of external electrical work and seem to have been in use, although unwittingly, some 2,500 years before the Christian era for the gold plating of women's necklaces[1]. The Daniell cell is a characteristic example. It consists of a copper electrode immersed in a solution of copper sulphate and a zinc electrode immersed in a solution of zinc sulphate. When the two solutions are brought into contact, without being allowed to mix, *e.g.* by means of a porous diaphragm, and the two metal electrodes are connected through an external circuit, containing a measuring instrument, a current will pass through this. Thus an electric tension has been set up between the two electrodes and gives rise to an electric current when the circuit is completed. At the same time as the current flows in the external circuit, certain chemical changes take place within the system: the weight of the copper electrode increases, the copper solution is diluted, the zinc electrode dissolves and the concentration of the zinc sulphate solution increases. In other words, the reaction

$$CuSO_4 + Zn \rightarrow Cu + ZnSO_4$$

occurs, which can be written in its ionic form as

$$Cu^{2+} + Zn \rightarrow Cu + Zn^{2+}$$

Other systems exist which can yield external electrical work from physico-chemical changes within a system, *e.g.* the passage of an ion from a more concentrated to a more dilute solution, changes in ionic charge, etc.

In general, a *galvanic cell* is defined as a series of conducting phases each in contact with the next, of which at least one is an electrolyte, and with the terminal phases excellent conductors (which are also called poles or electrodes) physically and chemically identical but not necessarily so from the point of view of internal electrical potential. In a narrow sense a cell is defined as a combination of this type, from which external electrical work can be

[1] *Cf.* H. Winkler, *Elektrie*, (1960) 2, 71.

obtained by means of physico-chemical changes in the system. The same system is called an electrolyser when the change in it involves the consumption of external electrical energy with a corresponding increase in its energy content (*cf.* Chapter I, 2, p. 3 *et seq.*). By convention a cell is symbolized by the series of phases of which it is composed, written with increasing numeration from left to right, *e.g.*

$$\underset{1}{Zn} \;/\; \underset{2}{Zn^{2+}} \;/\; \underset{3}{Cu^{2+}} \;/\; \underset{4}{Cu} \;/\; \underset{1'}{Zn} \qquad (1)$$

Its electric tension is equal to the internal electrical potential of the first electrode (1) minus the internal electrical potential of the second electrode (1′).

The two terminal phases which form the electrodes must be the same since it is not possible to measure, between internal points of different phases, any difference of potential and hence any galvani tension (*cf.* 4 of this chapter); but this is possible between two equal phases[1]. Cells may be reversible or irreversible. External electrical work may be obtained at the expense of the energy of the system (and sometimes of the environment) by the following electrochemical processes:

(a) The formation of ions from uncharged molecules or atoms and *vice versa*

$$Cu^+ + e^- \rightarrow Cu \quad {}^{2}$$

$$Cl_2 + 2\,e^- \rightarrow 2\,Cl^-$$

(b) A change in the charge on an ion

$$Fe^{3+} + e^- \rightarrow Fe^{2+}$$

(c) The formation of new ions by the conversion to the ionic state of neutral molecules formed by the decomposition of more complex ions and *vice versa.*

$$2\,MnO_4^- \rightarrow 2\,Mn^{2+} + 3\,O^{2-} + \tfrac{5}{2}\,O_2$$

$$\tfrac{5}{2}\,O_2 + 10\,e^- \rightarrow 5\,O^{2-}$$

(d) Changes in concentration at the individual electrodes without changes in the overall composition of the system; in such cases, the chemical

[1] See E. A. Guggenheim, *J. Phys. Chem.*, 33 (1929) 842; 34 (1930) 1540; J. W. Gibbs, *Collected Works*, Longmans, Green and Co., New York, (1949), p. 429.

[2] e^- represents the negative electrical charge, *i.e.* an electron.

reaction at one electrode occurs in the opposite sense and in equivalent concentration to that at the other.

(e) Electrokinetic phenomena (*cf.* Chapter XI, p. 395 *et seq.*).

However, for the free energy released in all these changes to be converted into external electrical work, it is essential in every case that the reactants do not mix but nevertheless remain in electrical and material contact with each other, through, for example, the solutions within a tube of narrow bore, through a porous diaphragm or by contact with an inert electrolyte.

The Energy of a Galvanic Cell and the Measurement of Electric Tension

The energy obtained from a galvanic cell is defined as the product of the quantity of electricity Q which circulates in the circuit, and the electric tension U existing between the electrodes, following the enumeration of the phases, *i.e.* it is QU. The measurement of the quantity of electricity is easy since each g equivalent of substance in the ionic state is linked with 96,500 C (1 F) so that it suffices to measure the quantity of substance transformed, expressed in equivalents, to obtain directly the amount of electricity ($Q = n\mathbf{F}$) involved in the transformation. By convention the quantity of electricity $n\mathbf{F}$ is given a positive sign when it moves within the cell from left to right, *i.e.* from phase 1 to phase 2 to phase 3 etc. (see diagram (1) on p. 97). The product $n\mathbf{F}$ is called the *reactional charge* when one stoichiometric unit of reaction, or *unit of advancement*, occurs within the cell whilst n is the number of electrical charges carried. The value of n will be positive if there is a movement of a positive charge within the cell in the sense of the phase enumeration and negative in the reverse case.

The system of conventions adopted leads to a convenient agreement between the thermodynamic and electrical signs which can be well illustrated by considering the Daniell cell. This can be represented as

$$\underset{1}{Zn} \ / \ \underset{2}{Zn^{2+}} \ / \ \underset{3}{Cu^{2+}} \ / \ \underset{4}{Cu} \ / \ \underset{1'}{Zn} \qquad (1)$$

The spontaneous reaction in such a cell is

$$Zn + Cu^{2+} \rightarrow Cu + Zn^{2+}$$

which is equivalent to the movement of a positive electrical charge within the cell from left to right, so that the reaction charge $n\mathbf{F}$ is positive.

Corresponding to the movement of charge within the cell, there will be a movement of positive current, in the classical sense, from phase Zn (1′) to phase Zn (1) through the external circuit. This means that the Zn (1′) electrode is at a more positive potential than the Zn (1) electrode. However, following the definition and convention adopted, the electric tension of this cell is negative so that $n\mathrm{F}U$ is also negative. This electrical work is thus usable work given up by the system to the environment. When the process is carried out reversibly, the work must be equal to the change in free enthalpy of the system and if the reaction is spontaneous $\Delta G < 0$, so that

$$\Delta G = n\mathrm{F}U$$

$$\frac{\Delta G}{n\mathrm{F}} = U = -\frac{A}{n\mathrm{F}} \quad (2)$$

since the affinity is equal, but of opposite sign, to the change in free enthalpy. Reversing the reaction but maintaining the phase order makes ΔG positive and A negative, but also changes the direction of movement of the positive electrical charge within the cell, so that the sign of $n\mathrm{F}$ also changes. As a result the signs of $\Delta G/n\mathrm{F}$ and $A/n\mathrm{F}$ are invariant once the phase order of the cell has been fixed; the former corresponds to the sign of the electric tension of the cell. If the electric tension U (which is unchanged in size and magnitude) is multiplied by the quantity of electricity – which is given a negative sign since the charge moves in the reverse direction within the system – the product $(-\ n\mathrm{F}U)$ assumes a positive value. This corresponds to an increase in the energy content of the system and thus to work performed by the environment on the system. The electric tension U, defined in (2), for a cell working reversibly, is called the reversible electric tension and is equal in absolute value, but of opposite sign, to the ratio of the chemical affinity of the reaction to the reactional charge. The electric tension of a galvanic cell may be measured experimentally or calculated thermodynamically.

The method of measurement most commonly used is that of compensation, first suggested by Poggendorf, and later perfected, in which the electric tension to be measured U_x is compared with a known tension U_n. Fig. III, 1 illustrates the principle of the method. A known electric tension U_n is applied to the ends of a calibrated homogeneous wire AB. This tension must remain constant for at least the time required to make the measurements and must be greater than that to be measured; it should

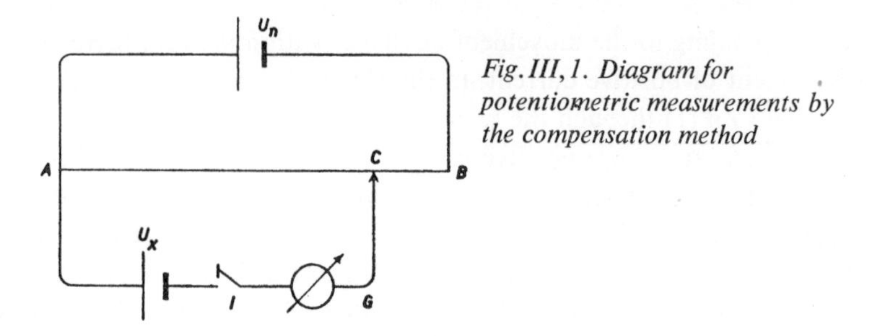

Fig. III, 1. Diagram for potentiometric measurements by the compensation method

preferably be the electric tension of an accumulator previously calibrated against a standard cell. The connections between the ends of the calibrated wire and the terminal of the source of the known tension, and the internal resistance of the cell of known electric tension, must both be negligible so that it can be assumed that the electric tension between points A and B is that of the calibrated accumulator, without any appreciable error. If the wire is homogeneous and uniformly calibrated, the electric tension will be uniformly divided along the wire AB so that across any length AC the tension will be proportional to this length. Similar considerations will apply to the relationship of the tension to the length CB.

The cell whose electric tension U_x is to be measured, is connected in opposition to the cell of known tension U_n with one pole connected to A and to the pole of the same sign of the latter. The other pole is connected through a key I to a zero instrument G and thence to the sliding contact C. The zero instrument may be a capillary electrometer or a sufficiently sensitive galvanometer. The contact is moved until no current passes in the circuit AU_xGC. This implies that the electric tension between points A and C, induced by the passage through the wire of the current supplied

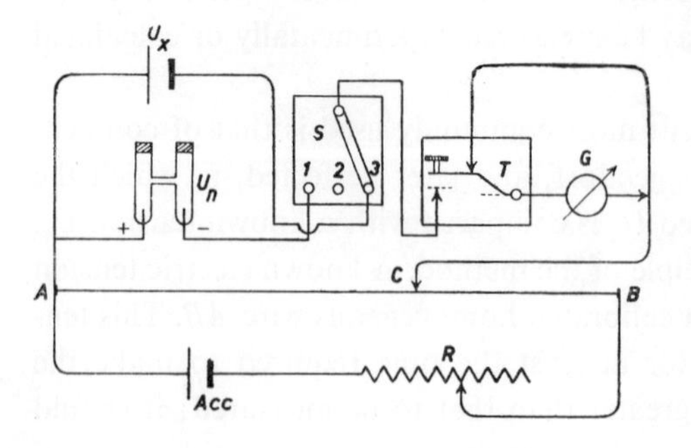

Fig. III, 2. Wiring for potentiometric measurements

by the cell U_n is equal and opposite to the electric tension of the cell U_x. Since the tension is proportional to the length of the wire

$$U_x = U_n \frac{AC}{AB}$$

Fig. III, 2 shows the type of circuit used for measuring electric tension with an accumulator *Acc* (which has been calibrated against the standard cell U_n) as the source of tension between A and B, and a galvanometer G as the zero instrument. It is possible to measure electric tensions greater than that of the standard cell by using an accumulator which itself has a greater tension. The measurement is made by placing the switch S in position 1, to put the standard cell in opposition to the accumulator *Acc*, and then moving the sliding contact C until the galvanometer shows that there is no current when the key T is closed.

Then the electric tension U between A and B is given by

$$U = U_n \frac{AB}{AC}$$

The switch S is then moved to position 3, to exclude the standard cell and insert the cell whose electric tension is to be measured. The sliding contact C is once more brought to an equilibrium position C'. The unknown electric tension U_x is thus

$$U_x = U \frac{AC'}{AB} = U_n \frac{AB}{AC} \cdot \frac{AC'}{AB} = U_n \frac{AC'}{AC}$$

The resistance R may be used to reduce the electric tension of the accumulator to a value which is more suitable for the measurement, should this be necessary.

More precise measurements may be made by this method, by replacing the wire by one or two accurately calibrated decade resistances of at least 1,000 or preferably 10,000 Ω total resistance, which can be exactly adjusted to within 1 Ω of any resistance. The circuit is shown in Fig. III, 3. The electric tension of the accumulator is applied across the resistances connected in series. Initially, the whole of resistance R_2 is inserted and the whole of resistance R_1 excluded. The switch is placed in position 1 to place the standard cell in the circuit. The resistance R_1 is gradually inserted whilst the resistance R_2 is excluded to the same amount, so that the total resistance of the accumulator circuit remains constant at the

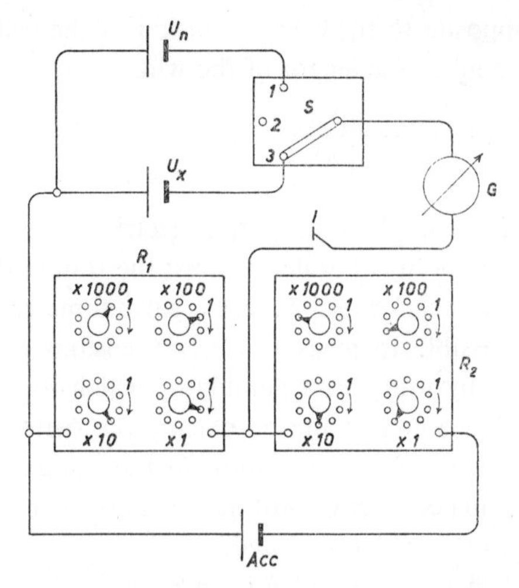

Fig. III,3. Wiring for precision potentiometric measurements

value of 1,000 or 10,000 Ω. If the resistances are 10,000 Ω each, the electric tension of the accumulator at the equilibrium position is given by

$$U_{Acc} = U_n \frac{10{,}000}{a}$$

where a is the value of the resistance inserted in the box R_1. The switch is moved to position 3 and a new value b is found for the resistance R_1, with the unknown electric tension U_x, which is given by

$$U_x = U_{Acc} \frac{b}{10{,}000} = U_n \frac{10{,}000}{a} \cdot \frac{b}{10{,}000} = U_n \frac{b}{a}$$

Both the calibrated wire and the resistance box types of apparatus are called *potentiometers*.

A standard cell is used as the source of known electric tension. Its electric tension and its variation with temperature are known exactly. These cells are not usually employed directly in opposition to the cell whose electric tension is to be measured, because when they supply current they become strongly polarized (*cf.* Chapter X) and their electric tension varies[1]. In general they are used to calibrate the electric tension (supplied by an accumulator) between A and B, or brought to the terminals of the decade boxes. The following standard cells are most frequently used.

[1] For the correct use of standard cells, particularly for measurements of very high precision, see H. H. UHLIG, *J. Electrochem. Soc.*, 100 (1953) 173; L. HARTSHORN and F. A. MANNING, *J. Sci. Instr.*, 31 (1954) 115.

(1) *The international Weston cell*

+ Cu	Hg	Hg_2SO_4 solid	$CdSO_4$ saturated solution	$CdSO_4 \cdot \frac{8}{3} H_2O$ solid	Cd amalgam 12.5% Cd	Hg −
1						1′

The electric tension of this cell in the direction of the phases from left to right (1 . . . . 1′) is given by the equation

$$U = 1.01830 - 4.075 \cdot 10^{-5}\,(t-20) - 9.444 \cdot 10^{-7}\,(t-20)^2 + 9.8 \cdot 10^{-9}\,(t-20)^3 \text{ V}$$

(2) *The standard Weston cell*

+ Cu	Hg	Hg_2SO_4 solid	$CdSO_4$ saturated solution at 4° C	Cd amalgam 12.5% Cd	Cu −
1					1′

The electric tension of this cell is virtually constant between 10° and 30° C at 1.01875 V.

(3) *The standard Clark cell*

+ Cu	Hg	Hg_2SO_4 solid	$ZnSO_4$ saturated solution	$ZnSO_4 \cdot 7\,H_2O$ solid	Zn amalgam 10% Zn	Cu −
1						1′

The electric tension of this cell is given by the equation

$$U = 1.4325 - 1.119 \cdot 10^{-3}\,(t - 15) - 7 \cdot 10^{-6}\,(t - 15)^2 \text{ V}$$

Another arrangement for measuring electric tensions without actually drawing any current from the cell is the electronic voltmeter. The elec-

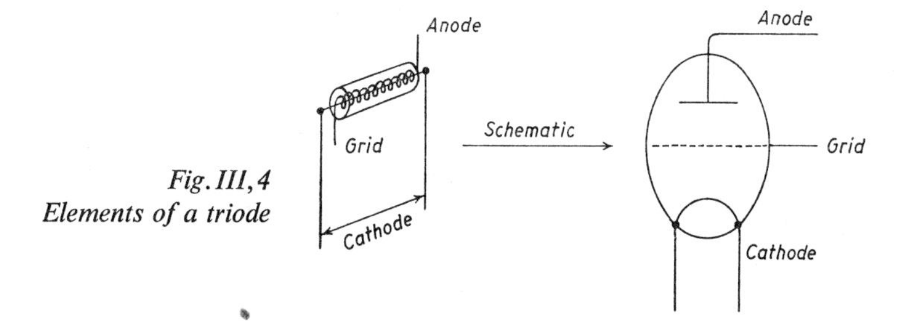

Fig. III, 4
Elements of a triode

tronic voltmeter allows the electric tension to be measured using a three electrode valve (a triode) without requiring a calibrated wire or resistance. Only a brief account will be given of this and reference should be made to a specialized text for the details of the technique of the electronic measurements. A simple triode (Fig. III,4) consists of a cathode (usually a metallic filament coated with a layer of alkaline earth metal oxide) around which is a second electrode, usually coiled into a spiral, which is called the grid; the cathode and grid in turn are within a metallic cylinder which forms the third electrode and is called the anode. The whole assembly is enclosed in a glass envelope under a very high vacuum. When the cathode (Fig. III,5) is heated by means of the heating battery *BR* to a sufficient temperature, it emits electrons. When a sufficiently high positive potential, with respect to the cathode, is applied to the anode by means of the high tension battery *BP*, the electrons are attracted towards it and flow through the circuit: anode → high tension battery → galvanometer → cathode. The galvanometer *G* then shows a certain current intensity which is called the anode current. If the grid is subjected to a variable potential which is negative with respect to that of the cathode, by means of the grid battery *BG* and the resistance *R*, the electrons emitted from the cathode are blocked in their passage towards the anode and the anode current diminishes.

If the grid potentials, referred to the cathode potential considered as zero, are plotted as the abscissae on a graph against the corresponding

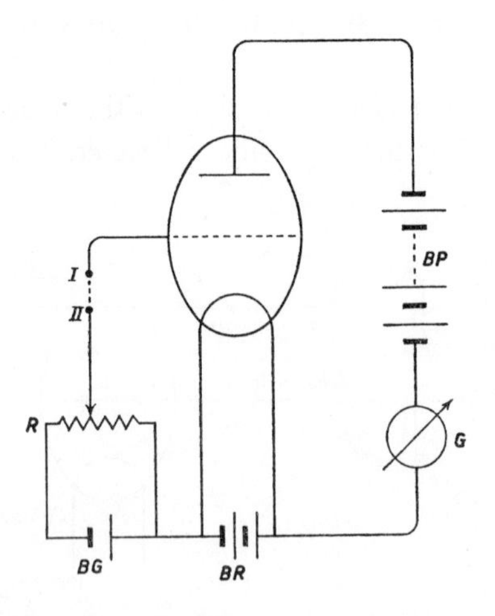

Fig. III, 5. Principle wiring of a valve voltmeter

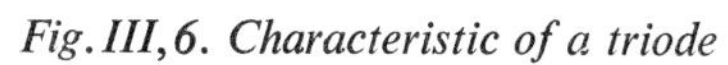
Fig. III, 6. Characteristic of a triode

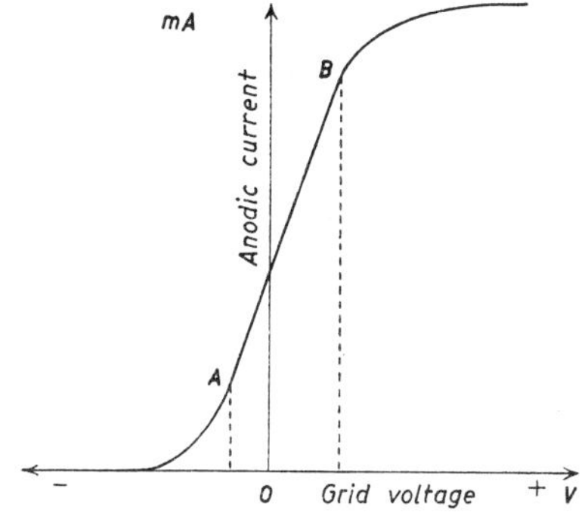

anode current intensity as ordinates (Fig. III, 6) a characteristic curve is obtained. The shape of the curve depends on the constructional characteristics of the valve, on the anode tension and on the cathode temperature *i.e.* indirectly on the heating battery tension. This graph is called the characteristic curve of the electronic valve. It always contains a straight line portion *AB*, where the change in the anode current is proportional to the change in the grid potential. This region of the curve may be used for the measurement of electric tension. The value of the tension U_x may be calculated graphically from the intensities of the anode current and the characteristic curve of the valve. The tube is checked from time to time by measurements of known electric tensions.

Electronic voltmeters have been much developed in recent years and at the present time many circuits exist, including bridge arrangements, for precise measurements. Reference may be made to specialized texts for details of these.

Very high precision measurements of electric tensions greater than about 0.5 V are made with an accuracy of up to 0.05%, using electrostatic electrometer techniques. The electrometers are advantageous not only for their precision, but also because they do not consume current and thus show the true electric tension. With such instruments it is possible under suitable conditions to measure tensions of as little as 10^{-5} V, although the precision also falls with the decreasing tension.

Every day measurements are made, more approximately, with voltmeters, which are direct reading instruments based on the motor action of an electric current, and are generally scaled directly in volts. For these instruments to work it is necessary that a certain current should pass through them so that they are inconvenient for measurements of the so-called clamp tension V. The difference between the equilibrium and clamp electric tensions is zero with a broken circuit but may become marked with a closed circuit, *i.e.* when the cell is supplying current. Ohm's Law applied to a closed circuit gives the equation

$$I = \frac{U}{R_i + R_e} = \frac{V}{R_e}$$

where I = the current intensity; R_i = the internal resistance; R_e = the external resistance. This leads to

$$V = U \frac{R_e}{R_i + R_e}$$

The clamp tension more closely approaches the tension of the cell as the internal resistance R_i of the cell becomes small in comparison with the external resistance R_e of the circuit.

Measurements with instruments which consume current have the further inconvenience of being liable to errors due to the polarization effect (*cf.* Chapter X, p. 634).

3. The Dependence of the Electric Tension of a Galvanic Cell on Temperature, Concentration and Pressure

A galvanic cell enables the variations in free enthalpy of a chemical or physico-chemical change in a system to be used as external electrical work. Thus, the cell changes into external electrical work not only internal energy of the system but possibly a certain amount of environmental thermal energy. It must therefore be possible to calculate thermodynamically the electric tension of a cell, whenever the process can be carried out reversibly.

In such a case the change in free enthalpy ΔG is independent of the route followed to achieve the change and if this is carried out within a cell, the change in free enthalpy – which is equal to the usable external work – becomes equal to the electrical work $n\mathrm{F}U$ which can be obtained from the circuit. It is necessary to bear in mind the signs of n and U, which follow the convention adopted, the phase enumeration and the direction of movement of the positive charge within the cell. Thus, the electric tension of a cell can be calculated thermodynamically[1] provided that the conditions under which the change occurs are rigorously established. By analogy with practical cases, it is assumed that the reaction takes place at constant pressure. The thermodynamic relationship which links the heat of reaction at constant pressure (the change in enthalpy ΔH) and the usable external electrical work (the change in free enthalpy ΔG) is the Gibbs–Helmholtz equation.

$$\Delta G = \Delta H + T\left(\frac{\partial(\Delta G)}{\partial T}\right)_p \tag{1}$$

The more generalized form of the Gibbs–Helmholtz equation may be applied to a fuller treatment with simultaneous changes of temperature and concentration

$$\mathrm{d}(\Delta G) = -\,\Delta S\mathrm{d}T + \Sigma_i\left(\frac{\partial \Delta G}{\partial c_i}\right)_{T,\,c_j}\frac{\partial c_i}{\partial T}\mathrm{d}T$$

See, for example, F.G.Brickwedde and L.H.Brickwedde, *Phys. Rev.*, 60 (1941) 172; R.E.Barieau, *J. Am. Chem. Soc.*, 72 (1950) 4024; *J. Chem. Phys.* 21 (1953) 1827; J.W.Stout, *J. Chem. Phys.*, 21 (1953) 1829; J.C.M.Li, *J. Chem. Phys.*, 19 (1951) 1059; 23 (1955) 2012; H.Mauser and G.Kortüm, *Z. Naturforsch.*, 11a (1956) 196.

Since the external electrical work $n\mathrm{F}U$ (where $n\mathrm{F}$ = the reactional charge, *cf.* 2 of this chapter) is usually expressed in Volt–Coulombs, it is multiplied by 0.239 which is the conversion factor to calories, the unit normally used for ΔH or ΔG. Equation (1) then becomes

$$0.239\, n\mathrm{F}U = \Delta H + T\left(\frac{\partial(0.239\, n\mathrm{F}U)}{\partial T}\right)_p \tag{2}$$

$$U = \frac{\Delta H}{0.239\, n\mathrm{F}} + T\left(\frac{\partial U}{\partial T}\right)_p$$

[1] It may be noted here that in electrochemistry in general, and in particular in the calculation of electric tensions and electrode processes, the methods of irreversible thermodynamics should strictly be used. Only as a limiting case can these processes be treated by the thermodynamics of reversible processes. However, the greater simplicity and clarity of reversible processes justifies their use in teaching, for a better understanding of the fundamental phenomena. Accounts of the use of the thermodynamics of irreversible processes are given by J.Prigogine, *Etudes Thermodynamiques des Phénomènes Irréversibles*, Dunod, Paris; Desoer, Liège (1947); S.R.de Groot, *Thermodynamics of Irreversible Processes*, Interscience, New York (1951); P.v.Rysselberghe, *Electrochemical Affinity*, Hermann, Paris (1955).

The term $(\partial(0.239\, n\mathrm{F}U)/\partial T)_p$ of (2) is the derivative of the variation in free enthalpy with respect to temperature, at constant pressure, and is thus a measure of the change in entropy $-\Delta S$, so that (*cf.* (40), p. 17)

$$T\left(\frac{\partial(0.239\, n\mathrm{F}U)}{\partial T}\right)_p = -\,T\Delta S = -\,q$$

This is thus a quantity of heat and represents its amount effectively involved in the transformation.

Table III, 1 shows the good agreement between the experimentally determined values for ΔH and those calculated by the Gibbs–Helmholtz equation from the electric tension and its temperature coefficient $\Delta U/\Delta T$, which may replace $(\partial U/\partial T)_p$ within the required accuracy.

TABLE III, 1

COMPARISON OF THE VALUES OBTAINED FOR ΔH BY MEASUREMENT AND CALCULATION

Cell	$C°$	$\frac{U}{\text{volt}}$	$\frac{\Delta U}{\Delta T}\cdot 10^4$	ΔH kcal *calc.*	ΔH kcal *measured*	ΔG
$-Pb/Pb$ acetate aq.//Cu acetate aq./Cu/Pb^+	0	-0.470	-3.85	$-16{,}830$	$-17{,}532$	$-21{,}68$
$-Zn/ZnCl_2+100\ H_2O/AgCl/Ag/Zn^+$	0	-1.015	$+4.02$	$-51{,}880$	$-52{,}046$	$-46{,}82$
$-Zn/ZnSO_4$aq. sat.//$CuSO_4$aq. sat./Cu/Zn^+	15	-1.0934	$+4.29$	$-56{,}140$	$-55{,}189$	$-50{,}44$
$-Ag/AgCl/HCl/Hg_2Cl_2/Hg/Ag^+$	25	-0.0455	-3.38	$+\ 1{,}270$	$+\ 1{,}900$	$-\ 1{,}05$

The sign // indicates that the diffusion tension (*cf.* p. 138 *et seq.*) has been eliminated.

The fourth cell is particularly interesting in that the application of the Gibbs–Helmholtz equation has given exact values for the changes in free enthalpy and in enthalpy, although the latter is positive (an endothermic reaction).

The dependence of the electric tension on the concentration may be easily deduced using the chemical potential (*cf.* (30), p. 11). At constant temperature with one unit of advancement, the chemical affinity is given by the equation

$$A = \Sigma \nu_i \mu_i \tag{3}$$

where the values ν_i are the stoichiometric coefficients of reaction for each of the substances involved and are given negative signs if the chemical species to which they refer are products of the reaction and positive signs if they are reactants. Keeping in mind the dependence of the chemical potential on the concentration, expressed in molality, equation (3) becomes

$$A = \Sigma \nu_i(\mu_{0i} + RT \ln a_i) \tag{4}$$

For any generalized reaction, where the electrical charges of the individual components A, B, C, D are omitted, for the sake of brevity,

$$pA + qB \rightarrow rC + sD$$

equation (4) becomes

$$A = p(\mu_{0_A} + RT \ln a_A) + q(\mu_{0_B} + RT \ln a_B) - r(\mu_{0_C} + RT \ln a_C) - s(\mu_{0_D} + RT \ln a_D)$$

$$= p\mu_{0_A} + q\mu_{0_B} - r\mu_{0_C} - s\mu_{0_D} + RT \ln \frac{a_A{}^p \, a_B{}^q}{a_C{}^r \, a_D{}^s} \tag{5}$$

The first four terms of the right-hand side of equation (5) are constant since they represent the standard chemical potential, and may be subsumed in a single constant. Considering the reaction at equilibrium where $\Delta G = -A = 0$.

$$A = 0 = \Sigma \nu_i \mu_{0i} - RT \ln \frac{a_C{}^r \, a_D{}^s}{a_A{}^p \, a_B{}^q} \tag{6}$$

where the activity is now that at equilibrium, so that equation (6) becomes

$$RT \ln \frac{a_C{}^r \, a_D{}^s}{a_A{}^p \, a_B{}^q} = RT \ln K = \Sigma \nu_i \mu_{0i}$$

where K is the equilibrium constant of the reaction. In general, then, equation (5) may be written as

$$A = RT \ln K - RT \ln \frac{a_C{}^r \, a_D{}^s}{a_A{}^p \, a_B{}^q} \tag{7}$$

Transforming (7) into electrical units gives

$$-\,0.239\, n\mathrm{F}U = RT \ln K - RT \ln \frac{a_C{}^r \, a_D{}^s}{a_A{}^p \, a_B{}^q}$$

whence

$$U = -\frac{RT}{0.239\, n\mathrm{F}} \ln K + \frac{RT}{0.239\, n\mathrm{F}} \ln \frac{a_C{}^r \, a_D{}^s}{a_A{}^p \, a_B{}^q} = U_0 + \frac{RT}{0.239\, n\mathrm{F}} \ln \frac{a_C{}^r \, a_D{}^s}{a_A{}^p \, a_B{}^q}$$

Introducing the numerical values of R and F and converting to decimal logarithms, at 25° C (298.16° K), gives the factor $(RT/0.239\ n\text{F})\ \ln \ldots$ the value $(0.059155/n)\ \log \ldots$

In the following discussion the temperature is considered to be maintained at 25° C throughout the transformations. To calculate the electric tensions of particular processes at other temperatures it suffices to apply the considerations previously given on the dependence of the electric tension upon the temperature.

As an example the H_2/Cl_2 gas cell may be considered. This consists of a hydrogen and a chlorine electrode immersed in a solution of hydrochloric acid (*cf.* p. 123 *et seq.*),

$$\underset{1}{\text{Pt}-\text{H}_2}\ /\ \underset{2}{\text{H}^+\text{Cl}^-\text{aq.}}\ /\ \underset{3}{\text{Pt}-\text{Cl}_2}$$

The chemical reaction in this cell is:

$$\text{H}_2 + \text{Cl}_2 \rightarrow 2\ \text{HCl}$$

The change in free enthalpy for one unit of advancement under reversible conditions, when the hydrochloric acid is considered as dissociated into its ions, is

$$\Delta G = -\mu_{\text{H}_2} - \mu_{\text{Cl}_2} + 2\ \mu_{\text{H}^+} + 2\ \mu_{\text{Cl}^-} = -A$$

If the gases are taken as being in their standard state of 25° C and 1 atmosphere

$$\Delta G = -\mu_{0\text{H}_2} - \mu_{0\text{Cl}_2} + 2(\mu_{0\text{H}^+} + RT \ln a_{\text{H}^+} + \mu_{0\text{Cl}^-} + RT \ln a_{\text{Cl}^-})$$

If the constants are extracted and the units are changed into electrical ones, then, since $n = +2$

$$-\frac{A}{0.239 \cdot 2\ \text{F}} = U = \frac{2(\mu_{0\text{H}^+} + \mu_{0\text{Cl}^-}) - \mu_{0\text{H}_2} - \mu_{0\text{Cl}_2}}{0.239 \cdot 2\ \text{F}} + 2\frac{RT}{0.239 \cdot 2\ \text{F}}(\ln a_{\text{H}^+}\, a_{\text{Cl}^-})$$

and from this, at 25° C,

$$U = U_0 + 0.05915 \log (a_{\text{H}^+} a_{\text{Cl}^-})$$
$$= U_0 + 0.05915 \log (a_{\pm\text{H}^+} a_{\pm\text{Cl}^-})$$

where $a_{\pm\text{H}^+}$ and $a_{\pm\text{Cl}^-}$ represent the mean activities of the two ionic species, which are equal. Thus,

$$U = U_0 + 0.05915 \log a^2_{\pm\text{HCl}}$$
$$= U_0 + 0.05915 \cdot 2 \log a_{\pm\text{HCl}}$$

When $a_{HCl} = 1$ the second term of the right hand side of this equation disappears and

$$U = U_0$$

The constant U_0 is called the standard electric tension of the cell. For the Daniell cell, at 25° C, written in the phase order already used in 1 and 2 of this chapter where the reaction is spontaneous,

$$Zn + Cu^{2+} \rightarrow Zn^{2+} + Cu$$

With the metallic copper and zinc in their standard state, the electric tension will similarly be given by

$$U = U_0 + \frac{0.05915}{2} \log \frac{a_{Zn^{2+}}}{a_{Cu^{2+}}}$$

The dependence of the reversible electric tension on pressure can be derived from the following simple argument by applying the principle of the equivalence of the second mixed derivatives, *i.e.* if a quantity is a function of two independent variables, the value of the second mixed derivative is independent of the order of derivation so long as the function may be derived within the interval considered. Thus, for $f(x, y)$

$$\frac{\partial}{\partial x}\left(\frac{\partial f}{\partial y}\right) = \frac{\partial}{\partial y}\left(\frac{\partial f}{\partial x}\right)$$

This is true for the free enthalpy function at constant temperature,

$$\frac{\partial}{\partial p}\left(\frac{\partial G}{\partial n_i}\right)_T = \frac{\partial}{\partial n_i}\left(\frac{\partial G}{\partial p}\right)_T \tag{8}$$

Equation (8) is valid for all of the components of the system and hence also for the algebraic sum of the individual expressions

$$\Sigma\nu_i \frac{\partial}{\partial p}\left(\frac{\partial G}{\partial n_i}\right)_T = \Sigma\nu_i \frac{\partial}{\partial n_i}\left(\frac{\partial G}{\partial p}\right)_T \tag{9}$$

The factors $(\partial G/\partial n_i)_T$ are the chemical potentials (*cf.* Chapter I, 2) of each component of the system. Their sum is thus the total useful work of the reaction, which in this case is electrical work $(-0.239\ nFU)$[1]; since, moreover, $(\partial G/\partial p)_T = v$ (see (26) on p. 10).

$$\frac{\partial}{\partial n_i}\left(\frac{\partial G}{\partial p}\right)_T = \left(\frac{\partial v}{\partial n_i}\right)_T$$

[1] Note that the same symbol is used for the number of moles in the derivatives (8) and (9) and for the number of equivalents of electricity.

The summation of these terms is thus the change in volume Δv produced by the reaction, so that (9) becomes

$$\left(\frac{\partial(0.239\, n\mathbf{F}U)}{\partial p}\right)_T = 0.239\, n\mathbf{F}\left(\frac{\partial U}{\partial p}\right)_T = \Delta v$$

i.e.

$$\left(\frac{\partial U}{\partial p}\right)_{T,n_i} = \frac{\Delta v}{0.239\, n\mathbf{F}} \text{ and } \mathrm{d}U = \frac{1}{0.239\, n\mathbf{F}} \Delta v \mathrm{d}p$$

Integrating between the limits p_1 and p_2 gives

$$U_{p_2} = U_{p_1} + \frac{1}{0.239\, n\mathbf{F}} \int_{p_1}^{p_2} \Delta v \mathrm{d}p$$

It can readily be seen from this that for reactions between condensed phases where the change in volume is negligibly small the electric tension is virtually independent of the pressure, but becomes markedly dependent in reactions involving the formation or disappearance of gases.

Thus, two kinds of forces are at work simultaneously in a galvanic cell. There are chemical forces which as a consequence of the reaction tend to separate the electrical charges within the cell and tend to confer different polarities on the two electrodes; on the other hand, there are electrostatic forces derived from the difference in potential between the two ends of the galvanic cell, and these tend to annul the existing potential difference and thus to oppose the first kind of force. The size of the chemical forces is given by the ratio of the affinity to the reactional charge and is called the *chemical tension* or electromotive force **E** of the cell. As a result of the conventions adopted earlier a positive electromotive force is the manifestation of chemical, *i.e.* non-electrostatic, forces, tending to make positive charges circulate within the cell in the phase order 1, 2, 3, ... 1′. The size of the electrostatic force is given by the electric tension of the cell. In a cell working reversibly, where the current intensity is infinitely small and tends towards zero, the electric tension is equal in absolute value but of opposite sign to the ratio of the affinity to the reactional charge. Thus, under equilibrium conditions, the reversible electric tension is equal in absolute value but of opposite sign to the electromotive force.

To illustrate this, a Daniell cell may be again considered. This can be arbitrarily written according to one of the two following schemes[1].

$$\text{(I)}\ \underset{1}{Zn}\ /\ \underset{2}{Zn^{2+}}\ //\ \underset{3}{Cu^{2+}}\ /\ \underset{4}{Cu}\ /\ \underset{1'}{Zn}$$

$$\text{(II)}\ \underset{1}{Cu}\ /\ \underset{2}{Cu^{2+}}\ //\ \underset{3}{Zn^{2+}}\ /\ \underset{4}{Zn}\ /\ \underset{1'}{Cu}$$

The reaction involved may also be written arbitrarily – ignoring the direction of its spontaneous progress – according to one of the two following schemes

$$(a)\ Cu^{2+} + Zn \rightarrow Cu + Zn^{2+}$$

$$(b)\ Cu + Zn^{2+} \rightarrow Cu^{2+} + Zn$$

It is possible to couple reaction (*a*) or (*b*) with either of the schemes (I) and (II) to give four different cases. It is apparent that the reactional charge for each of the four cases is as follows.

Case	*Reactional charge, n*
I *a*	+ 2
II *a*	− 2
I *b*	− 2
II *b*	+ 2

The chemical affinities A and the respective changes in free enthalpy ΔG of the reactions (*a*) and (*b*) satisfy the equations

$$A_a = -\ \Delta G_a = -\ A_b = +\ \Delta G_b$$

The electromotive force corresponding to scheme (I) is given by the equations

$$\mathbf{E} = \frac{+A_a}{+2\mathbf{F}} = \frac{-\Delta G_a}{+2\mathbf{F}} = \frac{+A_b}{-2\mathbf{F}} = \frac{-\Delta G_b}{-2\mathbf{F}}$$

and the reversible electric tension U_{rev} for the same scheme is given by the four equations

$$U_{rev} = \frac{-A_a}{+2\mathbf{F}} = \frac{+\Delta G_a}{+2\mathbf{F}} = \frac{-A_b}{-2\mathbf{F}} = \frac{+\Delta G_b}{-2\mathbf{F}}$$

[1] The // sign between phases 2 and 3 indicates that diffusion phenomena are for the moment being ignored (*cf.* p. 138 *et seq.*)

The invariance of these two sets of relationships is readily appreciated in that once the phase order has been established, inverting the reaction changes the signs of both the affinity and the reactional charge. Identical considerations apply for scheme (II) so that in general at equilibrium, *i.e.* under reversible conditions, the equations

$$U_{rev} = -\mathbf{E}$$

$$U_{rev} + \mathbf{E} = 0$$

hold true. Under non-equilibrium conditions the electromotive force remains constant whilst the electric tension varies from the value under reversible conditions, becoming greater or lesser depending on the direction in which the current flows through the cell; this difference becomes greater as the current intensity increases.

4. Electrode-Solution Electric Tensions; Simple and Multiple Electrodes; Electrodes of the First Kind

The thermodynamic theory of cells, like every thermodynamic treatment, is independent of any special hypothesis of the mechanism of the process which may be interpreted by widely varying ways. It is open to discussion, however, whether it is of any value to use hypotheses which are difficult to prove in order to achieve results which can be arrived at more simply thermodynamically.

For this reason Nernst's theory of the electrolytic tension of solutions, which he proposed in 1889, will not be considered and the calculation of the electric tensions of individual electrodes will be carried out exclusively by a thermodynamic treatment.

It has been shown how a galvanic cell permits the transformation into externally useful work of the change in free enthalpy involved in the transformation which occurs within that cell. This transformation which is usually a chemical reaction, is split by the cell into partial processes which occur at the individual electrodes and induce at each of these a definite chemical tension or electromotive force, and thus an electrode-solution electric tension which at equilibrium is equal, but of opposite sign, to the electromotive force. The total electric tension of the cell is given by the algebraic sum of the electric tensions at each of the interfaces and especially at the electrode-solution interfaces. It is thus useful to know exactly the electric tensions of each half-cell.

For a concise treatment of electric tensions it will be convenient to define certain quantities relative to single phase systems. Each phase has its own electrical potential in the sense given in Chapter I, 3 and defined by the work involved in moving a unit electrical charge[1] from infinity to a distance away from the surface of the phase which is rather greater than the action radius of the so-called image forces[2]. This potential which is given the symbol ψ, is called the external electrical potential or the volta potential and its magnitude and sign are determined by the amount and sign of the excess electrical charge in the conductor with respect to electroneutrality.

At the surface there is also always present a double electrochemical layer composed of orientated dipoles whose presence makes it necessary for work to be performed in order for an electrical charge to pass through them. In other words, there is a second potential, given the symbol χ, and called the electric surface potential[3] which is independent of the amount of charge in excess in the conductor. To carry a unit charge from infinity to the interior of the phase involves the performance of a quantity of work defined by the sum of the external and surface potentials; this quantity is called the internal electrical potential or galvani potential and is given the symbol φ.

Thus, the volta potential (ψ), the electric surface potential (χ) and the galvani potential (φ) are connected by the relationship

$$\varphi = \psi + \chi$$

The volta potential, and hence the volta tension, between two phases is directly susceptible to measurement whilst the electric surface and galvani potentials are not. The galvani tensions between two different phases are also not susceptible to experiment whilst the galvani tensions between two equal phases are identical to the volta tensions, since in this case the two electric surface potentials are identical and thus cancel each other out.

[1] In the metric sense of one unit of charge relative to the system of measurements chosen.

[2] The image forces are the electrostatic forces existing between a charge and a conductor as a consequence of the electrostatic charge induced in the conductor by the given external charge. Their action radius is about 10^{-4} cm.

[3] In fact, the electric surface potential is apparently a rise in potential rather than a single potential, and is thus a tension.

The electric tensions of cells which have been discussed so far have all been galvani tensions.

Moving a charge from the interior of one phase to that of another is thermodynamically equivalent to carrying the same charge from the interior of the first phase to infinity, and then back again to the interior of the second phase. The two partial works involved are given by the respective galvani potentials and thus the total work is given by the galvani tension existing between the two phases. The isolated electrochemical process at an electrode by virtue of which the charge, whether or not linked to a material particle, passes from the interior of one phase to that of another thus gives rise to a galvani tension between the two phases; this corresponds to the absolute electric tension of the electrode (U_{abs}). It is given by the equation

$$U_{abs} = {}^1\varphi - {}^2\varphi \quad [1]$$

It is convenient to base the thermodynamic calculation of the electric tension of an electrode on the electrochemical potentials. The chemical potential of a component in a system is the derivative of the free energy of the system with respect to the number of moles of the component under consideration, when all the other variables – temperature, pressure and the numbers of moles of other components – are kept constant. For components having an electrical charge (ions and electrons) this charge and the internal electrical potential of the phase, which contribute to the energy of the system, must also be considered. The *electrochemical potential* $\tilde{\mu}$ of a component is defined as the sum of its chemical potential and the product of the molar charge $z\mathbf{F}$ with the internal electrical potential of the phase, φ

$$\tilde{\mu}_i = \mu_i + z\mathbf{F}\varphi$$

where z is the number of elementary charges on the ions and may be either positive or negative. An isolated electrode consists of two phases in contact with each other; one of these is a solid, electronic conducting phase (*i.e.* of the first class: metals or graphite) and the other an ionic conducting phase (*i.e.* of the second class and usually an electrolytic solution but sometimes a pure electrolyte). Electrodes may be classified on the basis of two criteria. One of these is rigorously electrochemical, based on the

[1] The raised number on the left indicates the succession order of the phases which are considered.

number of electrically charged species which are able to traverse the surface between the conductor and the solution, and on the number of electrochemical reactions occurring at the electrode. The other criterion, which has historical precedence, is the composition of the electrodes.

According to the first system of classification an electrode is *simple* when only a *single* charged particle (an ion or an electron) can pass through the conductor-electrolyte interface, giving rise to only a *single* electrode reaction; an electrode is *double*, or more generally *multiple*, when two or more charged particles can pass through the interface or when there are two or more contemporaneous electrode reactions. On the basis of the second scheme of classification, electrodes are described as being of the 1*st*, 2*nd*, or 3*rd* kind, with amalgam, gas, and oxido-reduction or redox electrodes, respectively; in the calculation of their electric tensions they are all considered as simple electrodes. Although the first criterion is the more rigorous, electrodes will be described here on the basis of the second, both because the electrodes are the more easily recognized and, secondarily, because from a theoretical thermodynamic standpoint, multiple electrodes have mixed tensions which are readily obtained from the equilibrium conditions of the coupled reactions of the corresponding simple electrodes.

Electrodes of the 1*st* kind consist of a metal immersed in a solution of one of its salts and are reversible with respect to the cation, *e.g.* copper in copper sulphate solution, silver in silver nitrate solution. The reaction occurring with such electrodes is in general

$$^{1}[Me^{z+}] \rightleftharpoons {}^{2}Me^{z+}_{sol} \quad {}^{1} \tag{1}$$

Thus, it consists in the passage of the charged particle or 'metallic ion' from the electrolytes to the metal, or *vice versa;* the metal is considered as being formed of an ionic lattice in which the valency electrons are circulating more or less freely. If as a consequence of the electrochemical reaction the ions pass from the solid electrode to the solution, leaving the valency electrons in the solid phase, the electrode becomes negatively charged with respect to the solution; whilst for the reverse reaction the electrode assumes a positive sign.

The thermodynamic calculation of the electric tension of such an electrode is based on the study of such a system, determining the electric ten-

[1] A symbol within square brackets indicates that the corresponding substance is in the solid state, and the subscript *sol* indicates a solution.

sion existing at electrochemical equilibrium at the electrode solution interface. At constant temperature and pressure under conditions of electrochemical equilibrium, the following relationship must hold[1]

$$^{1}\tilde{\mu}_{\mathrm{Me}^{z+}{}_{el}} = {}^{2}\tilde{\mu}_{\mathrm{Me}^{z+}{}_{sol}}$$

so that

$$^{1}(\mu + z\mathbf{F}\varphi_1)_{\mathrm{Me}^{z+}{}_{el}} = {}^{2}(\mu + z\mathbf{F}\varphi_2)_{\mathrm{Me}^{z+}{}_{sol}} \tag{2}$$

Expressing the electrochemical potentials in an explicit form as functions of the standard chemical potential μ_0 and of the activity a and bearing in mind that the chemical potential of the solid phase is equal to the standard chemical potential and writing the phases of the electrode in the order

$$\underset{1}{\mathrm{Me}} \ / \ \underset{2}{\mathrm{Me}^{z+}{}_{sol}} \tag{3}$$

gives directly from (2) the absolute electrode-solution tensions[2] U_{abs}

$$U_{abs} = {}^{1}\varphi - {}^{2}\varphi = \frac{{}^{2}\mu_{0\mathrm{Me}^{z+}} - {}^{1}\mu_{0\mathrm{Me}^{z+}}}{0.239\ z\mathbf{F}} + \frac{RT}{0.239\ z\mathbf{F}} \ln a_{\mathrm{Me}^{z+}} \tag{4}$$

$$= U_{0abs} + \frac{RT}{0.239\ z\mathbf{F}} \ln a_{\mathrm{Me}^{z+}} \tag{5}$$

Equation (4) is identical to equation (2) on page 99 since $({}^{2}\mu_{0\mathrm{Me}^{z+}} + RT \ln a_{\mathrm{Me}^{z+}} - {}^{1}\mu_{0\mathrm{Me}^{z+}})$ is the free enthalpy of reaction (1) in the direction left to right and $z\mathbf{F}$ is the corresponding reactional charge. Equation (5), which is also called Nernst's equation, gives the sign of the electric tension and its dependence on the concentration. For ${}^{2}\mu_{0\mathrm{Me}^{z+}} \ll {}^{1}\mu_{0\mathrm{Me}^{z+}}$ *i.e.* for a reaction which is spontaneous for the passage of the ions from the electrode to the solution, $({}^{2}\mu_{0\mathrm{Me}^{z+}} - {}^{1}\mu_{0\mathrm{Me}^{z+}}) < 0$ and outweighs the second term of equation (5), so that $U_{abs} < 0$. This is true for electrodes of the base metals. On the other hand $({}^{2}\mu_{0\mathrm{Me}^{z+}} - {}^{1}\mu_{0\mathrm{Me}^{z+}}) = \Delta G$ for a unit of advancement under standard concentrations, for reaction (1) from left to right, and since $\Delta G < 0$ in this case, the reaction is in fact spontaneous in the direction shown. For ${}^{2}\mu_{0\mathrm{Me}^{z+}} \gg {}^{1}\mu_{0\mathrm{Me}^{z+}}$ this difference is positive and outweighs the second term of equation (5) so that $U_{abs} > 0$; this is true for the noble metals where ΔG $(= {}^{2}\mu_{0\mathrm{Me}^{z+}} - {}^{1}\mu_{0\mathrm{Me}^{z+}}$, once again for reaction (1) in the direction left to right, so that $z\mathbf{F} > 0)$ has a

[1] The subscript *el* indicates the metallic phase of the electrode

[2] In some texts the electrode tension is called the electrode potential.

positive value. In other words the reverse reaction is spontaneous. If the electric tension is now calculated from the free enthalpy ΔG of the spontaneous reaction of a noble metal — *i.e.* from right to left in reaction (1) — it will be negative but at the same time the sign of the reaction charge zF will be inverted, following the convention adopted for phase enumeration (*cf.* p. 98). Thus, once the order of the phases has been fixed, the sign of U_{abs} becomes independent of the way the reaction is written. If the electrolyte involved in the reaction has an activity $a = 1$, $U = U_0$ and is called the standard absolute electrode tension. For ${}^2\mu_{0\mathrm{Me}^{z+}} \cong {}^1\mu_{0\mathrm{Me}^{z+}}$ and thus for $U_0 \cong 0$ the sign of the electric tension is mainly determined by the second term of equation (5) and may be either positive or negative. In this case again it must be emphasized that the choice of the convention[1] adopted leads to an equivalence of the thermodynamic and electrical signs which is however purely coincidental in that there is no connection between thermodynamic and electrical signs. It is apparent that the case dealt with above refers to the passage of an ion from one phase to another and that the value of U_{abs} thus corresponds to a galvani tension between different phases and hence is not susceptible to direct measurement.

The electric tension which is thus established between the electrode and the solution of the electrolyte as a consequence of the difference in their chemical potentials, actually appears as a double electrochemical layer, one part of which is ionic and one dipolar. With a zinc electrode, for example, immersed in zinc sulphate solution the reaction proceeds spontaneously towards the formation of zinc ions in solution, since $\mu_{0[\mathrm{Zn}^{2+}]} \gg \mu_{0\mathrm{Zn}^{2+}{}_{sol}}$ so that the electrode becomes negatively charged. However, the amount of zinc ions which actually passes into solution with open circuit, is very small — of the order of 10^{-9} g of zinc for each cm^2 of electrode surface — and does not pass into the bulk of the solution because it is electrostatically attracted by the charge of opposite sign which remains on the electrode, made up of the valency electrons left there.

[1] The convention for electrical signs adopted in this book is the European one. Many texts, particularly American ones, follow a different sign convention, giving electrodes negative signs when the electrolytic solution assumes a negative charge with respect to the electrode. For the history of the sign conventions of electrode tensions *cf.* T. S. Licht and A. J. de Bethune, *J. Chem. Educ.*, 34 (1957) 433.

5. Electrode-Solution Electric Tensions; Electrodes of the Second and Third Kind

Electrodes of the second kind are also metallic but consist of a metal covered by one of its slightly soluble salts, immersed in a solution of an electrolyte which has the same anion as the slightly soluble salt. With electrodes of the first kind the passage of electricity from the electrode to the solution, and *vice versa*, is brought about exclusively by the discharge or formation of cations, whilst with electrodes of the second kind it may be assumed that the passage of electricity is mediated by the anions of the slightly soluble salt. An example of an electrode of the second kind consists of a mercury layer covered with solid mercurous chloride (calomel), in a solution of potassium chloride. Here, the passage of current from the electrode to the solution, with the consequent generation of a negative charge on the electrode is brought about by the reaction of Cl^- ions from the solution with the metallic mercury of the electrode to form solid calomel. The electrode becomes positively charged, however, when calomel disappears giving rise to metallic mercury and Cl^- ions in solution. Here the electricity passes from the solution to the electrode. In other words, the process is analogous to that in electrodes of the first kind, with the difference that the electrochemically active ion is the Cl^-. This is equivalent to saying that electrodes of the second kind behave as if they were formed of an anion in solution and its elementary form in contact with the solution and were reversible with respect to the anion. In the case illustrated, the calomel electrode behaves like a true chlorine electrode (which also exists *cf.* p. 125 *et seq.*) but is more convenient in use.

A generalized electrode of the second kind may be expressed as

$$\underset{1}{[Me^{mz+}]} \ / \ \underset{2}{[MeA_m]} \ / \ \underset{3}{A^{z-}_{sol}}$$

with the electrode reaction

$$[Me^{mz+}] + mA^{z-} \rightleftharpoons MeA_m \tag{1}$$

where $[Me^{mz+}]$ represents a metal of valency mz; A^{z-} represents the anion of valency z and m is the number of anions combined with one metal ion (which may contain more than one atom *e.g.* Hg_2^{2+}) to form the slightly soluble compound MeA_m. The cation of the soluble electrolyte has been omitted from reaction (1) since it does not take part in the reaction. The electric tension of such an electrode may be calculated starting from the general equation for electrodes of the first kind (*cf.* p. 114 *et seq.*). The actual activity of the cations of the slightly soluble salt is connected to the actual activity of the anions a_- through the solubility product (L)

$$a_+ a_-^m = L$$

whence

$$a_+ = L/a_-^m$$

so that

$$U_{abs} = U_{0\,abs} + \frac{RT}{0.239\ mz\mathrm{F}} \ln \frac{L}{a_-^m}$$

Subsuming the constant L into the constant $U_{0\,abs}$, which becomes $U_0{}'_{abs}$, gives

$$U_{abs} = U_0{}'_{abs} + \frac{RT}{0.239\ mz\mathrm{F}} \ln \frac{1}{a_-^m}$$

With the calomel electrode, previously discussed, the electrochemical reaction is

$$2\,[Hg^+] + 2\,Cl^- \rightleftharpoons Hg_2Cl_2$$

where $m = 2$ and $z = 1$. At 25° C this gives

$$U_{abs} = U_0{}'_{abs} + \frac{0.05915}{2} \log \frac{1}{a_{Cl^-}{}^2}$$

For the silver chloride electrode the electrochemical reaction is

$$[Ag^+] + Cl^- \rightleftharpoons AgCl$$

where $m = 1$ and $z = 1$. Hence,

$$U_{abs} = U_0{}'_{abs} + 0.05915 \log \frac{1}{a_{Cl^-}}$$

Here too, making the activity of the ions in solution equal to 1, gives $U_{abs} = U_0{}'_{abs}$ *i.e.* the standard absolute electric tension.

Electrodes of the third kind are of relatively less importance than those of the first and second kind and have been developed principally to provide electrodes which are reversible with respect to metals which decompose water but can form slightly soluble salts. In view of their minor theoretical and practical importance they will be briefly illustrated by an example rather than by a generalized statement. A reversible calcium electrode can be formed by the following arrangement.

$$[Zn^{2+}]\ /\ [ZnC_2O_4];\ [CaC_2O_4]\ /\ Ca^{2+}{}_{sol} \qquad (2)$$

i.e. electrode metal [Zn], slightly soluble salt of the electrode metal (solid phase I); slightly soluble salt of the anion of solid phase I and the cation being tested (solid phase II), the cation for which the electrode must be reversible. The electric tension of the electrode (2) can be obtained from the general equation for electrodes of the first kind referred to zinc in a solution of Zn^{2+} ions.

$$U_{abs} = U_{0\,abs} + \frac{RT}{0.239 \cdot 2\,\mathrm{F}} \ln a_{Zn^{2+}} \tag{3}$$

Bearing in mind the two equations for the solubility products of the two solid phases

$$a_{Zn^{2+}} \cdot a_{C_2O_4^{2-}} = L_1 \tag{4}$$

$$a_{Ca^{2+}} \cdot a_{C_2O_4^{2-}} = L_2 \tag{5}$$

and that the activity of the $C_2O_4^{2-}$ in a solution saturated with the two solid phases is the same in both of these equations, gives

$$\frac{a_{Zn^{2+}}}{a_{Ca^{2+}}} = \frac{L_1}{L_2}$$

$$a_{Zn^{2+}} = a_{Ca^{2+}} \frac{L_1}{L_2} \tag{6}$$

Substituting the values obtained from equation (6) into equation (3) gives

$$U_{abs} = U_{0\,abs} + \frac{RT}{0.239 \cdot 2\,\mathrm{F}} \ln \left(\frac{L_1}{L_2} a_{Ca^{2+}}\right)$$

$$= U_0{}'_{abs} + \frac{RT}{0.239 \cdot 2\,\mathrm{F}} \ln a_{Ca^{2+}} \tag{7}$$

In equation (7) the value of $U_0{}'_{abs}$ is obtained by combining the constant $U_{0\,abs}$ with $(RT / 0.239 \cdot 2\mathrm{F}) \ln (L_1 / L_2)$.

Such electrodes are subject to a series of limitations which stem essentially from the relative values of the solubility products of the two solid phases, from the possibility of secondary reactions of the two solid phases with each other, with the electrode metal or with the solution, and from the possibility of forming double salts or isomorphic crystals between the solid phases. Le Blanc and Harnapp have published a detailed discussion of these electrodes and their limitations in use[1].

[1] M. Le Blanc and O. Harnapp, *Z. physik. Chem.*, A 166 (1933) 321.

6. Electrode-Solution Electric Tensions; Gas Electrodes and Amalgam Electrodes

Active electrochemical processes of the kind illustrated with electrodes of the first kind can be obtained not only from metals but also from other substances in the gaseous or liquid state, when they are provided with a suitable inert support which allows an equilibrium to be obtained between the elementary molecules and the ions derived from them. This support often consists of an electrode of platinized platinum in simultaneous contact with the solution and with the elements in the gaseous or liquid state. Hydrogen, chlorine etc. are soluble in the platinum black which forms the coating of a platinized platinum electrode and can pass from it into solution as ions, *e.g.*

$$\text{Pt–H} \rightleftharpoons \text{H}^+ + \text{e}^- \quad [1]$$

$$\text{e}^- + \text{Pt–Cl} \rightleftharpoons \text{Cl}^-$$

The symbols Pt–H and Pt–Cl represent respectively hydrogen and chlorine dissolved in the platinum.

In fact, a single equilibrium does not exist but rather a number of different equilibria corresponding to the various intermediary steps of the reaction converting elementary molecules to ions. The following equilibria coexist in the case of hydrogen

$$\text{H}_{2\,gas} \rightleftharpoons \text{Pt–H}_2 \qquad (1)$$

$$\text{Pt–H}_2 \rightleftharpoons \text{Pt–2H} \qquad (2)$$

$$\text{Pt–2H} \rightleftharpoons 2\,\text{H}^+ + 2\,\text{e}^- \qquad (3)$$

Equation (1) represents the equilibrium between gaseous molecular hydrogen and hydrogen dissolved in the platinum, to which Henry's Law applies; equation (2) represents the equilibrium between molecular hydrogen and atomic hydrogen both dissolved in the platinum, to which the Law of Mass Action applies; whilst equation (3) represents the true electrochemical process which determines the electric tension. This implies that the hydrogen, chlorine, etc. behave as if they were existing in a metallic state. In electrodes of this type the physical function of a true metallic electrode which serves as a conductor into the solution is split from

[1] The hydrogen ion is written as H^+ to avoid the question of whether it is actually linked to a single water molecule in the form H_3O^+ or to more than one molecule of water in the form $H^+ \cdot n\,H_2O$; this point is still controversial.

the electrochemical function which is carried out by another substance partly dissolved in the electrode metal and partly in contact with it as an adsorbed layer at the surface in equilibrium with the solution. It is the substance in the adsorbed layer that is electrochemically active. The electric tension of such electrodes is defined if the following conditions are fulfilled:

(a) The equilibrium (1) must obey Henry's Law.

(b) The equilibrium (2) must obey the Law of Mass Action.

(c) The material forming the electrode must not give up or receive ions from the electrolyte, *i.e.* it must be inert with respect to the electrolyte so that the electrode functions as a single electrode, with a single electrode reaction.

The electric tension of these electrodes may also be derived from the general equation (5) on page 118. In gas electrodes the charged particles which pass through the interface and to which the electric tension refers, are electrons which react with the corresponding atoms of the gas in the adsorbed layer. With hydrogen the reaction is

$$H \rightleftharpoons H^+ + e^- \tag{4}$$

which gives rise to an equilibrium whose constant K may be written in the form

$$K = \frac{a_{e^-} \cdot a_{H^+}}{a_H} \tag{5}$$

With hydrogen electrodes written in the phase order

$$\underset{1}{Pt\text{–}H_2} \;/\; \underset{2}{H^+_{sol}}$$

the electrochemical potential of the electrons at electrochemical equilibrium must be the same inside and outside the crystal lattice of the supporting metal — in this case platinum — so that the following equation must apply.

$${}^1\tilde{\mu}_{e^-} = {}^2\tilde{\mu}_{e^-} \tag{6}$$

The electrochemical potentials may be written in their explicit forms and the activity of the electrons a_{e^-} derived from the equation (5); then, bearing in mind that the electron charge is negative (so that its electrochemical potential is $\tilde{\mu}_{e^-} = \mu_{0\,e^-} - \mathbf{F}\varphi$)

$$^1(\mu_{e^-} - \mathbf{F}\varphi) = {}^2(\mu_{e^-} - \mathbf{F}\varphi)$$
$$= {}^2(\mu_{0\,e^-} + RT \ln a_{e^-} - \mathbf{F}\varphi)$$

whence

$$U_{abs} = {}^1\varphi - {}^2\varphi = \frac{^1\mu_{0\,e^-} - {}^2\mu_{0\,e^-}}{0.239\,\mathbf{F}} - \frac{RT}{0.239\,\mathbf{F}} \ln \frac{Ka_{\mathrm{H}}}{a_{\mathrm{H}^+}} \tag{7}$$

Combining all the constants and making the activity of the atomic hydrogen dissolved within the platinum proportional to the square root of the pressure of the molecular hydrogen — as a consequence of equilibria (1) and (2) — gives

$$U_{abs} = U_{0\,abs} + \frac{RT}{0.239\,\mathbf{F}} \ln \frac{a_{\mathrm{H}^+}}{p_{\mathrm{H}_2}^{\frac{1}{2}}} \tag{8}$$

For a chlorine electrode written in the phase order

$$\underset{1}{\mathrm{Pt{-}Cl_2}} \;/\; \underset{2}{\mathrm{Cl^-}_{sol}} \tag{9}$$

the reasoning is perfectly analogous for the reaction

$$\mathrm{Cl} + e^- \rightleftharpoons \mathrm{Cl^-}_{sol} \tag{10}$$

whose equilibrium constant is

$$K = \frac{a_{\mathrm{Cl}^-}}{a_{\mathrm{Cl}}a_{e^-}}$$

Repeating the calculation performed for the hydrogen electrode gives finally

$$U_{abs} = U_{0\,abs} + \frac{RT}{0.239\,\mathbf{F}} \ln \frac{p_{\mathrm{Cl}_2}^{\frac{1}{2}}}{a_{\mathrm{Cl}^-}} \tag{11}$$

Equation (11) illustrates how, apart from the constant value for the standard tension $U_{0\,abs}$, a calomel electrode or a silver chloride electrode (*cf.* p. 121) behaves as a true chlorine electrode at a constant pressure p_{Cl_2}.

Equations (8) and (11) show how the activity and hence the pressure of the gaseous components must also be borne in mind with gas electrodes. The relationship that once the phase order is fixed the sign of the electric tension is also fixed independently of the direction in which the electrode reaction is considered to occur, is also valid for these electrodes.

Making both the gas pressure (usually expressed in atmospheres) and the activity of the corresponding ions equal to 1 gives the absolute standard electric tension.

Amongst those for which the concentration of the electrode has to be considered, must be included metallic ones which are not pure metals but a solution in another metal, *i.e.* an alloy. Of these, the most important are the amalgams. In these cases the calculations made for electrodes of the first kind may be repeated considering the electrode as a simple one. The chemical potential of an electrochemically active metal, dissolved in mercury or another metal, is no longer equal to its standard chemical potential but depends also on its activity in the metallic phase. Bearing this in mind leads to

$$U_{abs\ amalg.} = U_{0\ abs\ amalg.} + \frac{RT}{0.239\ z\mathbf{F}} \ln \frac{a_{\mathrm{Me}^{z+}\,sol}}{a_{\mathrm{Me}^{z+}\,amalg.}}$$

provided that the metallic system behaving as an electrode (amalgam or solid solution) is monophasic.

Amalgam electrodes provide a simple approach to the functioning of multiple electrodes. Considering a zinc amalgam electrode in contact with a solution containing both Zn^{2+} and Hg^{2+} ions, it is evident that it is a double electrode with charged particles of two different species able to traverse the metal-solution interface. For the complete system to be in electrochemical equilibrium, the electrochemical equilibrium requirements of each of the two ionic species must be fulfilled, which implies that if 1 is the metallic phase and 2 the solution phase,

$$^{1}\tilde{\mu}_{\mathrm{Hg}^{2+}} = {}^{2}\tilde{\mu}_{\mathrm{Hg}^{2+}}$$

$$^{1}\tilde{\mu}_{\mathrm{Zn}^{2+}} = {}^{2}\tilde{\mu}_{\mathrm{Zn}^{2+}}$$

Neither the zinc nor the mercury are in their standard state in the amalgam and hence in writing the electrochemical potentials in their explicit forms, their respective activities in the metallic phase must be included

$$\left.\begin{aligned} {}^{1}(\mu_0 + RT\ln a + 2\,\mathbf{F}\varphi)_{\mathrm{Zn}^{2+}} &= {}^{2}(\mu_0 + RT\ln a + 2\,\mathbf{F}\varphi)_{\mathrm{Zn}^{2+}} \\ {}^{1}(\mu_0 + RT\ln a + 2\,\mathbf{F}\varphi)_{\mathrm{Hg}^{2+}} &= {}^{2}(\mu_0 + RT\ln a + 2\,\mathbf{F}\varphi)_{\mathrm{Hg}^{2+}} \end{aligned}\right\} \quad (12)$$

Since the galvani tension ${}^1\varphi - {}^2\varphi$ is the same for both processes, equation (12) leads to

$$
{}^1\varphi - {}^2\varphi = \frac{{}^2\mu_{0\,\mathrm{Zn}^{2+}} - {}^1\mu_{0\,\mathrm{Zn}^{2+}}}{0.239 \cdot 2\,\mathbf{F}} + \frac{RT}{0.239 \cdot 2\,\mathbf{F}} \ln \frac{{}^2a_{\mathrm{Zn}^{2+}}}{{}^1a_{\mathrm{Zn}^{2+}}}
$$

$$
= \frac{{}^2\mu_{0\,\mathrm{Hg}^{2+}} - {}^1\mu_{0\,\mathrm{Hg}^{2+}}}{0.239 \cdot 2\,\mathbf{F}} + \frac{RT}{0.239 \cdot 2\,\mathbf{F}} \ln \frac{{}^2a_{\mathrm{Hg}^{2+}}}{{}^1a_{\mathrm{Hg}^{2+}}}
$$

whence

$$
U_{0\,abs_{\mathrm{Hg}}} - U_{0\,abs_{\mathrm{Zn}}} = \frac{RT}{0.239 \cdot 2\,\mathbf{F}} \left(\ln \frac{{}^2a_{\mathrm{Zn}^{2+}}}{{}^1a_{\mathrm{Zn}^{2+}}} - \ln \frac{{}^2a_{\mathrm{Hg}^{2+}}}{{}^1a_{\mathrm{Hg}^{2+}}} \right)
$$

The difference between the absolute standard electric tensions is equal to that between the relative standard electric tensions (*cf.* p. 133 *et seq.*). Then converting to decimal logarithms and bearing in mind the relative standard electric tensions gives (Table III, 3)

$$
\log \frac{{}^1a_{\mathrm{Hg}^{2+}} \cdot {}^2a_{\mathrm{Zn}^{2+}}}{{}^1a_{\mathrm{Zn}^{2+}} \cdot {}^2a_{\mathrm{Hg}^{2+}}} = \frac{1.61}{0.03} \cong 54
$$

$$
\frac{{}^1a_{\mathrm{Hg}^{2+}}}{{}^2a_{\mathrm{Zn}^{2+}}} \cong 10^{54} \frac{{}^2a_{\mathrm{Hg}^{2+}}}{{}^2a_{\mathrm{Zn}^{2+}}}
$$

In other words, the activity of the Zn^{2+} ions in solution is at equilibrium so preponderant that over the whole range of amalgam composition, right down to the smallest possible zinc activity which can still be measured, it is the zinc which determines the electric tension. This holds true whenever the difference between the U_0 values is sufficiently high. With lower values of this difference however, such as with silver amalgam, the resolution of system (12) gives the mixed electric tension of the equilibrium which is established.

7. Electrode-Solution Electric Tensions; Oxido-Reduction (Redox) Electrodes

It was mentioned (*cf.* p. 97) that the electrochemical processes which can transform the free enthalpy of a reaction into external electrical work include chemical reactions consisting of changes in the charge on an ion, *e.g.*

$$
\mathrm{Fe}^{2+} \rightarrow \mathrm{Fe}^{3+} + \mathrm{e}^-
$$

and those in which the reaction consists in the formation of new ions by the conversion of neutral molecules into the ionic state, by the decomposition of other ions, or *vice versa*, *e.g.*

$$2\ MnO_4^- + 16\ H^+ + 10\ e^- \rightarrow 2\ Mn^{2+} + 8\ H_2O$$

This latter may be schematically divided into partial reactions.

$$2\ MnO_4^- + 6\ H^+ \rightarrow 2\ Mn^{2+} + 3\ H_2O + 5\ O$$

$$5\ O + 10\ H^+ + 10\ e^- \rightarrow 5\ H_2O$$

The first example given above is of the oxidation of a ferrous ion to a ferric one. The second process is one of reduction; the heptavalent manganese of the permanganate ion is reduced to bivalent manganese. In both of these examples and indeed, for all the many others which could be given, electrons appear or disappear in the corresponding chemical equations, which shows that when suitably performed oxidation and reduction can also give rise to external electrical work.

The corresponding half cells are therefore called *oxido-reduction electrodes*, or briefly, *redox* electrodes.

Before continuing with the calculation of the electric tension of redox electrodes, it will be convenient to define exactly the concept of oxidation. Those processes are defined as oxidative in which the substance involved gives up electrons, and reductive those in which it receives electrons. In the permanganate example, there is a reduction of the permanganate ion by the addition of electrons and chemical transformations of the ions. Actually all electrochemical processes based on chemical reactions originate in an oxidation or reduction. When metallic silver separates from silver nitrate solution it is said to be *reduced* and the reaction effectively consists in the passage of silver from the ionic to the elementary state by the acquisition of electrons. These are in fact the reactions underlying electrodes of the first kind, considered from another point of view.

The electrodes now being considered differ from those of the first and second kinds and from gas electrodes in that both the oxidized and reduced forms of the system remain in the solution during and after the reaction. If an inert electrode is immersed in a solution containing both the oxidized and reduced forms of an ion, it will assume a well-defined electric tension, dependent on the ionic activity of the substances taking part in the electrochemical reaction. The most suitable metals for the inert electrodes in oxido-reduction processes are platinum, iridium, rhodium and gold. With one of these metals, which must however be chosen in accordance with the composition of the system, the electric tension is definite and constant. In some cases, other metals such as molybdenum or tungsten, may be used. However, these do not always give constant electric tensions because, dependent on the composition of the electrolyte, they may be attacked and pass into solution. They thus play a part in the electrochemical process and produce electric tensions in virtue of the presence of their ions in the solution. In other words they behave as multiple electrodes.

Also in this redox type of half-cell the electric tension is located at the boundary between the electrode and the solution. Since, by definition, the reaction always involves the movement of electrical charges, the electrode which has lost any chemical function exists only as a metallic conductor, taking up or supplying electrons. The mechanism of the electrochemical reaction which takes place at the boundary between electrode and electrolyte may be interpreted as follows. The electrons released in an oxidation are eliminated from the reacting substances through an inert electrode into an external circuit, whilst the electrons necessary for a reduction are given to the system by an external circuit. An equilibrium is thus established characterized by the fact that the electrode surface becomes charged with a certain quantity of electrons or else becomes deficient in electrons with respect to the electro-neutral state.

From this it is easy to deduce the sign of an inert electrode immersed in an oxidizing or reducing system. The sign is positive if the system tends to fix electrons *i.e.* if it is oxidizing, and is negative in the reverse case.

Naturally, the same general rule applies to these electrodes, that it is possible to convert all the free enthalpy of the reaction into external electrical work only if the change is reversible; and if the oxidizing system is separated from the reducing system but is in material and electrical contact with it, so that the passage of electrons from one system to the other occurs through an external metallic circuit and not by direct passage from the reducing to the oxidizing substance. Mixing the oxidizing and the reducing systems would lead to what might be called a 'chemical short circuit' and the free enthalpy of the reaction would be converted into heat. The calculation of the electric tension is carried out — remembering that the charged particle which traverses the electrode solution interface is the electron and that, as in the case of gas electrodes, the electrochemical equilibrium must be achieved for this particle also — according to the equation

$$^{1}\tilde{\mu}_{e^-} = {}^{2}\tilde{\mu}_{e^-} \tag{1}$$

Here, phase 1 is the inert metal electrode and phase 2 is the electrolyte solution. With the electrode

$$\underset{1}{\text{Pt}} \;/\; \underset{2}{Fe^{2+}{}_{sol};\; Fe^{3+}{}_{sol}}$$

the electrode reaction is

$$Fe^{3+} + e^- \rightleftharpoons Fe^{2+} \tag{2}$$

whose equilibrium constant may be written

$$K = \frac{a_{Fe^{2+}}}{a_{Fe^{3+}} a_{e^-}}$$

The activity of the electrons in phase 2 may be deduced from this:

$${}^2a_{e^-} = \frac{a_{Fe^{2+}}}{a_{Fe^{3+}}\, K} \tag{3}$$

Writing equation (1) in the explicit form, bearing in mind that the electron charge is negative and that the chemical potential of the electrons in the metallic phase 1 is equal to their standard potential, gives

$${}^1(\mu_{0\,e^-} - \mathbf{F}\varphi) = {}^2(\mu_{0\,e^-} + RT \ln a_{e^-} - \mathbf{F}\varphi) \tag{4}$$

Combining this with equation (3) gives finally

$$U_{abs} = {}^1\varphi - {}^2\varphi = \frac{{}^1\mu_{0\,e^-} - {}^2\mu_{0\,e^-} + RT \ln K}{0.239\,\mathbf{F}} + \frac{RT}{0.239\,\mathbf{F}} \ln \frac{a_{Fe^{3+}}}{a_{Fe^{2+}}}$$

$$= U_{0\,abs} + \frac{RT}{0.239\,\mathbf{F}} \ln \frac{a_{Fe^{3+}}}{a_{Fe^{2+}}} \tag{5}$$

If more than one charge is involved, *e.g.*

$$Sn^{4+} + 2\,e^- \rightleftharpoons Sn^{2+}$$

the equilibrium constant may be written in the form

$$K^{\frac{1}{2}} = \frac{a^{\frac{1}{2}}{}_{Sn^{2+}}}{a^{\frac{1}{2}}{}_{Sn^{4+}}\, a_{e^-}}$$

and equation (5) becomes

$$U_{abs} = U_{0\,abs} + \frac{RT}{0.239 \cdot 2\,\mathbf{F}} \ln \frac{a_{Sn^{4+}}}{a_{Sn^{2+}}}$$

Similar equations may be derived for even more complex reactions. For a generalized chemical reaction written in the form

$$r\,Ox + ze^- \rightleftharpoons s\,Red$$

(where *Red* indicates a reduced system, *Ox* an oxidized system and *r* and *s* represent the coefficients of the chemical reaction), one obtains

$$U_{abs} = U_{0\,abs} + \frac{RT}{0.239 \cdot z\mathbf{F}} \ln \frac{a^r{}_{Ox}}{a^s{}_{Red}}$$

With, for example, a permanganate electrode in acid solution, the electrode reaction is

$$MnO_4^- + 8\,H^+ + 5\,e^- \rightleftharpoons Mn^{2+} + 4\,H_2O \qquad (6)$$

and the electric tension of this electrode at 25° C is

$$U_{abs} = U_{0\,abs} + \frac{0.05915}{5} \log \frac{a_{MnO_4^-} \cdot a^8{}_{H^+}}{a_{Mn^{2+}}}$$

In equation (6) the activity of the water has been considered as constant and subsumed into the constant U_0. As with other electrodes when $a_{ox} = a_{red} = 1$, $U_{abs} = U_{0\,abs}$, *i.e.* the absolute standard electric tension. From the preceding it appears that the quantity which actually defines the oxidizing or reducing power of a system is the redox tension.

In some papers and books the *r*H notation is found. This was introduced by Clark to indicate numerically the oxidizing or reducing power of a system under certain conditions. However, the *r*H notation is gradually being abandoned since it has caused more confusion than simplification. Since, however, the *r*H notation will be frequently encountered in the literature, it will be as well to outline its significance. An inert electrode immersed in an oxido-reduction system assumes an electric tension which is perfectly defined by the relationships discussed above. It is always possible to construct a hydrogen electrode in a solution of the same *p*H as the oxido-reduction solution and having this same electric tension. Naturally, the pressure of gaseous hydrogen in this particular hydrogen electrode, cannot be one atmosphere, but results from the general equation which gives the electric tension of a hydrogen electrode as a function of the activity of the H^+ ion and of the pressure of the molecular hydrogen pH_2; the latter is unknown but the tension is known

$$U_{abs} = -\frac{RT}{0.239\,\mathbf{F}} \ln \frac{p^{\frac{1}{2}}{}_{H_2}}{a_{H^+}}$$

which at room temperature becomes

$$U_{abs} = -\,0.029 \log pH_2 - 0.059\,pH$$

According to Clark the reciprocal of the logarithm of the pressure of the gaseous hydrogen is indicated with the symbol rH, by analogy with pH. Thus, the preceding equation becomes

$$U_{abs} = 0.029\ rH - 0.059\ pH$$

$$rH = \frac{U_{abs} + 0.059\ pH}{0.029}$$

$$= \frac{U_{abs}}{0.029} + 2\ pH$$

Keeping the pH value constant, the value of rH is a linear function of the redox tension and thus can serve to indicate the oxidizing or reducing power of a system. It is necessary, however, to emphasize that the rH value of a given system has no significance if the pH value is not specified, that for certain kinds of electrodes (*e.g.* the Fe^{2+}; Fe^{3+} electrode) it does not entirely indicate their oxidizing power, that within certain limits it is independent of the pH and that finally two measurements are necessary (U and pH) for expressing a quantity which is actually unequivocally defined by the tension U alone (*cf.* G. Milazzo, *Ann. chim.* (*Rome*), 38 (1948) 714).

Moreover, in a solution neither *pure* oxidizing nor *pure* reducing substances can exist, since the electric tension of an inert electrode would then be $+$ or $-\ \infty$ and thus the substance would oxidize or reduce the solvent itself. Even if the amounts of substance which react are extremely small and not detectable by analysis they are sufficient to bring the electric tension from $\pm\infty$ to within the range of reversible oxido-reduction systems. Such a tension does not depend upon an individual activity, but on the ratio of the activities of the substances representing the oxidized and the reduced states of the system. It is defined only when the two systems co-exist in a perfectly defined ratio of activities. Again, in this case as in the preceding ones, the sign of the electric tension is unequivocally defined by the order of enumeration of the phases and does not depend upon the particular direction, direct or inverse, considered for the reaction.

8. Absolute and Relative Electric Tensions[1]

The electric tensions of the electrodes so far described could be calculated if the absolute standard chemical potentials and the activities of the individual ionic species involved in the electrode reaction under the experimental conditions of temperature, ionic strength, etc. were known. Such tensions are called absolute because they represent the galvani tensions which actually exist at the interfaces of the electrodes. Such absolute values cannot in any way be calculated in the present state of our knowledge, so that absolute electric tensions cannot be derived *a priori.* They could be experimentally measured if the absolute electric tension of at least one electrode were known. It would then be possible, by measuring the electric tensions of a number of galvanic cells produced by suitable pairings of selected electrodes to determine and eliminate the diffusion electric tensions (*cf.* p. 138 *et seq.*), and the electric tensions existing between the two terminal metallic phases. Thus, the desired absolute electric tensions of each electrode could be obtained. Various ways have been tried to calculate the absolute electric tensions by means of thermodynamic cycles. These could, for example, be based on the sublimation of a metal, ionization of the gaseous atoms, solution and hydration of the gaseous ions, return to the metal of the electrons released during the ionization and finally discharge of the hydrated ions. Another approach might be to construct an electrode of known absolute electric tension: possibly of zero electric tension (dropping mercury electrode, rapid immersion microelectrode, mercury electrode at the maximum of its electrocapillary curve or a scraped electrode etc.). But none of these methods have given satisfactory results, so that, following a suggestion by Nernst, all electric tensions are referred to that of the standard hydrogen electrode. This consists of a sheet of platinized platinum immersed in a solution of H^+ ions, at an activity of 1 g/kg of water, through which bubbles gaseous hydrogen at 1 atmosphere pressure[2]. Electric tensions referred to this electrode are called the *relative electric tensions* with respect to hydrogen

[1] For a fuller treatment of the difficulties and uncertainties which affect absolute and relative electric tensions and for the attempts which are at present being made to resolve these difficulties *cf.* G. Milazzo and G. Bombara, *J. Electroanal. Chem.*, 1 (1960) 265.

[2] The pressure of the gaseous hydrogen is referred to 760 mm of mercury and has to be corrected for the vapour pressure of water and the hydrostatic pressure of the water column standing above the hydrogen exit in the bulk of the solution.

and should be given the symbol U_H; however, at the present time, the symbol U itself represents the electric tensions referred to the hydrogen electrode. Standard electric tensions referred to this electrode are known as standard electric tensions with respect to hydrogen or more shortly as standard electric tensions, with the symbol U_0.

In other words, relative electric tensions with respect to hydrogen, or more briefly electrode electric tensions, are simply the electric tensions of galvanic cells which consist of the electrode under examination, a standard hydrogen electrode and finally of the same metal as the first electrode, assembled in this order. Such electric tensions are actually rigorously and exactly defined only with galvanic cells with a single electrolyte, which are thus without diffusion electric tensions (*cf.* p. 138 *et seq.*). For electrodes with electrolytes differing from that of the hydrogen electrode, the electric tension is measured after reducing the diffusion electric tension[1] as far as possible to a few millivolts. The standard electric tension of a chlorine electrode, for example, is rigorously given by the electric tension of the cell

$$\begin{array}{c} Cl_2\text{–}Pt \ / \ Cl^-H^+ \ / \ Pt\text{–}H_2 \\ p_{Cl_2} = 1 \text{ Atm.} \ / \ a_{\pm \, HCl} = 1 \ / \ p_{H_2} = 1 \text{ Atm.} \end{array}$$

and the standard electric tension of a silver chloride electrode is given by the tension of the cell

$$\begin{array}{c} Ag \ / \ AgCl \ / \ Cl^-H^+ \ / \ Pt\text{–}H_2 \ / \ Ag \\ a_{HCl} = 1 \ / \ p_{H_2} = 1 \text{ Atm.} \end{array}$$

However, the electric tension of a copper electrode is given only approximately by the electric tension of the cell [1]

$$\begin{array}{c} Cu \ / \ Cu^{2+} \ // \ H^+ \ / \ Pt\text{–}H_2 \ / \ Cu \\ a_{H^+} = 1 \ / \ p_{H_2} = 1 \text{ Atm.} \end{array}$$

Ions not involved in the electrode reactions have been omitted for the sake of brevity.

The hydrogen electrode is rather inconvenient in use and is often replaced by other comparison electrodes whose electric tensions are known with sufficient accuracy. Table III, 2 collects the values of the electric ten-

[1] The symbol // indicates that the diffusion electric tension between the electrodes which are in contact has been reduced as far as possible.

sions of some of the most frequently used comparison electrodes[1], referred to the standard hydrogen electrode.

TABLE III,2

ELECTRIC TENSIONS OF COMPARISON ELECTRODES AT 25° C

Electrode	*Electric tension* *
$Hg/Hg_2Cl_2/KCl$ sat.	+ 0.2412
$Hg/Hg_2Cl_2/KCl$ $c_{KCl} = 1\ m$	+ 0.2810
$Hg/Hg_2Cl_2/KCl$ $c_{KCl} = 1\ N$	+ 0.2801
$Hg/Hg_2Cl_2/KCl$ $c_{KCl} = 0.1\ N$	+ 0.3337
$Hg/Hg_2Cl_2/KCl$ $a_{KCl} = 1$	+ 0.26796
$Hg/HgO/Ba(OH)_2$ sat.	+ 0.1462
$Hg/HgO/Ca(OH)_2$ sat.	+ 0.1923
$Ag/AgCl/HCl$ $a_{HCl} = 1$	+ 0.2224
$Hg/HgSO_4/H_2SO_4$ $a_{H_2SO_4} = 1$	+ 0.6151

* The positive sign indicates that the electrode is the positive pole of the cell formed by it and the hydrogen electrode.

The electric tensions of electrodes are usually referred to that of the hydrogen electrode even at temperatures different from 25 °C. The simple reference of an electrode at any temperature to the hydrogen electrode at the same temperature would be acceptable if the free enthalpy of the reaction of the hydrogen electrode were completely independent of the temperature. However, there is experimental evidence to the contrary, which indicates that the free enthalpy of the reaction of the hydrogen electrode varies with temperature and that the absolute electric tension of the hydrogen electrode also varies with temperature. In fact, the overall electric tension of the galvanic cell

$$\underset{t^\circ}{Pt} \;/\; \underset{t^\circ}{Pt\text{–}H_2} \;/\; H^+A^- \;/\; \underset{(t^\circ+\Delta t^\circ)}{Pt\text{–}H_2} \;/\; \underset{t^\circ}{Pt} \qquad (1)$$

is not zero.

Thus, the conventional measurement of the electric tension of an elec-

[1] For the correct preparation and use of comparison electrodes *cf.* G.J. HILLS and D.J.G. IVES, *J. Chem. Soc.*, (1951) 305, for hydrogen electrodes; G.J. JANZ and H. TANIGUCHI, *Chem. Revs.*, 53 (1953) 397, *J. Electrochem. Soc.*, 104 (1957) 123; E.A. GUGGENHEIM and J.E. PRUE, *Trans. Faraday Soc.*, 50 (1954) 231; R.G. BATES *et al.*, *J. Chem. Phys.*, 25 (1956) 361 for silver chloride electrodes; G.J. HILLS and D.J.G. IVES, *J. Chem. Soc.*, (1951) 311, 381 for calomel electrodes.

trode by reference to that of the hydrogen electrode without allowing for the change in electric tension of the latter, attributes to the test electrode not only changes in electric tension due to the variation of the free enthalpy of its own electrode reaction with temperature, but also those due to the variation in electric tension of the hydrogen electrode. Thus, it would be more exact to make the electric tension of a standard hydrogen electrode at a particular temperature, such as 25° C, the conventional zero; to determine the non-isothermal temperature coefficient of the electric tension of the standard hydrogen electrode, and hence to calculate the electric tension of the standard hydrogen electrode at various temperatures, referred to that of the same electrode at 25° C. All electric tensions could then be measured allowing for that of the standard hydrogen electrode at the experimental temperature. Making the temperature t of the reference electrode 25° C [1] within the cell (1), the non-isothermal temperature coefficient of the electric tension of the hydrogen electrode is given by the limiting value of the ratio of the measured electric tension ΔU to the temperature difference Δt between the two elements, as Δt tends towards zero. Measurements at temperatures removed from 25° C or for rather wide temperature differences must also take account of a possible dependence of $\Delta U/\Delta t$ on temperature and any necessary correction terms must be calculated or measured.

Many recent studies have investigated the non-isothermal temperature coefficient [1]. Fairly concordant results were obtained and can be summarized as follows. For a galvanic cell composed as described (1) of two equal electrodes with only a single ion determining the electrode tension, the overall temperature coefficient $\mathrm{d}U/\mathrm{d}t$, bearing in mind various factors which can affect it [2], may be expressed by

$$\frac{\mathrm{d}U}{\mathrm{d}t} = \frac{\mathrm{d}U_0}{\mathrm{d}t} + \frac{R}{z\mathrm{F}} \ln m + \frac{R}{z\mathrm{F}} \ln f + \frac{RT}{z\mathrm{F}} \frac{\mathrm{d} \ln f}{\mathrm{d}t} - \Sigma_i \frac{t_i q_i}{z_i \mathrm{F} T} - \Sigma_i \frac{RT}{z_i \mathrm{F}} t_i \frac{\mathrm{d} \ln m_i}{\mathrm{d}t} \tag{2}$$

where $\mathrm{d}U/\mathrm{d}t$ = the measured, overall temperature coefficient; $\mathrm{d}U_0/\mathrm{d}t$ = the true temperature coefficient for the standard electrode ($a = 1$); z = the

[1] *Cf.* G. MILAZZO and R. DEFAY, *J. Electroanal. Chem.*, 2 (1961) 419.

[2] The contribution by the homogeneous electric tension developed in platinum wire under a temperature gradient (Thomson effect) has been neglected in this case because of its much smaller order of magnitude.

valency of the ions producing the electrode tension; m = their molality; f = their activity coefficient; t_i = the transference number [1] of the ions i; q_i = the heat of transfer of the ions i from the region where they disappear to that where they appear.

The first term of the right-hand side of equation (2) is due to the change in free enthalpy of the electrode reaction for the standard electrode. The second, third and fourth terms give the dependence on concentration. The fifth is due to the change in entropy as a consequence of thermal diffusion and also allows for the variations in entropy produced by the transfer of electrons to regions at different temperatures. The last term on this side, is due to the change in concentration caused by the thermal diffusion (Soret effect). On extrapolation to zero time, or on preventing the Soret effect, this last term disappears; with sufficiently dilute solutions ($\ln f$) and ($d \ln f$) may be fairly readily calculated. Again, if an attempt is made to estimate the term

$$\Sigma_i \frac{t_i q_i}{z_i \mathbf{F} T}$$

using experimental data on the Soret effect, a fairly good agreement is found between the values of the temperature coefficient and the value predicted from equation (2). This shows the validity of this approach.

However, even today, the use of the so-called isothermal temperature coefficient of electrode tension (defined by the ratio dU/dt) is still widespread. This ratio is obtained from measurements on two equal galvanic cells, each of them isothermal, but at different temperatures, formed of the test electrode and a hydrogen electrode. With a procedure similar to that described for comparison electrodes, the electric tensions of the two cells are found with reference to that of the standard hydrogen electrode. These electric tensions are, by definition, made equal to that of the test electrode at various temperatures, so that the temperature coefficient of the electric tension of the standard hydrogen electrode is also attributed to the test electrode. The most reliable data on the isothermal temperature coefficients are those published by R. G. Bates and his colleagues of the National Bureau of Standards [2].

[1] Attention is drawn to the sometimes inevitable confusion of symbols; in this case both the centigrade temperature and the transference number have the same symbol t, but the latter is distinguished by the subscript i.

[2] See the publications of R. G. Bates and his colleagues in the collection of the *J. Research Nat. Bur. Standards*, and D. J. G. Ives and G. J. Janz, *Reference Electrodes*, Academic Press, New York & London (1961).

9. Diffusion Electric Tensions between Solutions of Electrolytes

The boundary between two electrolyte solutions can also be the location of an electric tension, not only when the two electrolytes are different, but also when they are qualitatively the same but of different concentrations. Although these electric tensions are small they often cannot be ignored when considering the electric tensions of the electrodes so far studied. Hence, in measuring electric tensions with cells composed of a test electrode and a comparison electrode, it is necessary either to use some device to diminish these diffusion tensions as far as possible, or to calculate them with sufficient accuracy.

To illustrate the principle of such a calculation, consider two solutions of hydrochloric acid at concentrations c_1 and c_2 ($c_1 < c_2$); these solutions are in contact along an ideal boundary $A—B$ (Fig. III, 7). The electrolytes tend to diffuse from the solution at concentration c_2, which has the higher chemical potential, to that at concentration c_1 which has the lesser chemical potential. However, each of the two ionic species present diffuses independently of the other under the influence of the chemical potential, with a diffusion velocity which is a characteristic of the species and is proportional to their respective electrical mobilities, as well as to the concentration difference. With hydrochloric acid, the H^+ ions diffuse more rapidly than the Cl^- ions so that an excess of H^+ ions is produced in the more dilute solution, and is not compensated by a corresponding quantity of Cl^- ions. Since the H^+ ions are positively charged the more dilute solution also becomes positively charged with respect to the more concentrated and hence an electric tension is produced between the two sides of the boundary $A—B$. Thus, an electrochemical double layer is immediately formed, whose electric tension opposes any further diffusion of H^+ ions, whilst aiding that of the Cl^- ions until the diffusion velocities of the two species become equal.

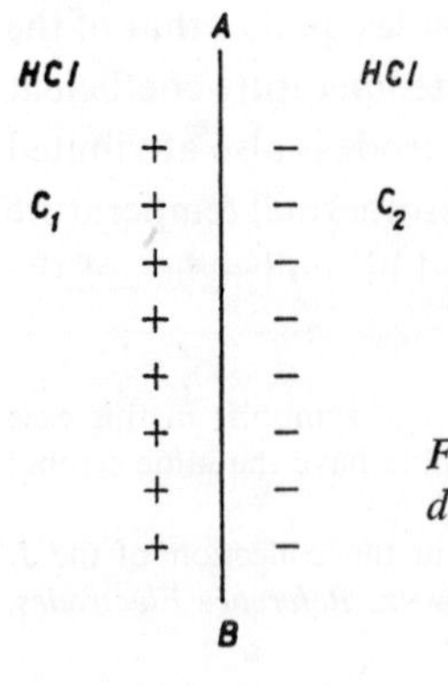

Fig. III, 7. Representation of diffusion electric tension

This diffusion is irreversible and always occurs when a solution of an electrolyte is in contact with another solution which is qualitatively the same but quantitatively different. When, for example, the two solutions are the electrolytes of two half cells in open electrical circuit, the two ionic species will migrate in the same direction. This does not give rise to external electrical work, because in the example quoted for instance, the electrical work necessary to move the H^+ ions through the double layer is exactly equal but of opposite sign to that produced by the passage of the Cl^- ions through the same double layer in the same direction. When the external circuit is closed however, and the cell is allowed to function spontaneously, the process becomes reversible and the ions migrate in opposite directions.

If the quantity of electricity which passes is 1 **F**, it will be divided between the two ionic species in the proportion of t_- for the Cl^- anions and $t_+ = (1 - t_-)$ for the H^+ cations. Since the process is reversible the free enthalpy of the transformation is equal to the external work, calculated electrically. This is $U\mathbf{F}$ (when U is the electric tension of the electrochemical double layer at the boundary between the two electrolyte solutions). The change in free enthalpy for each of the two ionic species is given by the difference between the chemical potentials of each, within the two solutions, multiplied by the number of ions which effectively move. Thus, for the anions

$$\Delta G = t_- \left[(\mu_{0\,an} + RT \ln a_{an.\,final}) - (\mu_{0\,an} + RT \ln a_{an.\,initial})\right]$$

$$= -\, t_- RT \ln \frac{a_{an.\,initial}}{a_{an.\,final}} \qquad (1)$$

and for the cations

$$\Delta G = -\,(1 - t_-)\, RT \ln \frac{a_{cat.\,initial}}{a_{cat.\,final}} \qquad (2)$$

Making the electrical work equal to the total change in free enthalpy, which is itself equal to the sum of (1) and (2), gives

$$0.239\, U_{diff.}\mathbf{F} = -\,(1 - t_-)\, RT \ln \frac{a_{cat.\,initial}}{a_{cat.\,final}} - t_- RT \ln \frac{a_{an.\,initial}}{a_{an.\,final}}$$

$$U_{diff.} = -\,(1 - t_-) \frac{RT}{0.239\,\mathbf{F}} \ln \frac{a_{cat.\,initial}}{a_{cat.\,final}} - t_- \frac{RT}{0.239\,\mathbf{F}} \ln \frac{a_{an.\,initial}}{a_{an.\,final}}$$

$$= (1 - t_-) \frac{RT}{0.239\,\mathbf{F}} \ln \frac{a_{cat.\,final}}{a_{cat.\,initial}} - t_- \frac{RT}{0.239\,\mathbf{F}} \ln \frac{a_{an.\,initial}}{a_{an.\,final}}$$

Since the two ionic species migrate in opposite directions, the final concentration and hence very approximately the final activity, of one is equal to the initial concentration, and hence activity, of the other; so that

$$U_{diff.} = (1 - t_{-})\frac{RT}{0.239\ \mathbf{F}} \ln \frac{a_{an.\ initial}}{a_{an.\ final}} - t_{-} \frac{RT}{0.239\ \mathbf{F}} \ln \frac{a_{an.\ initial}}{a_{an.\ final}}$$

$$= (1 - 2\,t_{-})\frac{RT}{0.239\ \mathbf{F}} \ln \frac{a_{an.\ initial}}{a_{an.\ final}}$$

$$= \frac{u_{+} - u_{-}}{u_{+} + u_{-}} \cdot \frac{RT}{0.239\ \mathbf{F}} \ln \frac{a_{an.\ initial}}{a_{an.\ final}}$$

and at 25° C

$$U = \frac{u_{+} - u_{-}}{u_{+} + u_{-}} \cdot 0.05915 \log \frac{a_{an.\ initial}}{a_{an.\ final}}$$

This calculation is valid for two solutions at different concentrations, of the same uni-univalent binary electrolyte. Following the conventions adopted, the passage of the anions from the more concentrated solution (phase 2) to the more dilute (phase 1) is equivalent to the passage of a corresponding positive charge in the reverse direction (1 → 2) and the sign of $RT/0.239\ \mathbf{F}$ is positive. This leads to a positive electric tension for $u_{+} > u_{-}$ (as with hydrochloric acid), *i.e.* the dilute solution (phase 1) is positive with respect to the concentrated solution; whilst it leads to a negative electric tension for $u_{+} < u_{-}$ (as with sodium hydroxide), *i.e.* the dilute solution (phase 1) is negative with respect to the concentrated solution (phase 2). The diffusion electric tension thus obtained, for the diffusion from the more concentrated to the more dilute phase, is algebraically additive with that derived from the electric tensions of the individual electrodes [1].

When the electrolyte is not uni-univalent, but gives rise to z-valent ions, the expression for the diffusion tension assumes the approximate form

$$U_{diff.} = \frac{RT}{0.239\ \mathbf{F}} \cdot \frac{\frac{u_{+}}{z_{+}} - \frac{u_{-}}{z_{-}}}{u_{+} + u_{-}} \ln \frac{a_{initial}}{a_{final}}$$

[1] This calculation, whilst being a clear illustration for teaching purposes of the origin of diffusion electric tensions, is not rigorous since it considers the rates of migration and hence the transference numbers of various ionic species, to be constant at different concentrations. It also requires a knowledge of the activity of individual ionic species which is in fact unknown. For an accurate analysis of diffusion tensions see, amongst others, G. Maronny and G. Valensi, *C.R. VII Réunion CITCE*, Lindau 1955, Butterworth, London, 1957, p. 38.

in which the absolute values of the respective valencies are used in place of z_+ and z_-. The equation becomes extremely complex for the general case of any electrolyte at any concentration.

Henderson[1] derived an equation based on certain simplifying hypotheses, which gives a general idea of the magnitude of diffusion tensions. He assumed that the diffusion layer was such that the change in concentration was linear in the direction of the diffusion, and that the activities were equal to the concentrations. Then,

$$U_{diff.} = \frac{RT}{0.239\,\mathbf{F}} \frac{(U_1 - V_1) - (U_2 - V_2)}{(U_1' + V_1') - (U_2' + V_2')} \ln \frac{U_1' + V_1'}{U_2' + V_2'}$$

$U_1 = u_{1+}c_{1+} + u_{2+}c_{2+} + u_{3+}c_{3+} \ldots$ is the sum of the products of the electrical mobilities $u_{1+}, u_{2+}, u_{3+} \ldots$ of the cations 1, 2, 3, ... with their respective ionic concentrations $c_{1+}, c_{2+}, c_{3+} \ldots$;

$V_1 = u_{1-}c_{1-} + u_{2-}c_{2-} + u_{3-}c_{3-} \ldots$ is the sum of the products of the electrical mobilities $u_{1-}, u_{2-}, u_{3-} \ldots$ of the anions 1, 2, 3, ... with their respective ionic concentrations $c_{1-}, c_{2-}, c_{3-} \ldots$;

$U_1' = u_{1+}c_{1+}z_{1+} + u_{2+}c_{2+}z_{2+} + u_{3+}c_{3+}z_{3+} \ldots$ is the sum of the products of the electrical mobilities $u_{1+}, u_{2+}, u_{3+} \ldots$ of the cations 1, 2, 3, with their respective ionic concentrations $c_{1+}, c_{2+}, c_{3+} \ldots$ and with their respective valencies $z_{1+}, z_{2+}, z_{3+} \ldots$;

$V_1' = u_{1-}c_{1-}z_{1-} + u_{2-}c_{2-}z_{2-} + u_{3-}c_{3-}z_{3-} \ldots$ is the sum of the products of the electrical mobilities $u_{1-}, u_{2-}, u_{3-} \ldots$ of the anions 1, 2, 3 ... with their respective ionic concentrations $c_{1-}, c_{2-}, c_{3-} \ldots$ and with their respective valencies $z_{1-}, z_{2-}, z_{3-} \ldots$. All these apply to the first solution, and U_2, V_2, U_2', V_2' are the analogous expressions for the second solution. The values of z_+ and z_- are absolute in Henderson's equation also.

When the two solutions containing ions with different migration velocities come into contact, an electric tension is thus always produced at the boundary. The exact calculation of this electric tension is impossible and so is its experimental determination, since it is a galvani tension between two different phases. It is thus preferable to attempt to eliminate or to diminish this tension as far as possible by a suitable experimental arrangement. One such arrangement is to connect the two solutions with a salt

[1] P. HENDERSON, *Z. physik. Chem.*, 59 (1907) 118; 63 (1908) 325.

bridge composed of a highly concentrated, or even saturated, solution of a uni-univalent binary electrolyte in which the migration velocity of the cation is as close as possible to that of the anion. Diffusion occurs from the concentrated to the dilute solution and $u_+ \cong u_-$, so that $u_+ - u_- \cong 0$ and hence $U_{diff} \cong 0$. The electrolytes which have so far proved to be most suitable for this purpose are potassium chloride, sodium nitrate and ammonium nitrate. Diffusion electric tensions are in this way diminished but not completely eliminated[1].

10. Concentration Cells[2]

The process described in the preceding paragraph allows the transformation into external electrical work of the work obtained when an ion passes from an initial activity a_2 to a final activity a_1, *i.e.* when the ionic concentration changes. This process is utilized in the construction of concentration cells. These are formed of two electrodes of the first or second kind or of gas electrodes, which are qualitatively the same and differ only in their electrolytic concentration. Consider the following cell

$$\begin{array}{ccccccc} & a & & b & & c & \\ \text{Pt–H}_2 & / & \text{HCl} & / & \text{HCl} & / & \text{Pt–H}_2 \\ & & a_1 & & a_2 & & \\ 1 & & 2 & & 3 & & 4 \end{array}$$

formed of two hydrogen electrodes, in both of which the gaseous hydrogen has the same pressure whilst the concentration of hydrochloric acid differs, such that $a_1 < a_2$. The spontaneous process is such that the more concentrated solution tends to become diluted. By carrying out the process in a suitable manner it is possible to transform the dilution work into external electrical work. The spontaneous processes which occur when 1 F of electricity is passed are as follows.

[1] *Cf.* G. Kortüm, *Lehrbuch der Elektrochemie*, Verlag Chemie, Weinheim (1957) p. 254 *et seq.*; G. Maronny and G. Valensi, *C.R. VII Réunion CITCE*, Lindau 1955, Butterworth, London (1957) p. 38; G. Milazzo, *Rend. ist. super. sanità*, 20 (1957) 379; G. Milazzo and G. Bombara, *J. Electroanal. Chem.*, 1 (1960) 265; N. P. Finkelstein and E. T. Verdier, *Trans. Faraday Soc.*, 53 (1957) 1618.

[2] It should be noted that all the diagrams of concentration cells in this section are written with a phase order such that the positively charged particles move within the cell in the direction of the phase enumeration when the cell works spontaneously; so that the electrical charge nF enters the calculations with a positive sign following the conventions adopted.

(a) One g equivalent of hydrogen passes from the elementary to the ionic state at the boundary *a* between the electrode and the solution.

(b) The process described in the preceding section takes place at the boundary *b* between the two solutions of electrolyte.

(c) One g ion of hydrogen is discharged at the boundary *c* between the solution and the electrode.

Within the cell the positive current, represented by H^+ ions, goes from the dilute to the concentrated solution, *i.e.* in the direction of the phase enumeration, since only thus can the two concentrations become equal. In fact, on the passage of current within the half cell on the left, 1 g equivalent of H^+ ion is formed and at the same time $t_+ = (1 - t_-)$ equivalents migrate towards the half cell on the right traversing the boundary *b*, so that t_- equivalents of H^+ ion remain in excess in the left-hand half cell. In the half cell on the right 1 equivalent of H^+ ions are discharged and at the same time $(1 - t_-)$ equivalents enter the cell from the half cell on the left, so that there is a net loss of t_- equivalents of H^+ ions. Simultaneously, t_- equivalents of Cl^- ions migrate from the half cell on the right towards that on the left. The overall effect is thus equivalent to a passage of t_- equivalents of hydrochloric acid from the half cell at higher concentration to that at lower concentration.

The change in free enthalpy ΔG for each of the two ionic species[1] is

$$- t_- RT \ln (a_2/a_1)$$

so that the total is

$$- 2\, t_- RT \ln (a_2/a_1)$$

where the activity is the mean activity of the electrolyte (*cf.* p. 148 *et seq.*).

Putting 0.239 $U\mathbf{F}$ for the electrical work gives

$$0.239\, U\mathbf{F} = - 2\, t_- RT \ln (a_2/a_1)$$

$$U_{cat} = 2\, t_- \frac{RT}{0.239\, \mathbf{F}} \ln (a_1/a_2)$$

In the final equation U_{cat} indicates that the cell is reversible with respect to the cations. If the cell were, however, reversible with respect to the

[1] To a first approximation the activity of the cations is considered to be equal to that of the anions in each of the two solutions.

anions, a similar argument would give the equation for the electric tension

$$U_{an} = -\,2\,t_+ \frac{RT}{0.239\ \mathrm{F}} \ln\,(a_1/a_2)$$

For the general case of an electrolyte with z-valent ions, the equations assume the form

$$U_{cat} = \left(\frac{1}{z_+} + \frac{1}{z_-}\right) \frac{RT}{0.239\ \mathrm{F}}\ t_-\ln \frac{a_1}{a_2}$$

$$U_{an} = -\left(\frac{1}{z_+} + \frac{1}{z_-}\right) \frac{RT}{0.239\ \mathrm{F}}\ t_+\ln \frac{a_1}{a_2}$$

These are readily derived if it is borne in mind that with the passage of 1 F of electricity, the migrating cations are given by the fraction t_+/z_+ and the anions by t_-/z_-. The last two equations lead to a simple method of determining transference numbers by measuring the electric tensions of concentration cells with transference, working reversibly, assuming that the values of the transference numbers do not change between the concentrations c_1 and c_2.

In a concentration cell the positive pole is in the more concentrated solution if the cations are active in the electrochemical process, whilst it is in the more dilute solution if the anions are active in the electrochemical process, *e.g.* two chlorine electrodes in hydrochloric acid rather than two hydrogen electrodes in hydrochloric acid.

It should be noted that in fact, cells of the type described do not transform chemical energy into external electrical work, since the isothermal and reversible diminution of the osmotic pressure of an ideal solution (analogous to the isothermal and reversible expansion of an ideal gas) does not change its internal energy. The external electrical work is produced exclusively at the expense of the thermal energy of the environment. Cells of the type described are called *concentration cells with transference*.

Another type of concentration cell exists which is called a *concentration cell without transference*. This consists of two equal cells with the same electrolyte, but of different concentrations, connected in opposition. The cell

$$\begin{array}{ccccccc} & & & \mathrm{Cu} & & & \\ {}^-\mathrm{Ag}\ / & \mathrm{AgCl\text{-}HCl}\ / & \mathrm{Pt\text{-}H_2} & \text{—} & \mathrm{Pt\text{-}H_2}\ / & \mathrm{HCl\text{-}AgCl}\ / & \mathrm{Ag}^+ \\ & \text{solid}\quad a_2 & & & & a_1\quad \text{solid} & \\ 1 & 2 & 3 & 4 & 5 & 6 & 7 \end{array}$$

consists inherently of two cells

$$\begin{array}{ccccccc} & \text{II} & & & & \text{I} & \\ {}^{+}\text{Ag} / & \text{AgCl–HCl} / & \text{Pt–H}_2{}^{-} & \text{and} & {}^{-}\text{Pt–H}_2 / & \text{HCl–AgCl} / & \text{Ag}^{+} \\ & \text{solid} \quad a_2 & & & & a_1 \quad \text{solid} & \\ \text{electrode 1} & & \text{electrode 2} & & \text{electrode 3} & & \text{electrode 4} \end{array}$$

in which the same electrolyte — hydrochloric acid — has activities a_1 and a_2 ($a_1 < a_2$). Of these two, cell I has an electric tension which in absolute value is greater than that of cell II, so that the spontaneous electrochemical process of cell I makes the same reaction occur in the inverse direction in cell II. Thus, cell II functions as an electrolytic, rather than a galvanic, cell. For the passage of 1**F** of electricity in the direction of the phase enumeration, 1 mole of silver chloride disappears at the silver electrode of cell I, depositing 1 equivalent of metallic silver on the electrode and liberating 1 equivalent of Cl^- ion into solution; at the hydrogen electrode 1 equivalent of hydrogen passes into solution as ions, so that in cell I the overall concentration of the electrolyte increases by 1 equivalent. In cell II the inverse process occurs. At the hydrogen electrode 1 equivalent of H^+ ions is discharged, whilst at the silver electrode 1 mole of chloride is formed by the passage into the ionic state of 1 equivalent of silver, which immediately reacts with an equivalent of Cl^- ions released by the discharge of the H^+ ion. Thus, in cell II, 1 equivalent of electrolyte disappears. The total process is the sum of the reactions taking place in the two cells and consists in the reversible and isothermal passage of 1 mole of electrolyte from the solution at higher activity a_2 to the solution at lower activity a_1. Since the process is isothermal and reversible it is possible to equate its free enthalpy with the external electrical work considering, as a first approximation, that the electrolyte is completely dissociated and that the activity of the anions is equal to that of the cations; then the change in free enthalpy for each of the two ionic species is $-RT\ln(a_2/a_1)$. The sum of the two is equal to the electrical work, so that

$$0.239\ U\mathbf{F} = -2\ RT \ln (a_2/a_1)$$

$$U = 2\frac{RT}{0.239\ \mathbf{F}} \ln \frac{a_1}{a_2}$$

The same result may be achieved by calculating the total electric tension as the algebraic sum of the electric tensions of cells I and II. In turn, these electric tensions are in each case the differences between the electric tensions of each of the individual electrodes. Applying the customary conventions to calculate each of these, gives

$$U = (U_1 - U_3) + (U_5 - U_7) \tag{1}$$

The electric tensions of the individual electrodes are

$$U_1 = U_0 - \frac{RT}{0.239\ \mathbf{F}} \ln a_{2\ \mathrm{Cl^-}}$$

$$U_3 = \frac{RT}{0.239\ \mathbf{F}} \ln a_{2\ \mathrm{H^+}}$$

$$U_5 = \frac{RT}{0.239\ \mathbf{F}} \ln a_{1\ \mathrm{H^+}}$$

$$U_7 = U_0 - \frac{RT}{0.239\ \mathbf{F}} \ln a_{1\ \mathrm{Cl^-}}$$

So that, considering the activities of each of the two ionic species as equal to each other and to the mean activity $a_{\pm}$ (*cf.* p. 148 *et seq.*), equation (1) becomes

$$U = 2\frac{RT}{0.239\ \mathbf{F}} \ln \frac{a_{1\pm}}{a_{2\pm}} \tag{2}$$

The electric tension of any concentration cell without transference, with any electrodes, is given by the general equation which is easily deduced by a similar reasoning.

$$U = \frac{p}{q}\,\frac{RT}{0.239\ z\mathbf{F}} \ln \frac{a_1}{a_2}$$

where p is the total number of ions produced by the dissociation of one molecule of electrolyte, q is the number of ions (anions or cations) for which the terminal electrodes are reversible produced by the dissociation of 1 molecule of electrolyte and z is the number of electric charges involved, corresponding to the valency of the ions with respect to which the terminal electrodes are reversible.

Concentration cells are used especially for measurements of the activities of electrolytes (*cf.* p. 148 *et seq.*).

A final type of concentration cell consists of two electrodes of the same metal but with different activities, in a solution of its ions, *e.g.* two amalgam electrodes (*cf.* p. 126 *et seq.*) at different concentrations. A cell composed of amalgams of a metal ($\mathrm{Me}_{amalg.}$) at two different concentrations ($c_2 > c_1$) in a solution of its ions of valency z (Me^{z+}), according to the scheme

$$\underset{c_2}{Me_{amalg.}} \;\Big|_{1}\; Me^{z+}_{sol} \;\Big|_{2}\; \underset{c_1}{Me_{amalg.}}\;{}_{3}$$

gives the following spontaneous process on closed circuit. The electrode at the higher concentration has a greater chemical potential than the other and hence metal passes from this into solution as ions. On the other hand, at the other electrode an equal quantity of ions are discharged and form metal dissolved in the amalgam. Thus the metal passes from the more concentrated amalgam (phase 1) to the less concentrated (phase 3) whilst the concentration of the electrolyte is unchanged. If 1 g atom of metal passes from concentration c_2 to concentration c_1, the change in free enthalpy is $-RT\ln(a_2/a_1)$; obviously this is recovered as external electrical work 0.239 UFz. Equating these two expressions gives

$$U = \frac{RT}{0.239\, z\mathrm{F}} \ln \frac{a_1}{a_2}$$

This equation is perfectly analogous to that expressing the electric tension of a concentration cell without transport.

Such equations have proved to be completely valid for very dilute amalgams, to which the laws of ideal solutions apply. Increasing the concentration leads to a less satisfactory agreement between the calculated and measured values for the electric tension. This may be attributed to several causes:

(a) The laws of ideal solutions are of limited validity for concentrated amalgams.

(b) Marked thermal effects appear with the dilution and concentration of the respective amalgams that do not cancel each other out.

(c) There may be variations in the state of the dissolved metal (associations of atoms of the dissolved metal to give polyatomic molecules, formation of compounds between atoms of the dissolved metal and the mercury, etc.).

(d) Finally, above certain limits of concentration, corresponding to saturation, new metallic phases appear in the electrode with abrupt changes of composition. It is evident that the validity of these equations presupposes that the electrodes are monophasic so that the equilibrium between the electrode and the solution may be defined on the basis of the thermodynamic activity.

The equation giving the electric tension of a concentration cell with amalgam electrodes is applicable not only to these but to all others consisting of a solution of one metal in another so long as this solution is monophasic. The electric tension of such cells does not depend on the particular electrolyte used, *i.e.* it does not matter if the electrolyte is dissolved in water, in a non-aqueous solvent or even in a molten electrolyte (*cf.* p. 177 *et seq.*).

11. Ionic Activities

In all physico-chemical, and hence electrochemical, calculations carried out during the last century and the first few years of the present one, the ionic concentration was used. This was tacitly deduced from the total concentration and the degree of dissociation. However, substances taking part in electrochemical processes in cells generally belong to the group of so-called strong electrolytes; these are assumed to dissociate completely so that the classical Arrhenius theory of dissociation equilibria is not applicable. It is thus not correct to calculate the effective ionic concentration from the total concentration of electrolytes using a degree of dissociation, derived for example, from cryoscopic or conductance measurements. With strong electrolytes, in particular, the degree of dissociation cannot be obtained from such measurements (*cf.* Chapter II, p. 85 *et seq.*). Even if it were possible to determine exactly the true degree of dissociation and to calculate the actual ionic concentration, this would not be sufficient for calculations of electrode tension since these actually depend on other parameters which do not enter into the dissociation equilibrium.

It has already been noted when dealing with the state of strong electrolytes (Chapter II, p. 85 *et seq.*) that although these had to be considered as completely dissociated, yet they had to be treated as if they were only partially active because of the inter-ionic forces of an electrostatic nature; and that the ratio Λ_v/Λ_0 serves to indicate rather a variation in migration velocity than a degree of dissociation. For this reason Lewis in 1900 introduced the concept of *activity* (*cf.* Chapter I, on page 13). The activity is that quantity which must be used in place of the actual concentration and represents the active mass which satisfies, for example, the law of mass action for the equilibria of strong electrolytes, and explains the variations of their cryoscopic and osmotic behaviour from the ideal, etc. The value of the activity differs from that of the concentration, but the difference diminishes as the solution becomes more dilute. Hence, of necessity, the equations so far derived can not be in accordance with theory when the concentrations are such that the solutions can no longer be considered as ideally dilute. In fact, only when they are very dilute, do the activity and the concentration coincide numerically, and the equations give the same results on introducing concentration as with activity.

If a represents the activity and c the concentration,

$$a = f_a c$$

where f_a is a factor, called the *stoichiometric coefficient of activity*[1], which in general is less than 1 and reaches the value of 1 only with ideally dilute solutions. It is not, however, identical with the degree of dissociation, although it has qualitatively the same behaviour as a function of concentration (with the exception of very concentrated solutions[2]). The origin of the degree of dissociation is different from that of the activity coefficient. The former is a result of the equilibrium for the dissociation of the molecules into ions, whilst the latter although including the effect of any partial dissociation, originates in the interionic forces of an electrostatic nature[3], between ions which are already dissociated. It also takes account of any possible deviations from ideal thermodynamic behaviour. Whilst the degree of dissociation is independent of the presence of ions of other chemical species the activity coefficient is a function, not only of its own concentration, but also of the concentrations and charges of all the other ions present, independently of their chemical nature.

A further activity coefficient has been defined which is called the *rational activity coefficient*. This second way of expressing activity as a function of the concentration refers not to the total concentration of electrolyte, but to the actual ionic concentration. Thus, whilst the stoichiometric activity coefficient takes account of all the causes of deviation

[1] Some texts indicate the stoichiometric activity coefficient by the symbol γ.

[2] Values for the activity coefficient greater than 1 are possible in very concentrated solutions and have in fact been found.

[3] The concept of activity is not limited to electrolytes alone, it can be applied also to neutral molecules and in this case its origin is to be found in various types of intermolecular forces. With electrolytes, however, these latter types of forces are so small as to be negligible in comparison with the interionic electrostatic forces.

from ideal behaviour and in particular includes the effect of any incomplete dissociation or any formation of associated ion pairs (*cf.* Chapter II, p. 91), the rational activity coefficient takes account only of those causes of deviations from ideal behaviour which can be referred to the actual ionic concentration.

The activity coefficient was first introduced by Lewis as a purely empirical factor to maintain the validity and the formal expressions of the relationships of classical thermodynamics such as the Law of Mass Action. Following recent studies on strong electrolytes, it has now become possible in some cases (*i.e.* with dilute solutions within the field of validity of the Debye and Hückel theory) to calculate with certain approximations, the rational activity coefficients on a purely theoretical basis. Thus, a definite and real physical significance must be attributed to this quantity, as can be seen from the following. For an ionic species which behaves ideally in solution,

$$\mu_1 = \mu_0 + RT \ln c \tag{1}$$

where c is the actual ionic concentration. If the behaviour of the system is not ideal, equation (1) must be written

$$\mu_2 = \mu_0 + RT \ln (fc)$$

$$\mu_2 = \mu_0 + RT \ln c + RT \ln f \tag{2}$$

where f is the rational activity coefficient. Subtracting (1) from (2) gives

$$\mu_2 - \mu_1 = RT \ln f \tag{3}$$

Equation (3) gives the difference between the chemical potentials of an ionic species in a real and in an ideal solution. This difference corresponds to a variation in energy and may be considered as equivalent to the additional energy of the ion due to the presence of the ionic atmosphere. It is possible to calculate this energy, using the Debye and Hückel theory[1],

[1] In view of the purpose and size of this text it is not possible to develop in detail the Debye and Hückel theory of electrolytes which today represents the only method available for the calculation of the activity coefficients of individual ionic species; and even this method is only approximate. Moreover, it falls far outside the scope outlined in the introduction since it is not a theory concerned with electrochemical phenomena as there defined and delimited. Hence, only the final results of this theory are given and reference should be made to a specialized work to deepen these concepts and for the mathematical development of the calculations.

which depends on the so-called *ionic strength* of the solution. This is defined by the equation

$$I = \tfrac{1}{2}\Sigma c_i z_i^2$$

where c_i is the actual concentration of each ionic species and z_i is its respective valency. This quantity is a measure of the electric field, existing in the solution, and acting on the individual ions. More precisely, it may be calculated that

$$-\log f_i = \frac{A'}{(\varepsilon T)^{\frac{3}{2}}} z_i^2 \sqrt{I} \quad (4)$$

where ε is the dielectric constant of the solvent and A' is a constant made up of universal constants. For a particular solvent at a particular temperature both the dielectric constant and the temperature are fixed so that equation (4) becomes

$$-\log f_i = A z_i^2 \sqrt{I}$$

For water at 25° C the value of A is 0.51 to a sufficient approximation.

Thus, in all physicochemical calculations the activity must be used in place of the concentration and this is especially true in the calculation of electric tensions. There is no need here to expand the concepts and methods of measurement which are fully dealt with in textbooks of physical chemistry. It is, however, convenient to indicate the principles underlying some of the methods which permit the experimental determination of stoichiometrical activity coefficients so as to determine the activities which are involved in electrochemical phenomena; these coefficients cannot be calculated *a priori* like the rational activity coefficients. Above all, it must be emphasized that every ionic species has its own activity and thus its own activity coefficient, which is given by the equation

$$f_a = \frac{a}{c}$$

However, no method is yet available, either experimental or theoretical, for the determination of activity coefficients of individual ionic species in a rigorous and absolute sense. For this reason use is made of the so-called *mean coefficients of activity*. These are formally defined as the geometric means of the individual coefficients

$$f_{a\pm} = \sqrt[(p+q)]{f_{a+}{}^p + f_{a-}{}^q} \quad (5)$$

where p is the number of cations and q the number of anions formed by the dissociation of one molecule of electrolyte. The mean activity coefficient is susceptible to experiment.

A first method is based on the determination of the solubilities of poorly soluble salts. In a saturated solution of a poorly soluble salt A_pB_q, the salt in solution is in equilibrium with the undissolved salt so that the chemical potential of this undissolved salt must be equal to that of the undissociated electrolyte in solution and also equal to the sum of the chemical potentials of the ions resulting from the dissociation

$$A_pB_q \rightleftharpoons pA^+ + qB^-$$

So that, (*cf.* Chapter I, p. 15 *et seq.*).

$$\begin{aligned}\mu_{[A_pB_q]} &= \mu_{0\,[A_pB_q]} + RT \ln a_{A_pB_q} \\ &= p(\mu_{0\,A^+} + RT \ln a_{A^+}) + q(\mu_{0\,B^-} + RT \ln a_{B^-})\end{aligned}$$

Since the values of μ_0, the coefficients p and q, and R are all constants at constant temperature

$$(a_{A^+})^p \times (a_{B^-})^q = \text{constant} = L \tag{6}$$

Where L is the solubility product referred to the activity. Hence, from the relationship of activity to concentration, equation (6) becomes

$$(c_{A^+})^p\ (c_{B^-})^q\ (f_{A^+})^p\ (f_{B^-})^q = L \tag{7}$$

If S is the solubility of the electrolyte in moles per litre, the individual ionic concentrations become

$$c_{A^+} = pS$$

$$c_{B^-} = qS$$

So that, bearing in mind the definition of the mean activity coefficient (5), equation (7) becomes

$$(p^p + q^q)\ S^{(p+q)}\ f_{a\pm}{}^{(p+q)} = L$$

whence

$$Sf_{\pm} = \text{constant}$$

This is valid for any solution, provided that undissolved salt is present, independently of the presence or absence of other electrolytes. Measuring the solubility S in two solutions I and II, of different ionic strengths gives

$$\frac{S_I}{S_{II}} = \frac{f_{\pm II}}{f_{\pm I}}$$

By extrapolation of the value of S_I for the solution I to zero ionic strength, so that the mean activity coefficient in this solution becomes equal to 1, it is possible to deduce the mean activity coefficient of solution II at finite ionic strengths.

A second method permitting the exact determination of mean activity coefficients is based on the measurement of the electric tensions of galvanic cells with only one electrolyte in solution. If, for example, it is wished to determine the mean activity coefficient of hydrochloric acid, the electric tension of the following cell may be measured

$$\underset{1}{Pt–Cl_2} \ / \ Cl^-H^+ \ / \ \underset{2}{Pt–H_2}$$

The electric tension is given by that of the first electrode less that of the second (*cf.* p. 157 *et seq.*) *i.e.* that of the chlorine electrode less that of the hydrogen electrode.

$$U = U_{Cl^-} - U_{H^+}$$

Writing these two electric tensions in their explicit forms gives

$$U = U_{0\,Cl^-} - \frac{RT}{0.239\ \mathbf{F}} \ln a_{Cl^-} - \frac{RT}{0.239\ \mathbf{F}} \ln a_{H^+}$$

$$= U_{0\,Cl^-} - \frac{RT}{0.239\ \mathbf{F}} \ln (a_{Cl^-} \cdot a_{H^+})$$

$$= U_{0\,Cl^-} - \frac{RT}{0.239\ \mathbf{F}} \ln (c_{Cl^-} f_{Cl^-} c_{H^+} f_{H^+})$$

Since the concentration of the Cl^- ions is equal to that of the H^+ ions and both are equal to that of the hydrochloric acid, and since the mean activity coefficient $f_\pm$ is given by equation (5) *i.e.*

$$f_\pm = \sqrt{f_{Cl^-} f_{H^+}}$$

then

$$U = U_{0\,Cl^-} - \frac{RT}{0.239\ \mathbf{F}} \ln (c^2_{HCl} f^2_\pm)$$

$$= U_{0\,Cl^-} - \frac{2\,RT}{0.239\ \mathbf{F}} \ln c_{HCl} - \frac{2\,RT}{0.239\ \mathbf{F}} \ln f_\pm \qquad (8)$$

Equation (8) permits the unequivocal determination of the mean activity coefficient of hydrochloric acid if U_0 is kwown and U measured.

A third method consists in the use of a concentration cell without transference, with only one electrolyte and identical terminal electrodes. If, for example, it is wished to determine the mean activity coefficient of a solution of sodium chloride, the following concentration cell without transference could be used

$$\underbrace{\text{Na–Hg / NaCl / AgCl / Ag}}_{1}\underbrace{\text{ / AgCl / NaCl / Na–Hg}}_{2}$$

where the electric tension is given by the equation (2) on page 146, *i.e.*

$$U = \frac{2\,RT}{0.239\,\mathbf{F}} \ln \frac{c_1 f_1}{c_2 f_2}$$

By making a series of measurements keeping one of the two solutions constant and progressively diluting the other until, by extrapolation, the mean activity coefficient can be made equal to 1, the mean activity coefficient of the solution whose concentration was maintained, can be unequivocally determined.

A noteworthy attempt to resolve the problem of activity coefficients was made by Valensi[1] by the introduction of ionic activity coefficients relative to an electrolyte AB, which gives a more rigorous significance to the ionic activity coefficient. The choice of potassium chloride as a reference electrolyte signifies making $f_{Cl^-} = f_{K^+} = f_{\pm}$, as had already been done by MacInnes[2], this time however on a logical basis and not simply *a priori* as a hypothesis, no matter how plausible. With the introduction of relative activity coefficients a more rigorous significance is given to measurements of ionic activity such as pH (*cf.* Chapter V, p. 289).

Other workers have followed a different route which might lead to an experimental resolution of the probleem of ionic activities. The fundamental idea underlying these studies is that certain ion exchange resins behave almost ideally as semipermeable membranes for a single ionic species, either cationic or anionic, provided that the concentration of fixed ions is sufficiently high in relation to the concentration in the solution of the electrolyte. Under such conditions, the semipermeability holds for those ionic species for which the membrane functions as an

[1] G. Valensi, *C.R. III Réunion CITCE*, Berne 1951 (1952), Manfredi, Milan, p. 438.
[2] D. A. MacInnes, *J. Am. Chem. Soc.*, 41 (1919) 1086.

ion exchanger. If, for example, such a cationic membrane separates two solutions at different concentrations, the electrolyte cannot diffuse since in order to maintain electroneutrality, the cations must diffuse with the anions, which is not possible. But since the water can diffuse, there is a tendency for the concentrations on each side of the membrane to become equal, and hence an electric tension develops whose value is given by $RT \ln (a_1/a_2)/(0.239\ z\mathrm{F})$ where z is the valency of the diffusable ion and a_1 and a_2 are the activities on the two sides of the membrane. With a cell of the type

$$\mathrm{Hg}\ /\ \mathrm{Hg_2Cl_2};\underset{\text{sat}}{\mathrm{KCl}}\ /\ \underset{c_1}{\mathrm{MeX}}\ /\ \text{Membrane}\ /\ \underset{c_2}{\mathrm{MeX}}\ /\ \underset{\text{sat}}{\mathrm{KCl}};\mathrm{Hg_2Cl_2}\ /\ \mathrm{Hg}$$

with two constant tension electrodes, theoretically the actual size of $RT \ln (a_1/a_2)/(0.239\ z\mathrm{F})$ is measured. There is thus the possibility of measuring directly, at least under certain conditions, the ratio of the ionic activities and thus of the true activity coefficients, on the hypothesis that the diffusion electric tensions at the point of liquid contact of the electrodes (KCl/MeX) are eliminated or at least reduced below the limits of experimental error. This hypothesis is probably true. It is also necessary that one at least of the two ionic activities is known, *e.g.* in a system of such a dilution that the activity coefficient calculated according to Debye and Hückel is sufficiently exact[1].

Activities can also be obtained by cryoscopic, osmotic, vapour pressure measurements, etc. Many values of activities have already been collected and tabulated[2].

12. The Series of Electric Tensions

Every chemical process which takes place in solution with a change of the charge of one or more of the ions present can give rise to a galvanic cell characterized by an electric tension. In general also, the formation or discharge of an ion where the charge changes from zero to the final value or *vice versa* must be considered as a change in charge. For every process,

[1] For the relevant literature see G. MILAZZO, *Rend. ist. super. sanità*, 20 (1957) 739; G. MILAZZO and G. BOMBARA, *J. Electroanal. Chem.*, 1 (1960) 265.

[2] G.N. LEWIS and M. RANDALL, *Thermodynamics and the Free Energy of Chemical Substances*, McGraw Hill, New York (1923). R.R. ROBINSON and H.S. HARNED, *Chem. Revs.*, 28 (1941) 419; H.S. HARNED and B.B. OWEN, *The Physical Chemistry of Electrolytic Solutions*, 3rd Edition, Reinhold, New York, 1958.

TABLE III, 3

SERIES OF ELECTRIC TENSIONS FOR CATIONS

($t = 25°$ C, $a = 1$ N)

Element	U_0	*Note*	*Element*	U_0	*Note*
Li/Li^{+}	− 3.045	3	Tl/Tl^{+}	− 0.335	1
Cs/Cs^{+}	− 2.923	3	Co/Co^{2+}	− 0.30	3
Rb/Rb^{+}	− 2.925	3	Ni/Ni^{2+}	− 0.25	3
K/K^{+}	− 2.925	3	Mo/Mo^{3+}	− 0.2	2
Ra/Ra^{2+}	− 2.92	3	In/In^{+}	− 0.14	3
Ba/Ba^{2+}	− 2.90	3	Sn/Sn^{2+}	− 0.140	3
Sr/Sr^{2+}	− 2.89	1	Pb/Pb^{2+}	− 0.126	1
Ca/Ca^{2+}	− 2.87	3	Fe/Fe^{3+}	− 0.036	2
Na/Na^{+}	− 2.713	1	$D_2/2\,D^{+}$	− 0.003	1
La/La^{3+}	− 2.52	3	$H_2/2\,H^{+}$	0.000	
Ce/Ce^{3+}	− 2.48	3	Sb/Sb^{3+}	+ 0.1	4
Mg/Mg^{2+}	− 2.37	3	Bi/Bi^{3+}	+ 0.2	4
Y/Y^{3+}	− 2.37	3	As/As^{3+}	+ 0.3	4
Sc/Sc^{3+}	− 2.08	3	Cu/Cu^{2+}	+ 0.337	3
Th/Th^{4+}	− 1.90	3	Co/Co^{3+}	+ 0.4	4
Be/Be^{2+}	− 1.85	3	Ru/Ru^{2+}	+ 0.45	2
U/U^{3+}	− 1.80	3	Cu/Cu^{+}	+ 0.52	1
Hf/Hf^{4+}	− 1.70	3	Te/Te^{4+}	+ 0.56	1
Al/Al^{3+}	− 1.66	1	Tl/Tl^{3+}	+ 0.71	2
Ti/Ti^{2+}	− 1.63	3	$2\,Hg/Hg_2^{2+}$	+ 0.792	3
Zr/Zr^{4+}	− 1.53	3	Ag/Ag^{+}	+ 0.800	3
U/U^{4+}	− 1.4	3	Rh/Rh^{3+}	+ 0.8	4
Mn/Mn^{2+}	− 1.19	4	Pb/Pb^{4+}	+ 0.80	4
V/V^{2+}	− 1.18	3	Os/Os^{2+}	+ 0.85	3
Cb/Cb^{3+}	− 1.1	2	Hg/Hg^{2+}	+ 0.854	3
Cr/Cr^{2+}	− 0.86	2	Pd/Pd^{2+}	+ 0.987	3
Zn/Zn^{2+}	− 0.763	1	Ir/Ir^{3+}	+ 1.15	3
Cr/Cr^{3+}	− 0.74	3	Pt/Pt^{2+}	+ 1.2	2
Ga/Ga^{3+}	− 0.53	3	Ag/Ag^{2+}	+ 1.369	
Ga/Ga^{2+}	− 0.45	2	Au/Au^{3+}	+ 1.50	3
Fe/Fe^{2+}	− 0.44	1	Ce/Ce^{4+}	+ 1.68	5
Cd/Cd^{2+}	− 0.402	1	Au/Au^{+}	+ 1.68	3
In/In^{3+}	− 0.335	3			

[1] Values critically reviewed by J. O'M. BOCKRIS and J. F. HERRINGSHAW, *Discussions of the Faraday Society No.1*, Electrode Processes, 1947.

[2] From W. M. LATIMER, *The Oxidation States of the Elements and their Potentials in Aqueous Solutions*, Prentice Hall, New York, 1938.

[3] Figures taken from recent literature.

[4] Older figures.

[5] This value is probably in error.

For other figures of less common elements *cf.* G. CHARLOT, D. BÉZIER and J. COURTOT, *Tables of Constants and Numerical Data No.8*, Selected Constants, Oxido-Reduction Potentials, Pergamon Press, London, 1958.

TABLE III,4

SERIES OF ELECTRIC TENSIONS FOR ANIONS
($t = 25°$ C, $a = 1\ N$)

Element	U_0	*Note*	*Element*	U_0	*Note*
Te^{2-}/Te	− 1.14	2	$2\ CNS^-/(CNS)_2$	+ 0.77	2
$Te_2^{2-}/2\ Te$	− 0.84	2	$2\ Br^-/Br_2$ *l*	+ 1.066	1
Se^{2-}/Se	− 0.92	1	ClO_2^-/ClO_2 *g*	+ 1.16	3
S^{2-}/S	− 0.52	3	$2\ Cl^-/Cl_2$ *g*	+ 1.358	1
$4\ OH^-/O_2+2\ H_2O$	− 0.401	2	OH^-/OH	+ 2.0	3
Re^-/Re	− 0.4	2	$2\ F^-/F_2$ *g*	+ 2.65	3
$2\ I^-/I_2$ *s*	+ 0.535	3			

Abbreviations: *s* = solid, *g* = gas and *l* = liquid.
[1] Values critically reviewed by J. O'M. Bockris and J. F. Herringshaw, *loc. cit.*
[2] From W. M. Latimer, *loc. cit.*
[3] Figures taken from recent literature.

it is possible to calculate or measure the electric tension relative to that of hydrogen and a standard electric tension[1] has also been defined. The standard tensions are of fundamental importance not only for the calculation of the electric tension of a given electrode as a function of the concentration and for use in a galvanic cell, but also for the correct interpretation of the chemical behaviour of all the substances taking part in the reactions of galvanic cells, for predicting the possibility of a particular reaction occurring, for the calculation of affinities and equilibrium constants and finally for studies on the corrosion and protection of metals.

If all the standard relative tensions are arranged in ascending order starting from the most negative, passing through zero and going on to the most positive, the *series of electric tensions* are obtained (Tables III, 3 to III, 8). This deserves some comment.

If a cell is constructed with two different electrodes, *i.e.* a so-called chemical cell, in which the electrolytes have unit activity, its electric tension in absolute value is given directly by the standard tension of the first electrode minus that of the second, when the diffusion electric tension[2] is eliminated. The positive electrode of the cell is that which has the more positive standard tension so that here the cations are discharged or the anions formed, whilst the negative electrode is that with the lowest

[1] From now on this will be called simply the 'standard tension'.
[2] *Cf.* p. 138 *et seq.*

standard tension and this functions as a soluble electrode, if it is metallic, giving up cations to the solution or alternatively discharging anions. If the cell consists of two metallic electrodes, the metal with the more negative standard tension displaces from the solution that with the more positive standard tension and this also occurs directly if a metal with a more negative electric tension is immersed in a solution of a salt of one with a more positive tension. Since metals with the higher electric tensions are the so-called noble metals, it is said that an electric tension is more or less noble depending on whether it is more or less positive. In particular, all the metals with negative standard tensions displace the H^+ ion from a 1 *N* solution of acid, liberating hydrogen in the elementary state and themselves passing into the ionic state; metals with a positive standard tension cannot give this reaction. In other words, metals with negative standard tensions are attacked by acids and to a greater extent the more negative is their standard tension, whilst those with positive tensions cannot be attacked. If the concentrations, or activities, are not unitary it is first necessary to calculate the respective electric tensions, using the generalized equations, to determine which will be the positive pole of the combination. The same conclusions are valid for anions. In this case, however, it must be remembered that in view of the inversion of the electrochemical process by the opposite sign of the anionic charge, it is those anions with a more positive tension which will displace those with a more negative.

Similar conclusions can be drawn from the series of electric tension for redox electrodes (Tables III, 5 and III, 6) where a high positive electric tension corresponds to a strong oxidizing power whilst a high negative electric tension indicates a strong reducing power. By connecting two half cells of this type into a cell, the system with the more positive tension oxidizes that with the more negative. For example, the system Fe^{3+}/Fe^{2+} oxidizes Cu^{2+}/Cu^{+}, but will be oxidized by the system $MnO_4^- + 8\,H^+/Mn^{2+} + 4\,H_2O$. In general, the more negative the electric tension, the greater is the tendency of the system to pass from the reduced to the oxidized state, no matter whether it involves the formation of cations, the discharge of anions or redox reactions. In other words, the more negative the electric tension the greater is the system's reducing power. At the same time as the electric tensions become more positive the reducing power diminishes, or the oxidizing power increases, until a strongly oxidizing system is obtained.

As was mentioned for redox tensions, allowance must be made for the

TABLE III, 5

SERIES OF ELECTRIC TENSIONS FOR REDOX REACTIONS

Changes of charge

Element	U_0	*Element*	U_0
$Cr(CN)_6^{4-}/Cr(CN)_6^{3-}$	− 1.28	RuO_4^{2-}/RuO_4^- (alkaline solution)	+ 0.6
$Co(CN)_6^{4-}/Co(CN)_6^{3-}$	− 0.83	$Mo(CN)_8^{4-}/Mo(CN)_8^{3-}$	+ 0.73
FeO_2^{2-}/FeO_2^- (40% NaOH; 80°)	− 0.68	Fe^{2+}/Fe^{3+}	+ 0.771
Ga^{2+}/Ga^{3+}	− 0.65	$OsCl_6^{3-}/OsCl_6^{2-}$	+ 0.85
$2\,S^{2-}/S_2^{2-}$	− 0.52	Ru^{3+}/Ru^{4+} (in 2 *N*HCl; $c = 1$)	+ 0.86
In^{2+}/In^{3+}	− 0.49	$IrCl_6^{3-}/IrCl_6^{2-}$	+ 0.87
Cr^{2+}/Cr^{3+}	− 0.41	$Hg_2^{2+}/2Hg^{2+}$	+ 0.920
In^+/In^{3+}	− 0.404	$IrBr_6^{4-}/IrBr_6^{3-}$	+ 0.99
WCl_5^{2-}/WCl_5^-	− 0.4	RuO_4^-/RuO_4	+ 1.00
In^+/In^{2+}	− 0.40	$3\,Br^-/Br_3^-$	+ 1.06
Ti^{2+}/Ti^{3+}	− 0.37	Fe^{2+}*o*-phenanthroline/ Fe^{3+}*o*-phenanthroline	+ 1.14
V^{2+}/V^{3+}	− 0.25	Fe^{2+}nitrophenanthroline/ Fe^{3+}nitrophenanthroline	+ 1.25
$Mn(CN)_6^{4-}/Mn(CN)_6^{3-}$	− 0.24	Tl^+/Tl^{3+}	+ 1.25
UO_2^+/UO_2^{2+}	− 0.06	Au^+/Au^{3+}	+ 1.29
$Co(NH_3)_6^{2+}/Co(NH_3)_6^{3+}$	+ 0.1	$Ce^{3+}/Ce^{4+}(1\,N\,H_2SO_4)$	+ 1.44
Mo^{3+}(red)/Mo^{5+}	+ 0.11	$Ce^{3+}/Ce^{4+}(1\,N\,HNO_3)$	+ 1.60
Sn^{2+}/Sn^{4+}	+ 0.154	$Mn^{2+}/Mn^{3+}(15\,N\,H_2SO_4)$	+ 1.51
Cu^+/Cu^{2+}	+ 0.153	Mn^{3+}/Mn^{4+}(in H_2SO_4)	+ 1.65
$Fe(CN)_6^{4-}/Fe(CN)_6^{3-}$	+ 0.356	Pb^{2+}/Pb^{4+}	+ 1.69
U^{4+}/U^{6+}	+ 0.4	$Co^{2+}/Co^{3+}(3\,N\,HNO_3)$	+ 1.842
$W(CN)_8^{4-}/W(CN)_8^{3-}$	+ 0.457	Ag^+/Ag^{2+}	+ 1.939
Mo^{5+}/Mo^{6+}	+ 0.53		
$3\,I^-/I_3^-$	+ 0.535		
MnO_4^{2-}/MnO_4^-	+ 0.564		

environmental medium in evaluating the stability of an oxidizing or reducing system. A hydrogen electrode in a neutral solution (*i.e.* one containing 10^{-7} g ions/1) has a relative electric tension of −0.414 V so that all systems which in a neutral environment have electric tensions more negative than −0.414 V reduce H^+ ions to elementary hydrogen; in other words, they are capable of decomposing water and this in fact happens with all the metals from lithium to iron. If, however, the environment is 1 *N* alkaline, the hydrogen electrode tension is −0.83 V and thus a reducing agent in 1 *N* alkali is stable until it reaches this electric tension; it can decompose water only if its electric tension is more negative.

[Text continued p. 172]

TABLE III,6

SERIES OF ELECTRIC TENSIONS

Miscellaneous reactions*

Reduced state			*Oxidized state*			n**	U_0
Ag	*s*	$+ Br^-$	AgBr	*s*		1	+ 0.071
Ag	*s*	$+ BrO_3^-$	$AgBrO_3$	*s*		1	+ 0.55
Ag	*s*	$+ CH_3COO^-$	$AgCH_3COO$	*s*		1	+ 0.64
Ag	*s*	$+ Cl^-$	AgCl	*s*		1	+ 0.222
Ag	*s*	$+ CN^-$	AgCN	*s*		1	− 0.017
Ag	*s*	$+ 2\,CN^-$	$[Ag(CN)_2]^-$			1	− 0.31
Ag	*s*	$+ 3\,CN^-$	$[Ag(CN)_3]^{2-}$			1	− 0.51
Ag	*s*	$+ CNO^-$	AgCNO	*s*		1	+ 0.41
Ag	*s*	$+ CNS^-$	AgCNS	*s*		1	+ 0.09
2 Ag	*s*	$+ C_2O_4^{2-}$	$Ag_2C_2O_4$	*s*		2	+ 0.47
2 Ag	*s*	$+ CO_3^{2-}$	Ag_2CO_3	*s*		2	+ 0.47
2 Ag	*s*	$+ CrO_4^{2-}$	Ag_2CrO_4	*s*		2	+ 0.445
4 Ag	*s*	$+ [Fe(CN)_6]^{4-}$	$Ag_4[Fe(CN)_6]$	*s*		4	+ 0.194
Ag	*s*	$+ I^-$	AgI	*s*		1	− 0.152
Ag	*s*	$+ IO_3^-$	$AgIO_3$	*s*		1	+ 0.355
2 Ag	*s*	$+ MoO_4^{2-}$	Ag_2MoO_4	*s*		2	+ 0.49
Ag	*s*	$+ 2\,NH_3$ *sol*	$[Ag(NH_3)_2]^+$			1	+ 0.373
Ag	*s*	$+ NO_2^-$	$AgNO_2$	*s*		1	+ 0.56
2 Ag	*s*	$+ 2\,OH^-$	Ag_2O	*s*	$+ H_2O$	2	+ 0.344
2 Ag	*s*	$+ OH^- + SH^-$	Ag_2S	*s*	$+ H_2O$	2	− 0.67
2 Ag	*s*	$+ S^{2-}$	Ag_2S	*s*		2	− 0.71
Ag	*s*	$+ 2\,SO_3^{2-}$	$[Ag(SO_3)_2]^{3-}$			1	+ 0.30
Ag	*s*	$+ 2\,S_2O_3^{2-}$	$[Ag(S_2O_3)_2]^{3-}$			1	+ 0.01
2 Ag	*s*	$+ SO_4^{2-}$	Ag_2SO_4	*s*		2	+ 0.653
2 Ag	*s*	$+ H_2S$ *g*	Ag_2S	*s*	$+ 2\,H^+$	2	− 0.036
2 Ag	*s*	$+ WO_4^{2-}$	Ag_2WO_4	*s*		2	+ 0.53
Ag_2O	*s*	$+ 2\,OH^-$	2 AgO	*s*	$+ H_2O$	2	+ 0.57
Al	*s*	$+ 6\,F^-$	AlF_6^{3-}			3	− 2.07
Al	*s*	$+ 3\,OH^-$	$Al(OH)_3$	*s*		3	− 2.31
Al	*s*	$+ 4\,OH^-$	$H_2AlO_3^-$		$+ H_2O$	3	− 2.35
As	*s*	$+ 2\,H_2O$	$HAsO_2$	*sol*	$+ 3\,H^+$	3	+ 0.25
As	*s*	$+ 3\,H_2O$	H_3AsO_3	*sol*	$+ 3\,H^+$	3	+ 0.24
2 As	*s*	$+ 3\,H_2O$	As_2O_3	*s*	$+ 6\,H^+$	6	+ 0.234
As	*s*	$+ 4\,OH^-$	AsO_2^-		$+ 2\,H_2O$	3	− 0.68
As	*s*	$+ 2\,S^{2-}$	AsS_2^-			3	− 0.75
AsH_3	*g*		As	*s*	$+ 3\,H^+$	3	− 0.60
AsH_3	*g*	$+ 3\,OH^-$	As	*s*	$+ 3\,H_2O$	3	− 1.37

Reduced state			*Oxidized state*			n^{**}	U_0
AsS_2^-		$+ 2\,S^{2-}$	AsS_4^{3-}			2	− 0.6
$HAsO_2$	*sol*	$+ 2\,H_2O$	H_3AsO_4	*sol*	$+ 2\,H^+$	2	+ 0.559
AsO_2^-		$+ 4\,OH^-$	AsO_4^{3-}		$+ 2\,H_2O$	2	− 0.67
H_3AsO_3	*sol*	$+ H_2O$	H_3AsO_4	*sol*	$+ 2\,H^+$	2	+ 0.559
Au	*s*	$+ 2\,Br^-$	$AuBr_2^-$			1	+ 0.96
Au	*s*	$+ 4\,Br^-$	$AuBr_4^-$			3	+ 0.87
Au	*s*	$+ 2\,Cl^-$	$AuCl_2^-$			1	+ 1.11
Au	*s*	$+ 4\,Cl^-$	$AuCl_4^-$			3	+ 0.99
Au	*s*	$+ 2\,CN^-$	$[Au(CN)_2]^-$			1	− 0.60
Au	*s*	$+ 2\,CNS^-$	$[Au(CNS)_2]^-$			1	+ 0.69
Au	*s*	$+ 4\,CNS^-$	$[Au(CNS)_4]^-$			3	+ 0.66
Au	*s*	$+ 3\,H_2O$	$Au(OH)_3$		$+ 3\,H^+$	3	+ 1.45
2 Au	*s*	$+ 3\,H_2O$	Au_2O_3	*s*	$+ 6\,H^+$	6	+ 1.363
Au	*s*	$+ I^-$	AuI	*s*		1	+ 0.50
Au	*s*	$+ 4\,OH^-$	AuO_2^-		$+ 2\,H_2O$	3	+ 0.5
$AuBr_2^-$		$+ 2\,Br^-$	$AuBr_4^-$			2	+ 0.82
$AuCl_2^-$		$+ 2\,Cl^-$	$AuCl_4^-$			2	+ 0.94
$[Au(CNS)_2]^-$		$+ 2\,CNS^-$	$[Au(CNS)_4]^-$			2	+ 0.645
B	*s*	$+ 4\,F^-$	BF_4^-			3	− 1.06
B	*s*	$+ 3\,H_2O$	H_3BO_3	*sol*	$+ 3\,H^+$	3	− 0.87
B	*s*	$+ 4\,OH^-$	$H_2BO_3^-$		$+ H_2O$	3	− 1.79
Ba	*s*	$+ 2\,OH^- + 8\,H_2O$	$Ba(OH)_2 \cdot 8\,H_2O$	*s*		2	− 2.97
2 Be	*s*	$+ 6\,OH^-$	$Be_2O_3^{2-}$		$+ 3\,H_2O$	4	− 2.62
Bi	*s*	$+ Cl^- + H_2O$	BiOCl	*s*	$+ 2\,H^+$	3	− 0.16
Bi	*s*	$+ 4\,Cl^-$	$BiCl_4^-$			3	+ 0.167
Bi	*s*	$+ H_2O$	BiO^+		$+ 2\,H^+$	3	+ 0.32
Bi	*s*	$+ 3\,OH^-$	BiOOH	*s*	$+ H_2O$	3	− 0.46
2 Bi	*s*	$+ 6\,OH^-$	Bi_2O_3	*s*	$+ 3\,H_2O$	6	− 0.44
Bi^{3+}		$+ 3\,H_2O$	$HBiO_3$		$+ 5\,H^+$	2	+ 1.7(?)
BiH_3	*g*		Bi	*s*	$+ 3\,H^+$	3	− 0.8
BiO	*s*		BiO^+			1	+ 0.38
2 BiO^+		$+ 2\,H_2O$	Bi_2O_4	*s*	$+ 4\,H^+$	2	+ 1.59
Bi_2O_3	*s*	$+ 2\,OH^-$	Bi_2O_4	*s*	$+ H_2O$	2	+ 0.56
2 Bi_2O_3	*s*	$+ 2\,OH^-$	Bi_4O_7	*s*	$+ H_2O$	2	+ 0.51
Bi_4O_7	*s*	$+ 2\,OH^-$	2 Bi_2O_4	*s*	$+ H_2O$	2	+ 0.62
Br_2	*l*	$+ 2\,H_2O$	2 HBrO		$+ 2\,H^+$	2	+ 1.59
Br_2	*l*	$+ 6\,H_2O$	2 BrO_3^-		$+ 12\,H^+$	10	+ 1.52

Reduced state			*Oxidized state*			n^{**}	U_0
Br_2		$+ 4\,OH^-$	$2\,BrO^-$		$+ 2\,H_2O$	2	+ 0.45
Br^-		$+ Cl^-$	$BrCl$			2	+ 1.20
Br^-		$+ H_2O$	$HBrO$		$+ H^+$	2	+ 1.33
Br^-		$+ 3\,H_2O$	BrO_3^-		$+ 6\,H^+$	6	+ 1.42
Br^-		$+ 2\,OH^-$	BrO^-		$+ H_2O$	2	+ 0.76
Br^-		$+ 6\,OH^-$	BrO_3^-		$+ 3\,H_2O$	6	+ 0.61
BrO^-		$+ 4\,OH^-$	BrO_3^-		$+ 2\,H_2O$	4	+ 0.54
$HBrO$		$+ 2\,H_2O$	BrO_3^-		$+ 5\,H^+$	4	+ 1.49
Ca	*s*	$+ 2\,OH^-$	$Ca(OH)_2$	*s*		2	− 3.02
$2\,Cb$		$+ 5\,H_2O$	Cb_2O_5		$+ 10\,H^+$	10	− 0.62
Cd	*s*	$+ 4\,CN^-$	$[Cd(CN)_4]^{2-}$			2	− 1.03
Cd	*s*	$+ CO_3^{2-}$	$CdCO_3$	*s*		2	− 0.74
Cd	*s*	$+ 4\,NH_3$ *sol*	$[Cd(NH_3)_4]^{2-}$			2	− 0.597
Cd	*s*	$+ 2\,OH^-$	$Cd(OH)_2$	*s*		2	− 0.81
Cd	*s*	$+ S^{2-}$	CdS	*s*		2	− 1.23
$Cd_{amalg.}$ $CdSO_4 \cdot \frac{8}{3}\,H_2O$ sat.		$+ SO_4^{2-}$ in	$CdSO_4 \cdot \frac{8}{3}\,H_2O$	*s*		2	− 0.435
Ce^{3+}		$+ 2\,H_2O$	CeO_2		$+ 4\,H^+$	1	+ 1.5
Cl_2	*g*	$+ 2\,H_2O$	$2\,HClO$		$+ 2\,H^+$	2	+ 1.63
Cl_2	*g*	$+ 4\,H_2O$	$2\,HClO_2$		$+ 6\,H^+$	6	+ 1.63
Cl_2	*g*	$+ 4\,H_2O$	$2\,ClO_2$	*g*	$+ 8\,H^+$	8	+ 1.53
Cl_2	*g*	$+ 6\,H_2O$	$2\,ClO_3^-$		$+ 12\,H^+$	10	+ 1.47
Cl_2	*g*	$+ 8\,H_2O$	$2\,ClO_4^-$		$+ 16\,H^+$	14	+ 1.34
Cl_2	*g*	$+ 4\,OH^-$	$2\,ClO^-$		$+ 2\,H_2O$	2	+ 0.52
Cl^-		$+ H_2O$	$HClO$		$+ H^+$	2	+ 1.50
Cl^-		$+ 2\,H_2O$	$HClO_2$		$+ 3\,H^+$	4	+ 1.56
Cl^-		$+ 2\,H_2O$	ClO_2	*g*	$+ 4\,H^+$	5	+ 1.50
Cl^-		$+ 3\,H_2O$	ClO_3^-		$+ 6\,H^+$	6	+ 1.45
Cl^-		$+ 4\,H_2O$	ClO_4^-		$+ 8\,H^+$	8	+ 1.35
Cl^-		$+ 2\,OH^-$	ClO^-		$+ H_2O$	2	+ 0.89
Cl^-		$+ 4\,OH^-$	ClO_2^-		$+ 2\,H_2O$	4	+ 0.76
Cl^-		$+ 4\,OH^-$	ClO_2	*g*	$+ 2\,H_2O$	5	+ 0.76
Cl^-		$+ 6\,OH^-$	ClO_3^-		$+ 3\,H_2O$	6	+ 0.62
Cl^-		$+ 8\,OH^-$	ClO_4^-		$+ 4\,H_2O$	8	+ 0.51
ClO^-		$+ 2\,OH^-$	ClO_2^-		$+ H_2O$	2	+ 0.66
ClO_2^-		$+ 2\,OH^-$	ClO_3^-		$+ H_2O$	2	+ 0.33
ClO_2	*g*	$+ 2\,OH^-$	ClO_3^-		$+ H_2O$	1	− 0.45
ClO_2	*g*	$+ H_2O$	ClO_3^-		$+ 2\,H^+$	1	+ 1.21

Reduced state			*Oxidized state*			*n***	U_0
ClO_3^-		$+ H_2O$	ClO_4^-		$+ 2 H^+$	2	+ 1.19
ClO_3^-		$+ 2 OH^-$	ClO_4^-		$+ H_2O$	2	+ 0.36
HClO		$+ H_2O$	$HClO_2$		$+ 2 H^+$	2	+ 1.63
$HClO_2$			ClO_2	*g*	$+ H^+$	1	+ 1.27
$HClO_2$		$+ H_2O$	ClO_3^-		$+ 3 H^+$	2	+ 1.21
$(CN)_2$	*g*	$+ 2 H_2O$	2 HCNO		$+ 2 H^+$	2	+ 0.33
CN^-		$+ 2 OH^-$	CNO^-		$+ H_2O$	2	− 0.97
HCN		$+ H_2O$	HCNO		$+ 2 H^+$	2	0.0
2 HCN			$(CN)_2$	*g*	$+ 2 H^+$	2	+ 0.37
HCNS		$+ 3 H_2O$	HCN		$+ H_2SO_3 + 4H^+$	4	− 0.55
$HCOO^-$		$+ 3 OH^-$	CO_3^{2-}		$+ 2 H_2O$	2	− 0.95
$2 H_2CO_3$			$C_2O_6^{2-}$		$+ 4 H^+$	2	+ 1.7
Co	*s*	$+ CO_3^{2-}$	$CoCO_3$	*s*		2	− 0.632
Co	*s*	$+ 6 NH_3$ *sol*	$[Co(NH_3)_6]^{2+}$			2	− 0.422
Co	*s*	$+ 2 OH^-$	$Co(OH)_2$			2	− 0.73
Co	*s*	$+ S^{2-}$	CoS	*sα*		2	− 0.93
Co	*s*	$+ S^{2-}$	CoS	*sβ*		2	− 1.07
CoO		$+ 2 OH^-$	CoO_2		$+ H_2O$	2	+ 0.9
$Co(OH)_2$		$+ OH^-$	$Co(OH)_3$			1	+ 0.17
Cr	*s*	$+ 2 Cl^-$	$CrCl_2^+$			3	− 0.74
Cr	*s*	$+ 3 OH^-$	$Cr(OH)_3$	*s*		3	− 1.3
Cr	*s*	$+ 4 OH^-$	CrO_2^-		$+ 2 H_2O$	3	− 1.2
$2 Cr^{3+}$		$+ 7 H_2O$	$Cr_2O_7^{2-}$		$+ 14 H^+$	6	+ 1.33
$Cr(OH)_3$		$+ 5 OH^-$	CrO_4^{2-}		$+ 4 H_2O$	3	− 0.13
Cu	*s*	$+ Br^-$	CuBr	*s*		1	+ 0.033
Cu	*s*	$+ 2 Br^-$	$CuBr_2^-$			1	+ 0.05
Cu	*s*	$+ Cl^-$	CuCl	*s*		1	+ 0.137
Cu	*s*	$+ 2 Cl^-$	$CuCl_2^-$			1	+ 0.19
Cu	*s*	$+ 2 CN^-$	$[Cu(CN)_2]^-$			1	− 0.43
Cu	*s*	$+ 2 CNS^-$	$[Cu(CNS)_2]^-$			1	− 0.27
Cu	*s*	$+ CO_3^{2-}$	$CuCO_3$	*s*		2	+ 0.053
Cu	*s*	$+ H_2S$ *g*	CuS	*s*	$+ 2 H^+$	2	− 0.259
Cu	*s*	$+ I^-$	CuI	*s*		1	− 0.18
Cu	*s*	$+ 2 I^-$	CuI_2^-			1	0.00
Cu	*s*	$+ 2 NH_3$ *sol*	$[Cu(NH_3)_2]^+$			1	− 0.12
Cu	*s*	$+ 4 NH_3$ *sol*	$[Cu(NH_3)_4]^{2+}$			2	− 0.05
Cu	*s*	$+ 2 OH^-$	CuO	*s*	$+ H_2O$	2	− 0.258
Cu	*s*	$+ 2 OH^-$	$Cu(OH)_2$	*s*		2	− 0.244
2 Cu	*s*	$+ 2 OH^-$	Cu_2O		$+ H_2O$	2	− 0.36

Reduced state			*Oxidized state*			n^{**}	U_0
Cu	*s*	$+ S^{2-}$	CuS	*s*		2	− 0.76
2 Cu	*s*	$+ S^{2-}$	Cu_2S			2	− 0.54
CuBr	*s*		Cu^{2+}		$+ Br^-$	1	+ 0.64
CuCl	*s*		Cu^{2+}		$+ Cl^-$	1	+ 0.54
$[Cu(CN)_2]^-$			Cu^{2+}		$+ 2\ CN^-$	1	+ 1.1
CuI	*s*		Cu^{2+}		$+ I^-$	1	+ 0.85
CuI_2^-			Cu^{2+}		$+ 2\ I^-$	1	+ 0.690
$[Cu(NH_3)_2]^+ + 2\ NH_3$ *sol*			$[Cu(NH_3)_4]^{2+}$			1	0.0
Cu_2O	*s*	$+ 2\ OH^- + H_2O$	$2\ Cu(OH)_2$			2	− 0.08
Cu_2S	*s*	$+ S^{2-}$	2 CuS			2	− 0.58
$2\ F^-$		$+ H_2O$	F_2O		$+ 2\ H^+$	4	+ 2.1
2 HF	*sol*		F_2	*g*	$+ 2\ H^+$	2	+ 3.06
Fe	*s*	$+ 6\ CN^-$	$[Fe(CN)_6]^{4-}$			2	− 1.5
Fe	*s*	$+ CO_3^{2-}$	$FeCO_3$	*s*		2	− 0.755
Fe	*s*	$+ 2\ OH^-$	$Fe(OH)_2$	*s*		2	− 0.87
Fe	*s*	$+ 3\ OH^-$	$Fe(OH)_3$	*s*		3	− 0.56
Fe	*s*	$+ S^{2-}$	FeS	*s*		2	− 1.00
Fe^{2+}		$+ 6\ F^-$	FeF_6^{3-}			1	+ 0.4
Fe^{3+}		$+ 4\ H_2O$	FeO_4^{2-}		$+ 8\ H^+$	3	+ 1.9
$[Fe(C_2O_4)_2]^{2-} + C_2O_4^{2-}$			$[Fe(C_2O_4)_3]^{3-}$			1	+ 0.02
FeO_2^-		$+ 4\ OH^-$	FeO_4^{2-}		$+ 2\ H_2O$	3	+ 0.9
$Fe(OH)_2$	*s*	$+ OH^-$	$Fe(OH)_3$	*s*		1	− 0.56
2 FeS	*s*	$+ S^{2-}$	Fe_2S_3	*s*		2	− 0.7
2 Ga	*s*	$+ H_2O$	Ga_2O		$+ 2\ H^+$	2	− 0.4
Ga	*s*	$+ 4\ OH^-$	$H_2GaO_3^-$		$+ H_2O$	3	− 1.22
Ga_2O		$+ 2\ H_2O$	Ga_2O_3		$+ 4\ H^+$	4	− 0.5
Ge	*s*	$+ 2\ H_2O$	GeO_2		$+ 4\ H^+$	4	− 0.15
Ge^{2+}		$+ 2\ H_2O$	GeO_2		$+ 4\ H^+$	2	− 0.15
$HGeO_2^-$		$+ 2\ OH^-$	$HGeO_3^-$		$+ H_2O$	2	− 1.4
H_2	*g*		$2\ H^+(10^{-7} m)$			2	− 0.414
H_2	*g*	$+ 2\ OH^-$	$2\ H_2O$			2	− 0.828
Hf	*s*	$+ H_2O$	HfO^{2+}		$+ 2\ H^+$	4	− 1.68
Hf	*s*	$+ 2\ H_2O$	HfO_2	*s*	$+ 4\ H^+$	4	− 1.57
Hf	*s*	$+ 4\ OH^-$	$HfO(OH)_2$	*s*	$+ H_2O$	4	− 2.50
Hg	*l*	$+ 4\ Br^-$	$HgBr_4^{2-}$			2	+ 0.21
2 Hg	*l*	$+ 2\ Br^-$	Hg_2Br_2	*s*		2	+ 0.139
2 Hg	*l*	$+ 2\ CH_3COO^-$	$Hg_2(CH_3COO)_2$			2	+ 0.51

Reduced state			*Oxidized state*			n^{**}	U_0
Hg	*l*	$+ 4\,Cl^-$	$HgCl_4^{2-}$			2	+ 0.38
2 Hg	*l*	$+ 2\,Cl^-$	Hg_2Cl_2	*s*		2	+ 0.268
Hg	*l*	$+ 4\,CN^-$	$[Hg(CN)_4]^{2-}$			2	− 0.37
2 Hg	*l*	$+ 2\,CN^-$	$Hg_2(CN)_2$	*s*		2	− 0.36
2 Hg	*l*	$+ 2\,CNS^-$	$Hg_2(CNS)_2$	*s*		2	+ 0.22
2 Hg	*l*	$+ CO_3^{2-}$	Hg_2CO_3	*s*		2	+ 0.32
2 Hg	*l*	$+ C_2O_4^{2-}$	$Hg_2C_2O_4$	*s*		2	+ 0.417
2 Hg	*l*	$+ CrO_4^{2-}$	Hg_2CrO_4	*s*		2	+ 0.54
Hg	*l*	$+ 4\,I^-$	HgI_4^{2-}			2	− 0.04
2 Hg	*l*	$+ 2\,I^-$	Hg_2I_2	*s*		2	− 0.041
Hg	*l*	$+ 2\,IO_3^-$	$Hg(IO_3)_2$	*s*		2	+ 0.40
2 Hg	*l*	$+ 2\,IO_3^-$	$Hg_2(IO_3)_2$	*s*		2	+ 0.394
Hg	*l*	$+ 2\,OH^-$	HgO	*s*	$+ H_2O$	2	+ 0.098
2 Hg	*l*	$+ 2\,OH^-$	Hg_2O	*s*	$+ H_2O$	2	+ 0.123
Hg	*l*	$+ OH^- + SH^-$	HgS	*s*	$+ H_2O$	2	− 0.77
Hg	*l*	$+ S^{2-}$	HgS	*s*		2	− 0.72 (?)
2 Hg	*l*	$+ S^{2-}$	Hg_2S	*s*		2	− 0.53
2 Hg	*l*	$+ SO_4^{2-}$	Hg_2SO_4	*s*		2	+ 0.615
Hg_2Cl_2	*s*	$+ 2\,Cl^-$	$2\,HgCl_2$			2	+ 0.63
I_2	*s*	$+ 2\,Br^-$	2 IBr	*sol*		2	+ 1.02
I_2	*s*	$+ 4\,Br^-$	$2\,IBr_2^-$			2	+ 0.87
I_2	*s*	$+ 2\,Cl^-$	2 ICl	*sol*		2	+ 1.19
I_2	*s*	$+ 4\,Cl^-$	$2\,ICl_2^-$			2	+ 1.06
I_2	*s*	$+ 6\,Cl^-$	$2\,ICl_3$	*s*		6	+ 1.28
I_2	*s*	+ 2 HCN	2 ICN		$+ 2\,H^+$	2	+ 0.625
I_2	*s*	$+ 2\,H_2O$	2 HIO		$+ 2\,H^+$	2	+ 1.45
I_2	*s*	$+ 6\,H_2O$	$2\,IO_3^-$		$+ 12\,H^+$	10	+ 1.195
I^-		$+ H_2O$	HIO		$+ H^+$	2	+ 0.99
I^-		$+ 3\,H_2O$	IO_3^-		$+ 6\,H^+$	6	+ 1.085
I^-		$+ 4\,H_2O$	IO_4^-		$+ 8\,H^+$	8	+ 1.4
I^-		$+ 2\,OH^-$	IO^-		$+ H_2O$	2	+ 0.49
I^-		$+ 6\,OH^-$	IO_3^-		$+ 3\,H_2O$	6	+ 0.26
ICl	*sol*	$+ 2\,Cl^-$	ICl_3	*s*		2	+ 0.99
ICl_2^-		$+ 3\,H_2O$	IO_3^-		$+ 6H^+ + 2Cl^-$	4	+ 1.23
IO^-		$+ 4\,OH^-$	IO_3^-		$+ 2\,H_2O$	4	+ 0.56
IO_3^-		$+ 3\,H_2O$	H_5IO_6		$+ H^+$	2	+ 1.6
IO_3^-		$+ 3\,OH^-$	$H_3IO_6^{2-}$			2	+ 0.7
HIO		$+ 2\,H_2O$	IO_3^-		$+ 5\,H^+$	4	+ 1.13
In	*s*	$+ Cl^-$	InCl	*s*		1	− 0.34
In	*s*	$+ 3\,OH^-$	$In(OH)_3$	*s*		3	− 1.0
2 In	*s*	$+ 6\,OH^-$	In_2O_3	*s*	$+ 3\,H_2O$	6	− 1.18

Reduced state			*Oxidized state*			n**	U_0
Ir	*s*	$+ 6\,Cl^-$	$IrCl_6^{3-}$			3	+ 0.77
2 Ir	*s*	$+ 6\,OH^-$	Ir_2O_3		$+ 3\,H_2O$	6	+ 0.1
Ir^{3+}		$+ 2\,H_2O$	IrO_2	*s*	$+ 4\,H^+$	1	+ 0.7
Ir_2O_3		$+ 2\,OH^-$	$2\,IrO_2$		$+ H_2O$	2	+ 0.1
La	*s*	$+ 3\,OH^-$	$La(OH)_3$	*s*		3	− 2.90
Mg	*s*	$+ 2\,OH^-$	$Mg(OH)_2$	*s*		2	− 2.67
Mn	*s*	$+ CO_3^{2-}$	$MnCO_3$	*s*		2	− 1.48
Mn	*s*	$+ 2\,OH^-$	$Mn(OH)_2$			2	− 1.55
Mn^{2+}		$+ 2\,H_2O$	MnO_2	β	$+ 4\,H^+$	2	+ 1.22
Mn^{2+}		$+ 4\,H_2O$	MnO_4^-		$+ 8\,H^+$	5	+ 1.52
$[Mn(CN)_4]^{2-}$		$+ 2\,CN^-$	$[Mn(CN)_6]^{3-}$			1	− 0.7
MnO_2	*s*	$+ 2\,H_2O$	MnO_4^-		$+ 4\,H^+$	3	+ 1.69
MnO_2	*s*	$+ 4\,OH^-$	MnO_4^{2-}		$+ 2\,H_2O$	2	+ 0.60
MnO_2	*s*	$+ 4\,OH^-$	MnO_4^-		$+ 2\,H_2O$	3	+ 0.587
$Mn(OH)_2$		$+ OH^-$	$Mn(OH)_3$	*s*		1	− 0.4
$Mn(OH)_2$		$+ 2\,OH^-$	MnO_2		$+ 2\,H_2O$	2	− 0.05
Mo	*s*	$+ 3\,H_2O$	MoO_3	*s*	$+ 6\,H^+$	6	+ 0.25
Mo	*s*	$+ 4\,H_2O$	H_2MoO_4	*sol*	$+ 6\,H^+$	6	0.0
Mo	*s*	$+ 8\,OH^-$	MoO_4^{2-}		$+ 4\,H_2O$	6	− 1.05
MoO^{3+}		$+ 2\,H_2O$	MoO_3	*s*	$+ 4\,H^+$	1	+ 0.48
$2\,NH_3$	*sol*	$+ H_2$	$2\,NH_4^+$			2	− 0.55
NH_3	*sol*	$+ 9\,OH^-$	NO_3^-		$+ 6\,H_2O$	8	− 0.12
$2\,NH_4^+$			$N_2H_5^+$		$+ 3\,H^+$	2	+ 1.27
$3\,NH_4^+$			HN_3		$+ 11\,H^+$	8	+ 0.69
NH_4^+		$+ 2\,H_2O$	HNO_2		$+ 7\,H^+$	6	+ 0.86
NH_4^+		$+ 3\,H_2O$	NO_3^-		$+ 10\,H^+$	8	+ 0.87
N_2H_4		$+ 2\,OH^-$	$2\,NH_2OH$			2	+ 0.74
N_2H_4		$+ 4\,OH^-$	N_2	*g*	$+ 4\,H_2O$	4	− 1.15
N_2H_4		$+ 8\,OH^-$	$2\,NO_2^-$		$+ 4\,H_2O$	6	− 0.21
N_2H_4		$+ 16\,OH^-$	$2\,NO_3^-$		$+ 10\,H_2O$	14	− 0.23
$N_2H_5^+$			N_2	*g*	$+ 5\,H^+$	4	− 0.23
$N_2H_5^+$		$+ 4\,H_2O$	$2\,HNO_2$		$+ 11\,H^+$	10	+ 0.79
$N_2H_5^+$		$+ 6\,H_2O$	$2\,NO_3^-$		$+ 17\,H^+$	14	+ 0.84
NH_2OH		$+ 5\,OH^-$	NO_2^-		$+ 4\,H_2O$	4	+ 0.45
NH_2OH		$+ 7\,OH^-$	NO_3^-		$+ 5\,H_2O$	6	− 0.30
NH_3OH^+		$+ H_2O$	HNO_2		$+ 5\,H^+$	4	+ 0.62
NH_3OH^+		$+ 2\,H_2O$	NO_3^-		$+ 8\,H^+$	6	+ 0.73
NH_4OH		$+ 2\,OH^-$	NH_2OH		$+ 2\,H_2O$	2	+ 0.42

Reduced state			*Oxidized state*			n^{**}	U_0
$2\,NH_4OH$		$+ 2\,OH^-$	N_2H_4		$+ 4\,H_2O$	2	+ 0.1
NH_4OH		$+ 9\,OH^-$	NO_3^-		$+ 7\,H_2O$	8	− 0.10
NO	*g*	$+ H_2O$	HNO_2		$+ H^+$	1	+ 0.99
NO	*g*	$+ 2\,H_2O$	NO_3^-		$+ 4\,H^+$	3	+ 0.96
2 NO	*g*	$+ 2\,H_2O$	N_2O_4		$+ 4\,H^+$	4	+ 1.03
N_2O		$+ 3\,H_2O$	$2\,HNO_2$		$+ 4\,H^+$	4	+ 1.29
N_2O_4	*g*	$+ 2\,H_2O$	$2\,NO_3^-$		$+ 4\,H^+$	2	+ 0.81
$2\,NO_2^-$			N_2O_4	*g*		2	+ 0.88
NO_2^-		$+ 2\,OH^-$	NO_3^-		$+ H_2O$	2	+ 0.01
$2\,HNO_2$			N_2O_4	*g*	$+ 2\,H^+$	2	+ 1.07
HNO_2		$+ H_2O$	NO_3^-		$+ 3\,H^+$	2	+ 0.94
$H_2N_2O_2$			2 NO		$+ 2\,H^+$	2	+ 0.71
$H_2N_2O_2$		$+ 2\,H_2O$	$2\,HNO_2$		$+ 4\,H^+$	4	+ 0.86
2 Nb	*s*	$+ 5\,H_2O$	Nb_2O_5	*s*	$+ 10\,H^+$	10	− 0.65
Ni	*s*	$+ CO_3^{2-}$	$NiCO_3$	*s*		2	− 0.45
Ni	*s*	$+ 6\,NH_3$ *sol*	$[Ni(NH_3)_6]^{2+}$			2	− 0.48
Ni	*s*	$+ 2\,OH^-$	$Ni(OH)_2$ *s*			2	− 0.72
Ni	*s*	$+ S^{2-}$	NiS	*sα*		2	− 0.83
Ni	*s*	$+ S^{2-}$	NiS	*sγ*		2	− 1.07
Ni^{2+}		$+ 4\,H_2O$	$NiO_2 \cdot 2\,H_2O$ *s*		$+ 4\,H^+$	2	+ 1.75
Ni^{2+}		$+ 2\,OH^-$	NiO_2		$+ 2\,H^+$	2	+ 1.75
$[Ni(CN)_3]^{2-}$		$+ CN^-$	$[Ni(CN)_4]^{2-}$			1	− 0.82
$Ni(OH)_2$*s*		$+ 2\,OH^-$	NiO_2	*s*	$+ 2\,H_2O$	2	+ 0.49
O_2	*g*	$+ H_2O$	O_3	*g*	$+ 2\,H^+$	2	+ 2.07
O_2	*g*	$+ 2\,OH^-$	O_3	*g*	$+ H_2O$	2	+ 1.24
OH		$+ H_2O$	H_2O_2		$+ H^+$	1	+ 0.72
OH^-		$+ HO_2^-$	O_2		$+ H_2O$	2	− 0.042
$3\,OH^-$			HO_2^-		$+ H_2O$	2	+ 0.87
$4\,OH^-$			O_2	*g*	$+ 2\,H_2O$	4	− 0.401
$2\,H_2O$			O_2	*g*	$+ 4\,H^+$	4	+ 1.229
$2\,H_2O$			O_2	*g*	$+ 4\,H^+$ $(10^{-7}m)$	4	+ 0.815
$2\,H_2O$			H_2O_2		$+ 2\,H^+$	2	+ 1.77
H_2O_2			O_2	*g*	$+ 2\,H^+$	2	+ 0.69
Os	*s*	$+ 6\,Cl^-$	$OsCl_6^{3-}$			3	+ 0.6
Os	*s*	$+ 4\,H_2O$	OsO_4	*s*	$+ 8\,H^+$	8	+ 0.85
Os	*s*	$+ 4\,OH^-$	OsO_2	*s*	$+ 2\,H_2O$	4	− 0.15
Os	*s*	$+ 9\,OH^-$	$HOsO_5^-$		$+ 4\,H_2O$	8	+ 0.02
Os^{2+}		$+ 6\,Cl^-$	$OsCl_6^{3-}$			1	+ 0.3
$OsCl_6^{2-}$		$+ 4\,H_2O$	OsO_4	*s*	$+ 6\,Cl^- + 8\,H^+$	4	+ 1.0

Reduced state			Oxidized state			n**	U_0
OsO_2	*s*	$+ 4\ OH^-$	OsO_4^{2-}		$+ 2\ H_2O$	2	+ 0.1
OsO_2	*s*	$+ 5\ OH^-$	$HOsO_5^-$		$+ 2\ H_2O$	4	+ 0.2
OsO_4^{2-}		$+ OH^-$	$HOsO_5^-$			2	+ 0.3
$OsO_2Cl_4^{2-}$		$+ 2\ H_2O$	OsO_4		$+ 4\ H^+ + 4\ Cl^-$	2	+ 1.0 (?)
P	*s*	$+ 2\ H_2O$	H_3PO_2		$+ H^+$	1	− 0.51
P	*s*	$+ 3\ H_2O$	H_3PO_3		$+ 3\ H^+$	3	− 0.49
P	*s*	$+ 4\ H_2O$	H_3PO_4		$+ 5\ H^+$	5	− 0.3
P	*s*	$+ 2\ OH^-$	$H_2PO_2^-$			1	− 2.05
P	*s*	$+ 5\ OH^-$	HPO_3^{2-}		$+ 2\ H_2O$	3	− 1.71
PH_3	*g*		P	*s*	$+ 3\ H^+$	3	− 0.06
PH_3	*g*	$+ 3\ OH^-$	P	*s*	$+ 3\ H_2O$	3	− 0.87
HPO_3^{2-}		$+ 3\ OH^-$	PO_4^{3-}		$+ 2\ H_2O$	2	− 1.05
H_3PO_2		$+ H_2O$	H_3PO_3		$+ 2\ H^+$	2	− 0.59
$H_2PO_2^-$		$+ 3\ OH^-$	HPO_3^{2-}		$+ 2\ H_2O$	2	− 1.57
H_3PO_2		$+ H_2O$	H_3PO_3		$+ 2\ H^+$	2	− 0.59
H_3PO_3		$+ H_2O$	H_3PO_4		$+ 2\ H^+$	2	− 0.276
Pb	*s*	$+ 2\ Br^-$	$PbBr_2$	*s*		2	− 0.280
Pb	*s*	$+ 2\ Cl^-$	$PbCl_2$	*s*		2	− 0.268
Pb	*s*	$+ CO_3^{2-}$	$PbCO_3$	*s*		2	− 0.506
Pb	*s*	$+ HPO_4^{2-}$	$PbHPO_4$	*s*		2	− 0.251
Pb	*s*	$+ H_2S$ *g*	PbS	*s*	$+ 2\ H^+$	2	− 0.07
Pb	*s*	$+ 2\ I^-$	PbI_2	*s*		2	− 0.365
Pb	*s*	$+ 2\ OH^-$	PbO	*s* red	$+ H_2O$	2	− 0.578
Pb	*s*	$+ 2\ OH^-$	PbO	*s* yellow	$+ H_2O$	2	− 0.575
Pb	*s*	$+ 3\ OH^-$	$HPbO_2^-$		$+ H_2O$	2	− 0.54
Pb	*s*	$+ 4\ OH^-$	PbO_2	*s*	$+ 2\ H_2O$	4	− 0.16
3 Pb	*s*	$+ 2\ OH^- + 2\ CO_3^{2-}$	$Pb_3(CO_3)_2(OH)_2$	*s*		6	− 0.59
Pb	*s*	$+ OH^- + SH^-$	PbS	*s*	$+ H_2O$	2	− 0.56
Pb	*s*	$+ S^{2-}$	PbS	*s*		2	− 0.98
Pb	*s*	$+ SO_4^{2-}$	$PbSO_4$	*s*		2	− 0.356
Pb^{++}		$+ 2\ H_2O$	PbO_2	*s*	$+ 4\ H^+$	2	+ 1.45
PbO	*s*	$+ 2\ OH^-$	PbO_2	*s*	$+ H_2O$	2	+ 0.28
3 PbO	*s*	$+ 2\ OH^-$	Pb_3O_4	*s*	$+ H_2O$	2	+ 0.25
$PbSO_4$	*s*	$+ 2\ H_2O$	PbO_2	*s*	$+ 4\ H^+ + SO_4^{2-}$	2	+ 1.680
Pd	*s*	$+ 4\ Br^-$	$PdBr_4^{2-}$			2	+ 0.6
Pd	*s*	$+ 4\ Cl^-$	$PdCl_4^{2-}$			2	+ 0.62
Pd	*s*	$+ 2\ OH^-$	$Pd(OH)_2$ *s*			2	+ 0.07
$PdBr_4^{2-}$		$+ 2\ Br^-$	$PdBr_6^{2-}$	in NaBr 1 *N*		2	+ 0.99
$PdCl_4^{2-}$		$+ 2\ Cl^-$	$PdCl_6^{2-}$	in HCl 1 *N*		2	+ 1.29
PdI_4^{2-}		$+ 2\ I^-$	PdI_6^{2-}	in KI 1 *N*		2	+ 0.48

Reduced state			*Oxidized state*			n**	U_0
PdO_2	*s*	+ 2 OH^-	PdO_3	*s*	+ H_2O	2	+ 1.2
$Pd(OH)_2$	*s*	+ 2 OH^-	$Pd(OH)_4$	*s*		2	+ 0.8
Po		+ 6 OH^-	PoO_3^{2-}		+ 3 H_2O	4	− 0.5
Pt	*s*	+ 4 Br^-	$PtBr_4^{2-}$			2	+ 0.58
Pt	*s*	+ 4 Cl^-	$PtCl_4^{2-}$ in HCl 1 *N*			2	+ 0.76
Pt	*s*	+ 2 H_2O	$Pt(OH)_2$	*s*	+ 2 H^+	2	+ 0.99
Pt	*s*	+ H_2S *g*	PtS	*s*	+ 2 H^+	2	− 0.30
Pt	*s*	+ 2 OH^-	$Pt(OH)_2$	*s*		2	+ 0.16
Pt	*s*	+ S^{2-}	PtS	*s*		2	− 0.83
$PtBr_4^{2-}$		+ 2 Br^-	$PtBr_6^{2-}$			2	+ 0.63
$PtCl_4^{2-}$		+ 2 Cl^-	$PtCl_6^{2-}$			2	+ 0.68
$[Pt(CN)_4]^{2-}$		+ 2 Cl^-	$[PtCl_2(CN)_4]^{2-}$			2	+ 0.89
Re	*s*	+ 2 H_2O	ReO_2	*s*	+ 4 H^+	4	+ 0.252
Re	*s*	+ 4 H_2O	ReO_4^-		+ 8 H^+	7	+ 0.36
Re	*s*	+ 4 OH^-	ReO_2	*s*	+ 2 H_2O	4	− 0.576
ReO_2	*s*	+ 2 H_2O	ReO_4^-		+ 4 H^+	3	+ 0.51
ReO_2	*s*	+ 4 OH^-	ReO_4^-		+ 2 H_2O	3	− 0.594
ReO_3		+ H_2O	ReO_4^-		+ 2 H^+	1	+ 0.77
Rh	*s*	+ 6 Cl^-	$RhCl_6^{3-}$			3	+ 0.44
Rh	*s*	+ 6 OH^-	Rh_2O_3		+ 3 H_2O	6	+ 0.04
Rh^{3+}		+ H_2O	RhO^{2+}		+ 2 H^+	1	+ 1.40
RhO^{2+}		+ 3 H_2O	RhO_4^{2-}		+ 6 H^+	2	+ 1.46
Ru	*s*	+ 3 Cl^-	$RuCl_3$	*s*		3	+ 0.65
Ru	*s*	+ 5 Cl^-	$RuCl_5^{2-}$			3	+ 0.6
Ru	*s*	+ 5 Cl^- + H_2O	$RuCl_5OH^{2-}$		+ H^+	4	+ 0.6
Ru	*s*	+ 2 H_2O	RuO_2	*s*	+ 4 H^+	4	+ 0.79
Ru	*s*	+ 4 OH^-	RuO_2	*s*	+ 2 H_2O	4	− 0.04
Ru^{2+}		+ 5 Cl^-	$RuCl_5^{2-}$			1	+ 0.3
$RuCl_5^{2-}$		+ H_2O	$RuCl_5OH^{2-}$		+ H^+	1	+ 1.3
$RuCl_5OH^{2-}$		+ $3H_2O$	RuO_4	*s*	+ 5 Cl^- + 7 H^+	4	+ 1.5
S	*s*	+ 3 H_2O	H_2SO_3		+ 4 H^+	4	+ 0.45
2 S^{2-}			S_2^{2-}			2	− 0.51
S^{2-}		+ 6 OH^-	SO_3^{2-}		+ 3 H_2O	6	− 0.61
SH^-		+ OH^-	S	*s*	+ H_2O	2	− 0.478
SO_3^{2-}		+ 2 OH^-	SO_4^{2-}		+ H_2O	2	− 0.90
2 SO_4^{2-}			$S_2O_8^{2-}$			2	+ 2.01
2 $S_2O_3^{2-}$			$S_4O_6^{2-}$			2	+ 0.08
$S_2O_3^{2-}$		+ 3 H_2O	2 H_2SO_3		+ 2 H^+	4	+ 0.40

Reduced state			*Oxidized state*			n^{**}	U_0
$S_2O_3^{2-}$		$+ 6\,OH^-$	$2\,SO_3^{2-}$		$+ 3\,H_2O$	4	− 0.58
$S_2O_4^{2-}$		$+ 4\,OH^-$	$2\,SO_3^{2-}$		$+ 2\,H_2O$	2	− 1.12
$S_2O_6^{2-}$		$+ 2\,H_2O$	$2\,SO_4^{2-}$		$+ 4\,H^+$	2	+ 0.22
$S_3O_6^{2-}$		$+ 3\,H_2O$	$3\,H_2SO_3$			2	+ 0.68
$S_4O_6^{2-}$		$+ 6\,H_2O$	$4\,H_2SO_3$		$+ 4\,H^+$	6	+ 0.48
H_2S	*g*		S	*s*	$+ 2\,H^+$	2	+ 0.14
$2\,H_2SO_3$			$S_2O_6^{2-}$		$+ 4\,H^+$	2	+ 0.60
H_2SO_3		$+ H_2O$	SO_4^{2-}		$+ 4\,H^+$	2	+ 0.17
$HS_2O_4^-$		$+ 2\,H_2O$	$2\,H_2SO_3$		$+ H^+$	2	− 0.08
Sb	*s*	$+ H_2O$	SbO^+		$+ 2\,H^+$	3	+ 0.212
2 Sb	*s*	$+ 3\,H_2O$	Sb_2O_3		$+ 6\,H^+$	6	+ 0.150
Sb	*s*	$+ 4\,OH^-$	SbO_2^-		$+ 2\,H_2O$	3	− 0.67
Sb	*s*	$+ 2\,S^{2-}$	SbS_2^-			3	− 0.85
SbH_3	*g*		Sb	*s*	$+ 3\,H^+$	3	− 0.51
$2\,SbO^+$		$+ 3\,H_2O$	Sb_2O_5		$+ 6\,H^+$	4	+ 0.58
Sb_2O_3		$+ 2\,H_2O$	Sb_2O_5		$+ 4\,H^+$	4	+ 0.73
Sb_2O_4	*s*	$+ H_2O$	Sb_2O_5	*s*	$+ 2\,H^+$	2	+ 0.48
H_3SbO_3		$+ H_2O$	H_3SbO_4		$+ 2\,H^+$	2	+ 0.75
Sc	*s*	$+ 3\,OH^-$	$Sc(OH)_3$	*s*		3	− 2.6(?)
2 Se	*s*	$+ 2\,Cl^-$	Se_2Cl_2			2	+ 1.06
Se	*s*	$+ 3\,H_2O$	H_2SeO_3		$+ 4\,H^+$	4	+ 0.74
Se	*s*	$+ 6\,OH^-$	SeO_3^{2-}		$+ 3\,H_2O$	4	− 0.36
SeO_3^{2-}		$+ 2\,OH^-$	SeO_4^{2-}		$+ H_2O$	2	+ 0.05
H_2Se	*sol*		Se		$+ 2\,H^+$	2	+ 0.40
H_2SeO_3		$+ H_2O$	SeO_4^{2-}		$+ 4\,H^+$	2	+ 1.15
Sn	*s*	$+ 3\,OH^-$	$HSnO_2^-$		$+ H_2O$	2	− 0.79
Sn	*s*	$+ S^{2-}$	SnS	*s*		2	− 0.94
$HSnO_2^-$		$+ 3\,OH^- + H_2O$	$Sn(OH)_6^{2-}$			2	− 0.90
Sr		$+ 2\,OH^- + 8\,H_2O$	$Sr(OH)_2 \cdot 8\,H_2O$	*s*		2	− 2.99
2 Ta	*s*	$+ 5\,H_2O$	Ta_2O_5	*s*	$+ 10\,H^+$	10	− 0.81
Te	*s*	$+ 6\,Cl^-$	$TeCl_6^{2-}$			4	+ 0.55
Te	*s*	$+ 2\,H_2O$	TeO_2	*s*	$+ 4\,H^+$	4	+ 0.53
Te	*s*	$+ 2\,H_2O$	$TeO(OH)^+$		$+ 3\,H^+$	4	+ 0.559
Te	*s*	$+ 6\,OH^-$	TeO_3^{2-}		$+ 3\,H_2O$	4	− 0.57
TeO_2	*s*	$+ 4\,H_2O$	H_6TeO_6	*s*	$+ 2\,H^+$	2	+ 1.02
TeO_3^{2-}		$+ 2\,OH^-$	TeO_4^{2-}		$+ H_2O$	2	+ 0.4
H_2Te			Te		$+ 2\,H^+$	2	− 0.670

Reduced state			*Oxidized state*			n^{**}	U_0
Th	*s*	+ 4 OH^-	ThO_2	*s*	+ 2 H_2O	4	− 2.64
Th	*s*	+ 2 H_2O	ThO_2	*s*	+ 4 H^+	4	− 1.80
Ti	*s*	+ H_2O	TiO^{2+}		+ 2 H^+	4	− 0.89
Ti	*s*	+ 2 H_2O	TiO_2		+ 4 H^+	4	− 0.86
Ti^{3+}		+ H_2O	TiO^{2+}		+ 2 H^+	1	+ 0.1
Ti^{3+}		+ 2 SO_4^{2-}	$Ti(SO_4)_2$			1	+ 0.04
Tl	*s*	+ Br^-	TlBr	*s*		1	− 0.658
Tl	*s*	+ Cl^-	TlCl	*s*		1	− 0.557
Tl	*s*	+ I^-	TlI	*s*		1	− 0.753
Tl	*s*	+ OH^-	TlOH	*s*		1	− 0.344
2 Tl	*s*	+ S^{2-}	Tl_2S	*s*		2	− 0.96
2 Tl	*s*	+ SO_4^{2-}	Tl_2SO_4			2	− 0.436
TlCl	*s*		Tl^{3+}		+ Cl^-	2	+ 1.36
TlOH	*s*	+ 2 OH^-	$Tl(OH)_3$	*s*		2	− 0.05
U	*s*	+ 2 H_2O	UO_2	*s*	+ 4 H^+	4	− 1.40
U	*s*	+ 2 H_2O	UO_2^{2+}		+ 4 H^+	6	− 0.82
U	*s*	+ 3 OH^-	$U(OH)_3$			3	− 2.17
U	*s*	+ 4 OH^-	UO_2		+ 2 H_2O	4	− 2.39
UO_2	*s*		UO_2^{2+}			2	+ 0.33
$U(OH)_4$		+ 2 Na^+ + 4 OH^-	Na_2UO_4		+ 4 H_2O	2	− 1.61
$U(SO_4)_2$	*sol*	+ 2 H_2O	UO_2^{2+}		+ 2SO_4^{2-}+4H^+	2	+ 0.36
V	*s*	+ H_2O	VO^{2+}		+ 2 H^+	4	+ 0.3
V	*s*	+ 4 H_2O	$V(OH)_4^+$		+ 4 H^+	5	− 0.253
V^{3+}		+ H_2O	VO^{2+}		+ 2 H^+	1	+ 0.36
VO^{2+}		+ H_2O	VO_2^+		+ 2 H^+	1	+ 0.999
VO^{2+}		+ 2 H_2O	HVO_3		+ 3 H^+	1	+ 1.02
VO^{2+}		+ 3 H_2O	VO_4^{3-}		+ 6 H^+	1	+ 1.031
VO^{2+}		+ 3 H_2O	$V(OH)_4^+$		+ 2 H^+	1	+ 1.00
W	*s*	+ 2 H_2O	WO_2	*s*	+ 4 H^+	4	− 0.05
W	*s*	+ 3 H_2O	WO_3	*s*	+ 6 H^+	6	+ 0.09
W	*s*	+ 8 OH^-	WO_4^{2-}		+ 4 H_2O	6	− 1.05
WO^{3+}		+ 2 H_2O	WO_3	*s*	+ 4 H^+	1	0.0 (?)
2 WO_2	*s*	+ H_2O	W_2O_5	*s*	+ 2 H^+	2	0.00
W_2O_5	*s*	+ H_2O	2 WO_3	*s*	+ 2 H^+	2	+ 0.15
Zn	*s*	+ 4 CN^-	$[Zn(CN)_4]^{2-}$			2	− 1.26
Zn	*s*	+ CO_3^{2-}	$ZnCO_3$	*s*		2	− 1.07
Zn	*s*	+ 4 NH_3 *sol*	$[Zn(NH_3)_4]^{2+}$			2	− 1.03

Reduced state			Oxidized state			n**	U_0
Zn	s	$+ 2\,OH^-$	$Zn(OH)_2$			2	− 1.245
Zn	s	$+ 4\,OH^-$	ZnO_2^{2-}		$+ 2\,H_2O$	2	− 1.216
Zn	s	$+ S^{2-}$	ZnS	s		2	− 1.44
$Zn_{amalg.}$		$+ SO_4^{2-}$ in $ZnSO_4 \cdot 7\,H_2O$ sat.	$ZnSO_4 \cdot 7\,H_2O$	s		2	− 0.799
Zr	s	$+ 2\,H_2O$	ZrO_2	s	$+ 4\,H^+$	4	− 1.43
Zr	s	$+ 4\,OH^-$	H_2ZrO_3		$+ H_2O$	4	− 2.36

Abbreviations: s = solid; g = gas; *sol* = solution; l = liquid;
amalg = amalgam; α, β, γ, etc. = crystalline phases;
m = molar; N = normal.

* The values given in this table are, in general, considerably less accurate than those in Table III, 3 both because they are in part calculated from insufficiently exact thermal values, and because the original articles often do not rigorously define the experimental conditions or the precision of the measurements. They can therefore be considered only as indications, even although the original literature gives the values of the electric tensions in terms of millivolts or less.

To make it easier to find redox tensions in this table, the processes are arranged alphabetically according to the most important, electrochemically active, elements, rather than according to the tensions.

Other reactions are reported in: OSTWALD–LUTHER, *Handbuch der allgemeinen Chemie*; LANDOLT–BÖRNSTEIN, *Physikalisch–chemische Tabellen*; *Critical Tables*; W. M. LATIMER, *loc. cit.*; *Handbook of Chemistry and Physics*; R. PARSON, *Handbook of Electrochemical Constants*; Butterworth, London, 1959; G. CHARLOT, D. BÉZIER and J. COURTOT, see note Table III, 3.

** Number of electrical charges involved in the reaction.

Analogous considerations may be applied to oxidizing systems whose electric tensions are referred to that of the oxygen electrode. These decompose water if their electric tensions are more positive than those of the oxygen electrode in a solution of an equal OH^- ion concentration. In practice, it is possible to maintain such oxidizing and reducing systems under conditions of metastability thanks to the overtension of hydrogen and oxygen (*cf.* Chapter IV, p. 229 *et seq.*, 256 *et seq.*).

Recently, the so-called *formal tensions* have been introduced. These are analogous to standard tensions referred to unit concentration (and not unit activity) of the substances involved (*i.e.* 1 mole per litre). They therefore do not take account either of the degree of dissociation, which can be less than 1, nor of the effect of interionic forces, nor of any secondary reactions, such as association, hydrolysis etc. These formal tensions are sometimes more useful than standard tensions in many problems of analytical chemistry. Formal potentials are also collected in tables[1]. It must, however, be noted that every table of standard tensions in aqueous media is sometimes insufficient. The acidity or alkalinity of the medium has often in fact great importance, not only when one of the ionic components of water, H^+ or OH^-, is directly involved in the reaction, but also when the acid or alkaline medium can determine the actual existence of one or more participants in the reaction. The electric tension of the Fe^{2+}/Fe^{3+} electrode, for example, should be independent of the pH of the medium and is in fact so, but only within a certain range of pH values lying below 2. If the pH is raised above this value, hydrolysis and precipitation of ferric hydroxide occurs so that the activity of the Fe^{3+} species becomes solely a function of pH and is no longer dependent on the total concentration of trivalent iron. It would be possible to give many other examples to illustrate the sometimes overwhelming importance of the pH of the medium. To avoid this inadequacy of tables of electric tensions, Pourbaix and his colleagues have for some years been studying the construction of pH-tension diagrams, on a purely thermodynamic basis; by the use of these it is possible to obtain rapidly much information which sometimes cannot be got from tables of tensions[2, 3].

13. Galvanic Cells in Non-Aqueous Solvents and in Molten Electrolytes[3]

It is possible to construct Daniell, redox, concentration, etc. cells also in non-aqueous solvents and the same laws which were described above for electrodes in aqueous solvents can be applied here. The principal

[1] See, for example, C.S. Garner in E.H. Swift, *System of Chemical Analysis*, Prentice Hall, New York (1939); H.H. Willard and G.D. Manalo, *Anal. Chem.*, 19 (1947) 462.

[2] An excellent article which gives the theory of such diagrams and discusses in detail their construction and use has been published by P. Delahay, M. Pourbaix and P. von Rysselberghe, *J. Chem. Educ.*, 27 (1950) 683.

An atlas of those pH-tension diagrams which have so far been studied will shortly be published.

In this a number of articles describe the applications of such diagrams to a wide range of fields of research and technology: reactivity, analysis, corrosion etc.

[3] After the completion of the section of the text dealing with non-aqueous solvents and molten electrolytes, a review monograph was published by D.J.G. Ives and G.J. Janz (*Reference Electrodes*, Academic Press, New York and London, 1961), to which reference should be made for the most accurate numerical values.

difficulty in the study of electrodes in non-aqueous solvents is the determination of the activity, which is often somewhat uncertain and thus leads to a divergence between experiment and theory. A second cause of uncertainty comes from the indeterminacy of the solvation energy of ions in non-aqueous solutions and from the possibility of forming imperfectly defined complexes between ions and solvent molecules.

In every aqueous medium the reference point is the electric tension of the standard hydrogen electrode, taken as zero. Systems in non-aqueous solvents should not be considered as isolated cases but an attempt should be made to relate their electric tensions to that of the standard hydrogen electrode in water. Ignoring, for the moment, the fact that the electric tensions of electrodes in non-aqueous media are uncertain to within millivolts, and sometimes even centivolts, because of the uncertainty about the activity coefficient, it will nevertheless be interesting to examine the question from a theoretical standpoint. The standard tension of an electrode of, for example, a monovalent metal, depends on the free enthalpy of the general reaction.

$$\tfrac{1}{2}\,H_2 + Me^{+}{}_{solv.} \rightarrow H^{+}{}_{solv.} + Me$$

where the suffix *solv.* indicates that the relative ion is solvated. To connect together the standard tensions in different solvents it is necessary to choose a reference point such that the difference between the standard tensions of the same element in the two different solvents measures the free enthalpy of the transfer of the ions of this element from one solvent to the other, *i.e.* it measures the difference between the free enthalpies of the ionic solvations. This quantity is not directly susceptible to measurement because all that can actually be measured is the free enthalpy of the transfer of a whole electrolyte.

If by other means it is possible to obtain the free enthalpy for the transfer of a single ion then the possibility arises of developing a series of electric tensions. It may be assumed for this purpose that the free enthalpy of solvation of the Rb^+ ion is the same in every solvent, since it is of large size, not polarizable to any extent and does not tend to form complexes. On this hypothesis Pleskow[1] obtained series of tensions referred to a Rb^+ electrode considered as zero in every solvent.

These series follow more or less the same pattern but there are, however, marked absolute differences between the electric tensions, and certain

[1] V. A. Pleskow, *Acta Physicochim. U.R.S.S.*, 13 (1940) 662; 21 (1946) 41.

inversions which are probably due to differences in the free enthalpy of solvation and to uncertainties in the determinations of activity. Pleskow's hypothesis, however, can only be considered as a first approximation, since the constancy of the solvation energy of the Rb^+ ion in different solvents is true only within certain limits. To improve this approximation, it is necessary to evaluate the small differences between the solvation energies of the Rb^+ ions. Following the procedure of Latimer, Pitzer and Slansky[1], the corrected Born equation may be used.

$$\Delta G_{\pm H_2O} = -\frac{N(e^-)^2}{2}\left(1-\frac{1}{\varepsilon}\right)\left(\frac{1}{r^+ + R^+} + \frac{1}{r^- + R^-}\right) \qquad (1)$$

Where $\Delta G_{\pm H_2O}$ is the free enthalpy of hydration of the electrolyte, N is Avogadro's number, e^- is the charge of an electron, ε is the dielectric constant and r^+ and r^- are the radii of the cation and anion in the crystalline state. R^+ and R^- are empirical correction terms which must be added to the ionic radii in the crystalline state so as to obtain an experimental free enthalpy which is equal to that calculated with equation (1) for each pair of ions. Thus, the free enthalpy of solvation $\Delta G_{\pm solv.}$ in non-aqueous solvents is obtained using the equation

$$\Delta G_{\pm solv.} = \Delta G_{\pm H_2O} + 2\,RT \ln \frac{a_{sat.\ H_2O}}{a_{sat.\ solv.}}$$

where $a_{sat.}$ is the activity of the electrolyte in the respective saturated solutions. Applying equation (1) referred to the free enthalpy of solvation instead of to that of hydration, correction terms R^+ and R^- are sought such that the values calculated for $\Delta G_{solv.}$ agree with the measured values for each pair of ions of opposite sign. Having once found the constant terms $(r + R)$ the value of the ionic $\Delta G_{solv.}$ for the different solvents can be calculated. Finally, from these values the shift of the standard tensions of rubidium in different solvents can be calculated in relation to its standard tension in water. From these values and from the series of electric tensions in various solvents it is possible to recalculate the standard tensions of various electrodes in these different solvents all referred to the electric tension of the standard hydrogen electrode in water[2]. The results of this work need further confirmation. By recalculating Pleskow's results[3] with reference to the standard hydrogen electrode, rather than the rubi-

[1] W. L. LATIMER, K. S. PITZER and C. M. SLANSKY, *J. Chem. Phys.*, 7 (1939) 108.
[2] H. STREHLOW, *Z. Elektrochem.*, 56 (1952) 827.
[3] V. A. PLESKOW, *Acta Physicochim. U.R.S.S.*, 13 (1940) 662; 21 (1946) 41.

TABLE III, 7

SERIES OF ELECTRIC TENSIONS IN NON-AQUEOUS SOLVENTS

(referred to the electric tension of the standard hydrogen electrode in water)

Element	H_2O	CH_3OH	CH_3CN	HCOOH	$HCONH_2$	N_2H_4	NH_3
Li/Li^+	− 3.04	− 3.13	− 3.09	− 3.01	—	− 3.11	− 3.23
Cs/Cs^+	− 2.92	—	− 3.02	− 2.97	—	—	− 2.94
Rb/Rb^+	− 2.92	− 2.97	− 3.03	− 2.98	− 2.92	− 2.92	− 2.92
K/K^+	− 2.92	—	− 3.02	− 2.89	− 2.94	− 2.93	− 2.97
Ca/Ca^{2+}	− 2.87	—	− 2.61	− 2.73	—	− 2.82	− 2.73
Na/Na^+	− 2.71	− 2.76	− 2.73	− 2.95	—	− 2.74	− 2.84
Zn/Zn^{2+}	− 0.76	-0.7_7	− 0.60	− 0.58	− 0.83	− 1.32	− 1.52
Cd/Cd^{2+}	− 0.40	-0.4_6	− 0.33	− 0.28	− 0.48	− 1.01	− 1.19
Tl/Tl^+	− 0.33	-0.4_1	—	—	− 0.41	—	—
Pb/Pb^{2+}	− 0.13	-0.2_3	+ 0.02	− 0.25	− 0.26	− 0.56	− 0.67
$H_2/2\,H^+$	0	-0.0_3	+ 0.14	+ 0.47	− 0.07	− 0.91	− 0.99
Cu/Cu^{2+}	+ 0.34	$+0.3_1$	− 0.24	+ 0.33	+ 0.21	—	− 0.56
Cu/Cu^+	+ 0.52	—	− 0.14	—	—	− 0.69	− 0.58
Hg_2/Hg_2^{2+}	+ 0.79	$+0.7_1$	—	+ 0.65	—	—	—
Ag/Ag^+	+ 0.80	$+0.7_3$	+ 0.37	+ 0.64	—	− 0.14	− 0.16
Hg/Hg^{2+}	+ 0.91	—	+ 0.39	—	—	—	− 0.24
$2\,Cl^-/Cl_2$	+ 1.36	$+1.0_9$	$+0.7_2$	+ 1.24	—	—	$+1.0_4$
$2\,Br^-/Br_2$	+ 1.07	$+0.8_6$	$+0.6_1$	+ 0.99	—	—	$+0.8_4$
$2\,I^-/I_2$	+ 0.54	$+0.3_3$	$+0.2_1$	+ 0.44	—	—	$+0.4_6$

dium electrode, and by utilizing the values published by Strehlow[1], and Pavlopoulos and Strehlow[2] also, the series of electric tensions reported in Table III, 7 were obtained.

The measurement of electrode tensions in non-aqueous solvents is obviously dependent on the availability of good comparison electrodes. In aqueous solvents the best comparison electrodes are of the second kind. In non-aqueous solvents these electrodes in general are neither stable nor reproducible because the halides of the electrode metals often form complexes with the alkaline halides or alkylammonium compounds used as electrolytes, or else they tend to form solvates with the solvent, in the solid state. It is thus very difficult to interpret the behaviour of these electrodes. There are exceptions, however, and some of these electrodes be-

[1] H. Strehlow, *Z. Elektrochem.*, 56 (1952) 827.
[2] T. Pavlopoulos and H. Strehlow, *Z. physik. Chem.* (*Frankfurt*), 2 (1954) 89.

TABLE III, 8

COMPARISON ELECTRODES IN NON-AQUEOUS SOLVENTS

Electrode	*Solvent*	U_0
$Ag/AgCl/Cl^-$	CH_3OH	+ 0.010
$Ag/AgCl/Cl^-$	C_2H_5OH	+ 0.081
$Ag/AgCl/Cl^-$	CH_3COCH_3	− 0.53
$Ag/AgCl/Cl^-$	HCOOH	− 0.120
$Ag/AgCl/Cl^-$	$HCONH_2$	+ 0.204
$Ag/AgCl/Cl^-$	CH_3COOH	− 0.618
$Ag/AgBr/Br^-$	CH_3OH	− 0.133
$Ag/AgBr/Br^-$	C_2H_5OH	− 0.0815
$Ag/AgI/I^-$	C_2H_5OH	− 0.253
$Cd/CdCl_2/Cl^-$	$HCONH_2$	− 0.617
$Hg/Hg_2(CH_3COO)_2/CH_3COO^-$	CH_3OH: $NaCH_3COO$ *sat.*	+ 0.423

have fairly well in various solvents. Table III, 8 collects values of the electric tensions of some comparison electrodes in non-aqueous solvents, referred to the conventional zero of the standard hydrogen electrode in the given solvent.

Many other values of electric tensions for electrodes of the second kind can be found in the literature and, in particular, values for the standard tensions of silver chloride electrodes in various solvents and mixtures of solvents with water. However, it is not always possible to rely upon the published figures and much care must be exercised in their use.

The measurement of electric tensions of galvanic cells with molten electrolytes has not so far given definitive results for various reasons although much valuable work has been done in this field. Above all, the state of molten electrolytes is not exactly known; it is not known if the classical theory of dissociation equilibria or the theory of total dissociation of strong electrolytes should be applied. Secondly, no comparison electrode is yet available for molten electrolytes whose electric tension is known for certain and which could be used for measuring the electric tensions of other electrolytes when connected with them to form a cell. Further, it has been found experimentally that at the boundary between two molten electrolytes exist marked diffusion electric tensions which it is not yet possible to determine exactly. Finally, major experimental difficulties are encountered which are not easily surmounted, due to the fact that many metallic electrodes in contact with molten electrolytes give rise to metallic clouds (*cf.* Chapter X, p. 592 *et seq.*), *i.e.* to solutions or dispersions of the metal in the electrolyte, which diffusing towards the other electrode cause variations in its electric tension. Some good results have, however, been obtained. For concentration cells in molten electrolytes the general equation of concentration cells is valid, particularly for low concentrations of an electrolyte in another as shown in Table III, 9.

TABLE III, 9

ELECTRIC TENSIONS OF CONCENTRATION CELLS IN MOLTEN ELECTROLYTES

Cell	*Solvent*	c_1*	c_2*	°*K*	$U_{obs.}$	$U_{calc.}$
$Cu/CuClc_1/CuClc_2/Cu$	KCl	0.400	0.0548	1097	0.1814	0.1828
$Cu/CuClc_1/CuClc_2/Cu$	KCl	0.605	0.474	1064	0.0225	0.0224
$Cu/CuClc_1/CuClc_2/Cu$	NaCl	0.833	0.265	1116	0.1075	0.1097
$Cu/CuClc_1/CuClc_2/Cu$	NaCl	1.010	0.497	1104	0.0690	0.0674
$Ag/AgClc_1/AgClc_2/Ag$	KCl	0.712	0.203	1089	0.1172	0.1174
$Ag/AgClc_1/AgClc_2/Ag$	KCl	0.712	0.410	1074	0.0490	0.0510

* c_1 and c_2 indicate the concentrations expressed in moles per 1,000 g of solvent.

These results favour the theory of a complete dissociation of electrolytes dissolved in other molten electrolytes, with activity coefficients equal for the two solutions. For chemical cells with molten electrolytes of the type $Ag/AgCl/Cl_2$ based on the reaction $Ag + 1/2\ Cl_2 \rightarrow AgCl$ or of the type $Pb/PbCl_2/Cl_2$ based on the reaction $Pb + Cl_2 \rightarrow PbCl_2$, where the negative pole consists of the molten metal in contact with its molten salt – preferably a halide – whilst the positive pole consists of a graphite rod immersed in this molten salt and around which is bubbled the gaseous halogen, it is possible to measure the electric tensions which arise from the formation of the halide.

Measuring the electric tensions of many cells in which the same halide of various metals is formed, gives a series of relative electric tensions for the various metals with respect to their ions, assuming that the activities of both the metallic and halogen ions in the pure molten halide are equal to one. In this case the electric tension of the halogen electrode is constant, provided that the gaseous pressure of the halogen and the temperature are also kept constant. If the electric tensions of cells of the type

$$Me/MeX_n/X_2$$

TABLE III, 10

SERIES OF ELECTRIC TENSIONS IN MOLTEN ELECTROLYTES

Metal	U_0 *in*			
	$MeCl_n$	$MeBr_n$	MeI_n	H_2O
Cs/Cs^+	− 2.66	—	—	− 2.92
Rb/Rb^+	− 2.60	− 2.10	—	− 2.92
Ba/Ba^{2+}	− 2.60	− 2.62	—	− 2.90
Sr/Sr^{2+}	− 2.52	− 2.41	—	− 2.89
K/K^+	− 2.51	− 2.53	− 2.41	− 2.92
Li/Li^+	− 2.39	− 2.40	− 2.41	− 3.04
Ca/Ca^{2+}	− 2.36	− 2.25	—	− 2.87
Na/Na^+	− 2.36	− 2.35	− 2.27	− 2.713
Mg/Mg^{2+}	− 1.59	− 1.58	− 1.47	− 2.37
Mn/Mn^{2+}	− 0.86	− 0.83(?)	− 0.91	− 1.19
Al/Al^{3+}	− 0.85(?)	− 0.81	—	− 1.66
Tl/Tl^+	− 0.45	− 0.69	− 0.87	− 0.335
Zn/Zn^{2+}	− 0.41	− 0.50	− 0.73	− 0.763
Cd/Cd^{2+}	− 0.26	− 0.46	− 0.65	− 0.402
Pb/Pb^{2+}	− 0.10	− 0.38	− 0.44	− 0.126
Sn/Sn^{2+}	− 0.06	− 0.13	− 0.49	− 0.140
Ni/Ni^{2+}	+ 0.01(?)	—	—	− 0.25
Co/Co^{2+}	+ 0.06(?)	− 0.05(?)	− 0.03(?)	− 0.30
Ag/Ag^+	+ 0.18	+ 0.10	− 0.43	+ 0.800
Cu/Cu^+	+ 0.28	− 0.06	− 0.29	+ 0.52
Bi/Bi^{3+}	+ 0.38	+ 0.19	− 0.13	+ 0.2

For further information concerning electric tensions in molten systems see IU. K. DELIMARSKII and B. F. MARKOW; *Electrochemistry of Fused Salts* (English translation from the Russian), Sigma Press, Washington, D.C., 1961.

where X represents the halogen, are arranged according to increasing values, for example, after eliminating the electric tension of the halogen electrode[1], a series is obtained whose order should be equal to that for

[1] The electric tension of the halogen electrode is calculated for a temperature of 700° C from the equilibrium

$$H_2 + X_2 \rightleftharpoons 2HX \qquad (X = \text{halogen})$$

If in the cell

$$X_2/HX/H_2$$

the electric tension of the hydrogen electrode is taken as zero, the electric tension of the cell calculated from the affinity of the above reaction is equal to that of the halogen electrode.

the same elements in the series of electric tensions in water. Table III, 10 reports the values for the electric tensions of metallic electrodes derived from cells forming respectively the chlorides, bromides and iodides at 700°; the electric tensions are referred to that of the hydrogen electrode taken as zero. It is easy to see from Table III, 10 that not only does the electric tension series at 700° C in molten electrolytes, not correspond with that at 25° C in aqueous solutions but also that the order depends upon the particular type of halogen involved. So long as it is not possible to construct a comparison electrode which is certain and independent, like those used for solutions, and so long as it is not possible to determine the activities of the various phases exactly, the standard tensions of the individual elements with respect to their ions in the molten state will remain indeterminate and, moreover, the orders of the corresponding series of electric tensions will remain uncertain. And even if these two conditions were fulfilled it would still be essential to achieve a third, that is the elimination of the diffusion electric tension between the different electrolytes in the molten state; these electric tensions are not known and are difficult to calculate since the electrical mobilities in the molten state are also unknown. These diffusion tensions are often not of negligible size. The comparison electrode which, at the present time appears to be the best for use with molten electrolytes, is the Ag/AgCl[1] electrode.

Finally, it must be emphasized that the positions of the various metals in the series of electric tensions are probably not constant but vary with the temperature and the solvent[2].

[1] S. Senderoff and A. Brenner, *J. Electrochem. Soc.*, 101 (1954) 31; M. Bonnemay and R. Pineau, *Compt. rend.* 240 (1955) 1774; *C.R. VIII Réunion CITCE*, Madrid, 1956 (1958) Butterworth, London, p. 190; S. N. Flengas and T. R. Ingraham, *Can. J. Chem.*, 35 (1957) 1139.

[2] *Cf.* for example K. Delimarskii, *C. A.*, 40 (1946) 1737; V. A. Plotnikov, E. J. Kirichenko and N. S. Furtunatov, *C. A.*, 35 (1941) 3530; K. Delimarskii and A. A. Kolotts, *Chem. Zentr.*, I (1950) 1947; K. Grjotheim, *Z. physik. Chem., Frankfurt*, 11 (1957) 150.

CHAPTER IV

ELECTROLYSIS AND ELECTROCHEMICAL KINETICS IN AQUEOUS SOLUTIONS[1, 2]

by

CLAUDIO FURLANI[3] and GIULIO MILAZZO

1. Faraday's Laws and Current Efficiency

If two electrodes are immersed in an aqueous solution of a salt, acid or base and connected to a source of continuous current of a sufficiently high tension, there will be a passage of electricity through the solution and at the same time various chemical reactions will occur at the electrodes. These may include the evolution of gas, the separation of substances, the dissolution of the electrode or the appearance of new substances in the solution. The passage of current through a solution of hydrochloric acid, for example, under suitable conditions leads to the evolution of gaseous chlorine on the anode and of gaseous hydrogen on the cathode; in a solution of copper sulphate with a copper anode, metallic copper will separate at the cathode and dissolve from the anode; in a solution containing ferrous salts there will be a production of ferric salts. Thus, a change is brought about in the system at the expense of the electrical energy furnished from outside the system. The chemical change in an electrolytic system where the final composition is not equal to the initial one, at the expense of externally supplied electrical energy, is therefore simply the inverse of the process of the production of external electrical work at the expense of changes in free energy produced by chemical reactions in the system, *i.e.* the process on which the galvanic cell is based (*cf.* Chapter III and Chapter XI).

[1] The theoretical part of electrolysis in molten electrolytes is treated in Chapter IX on page 592 *et seq.*

[2] Throughout this and the following chapters, where the activity occurs in equations it is assumed for the sake of simplicity that the activity coefficient is 1, so that the concentration can replace the activity. It is apparent that this is not valid for rigorous calculations. The constant 0.239 is also assumed as included in the value of the constant R.

[3] University of Rome.

Reactions at the electrode surface between the particles in solution and the electrical charges of the electrode are called primary reactions (*cf.* Chapter II, p. 24 *et seq.*) and may be accompanied by secondary reactions. Primary chemical reactions follow two fundamental laws of electrolysis which were discovered by Faraday in 1833. The first of these states that the mass of substance m formed at an electrode by the passage of a current through an electrolyte is proportional to the total quantity of electricity which passes. So that

$$m = kIt$$

where I is the current intensity and t is time, or more exactly, if the current varies

$$m = k \int_0^t I\mathrm{d}t$$

Faraday's Second Law states that the masses of various substances separated by the same amount of electricity are in the ratios of their respective chemical equivalents. Faraday's Laws are experimental. The First Law may be illustrated by the electrolysis of two solutions of silver nitrate with a current of constant and equal intensity. This current, however, flows through one solution for twice as long as it does through the other. Thus, twice the quantity of electricity is passed through one solution as through the other and it will be found that exactly twice the weight of silver will be deposited on the cathode in this solution as on that in the other. Similarly, if the electrolyses are carried out for the same length of time in two cells but with twice the current intensity in one compared to the other, there will again be double the amount of silver deposited in the cell which has had the larger current with respect to the other.

The Second Law may be illustrated by connecting three electrolytic cells in series so that the current through each of them is exactly the same and hence exactly the same quantity of electricity passes through each cell. Hydrochloric acid is placed in the first cell, silver nitrate solution in the second and nickel sulphate solution in the third, and electrolysis is continued until 1 g equiv. of hydrogen has collected in the first cell. It will then be found that exactly 107.88 g of silver and 58.69/2 g of nickel will have been deposited in the other two cells respectively, *i.e.* 1 g equiv. in each case. Moreover, 35.4 g of chlorine will be evolved at the anode of the first cell and 8 g of oxygen at the anodes of the other two, *i.e.* once more 1 g equiv. This shows that the quantity of electricity required to separate 1 g equiv. is always the same. It has been determined and is 96491.4 C;

if now 96491.4 C separate 1 g equiv. Ag, 1 C separates 107.88/96491.4 g of silver, *i.e.* 0.001118 g. This is the rational basis for the legally established figure used in the definition of the ampere. The ampere, being a current of 1 C/sec separates 0.001118 g of silver per second at the cathode of an electrolytic cell containing silver nitrate.

The *electrochemical equivalent* of any element is defined as the mass of that element which is separated electrochemically by one coulomb; for silver it has this value of 0.001118 g. For any given electrochemical process, the electrochemical equivalent is defined in general as the quantity of substance transformed by one coulomb. Table IV, 1 shows the electrochemical equivalents (g/C) of the more important elements together with their reciprocals (the number of coulombs required to separate 1 g of the element) and their multiples referred to Ah units (1 Ah = 3,600 C). Similar tables could be calculated for any other process.

The quantity of 96491.4 $\simeq$ 96500 C required to separate 1 equivalent of a substance is called a faraday (1 **F**).

In using Faraday's Law allowance must be made for valencies and for the reactions which occur. If, for example, a metal has various valency states its electrochemical equivalent for a deposition reaction will vary depending upon the particular valency which it has in the solution in which the electrolysis occurs, and hence the amount deposited by 1 **F** will also vary. These two laws have always proved to be rigorously valid even when apparent discords exist between theory and practice. In effect such divergencies can always be explained by the particular experimental conditions used and the way in which the reaction occurs so that there can be no doubt of the validity of Faraday's Laws. Above all, it must be realized that even although the greater part of the current flowing through a cell during electrolysis leads to a chemical reaction, yet a certain small fraction of the total current may be utilized for other purposes (*e.g.* dispersed in capacitive effects). That part of the current which is utilized in chemical phenomena is called the *faradic current* whereas the other, which is not so used, is called the *nonfaradic current*. Faraday's Laws always mean the faradic current when they refer to 'current'. Allowance must also be made for other contemporaneous primary processes or for further changes in the system due to secondary reactions. The electrolysis of a solution of zinc sulphate acidified with sulphuric acid, does not lead to the separation of one equivalent of zinc at the cathode for each **F** which passes through the cell, since H^+ is discharged at the same time as the Zn^{2+} ions. Hence, two

TABLE IV,1

ELECTROCHEMICAL EQUIVALENTS

Element	*Valency*	$(g/C) \cdot 10^3$	$(C/g) \cdot 10^{-3}$	g/Ah	Ah/g
Ag	1	1.11793	0.89451	4.02452	0.24848
Al	3	0.09316	10.73415	0.33538	2.98171
As	3	0.25876	3.86464	0.93152	1.07351
As	5	0.15254	6.44106	0.55891	1.78918
Au	1	2.04352	0.48935	7.35668	0.13593
Au	3	0.68117	1.46805	2.45223	0.40779
Ba	2	0.71171	1.40507	2.56216	0.39030
Be	2	0.04674	21.39688	0.16825	5.94358
Bi	3	0.72193	1.38517	2.59896	0.38477
Bi	5	0.43316	2.30861	1.55938	0.64128
Br	1	0.82815	1.20752	2.98132	0.33452
Ca	2	0.20767	4.81537	0.74761	1.33760
Cd	2	0.58244	1.71693	2.09677	0.47692
Ce	3	0.48404	2.06594	1.74255	0.57387
Cl	1	0.36743	2.72161	1.32275	0.75600
Co	2	0.30539	3.27452	1.09931	0.90966
Cr	3	0.17965	5.56624	0.64676	1.54618
Cr	6	0.08983	10.13247	0.32338	3.09235
Cs	1	1.37731	0.72606	4.95830	0.20168
Cu	1	0.65876	1.51801	2.37152	0.42167
Cu	2	0.32938	3.03602	1.18576	0.84334
Fe	2	0.28938	3.45568	1.04176	0.95991
Fe	3	0.19291	5.18353	0.69451	1.43987
H	1	0.010446	95.73321	0.037605	26.59256
Hg	1	2.07886	0.48103	7.48390	0.13362
Hg	2	1.03943	0.96207	3.74195	0.26724
I	1	1.31523	0.76032	4.73484	0.21120
Ir	4	0.50026	1.99896	1.80095	0.55546
K	1	0.40514	2.46828	1.45850	0.68563
Li	1	0.07192	13.90490	0.25890	3.86247
Mg	2	0.12601	7.93586	0.45364	2.20440
Mn	2	0.28461	3.51363	1.02458	0.97601
Mn	4	0.14230	7.02727	0.51229	1.95202
Mo	6	0.16580	6.03125	0.59689	1.67535
N	3	0.048387	20.66676	0.17419	5.74077
N	5	0.029032	34.44446	0.10452	9.56795
Na	1	0.23831	4.19620	0.85792	1.16561
Ni	2	0.30409	3.28846	1.09474	0.91346
O	2	0.082902	12.06250	0.29845	3.35069
Os	4	0.49611	2.01567	1.78601	0.55991

Element	*Valency*	(g/C) · 10^3	(C/g) · 10^{-3}	g/Ah	Ah/g
P	5	0.06421	15.57456	0.23115	4.32627
Pb	2	1.07363	0.93142	3.86506	0.25873
Pb	4	0.53681	1.86284	1.93253	0.51746
Pd	4	0.27642	3.61762	0.99513	1.00489
Pt	4	0.50578	1.97716	1.82080	0.54921
Rb	1	0.88580	1.12892	3.18889	0.31359
Re	7	0.27581	3.62568	0.99292	1.00713
Rh	4	0.26661	3.75085	0.95978	1.04190
Ru	4	0.26347	3.79548	0.94850	1.05430
S	2	0.16611	6.01996	0.59801	1.67221
S	4	0.08306	12.03993	0.29901	3.34442
S	6	0.05537	18.05989	0.19934	5.01664
Sb	3	0.42059	2.37763	1.51411	0.66045
Sb	5	0.25235	3.96272	0.90847	1.10075
Se	6	0.13637	7.33283	0.49094	2.03690
Sn	2	0.61503	1.62595	2.21409	0.45165
Sn	4	0.30751	3.25190	1.10705	0.90330
Sr	2	0.45404	2.20244	1.63455	0.61179
Te	6	0.22040	4.53726	0.79343	1.26037
Ti	4	0.12409	8.05846	0.44674	2.23846
Tl	3	0.70601	1.41641	2.54164	0.39345
U	6	0.41117	2.43206	1.48023	0.67557
V	5	0.10560	9.47007	0.38015	2.63057
W	6	0.31779	3.14674	1.14404	0.87409
Zn	2	0.33876	2.95197	1.21952	0.81999

contemporaneous primary reactions occur at the cathode; adding together the values of the equivalents of zinc and hydrogen separated at the cathode, gives a sum which is rigorously equal to the number of **F** which have passed through the cell. Similarly, in the electrolysis of sodium chloride, one equivalent of chlorine would be expected to separate for every **F** passing through the cell; but this does not occur. Here the anodic process is unique but is accompanied by secondary reactions such as the dissolution of a part of the evolved chlorine and reactions of the chlorine with OH^- ions from the cathode region to give hypochlorous acid and ClO^- ions etc. These reactions use up part of the chlorine separated by the primary discharge reaction and apparently invalidate Faraday's Law. If the secondary reactions are eliminated however the law proves to be perfectly valid.

These and other concomitant phenomena mean that the quantity of substance, produced at the electrode, rarely corresponds to the quantity of electricity which has passed through the cell according to Faraday's Laws. The ratio between the quantity of substance effectively obtained by electrolysis and the quantity theoretically obtainable according to these laws is called the *current efficiency* ($R_{cur.}$). Faraday's Laws are valid not only for electrolytic phenomena in general with any electrode reaction but also for the electrode reactions of galvanic cells and both for aqueous solutions, any other solvents and molten electrolytes. In this last case, the exact and unequivocal confirmation of the laws encounters gross experimental difficulties due to the particular electrochemical nature of these systems and to the generally high temperatures, at which salts melt, leading to significant amounts of secondary reactions (*cf.* Chapter X, p. 592 *et seq.*).

2. The Mechanism of Current Flow through Electrodes

An electrode is a heterogeneous system composed of at least one conductor of 1st class and at least one of 2nd class, *e.g.* a metal and an aqueous solution of one of its salts. There is thus the possibility of a spontaneous electrochemical reaction between the substances present in the different phases of the electrode itself. This may be expressed in another way, by saying that in the heterogeneous system a certain positive chemical affinity exists, which tends to make a reaction occur. An electrode reaction cannot take place if the electrode is not coupled with another one and with an external circuit to form a closed electrochemical system. To study the kinetics of individual electrochemical processes, the electrode should ideally be considered as coupled with a normal hydrogen electrode (*i.e.* a true reference electrode rather than simply a comparison electrode). This hydrogen electrode should be of such a size that the progress of the reactions and the magnitudes of the dynamic electrical quantities (the electrode current intensity and density, *cf.* p. 203 *et seq.*) may be neglected, and will not affect either its composition or its own electrical quantities. With such a reference electrode the electric tension of the system is, by definition, the true relative electric tension of the other electrode. In what follows, it will be implicitly assumed that the particular electrode studied functions in association with a reference electrode of constant characteristics independent of its own operation. Observations on such a system will, according to the conventions already adopted (Chapter III), be attributed solely to the test electrode. With electrode reactions there is a separation of charges consequential upon the reaction (*e.g.* $Cu \rightarrow Cu^{2+} + 2e^-$). The passage of charges across the boundary between the two conducting phases constitutes the electrical current. These charges, separated by the chemical reaction, also generate an electrical field and hence an *electric tension U* which tends to make the charges reunite electrostatically. The electric tension thus derives from chemical affinity. This, when expressed in electrical units and referred to unit charge, corresponds to a tension which is the *chemical tension* (Chapter III, 3 on p. 112 *et seq.*). At equilibrium this is equal but of opposite sign to the electric tension. The electric tension is not always a spontaneous factor arising as a consequence of the chemical reaction, but may depend upon the characteristics of the external electrical circuit in which the electrode is connected. Suppose that the external electrical circuit has an infinite resistance so that no current passes ($I = 0$); since the passage of current is a consequence of a chemical reaction, when $I = 0$ no reaction is taking place and hence an equilibrium state has been established. The affinity, and thus the chemical tension, always exist but the electric tension is such as to completely neutralize their effect. Hence, the equilibrium chemical tension E and the equilibrium electric tension U are equal but of opposite sign (Chapter III p. 112 *et seq.*) ($E = -U$).

When, however, equilibrium conditions do not apply $E \neq -U$. The chemical tension is constant for a system of constant chemical composition since it is a measure of the affinity of the reaction which in turn is defined by the chemical potentials of the substances present in the two phases. Hence, the displacement of the equilibrium can occur only by a change in the electric tension U such as may be obtained by closing the external circuit through a very low ohmic resistance. The electric tension now becomes very small in absolute value and also absolutely smaller than the chemical tension. In this condition the chemical tension overcomes the electric, and the chemical reaction at the electrode (the *electrode reaction*) proceeds in the natural sense (*i.e.* in the sense determined by the chemical affinity of the system). Current will flow through the electrode in a particular direction. This is what happens in a galvanic cell in which the electric tension arises spontaneously as a consequence of the chemical electrode reaction.

If, on the other hand, a source of electric tension is placed in the external circuit in opposition to the electric tension of the electrode so that $|U| > |E|$; the electric tension now overcomes the chemical and a current is passed inducing an electrode reaction in the opposite direction to that in the case above. Thus, the chemical reaction now runs in the direction corresponding to a negative affinity. In general, the passage of a current through an electrode requires an electric tension different from that which it would assume at equilibrium ($U \neq U_{rev} = -E$). Any changes in tension ($U - U_{rev}$) applied to an electrode to make a current pass through it are called *overtensions*, given the symbol η. From what has been said so far, an overtension may be positive or negative and hence the current may flow in either direction. All that has been said so far and all that will be added now, applies equally well to both cases, *i.e.* to spontaneous processes in galvanic cells where the chemical tension overcomes the electric, and to electrolytic processes induced under the action of an applied external electric tension which overcomes the chemical tension. For simplicity's sake, however, the argument will be developed solely for the case of an applied external electric tension. This is electrolysis in the classical sense which industrially is by far the most important case. It will be apparent, however, that the conclusions obtained could be equally well applied to galvanic cells.

When a flow of current is induced through an electrode, it may be assumed that an external electric tension is first applied equal to the equilibrium electric tension, followed by a suitable overtension. Before describing the phenomenon of overtension in quantitative detail it will be as well to discuss why an overtension is required and how the size of an overtension needed to give a particular current may be predicted. If the changes which occur at the electrode during an electrode reaction were perfectly reversible an infinitesimal shift in the position of the equilibrium would suffice to cause a change in the system and an infinitesimal overtension would lead to a finite current flow. In practice, however, electrode reactions like all real processes are not thermodynamically reversible even though they can be inverted. To overcome the causes of this irreversibility it is necessary to increase what might be called the 'motive power' of the electrode reaction, *i.e.* the predominance of the electric tension over the chemical: the overtension. Thus, to induce the passage of a finite current it is necessary to give the electrode a finite overtension of a size adapted to the size of the forces which make it necessary. These various forces which cause overtension may be conveniently used as a basis for classification into *transfer*, *diffusion*, *reaction* and *resistance* overtensions[1].

[1] In some cases of cathodic deposition of metals a particular type of overtension appears: the so-called crystalline overtension (*cf.* p. 253 *et seq.* and Chapter VIII).

The *transfer* overtension may be understood by considering the electrode process as a chemical reaction. As with all chemical reactions it will have a certain activation energy and hence will take place with a finite reaction velocity. In general, the reaction velocity is lower as the activation energy is higher. Since, morever the electrode reaction can usually be inverted it is necessary to consider two reactions: one the forward and the other the reverse reaction. Each of these will have its own activation energy and its own velocity constant. At equilibrium the two reactions proceed with equal velocity and hence there is no net chemical change. As the chemical reaction of the electrode is accompanied by a transfer of electrical charges from one phase to another in the electrode, it is possible to distinguish a *progressive* partial current $\vec{I}$ accompanying the forward electrode reaction and a *regressive* partial current $\overleftarrow{I}$ of opposite sign, accompanying the reverse reaction. These two partial currents may be cathodic or anodic; the *cathodic current* I_- carries positive charges from the solution phase to the solid phase of the electrode, whereas the *anodic current* I_+ carries negative charges from the solution phase to the solid phase. The individual partial currents cannot be measured experimentally in a direct fashion. However, it is their difference which gives the *total current* (or simply *current*) which can be measured simply by putting an ammeter in the external circuit. At electrochemical equilibrium the forward reaction exactly compensates the reverse reaction so that the partial cathodic current is equal, but of opposite sign, to the partial anodic current and the net current is zero. The transfer overtension is the increase (taken algebraically, *i.e.* positive or negative) in the electric tension which favours the passage of charges in the direction corresponding to one of the two partial electrode reactions and simultaneously prevents the transfer of charges corresponding to the other partial reaction. It therefore accelerates the transfer of charges in one direction and slows it down in the other, so that the electric tension is no longer equal to the chemical tension and a net current will pass. The overtension furnishes energy in a 'single direction' and can overcome the activation energy of only one of the two reactions: the forward or the reverse. If, for example, the forward reaction is cathodic, a negative overtension will overcome the activation energy of the forward reaction and accelerate the cathodic transfer of charges; the direct reaction will thus become faster than the reverse and the cathodic current will be of greater intensity than the anodic. A certain net cathodic current will result which can be detected with a measuring instrument. A positive overtension will produce the opposite effect giving an anodic current.

The *diffusion* overtension is caused by the limited rate at which the electrochemically active substance, which is consumed during electrolysis, can be replaced at the electrode in case this replacement occurs only by diffusion from the bulk of the solution. For an electrolysis to proceed with a constant current intensity, other conditions being equal, the concentration of the substance involved in the electrode reaction must remain constant. Since the reaction consumes the electrochemically active substance its concentration close to the electrode tends to diminish and must be renewed from the bulk of the solution. This renewal may occur by migration under the influence of the electrical field between the electrodes or by convection (*e.g.* in stirred solutions) or by *diffusion.* This last is the most usual case. Here the renewal is generally slow so that the layer of solution in immediate contact with the electrode metal becomes impoverished in electrochemically active substance. The equilibrium electric tension is therefore altered and to make the reaction proceed at the same velocity, the electrode must be given an increase in electric tension to bring the overtension back to the same value which existed before this change in concentration appeared. The diffusion overtension is thus due to the fact that the equilibrium electric tension to which the overtension is referred has been calculated from a concentration of electrochemically active substance which is not that of the substance actually determining the electrochemical equilibrium at the electrode – solution boundary but that of the bulk of the solution.

The *reaction* overtension is due to a similar mechanism to the diffusion overtension. It compensates for the effects of changes in concentration due to chemical reactions other than the electrodic ones, *i.e.* to secondary reactions, that consume or produce the electrochemically active substances. Supposing, for example, that the substance consumed during the electrode process is formed by a chemical reaction preceding the electrochemical process; then if this preceding reaction is sufficiently rapid, the electrochemically active substance will always have a relatively constant concentration near to the electrode. However, if this reaction is slow in comparison with the rate at which the electrochemically active substance is consumed by the electrode reaction, there will be a fall in its concentration with similar results to those described for diffusion overtension. If the slow reaction follows the electrochemical one and consumes the product of the electrode reaction, there will be an increase in the concentration of this product near to the electrode with a consequent change in the true equilibrium electric tension of the electrode in the same sense as that provoked by the lack of electrochemically active substance described above. If the external electric tension remains constant the reaction will slow down since the actual overtension between the electrodes will have been diminished. In this case too a supplementary overtension is required to keep the current intensity, and hence the reaction velocity, constant.

Finally, the *resistance* overtension is needed to overcome the ohmic resistance of the electrolyte and the other conducting parts of the electrolysis cell. The total applied electric tension of an electrolysis cell serves not only to bring about a chemical change at the electrode and to overcome the obstacles which accompany this, but also to move the electrical charges within the conducting phases overcoming their electrical resistance. If I is the current intensity and R is the total ohmic resistance of the electrolysis cell, the resistance overtension will be IR and the corresponding energy will be dissipated within the bulk of the conducting phases as heat, by the Joule effect.

In general all the various types of overtension, other than the resistance overtension, originate in a slow stage in the overall process occurring at the electrode.

Both the diffusion and the reaction overtensions depend upon changes in the concentration of the electrochemically active substance or substances and hence can both be considered as expressions of a 'concentration' overtension; this is sometimes also called concentration polarization. This latter term was, and still is, often used ambiguously since this overtension can have two distinct causes: diffusion or chemical reactions. It is therefore preferable to use the terms diffusion or reaction overtensions for the particular cases rather than the term concentration overtension, to avoid ambiguity.

The magnitude of the overtension required to pass a given finite current through an electrode is a measure of the magnitude of the phenomena which oppose the overall electrode reaction, *i.e.* its degree of thermodynamic irreversibility. The current which flows through the electrolytic cell becomes greater as the overtension increases so that the total current intensity is an increasing function of η. The mathematical form of this function may be theoretically predicted and explained in various ways (*cf.* p. 203 *et seq.*). It can also be derived experimentally and is known as *Tafel's equation*[1].

$$\eta = a + b \log I$$

(when η and I are sufficiently large) where a and b are constants to be determined for each particular case and may have either positive or negative values depending on whether the overtension is anodic or cathodic. The estimation of the degree of irreversibility of an electrode process requires a knowledge of the parameters a and b of Tafel's equation. These can be obtained by drawing a graph of the experimentally determined values of $\log I$ against η (see section 3 below for the measurement of overtensions).

In particular, in determining the transfer and diffusion overtensions, it is usual to select the extrapolated value of the current intensity at zero overtension. In fact, Tafel's equation is valid only for finite positive and negative overtensions which are not close to zero, because at electrical equilibrium $I = 0$ and hence $\log I \rightarrow -\infty$, and the graph of $\log I$ against η cannot be a straight line in this region (see Fig. IV, 8). It will be shown on page 203 *et seq.* that the value of the current intensity extrapolated from Tafel's equation to $\eta = 0$ is equal to the absolute value which the progressive and regressive partial currents assume at electrochemical equilibrium; it is called the *exchange current* and is a highly significant quantity for the characterization of electrochemical processes and the degree of their thermodynamic irreversibility. In fact with no applied overtension, when the cathodic and anodic reactions proceed with equal velocity, this velocity and hence the intensity of the exchange current depend solely on the size of the activation energies of the two electrode chemical reactions. This thus provides a method of deciding if the transfer of charges in both directions through the electrode surface is more or less free or hindered, *i.e.* if the electrode process is more or less thermodynamically reversible.

Once the transfer overtension has been evaluated, measurements of the total overtension can give the values of the reaction and diffusion overtensions after the elimination of the resistance overtension.

3. Polarization of Electrodes and Decomposition Tensions

Consider a particularly simple example of electrolysis consisting of a cell containing $1\,N$ $CuSO_4$ solution into which dip two metallic copper elec-

[1] J. TAFEL, *Z. physik. Chem.*, 50 (1905) 641.

trodes connected to an external circuit by conductors of the first class (metal wires):

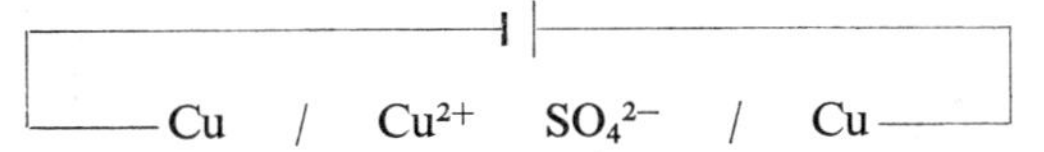

In this system the electrodes assume spontaneously (before electrolysis) the equilibrium electric tension which following the accepted conventions (*cf.* Chapter III) is +0.34 V for the right-hand electrode and −0.34 V for the left-hand electrode. Their chemical composition does not alter during electrolysis. Electrodes with this characteristic are called *nonpolarizable*[1]. In the present example, moreover, the electrolytic reaction occurring during electrolysis ($Cu \rightleftharpoons Cu^{2+} + 2e^-$) is the same at both electrodes but occurs in opposite directions. Since the equilibrium electric tension is +0.34 V and −0.34 V for the two electrodes respectively, the overall tension of the cell is zero and in the absence of a source of electric tension in the external circuit, no current will pass. Certain further conditions are required to make a current flow in the electrolysis circuit. Firstly, it is necessary to give each electrode a transfer overtension; the electrodes will then be displaced from the state of electrochemical equilibrium, and one of the two reactions — forward and reverse — which at equilibrium occur to an equal extent on both electrodes, will prevail. Thus, a current is enabled to pass through the electrode interfaces. However the transfer overtension is sufficient only to make the current flow across the boundary between the two electrode phases but not to overcome the ohmic resistance of the circuit and thus to make the current flow also through the other parts of the circuit in series with the electrodes. For this it is necessary to provide a further resistance overtension.

To provoke a net flow of current through the electrolysis cell it is thus necessary to apply to it an electric tension greater than that at equilibrium[2], *i.e.* an overtension or a *polarization*. In practice, this is obtained by placing a generator of electric tension in series in the external circuit. In this case, the electric tension at the ends of the cell is no longer zero but assumes a value U which is divided into three parts. One of these is

[1] The extreme cases of *ideally nonpolarizable* and *totally polarizable* electrodes are characterized by the fact that the charged particles can in the first case freely traverse the electrode surface without any hindrance whereas in the second case no transfer of charged particles through this surface is possible. Actual electrodes fall between these two extremes.

[2] In this particular case the overall equilibrium electric tension is zero.

utilized as the resistance overtension, whilst the other two create two transfer overtensions at the two electrodes so that at the cathode the electrode tension becomes less than that at equilibrium and at the anode it becomes greater. The electrochemical equilibrium at the interfaces of the two electrodes is thus destroyed. At the cathode, which has been given a negative overtension, occurs the cathode reaction (a deposition of solid copper on the metallic electrode) whereas at the anode the positive overtension leads to the anodic reaction (the dissolution of electrode metal). The reactions occur according to the equations

$$Cu^{2+} + 2\,e^- \rightarrow Cu \qquad \text{(at the cathode)}$$

$$Cu \rightarrow Cu^{2+} + 2\,e^- \qquad \text{(at the anode)}$$

With nonpolarizable metallic electrodes like that described, the transfer overtensions are usually small. In other words, the electrode reaction $Me \rightarrow Me^{z+} + ze^-$ requires a small activation energy and the exchange current intensity is fairly high. In such cases, the greater part of the applied electric tension is utilized as resistance overtension.

An electrolytic cell with polarizable electrodes presents quite a different picture. These electrodes do not correspond to any reversible galvanic half cell before electrolysis but function as such during electrolysis. It is necessary to give such electrodes an external electric tension, called an *electrolytic polarization* to obtain electrochemical equilibrium. An example of this would be a cell consisting of iron and carbon electrodes in a 1 *N* solution of zinc sulphate. When an increasing electric tension is applied to such a cell there would first be only a very small, almost zero, flow of current when the applied electric tension was low (nonfaradic current; see below). Above a certain value of the applied electric tension (the decomposition tension) however, a strong flow of current begins. Fig. IV, 1

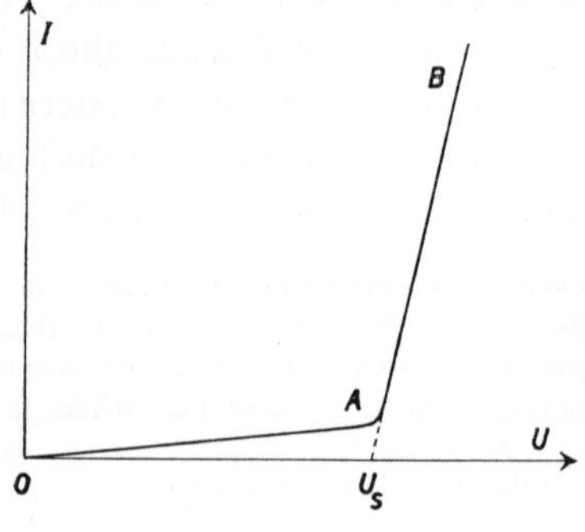

Fig. IV, 1. Diagram showing decomposition electric tension

shows the relationship of current to applied electric tension. Initially, there is no spontaneous electric tension at the interfaces of the two electrodes corresponding to that of a true galvanic half cell, but the two conducting phases of each electrode facing on to a single boundary are equivalent to the two plates of a still uncharged capacitor. Hence, they have potentially the structure of an electrochemical double layer. On applying externally an electric tension there will be a single initial current pulse which will give a certain charge to the metallic side of each of the two electrode interfaces, hence each will have a certain electrical potential. Electrostatic forces will augment the concentration in the solution side, of ions of opposite sign to the charge on the metal. If the electric tension thus established is less than that corresponding to the electrode reaction, the flow of current stops immediately; it would start again only if there were a change in the applied electric tension. In this state the charge given by the instantaneous current to the interface boundary of the electrodes, *i.e.* to the electrochemical double layer, may be related to the charge on a capacitor of capacitance $C = I\mathrm{d}t/\mathrm{d}U$. No chemical reaction occurs during the instantaneous flow of current and hence there is no transformation of substance; there has simply been a charging of a capacitor. The corresponding current, $I = C\mathrm{d}U/\mathrm{d}t$ is therefore called the *nonfaradic* current (in practice, at electric tensions below the decomposition tension it is possible, besides the instantaneous nonfaradic current, to have a small permanent, or *residual*, current due to various factors which will be discussed later). In Fig. IV, 1 the segment OA corresponds to the sum of the nonfaradic and residual currents.

The effect of the nonfaradic current is simply to modify the electric tension of the electrodes by charging the capacitors representing the electrochemical double layers of the two interfaces. The modification of the electric tensions of an electrode by the instantaneous nonfaradic current is called *electrolytic polarization*; in general any electric tension generated as a result of the passage of an electric current through a galvanic or electrolytic system (*e.g.* a system with polarizable electrodes), which is such as to oppose the current flow is called *polarization*. Electric tensions due to electrolytic polarization may be easily demonstrated, in polarizable electrode systems, by disconnecting the external applied tension and replacing it by a tension measuring instrument. Immediately after the external electric tension is removed the instrument will indicate an electric tension which starting from the value of the external tension will fall more or less rapidly to zero, or to the original value existing before the application of the external tension, as the capacitance of the electrochemical double layer is discharged and it returns to its initial state. Naturally, electrolytic polarization may involve only one or both of the electrodes depending on whether one or both are polarizable; in other words, the polarization may be anodic, cathodic, or both. As the external electric applied tension gradually increases the cathodic and anodic polarizations rise and their sum is always equal to the external electric tension. The nonfaradic current which corresponds to each increase of the external tension is only instantaneous because as soon as the capacitors of the double layers are charged the sum of their polarizations equals the external electric tension and no excess of it remains available to overcome the ohmic resistance of the cell and create a transfer overtension. For a permanent flow of current through the cell, the externally applied electric tension must be greater than the sum of the polarizations of the anode and cathode so that part of the applied electric tension remains in excess of that required to charge the interfacial capacitances. Now, the ions present in the double layer, or formed by the electrode reaction, will pass across the electrode–electrolyte boundary: true electrolysis will begin with a flow of current provoking chemical reactions in accordance with Faraday's Laws. This new current flow is therefore called *faradic*. In Fig. IV, 1 the faradic current corresponds to segment AB. The tension U_s at which the faradic current flow begins — allowing for the fall IR in electric tension due to the residual currents and the resistance of the cell — is called the *decomposition tension*. It has been seen that with each current impulse there is a charging of the electrochemical double layer and a consequent increase in the polarization of the electrodes. This polarization obviously cannot increase to infinity and stops when the electric tension of the electrode is equal to that of the corresponding galvanic half cell. All the electric tension applied from the outside above this value can be used to induce the flow of a permanent faradic current. In the present case of a cell containing $1N$ $ZnSO_4$ with iron and carbon electrodes

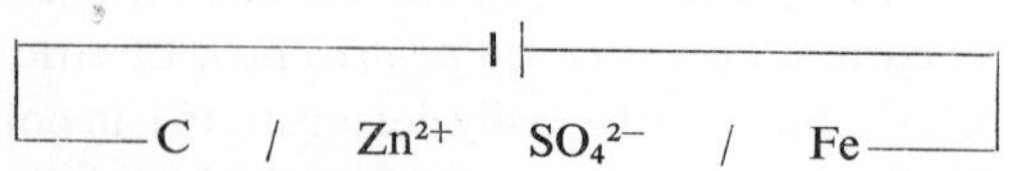

an increase in the external electric tension raises the electrolytic polarization of both the carbon anode and the iron cathode. Considering the behaviour of the latter in particular, it will be seen that the cathodic polarization which it is given, rises progressively with the external electric tension and merely increases the charge of the double layer, until a value of -0.763 V is reached. This is the equilibrium electric tension of the galvanic half-cell Zn/Zn^{2+} (1 *N*). The Zn^{2+} ion may be discharged to the metallic state and it requires only a few metallic zinc atoms to be deposited on the iron cathode for the electrode to start behaving as if it were a metallic zinc electrode immersed in a solution of Zn^{2+} *i.e.* as a Zn/Zn^{2+} (1 *N*) galvanic half cell. Once this half cell has been formed by the deposition of a layer of metallic zinc the role of the electrolytic polarization tension is finished. If the polarization of the cathode is made more negative than -0.763 V its cathodic excess does not form a further electrolytic polarization tension but is used to overcome the transfer and resistance overtensions *i.e.* to induce the faradic current flow through the electrode. Thus begins true electrolysis which will continue so long as the electric tension is maintained at below -0.763 V.

The behaviour of the anode is analogous. Here the electrode process *i.e.* the direct reaction caused by the passage of the current, is the discharge of OH^- ions present in the aqueous solution of the Zn salt forming oxygen; and the carbon, once electrolysis has begun, becomes solely the conducting support of an oxygen electrode without taking any further active part in the electrode process. Such electrodes are called *inert*. If the anode instead of being carbon were a base metal such as copper, the anodic process would be different. Initial polarization would first be obtained to start the dissolution of the metal; after this the anode would assume the characteristics of a nonpolarizable electrode (a metal in the presence of its ions) as described earlier.

In general, in any electrolytic cell two electrodes are required, one of which functions as the cathode and the other as the anode. Cations are discharged at the cathode (where, more generally, a reduction occurs) whereas anions are discharged, or cations formed by the dissolution of the anode, at the anode (where, more generally, an oxidation occurs). At both anode and cathode a galvanic half cell is formed, if it does not already exist, which corresponds to the substances involved in the respective electrode processes. It will be clear from what has been said that the electrodes of electrolysis cells may be classified into two groups, *i.e.* polarizable and nonpolarizable[1], depending on whether or not they spontaneously assume the electric tension of the galvanic half cell corresponding

[1] See note page 193.

to the electrode process. An electrolytic cell may contain electrodes of the same or of different types. In each case the sum of the anodic and cathodic polarizations equals the electric tension of the galvanic cell formed from the two half cells; this electric tension itself tends to circulate current in the direction opposite to that of the faradic electrolysis current. If, for example, water which has been made conducting by adding any base, or oxygenated acid or salt, is electrolyzed between platinized platinum electrodes there will be a formation of a hydrogen electrode at the cathode and an oxygen electrode at the anode. It is necessary to overcome the electric tensions of these two half cells before the electrolysis can take place and thus the electric tension of the galvanic cell formed as a consequence of electrolysis, is called the *counter tension.* It may easily be measured by interrupting the electrolysis and putting a measuring instrument into the external circuit in place of the external generator of electricity.

The principle of dividing the total polarization of an electrolytic cell — other than the resistance overtension — into two parts attributed to the cathode and anode respectively, may be extended to any other form of overtension. This treatment is useful in applying Ohm's Law to electrolytic cells. Consider the cathodic polarization U_c and the anodic polarization U_a as including all forms of overtension originating in the respective electrode interfaces. This, of course, excludes the resistance overtension which corresponds to a fall in electric potential distributed throughout all the conducting phases. The external applied tension U may thus be divided into three parts.

$$U = U_c + U_a + IR \tag{1}$$

whence

$$I = \frac{U - (U_c + U_a)}{R} \tag{2}$$

In applying Ohm's Law, allowance must therefore be made for the polarization of the electrodes, and in particular for the electrolytic polarization, in order to predict the value of the current I which will flow in a cell of resistance R under an applied electric tension U. The terms U_c and U_a must be taken as generalized, *i.e.* as including not only the electrolytic polarization even although this constitutes the preponderant part of the total polarization, but also any other forms of overtension such as transfer, diffusion or reaction. The Ohm's Law equation when thus applied to electrolytic processes serves not only to predict the intensity of the faradic current as a function of the total applied tension but is also useful for obtaining a better definition of the decomposition electric tension. Assuming that the intensity of the current through the cell tends towards zero, the significance of the various terms in equation (1) will be simplified since all the overtensions will also tend towards zero. (It will be remembered that by definition, overtensions arise only when a finite faradic current passes between the electrodes.) U_c and U_a will then become equal to the electrolytic polarizations of the respective electrodes. The term IR will also obviously tend towards zero so that equation (1) becomes

$$U = U_a + U_c = U_s \tag{3}$$

U_s is the *decomposition electric tension, i.e.* the electric tension at which the faradic current first begins to differ from zero; it is equal to the sum of the electrolytic polarizations of the two electrodes and it corresponds to the electric tension of the galvanic cell formed during the electrolysis or pre-existing if the electrodes are nonpolarizable. In practice, the decomposition tension does not always correspond exactly to the electric tension of this galvanic cell. This is true only when the electrolytic process is nearly reversible in the sense that the application of an external tension slightly greater than the equilibrium electric tension will lead to the flow of a finite current. Such nearly reversible behaviour is found in some cases but not in all; for example, the decomposition tensions of the halogen acids in 1 N solutions have been determined experimentally between platinum electrodes as 1.41 V, 1.07 V and 0.52 V for HCl, HBr and HI respectively. When the sums of the equilibrium electric tensions for the hydrogen electrode and the respective chlorine, bromine and iodine electrodes are compared they give values of 1.418 V, 1.14 V and 0.54 V respectively. The agreement must be considered as very good, particularly since such measurements are of low accuracy. If, however, the electrode process is irreversible, *i.e.* if other forms of overtension are involved, the passage of a detectable current occurs only when an electric tension significantly higher than the equilibrium value is applied. However, even if the electrode process itself is reversible because the decomposition tension coincides with the reversible equilibrium tension, the current intensity must be fairly low so as not to provoke changes in concentration near to the electrodes. In cases where this is not true, like those where particularly marked reaction or transfer overtensions are present, the concept of decomposition tensions has no longer any significance since the electric tension at which an appre-

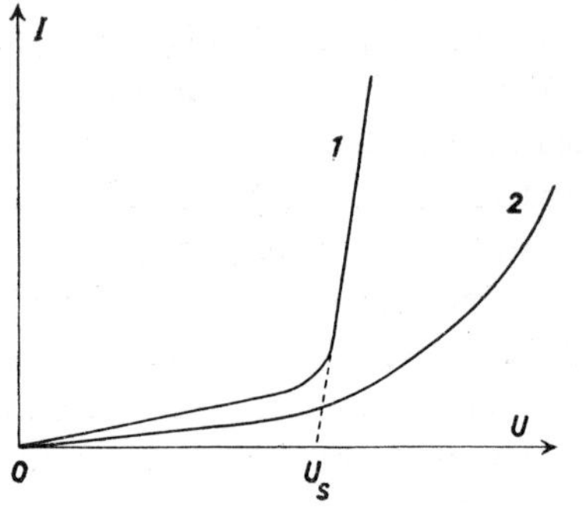

Fig. IV, 2. Different types of electric tension – current intensity diagrams

ciable current flow occurs is no longer a characteristic of the electrode system but rather of the experimental conditions. Moreover, it must not be overlooked that the possibility of measuring exactly the decomposition tension — taken as that above which the faradic current flow starts — is dependent on the existence of an abrupt change in the diagram current intensity–applied tension, like that shown in curve 1 of Fig. IV, 2. If, however, marked overtensions flatten the current–tension curve as in curve 2 of Fig. IV, 2 the determination of the decomposition electric tension becomes uncertain. For all these reasons the decomposition electric tension, although it is a quantity of great practical interest, does not have any well-defined theoretical significance at least for those electrode processes which are not perfectly reversible.

The experimental measurement of decomposition tensions may be carried out with the arrangement shown in Fig. IV, 3. A source *A* provides an electric tension which is drawn in variable amount from across the terminal *B* and the sliding contact *E* and is applied to the electrolysis cell *D*. An ammeter *G* is in series with this and *V* is a voltmeter of high internal resistance, which indicates the electric tension applied to the terminals of

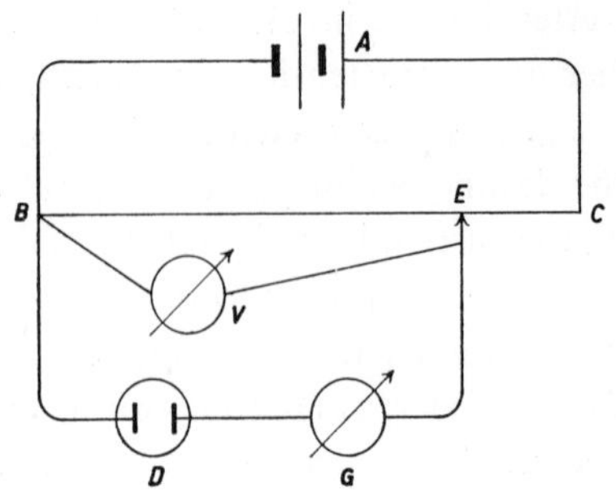

Fig. IV, 3. Wiring for measurements of electric tension of decomposition

TABLE IV, 2

DECOMPOSITION ELECTRIC TENSIONS U_s

Electrolyte	*Concn.* mol/l	U_s volt	*Electrolyte*	*Concn.* mol/l	U_s volt
CH_3COOH*	—	1.57	NH_4Cl*	—	1.76
$CH_2(COOH)_2$	0.5	1.69	NH_4Br*	—	1.46
$CH_2ClCOOH$	1.0	1.72	$AgNO_3$	0.1	0.84
CCl_3COOH	1.0	1.51	Ag_2SO_4*	—	0.80
$(COOH)_2$	0.5	0.95	$CuSO_4$	0.5	1.49
HNO_3	1.0	1.69	$CdBr_2$	0.5	1.53
H_3PO_4	0.33	1.70	$CdCl_2$	0.5	1.88
H_2SO_4	0.5	1.67	$CdSO_4$	0.5	2.03
$HClO_4$	1.0	1.65	$Cd(NO_3)_2$	0.5	1.98
HCl	0.17	1.41	$ZnBr_2$	0.5	1.80
HBr	0.1	1.07	$ZnCl_2$	1.0	2.28
HI	1.0	0.52	$ZnSO_4$	0.5	2.35
$NaOH$	1.0	1.69	$HgBr_2$	$\sim 2.7 \cdot 10^{-4}$	1.56
$NaNO_3$*	—	2.15	$Pb(NO_3)_2$	0.5	1.52
KOH	1.0	1.69	$CoCl_2$	0.5	1.78
KI*	—	1.16	$CoSO_4$	0.5	1.92
KBr*	—	1.74	$NiCl_2$	0.5	1.85
NH_4OH	1.0	1.74	$NiSO_4$	0.5	2.09
NH_4NO_3*	—	2.04			

* Experimental conditions not specified.

cell *D*. On moving the contact *E* so as to gradually increase the applied tension, it will be observed that with low electric tensions the current intensity is virtually zero. As the applied electric tension increases, however, beyond a certain point the passage of a faradic current begins and the ammeter shows a sudden deflection. The tension indicated by the voltmeter at this moment is the decomposition electric tension corresponding to the point U_s in Figs. IV, 1 and IV, 2. Table IV, 2 shows values of the decomposition electric tension for some of the more common electrolytes measured between smooth platinum electrodes.

It is also possible to determine the polarization of single electrodes by connecting them through a salt bridge or porous septum (Fig. IV, 4) with any comparison electrode to form a galvanic cell. By measuring the electric tension and subtracting the known electric tension of the comparison

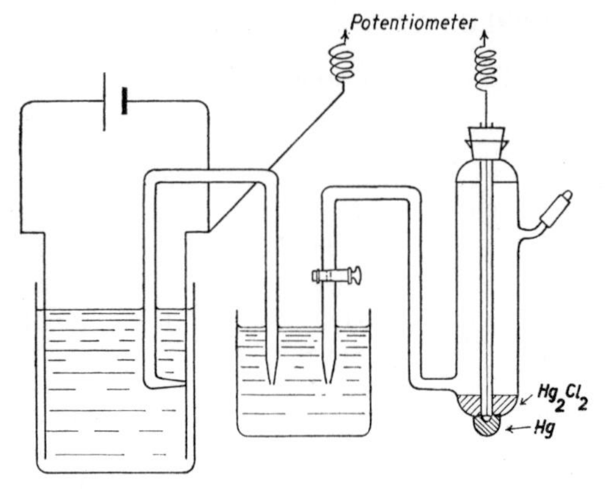

Fig. IV,4. Wiring for measurements of electric tension of individual electrodes

electrode it is easily possible to determine the actual electric tension of the test electrode even if it is working.

With such an arrangement, it can be shown for example, that the deposition of copper from a 1 *N* solution of $CuSO_4$ occurs virtually at $+0.34$ V, *i.e.* at the equilibrium electric tension. Thus, the electrochemical process $Cu \rightleftharpoons Cu^{2+} + 2e^-$ is practically reversible since its decomposition tension coincides with the theoretical value for the reversible equilibrium electric tension.

It will appear, from what has been said, that below the decomposition tension no permanent current should pass and that only the nonfaradic instantaneous current should occur, whereas even below the decomposition tension there is a small permanent current which is called the *residual current*. This corresponds to segment *OA* in Fig. IV, 1 and it is clear that if this should be of any significant size it would make the experimental determination of the decomposition electric tension inaccurate. There are two principal causes of residual currents. Firstly, the capacitor formed by the double layer whose charging gives rise to the nonfaradic current may, like all capacitors, not be perfect and lose charge. This implies that it has an equivalent ohmic resistance which although generally very high may nevertheless cause a continuous flow of a permanent component of the charging current of the capacitor. In electrolysis, the capacitor loss is represented by the thermal agitation of the particles in solution, which tends to disperse into the bulk of the solution the ions which have accumulated at the electrode interface forming the charge of one plate of the capacitor. Thus, to maintain the electric tension at a constant value, this dispersed charge must be continuously renewed, giving rise to a permanent residual current.

Secondly, the residual current may be due to the fact that the progressive electrode reaction is never completely zero even at applied electric tensions less than the decomposition tension. Even before this value is reached there will be, for example, a small discharge of cations at the cathode leading to a deposition of traces of metal which as yet are not organized into a continuous phase, or of gases in amounts too small to reach atmospheric pressure. The activity of the deposited product is under these conditions less than that corresponding to the state in an electrode with the electrolysis products in their final stable form. Hence, the countertension generated is also correspondingly less and is equal to the applied external tension. The substance deposited on the electrode in this low activity state tends to move away from the electrode surface either by returning to the ionic state or by diffusing into the interior of the electrode. To keep the countertension equal to the applied external tension it is necessary to continuously renew electrolytically the small amount of the end product of electrolysis which has moved away from the electrode surface. This mechanism too can give rise to a residual current.

4. The Quantitative Treatment of Overtensions. Current Density–Electric Tension Diagrams

The various types of overtension — transfer, diffusion, reaction and resistance — have been defined and their origins have been described qualitatively. The most important of these is the transfer overtension whose quantitative treatment is still not satisfactory despite the numerous studies which have been made of it.

One approach, which is not immune to certain criticisms particularly of its logic, has been however used by a number of authors. The quantitative treatment of transfer overtension is started by considering the double electrical and chemical aspect of the electrode reaction. An electric tension U always exists at the boundary between the two phases of the electrode. This electric tension is localized in the thickness of the electrochemical double layer and is a function of the electrochemical potentials of the electrochemically active components in the solid metal and the solution. The passage of a current through the interface of an electrode, *i.e.* the transport of electrical charges from one phase to the other, requires these charges to pass through the double layer and its electric tension leading to a performance of work $QU = z\mathbf{F}U$. The passage of a current through an electrode is always associated with a chemical electrode reaction which can usually be inverted. Thus, at the electrode forward and reverse reactions are possible, with which are associated the progressive and regressive partial currents. Like all chemical reactions, electrode reactions proceed with a particular velocity and require a certain activation energy; but the activation energy is not the same when the reaction occurs under an applied electric tension as it would be independently of an applied tension. In fact, the electric tension existing between the electrode phases alters to a different extent the differences between the energies of the final, or the initial, states and the activated state of the electrochemical process, *i.e.* it affects differently the activation energies of the forward and reverse reactions. The electrode reaction thus requires an overall activation energy E which is the resultant of a term representing the activation energy for the reaction in the absence of any electric tension at the interface — called the chemical contribution — and a term due to the presence of the electric tension, called the electrical contribution. The electrode reaction may be given diagrammatically as the transformation of a generalized oxidized form Ox^{z+} into a generalized reduced form Red by the addition of z electrons supplied by the electrode.

$$Ox^{z+} + ze^- \rightleftharpoons Red$$ [1]

The theory of the velocity of reactions supplies, for the reaction at the cathode (where negative charges are taken from the metal), the equation[2]

$$-\,d[Ox]/dt = \vec{k'}_-[Ox]e^{-E_-/(RT)} \tag{1}$$

and similarly for the partial anodic reaction

$$-\,d[Red]/dt = \vec{k'}_+[Red]e^{-E_+/(RT)} \tag{2}$$

where E_- and E_+ are the activation energies of the reduction and oxidation reactions respectively. The intensity of the partial anodic and cathodic currents is proportional to the number of charges transferred in each

[1] For simplicity in the following formula the electrical charges of the symbols Ox and Red will be omitted; it should also be noted that both the electron and the base for natural logarithms have the same letter but that italic type is used for the exponential.
[2] To distinguish kinetic constants from other proportionality constants such as equilibrium, the former will be indicated by an arrow above the symbol.

direction in unit time, and is thus also proportional to the reaction velocity.

$$\left.\begin{aligned} I_- &= \vec{k}''\mathrm{d[Ox]/d}t \\ I_+ &= -\vec{k}''\mathrm{d[Red]/d}t \end{aligned}\right\} \quad (3)$$

The overall activation energy thus contains a chemical contribution $E°$ and an electrical contribution derived from the presence of the double layer and its electric tension. For the anodic reaction with a certain electric tension U in the double layer, the chemical activation energy $E°_+$ will alter by a certain amount proportional, with a factor α, to the energy $zU\mathbf{F}$; and for the cathodic reaction the corresponding activation energy $E°_-$ will alter by an amount which is of opposite sign proportional, according to a factor β, to the energy $zU\mathbf{F}$. Hence, the activation energies for the cathodic and anodic processes are respectively

$$E_+ = E°_+ - \alpha zU\mathbf{F}$$

$$E_- = E°_- + \beta zU\mathbf{F}$$

At constant temperature and pressure $E°$ may be taken as constant so that equations (1) and (3) become

$$\begin{aligned} -\mathrm{d[Ox]/d}t &= \vec{k}'_-\,[\mathrm{Ox}]e^{-(E°_- + \beta zU\mathbf{F})/(RT)} \\ &= \vec{k}'_-\,[\mathrm{Ox}]e^{-E°_-/RT}e^{-(\beta zU\mathbf{F})/(RT)} \end{aligned}$$

$$\begin{aligned} I_- &= -\vec{k}''_-\vec{k}'_-\,[\mathrm{Ox}]e^{-E°_-/RT}e^{-(\beta zU\mathbf{F})/(RT)} \\ &= -\vec{k}_-\,[\mathrm{Ox}]e^{-(\beta zU\mathbf{F})/(RT)} \end{aligned} \quad (4)$$

where

$$\vec{k}_- = \vec{k}''_-\vec{k}'_-\,e^{-E°_-/(RT)}$$

Similar reasoning gives for the anodic reaction

$$I_+ = \vec{k}_+\,[\mathrm{Red}]e^{(\alpha zU\mathbf{F})/(RT)} \quad (5)$$

Since the overall electric tension is the sum of the reversible electric tensions and the overtensions, equations (4) and (5) now become

$$I_- = -\vec{k}_-\,[\mathrm{Ox}]e^{-(\beta\{U_{rev}+\eta\}z\mathbf{F})/(RT)} \quad (6)$$

$$I_+ = \vec{k}_+\,[\mathrm{Red}]e^{(\alpha\{U_{rev}+\eta\}z\mathbf{F})/(RT)} \quad (7)$$

As each of the two current intensities is a measure of the velocity of one of the two reactions — forward or reverse — the algebraic sum of the two current intensities will be a measure of the reaction velocity and this sum will be the net current detectable in the external circuit. At electrochemical equilibrium the electric tension existing at the electrode interface is the reversible equilibrium electric tension and is equal, in absolute value, to the chemical tension; and the forward and reverse partial velocities are equal and opposite. The overall reaction is thus zero and the two partial current intensities are also equal and opposite with their common absolute value equal to I_0 and with the net current intensity zero.

$$|I_+| = |I_-| = |I_0| \tag{8}$$

Equations (4), (5) and (8) give

$$\begin{aligned} I_0 &= |\vec{k}_- [\mathrm{Ox}] e^{-(\beta U_{rev} z\mathbf{F})/(RT)}| \\ &= |\vec{k}_+ [\mathrm{Red}] e^{(\alpha U_{rev} z\mathbf{F})/(RT)}| \end{aligned} \tag{8a}$$

and for $U \neq U_{rev}$, equations (6) and (7) become

$$\begin{aligned} I_- &= -\, I_0\, e^{-(\beta\eta z\mathbf{F})/(RT)} \\ I_+ &= I_0\, e^{(\alpha\eta z\mathbf{F})/(RT)} \end{aligned}$$

It is obvious that a positive overtension will favour the anodic reaction and a negative overtension will favour the opposite cathodic reaction. The overtension $\eta = U - U_{rev}$ is the transfer overtension which exists to a greater or lesser extent in all electrode processes. The intensity of the net current is given by the algebraic sum of the two partial current intensities

$$\begin{aligned} I &= I_+ + I_- \\ &= I_0 \left(e^{(\alpha\eta z\mathbf{F})/RT} - e^{-(\beta\eta z\mathbf{F})/(RT)}\right) \end{aligned} \tag{9}$$

For increasing absolute values of the overtension one of the two exponentials rapidly becomes negligible with respect to the other. For example, with increasing negative (cathodic) overtensions, the exponential relative to the partial anodic current rapidly becomes negligible and the intensity of the net current becomes practically equal to the intensity of the partial cathodic current.

$$I = -\, I_0\, e^{-(\beta z\eta\mathbf{F})/(RT)}$$

which logarithmically gives

$$\ln |I| = \ln I_0 - (\beta\eta z\mathbf{F})/(RT)$$

whence

$$\eta = (RT/\beta z\mathbf{F}) \ln I_0 - (RT/\beta z\mathbf{F}) \ln |I|$$
$$= a - b \log |I|$$

For increasing values of a positive overtension (anodic) an analogous reasoning leads to the equations

$$\eta = -(RT/\alpha z\mathbf{F}) \ln I_0 + (RT/\alpha z\mathbf{F}) \ln I$$
$$= a' + b' \log |I|$$

Thus, Tafel's experimental equation has been deduced theoretically. Various authors have attributed values to the transfer coefficients α and β such that their sum is unity

$$\alpha + \beta = 1$$

and

$$\beta = 1 - \alpha$$

Giving α and β the common value of 0.5 the diagram of I *vs.* η assumes a symmetrical form with respect to U_{rev} ($\eta = 0$) (Fig. IV, 5) whereas the symmetry is lost when $\alpha \neq 0.5$. The simple treatment of the processes of charge transfer at the electrode, which has been developed so far, is not entirely free from theoretical objections due to the simplicity of the hypotheses taken as the starting point. In fact, the procedure followed so far to interpret the kinetics of the electrode processes has consisted essentially of applying to them the laws of the kinetics of purely chemical proc-

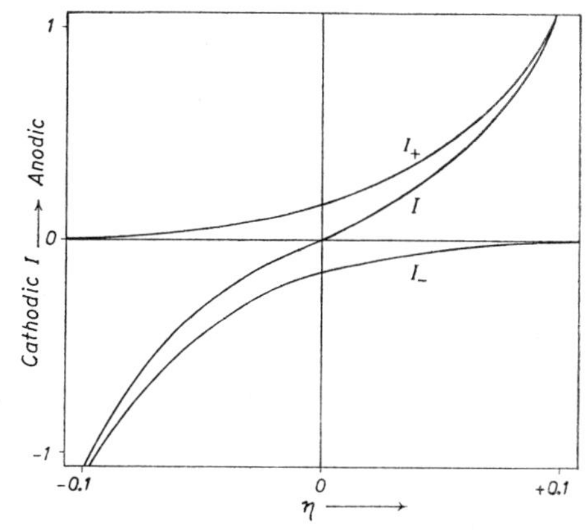

Fig. IV,5. Overtension – current intensity diagram for $\alpha = \beta = 0{,}5$

esses and introducing certain additional *ad hoc* considerations, developed for a simple model, to explain the effect of the electrode tension on the rate of the electrode reactions. In the last analysis, this procedure is empirical. Its principal justification is that it allows the deduction of Tafel's equation (which is an experimental relationship); however, it does not permit a definition of the transfer coefficients α and β other than as purely phenomenological parameters. Nor, moreover, does it allow any prediction of how or to what extent their value depends on the electrode tension and on the concentrations of the reactants. In fact, it is normal to assume that α and β are constant and to justify this hypothesis by the fact that experimentally they are often almost constant. However, other experimental results and theoretical considerations are available which show how in fact α and β may vary in certain cases for a particular electrode reaction, as the electric tension applied to the electrode and the concentrations of the electrochemically active species vary.

Another criticism of the simple treatment used is that it seems to suggest that the chemical affinity and the electrical affinity exert effects on the rate of an electrode reaction by two independent mechanisms whereas it would be more logical to admit that in reality only a single overall electrochemical affinity exists, which is responsible for the course of an electrode reaction.

Recently, attempts have been made to overcome this difficulty by developing a theoretically more consistent theory for the rate of electrode processes. In particular, van Rysselberghe[1] has pointed out the usefulness of basing the interpretation of the kinetics of electrode processes on the use of the electrochemical potentials of the reacting species of the rate-regulating step. Defining the electrochemical affinity of the progressive electrode reaction as $\vec{\tilde{A}}$ and that of the regressive electrode reaction as $\overleftarrow{\tilde{A}}$ and taking their common value at electrochemical equilibrium as $\tilde{A}_0$, van Rysselberghe proposed the following equation for the net rate V of an electrode reaction

$$V = V_0 \left(e^{(\vec{\tilde{A}} - \tilde{A}_0)/(RT)} - e^{(\overleftarrow{\tilde{A}} - \tilde{A}_0)/(RT)}\right) \tag{9a}$$

[1] P. VAN RYSSELBERGHE, *J. Chem. Phys.*, 29 (1958) 640 for the discussion of the purely chemical reactions. See also: P. VAN RYSSELBERGHE, *Electrochemical Affinity*, Hermann et Cie, Paris, 1955, (with references to earlier articles); *Proc. VIII Meeting CITCE*, Madrid 1956, Butterworth, London, 1958, p. 405.

where V_0 indicates the common value of the two rates of the partial reactions at equilibrium[1].

The net rate of the reaction is taken as the derivative of the advancement of the electrode reaction with respect to time ($d\xi/dt$); this is proportional to the net value of the current intensity at the electrode

$$I = I_+ + I_- = z\mathbf{F}V$$

The definition of the net current at the electrode which can be deduced from (9a) is similar to that given by equation (9) but has the advantage of having been derived in a way which is in better accord with the principles of the thermodynamics of irreversible processes, and of postulating an overall electrochemical affinity.

It also has the advantage of being open to further development on a more rigorously thermodynamic basis and especially for the definition and the explanation of the physical significance of the coefficients α and β.

The second type of overtension is the *diffusion overtension.* It will be useful to examine the electrolytic processes and other phenomena which occur at the electrode–electrolyte boundary during a particularly simple electrolytic process, in which the electrode reaction is not accompanied by either other chemical reactions or by secondary electrode reactions, *e.g.* the electrolysis of a solution of $CuSO_4$ between copper anode and cathode. This system is initially perfectly symmetrical and should theoretically remain so during the electrolysis since the anodic process — the dissolution of metallic copper — is simply the reverse of the cathodic process: the deposition of copper. Since the electrolytic polarization is zero, all the external applied tension should be available to overcome the transfer and resistance overtensions and no other causes of overtension should exist. In practice, however, during electrolysis with a constant current intensity the system becomes asymmetrical since concentration gradients are set up in the layers of solution about the electrodes. A primary effect is as follows. Supposing that the electrolysis proceeds with a constant finite current intensity in unagitated liquid with a current yield of 1 for both the anode and the cathode, and supposing also that diffusion phenomena may be ignored so that material is brought from the bulk of the electrolyte only by migration, then the passage of 1 **F** of charge will

[1] Care must be taken not to confuse V, the symbol for velocity in equations (9a) *et seq.*, with V, the symbol for volt.

deposit 1 equivalent of copper on the cathode whilst the same amount of copper will dissolve from the anode, so that 1 equiv. of copper ions pass through each interface. In the bulk of the solution, however, the current is carried not only by the cations Cu^{2+} but also by the anions SO_4^{2-} in amounts proportional to the respective transference numbers, *i.e.* the quantity of Cu^{2+} ions which passes through the solution is less than 1 equiv. As a result, there will be a loss of Cu^{2+} ions from the region near to the cathode and a gain of Cu^{2+} ions near to the anode. As the electrolysis proceeds, the concentration difference between the anode and the cathode increases. In practice diffusion phenomena cannot be completely eliminated and, tend to diminish the concentration difference to a value below the predicted one. In every case, the concentration difference between the cathode and the anode cannot increase indefinitely; after a while so many anions will have moved, under the influence of the electric field, towards the anode without being discharged that their accumulation raises the osmotic pressure about the anode and tends to carry them back again to the cathodic region where there is a lack of anions. This implies that the anions are urged by the force of the electric field in one direction and by the osmotic pressure difference in the other. When these two forces become of equal intensity the anions no longer move; this means that the transport of current in the bulk of the liquid is entrusted solely to the cations. The difference in concentration between the anode and cathode now reaches its maximum value and remains constant. The change in concentration thus produced in the region of the electrodes is directly proportional to the intensity of the electrolysis current I and is inversely proportional to the surface area S of the electrode, *i.e.* it is proportional to J where

$$J = I/S \tag{10}$$

J is called the *current density* and it is the significant quantity for studies of electrode processes rather than the current intensity I.

As a consequence of these variations in concentration, the equilibrium electric tensions of individual electrodes will also vary and a concentration cell will be produced. The applied external tension will not then be completely available to overcome the ohmic resistance and the transfer overtension since a part of it will be absorbed by the concentration cell which has been formed. To make the electrolysis continue with the same current density it will now be necessary to apply an electric tension greater than that used initially. In general, for any system which is asymmetrical at the start or becomes so during electrolysis, the electric tension required for the electrolysis to continue after a certain time becomes greater than that calculated from the concentration of the bulk of the solution, due to concentration changes occurring about the electrodes. This first effect is not, however, very important since the concentration differences produced may be simply eliminated by agitation.

A second effect of the concentration polarization is more important and can be obtained even in agitated solutions. It appears when the transport of electrochemically active substance is due not only to migration under the influence of the electrical field existing between the electrodes, but also to diffusion[1]. In the preceding case it was assumed that diffusion phenomena could be ignored. This is in general permissible when the migration under the influence of the electric field is performed in the main by species involved in the electrode reaction. If, however, other electrolytes are present in the solution which do not take part in the electrode reaction (inert electrolytes) the transport of charges in the bulk of the solution by migration is entrusted to all the electrolytes present; each acting in proportion to its concentration and transference number. As a consequence, particularly when there is a strong excess of inert electrolyte, the transport of the electrochemically active species by migration becomes very small or even negligible and a lack of active substance may be produced near to one of the electrodes where it is consumed during electrolysis. The renewal of this substance will then be primarily due to diffusion from the bulk of the solution and this diffusion will thus become more important than migration. For a single electrode, *e.g.* the cathode, the quantity of ions discharged with a current density J is $J/z\mathbf{F}$ equiv. per second for each cm^2 of electrode surface area. Under the conditions described this quantity will derive essentially by diffusion through a thin layer existing in the immediate neighbourhood of the cathode and having a thickness l (the diffusion layer, *i.e.* the layer in which a concentration gradient exists due to the cathodic consumption). The thickness of the layer will depend upon the state of agitation of the liquid. Within the layer the concentration varies gradually between the value c_0 which is the concentration in the bulk of the solution, and the value c which is the concen-

[1] Diffusion is defined as the movement of a solute in the bulk of a solution from a zone of higher concentration to one of lower as a consequence solely of the concentration difference.

tration on the liquid side of the interface immediately adherent to the cathode. Since the coefficient of diffusion D is defined as the quantity of solute which moves in one second through a section of 1 cm^2 area under the action of a concentration gradient of 1 unit of concentration/cm, then

$$\frac{J}{z\mathrm{F}} = \frac{D(c_0 - c)}{l} \tag{11}$$

If l and D are constant for a given solution equation (11) becomes

$$J = k(c_0 - c) \tag{12}$$

or

$$c = c_0 - \frac{J}{k} \tag{13}$$

The electric tension of the cathode actually depends upon the ionic concentration in the layer of solution immediately adherent to it, *i.e.* on c and not on c_0 so that at 25° C

$$\begin{aligned} U_1 &= U_0 + (0.059/z)\ \log\ c \\ &= U_0 + (0.059/z)\ \log\ \left((c_0 - \frac{J}{k}\right) \end{aligned} \tag{14}$$

whilst the initial equilibrium tension is given by

$$U_{eq} = U_0 + (0.059/z)\ \log\ c_0 \tag{15}$$

so that in this case a difference is produced between the electric tension of the electrode in the resting state and in the working state, *i.e.* during electrolysis. This difference in electric tension constitutes a form of concentration overtension. It is that which is more correctly called the diffusion overtension or diffusion polarization and is represented by the symbol η_d. Its value may be obtained from the difference between equations (14) and (15)

$$\begin{aligned} \eta_d &= U_1 - U_{eq} \\ &= (0.059/z)\ \log\ \frac{c_0 - (J/k)}{c_0} \\ &= (0.059/z)\ \log\ \left(1 - \frac{J}{kc_0}\right) \end{aligned} \tag{16}$$

It usually has a numerical value of the order of hundredths of a volt.

As in the preceding case where there is also a polarization concentration the size of the concentration overtension depends exclusively, for any given initial concentration, on the current density and on the rate at which the concentration difference induced by electrolysis is compensated for by the migration or diffusion of the ions. On one hand an increase in the current density tends to increase the concentration difference at the electrodes but on the other hand the phenomena of migration and diffusion of the ions tend to diminish it but to a lesser extent than that in which it is increased, even if the electrolyte is vigorously agitated to aid the equilibration of the concentrations. Consequently, an exponential increase in the diffusion overtension always accompanies an increase in the current density. In practice, migration and diffusion cannot be completely separated so that the movement of ions in the bulk of the liquid is due to both forces: the electrical field, and the difference of concentration (or rather of osmotic pressure). Referring once more to the case of a metallic copper cathode on which Cu^{2+} ions are discharged, an increase in the polarization of the electrode increases the intensity of the electrical field; this in turn increases the effective migration velocity of the cations and hence the number of ions reaching the cathode every second and being discharged there. There is thus an increase in the current density J but increasing the number of ions discharged per second will also increase the concentration difference between the immediate neighbourhood of the electrode and the bulk of the solution. Hence, the diffusion will increase. However, this process cannot continue indefinitely and equation (12) shows that the current density reaches a limiting value when $c_0 = 0$, *i.e.* when all the ions which reach the electrode under the action of the electric field and diffusion are immediately discharged. It is not possible, at constant temperature, to increase the number of ions arriving at the electrode per second beyond this limit. A limiting value of the current density is reached which is called the *limiting current* and remains constant even when the external tension applied to the electrode is increased. The process under these conditions becomes partially irreversible and the excess of energy supplied is trans-

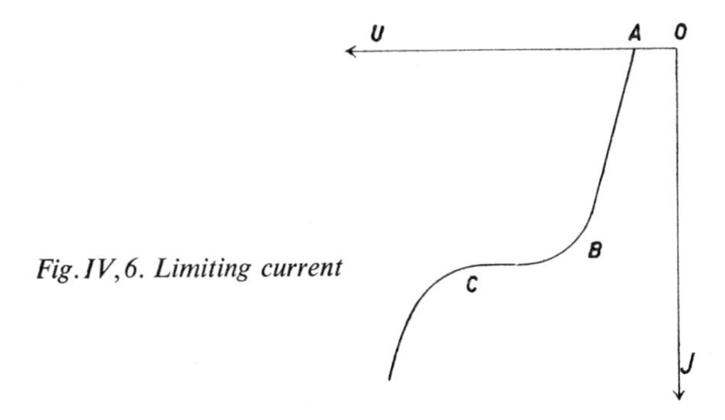

Fig. IV, 6. Limiting current

formed into heat. The progress of the phenomenon is shown in Fig. IV,6. The applied external tensions U are shown as abscissae and the resulting current densities J as ordinates. The segment AB corresponds to the increase in J, for increasing values of the electric tension and shows therefore the behaviour of diffusion overtension, and the segment BC corresponds to the limiting current. On continuing to increase the applied external tension, the current density remains constant until the point C is reached when a second ascending segment is obtained similar to the first. This is due to the fact that H^+ ions coming from the dissociation of water are always present in aqueous solutions of cations. If the discharge tension of the H^+ ions under the experimental conditions is less noble than the discharge tension of the cations, the segment ABC of the curve will correspond to the discharge of the cations and the discharge of the H^+ ions will begin at point C. If, however, the discharge tension of the H^+ ions is more noble than that of the cations, the H^+ ions are first discharged and then after segment BC, corresponding to the limiting current, the discharge of the other cations begins. If different ions of the same sign are present in the solution there will be a similar step for every ion and the J–U diagram will show as many steps as there are ions present. Each step will occur at a different electric tension characteristic of the relevant ion provided that the discharge tensions are sufficiently different. When they are close together there may be a simultaneous discharge of more than one ion (*cf.* p. 265 *et seq.*).

Fig. IV,6 shows the general features of a J–U diagram for an electrode process controlled by diffusion. In actual cases the characteristics of the diagram depend on the nature of the electrochemically active substance and on the particular conditions of the experiment, such as the concentration, temperature, agitation and the presence of inert electrolyte. Diffusion polarization is the only possible cause of overtension arising during the electrolysis for all reversible electrolytic processes having a simple course *i.e.* all those which are not followed by slow secondary, nor preceded by slow preliminary chemical reactions. In many cases the course of the reaction is complex and in particular the electrode reaction may be preceded or followed by one or more chemical reactions. An example is provided when the substance which is used in the electrode reaction is in turn the product of an earlier reaction. If the consumption of the electrochemically active substance is not compensated by a sufficiently rapid replacement by the preceding reaction, a new form of polarization will arise:

reaction polarization. This requires a corresponding increase in the applied tension as a *reaction overtension*. An overtension of this type may also arise if the electrode reaction is followed, rather than preceded, by a slow chemical reaction. In this case there will be an accumulation of substances leading to changes in the concentrations of the products of the electrode process. The treatment given here will refer principally to overtensions due to chemical reactions preceding the electrode process but analogous considerations will apply to reactions following the electrode process.

Reaction polarization may be encountered whenever the concentration of the electrochemically active substance consumed in the electrode reaction is not independent of the other substances present in solution, but is formed by successive balanced chemical reactions from other starting materials. If, for example, the electrode reaction is the reduction of a metallic ion Me^{z+} to a metal Me, this ion may not actually be present in a free form derived from the dissociation of a simple salt, but may be present as a complex $[MeX_p]^{(z-pq)+}$. The metal here is linked to p complexing groups, each bearing a charge q; it is assumed that the complex ion is not itself directly reducible. The cathodic discharge of the Me^{z+} ion must therefore be preceded by the reaction giving Me^{z+}:

$$[MeX_p]^{(z-pq)+} \rightleftharpoons Me^{z+} + pX^{q-} \tag{17}$$

Initially the active cation is present at a certain concentration $[Me^{z+}]$ in equilibrium with the complex ion. This concentration will depend upon the value of the dissociation constant of the complex and must be used in place of [Ox] in equations (1), (3), (4), (6), (8a) and (9). It will naturally determine the cathodic current intensity. Once electrolysis has begun the free metallic ions originally present, in equilibrium with the complex, will rapidly be used up about the cathode. Two responses may develop during the electrolysis depending on whether reaction (17) is slow or fast, *i.e.* on whether the rate of the reaction generating the electrochemically active species is low or high. If firstly, the rate of this reaction is very high, it can be assumed that the corresponding equilibrium is rapidly established so that the electrochemically active substance is continuously renewed and maintained at the equilibrium concentration. There will then be no change in concentration during electrolysis; no overtension will be generated apart from those of transfer and resistance, which are always present, provided that other factors do not intervene. Since however the substances involved in the preliminary reaction are themselves liable to suffer changes in concentration during electrolysis by the same mechanism which causes the diffusion overtension, overtensions due to diffusion may occur even when all the reactions preceding the electrode one are rapid. In general if the formation of the electrochemically active substance Ox is preceded by the equilibrium

$$aA_a + bA_b + cA_c \ldots \rightleftharpoons Ox + mA_m + nA_n \ldots \tag{18}$$

whose constant is K, the value of Ox required for equation (1) etc. may be obtained by solving the equilibrium equation *i.e.*

$$I_- = -\vec{k}_- K \frac{[A_a]^a[A_b]^b[A_c]^c \ldots}{[A_m]^m[A_n]^n \ldots} e^{-\{\beta z F(U_{rev}+\eta)\}/(RT)}$$

$$= -\vec{k}_- K\Pi\, [A_i]^i\, e^{-\{\beta z F(U_{rev}+\eta)\}/(RT)} \tag{19}$$

where Π is the product of the concentrations $[A_i]^i$.

The electric tension required to give a certain current density in such a cell will be more cathodic than that required for the same current density in a solution of pure Ox itself *i.e.* not complexed. In this case the overtension is actually a change in the equilibrium electric tension U_{rev}. Behaviour of this kind, where a reaction overtension is absent, is found with all labile metallic complexes *i.e.*, all those which can dissociate rapidly, liberating the metallic ion.

Secondly, if one of the preelectrodic reactions has a low rate, utilization of the electrochemically active substance present in the free form, will lead to its progressive loss from the region about the electrode; since the slow establishment of the preelectrodic equilibrium will not suffice to renew it rapidly. Under these conditions a new type of polarization — *reaction polarization* — will arise. There will be a corresponding reaction overtension, whose calculation will be demonstrated for the cathodic reduction of a substance Ox which is itself a product of the preceding general reaction (18). Under steady state conditions the rate of electrolytic consumption of Ox depends on the current density J and is equal to the rate of its formation by the chemical reaction (18). This in turn is the resultant of the forward and reverse reactions of (18) governed by the respective velocity constants $\vec{k}_1$ and $\overleftarrow{k}_2$. Thus

$$-\frac{1}{S}\mathrm{d}[Ox]/\mathrm{d}t = J = \frac{\vec{k}}{S}(\vec{k}_1[A_a]^a[A_b]^b[A_c]^c \ldots - \overleftarrow{k}_2[Ox]\,[A_m]^m[A_n]^n \ldots) \tag{20}$$

Moreover whether the current density is zero ($J = 0$) or finite ($J < 0$), the following two equations hold true.

$$U_{J=0} = U_0 + \frac{RT}{z\mathbf{F}} \ln [\mathrm{Ox}]_{equil} \tag{21}$$

$$U_{J<0} = U_0 + \frac{RT}{z\mathbf{F}} \ln [\mathrm{Ox}] \tag{22}$$

where [Ox] is the concentration, or more correctly the activity, in the layer of electrolyte in immediate contact with the electrode. Ignoring the transfer overtension, the total remaining overtension $\eta = U_{J<0} - U_{J=0}$ and will be equal to the reaction overtension, so that

$$\eta = \eta_r = \frac{RT}{z\mathbf{F}} \ln ([\mathrm{Ox}] / [\mathrm{Ox}]_{equil})$$

and hence

$$[\mathrm{Ox}] = [\mathrm{Ox}]_{equil}\, e^{(RT/z\mathbf{F})\eta_r} \tag{23}$$

If J is greatly increased, [Ox] will tend rapidly towards zero since its rate of formation is low and hence η_r will also tend towards $-\infty$ (23). Substituting equation (23) into equation (20) gives

$$J = \frac{\vec{k}}{S} (\vec{k}_1 [\mathrm{A}_a^-]^a [\mathrm{A}_b]^b [\mathrm{A}_c]^c \ldots - \overleftarrow{k}_2 [\mathrm{A}_m]^m [\mathrm{A}_n]^n [\mathrm{Ox}]_{equil}\, e^{(RT/z\mathbf{F})\eta_r}) \tag{24}$$

As J is increased the second term of the right hand side tends towards zero and J tends towards a constant value determined by both the diffusion of the substances producing Ox and the activation energy of the reaction which forms Ox. This latter is inherent in the term $\vec{k}_1$. A term may also appear due to the diffusion of any Ox present in the bulk of the electrolyte. The reaction overtension also shows characteristically a limiting current which, however, is normally much lower than that corresponding to the diffusion overtension alone, due to the presence of the exponential term $e^{-E/(RT)}$ (the energy of activation of the chemical reaction which forms Ox) contained in the constant $\vec{k}_1$, in addition to the diffusion terms themselves. Fig. IV,7 shows a typical pattern of a pure reaction overtension. In practice it is difficult to observe a pure reaction overtension, because the phenomena which tend to establish it are inevitably accompanied by diffusion polarization. It is also difficult to observe the limiting kinetic current, because the reaction overtension increases fairly rapidly with increasing current density and in particular very much more rapidly than

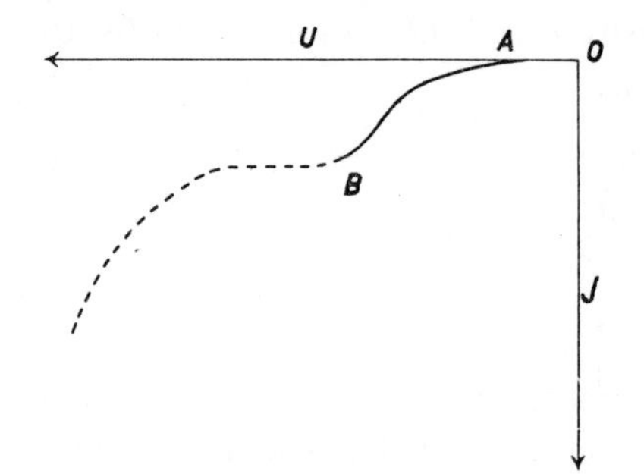

Fig. VI,7. Reaction overtension

in the case of diffusion polarization. This can be seen if the segment *AB* of Fig. IV,6 is compared with the segment *AB* of Fig. IV,7. Here, the limiting current can be observed only in the region of very high overtension which is difficult to use (the broken curve in Fig. IV,7).

Reaction polarization may also be called, quite correctly, chemical polarization or overtension. The common use of the terms 'activation overtension' or 'activation polarization' has less to commend it. This would logically refer to any overtension due to a high activation energy being required in any one of the series of reactions contributing to the overall electrochemical process and hence determining the velocity of this process. However, on this definition, the slow process could be not one of the preelectrodic chemical reactions but the electrode process itself, *i.e.* the charge transfer. This activation overtension would thus include not only reaction overtension but also transference overtension and the use of this term could be ambiguous.

Reaction polarization is, in general, caused by the slowness of one of the various, and possibly numerous, reactions preceding or following the electrode reaction of charge transfer. In general, however, it is difficult from the experimental data of applied tension and current density, to identify which of these reactions is actually determining the polarization. In fact, this difficulty pervades the whole field of overtension. It is rarely easy, and often impossible, to identify the cause of an observed overtension, or to determine the type of overtension (transfer, diffusion etc.) existing in a given electrolytic process from the experimental data of electric tension and current alone. The problem may be resolved by a careful examination of the shape of the electric tension–current density diagram. Comparing Figs. IV,5,6 and 7 shows that the *J–U* curves have rather different patterns for transfer, diffusion and reaction overtensions.

In practice, this method does not always give sure results because the experimentally measured *J–U* curves may not be sufficiently characteristic to decide to which example they belong. This is because in practice different forms of overtension are often present simultaneously. This makes it extremely difficult to decide which forms are present and to determine their quantitative relationships. However, other, more complex methods do exist which sometimes can aid studies of the mechanism more than do the simple data of current density and electric tension. These methods, however, require a knowledge of the intensity or density of partial currents and will be dealt with after a discussion of the methods of determining these.

5. Methods of Evaluating Currents and their Applications

Two types of parameter characterize the progress of an electrolysis: electric tension and current intensity or density. Methods for the determination of electrode tensions have already been discussed (Chapter III). Various forms of current intensity involved in electrolysis must now be considered, including the total current (I), the unidirectional, partial progressive ($\overrightarrow{I}$) and partial regressive ($\overleftarrow{I}$) currents one of which is cathodic (I_-) and the other anodic (I_+), and the exchange current (I_0). All these forms of current may be characterized by measurements of current intensities or densities. The only form directly accessible to experiment is the total current which corresponds to the algebraic sum of the partial progressive and regressive currents; it can be measured simply with an ammeter or galvanometer placed in series with the electrolysis circuit. The unidirectional partial currents and the exchange current are however not accessible to direct measurement and can only be determined by the elaboration of data from a number of different measurements. The determination of a unidirectional current at electric tensions far removed from equilibrium is relatively simple. Here, the other unidirectional current is practically zero and does not contribute to the total current. With a strong positive overtension, for example, the cathodic current may be neglected and the measured current is virtually equal to the anodic current. Naturally, however, the simplification cannot be used for the measurement of partial currents at low overtensions tending towards zero. In the particularly important case of zero overtension when the exchange current (I_0) is to be determined, recourse must be made to the following methods (a)–(d).

Sometimes, particularly in American literature, the situation with an electrode in electrochemical equilibrium is described not with the exchange current intensity I but by using the common value of the velocity constant of the anodic and cathodic processes; this is given the symbol k_0 defined by

$$I = z\mathbf{F}k_0[\mathrm{Ox}]^{\beta}_{equil}\,[\mathrm{Red}]^{\alpha}_{equil}\,S$$

where S is the surface area of the electrode and $[\mathrm{Ox}]_{equil}$ and $[\mathrm{Red}]_{equil}$ are the concentrations at the electrode surface under equilibrium conditions (*cf.* p. 203 *et seq.*). The quantity k_0 has the advantage over I_0 of being independent of the concentrations of the electrochemically active substances present in solutions. Still other authors have used a 'standard exchange current density' J_0 defined by

$$I = J_0 S[\mathrm{Ox}]^{\beta}_{equil}\,[\mathrm{Red}]^{\alpha}_{equil}$$

i.e.

$$J_0 = z\mathbf{F}k_0$$

The present text however will, for simplicity, continue to use I_0.

(a) Extrapolation of the experimental $I-\eta$ curve by *Tafel's equation.* It has been seen that when the overtension in one direction or the other is strong, only one of the unidirectional currents is of importance because the other tends to zero. Hence, equation (9) on page 206 may be simplified, for strongly positive overtensions

$$I \cong I_+ = I_0\, e^{(\alpha z\mathbf{F}\eta)/(RT)} \tag{1}$$

and for strongly negative overtensions

$$I \cong I_- = -\,I_0\, e^{-(\beta z\mathbf{F}\eta)/(RT)} \tag{2}$$

Multiplying I_- by -1 and taking logarithms of equations (1) and (2) gives

$$\log\,|\,I_+\,| = \log I_0 + (0.434\alpha z\mathbf{F}\eta)/(RT) \tag{3}$$

and

$$\log\,|\,I_-\,| = \log I_0 - (0.434\beta z\mathbf{F}\eta)/(RT) \tag{4}$$

which are both equations of the type $\eta = a + b \log I$ (*i.e.* Tafel's equation). If the overtensions and the logarithms of the measured current intensities at high overtensions are plotted on a graph, a straight line is obtained, which will have a slope of $(2.303RT)/(z\mathbf{F}\alpha)$ and $-(2.303RT)/(z\mathbf{F}\beta)$ in the two cases. The intensity at the intercept on the ordinate axis will be $\log I_0$ (Tafel's line shown broken in Fig. IV,8). It is thus possible to determine at the same time the values of α, β and I_0 starting from two easily meas-

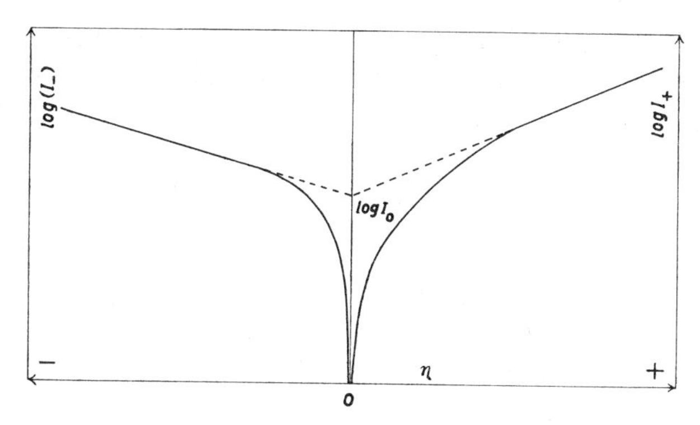

Fig. IV,8. Tafel's lines

ured quantities: the intensity of the total current and the corresponding overtension. It may also be noted that the intersection of the anodic and cathodic Tafel's lines shows the equilibrium electric tension U_{rev} which may not be easily measured by normal methods, as for example with irreversible processes. For small overtensions, the smaller of the two unidirectional currents cannot be considered as negligible and the experimental curve $\eta - \log |I|$ departs from the linear form predicted by Tafel's equation (3) or (4) (the solid curve in Fig. IV,8). The determination of α, β and I_0 may however be carried out by extrapolation, to $\eta = 0$, of the straight part of the curve corresponding to the high overtension region.

The method is completely valid only when the sole form of overtension present in the electrode process is that of transfer. In other cases, the relationship between $\log |I|$ and η may depart from linearity, but nevertheless if a linear segment exists, its extrapolation to $\eta = 0$ gives once more the correct value for the logarithm of the exchange current intensity, even if types of overtension other than the transfer appear at electric tensions away from the equilibrium.

(b) From measurements of the impedance of electrodes with alternating currents[1]. If a sinusoidal alternating tension of frequency ω is applied to

[1] P. Dolin and B. W. Ershler, *Acta Physicochim. U.R.S.S.*, 13 (1940) 747; J. E. B. Randles, *Discussions Faraday Soc.*, 1 (1947) 11; B. W. Ershler, *Discussions Faraday Soc.*, 1 (1947) 269; J. O'M. Bockris, *Modern Aspects of Electrochemistry*, Vol. 1, Butterworth, London, 1954, Chapter IV.

an electrode (so that $U' = U'_0 \cos \omega t$) already subjected to a continuous equilibrium tension U_{rev}, a pulsating current will flow which will be out of phase with the electric tension U', since the electrode will intermittently present a complex impedance to the passage of the alternating current. This impedance is partially ohmic and partially capacitive:

$$I = I_0 \cos (\omega t + \varphi) \tag{5}$$

It can be shown that the impedance Z at the interface of the electrode is related to the intensity of the exchange current I_0, to the capacity of the diffuse layer of the liquid phase C_d and to the frequency ω of the electric tension, by the equation

$$Z = \frac{RT}{z\mathbf{F}} \frac{1}{I_0} + \frac{1}{C_d \omega} \tag{6}$$

Hence, a graph of the experimentally determined values of Z at various frequencies, plotted against the frequency will on extrapolation to $\omega \to \infty$ give I_0.

(c) By measurements of the intial current with a constant tension[1]. If, when a constant electric tension is applied to an electrode the concentrations of the substances involved in the electrode reaction remain unaltered following the passage of the current, a current I will flow through the electrode; this current will be given by equation (9) on page 206. However, in practice, the concentration of the initial substance diminishes when the current flows and hence this also diminishes with time (Fig. IV,9). The initial current I_i which flows in the first instant of electrolysis, however, does not show any effect of the loss of the electrochemically active sub-

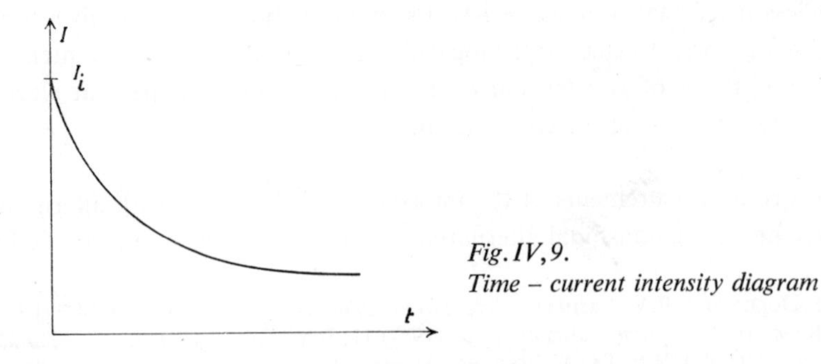

Fig. IV,9.
Time – current intensity diagram

[1] H. Gerischer and W. Vielstich, *Z. physik. Chem.*, N.F. 3 (1955) 16.

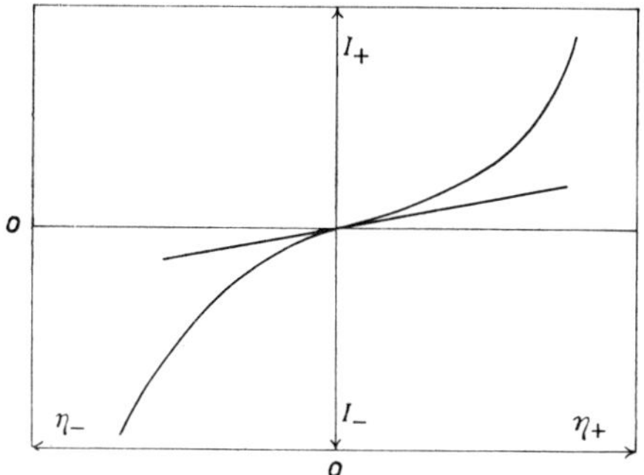

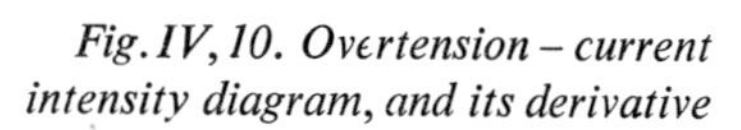
Fig. IV, 10. Overtension – current intensity diagram, and its derivative

stance and thus it is truly represented by equation (9) on page 206. In other words, the initial current I_i depends only on α, β and the transfer overtension η. From equation (9) on page 206 at the point $\eta = 0$, bearing in mind that $I = I_i$ and assuming that $\alpha + \beta = 1$

$$(\mathrm{d}I_i/\mathrm{d}\eta)_{\eta=0} = (I_0 z\mathbf{F})/(RT) \qquad (7)$$

To obtain I_0, I_i is measured for various positive and negative overtensions keeping the duration of each determination constant. The values obtained for I_i are plotted against the overtension η and the slope of the curve at the point $\eta = 0$ gives the factor $(\mathrm{d}I_i/\mathrm{d}\eta)_{\eta=0}$. I_0 can now be readily obtained using equation (7). The graph of I_i against η (Fig. IV, 10) may also be used for the determination of α and β on the basis of the asymmetry of the two limbs of the experimentally obtained curve. The method is in theory fairly simple but requires rather complicated apparatus to ensure the exact constancy of the applied electric tension throughout the whole period of measurement.

(d) From measurements of the initial electric tension with a constant current. This method is very similar to, but in some sense is the reverse of, the preceding. A constant current is passed through an electrode. This may be done for example, by supplying the electrolysis circuit from a continuous high tension source and placing a high external ohmic resistance in series in the circuit. Any changes in the resistance of the electrolysis circuit itself and any electrodic polarization (which may be considered as an increase in resistance) will be negligible in comparison with this high external resistance. Under these conditions, the intensity of the current flowing through the circuit will be virtually unchanged by variations in the resistance of the electrolytic part of the circuit and may be considered as constant. The overtension which is set up initially at the electrode will still be determined by equation (9) on page 206. The value of I, here, is fixed *a priori* even although the concentration of the electrochemically active substances may change as a result of the electrode reaction changing the actual overtensions. Knowing the values of I and η_i (the former being pre-selected and the latter measured) it is possible to construct an I–η_i graph analogous to that in Fig. IV, 10. Using the procedure described already, it is then possible to determine the exchange current with equation (8).

$$(\mathrm{d}I/\mathrm{d}\eta_i)_{\beta_i=0} = (I_0 z\mathrm{F})/(RT) \tag{8}$$

The importance of measurements of partial currents is principally that their knowledge is an important first step towards an understanding of the mechanism of electrode processes. In particular, the intensity of the exchange current is a characteristic parameter of the electrochemical equilibrium and is defined under conditions in which all the causes of overtension, other than that of transfer, are zero. This current thus characterizes the nature of the process of charge transfer which is the central step of the set of reactions making up the overall electrode process. Moreover, a knowledge of the size of the exchange current lies at the base of a method for the identification of the determining step in a succession of partial reactions forming a complex electrode process when reaction overtensions arise also. Since this method is one of the most characteristic and important applications of measurements of partial currents, it will be briefly described.

Consider an electrodic process, made up of several steps, corresponding to the overall reaction

$$a\mathrm{A} + b\mathrm{B} + c\mathrm{C} + \ldots + ze^- \rightleftharpoons u\mathrm{U} + v\mathrm{V} + w\mathrm{W}\,[1] \tag{9}$$

in which one of the chemical reactions preceding or following the charge transfer determines the kinetics of the overall process. Normally, this reaction will occur only once in the cycle of reactions (9) and may be written in the general form

$$h\mathrm{H} + k\mathrm{K} + \ldots \rightleftharpoons m\mathrm{M} + n\mathrm{N} + \ldots \tag{10}$$

[1] Each species A, B,...., U, V,.... may or may not be ionic, but for simplicity the indication of the charge will be omitted. Clearly equation (9) has to be a true equation with regard to electric charges also.

Assume also, for simplicity, that reaction (10) is the sole determinant of the overall kinetics; all the other steps of process (9), including the transfer of charge, are assumed to occur extremely rapidly. Thus, all the partial reactions preceding and following the determining step (10) may be considered to be virtually always in their equilibrium state. In practice, the electrode processes are often more complicated than is assumed here; the velocity of the charge transfer process may become sufficiently high not to affect the kinetics of the reaction, only if the applied overtension is also high and of the sign which favours the passage of current in the direction corresponding to the utilization of the products of the determining chemical reaction. Otherwise the charge transfer process could also affect the kinetics. But it will be convenient to develop the procedure used for the identification of the step determining the kinetics under ideally simplified conditions, so as to demonstrate it more clearly.

It will be assumed that the overall reaction (9) takes place through a series of intermediate steps as shown in Fig. IV,11. The substances A, B, C, ... U, V, W, ... in equation (9) are respectively those present in the initial (1) and final (z) states, and H, K ... and M, N ... are the substances consumed and formed respectively in the step from state (i) to state ($i + 1$). All the steps from the initial state (1) to the succeeding states (including the charge transfer $j \rightarrow j + 1$) require an energy barrier to be overcome, but this barrier is sufficiently low for the changes to occur rapidly so that equilibrium can be assumed to be always maintained. Only the step from state (i) to state ($i + 1$) – which is the rate determining reaction (10) – possesses an energy barrier much higher than that of the others, which makes it rate limiting. In this discussion, the velocity of the

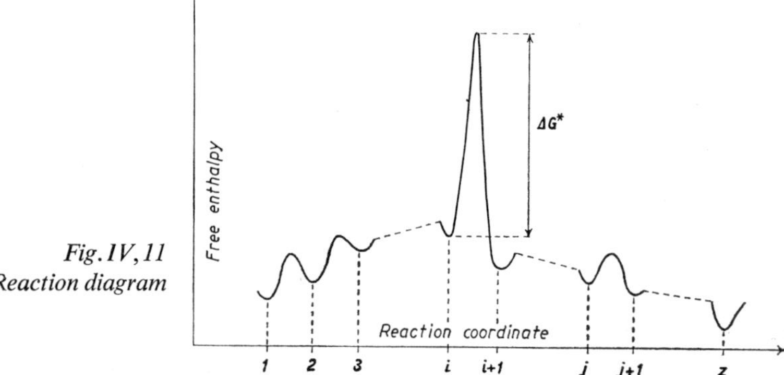

Fig. IV, 11
Reaction diagram

overall process $\vec{V}_{1\to z}$ (which is directly proportional to the current intensity $\vec{I}$) is equal to the velocity $\vec{V}_{i\to(i+1)}$ and hence is also proportional to the concentrations of the substances H, K, ... in state (*i*), each raised to the power of their stoichiometric coefficient in reaction (10), *i.e.*

$$\vec{I} \propto \vec{V}_{1\to z} = \vec{V}_{i\to(i+1)} = \vec{k}_{i\to(i+1)}\,[\mathrm{H}]^{h}[\mathrm{K}]^{k}... \tag{11}$$

The current intensity thus depends directly on the concentrations of the substances present in state (*i*). However, these concentrations are often not readily determined in practice as they may refer to unstable intermediates or other compounds which are difficult to measure or may not appear in the equation to the overall process. It is thus more convenient to relate the current intensity to the concentrations of the initial substances A, B, C ... This may be achieved by taking advantage of the fact that all the reactions leading from the initial state to the state (*i*) are in equilibrium. The overall equilibrium reaction for the passage from the initial state to state (*i*) may be represented by

$$a'\mathrm{A} + b'\mathrm{B} + c'\mathrm{C}\ldots \rightleftharpoons h'\mathrm{H} + k'\mathrm{K} + \ldots \tag{12}$$

where the coefficients a', b', c', ... may differ from the coefficients a, b, c, ... of equation (9) although in most cases they are the same. An overall equilibrium constant $K_{1\to i}$ will correspond to this reaction, so that

$$[\mathrm{H}]^{h'}[\mathrm{K}]^{k'}\ldots = K_{1\to i}\,[\mathrm{A}]^{a'}[\mathrm{B}]^{b'}[\mathrm{C}]^{c'}$$

Multiplication of this equation by $[\mathrm{H}]^{h}\,[\mathrm{K}]^{k}\ldots$ followed by division by $[\mathrm{H}]^{h'}\,[\mathrm{K}]^{k'}\ldots$ gives

$$[\mathrm{H}]^{h}[\mathrm{K}]^{k}\ldots. = K_{1\to i}\,[\mathrm{A}]^{a'}[\mathrm{B}]^{b'}[\mathrm{C}]^{c'}\ldots\,[\mathrm{H}]^{h-h'}[\mathrm{K}]^{k-k'}\ldots$$

Substitution of this expression in (11) gives

$$\vec{I} = z\mathbf{F}K_{i\to(i+1)}\,K_{1\to i}\,[\mathrm{A}]^{a'}[\mathrm{B}]^{b'}[\mathrm{C}]^{c'}\ldots\,[\mathrm{H}]^{h-h'}[\mathrm{K}]^{k-k'}\ldots = K\Pi[\mathrm{X}]^{x'} \tag{13}$$

where $\Pi[\mathrm{X}]^{x'}$ gives in abbreviated form, the product

$$[\mathrm{A}]^{a'}[\mathrm{B}]^{b'}[\mathrm{C}]^{c'}\ldots\,[\mathrm{H}]^{h-h'}[\mathrm{K}]^{k-k'}$$

Equation (13) shows that in the usual cases where $h = h'$, $k = k'$..., the value of $\vec{I}$ does not depend on the concentrations [H], [K]...

Taking logarithms of each term of equation (13) and differentiating each with respect to the logarithm of the concentration [X] of the initial substances whilst keeping the other terms constant, gives

$$(\partial \ln \vec{I} / \partial \ln [X])_{(Y) \neq (X)} = x'_i \tag{14}$$

Since in this case the progressive current intensity $\vec{I}$ is identical with that of the total current, it is only necessary to measure this at different concentrations of the starting substance to determine the values of the coefficients a', b', c',... from equation (14). These coefficients which appear in equation (12) are here called *stoichiometric factors.* The reaction determining the kinetics of the overall process can thus be recognized by using the stoichiometric factors. Once the left-hand side of equation (12) is known, the right-hand side will usually be determined unequivocally. It is this latter which indicates which substances react and in what amounts in the determining reaction.

The treatment developed so far for a slow chemical reaction which determines the kinetics of an overall electrode process may also be used when the slow reaction is that of the charge transfer itself, even when it is preceded or followed by rapid chemical reactions. The determination of the stoichiometric factors, which here assume the significance of the order of the electrochemical reaction, serves to identify the chemical species directly involved in the process of charge transfer. In this way, for example, it is possible to clarify the mechanism of the reduction of complex metallic ions which are not reduced as such but first undergo a partial or total dissociation of the complexing groups. It is possible to develop a perfectly analogous treatment for this, to that developed in the preceding case which led to equation (14). There are, however, some differences. The most important is that the velocity constant of the slow reaction will now depend upon the electric tension, so that in dealing with this problem allowance must be made for the electric tension applied to the electrode. Moreover, the intensity of the current which appears in equations of type (14) and from which the stoichiometric factors are determined, is always that of a partial unidirectional current. This can no longer be taken as equal to the total current since it is not now possible to assume that the value of the unidirectional current of opposite sign is negligible. Firstly, since only rarely does more than one chemical species take part in a charge transfer, not all the products of reaction (12) will be consumed during the determining step, so that equation (13) will have to contain concentration terms in the denominator. Consider, for example, the reduction of a complex metallic ion $(MeX_p)^{(z-pq)+}$ which dissociates before the reduction:

$$(MeX_p)^{(z-pq)+} \rightleftharpoons Me^{z+} + pX^{q-} \tag{15}$$

The expression for the current intensity is calculated as in the derivation of equations (13) and (14), and gives

$$\vec{I}_- = z\mathbf{F}\vec{k}(U)\,[Me^{z+}] = z\mathbf{F}\vec{k}(U)\,K_{(15)}\big[(MeX_p)^{z-pq+}\big]/[X^{q-}]^p \tag{16}$$

The expression $k(U)$ indicates that the constant k for the process is a function of the electric tension U; and $K_{(15)}$ indicates the equilibrium constant for reaction (15).

To interpret the effect of the electric tension of the electrode on the velocity of a rate determining charge transfer reaction, it is convenient to fix a value for the electric tension. Changes in $\vec{I}_-$ with changes in the concentration of the starting substance, may be measured whilst the electric tension is kept constant, so that the effect of the electric tension on the reaction velocity is constant. In this case, there will be complete identity with the preceding case for a slow chemical reaction, and equation (14) will still be valid. Obviously, the current intensity in equation (14) is not the overall net current but one of the two partial currents which must therefore be determined indirectly.

Another procedure which is sometimes used to determine the stoichiometric factors of electrochemical processes involving a slow transfer preceded or followed by rapid chemical reactions, consists in studying the velocity of the reactions at electrochemical equilibrium; the current intensity required is that of the exchange current and the following equations analogous to those of (13) and (16), can be written:

$$\vec{I}_0 = z\mathbf{F}\vec{k}_- e^{-(\beta z\mathbf{F}U_{rev})/(RT)}\,\Pi[X]^x \tag{17}$$

$$\vec{I}_0 = z\mathbf{F}\vec{k}_+ e^{(\alpha z\mathbf{F}U_{rev})/(RT)}\,\Pi[X]^x \tag{17'}$$

The relevant equation depends on whether the substance X which takes part in the reaction is the reduced or oxidized form of the redox couple directly linked with the charge transfer. Taking logarithms of (17) and (17′) and differentiating gives

$$(\partial \ln I_0/\partial \ln [X])_{(Y)\neq(X)} = x_i - (\beta z\mathbf{F}/RT)\,(\partial U_{rev}/\partial \ln [X])_{(Y)\neq(X)} \tag{18}$$

and

$$(\partial \ln I_0/\partial \ln [X])_{(Y)\neq(X)} = x_i + (\alpha z\mathbf{F}/RT)\,(\partial U_{rev}/\partial \ln [X])_{(Y)\neq(X)} \tag{18'}$$

respectively, where x_i are the stoichiometric factors. Vielstich and Gerischer[1] were able to determine with this procedure, for example, that the cathodic reduction of silver in the presence of a high concentration of cyanide ions occurs through $[Ag(CN)_2]^-$ ions, since the stoichiometric factor x_i for the $(CN)^-$ is $1.94 \cong 2$; in the presence of lower concentrations of $(CN)^-$ ions however the reaction occurs through the neutral complex $[Ag(CN)]$ since the same stoichiometric factor is here $1.06 \cong 1$. It should be noted that with this method it is possible to identify directly the chemical species responsible for the electrode reaction even although they do not appear in the overall reaction equation.

So far, methods have been described for identifying the kinetic determining steps in relatively simple cases where a single partial process which determines the overall kinetics may be either the charge transfer or a preceding or following chemical reaction. Naturally, actual cases are often more complicated. They may involve two fairly slow partial processes which both help to determine the overall kinetics, *e.g.* processes controlled both by a slow reaction and by the activation energy needed for the charge transfer. In other cases, the situation may be complicated by diffusion phenomena so that there may be processes controlled by activation and by diffusion, by a chemical reaction and by diffusion, or even by all three factors. Many of these cases have been examined polarographically by Brdička and Wiesner and by Delahay. The treatment follows the same principles used for the resolution of the more simple cases described here but leads to much more complicated expressions, for which reference may be made to the original literature.

6. Cathodic Processes; Discharge of the H^+ Ion

One of the most important cathodic processes is the discharge of the H^+ ion to form gaseous hydrogen H_2. Despite its apparent simplicity, this process actually follows a fairly complicated course which in some respects has still not been clarified. It has been the subject of very many studies since, in addition to its intrinsic importance, a knowledge of its mechanism may furnish a model for the interpretation of other electrode processes.

The electric tension at which hydrogen is discharged coincides with the equilibrium electric tension only on a platinized platinum electrode,

[1] W. Vielstich and H. Gerischer, *Z. physik. Chem.*, N.F. 4 (1955) 10.

TABLE IV, 3

CHARACTERISTICS OF THE HYDROGEN DISCHARGE PROCESS

*Electrode material**	*Overtensions (V) at room temp. in* N HCl *for various values of* J(A/cm²) *absolute values*				*Values of* J_0 *and* α *in solutions of various composition at* 25° C		
	$J=10^{-3}$	$J=10^{-2}$	$J=10^{-1}$	$J=1$	*Solution*	J_0(A/cm²)	α
Ag	0.44	0.66	0.76	—	HCl 0.1 *N*	10^{-6}	0.5–0.6
Ag	—	—	—	—	HCl 7 *N*	10^{-7}	0.53
Al	0.58	0.71	0.74	0.78	H_2SO_4 2 *N*	10^{-10}	0.6
Au elec.dep.	0.17	0.25	0.32	0.42	HCl 1 *N*	10^{-5}–10^{-6}	0.7–1.3
Be	0.63	0.73	—	—	HCl 1 *N*	10^{-9}	0.5
Bi	0.69	0.83	0.91	1.01	HCl 1 *N*	10^{-7}	0.4–0.6
C filaments	0.95	1.13	1.18	1.17	—	—	—
C graphite	0.47	0.76	0.99	1.03	—	—	—
C arc carb.	0.27	0.34	0.41	0.41	—	—	—
Cd	0.91	1.20	1.25	1.23	HCl 1 *N*	10^{-7}	0.3
Cr	—	—	0.67	0.77	—	—	—
Cu	0.60	0.75	0.82	0.84	HCl 0.1 *N*	$2 \cdot 10^{-7}$	0.50
Cu elec.dep.	0.50	0.62	0.74	0.80	—	—	—
Fe	0.40	0.53	0.64	0.77	H_2SO_4 2 *N*	10^{-6}	0.50
Ga	—	—	—	—	H_2SO_4 0.2 *N*	$2 \cdot 10^{-7}$ (87°)	0.50
Hg	1.04	1.15	1.21	1.24	HCl 0.1 *N*	5–$6 \cdot 10^{-13}$	0.50–0.
Hg	—	—	—	—	HCl 10 *N*	$2 \cdot 10^{-11}$	0.61
Hg	—	—	—	—	KOH 0.1 *N*	$4 \cdot 10^{-16}$	0.62
In	0.80	1.05	1.19	—	—	—	—
Mo	0.30	0.44	0.57	—	HCl 1 *N*	10^{-6}	1.5
Nb	0.65	0.74	0.82	—	HCl 1 *N*	10^{-7}	0.6
Ni	0.33	0.42	0.51	0.59	NaOH 0.12 *N*	$4 \cdot 10^{-7}$	0.58
Pb	0.67	0.97	1.12	1.08	HCl 0.1 *N*	$2 \cdot 10^{-13}$	0.48
Pb elec.dep.	0.91	1.24	1.26	1.22	—	—	—
Pd	—	—	—	—	HCl 0.6 *N*	$2 \cdot 10^{-4}$	2.0
Pd	—	—	—	—	NaOH 1 *N*	10^{-4}	0.45
Pt smooth	0.09	0.39	0.50	0.44	HCl 1 *N*	0.8–$1.0 \cdot 10^{-3}$	2.0
Pt elec.dep.	0.25	0.35	0.40	0.40	—	—	—
Pt platinized	0.01	0.03	0.05	0.07	—	—	—
Rh	0.08	0.22	0.33	0.34	—	—	—
Sb	—	—	—	—	H_2SO_4 2 *N*	10^{-9}	0.58
Sn	0.85	0.98	0.99	0.98	HCl 1 *N*	10^{-8}–10^{-10}	0.45
Ta	0.41	0.75	0.90	—	HCl 1 *N*	10^{-5}	0.7
Tl	1.05	1.13	1.15	—	—	—	—
W	0.27	0.35	0.47	0.54	HCl 5 *N*	10^{-5}	0.53

* Electrode metals are in the massive form unless otherwise stated.

whereas with electrodes of other materials, more or less elevated overtensions are required. These generally are variable and depend upon the environmental conditions and various other factors which cannot all be readily controlled and reproduced. The principal characteristics of the overtension of hydrogen are as follows.

1. It increases in absolute value as the current density J increases, following approximately Tafel's equation:

$$\eta_H = a + b \log |J| \tag{1}$$

2. It diminishes in absolute value with increasing temperature.

3. It depends on the chemical nature of the electrode material; in 1 N HCl, for $J = 1 \cdot 10^{-2}$ A/cm² at room temperature, the overtensions increase in the order: Pt (platinized), Rh, Au, W, Pt (smooth), Ni, Mo, Fe, Ag, Al, Be, Nb, Ta, Cu, C (graphite), Bi, Pb, Sn, In, Tl, Hg and Cd. Under other experimental conditions and depending on the purity of these materials, the series may undergo alterations.

4. Other factors which may effect the overtension of hydrogen are the surface state of the electrode (smooth, rough or spongey) its previous treatment, the pressure of gaseous hydrogen, the pH of the solution (in certain cases only) the nature of the solvent, the presence or absence of foreign electrolytes and above all the presence of even traces of certain substances which can act as catalytic poisons.

The electrochemical quantities which characterize the hydrogen reaction process are the intensity, or rather the density, of the exchange current J_0 and the transfer coefficient α, which can both be obtained from Tafel's equation. In this, the coefficient a allows J_0 to be deduced and is a constant depending primarily upon the nature of the electrode, and the coefficient b allows α to be deduced and depends not so much on the nature of the electrode as on the reduction process. Table IV, 3 presents experimental data which show the influence of the nature of the electrode material, the composition of the solution and the current density on the values of the overtension and the other characteristic parameters for the discharge of hydrogen, α and J_0.

The generally large overtensions observed in the discharge of hydrogen arise through the presence of one or more slow steps in the overall electrode process:

$$2\,H^+ + 2\,e^- \rightarrow H_2 \tag{2}$$

This has been shown by studying the effects of temperature. The partial process determining the observed overtension has not been identified for certain, but may be one of the following:

1. The flow of ions to the electrode.
2. The dehydration of the ions (free H^+ ions do not exist in aqueous solution but are solvated; the symbol H^+ is used for simplicity).
3. The discharge of the ions.
4. The formation of molecules.
5. The elimination of the gaseous hydrogen from the electrode.
6. The diffusion of the hydrogen away from the electrode, etc.

The transference phenomena (1), (5) and (6) become important only under particular experimental conditions. The equilibrium (2) is probably not rate determining, but the phenomena (3) and (4) become increasingly important in the succession of the following partial reactions:

(a) Reduction of the hydrogen ion to the hydrogen atom adsorbed on the metal M of the electrode

$$H^+ + e^- + M \rightarrow M(H) \tag{3}$$

M(H) represents the H atom adsorbed on the metal. In strongly alkaline media the reaction (3) may be replaced by

$$H_2O + e^- + M \rightarrow M(H) + OH^-$$

(b_1) The recombination of atomic hydrogen to form the H_2 molecule

$$M(H) + M(H) \rightarrow 2\,M + H_2 \tag{4}$$

or by

(b_2) the reduction of another H^+ ion followed by its combination with atomic H already present on the electrode

$$M(H) + H^+ + e^- \rightarrow M + H_2 \tag{5}$$

Equation (5) may in turn, be divided into two successive steps, *i.e.* the formation of a hydrogen ion-molecule H_2^+ adsorbed on the metal (5a) and its reduction (5b)

$$M(H) + H^+ \rightarrow M(H_2^+) \tag{5a}$$

$$M(H_2^+) + e^- \rightarrow M + H_2 \tag{5b}$$

Of the numerous theories which have been proposed to explain the overtension of hydrogen, three have proved to be most satisfactory and complete. All of these assume that the rate determining step is one or other of the partial reactions (3), (4), or (5), shown above. According to the *catalytic theory*, the determining reaction is the recombination (4) (the discharge (3) is supposed to be in equilibrium and (5) not to exist). According to the *slow discharge theory* the rate determining reaction is the discharge (3); and according to the *electrochemical theory* the rate determining reaction is the discharge of hydrogen through an intermediary ion molecule (5b).

(I) *The Slow Combination Theory*

According to this theory, which was originally developed by Tafel, the discharge of the hydrogen ion (3), is in equilibrium and is followed by the recombination reaction (4) which is rate determining. Thus, on this theory, the electrode would be covered with an adsorbed layer of atomic hydrogen and its electric tension during electrolysis will depend upon the concentration of this atomic hydrogen on the electrode.

$$U = (RT/\mathbf{F}) \ln ([\mathrm{H}^+]/k[\mathrm{H}]) \tag{6}$$

The reversible electric tension of the same hydrogen electrode at atmospheric pressure is given by

$$U_{rev} = (RT/\mathbf{F}) \ln [\mathrm{H}^+] \tag{7}$$

so that the overtension η is

$$\eta = U - U_{rev} = -(RT/\mathbf{F}) \ln (k[\mathrm{H}]) \tag{8}$$

i.e.

$$[\mathrm{H}] = k^{-1} e^{-(\eta\mathbf{F})/(RT)} \tag{9}$$

The velocity of the overall reaction of hydrogen discharge and hence the current density J_- would, according to Tafel, be controlled by the determining reaction (4):

$$J_- = \vec{k}'' \mathrm{d}[\mathrm{H}]/\mathrm{d}t = \vec{k}' [\mathrm{H}]^2 \tag{10}$$

which gives with equation (9)

$$J_- = \vec{k}' k^{-2} e^{-(2\mathbf{F}\eta)/(RT)} \tag{11}$$

Taking logarithms,

$$\ln |J_-| = \ln (\overrightarrow{k'}k^{-2}) - (2\eta\mathbf{F})/(RT) \tag{12}$$

or

$$\eta = \{(RT)/(2\,\mathbf{F})\} \ln (\overrightarrow{k'}k^{-2}) - \{(RT)/(2\,\mathbf{F})\} \ln |J_-| \tag{13}$$

which is of the type

$$\eta = \text{constant} + b \log |J| \tag{13a}$$

Equation (13a) is thus Tafel's equation with $b = -2.303\ RT/2\,\mathbf{F}$, *i.e.* $b \cong -0.0296$ V at room temperature.

The theory of slow combination has some advantages over other theories. It can account for the dependence of the overtension on the material of the electrode ($\overrightarrow{k'}$ in equations (10)... (13) is a characteristic of the electrode metal). Strong evidence in its support comes from the observation of Bonhoeffer[1] that the gaseous reaction $2\,H \rightarrow H_2$ is catalyzed by metallic surfaces and that the catalytic activity of various metals for this reaction diminishes in the order Pt, Pd, W, Fe, Ag, Cu, Pb, and Hg which is very similar to the order in which the overtensions for the discharge of the H^+ ion increase. Moreover, since a catalyst accelerates a reaction in both directions, if the slow combination theory is correct, the metals showing low overtensions for hydrogen should also catalyze the inverse reaction ($H_2 \rightarrow 2\,H$) and in fact Pt and Pd on which the hydrogen overtension is lowest, are excellent catalysts of hydrogenation. Nowadays, however, Bonhoeffer's observation has lost some of its value since experimental studies have shown that recombination is the rate determining phase on Pt and Pd, *i.e.* on metals which facilitate this reaction; but on metals like Pb and Hg which have little catalytic action, the rate determining step seems to be the discharge of the H^+ ion. The reason for this behaviour is probably that the more active a metal is catalytically, the more strongly will it adsorb the H atoms formed in the discharge and hence the more readily will the discharge occur; in these cases (Pt and Pd) the desorption will, however, be difficult and hence the combination reaction will be fairly slow. However, on metals which are only feebly catalytic the H atoms will be only weakly adsorbed and hence the discharge will not be facilitated and will become the rate determining step. The theory of slow combination predicts that $b = -0.029$ (or more, as absolute value when the phenomena of H adsorption are con-

[1] K. F. Bonhoeffer, *Z. physik. Chem.*, 113 (1924) 213.

sidered). Since the discharge reaction (3) is presumed to be rapid, this implies that the charge transfer will occur rapidly, *i.e.* that the exchange current intensity will be high and of the order of $10^{-3} - 10^{-4}$ A/cm². It is now recognized that the mechanism of the slow combination applies above all to H^+ discharge on Pt or Pd (*cf.* Tables IV,3 and IV,4). Two fundamental objections may be made to the theory of slow combination.

1. The coefficient b of Tafel's equation is predicted to be $-2.303\,RT/2\,\mathbf{F} = -0.029$ which coincides with the experimentally determined value in only a few cases (*e.g.* sometimes on Pt or Pd) whereas in most cases the experimental value is of the order of —0.118 or less. This difficulty may be partly overcome if it is assumed that the hydrogen atoms taking part in the combination reaction are actually adsorbed on the electrode metal so that the concentration of active hydrogen $[H]_1$ for reaction (4) may be expressed as a function of the total hydrogen concentration [H] using Freundlich's isotherm

$$[H]_1 = k[H]^{1/n} \tag{14}$$

Substituting $[H]_1$ for [H] in equation (10) finally gives a value for b in equation (13a) of $b = -2.303\,nRT/2\,\mathbf{F}$. Since, in general, $n > 1$, the new value of b will be closer to the experimental values.

2. It has been found experimentally[1] that the change in overtension as a function of time in the brief interval between the closing of the circuit and the moment at which the overtension reaches its stationary value, is linear for a constant current density, as shown in Fig.IV,12. It must,

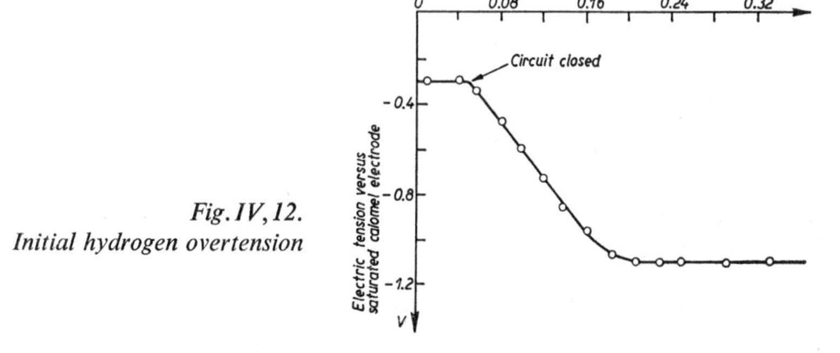

Fig.IV,12. Initial hydrogen overtension

[1] F.P.Bowden and E.K.Rideal, *Proc. Roy. Soc.*, A120 (1928) 59, 80; 125 (1929) 1446. H.Brandes, *Z. physik. Chem.*, A 142 (1929) 97.

therefore, also be a linear function of the concentration of atomic hydrogen which separates and accumulates during this time interval. Equation (8) however would predict a logarithmic relationship.

(II) *The Slow Discharge Theory*

This theory assumes that the rate determining reaction is the discharge of the hydrogen ion to form atomic hydrogen adsorbed on the metallic surface of the electrode (reaction 3); the following recombination (4) would then be in equilibrium. This theory was particularly developed by Erdey–Gruz and Volmer[1] but in its early form had the serious drawback of requiring the overtension to be dependent on the pH of the solution. This difficulty was eliminated by a treatment due to Frumkin employing Stern's hypothesis on the structure of the double layer. It is assumed that the concentration $[H^+]_{DL}$ of ions in the double layer is not that $[H^+]$ in the bulk of the solution. According to Stern

$$[H^+]_{DL} = [H^+]e^{-\psi \mathbf{F}/RT} \tag{15}$$

where ψ is the electric tension at a distance of one ionic radius from the electrode; this is given by

$$\psi = \psi_0 + (RT/\mathbf{F}) \ln [H^+] \tag{16}$$

which is valid for solutions which are not too concentrated, so that the cathode surface is not saturated.

Equation (4) on page 205 may be used for the cathodic discharge of H^+ ions bearing in mind that the ionic concentration required is that of the double layer, expressed by equation (15); in this equation [Ox] is replaced by $[H^+]_{DL}$, the general constant β by the transfer factor α,[2] and the electric tension U_{rev} by the fraction of the tension $(U_{rev} - \psi)$ relevant to the thickness of the double layer of one ionic radius, directly involved in the discharge process. Thus,

$$I_- = -\overrightarrow{k}[H^+]_{DL}e^{-\{(\alpha \mathbf{F})/(RT)\}(U_{rev}-\psi+\eta)} \tag{17}$$

multiplying by -1, re-arranging to give the current density and putting in logarithmic form gives

$$\ln|J_-| = \ln \overrightarrow{k} + \ln [H^+]_{DL} - \{(\alpha \mathbf{F})/(RT)\}U_{rev} + \{(\alpha \mathbf{F})/(RT)\}\psi - \{(\alpha \mathbf{F})/(RT)\}\eta - \ln S \tag{18}$$

[1] T. Erdey-Gruz and M. Volmer, *Z. physik. Chem.*, A 150 (1930) 203.

[2] These are merely changes of notation to agree with the original literature on this topic.

where S is the area of the electrode surface. Replacing U_{rev} by (RT/F) ln $[H^+]$ and $[H^+]_{DL}$ and ψ by the values given by equations (15) and (16) respectively gives

$$\ln |J_-| = \ln \vec{k} - \{(\mathrm{F}\psi_0)/(RT)\}(1 - \alpha) - (\alpha \mathrm{F}\eta)/(RT) - \ln S$$

$$\eta = \{(RT)/(\alpha\mathrm{F})\} \ln \vec{k} - (1/\alpha - 1)\psi_0 - \{(RT)/(\alpha\mathrm{F})\} \ln S - \{(RT)/(\alpha\mathrm{F})\} \ln |J_-| \quad (19)$$

which is of the type

$$\eta = \text{const.} + b \log |J_-| \quad (20)$$

where b is $(-\,2.303\,RT/\alpha\mathrm{F})$. If α is given the value 0.5, b will become 0.118 at 25° C.

The slow discharge theory leads to the correct value for the coefficient b for many metals (Table IV,4)[1] and describes correctly its behaviour as a function of temperature as was shown in a series of measurements with a mercury cathode (Table IV, 5). It can also interpret the results of Bowden and Rideal, and Brandes. In fact, if the discharge is slow when a constant current density is used there will first be an accumulation of the H^+ ions transported towards the cathode. Only when the electric tension reaches a certain value $(U_{rev} + \eta)$ will the discharge begin. If the double layer is considered as a capacitor of capacitance C, then as its charge — represented by the accumulation of H^+ ions — increases so will its electric tension, equal to that of the electrode, according to the equation

$$\Delta U = \Delta Q/C$$

TABLE IV,4

VALUES OF THE COEFFICIENT b FOR VARIOUS METALS

Metal	b	*Metal*	b	*Metal*	b
Ag	0.12	Fe	0.12	Pt	0.02–0.08
Al	0.12	Hg	0.12–0.15	(platinized)	
Au	0.08–0.12	In	0.25	Rh	0.14
Be	0.11	Mo	0.13	Sr	0.2
Bi	0.10	Nb	0.11	Ta	0.34
C	0.84	Ni	0.11	Tl	0.08
Cd	0.25	Pb	0.23–0.3	Zr	0.09
Cu	0.12–0.16	Pt smooth	0.19–0.3		

[1] J. O'M. Bockris, *Trans. Faraday Soc.*, 43 (1947) 417.

TABLE IV, 5

THE DEPENDENCE OF b ON TEMPERATURE, WITH A Hg CATHODE

T °K	*b (determined)*	*b (calculated with (20))*
273	0.108	0.107
309	0.123	0.122
345	0.141	0.136

Replacing ΔU by the change in overtension $\Delta\eta$, and ΔQ by the change in concentration of the $[H^+]$ ions (expressed in moles · **F**) gives

$$\Delta\eta = \mathbf{F}\Delta[H^+]/C$$

i.e. a linear, and not a logarithmic, relationship as observed experimentally.

The slow discharge theory has, however, the following weaknesses.

(a) If the cathodic discharge is the slow and difficult process this does not explain why such significant overtensions are encountered only with the H^+ ion and not with other cations. Similar overtensions would be expected for the discharge of other ions but are not in fact found.

(b) The value of 0.5 for the coefficient α, which gives b the value -0.118 is purely empirical and has no theoretical basis. The value of the coefficient should be exactly the same for all substances and should be constant in time; this is not found in practice (*cf.* Table IV, 4). The slow discharge theory predicts that b will have the value -0.118 and that J_0 will have a very small value (10^{-12} – 10^{-16} A/cm^2) which is in agreement with the supposed difficulty of charge transfer, but the theory does not explicitly predict any dependence of either η or b on the nature of the electrode material. The first two predictions are sufficiently well fulfilled particularly for the discharge of hydrogen on Hg, *i.e.* on a metal with a very low catalytic activity for the combination of atomic hydrogen. In fact, only on mercury can it be reasonably safely assumed that the mechanism of hydrogen discharge is that predicted by the slow discharge theory. For metals of intermediary catalytic activity (*i.e.* greater than that of Hg but less than that of Pt and Pd) the exact nature of the mechanism of the discharge of the H^+ ion is still uncertain.

(III) *The Electrochemical Theory*

The electrochemical theory, or better the theory of slow electrochemical processes, is due principally to Volmer and Horiuti. It has certain features in common with both the other two theories described above, but lies closer to the slow combination theory; to some extent it is an alternative to this. It assumes primarily, like the slow combination theory, that the discharge of the proton (3) is rapid. The electrode metal thus becomes covered with adsorbed atomic hydrogen and the successive reactions must be the desorption of the hydrogen and its liberation as H_2. The electrochemical theory assumes that with metals of high catalytic activity, like Pt and Pd, the most likely reaction to occur will be the recombination of the adsorbed H atoms to form molecular H_2 as suggested in the slow combination theory. However, with metals of lower catalytic activity there will be a possibility of another reaction occurring more easily than the recombination, and this other reaction will be rate determining. This reaction, which thus forms an alternative to the slow combination mechanism, is desorption following the electrochemical process of formation of an adsorbed ion-molecule of hydrogen and its electrochemical reduction to H_2 (the slow phase)

$$M(H) + H^+ \rightarrow M(H_2^+) \tag{21}$$

$$M(H_2^+) + e^- \rightarrow M + H_2 \tag{22}$$

i.e.

$$M(H) + H^+ + e^- \rightarrow M + H_2 \tag{23}$$

In common with the slow discharge theory, it supposes that the slow step in this case is the charge transfer and that the velocity of the corresponding reaction depends on the relative transfer coefficient α.

For this theory too, Stern's hypothesis of the structure of the electrochemical double layer may be used. It is assumed that the slow, rate determining reaction is (23), that the current density is proportional to the velocity of this reaction and that the concentration of atomic hydrogen adsorbed on the metal is stationary since as atomic hydrogen is formed by the discharge reaction (3) it is consumed by reaction (23) and by the reverse of (3). The treatment used for the slow discharge theory to obtain the concentration of $[H^+]$ ions in the double layer may be repeated. Then equation (17) may be rewritten with an additional factor to take account of the constant concentration of adsorbed atomic hydrogen, *i.e.*

$$I_- = -\vec{k}\,[H]_{ads}[H^+]_{DL}e^{-(\alpha F/RT)(U_{rev}-\psi+\eta)}$$

Developing this as before, and assuming that $\alpha = 0.5$ gives

$$J_- = -\overrightarrow{k'}\,[\mathrm{H}]e^{-(0.5\mathrm{F}/RT)\eta} \tag{24}$$

As mentioned above, the atomic hydrogen adsorbed on the electrode will be consumed in two reactions — the electrochemical desorption (23) and the reverse of reaction (3)

$$\mathrm{M(H)} \rightarrow \mathrm{H^+} + \mathrm{e^-} \tag{25}$$

The behaviour of the system may thus be reversed depending on which of the two reactions predominates. Two limiting cases may be distinguished.

(a) If reaction (25) is very slow relative to the direct reaction (3), the atomic hydrogen will be rapidly formed and only slowly used up. It can thus be presumed that the whole surface of the electrode will be covered by adsorbed atomic hydrogen; this implies that $[\mathrm{H}]_{ads}$ is constant, so that equation (24) will become equivalent to equation (20) with the same value for b of -0.118.

(b) If, however, reaction (25) is sufficiently rapid, not only will the electrode surface no longer be completely covered by adsorbed atomic hydrogen but the quantity of it used in the rate determining reaction (23) will no longer be equal to the total quantity of atomic hydrogen consumed. The quantity thus used up in the rate determining reaction (23) will be given by the ratio of the velocity $\overrightarrow{V}_{23}$ of reaction (23) to the sum of the velocities $(\overrightarrow{V}_{23} + \overrightarrow{V}_{25})$ of reactions (23) and (25). Equation (24) thus becomes

$$J_- = \{(\overrightarrow{k''}[\mathrm{H}]_{ads}\overrightarrow{V}_{23})/(\overrightarrow{V}_{23} + \overrightarrow{V}_{25})\}e^{-(\alpha\mathrm{F}/RT)\eta} \tag{26}$$

Clearly $\overrightarrow{V}_{23} = \overrightarrow{k}_{23}e^{-(\alpha\mathrm{F}/RT)\eta}$ and $\overrightarrow{V}_{25}$ for the partial anodic reaction will be given by

$$\overrightarrow{V}_{25} = \overrightarrow{k}_{25}e^{(\beta\mathrm{F}/RT)\eta}$$

where β is the transfer coefficient of reaction (25). Equation (26) now becomes

$$J_- = \overrightarrow{k''}[\mathrm{H}]_{ads}\frac{\overrightarrow{k}_{23}e^{-(\alpha\mathrm{F}\eta)/(RT)}}{\overrightarrow{k}_{23}e^{-(\alpha\mathrm{F}\eta)/(RT)} + \overrightarrow{k}_{25}e^{(\beta\mathrm{F}\eta)/(RT)}}\,e^{-(\alpha\mathrm{F}\eta)/(RT)} \tag{27}$$

In the other limiting case where $\overrightarrow{V}_{25} \gg \overrightarrow{V}_{23}$, the first term in the denominator of the right-hand side of equation (27) becomes negligible with respect to the second; giving α and β the common value 0.5 and considering $[H]_{ads}$ as constant, equation (27) will give finally

$$J_- = -\vec{k}\,\frac{e^{-(2\alpha F\eta)/(RT)}}{e^{(\alpha F\eta)/(RT)}} = -\vec{k}\,e^{-(3F\eta)/(2RT)}$$

i.e. once more:

$$\eta = \text{const.} + b \log |J_-| \tag{28}$$

where $b = -2.303 \cdot (2/3)(RT/F) = -0.039$ at 25° C. Clearly, in intermediary cases — where $\vec{V}_{25}$ lies between 0 and the limiting maximum predicted in case b — the factor b will have values such that $-0.039 \geqslant b \geqslant -0.118$. Moreover, b will in this case depend on η. The electrochemical theory thus predicts that b will lie between about -0.12 and -0.039 with values of J_0 in the range of magnitude of 10^{-6} and 10^{-10} A/cm^2.

The electrochemical theory subsumes some of the advantages of the other two. It is consistent with a correct value for b and it explains the effect of the electrode metal on the overtension; the velocity of reaction (23) will depend on how strongly the hydrogen atoms are adsorbed on the metal surface. However, it too does not explain the value of 0.5 for α and has to postulate it. The mechanism proposed by this theory is probably true for the discharge of hydrogen on metals having an intermediary catalytic activity.

(IV) *A comparison of the various theories*

These three theories are the most satisfactory, in that they interpret the greatest number of experimental facts, and are of the most general application, but many other theories have been proposed. These are now considered to be of less importance and will not be fully dealt with. One stated that the slow phase was the formation of bubbles of H_2 gas on the electrode surface[1]; another was an application of the theory of absolute reaction velocity to the process of the transfer of protons from the solution to the electrode[2] which is a development of the slow discharge theory; the theory of Knorr considered as rate determining the diffusion of discharged hydrogen towards the interior of the electrode metal. The possibility must not be overlooked that in individual cases other partic-

[1] D. A. McInnes and L. Adler, *J. Am. Chem. Soc.*, 41 (1919) 194.
[2] H. Eyring, S. Glastone and K. J. Laidler, *J. Chem. Phys.*, 7 (1939) 1053; *Trans. Electrochem. Soc.*, 76 (1939) 145.

TABLE IV,6

SUMMARIES OF VARIOUS THEORIES ON THE DISCHARGE OF HYDROGEN

Theory	*Presumed rate-determining reaction*	*b*	J_0 (A/cm²)	η *dependent on electrode metal?*	*Examples*
Slow combination (Tafel)	$M(H)+M(H) \rightarrow 2M+H_2$	− 0.029 or less	10^{-3}–10^{-4}	yes	Discharge c Pt and Pd electrodes
Slow discharge (Volmer, Erdey-Gruz, Frumkin)	$H^+ + M + e^- \rightarrow M(H)$	− 0.118	10^{-13}–10^{-16}	no	Discharge Hg electroc
Electrochemical (Volmer, Horiuti)	$M(H)+H^++e^- \rightarrow H_2+M$	between − 0.118 and − 0.039	10^{-6}–10^{-10}	yes	Discharge Pb and Cu electrodes

ular types of intermediary reactions may be concerned in the hydrogen discharge process. In the discharge of H^+ on arsenic it seems clear that arsine AsH_3 is first formed and then decomposes giving H_2 and regenerating metallic arsenic. In this case, the slow rate determining reaction could be either the formation or the decomposition of the AsH_3. Table IV,6 shows the principal characteristics of the three main theories.

Comparison of experimental data for the discharge of hydrogen with the values predicted by the various theories can show in general terms which of the electrode mechanisms is the most probable, but in fact it is still not possible to be certain which of them actually occurs in practice. There are various pieces of evidence to suggest that the discharge of hydrogen may occur in different ways on different electrode materials and under different experimental conditions; thus depending on the electrode material and the experimental conditions, one or other of the theories may best accord with the experimental observations.

7. Cathodic Processes; The Discharge of Metals

Another important type of cathode process is the discharge of a metallic cation with the deposition of the metal in the elementary crystalline state (electro-crystallization). It is characteristic of this type of cathodic process that in aqueous solution it is not the unique cathodic process possible but always occurs together with the discharge of H^+ ions of the water. The deposition of metal alone as the unique cathode process occurs only in the electrolysis of molten metallic salts or in certain cases of solutions in non-aqueous solvents. In this situation, a knowledge of the overtension of hydrogen and of other cations, under the various conditions of electrolysis in aqueous solutions, is of fundamental importance. This knowledge is necessary to decide whether an electrochemical process in an aqueous medium is possible or not, particularly for processes which occur at a reversible electric equilibrium tension less noble than that of hydrogen in a solution of the same acidity. In aqueous solutions of a metallic cation a certain quantity of H^+ ions are always present. The effective discharge tension U_{Me} of the cation is given by the equation

$$U_{Me} = U_{0\,Me} + \{(RT)/(z\mathbf{F})\} \ln [Me^{z+}] + \eta_{Me}$$

where the symbols have their usual significance and η_{Me} is an overtension for the cation and includes all its forms, and in particular diffusion and reaction overtensions. The effective discharge tension of the H^+ ion U_H is given by the analogous equation

$$U_H = (RT/\mathbf{F}) \ln [H^+] + \eta_H$$

Various cases may now be distinguished.

(a) The deposition tension for the cation U_{Me} may be more noble than the equilibrium electric tension of hydrogen under conditions of equivalent acidity. The relevant *J*–*U* curve of the electrode will be quite normal. In Fig. IV, 13 the process is illustrated by the two curves of current density–cathodic tension for the separation of the metal and of hydrogen. Starting from zero and polarizing the electrode cathodically, when the value *A* is reached (corresponding to the reversible equilibrium tension + any transfer overtension) the metal will begin to separate with a current efficiency of 1. As the current density increases, the cathodic polarization will rise as shown by the curve *AB*. At a density *B''* corresponding to the polarization *B'*, the value of the limiting kinetic, or diffusion, current is

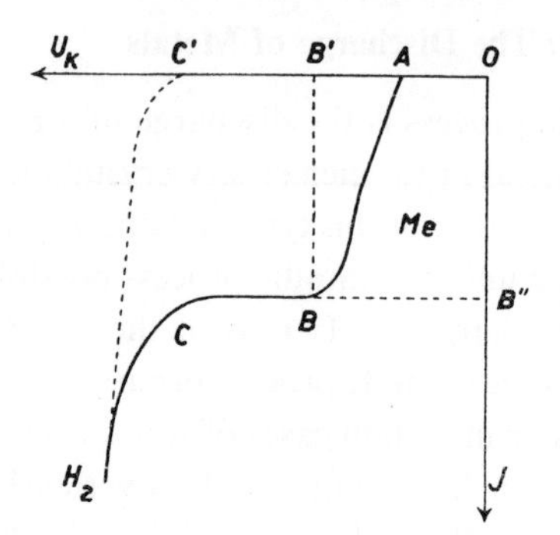

Fig. IV, 13.
Electric tension – current density diagram for discharge of cations

reached. This will remain constant until the polarization reaches the value C' which is the effective discharge tension of the H^+ ion. After point C a curve will be obtained for the discharge of the H^+ ion. The broken line beginning at point C' represents the discharge of the H^+ ion alone, in the absence of any more noble cation. Until the electric tension C' is reached, the current efficiency will be 1 for the discharge of metallic cations but after this point the quantity of electricity used for the discharge of these metallic cations will remain constant, whilst a certain quantity of electricity will be used for the discharge of hydrogen and this will increase with increasing polarization. Thus, the current efficiency which starts at 1 for the metallic cation gradually diminishes after the horizontal part of the curve has been reached.

(b) The deposition tension of the cation U_{Me} may be much less noble than the effective discharge tension of hydrogen U_H. A curve will now be obtained, perfectly analogous to that shown in Fig. IV, 13 but the first part *ABC* will refer to the discharge of hydrogen. After the limiting current has been reached for this ion, the discharge of the metallic cation will begin. It may or may not be possible to obtain a deposit of the metal on the cathode depending on the chemical properties of the metal itself. In general, all those metals which are less noble than hydrogen, spontaneously decompose water liberating gaseous hydrogen and forming the corresponding hydroxide. Only if the velocity of the discharge of the metallic ion is greater than the velocity of the reaction of the metal with water will it be possible to obtain a certain amount of the metal at the cathode.

In this second case, since the overtension of hydrogen depends on the chemical nature of the electrode, the simultaneous separation of metal with the discharge of the H^+ ion will begin sooner after the start of the faradic current, as the hydrogen overtension is greater on the original electrode metal. Once, however, the first layer of metal has been formed by discharge of cations present in the solution, the process will continue with the overtension characteristic of the separated metal which now functions as the electrode. This is illustrated in Fig. IV, 14 which shows the *J–U* curves for the separation of manganese from feebly acidified, concentrated solutions of manganese chloride with copper (curve I), platinum (curve II) and manganese (curve III) cathodes. The first portion of the curves is noticeably different for the three cathodes; but the three curves tend to run together into a single one after the copper and platinum electrodes have become covered with manganese by electrolysis. After this point the cathode metal is no longer the original but is manganese. All three curves correspond to that shown diagrammatically in the preceding figure; the first part which differs in each curve corresponds to the discharge of the H^+ ions on the different cathode metals; the second part, after a more or less pronounced plateau for the limiting current, corresponds to the discharge of the Mn^{2+} ions.

(c) The third case includes those cations whose effective discharge tension is close to that of the H^+ ion. Müller observed that the effective discharge tension of the H^+ ion on metals which were not much less noble than hydrogen was close to the effective discharge tensions of metallic cations in

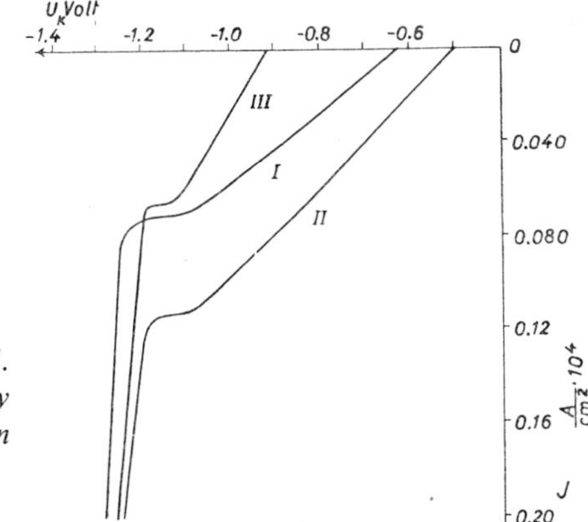

Fig. IV, 14.
Electric tension – current density diagram for Mn deposition

the same solution. This phenomenon controls the discharge of metals whose discharge tension is intermediary between that of hydrogen and that of much less noble metals. These metals lie in the tension series more or less between hydrogen and bivalent manganese (standard tensions corresponding to 0.000 and $-$ 1.1 V). If the effective discharge tension of the H^+ ion is more noble than that of the cation, the progress of the electrolysis will be that illustrated in Fig. IV, 15. Even at the relatively low current density A' corresponding to the polarization A, the discharge of the metallic cation will begin and proceed simultaneously with the discharge of the H^+ ion. The further progress of the electrolysis depends only on the relative slopes of the curves for the two individual processes.

The current is in part used for the discharge of the hydrogen and in part for the discharge of the metal. When the cathode is polarized at tension B, a current density B' is established on the electrode by the discharge of the metallic cations and one of B'' by the discharge of the hydrogen. The ratio of the current utilized for the discharge of the metallic cation, to that utilized for the discharge of the hydrogen will be as the ordinate a to the ordinate b, so that the current efficiency for the metallic cation will be

$$R_{cur.} = a/(a + b)$$

If, however, the effective discharge tension of the metallic ion is more noble than that of the H^+ ion, the progress of the reaction will be substantially like that shown in Fig. IV, 15 with the difference that curve I will refer to the discharge of the metallic cation whereas curve II will refer to the discharge of the H^+ ion. In this case, the current efficiency for the cation will be given by

$$R_{cur.} = b/(a + b)$$

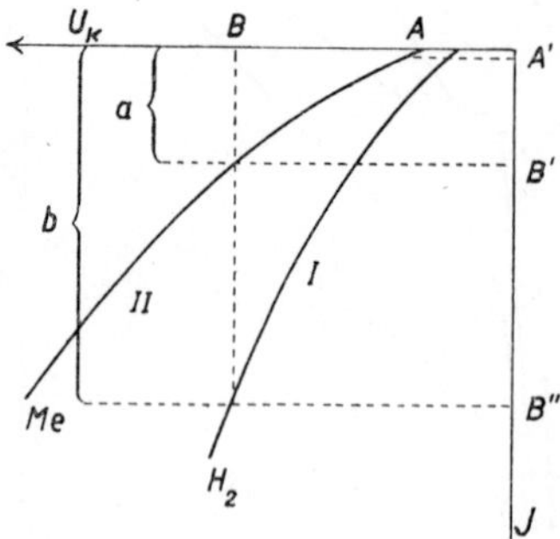

Fig. IV, 15. Electric tension – current density diagram for metals near hydrogen

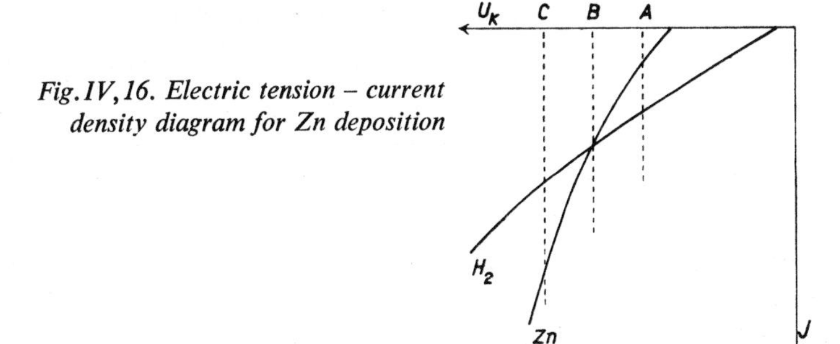

Fig. IV, 16. Electric tension – current density diagram for Zn deposition

A special case occurs when the curve for the discharge process whose tension is initially more noble has a lesser slope so that it occurs with an overall overtension which increases more rapidly, with increasing J, than that of the other process. A typical example is that of the electrolysis of zinc in a neutral aqueous solution of a simple salt. The two curves are shown diagrammatically in Fig. IV, 16. At a cathodic polarization A, corresponding to a low current density, there will be mainly an evolution of hydrogen; at the tension B at which the curves intersect, the current efficiency will be 0.5 for hydrogen and 0.5 for zinc; at the polarization C there will be mainly a deposition of zinc. As the current density is increased still further, the yield for the zinc will approach increasingly to 1 although it has a standard equilibrium electric tension (about − 0.76 V) which is considerably more negative than that of hydrogen in neutral solution (− 0.415 V).

All that has been said for the simultaneous discharge of metallic cations and H^+ ions is equally valid for cases where more than one metallic species is present in the solution. The first process to occur is invariably that requiring the lower cathodic polarization and from the progress of the current density–electric tension curves it is possible to judge whether under certain conditions one metal will separate exclusively from the solution or whether there will be a simultaneous discharge of two or more different cations. Whether this latter will occur depends not only on the values of the respective equilibrium electric tensions under the experimental conditions used but also, and sometimes predominantly, on the size of the diffusion and reaction overtensions for each of the metallic ions and on any depolarization effects for the simultaneous discharge of more than one cation (see below p. 265 *et seq.*).

8. Cathodic Processes; Overtensions of Metals

In the discharge of certain metallic cations such as those of nickel, iron and cobalt, overtensions are also important. Analogously to the discharge of hydrogen, these overtensions arise from a slow process forming part of the overall discharge reaction by which the cations pass into metallic atoms arranged in a crystal lattice. This overall process contains various steps: dehydration of the ions (or more generally any chemical reaction preceding the discharge such as the dissociation of complex ions), the discharge of the ions, the arrangement of the neutral atoms in the crystal lattice of the metal. As far as the effect of chemical reactions preceding the discharge is concerned it should be noted that at least one dissociation or dehydration reaction very often occurs before the discharge of a metallic cation; recent studies have led to the hypothesis[1] that the cathodic deposition of metals occurs preferentially through the discharge of complexes, (a term which here includes hydrated ions) with coordination numbers less than those of normal stable ions and sometimes even with overall charges which are lower. In the case of silver cyanide which has already been discussed (*cf.* p. 229), where the ion predominantly present in solution is $[Ag(CN)_3]^{2-}$, reduction occurs through the $[Ag(CN)_2]^-$ ion and the neutral species $[Ag(CN)]$. This hypothesis is supported by the fact that calculations of the concentration of free metallic ions in a solution of complex ions (in the narrow sense) on the basis of the actual reversible electric tension of the electrode and the standard reversible tension of the same electrode in solutions of a simple salt, will often give values corresponding to only a few free metallic ions per litre. If the electric tensions were truly defined by the activity of the free ions, this would imply that the electrode tension was unstable and variable from moment to moment, depending upon the temporarily variable distribution of these metallic cations in the liquid cathode film. Since, however, the electric tension is always well defined and constant, it must be concluded that it is not determined by the free metallic cations but by ionic species which are already complexed, and that the primary electrochemical process must occur directly through these complex ions which determine the reversible equilibrium electric tension of the electrode. As a consequence it follows that the discharge of a cation is preceded by a chemical reaction: either dehydration or the dissociation of the complex.

[1] E. H. Lyons, *J. Electrochem. Soc.*, 101 (1954) 363.

Two cases may arise. Firstly, the reaction forming the directly reducible ion may be rapid and at every instant its concentration may be in equilibrium with the concentration of hydrated ions. The electrode process will then occur as described for the normal discharge of a reducible substance with no overtension other than that of diffusion. If the complex ion predominantly present is not that which determines the reversible electric tension in solutions of simple salts but is nevertheless in equilibrium with it, there will be an actual concentration of electrochemically active, and directly reducible ions, less than the analytical net concentration. In this case it would be possible to speak of a concentration polarization, but the actual overtension (*i.e.* that calculated on the basis of the reversible electric tension corresponding to the equilibrium concentration of the directly reducible ions) which must be applied for the cathodic deposition of metal and its variation as a function of the logarithm of the current density are of the diffusion overtension type with a possible addition of a transfer overtension. Secondly, the reaction forming the reducible free ion by the dissociation of the hydrated or complex ion may be slow. Hence, once consumed during the electrolysis, the amount of reducible ion which will remain in solution in equilibrium with the complexed form must be renewed by the slow reaction which will hence lead to a constantly lower concentration than that of equilibrium. The slow reaction thus becomes rate determining for the renewal of electrochemically active substance at the electrode and causes a reaction overtension. An example of this is given by the deposition of cadmium from cyanide solutions. The ion predominantly present in solution is $[Cd(CN)_4]^{2-}$ whereas the reducible ion has a lower coordination number. The reaction determining the overall kinetics is

$$[Cd(CN)_4]^{2-} \rightarrow [Cd(CN)_3]^- + CN^-$$

which causes a reaction overtension. The existence of a slow dissociation reaction in the overall process of the discharge of a metallic ion may be very clearly demonstrated by measurements using a technique which has recently been developed: transitometry or chronopotentiometry. If electrolysis occurs with a constant current density in a nonagitated solution the renewal of the electrochemically active substance at the electrode surface will come about principally by diffusion. In the first moments of electrolysis the faradic current will consume the active electrochemical substance present in the immediate vicinity of the electrode and hence the concentration of this substance will fall below its initial value (c_0: the concentration existing in the bulk of the solution) to a value which after a certain period from the beginning of the electrolysis (the transition time) will tend towards zero. From this moment on, the electrochemically active substance will arrive at the electrode surface exclusively by diffusion. The electric tension with a mercury or platinum electrode, which at zero time tended towards a high positive value if only the metallic ions Me^{z+} (the oxidized form) were present on it, will assume values close to the equilibrium electric tension of the Me/Me^{z+} electrode as soon as electrolysis begins, *i.e.* as soon as an appreciable quantity of metal is deposited on the electrode surface. Once the transition time has been reached the electric tension of the electrode will tend towards a high negative value as the concentration of the oxidized form at the surface of the electrode tends towards zero (*cf.* Fig. IV, 17). There will thus be an abrupt change in the electrode tension which may be used to recognize the transition time. The theory of electrode processes of this type was developed by Delahay, Gierst, Juliard, Lingane and others who deduced that for an electrode process which occurs without kinetic complications the following equation holds

$$
\begin{aligned}
I_{-}\tau^{\frac{1}{2}} &= 2\, z\mathrm{F}D^{\frac{1}{2}}c_0/\sqrt{\pi} \\
&= kc_0
\end{aligned}
$$

Here D is the diffusion coefficient. This implies that for equal initial concentrations c_0 the product $I_{-}\tau^{\frac{1}{2}}$ is constant and independent of I_{-}, and is thus also independent of τ. In other words, this product does not vary even if τ does or if the disappearance of the electrochemically active substance from the electrode surface and its neighbourhood occurs rapidly

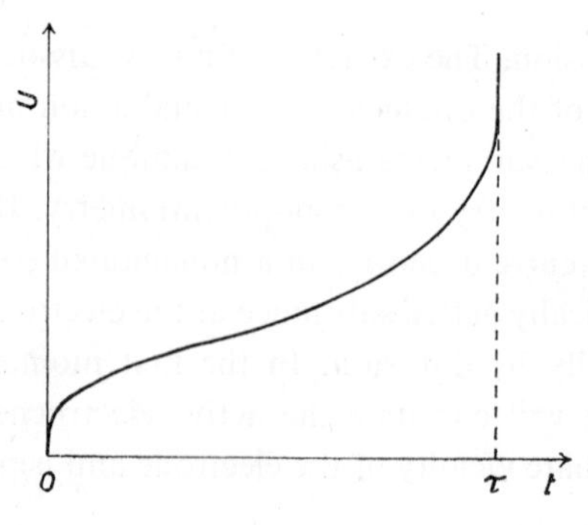

Fig. IV, 17.
Typical diagram of transitometry

or slowly, because the reducible substance present in solution is reduced as such or its reduction is preceded *only* by rapid reactions, *e.g.* rapid dehydration; c_0 indicates in this case the total analytical concentration of the metallic cation. Consider now the discharge of the cation preceded by a slow dissociation reaction of a hydrated or complexed ion. At low current intensity and hence at long transition time the reaction even although it is slow will still have time to occur and hence will not cause any appreciable delay. $I_{-}\tau^{\frac{1}{2}}$ will not differ significantly from the value which it would have if the dissociation reaction did not exist or were very rapid. However, with a high current density and hence a much lower transition time, the dissociation reaction will not have time to renew all the dissociated metallic ion consumed in the diffusion layer. This means that the effective quantity of the substance reduced is less and requires a smaller quantity of charge or that, with equal current intensities, requires a lesser reduction time. The value of τ will fall and so will that of the product $I_{-}\tau^{\frac{1}{2}}$. Whenever the discharge of a metallic cation is preceded by a slow chemical reaction, the value of $I_{-}\tau^{\frac{1}{2}}$ as determined by transitometric measurements, will not be constant throughout each experiment but will fall as the current density used in the electrolysis (constant for the whole duration of each individual experiment) is increased. The slope of the curve which gives the value of the characteristic product $I_{-}\tau^{\frac{1}{2}}$ as a function of I, allows the value of the velocity constant $\vec{k}$ of the rate determining reaction preceding the discharge, to be determined. Assuming that this rate determining reaction is of the first order, Gierst and Juliard deduced the equation

$$(I_{-}\tau^{\frac{1}{2}})_{el} = (I_{-}\tau^{\frac{1}{2}})_{I\to 0} - \vec{k}/(D^{\frac{1}{2}}\tau^{\frac{1}{2}})$$

where $(I_{-}\tau^{\frac{1}{2}})_{el}$ is the value of the characteristic product measured under the actual experimental conditions of electrolysis in the presence of kinetic effects and $(I_{-}\tau^{\frac{1}{2}})_{I\to 0}$ represents the value of $(I_{-}\tau^{\frac{1}{2}})$ which would have been obtained in their absence, or where in practice the current density is very low so that kinetic effects become negligible.

A third possibility exists that the product $I_{-}\tau^{\frac{1}{2}}$ for the reduction of a complex metallic ion may remain constant with changes in I_{-} even although a complex is involved which is known to be kinetically inert. This implies that it does not undergo rapid reactions and its dissociation, if it occurs, is slow. In this case, the only way of interpreting the experimental data is to assume that this complex does not in fact dissociate to

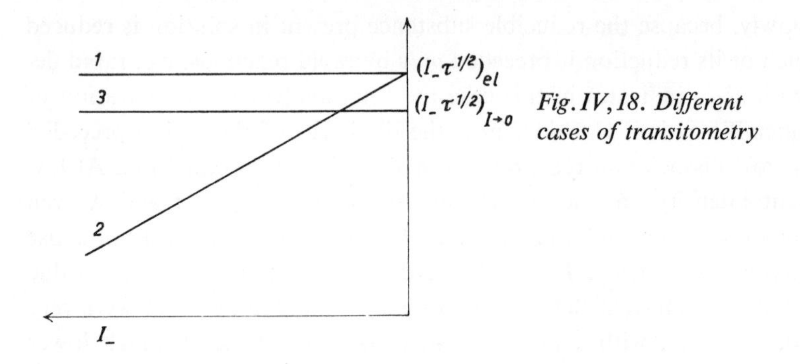

Fig. IV, 18. Different cases of transitometry

give a free metallic ion but is reduced as such. This case is not often encountered but does demonstrate that electrochemically active substances in the deposition of a metal may not necessarily be the free metallic ion but a complex ion (*e.g.* the deposition of silver from cyanide solutions dealt with on p. 229).

Many recent studies by various authors would seem to indicate that the true electrode process in these cases is the capture of electrons by the complex containing the metallic ion, without its destruction, thus forming the corresponding zerovalent metallic complex (these complexes contain metals in a zero-oxidation state and are not rare); $[Ni(CN)_4]^{4-}$ and $[Co(Dip)_3]$[1] are formed as the products of the electrolytic reduction of the complex ions $[Ni(CN)_4]^{2-}$ and $[Co(Dip)_3]^{3+}$ respectively; these zerovalent metallic complexes later decompose depositing the metal and liberating the complexing groups. Fig. IV, 18 illustrates the three cases so far discussed. In curve 1 the product $I_-\tau^{\frac{1}{2}}$ remains constant, which implies that the reduction is probably preceded by the dissociation of a complex but that this dissociation is rapid. In curve 2, which is typical of the deposition of Cd from cyanide solutions, the product $I_-\tau^{\frac{1}{2}}$ falls with increasing absolute values of the current intensity I_-; this implies that the deposition of metallic Cd from a solution containing the $[Cd(CN)_4]^{2-}$ ion is preceded by a slow reaction whose rate determining stage is

$$[Cd(CN)_4]^{2-} \rightarrow [Cd(CN)_3]^- + CN^-$$

Curve 3 which is typical for some copper complexes, shows that the reduction of the complex $[Cu(H_2NCH_2CH_2NH_2)_3]^{2+}$ occurs with $I_-\tau^{\frac{1}{2}}$ constant. As it is known that similar copper complexes are inert it cannot

[1] Dip = dipyridyl.

be assumed that a rapid dissociation occurs before the discharge reaction and since the constancy of $I\_\tau^{\frac{1}{2}}$ shows that a slow dissociation does not occur it must be assumed that this complex undergoes reduction directly without a previous dissociation. However, another interpretation of the situation is possible. It could be assumed that the primary electrode reaction is the discharge of the H^+ ion and that the deposition of the metal is due to a secondary reaction between the complex anion containing the metal and the atomic hydrogen produced by the primary cathodic reaction. This interpretation is supported by the observation that often during electrolysis of complex salts, hydrogen is evolved simultaneously with the deposition of metal at the cathode. If this secondary reaction following the electrochemical discharge is slow, there will be eventually an accumulation of atomic hydrogen with an increase in the cathode overtension. According to this interpretation also, the overtension is of the reaction type.

The second cause of overtension which must be considered in the cathodic deposition of metals, lies in the reaction of electron transfer. Analogously to the argument developed on p. 229 *et seq.* for the overtension of hydrogen, it can be assumed that a particularly high activation energy is required for the discharge of metallic cations and will determine the overtension. Particularly high transfer overtensions are present with metals of the transition groups and of sub-groups IV, V and VI of the periodic system. These give especially stable complexes (the so-called internal complexes) with covalent bonds between the central ion and the ligands. The ions of the other metals have the electronic structure of the rare gases and, giving rise to much less stable complexes, do not show particularly high overtensions. This situation, however, should rigorously be interpreted more as a difference of reaction energies than as a particularly high activation energy.

A third cause of overtensions of metals may be attributed to a possible delay in the arrangement of the metallic atoms, formed by the discharge, into the crystal lattice of the solid metal. It seems clear, at the present day, that the primary product of the reduction of a metallic ion is a relatively free or wandering atom on the surface of the underlying solid metal which forms the mass of the electrode. The insertion of this wandering atom in a suitable position in the ordered crystal lattice of the solid may be a slow process which would cause an overtension called the *crystallization overtension.* During the discharge of silver from perchloric solutions containing Ag^+ ions, Gerischer was able to show that the formation of the lattice was effectively a very slow reaction; its velocity was about 2,000-fold less than that of the actual electrochemical reaction, *i.e.* the charge transfer. The study of the ordered deposition of metallic atoms discharged in an electrolytic process, called electrocrystallization, is not only important for the overtensions of the deposition of metals but has an intrinsic interest. It has been studied by many workers and in particular by H. Fischer. According to this author, the deposition of the metal occurs on the surface of the electrode in what he calls a 'growth layer' in which the crystals of the metal grow by the addition of discharged metallic atoms. As a prime consequence, not all the faces nor all the sites of a metallic crystal are equally susceptible to growth. The deposition of the metal, *i.e.* the growth of the crystals will occur preferentially on certain faces or on a limited number of active sites. Fischer distinguishes experimentally four types of crystalline form or grain which may arise in the electrolytic deposition of a metal.

1. An isolated type orientated in the field. These are relatively gross crystals practically isolated from each other and growing in the direction of the lines of force of the electrical field about the cathode.

2. A reproductive type orientated on the base. These are also fairly large but compact crystals which reproduce and continue the crystalline structure of the metallic substratum of the electrode.

3. A texture structure type orientated in the field. These are minute and not clearly apparent crystals deposited in a compact form and aligned along the lines of force of the electrical field.

4. A disperse type. These are microcrystals with no apparent orientation of the crystallites.

There may be a continuous passage from type 1 through 2 and 3 to 4. In this order of succession they pass from the formation of a few large crystals (type 1), which have little tendency to form new crystalline nuclei and which themselves grow fairly easily, to the formation of numerous minute crystals (type 4), *i.e.* to the formation of numerous new crystalline nuclei rather than to the growth of pre-existing ones. As therefore the type of electrocrystallization changes from 1 through to 4 there is increasing inhibition, *i.e.* increasing difficulty of discharging the metallic ion and allowing it to reach a point in the growth layer suitable for the growth of the crystal. Thus, for example if the discharge of the cation is easy and if the metallic atom as soon as it is formed finds a site or a crystalline face on which growth readily occurs (case 1), or if the atom is reasonably free to migrate within the growth layer until it finds the most favourable site to insert itself into the crystal lattice (case 2), *i.e.* if the process requires only a small net overtension, the formation of a few large crystals will be favoured rather than the formation of many new crystalline nuclei which would require a considerable activation energy. If, however, the cation discharge process requires a high overtension and if the movement of the metallic atoms as soon as they are deposited within the growth layer is difficult (*e.g.* due to the presence of foreign substances adsorbed on the electrode surface which could block parts of the active centres) the activation energy for the creation of new crystalline nuclei may be less than the activation energy required to arrange the metallic atom on a pre-existing crystal. Many crystal nuclei will then be formed and will naturally be smaller and without an apparent preferred orientation (type 4). The type of electrocrystalline deposit which a metal will give depends to some extent on the conditions in which the electrolysis is carried out. The type of crystalline deposit may be changed from type 1 towards type 4 by adding complexing substances to the solution of the metallic salt or by adding foreign substances which will be adsorbed on the electrode, *i.e.* inhibitors, or even by lowering the temperature. The nature of the metals is a primary factor. Metals exist which will deposit readily and will preferentially crystallize in type 1 or perhaps type 2, *e.g.* Pb, Sn, Cd; there are 'inert' metals which require high overtensions for their discharge and usually deposit as types 3 or 4 *e.g.* Fe, Ni, Mn, Cr, and there are metals of intermediate characteristics *e.g.* Zn, Ag, Au. These will usually crystallize in types 2 and 3 but may change to type 4 if they are deposited from solutions in which they are highly complexed. Electrocrystallization is closely linked with the surface properties of the metals deposited electrically and its study is thus of technical importance in electrometallurgy (*cf.* Chapter VIII).

Piontelli has proposed a similar classification dividing metals into active and inert on the basis of their electrochemical behaviour in deposition and dissolution processes[1].

[1] For further details see: W. Lorenz, *Z. physik. Chem.*, 202 (1953) 275; *Z. Naturforsch.*, 9A (1954) 716; H. Fischer, *Elektrolytische Abscheidung und Elektrokristallisation von Metallen*, Springer Verlag, Berlin, 1954; and the papers of R. Piontelli and his colleagues.

9. Anodic Processes; the Discharge of Anions

In cathodic processes the discharge products are normally stable substances — metals or hydrogen — and the corresponding electrochemical processes are simple. Anodic processes of the discharge of anions are generally complex however, because the primary discharge products are usually unstable molecules or radicals. There are a few exceptions such as the halide ions Cl^-, Br^- and I^- but even the discharge of the OH^- ion takes place with a fairly complex mechanism. All the other anions are not normally discharged except in a peculiar way under exceptional conditions, and even then often not completely. This is because the discharge tension of the OH^- ion from aqueous solution is one of the lowest and hence this will be one of the first to be discharged when the electrode is anodically polarized. Only the Br^- and I^- ions have, under certain conditions, definitely lower discharge tensions than the OH^- ion. The process by which this is discharged is in itself less simple than that for the H^+ ion as can be seen even from the equation for the overall process:

$$4\,OH^- \rightarrow 2\,H_2O + O_2 + 4\,e^- \tag{1}$$

For this reaction, which should be that on which the oxygen electrode is based, it is not normally possible to measure any definite constant electric tension even with a platinized platinum electrode. In a cell composed of an oxygen and a hydrogen electrode it is apparent that the chemical reaction which gives rise to the electrode tension is the formation of H^+ and OH^- ions which then react to form undissociated water. Nernst and Wartemberg calculated thermodynamically from the equilibrium equation

$$2\,H_2 + O_2 \rightleftharpoons 2\,H_2O \tag{2}$$

that this cell would have an electric tension of 1.237 V at room temperature; from this the standard electric tension of an oxygen electrode at unit OH^- ion activity must be 0.401 V. Hence, the electric tension of an oxygen electrode at 25° C is given by

$$U = 0.401 + (1/4)0.0591\ \log\,(p_{O_2}/[OH^-]^4) \tag{3}$$

Since $[H^+]\,[OH^-] = 10^{-14}$,

$$U = 1.228 + 0.0296\ \log\,([H^+]^2\sqrt{p_{O_2}}) \tag{4}$$

Equation (4) gives the electric tension of an oxygen electrode as a function of the pH of the solution and it is assumed that this calculated value is the

TABLE IV,7

OVERTENSIONS OF OXYGEN

Overtensions (V) at given current densities J (A/cm²)

Material	10^{-5}	10^{-4}	10^{-3}	10^{-2}	10^{-1}	1
Ag*	0.41	0.45	0.60	0.71	0.94	1.06
Au*	0.73	0.93	0.96	1.05	1.53	1.63
C graphite	0.31	0.37	0.50	0.96	1.12	2.20
Cd**		0.67	0.80	0.96	1.21	1.21
Co**	0.27	0.32	0.39	0.46	0.54	0.61
Cr**	0.32	0.49	0.58	0.66	0.73	0.77
Fe**	0.35	0.37	0.41	0.48	0.56	0.63
Ni**	0.32	0.45	0.60	0.75	0.91	1.04
Pb**			0.80	0.97	1.02	1.04
Pd***	0.39	0.48	0.89	1.01	1.12	1.28
Pt***	0.52	0.80	1.11	1.32	1.50	1.55
Pt platinized	0.21	0.32	0.46	0.66	0.89	1.14

* electrolytically deposited, **precipitated powder, *** massive

reversible equilibrium electric tension of the OH^- ion. The difference between the discharge tension of the OH^- ion and this reversible equilibrium tension gives the overtension of oxygen.

Marked overtensions are observed in the discharge of the OH^- ion but they do not show the clear regularities encountered with discharge overtensions of the H^+ ion. Table IV,7 contains measurements by Hickling and Hill[1] which show that the overtension is so dependent on the actual current density that it is not possible to arrange anode materials in an order of overtensions. In general, Co, Fe and Cu have low overtensions and Pt and Au have high overtensions. It is clear that as in the discharge of H^+ ions, the discharge overtensions of the OH^- ions are to a large extent dependent on the nature and surface state of the electrode material. The dependence on the current density is difficult to observe and reproduce. Very careful measurements with specially prepared electrodes and with completely purified solutions subjected to prolonged preliminary electrolysis between auxiliary electrodes, have given the theoretical elec-

[1] A. Hickling and S. Hill, *Discussions Faraday Soc.*, No. 1 (1947) 236.

tric tension of the oxygen electrode, and a U–log J diagram which follows Tafel's equation

$$\eta = a + b \log J \tag{5}$$

where $b = 0.12$. For some materials such as Au, Pd and C, there are abrupt changes of the order of magnitude of tenths of a volt which can sometimes be related to changes in the surface characteristics of the anode material. This may sometimes undergo considerable attack visible even to the naked eye. An increase in the temperature also tends in this case to alter the overtension. There are not yet sufficient data available on the effect of pressure and electrolyte composition to determine their influence on overtensions.

A very particular characteristic of oxygen overtension is its dependence on the time of electrolysis as shown in Fig. IV, 19. It can be seen that with some metals (Ni, Fe, platinized Pt etc.) the overtension continues to increase, even although very slowly, with increasing electrolysis time and tends towards a limiting value which is difficult to determine. For other substances (Cu, Ag and Pb at low current density) sudden changes are present which are also of the order of tenths of a volt. At high current densities Pb shows a slight fall in the overtension.

In the interpretation of oxygen overtensions it should be noted that the overall reaction (1) probably occurs in several steps each of which may be slow and thus the origin of the overtension. Very probably the primary reaction in the discharge process of the OH^- ion is

$$OH^- \rightarrow OH + e^- \tag{6}$$

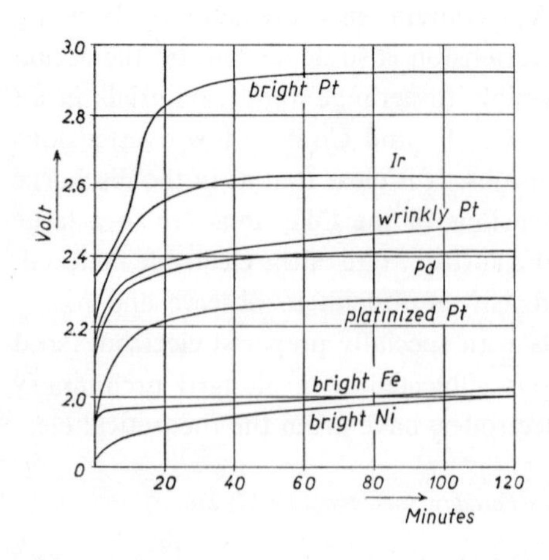

Fig. IV, 19. Oxygen overtension

forming adsorbed OH radicals which are unstable. There are at least two possible ways in which these can react further; firstly, by combining amongst themselves to form hydrogen peroxide

$$2\,OH \rightarrow H_2O_2 \tag{7a}$$

and secondly, by decomposing to form oxygen

$$4\,OH \rightarrow 2\,H_2O + O_2 \tag{7b}$$

The position is further complicated by the possibility that the primary oxidation product and the other intermediates of the overall electrode process may interact with the electrode metal to form true chemical compounds — more or less stable oxides and peroxides — or may be adsorbed on to the metal. There is experimental evidence of this; the dependence of the overtension of oxygen on the time of electrolysis shown in Fig. IV, 19, may be interpreted as due to a formation of more or less stable, pure or mixed oxides whose decomposition evolves gaseous oxygen and determines the electrode tension. It has been observed that an electrode of platinum covered with the oxide PtO_3 assumes an electric tension of 1.5 V when immersed in 2 *N* sulphuric acid; this value is close to that at which oxygen begins to be evolved when the same electrode acts as an anode in electrolysis with low current densities. This suggests that it is the oxide PtO_3 itself which determines the electrode tension. The overtension of oxygen on a platinized platinum electrode or on other electrodes of more or less readily oxidized materials, may be interpreted as due to the formation of such oxides; more or less stable oxides probably exist corresponding to higher oxidation states such as PtO_2, NiO_2, CoO_2, Cu_2O_3, CuO_3, FeO_3 etc. When these unstable oxides are formed on the electrode, not in a pure state but mixed with other oxides of the electrode material, the electric tension can no longer be defined. The problem of oxygen overtensions thus remains completely open, but this is not surprising since even the relatively simple problem of the hydrogen discharge mechanism has not yet been completely clarified. Another anodic discharge process of anions, of some importance, is that with halide ions. Here, too, the mechanism of the process is not well known, above all because of the difficulty of carrying out good experimental measurements to follow the exact overtensions accompanying the discharge. It should also be noted that there is often the possibility of a rather high residual current passing through halide solutions even at applied electric tensions much less than the reversible equilibrium value. This is due to the solubility in water of halogens in the elementary state; this solubility increases in the presence of halide ions particularly for bromine and iodine. The *J–U* curve thus shows a rather abnormal form.

It is also difficult for the surface characteristics of the electrodes to remain constant as electrolysis proceeds; after a certain time there will be an increase in the electrode polarization even though the current is kept constant. This new state is reached more readily with more positive equilibrium electric tensions, greater current densities and lower concentrations of halide and H^+ ions. All these conditions facilitate the discharge of OH^- ions with the consequent possibility of forming oxides on the electrode surface. It is thus necessary to distinguish between the relatively low, true overtensions of the halogens found at low current densities and electrode polarizations and the sudden increases in electrode polarization which are almost certainly connected with the discharge of OH^- ions. Changes in overtension with current density can sometimes be described, for these anions too, using Tafel's equation. Too little is known about other anodic partial discharge processes of anions to give a concise description.

10. Anodic Processes; Anodic Behaviour of Metals; Anodic Dissolution

A different type of anodic process is the oxidation of the metal forming the solid electrode, *i.e.* its anodic dissolution to form the corresponding cations in solution. Whether an anion is discharged or a cation formed at a particular electrode depends on the respective electric tensions necessary for the two processes (as necessary electric tension the total electrode tension is understood *i.e.* the equilibrium electric tension U_{rev} plus any overtension η). That anodic process will take place which needs the lower effective electrode tension, but if the two electrode tensions are very close, both processes may occcur simultaneously in such proportions that the total current efficiency is unity. Three cases may occur in practice.

1. The metal may have an equilibrium electric tension more positive than that corresponding to the discharge of the anions present in the solution. No metal will pass into solution and the electrochemical process will consist exclusively of the discharge of the anions with a current efficiency of 1. Such electrodes are called *unattackable* or *inert*. A platinum electrode

in an iodine solution is inert, since to make platinum go into solution as ions requires a markedly higher anodic polarization than is needed to discharge I^- ions. The most common inert electrodes are Pt, Au, arc carbon or graphite.

2. The metal, although having an equilibrium electric tension less positive than the discharge tensions of the anions present, nevertheless may have a sufficiently high overtension to reach the discharge tension of one of the anions present. The metal will pass into solution with a current efficiency less than 1, or may not be attacked. This overtension may become apparent either immediately the current starts to flow or else after a certain time depending upon the particular conditions of the experiment.

3. The effective discharge tensions of the anions present in solution may be more positive than the equilibrium electric tension of the metal and remain more positive at its effective dissolution tension even when this is accompanied by overtensions. The anodic process will now consist exclusively of the dissolution of the electrode metal, which passes into solution as its ions. Here too the overall anodic process will consist of a number of partial steps; any of these may be slow and cause overtensions. In particular, all the processes involved in the discharge of metallic cations may take place, in the reverse order and sense: destruction of the crystal lattice of the metal, charge transfer, hydration or complexing of the metallic ions formed and their removal from the electrode by convection, migration or diffusion. In this case, the hydration or complexing reactions to which the formed metallic ions are subject are rather less important in the kinetics of the overall process, since they are normally rapid, but the destruction of the crystal lattice of the metal becomes very important. The anodic dissolution of many metals, and in particular the current efficiency of the relative overtensions, often varies noticeably with changes in the surface state of the metal and this in itself is an indication of the importance of what happens at the solid surface which is attacked, before the charge transfer. The work of Gerischer, Fischer and Piontelli has led to a more precise and clear understanding of what happens in this step.

It is clear that in the anodic dissolution of a metal, the destruction of the crystal lattice of the solid and the charge transfer form two distinct steps in the overall process. It is in fact possible to demonstrate their existence separately by, for example, the analysis of the changes in overtension in the first few instants of electrolysis. The destruction of the lattice is analogous to a surface fusion of the metal and creates on the surface a certain number of metallic atoms wandering relatively freely over it. The amount of such atoms is often considerable; according to Gerischer, in the dissolution of Ag in perchloric acid solution, the concentration of wandering atoms on the electrode surface at room temperature is of the order of 1% of a monolayer. It is these wandering atoms which undergo charge transfer and are converted into ions. This is demonstrated, for example, by the fact that the exchange current density J_0 for the anodic dissolution of (solid) silver is very close to that for (liquid) mercury. This implies that the charge transfer comes about with the same facility in the two cases, so that atoms arranged in the lattice of the metallic silver are probably not involved; for the destruction of the lattice would slow the process and diminish the number of ions. The wandering silver atoms present in a relatively free state on the surface must therefore be involved; these atoms are in an almost liquid state equivalent to that of the atoms in liquid mercury. The destruction of the crystal lattice must thus necessarily precede the charge transfer. It may require a considerable activation energy and may thus be a cause of overtensions.

Once the charge transfer has occurred, *i.e.* once the cations are formed by the dissolution of the metal, the process will continue in different ways depending on whether the metal can form only one, or more, types of ions of different valencies. Three cases may be distinguished.

(a) In aqueous solutions the metal may form ions of only one valency (*e.g.* Ag, Cd, etc.). When an electrode of such a metal is anodically polarized, it will pass into solution with a current efficiency of 1 from the beginning of the electrolysis; the yield will remain unity provided sufficiently high overtensions do not arise during the electrolysis to make the effective dissolution tension more negative than that required for the discharge of the anions present in the solution. The J–U curve will be completely analogous to that for cathodic deposition. A diffusion overtension may occur due to the fact that not all the cations formed by the anodic dissolution of the metal succeed in migrating and diffusing from the layer about the anode towards the cathode at the same rate at which they are formed. Thus, the anodic concentration will rise and the electrode tension will correspondingly become more positive. A limiting current may also occur in the anodic dissolution of metals. As the metal passes into solution a certain number of anions will migrate at the same time towards the anode under the action of the electrical field and of the osmotic pressure difference, so as to compensate for the increased concentration of cations. When the solubility product of the electrolyte is reached, it will begin to crystallize and thus a certain number of anions will be removed from the solution. As soon as the number of anions removed by crystallization from the solution in the anodic region equals the number of anions which migrate towards the anodic region, a limiting current will be reached above which another anodic process will begin.

Metals requiring only a small overtension for their deposition, will generally go into solution with a correspondingly small overtension. However, metals which require a high overtension for their deposition will also require a similar one for the anodic dissolution; and the *J–U* curve will be flatter. Fig. IV, 20, for example, illustrates the case of nickel. The general effect of temperature on the overtension is also apparent.

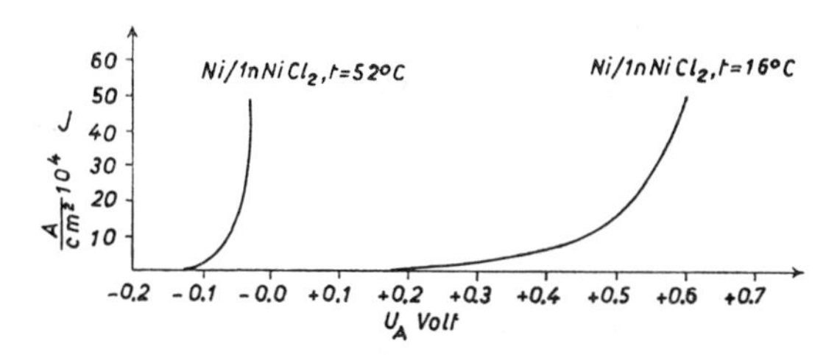

Fig. IV, 20. Anodic dissolution of nickel

(b) The equilibrium electric tension for the dissolution of the metal may be defined as in case (a), but the metal may form ions of different valencies (*e.g.* Cu^{+} and Cu^{2+}, Fe^{2+} and Fe^{3+}, etc.). The metal will pass into solution with a total current efficiency of 1, but the amounts of the different ions formed will depend on their respective equilibrium electric tensions with the metal, and on the redox tension. For example, in a system consisting of a metal such as iron which gives bi- and tri-valent ions (Fe^{2+} and Fe^{3+}) three processes are possible in a solution containing ions of the two degrees of oxidation.

$$Me \rightarrow Me^{2+} + 2\,e^{-} \quad (1)$$

$$Me \rightarrow Me^{3+} + 3\,e^{-} \quad (2)$$

$$Me^{2+} \rightarrow Me^{3+} + e^{-} \quad (3)$$

Their actual electric tensions may be indicated by U_1, U_2, and U_3. The work required to make one g atom of metal pass from the elementary to the trivalent state will be the same if the reaction occurs directly according to reaction (2) or by steps according to the reactions (1) and (3).

$$3\,U_2\mathrm{F} = 2\,U_1\mathrm{F} + U_3\mathrm{F}$$

whence

$$U_2 = \frac{2\,U_1 + U_3}{3}$$

The electric tension U_2 of reaction (2) is always intermediate between the electric tensions U_1 and U_3 of the processes (1) and (3). Which of these two tensions U_1 or U_3 is the more noble will depend upon the chemical properties of the metal and possibly on the ionic concentration. Luther's equation is generally true

$$U_{\mathrm{Me/Me}^{z+}} = (1/z_2)\,[z_1 U_{\mathrm{Me/Me}^{z_1+}} + (z_2 - z_1)U_{\mathrm{Me}^{z_1+}/\mathrm{Me}^{z_2+}}]$$

This is analogous to that above, in which the numbers 2 and 3 have been replaced by the general indices z_1 and z_2 ($z_1 < z_2$).

A metal immersed in a solution of its bi- and tri-valent ions is in equilibrium with the solution only if the electrode–electrolyte tension U is equal to the actual electrode tension of the two possible electrodes Me/Me^{2+} and Me/Me^{3+}, *i.e.* at 25° C

$$\begin{aligned} U &= U_1 = U_2 \\ &= U_{0\,1} + (0.059/2)\log\,[\mathrm{Me}^{2+}] \\ &= U_{0\,2} + (0.059/3)\log\,[\mathrm{Me}^{3+}] \end{aligned}$$

{$U_{0\,1}$ and $U_{0\,2}$ correspond to the respective standard electric tensions of reactions (1) and (2) and under these conditions (since $U_1 = U_2$) $U_1 = U_2 = U_3$}

In the case of iron at room temperature, equilibrium occurs when

$$-\,0.44 + (0.059/2)\log\,[\mathrm{Fe}^{2+}] = -\,0.64 + (0.059/3)\log\,[\mathrm{Fe}^{3+}]$$

so that at equilibrium

$$\sqrt{[\mathrm{Fe}^{2+}]}\,/\,\sqrt[3]{[\mathrm{Fe}^{3+}]} = 10^{6.9}$$

In the anodic dissolution of iron in a solution containing ferrous and ferric ions, ions are first formed in the valency state of the species whose concentration is less than that required for equilibrium. After this, the iron will pass into solution by forming bi- and tri-valent ions simultaneously, in the above proportion, *i.e.* predominantly Fe^{2+}. Analogous considerations for copper lead to the conclusion that at equilibrium

$$[Cu^{2+}] / [Cu^{+}]^2 \cong 10^{6.2}$$

i.e. copper dissolves anodically almost exclusively as Cu^{2+} ions. It can also be deduced from the three relative electric tensions and from the series of electric tensions, that a metallic electrode in which $U_1 < U_2$ as with iron, is unstable in the exclusive presence of ions of the higher oxidation state and that it will react by forming ions of the lower oxidation state. Iron will react according to the equation

$$2\,Fe^{3+} + Fe \rightarrow 3\,Fe^{2+}$$

However, a metallic electrode for which $U_1 > U_2$, as with copper, is unstable in the exclusive presence of ions of the lower valency and will react to form elementary metal and ions of the higher valency, *e.g.* for copper

$$2\,Cu^{+} \rightarrow Cu^{2+} + Cu$$

A special aspect of anodic overtensions of metals is the phenomenon of passivity. Since this is of primarily practical importance in relation to protection against corrosion it will be dealt with in Chapter VIII at the same time as the latter.

11. Depolarization

The theoretical value of the electrode tension at which an electrolytic process begins, is equal to the equilibrium electric tension of the galvanic half-cell which is formed; calculated using the actual concentrations of both the electrochemically active substance and the product of the electrode reaction. If, however, the product of the primary reaction is eliminated, or its concentration diminished, the equilibrium of the corresponding galvanic halfcell is either not established or will remain at an electric tension equivalent to a lower electrode polarization; and the electrolysis may continue despite this lower electrode polarization. The electrode is said to be *depolarized.*

Depolarization may occur under various conditions. If an anode on which gaseous chlorine should be evolved is treated so that it combines with a substance which fixes the chlorine as it is discharged, the anodic polarization cannot reach the value of the equilibrium electric tension at which chlorine is evolved as a gas at a pressure of 1 atmosphere. The elementary chlorine will be effectively present on the anode at a pressure corresponding to that of the dissociation equilibrium of the products of the reaction between the chlorine and the other substance. This pressure must be less than 1 atmosphere or the product would not be stable, and the discharge tension will therefore be less positive, *i.e.* the electrode will be depolarized.

All substances which act in this way to destroy the electrode polarization, by diminishing the electrolytic polarization or eliminating discharge overtensions of any type (diffusion, reaction etc.) are called depolarizers.

Two particular types of depolarization are interesting in the cathodic deposition of metals. These occur, firstly when the cations dissolve in the metal of the electrode to form an alloy or a compound which is soluble in excess of the electrode metal; and secondly, when there is a simultaneous discharge of two cations which can form an alloy of the solid solution type in the metallic state. In these cases, the J–U curve of the individual components cannot be used to interpret the progress of an electrolysis. In the former case, the activity of the metal corresponding to the cation which is discharged, will be diminished and hence the discharge tension will be depolarized. It is possible, for example, to discharge both Na^+ ions and K^+ ions from a *neutral* 1 N solution on a mercury cathode, even although the corresponding standard electric tensions are − 2.71 V and − 2.92 V respectively. Sodium and potassium are soluble in mercury to form alloys (amalgams) so that their discharge tensions remain depolarized. They will also form intermetallic compounds with mercury which are soluble in excess of mercury. This process is however made possible not only by the formation of the amalgams but also by the high overtensions of hydrogen on mercury. Similar examples are found in the discharge of zinc on palladium, of antimony on copper and of lead on platinum.

A brief discussion may be given of the method of calculating quantitatively the effect of depolarization on the discharge of a metal on mercury. The electrode tension $U_{amalg.}$ will be that of the part of the galvanic halfcell (allowing for the tension of the comparison electrode)

$$Me_{(Hg)} \,/\, Me^{z+} \,//\, \text{comparison electrode} \tag{1}$$

in which $Me_{(Hg)}$ indicates the metal Me deposited on mercury in the form of a solution (amalgam) or of intermetallic compounds with mercury which are soluble in it. Clearly,

$$U_{amalg.} = U_{0\,amalg.} + \{(RT)/(zF)\} \ln ([Me^{z+}] / [Me]_{(Hg)}) \tag{2}$$

where $[Me^{z+}]$ indicates the activity (and not the concentration) of the metallic ions in solution and $[Me]_{(Hg)}$ the activity of the metal dissolved in the amalgam or combined in the intermetallic compounds (assuming that the activity of the mercury in the amalgam is constant). For the cell with saturated amalgam

$$Me_{(Hg)sat} / Me^{z+} / Me \tag{3}$$

where $Me_{(Hg)sat}$ indicates the saturated amalgam of Me or the saturated solution of any intermetallic compounds in the mercury, the total electric tension U_{sat} of the galvanic cell (3) is independent of the activity $[Me^{z+}]$ in the solution and can be expressed by

$$U_{sat} = U_{0\,amalg.} - U_{0\,Me} - \{(RT)/(zF)\} \ln [Me]_{(Hg)sat} \tag{4}$$

From equation (4) $U_{0\,amalg.}$ becomes

$$U_{0\,amalg.} = U_{0\,me} + U_{sat} + \{(RT)/(zF)\} \ln [Me]_{(Hg)sat} \tag{5}$$

If the solubility of a metal in mercury and its standard electric tension are known, the determination of U_{sat} will allow $U_{0\,amalg.}$ to be predicted. Table IV, 8 shows that the values calculated in this way agree very well with the experimentally determined values of $U_{0\,amalg.}$.

TABLE IV, 8

CALCULATED AND DETERMINED VALUES (V) FOR $U_{0\,amalg.}$

Metal	$U_{0_{Me}}$	U_{sat}	*Solubility of metal in* Hg	*Activity coef. saturated amalgam*	*Activity* $[Me]_{(Hg)sat}$	$U_{0\,amalg.}$ *calc.*	$U_{0\,amalg.}$ *deter.*
Tl	− 0.336	+ 0.003	27.4	8.3	227	− 0.195	− 0.213
Na	− 2.714	+ 0.780	3.52	1.3	4.57	− 1.896	− 1.88
K	− 2.925	+ 1.001	1.69	5.6	9.46	− 1.866	− 1.90

It will be seen that the effect of depolarization is particularly marked in the discharge of sodium and potassium.

The second case of cathodic depolarization during the discharge of a metal occurs when two cations whose corresponding metals can form an alloy are present in the solution. It is apparent that for the simultaneous discharge of the two cations, their discharge tensions under the particular experimental conditions must not be too different; however, they need not be the same, since in general the formation of the alloy depolarizes the discharge of the more noble metal and modifies its overtension. It is not possible to predict the relative amounts of the two cations which will be discharged, from their respective *J–U* curves. When the electric tensions of the two metals, which are to form the alloy, are very different they may be brought closer together by forming suitably selected complex salts. In these, the cations no longer exist in a free state but are fixed in a complex ion of low dissociation constant; this will strongly diminish the concentrations of both cations but that of the more noble will be affected to the greater extent. In this way it is possible to obtain electrolytically, for example, brass from solutions of cyanide-complexed zinc and copper since the deposition tension of copper is made more negative by the formation of the complex whilst that of zinc is depolarized, by the formation of the alloy, with respect to the value which it would have in a solution of zinc alone.

A special depolarizing effect is caused by the imposition of an alternating current on to the continuous electrolyzing current. This effect is shown also for metals which tend to become passive, in that the appearance of anodic passivity is hindered.

12. Electrode Tensions in Redox Processes

Some electrode reactions do not involve chemically the material forming the solid conductors (first class conductors) of the electrode but only chemical species present in solution before and after the electrode reaction. These are usually called oxido-reduction or redox reactions. The first class conductors function exclusively as supports or inert conductors of electricity. The reactions may be largely classified into two groups. The first is characterized by an electrochemical process which consists exclusively of a change in the charge on an ion. The second is characterized by a change in the chemical composition of the substance subjected to the oxidation or reduction process, where the term chemical change includes polymerization processes also. It must be borne in mind as well that some processes which are apparently of the first type actually take place through a series of reactions and effectively also belong to the second type. A typical example is the reduction of tetravalent manganese to trivalent, according to the overall reaction

$$Mn^{4+} + e^- \rightarrow Mn^{3+}$$

This actually occurs through a transfer reaction

$$Mn^{3+} + e^- \rightarrow Mn^{2+}$$

followed by a dismutation

$$Mn^{4+} + Mn^{2+} \rightarrow 2\ Mn^{3+}$$

Processes of the first group are often reversible whereas those of the second are more commonly irreversible. The processes of the second group also occur in various steps consisting of a primary reaction followed by one or more secondary ones. Thus, the cathodic reduction in aqueous media may sometimes be considered as deriving from the primary process of discharge of the H^+ ion

$$H^+ + e^- \rightarrow H \tag{1}$$

followed by its reaction with the reducible substance. This will consequently lower the concentration of atomic hydrogen and act as a depolarizer. Similarly, in anodic oxidation, the primary reaction can sometimes be taken as the discharge of the OH^- ion

$$OH^- \rightarrow OH + e^- \tag{2}$$

followed by one of two possible secondary reactions

$$2\ OH \rightarrow H_2O_2 \rightarrow O + H_2O \tag{3}$$

or

$$2\ OH \rightarrow O + H_2O \tag{4}$$

These form atomic oxygen or hydrogen peroxide which can then react with the oxidizable substance. In this case, too, the oxidizable substance will act as a depolarizer. In both cases the depolarizer may be either an electrolyte or a nonelectrolyte. In other cases, as for example in the reduction of many organic substances, the overall reaction will occur through two partial reactions

$$R + e^- \rightarrow R^-$$

$$R^- + H^+ \rightarrow RH$$

The considerations discussed for the dissolution or deposition of metals are certainly applicable to processes of the first group. Whenever the process has an equilibrium electric tension which is more positive than that for the discharge of the H^+ ion (under the conditions of acidity in the solution) if it is a cathodic process, or more negative than that for the discharge of the OH^- ion if it is an anodic oxidation, and if it is perfectly reversible, it will take place with a current efficiency of 1. By making use of the overtensions of hydrogen and oxygen on different electrode materials, it is even possible to carry out processes which should not theoretically take place in view of their equilibrium electric tensions.

The *J–U* curve shows the sole peculiarity of a diminished slope, due to more marked diffusion overtensions, than in cathodic deposition or in anodic dissolution of a metal. This is due to the fact that in reduction processes there will be a diminution in concentration of the oxidized phase at the same time as the increase in concentration of the reduced phase. These two variations in concentration will act in the same sense. Similarly, in oxidation processes, there will be an increase in concentration of the oxidized phase and a diminution in concentration of the reduced phase. In some cases, a yet lower slope is found due to transfer overtensions.

Processes of the second group in which all the primary and secondary reactions are reversible and rapid, give *J–U* diagrams which show only diffusion overtensions. If, however, one of the secondary reactions is slow the whole process will be slowed down and a flatter curve will be obtained corresponding to a reaction overtension. This is due essentially to an increase in the concentration of atomic hydrogen in reduction processes and hence to a shift in the electrode tension towards more negative values. Similar considerations apply to anodic oxidations which are naturally referred to an oxygen electrode.

In the electrochemical processes of oxidation or reduction, and particularly in those of the second type in which the overall process actually consists of a primary reaction followed by one or more secondary ones, various factors may have an accelerating or retarding action which will be reflected in the shape of the *J–U* curve.

Above all, the material of the electrode may act in a purely electrochemical sense depending on the hydrogen or oxygen overtensions on it. The process will only occur if it takes place under the actual experimental conditions at a less negative electric tension (cathodic) or less positive electric tension (anodic) than that necessary for the discharge of hydrogen or oxygen respectively. This explains why some oxidations which are difficult to bring about by purely chemical routes will take place relatively easily by electrochemical ones. Under suitable conditions, it is possible to obtain notably more positive electric anode tensions than those corresponding to strong oxidizing agents. Metals on which the evolution of oxygen is accompanied by strong overtensions make possible oxidation processes requiring an elevated anodic tension and the same is true for cathodic reductions of difficultly reducible substances such as ketones and oximes. In general, electrodes for oxidation or reduction processes are selected from those metals which have a high overtension for oxygen or hydrogen, together with sufficient chemical resistance to attack by the electrolyte.

The electrode material may also act catalytically and this explains why no rigorous parallel exists between overtensions and reducing or oxidizing activity. In fact, a catalytic action which accelerates the process will simply eliminate the reaction overtension and thus diminish the polarization required for the process to occur.

The pH of the medium is important both as a factor determining the equilibrium electric tension and as a factor determining the constitution of the depolarizer. If, for example, a weak acid is subjected to oxidation, it will be present as the undissociated molecule in an acid medium but as the anion in an alkaline medium. The end product may be different depending on whether the electrochemical action is applied to the undissociated molecule or the dissociated anion.

Finally, the addition of various substances may influence the reaction both catalytically and by altering the alkalinity of the medium, etc., and by functioning as carriers of charge. A small quantity of a cerous salt will be readily oxidized anodically to the ceric state. The ceric ion will in turn react with the depolarizer to oxidize it and will itself be reduced to the cerous; this will then be rapidly reoxidized at the anode. Similarly, a titanic salt can be readily reduced cathodically to the titanous state which will reduce a depolarizer and be oxidized to the titanic state; this will then be once more reduced at the cathode. The Ce^{3+} and Ti^{4+} ions will thus accelerate oxidation and reduction processes respectively. Various other ions will act like those of cerium and titanium. Naturally, the action of carriers is to some extent specific with regard both to the depolarizer and the range of electric tensions within which they will work.

A theory developed by Glasstone and Hickling[1] should be mentioned in connection with the particular case of anodic oxidation. According to this, anodic oxidation is due to the hydrogen peroxide formed in one of the two secondary reactions between the OH radicals produced by the primary discharge of the OH^- ions. Many details of anodic oxidations are in agreement with the predictions of this theory. At the present time, however, certain anomalies exist which cannot be explained by any existing theory. For example, Kolbe's reaction of the formation of ethane by the anodic oxidation of acetate ions, occurs under particular conditions at an anodic tension which is 0.4 V more positive than that for the evolution of oxygen under the same conditions in the absence of acetate ions[2].

13. Energy Efficiency

If the reversible equilibrium electric tension of the galvanic element which is formed during an electrolysis between the anode and cathode is U_{rev}, the minimum work which should theoretically be expended to bring about the reaction of one mole of substance at each electrode will be $z\mathbf{F}U_{rev}$. However, this minimum work is sufficient to bring about this amount of electrode reaction only if the electrode processes occur under perfectly reversible conditions, *i.e.* when the effective electrode tension is equal to the reversible equilibrium electric tension. This implies that the current density must be very small, tending towards zero; but from what has been said above, it will be apparent that in general it is not possible to carry out electrolytic processes with a finite current density which does not tend towards zero, without causing overtensions. These then have to be overcome in addition to the resistance of the electrolytic cell. This, in turn, implies that the electric tension applied to the electrodes has to be greater than that which would theoretically be needed for a reversible electrochemical process carried out with an infinitely small current density. This leads to a greater consumption of energy than is theoretically required.

The *energy yield* (R_{en}) of a process is defined as the ratio of the quantity of energy theoretically required to that actually consumed under the par-

[1] S. Glasstone and A. Hickling, *Chem. Revs.*, 25 (1939) 407; see also A. Klemenč, *Z. physik. Chem.*, A 185 (1939) 1.

[2] A. Hickling, *Discussions Faraday Soc.*, No. 1 (1947) 227.

ticular conditions of electrolysis. The electrical energy is the product of the quantity of electricity and the applied electric tension. If z is the number of **F** of electricity theoretically required for the reaction of 1 mole of substance and p is the total number of moles of substance transformed, then

$$R_{en} = (U_{rev}/U)(z\mathbf{F}p/Q) \tag{1}$$

U_{rev} is the electric tension theoretically required for the reversible process and is equal to the sum of the anodic and cathodic polarizations at equilibrium; U is the actual electric tension applied to the cell; $z\mathbf{F}p$ is the theoretical quantity of electricity and Q is the quantity of electricity which actually passes through the cell. Using the definition of current efficiency (*cf.* p. 186) gives

$$R_{en} = (U_{rev}/U)R_{cur}$$

The actual consumption of energy E may be calculated as follows. The actual electric tension U applied to the terminals of the cell is the sum of the electrode polarizations P (decomposition tension + any overtensions) and the tension IR necessary to overcome the ohmic resistance of the cell, *i.e.*

$$U = P + IR$$

The energy consumed E is the product of the applied tension U and the quantity of electricity which has passed through the cell $\int_0^t I\mathrm{d}t$ (t = time). So that, if I is constant

$$E = PIt + I^2Rt$$

If the current efficiency is R_{cur} and the electrochemical equivalent in g/Ah is θ, then the mass of substance m which separates in unit time is

$$m = \theta ItR_{cur}$$

The energy consumption E' for each g of substance is given by

$$E' = E/m$$

$$= \frac{PIt + I^2Rt}{\theta ItR_{cur}}$$

$$= \frac{P + IR}{\theta R_{cur}} \tag{2}$$

since

$$R = \varrho l/S$$

and

$$I = JS$$

where ϱ = the specific resistance, l = the distance between the electrodes, S = the cross sectional area of the cell (taken as approximately equal to the surface area of the electrodes) and J = the current density; equation (2) will become

$$E' = (P + \varrho lJ) / (\theta R_{cur})$$

This equation permits the ready calculation of the energy consumed by each unit of mass of an electrolytic preparation where P includes the sum of the reversible electric tensions and all the overtensions applied to the electrodes but excludes the ohmic resistance overtension.

CHAPTER V

ANALYTICAL APPLICATIONS

by

ARNALDO LIBERTI[1] and GIULIO MILAZZO

The general term 'analytical applications' indicates all the techniques which use any electrical parameter to identify the presence or to establish the concentration in a solution of ionic species, either by direct measurement or by titration.

The possibility of expressing changes in any quantity of a system as a function of an electrical measurement makes electrochemical methods particularly interesting and gives them a wide field of application. Classical determinations are based on chemical changes resulting from the passage of an electric current but other experimental techniques have now been successfully applied. The analytical applications of electrochemistry may be conveniently split into the following methodological divisions:

1. Electrolytic analyses
2. Potentiometry
3. Conductometric methods, including high frequency conductometry
4. Polarographic and voltametric methods
5. Electrophoresis[2]

1. Electrolytic Analyses

The term *electrolytic analyses* includes all those analytical procedures which use electrolysis for the determination of one or more ionic species. Since Faraday's Law underlies all electrolytic analytical methods, the various processes of oxidation or reduction which occur respectively at the anode and the cathode, may be carried out with widely differing experimental provisions to give a large range of techniques. These may be classified according to the most important electrical parameter in the electrolytic process considered: into electrolysis with a constant current, electrolyses with controlled electric tensions, internal electrolysis, coulometry at controlled electric tension and coulometric titration.

[1] University of Naples.

[2] In view of its distinctive nature, electrophoresis will be dealt with in detail in the chapter on colloids.

(I). *Electrolysis with a Constant Current*

This is the classical method of electrolytic analysis and is also known as electrogravimetric analysis from its resemblance to gravimetric analysis. In this latter, an excess of a reagent is added to precipitate a given compound quantitatively; the passage of a quantity of electricity in excess of the theoretical requirement, in electrogravimetric analysis, leads to the deposition on an electrode of one ionic species which is then determined by weighing. The experimental arrangement is shown in Fig. V, 1.

The tension divider R allows the electric tension which is applied to the electrodes immersed in the solution, to be varied and brought to the value corresponding to the individual discharge tension (*cf.* p. 192 *et seq.*) of every ionic species. For the electrolysis to occur it is necessary to overcome all the ohmic resistances of the circuit by applying to the electrodes a polarization which is about 0.1–0.2 volts greater than the sum of the discharge tensions, taking account of any overtensions. There is thus a passage of an electrolytic current which remains virtually constant since the vigorous stirring of the solution eliminates any polarization of the concentration.

The electric tension which must be applied to the electrodes (U_A) may, for practical purposes, be considered as composed of three terms,

$$U_A = U_B + U_S + U_R$$

where U_B is the electric tension required to overcome that of the galvanic cell formed and is given by the equation

$$U_B = U_{anode} + U_{cathode}$$

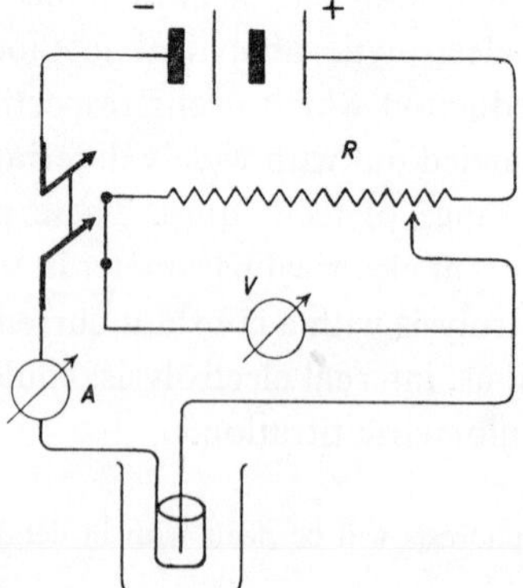

Fig. V, 1. Electrolytic analysis

where U_{anode} and $U_{cathode}$ are calculated by Nernst's equation, following the conventions established in Chapter III. U_S is the total overtension resulting from the sum of the anodic and cathodic overtensions (*cf.* Chapter IV) and U_R is the electric tension required to overcome the ohmic resistance of the solution ($U_R = IR$). The value of the resistance depends on the surface area of the electrodes, their distance apart and the composition of the solution.

A practical example will illustrate the use of this equation; and the calculation of the electric tension which must be applied for the electrolysis of a 0.1 *m* solution of copper sulphate, between platinum electrodes in 1 *N* acid solution, will be performed. U_B is calculated as the electric tension of a galvanic cell formed from an oxygen electrode ($\frac{1}{2}O_2 + H_2O + 2e^- = 2OH^-$) and a copper electrode, arranged in the following phase order.

$$Pt\text{–}O_2/H_2O,\ H^+,\ Cu^{2+}/Cu/Pt$$

The equation

$$U_B = U_{O_2} - U_{Cu} = 1.229 + 0.059 \log [H^+] - (0.345 + 0.029 \log [Cu^{2+}])$$

applies to this at 25° C. Substituting the analytical concentrations for $[H^+]$ and $[Cu^{2+}]$ gives a value for the electric tension of 0.913 volts. Since copper is deposited and oxygen evolved it is necessary to apply externally, this tension increased by the value of the anodic overtension — which is about 0.40 V for oxygen on smooth platinum in acid solution — whilst the overtension of copper on platinum is virtually negligible. Electrolysis will thus occur if an electric tension is applied sufficiently greater than 1.313 V to overcome the electrical resistance of the solution, *i.e.* *IR*. The electric tension of the electrodes will vary during the electrolysis according to Nernst's equation, but whilst the cathodic tension varies by about 30 mV for a 10-fold variation in copper concentration, the variation of the anodic tension may be ignored because the concentration of dissolved oxygen is virtually constant since it is in equilibrium with that of the atmosphere; this electric tension increases by only a few millivolts with the increase of the hydrogen ion concentration.

These considerations enable the calculation of the electric tension which has to be applied in order that an electrogravimetric determination may be considered analytically complete, or in order that the concentration of an ion may be diminished to a predetermined value. If, for example, the final concentration of copper in this example must be 10^{-10} *m* — a value more than sufficient to guarantee a quantitative determination of copper — it can be calculated that the electric tension of the copper electrode becomes 0.3 V more negative than the value of 1.313 V calculated above, and thus the tension U_A which must be applied to the electrodes is 1.608 V. In practice, in any case, in view of the uncertainties and errors surrounding certain values (*i.e.* the overtensions and the difference between concentrations and activities) and bearing in mind the fraction of the tension necessary to overcome the ohmic resistance of the electrolytic cell, these determinations are always applied with a tension U of about 2 V in order to have a margin of safety.

As was noted above, in a cathodic process the electric tension of the electrode used as the cathode, does not remain constant during the electrolysis because the concentration of the ion which is deposited, diminishes; the electrode tension thus becomes more negative according to Nernst's equation. If the electrolytic current were kept constant it would be possible to deposit any other ions which might be present. When more than one cation is present in the solution it is possible to separate them electrolytically, and to determine them quantitatively, provided that the discharge tensions of the various cations present in solution are sufficiently different so that the maximum polarization required for the quantitative separation of a more noble cation is always less than the minimum polarization necessary to start the separation of the next less noble cation. For example, in the electrolysis of copper sulphate in acid solution, the evolution of hydrogen forms a marked interference because it makes the deposit porous, spongey and poorly adherent to the electrode. For hydrogen to be evolved, an electric tension of about 2.2 V must be applied to the electrode, bearing in mind that the overtension of hydrogen on copper is about 0.60 V. With an electric tension of this value, or more, hydrogen and copper are discharged simultaneously. If the electrolysis is performed with a lead accumulator giving an electric tension of about two volts then, since $U_A = U + IR$ (where U is the sum of U_B and U_S), when U becomes equal to U_A, I becomes zero and the electrolysis stops without any loss of current.

Electrolytic analyses with constant current are of particular use in the separation of cations with very different decomposition tensions and may be used when the electrodic reaction leads to the complete discharge of the cation to give rise to a deposit, which must be well adherent to the cathode and must not occlude impurities from the solution. Electrolyses with constant current may be used, not only for reduction processes, but also for anodic oxidation processes. This procedure is often followed when the ions concerned cannot be quantitatively separated in a suitable form at the cathode whilst they will give a compact deposit of oxide at the anode. This is so, for example, with manganese which has a very negative discharge tension so that its cathodic separation — which is made possible by the overtension of hydrogen on the metallic manganese electrode — is always accompanied by a strong evolution of hydrogen and thus is not quantitative. It is, however, possible by anodic oxidation under suitable conditions, to obtain a compact and adherent deposit of manganese dioxide, which is easily washed and hence readily weighed.

It is opportune to record that the alkaline and alkaline earth metals may also be quantitatively determined electrolytically, using a mercury cathode, on which hydrogen has a fairly high overtension. The use of a mercury cathode is particularly helpful for separating metallic ions of differing electronegativity[1]; for example, in the analyses of steel and ferrous alloys with a mercury cathode, it is possible to separate the most electronegative elements such as titanium, tungsten, aluminium and molybdenum.

A fairly large number of electrodes have been proposed. A very common form consists of platinum foil rolled into a cylinder and used as the cathode, with a thick platinum wire which is concentric within the foil used as the anode (Winkler's electrode). In certain cases the platinum may be replaced by less costly metals. As far as current intensity is concerned it must be remembered that throughout the analysis, it must always be well below the limiting current density, so that the surface area of the electrode must be known or be readily calculable. From what has been said above (*cf.* Chapter IV) it is apparent that a vigorous stirring allows the value of the current intensity to be markedly increased so that the analysis may be carried out in much less time. In every case in order to get satisfactory results, particularly with separations, the greatest attention must be given to obtaining the optimum conditions, either those discovered personally or taken from the literature[2].

[1] Electronegativity is here, and in following sections, referred to the values of the standard electric tensions, but not in the sense given by Pauling and others as electronegativity of the elements.

[2] For a systematic description of the conditions for electrolysis of various systems consult H.J.S. Sand, *Electrochemistry and Electrolytical Analysis*, Vol. II and III, Blackie, London, 1946; A. Classen, *Quantitative Analyse durch Elektrolyse*, J. Springer, Berlin, 1927.

(II). *Electrolyses with Controlled Electric Tensions*

It has been seen that the fundamental factor which determines an electrodic process is neither the current intensity within the cell nor the overall electric tension applied to the electrodes, but the electric tension of the electrode at which the reaction occurs. From what has been said above, about electrogravimetry, it will be apparent that if a solution contains many metals of only slightly different discharge tensions they may be deposited simultaneously. This difficulty may be overcome by adding suitable agents which can form complexes of varying stability with the different cations and which can be reduced at different electric tensions. Another method, of a more general nature, consists in the control of the electric tension of the electrode at which the particular reaction occurs. This method which was developed by Sand and Fischer, has recently been widely applied following Hickling's electronic arrangement for automatically controlling the electrode tension. The electrolysis circuit (Fig. V, 2) is analogous to that used for the classic electrogravimetry and the essential difference consists in the insertion into the solution of a third electrode in addition to those to which the electrolysis tension is applied. This additional electrode has a definite electric tension (*e.g.* a saturated calomel electrode or a silver chloride electrode) with respect to which the electric tension of the electrode under consideration may be measured, using a potentiometer and an electronic voltmeter V_2. Thus, three electrodes are inserted into the cell in which electrolysis at constant electric tension is performed: the electrode maintained at a constant electric ten-

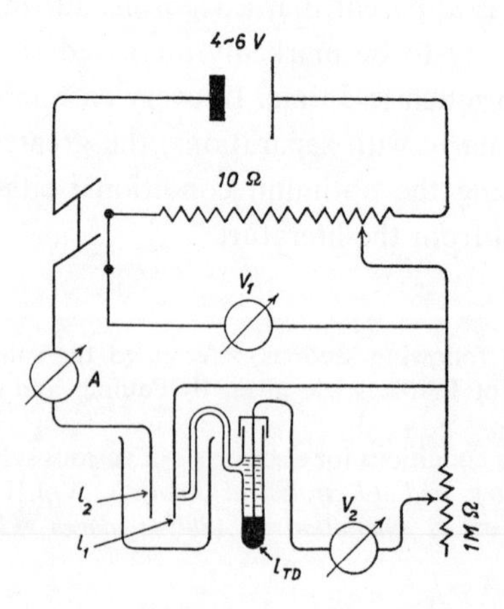

Fig. V, 2. Electrogravimetry with controlled electric tension

sion l_1, which will be called the working electrode; the second electrode l_2, which serves merely to conduct the electrolysis current; and the third electrode $l_{T.D.}$ which permits the continuous control of the electric tension of the working electrode. If a solution contains ions A_1 and A_2, whose discharge tensions are U_1 and U_2 at the respective initial concentrations c_1 and c_2, the determination of A_1 (which is the more readily reducible ion) can be carried out if a definite value U can be obtained for the cathodic electrode tension, such that A_1 is deposited uniquely and quantitatively, *i.e.* at least 99.99% of the initial quantity, from the initial concentration c_1. At 25° C the electric tension U must then satisfy the equation

$$U_{0\,2} + (0.0591/z_2) \log c_2 < U < U_{0\,1} + (0.0591/z_1) \log (c_1 \cdot 10^{-4})$$

where z_1 and z_2 are the numbers of electrons involved in the respective electrode processes. The cathodic tension is thus fixed at a value, which may be deduced from the polarization curve, for which the concentration of the ion to be reduced is virtually zero at the end of the electrolysis.

If the electrode reaction of the particular process takes place with a yield of 100%, the intensity of the current diminishes exponentially with time according to the equation

$$I_t = I_0\, 10^{-Kt}$$

where I_0 is the initial current intensity and I_t is the current intensity at time t, and K is a constant depending on the electrode area, the volume of the solution, the rate of agitation and the temperature, but is independent of the progress of the electrode reaction. The applied electric tension must be frequently adjusted to the predetermined value whilst the current intensity diminishes, and tends towards zero after a very long time. The electrolysis is, however, normally stopped when the current intensity becomes less than 0.1% of the initial value. The control of the electrode tension may be carried out manually with the apparatus indicated, or else automatically using an instrumental arrangement which is known as a potentiostat. The principle of the potentiostat is to convert the difference between the pre-selected and measured electric tensions at any moment to an alternating tension which is suitably amplified to operate, by means of relays servomotors and other arrangements, the sliding contact of the potential divider. The cathodic tension is thus kept constant at the pre-selected value despite the changes in the electrical quantities due to the progress of the electrolytic reaction.

Electrolysis at controlled electric tensions permits the electrogravimetric determination of a given ionic species without co-deposition, *i.e.* the simultaneous deposition of other ions present in the solution. The use of this technique is particularly important for the analytical determination of metals in various alloys such as copper, bismuth and lead in alloys of copper, and tin, lead, cadmium and zinc in zinc dust; many of these separations can be carried out for preparative purposes in addition to analytical ones.

The accuracy of this technique is the same as that of the classical electrogravimetric method but its principal advantage lies in its exceptional selectivity obtained by the control of the electric tension. For a complete description of the electrical circuits and the possibilities which can be achieved with this analytical technique, reference should be made to the specialized literature on this topic[1].

(III). *Internal Electrolysis*

The term internal electrolysis, or spontaneous electrogravimetric analysis, indicates that analytical procedure by which the electrolysis of a solution is obtained without the application of an external electric tension. This is achieved at the expense of the energy generated by a cell which has a platinum cathode and an anode of a baser metal than that which is to be deposited from the solution. It is thus an electrolysis with controlled electric tension where the determining factor is the tension of the electrode which forms the anode. This, on passing into solution, displaces as in a 'displacement' or 'cementation' reaction, the more electronegative[2] cations which are then deposited on the cathode. It is obvious that in any experimental arrangement which utilizes this method it is necessary for the two electrodes to be separated by a diaphragm and for the anode to be immersed in a solution of an inert electrolyte to avoid the direct deposition upon the anode. If, for example, the cadmium content of a solution is to be determined, a cell is constructed consisting of a platinum electrode immersed in the test solution in which is also immersed a vessel with a porous diaphragm containing a 0.1 *m* solution of zinc sulphate into which in turn,

[1] H. DIEHL, *Electrochemical Analysis with Graded Cathode Potential Control*, G.F. SMITH, Chemical Co. Columbus, 1948. J.J. LINGANE, *Electroanalytical Chemistry*, Chapter XI, Interscience Publ., New York, 1953.

[2] See footnote 2, page 279.

dips the zinc electrode. Short circuiting the electrodes leads to the reaction

$$Zn + Cd^{2+} = Zn^{2+} + Cd$$

For which at 25° C

$$U_{0\,Zn} + \frac{0.0591}{2} \log a_{Zn^{2+}} = U_{0\,Cd} + \frac{0.0591}{2} \log a_{Cd^{2+}}$$

and thus, ignoring the difference between activity and concentration,

$$-0.763 + \frac{0.0591}{2} \log 0.1 = -0.419 + \frac{0.0591}{2} \log [Cd^{2+}]$$

Resolving these equations shows that at the end of the electrolysis $[Cd^{2+}] = 10^{-13.9}$ so that the deposition of cadmium may be considered complete. This analytical procedure is very simple particularly as regards the experimental requirements, but it presents certain practical limitations in actual use, of which the most important is the long time required for electrolysis. It has been used for very many determinations and especially for the analysis of small quantities of impurities of relatively noble metals in various alloys. An excellent treatment of this subject has been given by Schleicher[1] with many references.

(IV). *Coulometry at Controlled Electric Tension*

The theoretical basis of this analytical procedure is Faraday's Law, according to which the total quantity of electricity, Q, expressed in faradays, consumed in a particular electrode reaction is equal to the number of equivalents involved in the reaction. If the current intensity is kept strictly constant during an electrolysis, the quantity of electricity is given by the product of the intensity and the time, but if the electrolysis is carried out with controlled tension the intensity of the current varies exponentially so that

$$I_t = I_0 \cdot 10^{-Kt}$$

and the quantity of current is then given by the equation

$$Q = \int_0^t I_t \mathrm{d}t = \frac{I_0}{2.3\,K}(1 - 10^{-Kt})$$

[1] A. SCHLEICHER, *Electroanalytische Schnellmethoden*, F. Enke, Stuttgart, 1947.

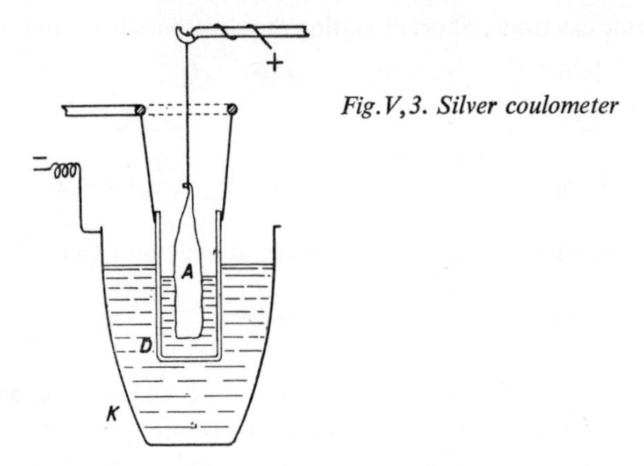

Fig. V, 3. Silver coulometer

If an electrodic process takes place with a current efficiency of 100% the calculation of the quantity of current can be performed either by integrating the current–time curve, or by placing a coulometer[1] in series in the electrolysis circuit. This appliance consists of an electrolytic cell in which a suitable process takes place with a yield of 100% without concomitant secondary reactions of any type; it is so constructed that the product of the electrolytic reaction may be accurately weighed or otherwise directly measured. One which gives very accurate results is the silver coulometer or voltameter shown in Fig. V, 3.

It consists of a platinum crucible *K* which functions as the cathode and contains a solution of pure silver nitrate with 20–40 parts of salt in 100 parts by weight of water. The anode, *A*, consists of a very pure silver rod dipping into a solution of silver nitrate of the same concentration contained in a cylindrical porous pot *D*. This in turn is immersed in the solution in the crucible. The porous pot functions as a diaphragm to prevent particles of silver falling from the cathode and becoming attached to the anode. The cathodic current density must not exceed 0.02 A/cm^2 and the anodic 0.2 A/cm^2. By weighing the crucible before and after the electrolysis, the quantity of electricity which has passed may be calculated knowing that one coulomb separates 0.001118 g of silver.

Another type which is widely used and is very simple to construct is the gas coulometer. It consists of a burette filled with electrolytes (generally 0.5 *m* K_2SO_4 or Na_2SO_4) with two electrodes inserted at its bottom. The

[1] In certain older texts, coulometers are called voltameters.

passage of current through these electrodes leads to the evolution of hydrogen and oxygen within the burette, according to the reactions

$$2\,H_2O = 4\,H^+ + O_2 + 4\,e^-$$

$$4\,H_2O + 4\,e^- = 2\,H_2 + 4\,OH^-$$

The gas is collected in the burette and its volume measured, brought to normal conditions and corrected for the vapour pressure of the solution. One coulomb releases 0.1733 cm^3 of gas at 0° C and 760 mm Hg.

Other types of coulometer have been proposed, such as the copper coulometer (which is analogous to the silver one) and titration coulometers such as that with silver in which the quantity of silver dissolved in the anodic region, by the passage of current in an electrolytic cell with a silver anode, is measured by titration. Another coulometer of this type is that with sodium iodide where the liberated iodine is titrated with thiosulphate.

Coulometers give very accurate results but are inconvenient in that they are not generally direct reading and cannot be used for the measurement of very small quantities of electricity. Alternatively, since the number of coulombs is equal to the integral of the current–time curve, any mechanical, electromechanical or electrical integrator yielding the value of $\int_0^t I\mathrm{d}t$ may be used as a coulometer. This type of integrator has the disadvantage, however, of being less accurate than the chemical coulometer.

This method may be used successfully for the determination of substances which give rise to soluble products on electrolysis. The degree of accuracy is equal to that obtained with the classical methods of electrogravimetry. An interesting application is the possibility of determining fairly simply the number of electrons involved in a particular electrodic process, which takes place with a yield of 100%, when the quantity of substance electrolyzed is known.

(V). *Coulometric Titration or Coulometry at Constant Current*

This method is analogous to that in the preceding section, consisting in the utilization of the principle of electrochemical equivalents expressed by Faraday's Law. It differs from the above however, in that whilst with coulometry at controlled electric tension, the current diminishes continuously throughout the electrolysis so that the quantity of current used has to be obtained by an integration process, in the present method the current is kept constant throughout the electrolysis so that the quantity of current consumed is given by the product of the constant current intensity and the time. If I is expressed in milliamperes and t in seconds,

$$It/96{,}500 = \text{the number of milliequivalents}$$

The current which passes through the cell can cause reduction or oxidation of the species in solution at one of the electrodes of the cell, or can generate a reagent. The first case is a primary coulometric titration and the other a secondary. This latter procedure is of more general interest since it allows the determination of substances which are not directly reducible or oxidizable. The passage of the constant current is stopped when a suitable indicator shows that the analytical reaction is finished. The following requirements are needed for the performance of this method.

1. A source of current of constant intensity (galvanostat)
2. The achievement of experimental conditions such that the desired reaction takes place with a yield of 100%
3. An electric chronometer
4. A device to indicate the end of the reaction

The current source may consist of a number of accumulators connected in series or of a suitable electronic apparatus. The 100% yield can be achieved by placing an excess of a suitable electrolyte in the cell, and the end of the reaction may be detected, not only by visual indicators, but also by potentiometry or amperometry. With these electrical indicator systems it is possible to build automatic appliances to perform the titration.

Coulometric titrations have the advantage of not requiring standardized solutions for the determination of a particular component since this is obtained using Faraday's Law. The errors common to microchemical determinations or determinations on very dilute solutions, are eliminated, since quantities of current corresponding to minute amounts of reagent can be measured with great precision. For such titrations also, unstable reagents such as Br_2, HBrO, Cu^+, Cl_2 may be used by being generated within the solution. The technique offers the advantage that it is readily made automatic and if necessary can be remotely controlled so that it can be used for dangerous substances such as radioactive materials. The

TABLE V,1

COULOMETRICALLY GENERATED REACTANTS AND THEIR ANALYTICAL USE

Reactant	*Titrated substances*
H^+	Bases
OH^-	Acids (strong and weak); S and C (in steel and in organic compounds after combustion)
Cu(I) ($CuCl_3^{2-}$)	Cr(VI), V(V)
Fe(II)	Mn(VII), V(V), Ce(IV), Cr(VI), Pu(VI)
Ti(III)	Fe(III), V(V), V(IV), U(VI), Ce(IV); (various organic compounds)
Br_2	Tl^+, CNS^-, N_2H_4, NH_2OH, As(III), Sb(III), I^-, Cu(I), U(IV), Hydroxyquinoline, Salicylic Acid, Aniline; Ascorbic Acid, Olefines, Thiodiglycol and other organic compounds
Cl_2	Tl^+, I^-, As(III), Fe(II)
I_2	Sb(III), As(III), $S_2O_3^{2-}$, IO_3^-, Thiamine, Boranes, Hydroquinone
BrO^-, ClO^-	NH_3, AsO_2^-, CNS^-, SbO_2^+, Aminoacids, various organic compounds
IO^-	BH_4^-
Mn(VI)	Fe(II), As(III), $H_2C_2O_4$
$Cr_2O_7^{2-}$	Fe(II)
Sn(II)	I_2, Br_2, Au(III), V(V), Cu(II), Pt(IV)
Ce(IV)	Fe(II), $[Fe(CN)_6]^{4-}$, Ti(III), U(IV), As(III), Hydroquinone
Hg_2^{2+}	Cl^-, Br^-, I^-, S^{2-}, SCN^-, $C_2O_4^{2-}$
Ag^+	Cl^-, Br^-, I^-, Thiourea, Mercaptans
Pb^{2+}	$C_2O_4^{2-}$, SO_4^{2-}, $[Fe(CN)_6]^{4-}$, $Cr_2O_7^{2-}$, WO_4^{2-}, MoO_4^{2-}
EDTA	Cu(II), Zn(II), Pb(II), Ca(II), Cd(II), Mn(II), Ni(II), Co(II)

principal reagents which can be generated coulometrically, together with the ionic species which they titrate, are shown in Table V,1. Reference may be made to textbooks for a more complete treatment of this subject[1].

2. Potentiometry

The measurement of the electric tension of a galvanic cell is a versatile analytical technique for the determination of many parameters: activities, activity coefficients, solubility products, dissociation constants, equilibrium constants, transference numbers, oxidation states etc.

[1] J. J. LINGANE, *Electroanalytical Chemistry*, Interscience Publ., New York, 1953; P. DELAHAY, *New Instrumental Methods in Electrochemistry*, Interscience Publ., New York, 1954.

One of its most important applications is analytical and is based on the principle that a galvanic cell composed of a suitable indicator electrode and a comparison electrode arranged in the following phase order (*cf.* Chapter III):

$$\underbrace{X \;/\; X_{ion}}_{\text{indicator electrode}} \;//\; \underbrace{Y_{ion} \;/\; Y \;/\; X'}_{\text{comparison electrode}} \qquad (\text{metal } X = X')$$

gives an electric tension which has a simple relation to the activity of one or more of the electrochemically active ionic species functioning in the electrode reaction of the indicator electrode. This tension may be utilized analytically either directly, or as an indicator for the identification of the end point of an analytical reaction.

The potentiometric determination of a given ionic species requires the construction of a suitable galvanic cell whose overall electric tension is then measured. Knowing the electric tension of the comparison electrode it is possible to obtain immediately the electric tension of the indicator electrode, and hence to deduce the activity of the given ionic species. If the determination is carried out in dilute solution the activity may, as a first approximation, be taken as equal to the concentration; so that from a practical standpoint the measurement of the electric tension leads to the determination of the concentration of an ionic species. This is applicable to any electrode which is reversible with respect to a single ionic species. This application is called direct potentiometry.

Sometimes, however, as for example with oxido–reductive systems, a simple measurement of the electric tension gives a value which cannot be used to establish the concentration of an ionic species, and thus it is necessary to determine the electric tension, between the indicator electrode and the comparison electrode, after successive additions of a titrating reagent. From the quantity of this reagent consumed at the end of the analytical reaction — as shown by the indicator electrode — the concentration of the test substance may be calculated. This second application is called potentiometric titration. The selected reaction must naturally be such as to vary the activity of the electrochemically active ionic species. It is not absolutely necessary, however, that both the substance to be analysed and the titrating reagent are electrolytes; it is sufficient that the activity of at least one of the ionic species varies as a consequence of the analytical procedure.

(I). *Direct Potentiometry; Measurement of* pH

The most important application of direct potentiometry is the determination of the pH of a solution, which is carried out by measuring the electric tension between an electrode, reversible with respect to the H^+ ion, and a comparison electrode immersed in the test solution. If U is the electric tension of this cell and U_0 is the tension of the same cell with the indicator electrode at pH = 0, then from the definition

$$\text{pH} = -\log a_{\text{H}^+}$$

at 25° C

$$\text{pH} = \frac{U_0 - U}{0.05916}$$

The numerical value obtained from this relationship does not, however, actually indicate either the activity or the concentration of hydrogen ion; it represents only an approximation to this which is indeterminate because the diffusion electric tension of the cell is unknown and the activity coefficients of the individual ionic species are also not known. To eliminate these uncertainties, and in view of the need for a very reliable and reproducible measure of acidity, the pH is defined by a 'conventional practical criterion' using the following equation

$$\text{pH} = \text{pH}_S + \frac{U_S - U}{0.059}$$

In this expression U is the measured electric tension of a cell composed of an electrode, which is reversible with respect to the H^+ ion and contains the test solution, and of a comparison electrode; U_S is the electric tension of the same cell with the electrode which is reversible with respect to the H^+ ion containing a solution whose pH is pH_S. In order to extend the range over which this criterion is valid the value of pH_S has been determined with four solutions which form the primary standards. These are

(a) 0.05 *m* potassium hydrogen phthalate
(b) 0.025 *m* monopotassium dihydrogen phosphate + 0.025 *m* disodium hydrogen phosphate
(c) 0.01 *m* borax
(d) a saturated solution, at 25°, of potassium hydrogen tartrate.

Two secondary standards for strongly acid and basic solutions are respectively, 0.05 *m* potassium tetraoxalate and a saturated solution at 25° C, of

TABLE V,2

VALUES OF pH_S FOR STANDARD REFERENCE SOLUTIONS

$t°$ C	*0.05 m* $KH_3(C_2O_4)_2 \cdot 2\ H_2O$	(*sat. 25°* C) $KHC_4H_4O_6$	*0.05 m* $KHC_8H_4O_4$	*0.025 m* KH_2PO_4 Na_2HPO_4	*0.01 m* $Na_2B_4O_7 \cdot 10\ H_2O$	(*sat. 25°* [illegible] Ca(O[illegible]
0	1.67		4.01	6.98	9.46	13.43
5	1.67		4.01	6.95	9.39	13.21
10	1.67		4.00	6.92	9.33	13.0[illegible]
15	1.67		4.00	6.90	9.27	12.81
20	1.68		4.00	6.88	9.22	12.6[illegible]
25	1.68	3.56	4.01	6.86	9.18	12.45
30	1.69	3.55	4.01	6.85	9.14	12.3[illegible]
35	1.69	3.55	4.02	6.84	9.10	12.1[illegible]
40	1.70	3.54	4.03	6.84	9.07	11.9[illegible]
45	1.70	3.55	4.04	6.83	9.04	11.8[illegible]
50	1.71	3.55	4.06	6.83	9.01	11.7[illegible]
55	1.72	3.56	4.07	6.84	8.99	11.5[illegible]
60	1.72	3.56	4.09	6.84	8.96	11.4[illegible]
70	1.74	3.58	4.12	6.85	8.93	
80	1.77	3.61	4.16	6.86	8.89	
90	1.80	3.65	4.20	6.88	8.85	
95	1.81	3.68	4.23	6.89	8.83	
dB/dpH*	0.070	0.027	0.016	0.029	0.020	0.[illegible]
dilution** effect	+ 0.186	+ 0.049	+ 0.052	+ 0.080	+ 0.01	+ 0.[illegible]
dpH/dt***	+ 0.001	+ 0.001	+ 0.01	− 0.003	− 0.008	− 0.[illegible]

* Buffer capacity expressed as the number of equivalents of strong base, B, which added to 1 litre of buffer solution cause a change of 1 unit of pH.
** The change in pH produced by diluting 1 volume of the buffer solution with 1 volume of water.
*** The change in pH for each degree of temperature.

calcium hydroxide. The values of pH_S determined by the National Bureau of Standards, U.S.A. are reported in Table V,2[1]; these were obtained by measuring the electric tension of a galvanic cell formed of a hydrogen

[1] V.E. Bower and R.G. Bates, *J. Research Natl. Bur. Standards*, 59 (1957) 261; R.G. Bates and V.E. Bower, *J. Research Natl. Bur. Standards*, 56 (1956) 305.

electrode containing the test solution and of a silver–silver chloride electrode, arranged in the phase order

$$Cu/Pt\text{–}H_2/H^+//Cl^-/AgCl/Ag/Cu$$

The electric tension of such a cell is given by

$$U = U_0 + RT/\mathbf{F}(\ln m_{H^+} + \ln f_{H^+} + \ln m_{Cl^-} + \ln f_{Cl^-})$$

Since the values of U_0 and m_{Cl^-} (molarity of Cl^-) are known, a value can be obtained, known as pwH, which is expressed by

$$\begin{aligned} \text{pwH} &= (U_0 - U)\mathbf{F}/RT + \ln m_{Cl^-} \\ &= -(\ln m_{H^+} + \ln f_{H^+} + \ln f_{Cl^-}) \end{aligned}$$

whence

$$\begin{aligned} \text{pwH} &= (U_0 - U)\mathbf{F}/RT\,2.3026 + \log m_{Cl^-} \\ &= -(\log m_{H^+} + \log f_{H^+} + \log f_{Cl^-}) \end{aligned}$$

The value of pH_S is thus given by

$$pH_S = \text{pwH} + \log f_{Cl^-}$$

This equation shows that the value of pwH differs from the 'standard' pH (pH_S) only by the term containing the logarithm of the activity coefficient of the chlorine ion. The sole arbitrary criterion in this definition of pH, is the value attributed to this activity coefficient. Since no theoretical or experimental technique exists for the determination of activity coefficients of individual species, certain practical criteria can be applied to calculate the value to be given to f_{Cl^-}, *e.g.* the value of a mean activity coefficient of an alkali chloride at the ionic strength of the buffer solution considered, or that of a solution of hydrochloric acid. For values of the ionic strength less than 0.2 these criteria give the same values as those obtained with the Debye–Hückel equation.

On this basis, the measurement of pH consists in the determination of the electric tension of a galvanic cell but for the correct interpretation of the experimental data it is necessary that the test solution has a composition and an ionic strength not too different from those of the standard solution, since the diffusion electric tension could have a decisive influence on the overall value of the measured electric tension. In view of the abnormal diffusion electric tensions which are encountered in strongly acid or alkaline conditions, it may be concluded that only the pH values of aqueous buffer solutions of simple salts at ionic strengths between 0.01

and 0.2 can be determined potentiometrically, and only within the pH range of 2 to 12.

Under conditions different from these, the pH of a solution has a purely indicative value. Whilst this is sufficient for industrial purposes, in the study of equilibria involving changes in hydrogen ion activity measurements of pH can have little significance for the above reasons, and especially for the uncertainties surrounding both the values of the activity coefficients of the ionic species involved in the equilibrium, and the diffusion electric tension. In reactions such as the hydrolysis of an ion or the formation of complexes, it is more important to know the real variations of the hydrogen ion concentration consequent on variations in the system; so that the measurements of electric tension may express such variations, the study of the system must be carried out in a medium of constant ionic strength. Equilibria are thus studied in solutions containing the ionic species under examination in the presence of other inert ions (ClO_4^-, Na^+, etc.) added in such concentration that the activity coefficients of the species under examination may be derived directly from the electric tension measurements. If the reaction is performed with the cell:

Electrode reversible with respect to H^+ or Me^{z+}	Solution S in equilibrium ClO_4^- 3 m; H^+	3 m $NaClO_4$	C.E. comparison electrode

the measured electric tension U may, from a practical standpoint, be considered as the sum of three terms: a constant value U_0 which is different for every type of half cell, a term due to the variation in U with the ionic concentration (which can be calculated with Nernst's equation) and a term U_j' which includes the diffusion electric tension (3 m $NaClO_4/S$) and a term due to the variation of the activity with H^+. For a cell formed by a comparison electrode (C.E.) and a hydrogen electrode

$$\text{Pt–H}_2/\text{H}^+/\text{C.E}$$

at 25° C

$$U = U_0 + 0.05916 \log [\text{H}^+] - (1/2) \cdot 0.05916 \log p_{\text{H}_2} + U_j'$$

$$U_j' = U_j + 0.05916 \log f_{\text{H}^+}$$

As the solution S tends towards the composition 3 m $NaClO_4$, U_j and hence U_j' tend to zero. Using solutions of known H^+ content and subtracting from the value of U obtained the term calculated with Nernst's equation, gives $(U_0 + U_j')$ which by extrapolation to $H^+ = 0$, yields the value of U_0[1].

The principal electrodes for the determination of pH are as follows.

Hydrogen electrode

This has been for a long time the most accurate electrode for the determination of pH and has been used both for the measurement of the pH of a solution and for the calibration of various electrodes which are reversible with respect to the hydrogen ion. In its most simple form it consists of a sheet of platinized platinum immersed in the solution whose pH is to be measured and through which bubbles pure gaseous hydrogen at a pressure of 1 atmosphere. Many forms of the hydrogen electrode have been described. A fairly useful type is shown in Fig. V, 4.

Platinization[2] of the platinum electrode consists in covering it with platinum black which acts as a catalyst for the overall electrode reaction

$$2\,H^+ + 2\,e^- \rightleftharpoons H_2\,(gas)$$

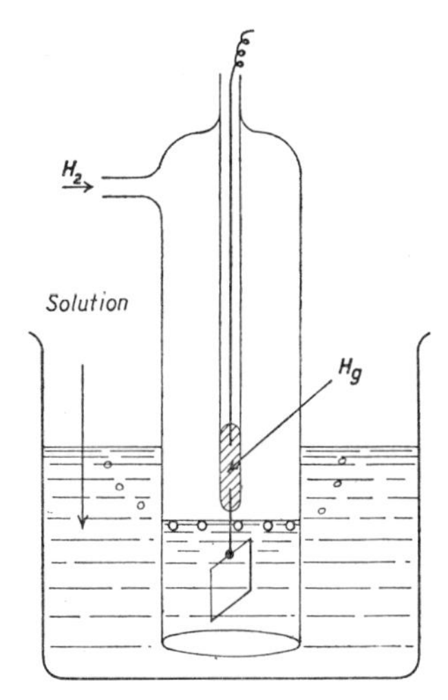

Fig. V, 4. Hydrogen electrode

[1] G. Biedermann and L. G. Sillén, *Arkiv Kemi*, 5 (1953) 425.

[2] Platinization is obtained by electrolytic reduction, with a current intensity of 200–400 mA for 1–3 minutes, of a 1–3% solution of chloroplatinic acid containing a small quantity of lead acetate trihydrate (80 mg per 100 ml of solution). The deposition of palladium black is sometimes used instead of platinization of the electrode.

The electric tension of the electrode is thus a function of the hydrogen ion activity and of the partial pressure of hydrogen, expressed in atmospheres; under working conditions the partial pressure of the hydrogen is not normally exactly equal to 760 mm Hg. If B indicates the atmospheric pressure and p_w the vapour pressure of water, then the partial pressure of the hydrogen is $(B - p_w)$ and the electric tension of the resulting hydrogen electrode is given by

$$U = (RT/\mathrm{F}) \ln (a_{\mathrm{H}^+}/\sqrt{p_{\mathrm{H}_2}})$$

$$= \frac{RT}{\mathrm{F}} \ln a_{\mathrm{H}^+} - \frac{RT}{2\mathrm{F}} \ln \frac{760}{B - p_{\mathrm{w}}}$$

Any substance which can effect the equilibrium of the electrode reaction makes the electrode give an incorrect value and many substances will 'poison' it permanently. Traces of oxygen interfere, since the platinum black acts as a catalyst for the combination of hydrogen and oxygen. Many organic substances also interfere because they can be hydrogenated when in contact with the platinum electrode; this is especially true of aromatic compounds so that buffers based on phthalate and benzoate cannot be used with this electrode. It cannot be used also in the presence of strongly oxidizing ($Cr_2O_7^{2-}$, MnO_4^-, etc.) or reducing (Sn^{2+}, Cr^{2+}, H_2SO_3) substances or metals whose standard electric tension is more positive than that of hydrogen (Cu^{2+}, Ag^+, Hg^{2+}) and which would hence be reduced. Equilibrium is often reached only slowly which makes it difficult to employ this electrode for everyday purposes; but notwithstanding all these limitations the hydrogen electrode in general gives the most accurate results of all the available electrodes.

Quinhydrone electrode

This is a redox electrode based on the reaction

$$\text{HO–C}_6\text{H}_4\text{–OH} \rightleftharpoons \text{O=C}_6\text{H}_4\text{=O} + 2\,\mathrm{H}^+ + 2\,\mathrm{e}^- \tag{1}$$

The electric tension of an inert electrode (gold or platinum) immersed in this system is given, at 25° C, by

$$U = U_0 + \frac{0.05916}{2} \log \frac{a_{\mathrm{Q}} \cdot a_{\mathrm{H}^+}}{a_{\mathrm{QH}_2}} \tag{2}$$

Quinhydrone is an addition compound of quinone (Q) and hydroquinone (QH_2) in equimolecular proportions; it is poorly soluble and in solution dissociates liberating equal parts of quinone and hydroquinone. In a solution containing an excess of quinhydrone the activities of the quinone and hydroquinone, produced by the dissociation of the quinhydrone, have virtually the same values and thus cancel out, so that equation (2) becomes

$$U = U_0 + \frac{0.05916}{2} \log (a_{H^+})^2$$

$$= U_0 - 0.05916 \text{ pH}$$

In other words, the electric tension of a quinhydrone electrode is, apart from a constant, equal to that of a hydrogen electrode immersed in the same solution. The value of the standard electric tension of the quinhydrone electrode varies with temperature and is reported in Table V,3

TABLE V,3

STANDARD ELECTRIC TENSIONS OF THE QUINHYDRONE ELECTRODE

$t°$ C	U_0	$t°$ C	U_0
0	0.7179	25	0.6995
5	0.7144	30	0.6954
10	0.7107	35	0.6920
20	0.7034	40	0.6886

The quinhydrone electrode is easily constructed since it is simply a bright (not platinized) platinum wire, immersed in the solution to which sufficient solid quinhydrone has been added to leave some undissolved (*i.e.* about the amount which will cover the tip of a spatula).

The quinhydrone electrode has the great advantage of simplicity; it is as accurate as the hydrogen electrode, assumes almost immediately the equilibrium electric tension, is not readily poisoned and may be used even in the presence of substances which can be reduced by hydrogen. Its main inconveniences are its limited range, which permits measurements only in solutions with a pH less than 8, and the inaccurate results which are obtained in the presence of high concentrations of electrolytes, oxidizing agents, proteins or complexing substances. In this case also the presence of any agent which alters the indicated electrode reaction equation, leads to an erroneous measure of pH. The two principal limitations of the quinhydrone electrode, *i.e.* high salt concentrations and alkaline conditions, can be explained. The former is due to different variations of the activity coefficients of quinone and hydroquinone so that their ratio does not remain unity. The second is due to acid dissociation of hydroquinone at pH values greater than 8. Moreover, in an alkaline medium, hydroquinone is readily oxidized by atmospheric oxygen.

Antimony electrode

The antimony–antimony oxide electrode has been much studied with the object of using it in place of the hydrogen electrode. It consists of a fused rod of antimony immersed in the liquid whose pH is to be determined; the addition of a little antimony trioxide to the solution is advantageous. The reaction which determines the electrode tension is not known for sure; it is believed to be due to a redox process between the antimony and the thin layer of antimony trioxide which is always present on the surface of the metal. This electrode reaction may be formulated as follows.

$$2\,\mathrm{Sb}(\mathit{solid}) + 3\,\mathrm{H_2O}(\mathit{liquid}) \rightleftharpoons \mathrm{Sb_2O_3}(\mathit{solid}) + 6\,\mathrm{H^+} + 6\,\mathrm{e^-}$$

Since antimony and antimony oxide are solids they may be considered to have unitary activity and so may the water, provided that the solution in use is sufficiently dilute. The electric tension of the electrode is thus a function of the hydrogen ion activity, which at 25° C is given by

$$U = U_0 + 0.05916 \log a_{\mathrm{H^+}}$$

The antimony electrode does not give very accurate results; it is not perfectly reversible; its electric tension depends not only on the hydrogen ion activity but also on the concentration of dissolved oxygen; and it may be further affected by the composition of the solution. Despite these limitations, the antimony electrode is widely used because it is robust, has a rapid response, and may be used at high temperatures and in the presence of cyanides, sulphites, reducing substances and alkaloids. However, oxidizing substances, complexing agents such as tartrates, citrates, phosphates and oxalates, and traces of some cations such as Ag^+, Cu^{2+}, and Cd^{2+}, interfere. It may be used over a range of pH values from 2 to 12, but the $\mathrm{d}U/\mathrm{dpH}$ function is not rigorously constant. One of the main causes of the deviation from complete reversibility with this electrode, is due to the possible formation of oxides other than Sb_2O_3, such as Sb_2O_4 and Sb_2O_5, so that collateral redox equilibria can be established besides that of the indicated electrode reaction. Such equilibria which, being in the solid phase are slow, give this the character of a mixed electrode so that its electric tension cannot be constant.

Other types of metal–metal oxide electrodes have been proposed, whose behaviour is identical to that of the antimony one. They may be used only in a restricted number of cases in that it is necessary for the oxide to have a very low solubility in the solution used and not to react with the ions present in the solution. Mention may be made of the mercury–mercurous oxide and silver–silver oxide electrodes which can be used in strongly alkaline ($pH > 9$) solutions. The electric tensions of these electrodes are strongly influenced by the presence of halides and ions which readily form complexes with the respective metals.

Glass electrode

Although not yet fully understood, the functioning of the glass electrode must be based on the existence of an electric tension between one surface of the glass and an aqueous solution, and varying regularly as a function of pH of the solution. Haber and Klemensiewicz[1] showed that the electric tension of the galvanic cell

$$Cu/Hg/Hg_2Cl_2\ KCl\,sat./Soln\ X/glass/0.1\ N\ HCl\ AgCl/Ag/Cu \qquad I$$

where a thin glass membrane separates the solution X of unknown hydrogen ion concentration from the 0.1 *N* solution of hydrochloric acid, varies with the hydrogen ion activity of solution X in the same way as in the following cell where the glass membrane has been replaced by a system of hydrogen electrodes.

$$Cu/Hg/Hg_2Cl_2\ KCl\,sat./Soln\ X/Pt,H_2,Pt/0.1\ N\ HCl\ AgCl/Ag/Cu \qquad II$$

A glass electrode thus consists essentially of a glass membrane in the form of a closed vessel containing a comparison electrode immersed in a solution of constant pH and containing the ion with respect to which the comparison electrode is reversible. The glass membrane is made from a special glass of low melting point and highly hygroscopic and conductive. Fig. V,5

[1] F. HABER and Z. KLEMENSIEWICZ, *Z. physik. Chem.*, 67 (1909) 385.

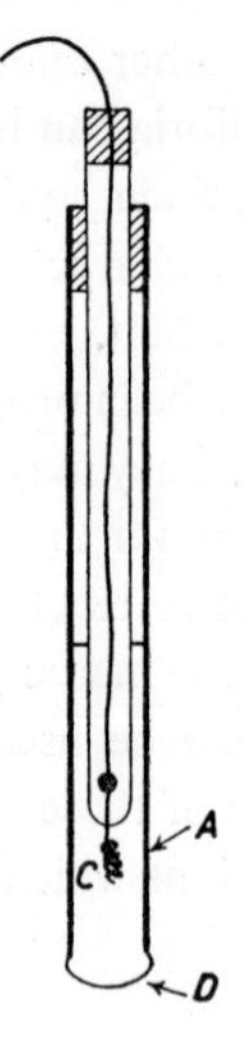

Fig. V,5. Glass electrode

shows the MacInnes and Dole[1] electrode obtained by fusing a membrane (D) of special glass[2] to the end of a tube (A); C is a silver–silver chloride or calomel, comparison electrode in hydrochloric acid or in a chloride buffer solution.

On the basis of cell diagrams I and II — which are thermodynamically identical — the behaviour of a glass electrode may be reduced to that of a concentration cell in that the work of these cells corresponds to the transfer of hydrogen ions from one cell to another. In cell II H^+ is reduced to gaseous hydrogen and then re-oxidized to H^+, whilst in cell I there is a simple reversible transfer of hydrogen ions. This diagram shows the behaviour of the glass electrode as it really occurs. It is not influenced by the presence of reducing or oxidizing substances and is the only electrode with this property. The specific behaviour of the glass electrode is interpreted, according to the most modern viewpoint, as a combination of ion exchange and proton transfer dependent on the special nature of the membrane. The hydrogen ions of the solution are exchanged with alkali ions of the glass, forming zones of enhanced stability. The glass electrode, when thus conditioned, forms a site where there is an exchange between the protons of the solution and those fixed on the surface of the glass; the potential of the surface varies with the loss or gain of protons.

The electric tension U_V of the glass electrode is given, at 25° C, by

$$U_V = U_{0V} + 0.05916 \log a_{H^+}$$

[1] D. A. MacInnes and M. Dole, *J. Am. Chem. Soc.*, 52 (1930) 29.
[2] The most commonly used glass is Corning 015 which has the molar composition: 72.2% SiO_2, 6.4% CaO, 21.4% Na_2O.

This is only valid for a limited range of pH values, which depends on the type of glass used for the membrane. Corning 015 glass gives a range of pH values from 1 to 9. A glass electrode of this type shows a negative error in strongly acid solutions and a positive error in alkaline solutions, *i.e.* it gives too high a pH value in acid solution and too low a value in alkaline solution. The first error is due to a variation of the activity of the water in the solution, whilst the second is essentially due to the presence of sodium ions which diffuse through the membrane. The measurement of the pH of strongly alkaline solutions may however be performed with electrodes whose membranes are made of special glasses (*e.g.* lithium glass). The equation given above shows that every glass electrode has a standard electric tension which may differ from electrode to electrode; thus each electrode must be calibrated with a solution of known hydrogen ion concentration, which is usually a buffer. If the glass electrode is placed in a solution of hydrogen ion activity identical to that within the membrane, no electric tension should be shown between the inside and the outside; but in fact a very small tension (1–2 mV) is found due to asymmetry of the system, *i.e.* to the fact that the two surfaces of the membrane are differently conditioned both in manufacture and by use, and so are not strictly equivalent. The asymmetry tension of a glass electrode is not constant and thus the electrode must be frequently controlled in use, with a buffer solution.

The glass electrode has an extremely high internal resistance of the order of 10^6–10^8 Ω and thus the electric tension of a cell which contains such an electrode cannot be measured with a normal potentiometer since a galvanometer of extremely high sensitivity would be needed; the measurement is normally made with an electronic voltmeter. The present day tendency is to use a very sensitive electronic arrangement with electrodes of very high internal resistance so as to make them as mechanically resistant as possible.

Glass electrodes have various advantages. They have a very wide field of application, rapidly reach the equilibrium electric tension, can be used in strongly oxidizing or reducing media and in the presence of organic substances, eliminate the need for platinization and for a supply of gaseous hydrogen and finally make it possible to carry out the measurements on very small amounts of liquid. Their disadvantages are the delicacy of the glass membrane and inaccurate results obtained in the presence of large quantities of ions of small diameter, *e.g.* Li^+ and Na^+.

TABLE V,4

CHARACTERISTICS OF ELECTRODES USED FOR THE MEASUREMENT OF pH

Electrode	pH	*Substances interfering with the electrode*	*Causes of errors*	*Repro-ducibility* (mV)	*Notes*
Hydrogen	0–14	Reducing and oxidizing agents, heavy metals, air	Incomplete saturation, O_2 in H_2, poisons, viscous solutions	0.1	Follows theoret-ical relationshi slow equilibriun strong reducing action
Quinhydrone	1–8	Alkalis, reducing and oxidizing agents, complexing agents, proteins	Poisons, salts	0.1	Follows theoret-ical relationship rapid equilibriu
Antimony	3–10	Strong alkalis and acids, H_2S, Cu^{2+}	Oxidizing agents, organic com-pounds, salts, Cl_2, H_2S	5	Does not follo exact theoretic relationships; must be calibra
Glass	0–12	Dehydrating solu-tions, colloids, surface deposits	High concentra-tions of alkali ions	0.5	Follows theore ical relationshi over certain pH ranges depend on the type of glass; must be calibrated

The mean error found with the glass electrode is ± 0.05 pH but this can be reduced to ± 0.02 pH if the electrode is calibrated immediately before and after each determination. Table V,4 shows the characteristics of the most important electrodes used for pH measurements.

The determination of pH *in nonaqueous solutions*

The above considerations are also valid for electrometric determinations of acidity in nonaqueous media. It must, however, be remembered that the pH of these solutions cannot be considered as a true value of the hydrogen ion concentration and so cannot be related directly with the equilibrium conditions of the solution itself. Glacial acetic acid is an exception to this and Kolthoff and Bruckenstein[1] developed a rigorous treatment for it. The values of pH obtained from electrometric determinations for the nonaqueous solutions and for gels, tissues, soft solids and similar systems must be considered simply as numerical values which are of practical use for their reproducibility. The equations derived for aqueous solutions are not applicable to nonaqueous ones, for several reasons. The most important of these is the high value of the electric tension which is established at the interface between different solvents of disparate physical and chemical properties as, for example, between the nonaqueous medium and the aqueous solution of potassium chloride in the comparison electrode. The value of this electric tension becomes less reproducible and less stable as the concentration of water diminishes. The solutions also have a high electrical resistance and so the potentiometric measurements are less sensitive. A further difficulty is the choice of a suitable standard solution and a suitable reference scale of acidity. In view of these difficulties, and above all the uncertainty about the contact electric tension, it is not worth while to seek to correlate the measured electric tensions under these conditions.

At the present time the only satisfactory method for dealing with electrometric determinations of acidity in nonaqueous media, is to consider each solvent as an absolutely independent system, without making any direct reference to values obtained in aqueous solutions.

The hydrogen electrode in any solvent is always the reference electrode and the comparison electrode must always contain the same nonaqueous solvent which is being used. Many values of standard electric tensions for electrolytes in various solvents (methanol, ethanol, ammonia etc.) have been obtained on this basis. These values have a purely conventional significance in that the electric tension of the standard hydrogen electrode is arbitrarily taken as zero in every solvent whereas in fact it almost certainly has different values in the different solvents. The other electrodes described (glass, hydroquinone, antimony) may be used as well as the hydrogen electrode, bearing in mind the limitations mentioned. The most convenient electrode for the determination of acidity in glacial acetic acid is that with chloranile — an aquimolecular mixture of tetrachloroquinone ($C_6Cl_4O_2$) and the corresponding hydroquinone ($C_6Cl_4(OH)_2$) — connected to the comparison electrode with a bridge composed of a supersaturated

[1] I. M. Kolthoff and S. Bruckenstein, *J. Am. Chem. Soc.*, 78 (1956) 1; 79 (1957) 1.

solution in acetic acid containing a little gelatine. The diffusion electric tension of this bridge is practically constant so that the measured electric tension may be considered as a direct measure of the hydrogen ion activity. The saturated calomel electrode is most generally used for comparison; the silver–silver chloride electrode has had more limited use.

(II). *Potentiometric Titrations; Theory*

This method makes possible determinations of the total concentration of a substance, independently of any incomplete dissociation, so long as certain conditions are fulfilled; and it is sometimes possible with a single estimation to determine a number of ionic species present together in the solution. The method is based on the examination of the changes in the electric tension of a suitable electrode — immersed in the solution of the electrolyte which is to be analysed — on the addition of a titrating reagent. This electric tension is dependent on the activity of the electrochemically active ions and, at constant temperature, varies as the activities of the ions vary as described in Chapter III.

During an analytical reaction the environmental conditions (ionic strength, temperature, activity of the solvents, etc.) may be considered as constant and hence the activity coefficients of the various species may also be considered as constant. In such conditions then, the changes in electric tension are actually uniquely determined by changes in the concentration of the electrochemically active ions so that their determination is possible. If they react with analytical titrating reagents or are produced by the reaction of the initial substance with the reagent, their concentration must be a function of the volume of reagent added so that the electric tension

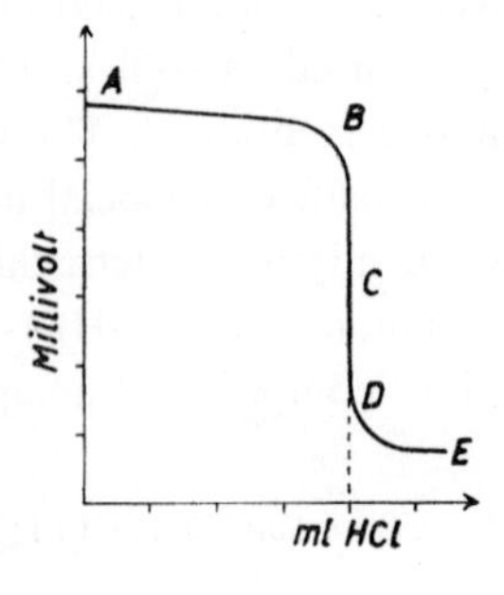

Fig. V,6. Potentiometric titration

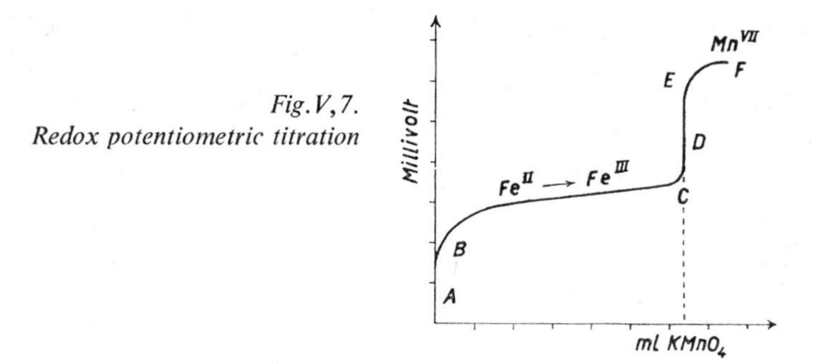

Fig. V, 7.
Redox potentiometric titration

of the electrode is also a function of this volume. If the values of the measured electric tension, or quantities proportional to it such as pH, pA[1], or pM, are plotted on a graph as ordinates against the added volume of reagent, titration curves are obtained of the kind showed in Fig. V, 6 and Fig. V, 7.

The two figures refer respectively to titrations of solutions of silver nitrate with hydrochloric acid and of ferrous sulphate with permanganate. The curves show points of inflexion which virtually coincide with the equivalence points of the particular reactions.

Potentiometric titrations may be applied to all types of volumetric determinations — neutralizations, precipitations, oxido–reductions and complex formation — provided that a convenient indicator electrode is available. The progress of a potentiometric titration may be easily followed by considering the reaction between a cation C and an anion A to form a poorly soluble salt AC. Consider 100 ml of the cation solution C to which is added a standardized solution of the anion A. This solution may be taken as having a normality 10-fold greater than that of C. A pair of electrodes is placed in the solution, one being the comparison electrode and the other the indicator of the activity of the cation C. The addition of the first 9 ml of the standard solution lowers the amount of cation C present in solution in the ratio 100/10 *i.e.* 10 because it precipitates the poorly soluble compound AC. Hence, ignoring for the moment the increase in volume of the solution and the solubility of the precipitate, the electric tension of an electrode which is reversible for cation C diminishes

[1] In a reaction which forms a complex, $M^{z+} + nH^- = MA_n^{(z-n)+}$, where $pA = -\log [A^-]$, and $pM = -\log [M^{z+}]$.

at room temperature, by $0.059/z$ V (where z is the cation charge). The addition of a further 0.9 ml of the solution of anion A lowers the quantity of cation C present still further in the same ratio of 10/1. Hence, the change in the electric tension of the electrode is the same, but for the addition of a much smaller volume. Adding a further 0.09 ml of the anion solution diminishes the amount of cation C present, in the same ratio 1/0.1 *i.e.* 10, and the change in electric tension is still the same. The addition of a further 0.01 ml brings the reaction to its end point when A has been added in an amount equivalent to C and the concentration of C corresponds to that of a saturated solution of AC. On adding still more of the anion solution, the concentration of cation C diminishes further since it is linked with the concentration of A by the solubility product relationship.

The progress of the titration curve described is shown diagrammatically in Fig. V, 6. It can be seen that the change in the electric tension is greatest at the end point which is located at point C and that the portion CDE is symmetrical with respect to the portion ABC since the charge on the cation is equal to that on the anion. However, the titration curve is not symmetrical in the general case where the charges on the ions involved in the reaction, are different.

A similar type of titration curve is observed in the potentiometric determination of an acid–base system. In this case a poorly dissociated compound is formed in place of the poorly soluble precipitate: $H^+ + OH^- \rightarrow H_2O$. A similar type of electric tension–volume diagram is found also for oxidimetric reactions, *e.g.* the titration of a ferrous salt with a suitable oxidant ($K_2Cr_2O_7$, $KMnO_4$ etc.), as shown in Fig. V, 7.

The electric tension of an inert electrode immersed in the initial solution of the ferrous salt, at room temperature, is given by the equation

$$U = U_0 + 0.059 \log (a_{Fe^{3+}}/a_{Fe^{2+}})$$

As was shown above, the ratio of the activities may be replaced by the ratio of the concentrations $[Fe^{3+}] / [Fe^{2+}]$, which initially has a very small value since the iron is almost exclusively in the ferrous state at equilibrium, with only analytically negligible traces of the Fe^{3+} ion. The first drop of oxidant thus leads to a marked percentage change in the concentration ratio $[Fe^{3+}] / [Fe^{2+}]$, and thus to a marked change in the electric tension of the electrode (the portion *AB* of the diagram). As the titration proceeds, the change in the concentration ratio $[Fe^{3+}] / [Fe^{2+}]$, for a constant addition of oxidant, diminishes and the change in the electric tension thus becomes smaller (the part *BC* of the diagram). As the end point is approached the concentration of ferric ion becomes preponderant over the concentration of ferrous ion and the same quantity of oxidant causes steadily increasing changes in the concentration ratio. The graph once more shows a rapid change (portion *CDE*). At the end of the reaction a certain, very small, amount of ferrous ion is always present since the oxidation is an equilibrium reaction. The further addition of oxidant thus makes only small changes in both concentrations and thus the ratio changes little and the consequent variations in the electric tension of the electrode are also small (portion *EF*). Other oxidation or reduction reactions give similar diagrams.

For all cases, it is clear that on adding small and constant amounts of the titrant and measuring the change in electrode tension for each addition, the end point is reached when the change becomes a maximum. The value of the first derivative dU/dv, and hence the slope of the curve here reaches a maximum and, moreover, since this is a point of inflexion, the second derivative d^2U/dv^2 becomes zero. For reactions of a simple type as for example between uni–univalent electrolytes

$$Ag^+ + Cl^- \rightarrow AgCl \ (solid)$$

starting with equivalent concentrations, d^2U/dv^2 becomes zero when the volume of titrant, containing the anion, added to the solution containing the cation is equal to the initial volume of this latter solution, *i.e.* this condition is satisfied at equivalence. When the initial concentrations are not equal, or when the reactions are more complex (such as poorly soluble precipitates of the formula A_mC_n when $m \neq n$ and both are greater than 1; more or less complex redox reactions, etc.), the point of inflexion does not coincide exactly with the equivalence point due to the dissymmetry which the diagram assumes about the equivalence point in these cases. If, however, the solubility product for the precipitation reactions is fairly small, or the equilibrium constant for the redox reactions is fairly large, the difference between the points of inflexion and equivalence is quite negligible.

In the case of acidometric titrations the point of inflexion coincides with the point of equivalence only in reactions between strong acids and strong bases. They no longer coincide in titrations of weak acids or weak bases. The difference is, however, negligible provided that $cK > 10^{-10}$ where c is the initial concentration of the weak acid or base and K is its dissociation constant. It is evident that in the special case of acid–base titrations, polyprotic acids or polyacidic bases can also be titrated potentiometrically. For a diprotic acid the two steps in electric tension are theoretically detectable if the ratio of the dissociation constants is greater than 16. In practice, however, it must be very much more than this to obtain accurate results.

Four conditions must be fulfilled by an analytical reaction which is to be used potentiometrically.

1. The substance to be titrated and the titrant must react with each other in a fixed, known, stoichiometric ratio.

2. An indicator electrode must be available whose electric tension is uniquely and reversibly determined by the concentration of at least one of the ions involved in the analytical reaction.

3. The analytical reaction must be as complete as possible, *i.e.* precipitation reactions giving a highly insoluble product; redox reactions with a high equilibrium constant; and complexometric reactions with a reasonably high formation constant; the conditions required for acidimetry have been described above.

4. A rapid equilibrium must be established between the reacting components and the indicator electrode.

(III). *Methods for the Detection of the End Point of a Potentiometric Titration*

The detection of end points may be carried out graphically or by calculation, or else the end point may be determined experimentally. Of the many methods which have been proposed, only a few of the most commonly used will be mentioned here.

Graphical and calculation methods

These methods are based on the pattern of the measured electric tension changes as a function of the volume of reagent added. For their use it is necessary to know the value of the electric tension which will be indicated by the electrode near the end point.

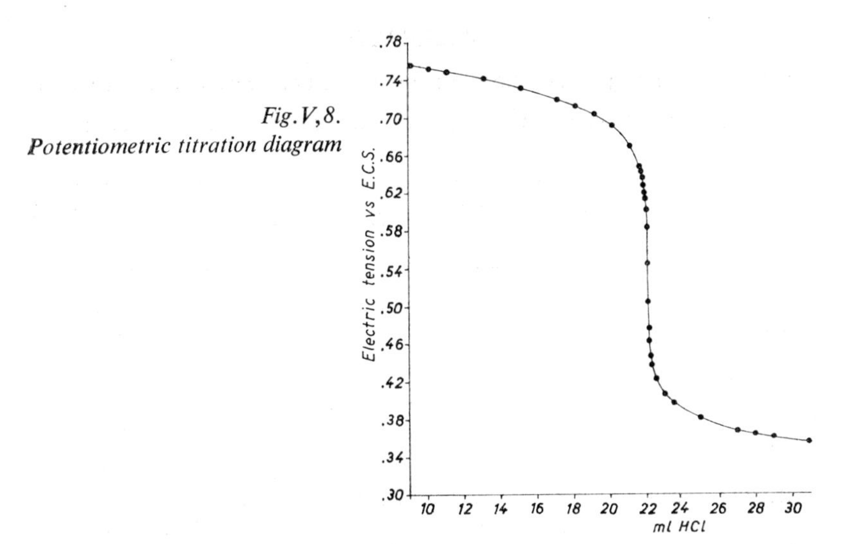

Fig. V, 8. Potentiometric titration diagram

One of the graphical procedures consists in drawing the titration curve with the measured electric tensions as ordinates and the added volumes of reagent as abscissae. If a marked change in electric tension is found, as in the graphs of Fig. V, 6 and Fig. V, 7, the point of inflexion is immediately visible and the corresponding abscissa gives the volume of reagent corresponding to the equivalence.

A more exact method is to draw a graph with the volumes as abscissae and the function $\Delta U/\Delta v$, or $\Delta^2 U/\Delta v^2$ or even, as Gran[1] recently proposed $(\Delta v/\Delta U)/\bar{v}$ where $\bar{v}$ is the mean volume at each point. This procedure is illustrated for the determination of the end point of a titration of 30 ml of 0.0732 *N* borax with 0.100 *N* HCl using the hydrogen electrode as the indicator and the saturated calomel electrode as the comparison electrodes (Fig. V, 8). The experimental values are reported in tables together with the calculaed factors required for determining the end point (Table V, 5).

Fig. V, 9 shows the three graphical procedures: (a) the method with the first derivative, (b) the method with the second derivative and (c) Gran's method.

All three procedures when correctly applied, give the same result but that of Gran is the easiest to use. In this method the calculated points

[1] G. Gran, *Acta Chem. Scand.*, 4 (1950) 559.

TABLE V,5

THE TITRATION OF 30 ML OF 0.0732 N BORAX WITH 0.1 N HYDROCHLORIC ACID

(Indicator electrode: hydrogen electrode; comparison electrode: saturated calomel electrode)

v ml	U volt	$\bar{v}$	$\dfrac{\Delta U}{\Delta \bar{v}}$	$\bar{v}_1$	$\dfrac{\Delta^2 U}{\Delta v^2}$	$\dfrac{\Delta v}{\Delta U}$	$\left(\dfrac{\Delta v}{\Delta U}\right)\dfrac{1}{\bar{v}}$
0.00	0.7893						
10.00	0.7538						
15.00	0.7328						
20.00	0.6932						
21.00	0.6725						
		21.25	− 0.042			23.81	1.11
21.50	0.6514			21.40	− 0.08		
		21.54	− 0.065			15.38	0.714
21.58	0.6462			21.58	− 0.27		
		21.63	− 0.089			11.24	0.520
21.67	0.6391			21.67	− 0.31		
		21.71	− 0.114			0.77	0.404
21.75	0.6300			21.75	− 0.51		
		21.78	− 0.155			6.45	0.296
21.80	0.6238			21.80	− 1.13		
		21.82	− 0.200			5.00	0.229
21.84	0.6158			21.84	− 2.58		
		21.86	− 0.303			3.30	0.151
21.88	0.6037			21.88	− 3.38		
		21.90	− 0.438			2.28	0.104
21.92	0.5852			21.92	− 12.57		
		21.94	− 0.948			1.05	0.047
21.96	0.5483			21.96	− 3.05		
		21.98	− 1.070			0.93	0.048
22.00	0.5055			22.00	10.05		
		22.02	− 0.668			1.50	0.068
22.04	0.4788			22.04	8.20		
		22.06	− 0.340			2.94	0.197
22.08	0.4652			22.10	2.40		
		22.13	− 0.196			5.10	0.230
22.17	0.4495			22.18	0.98		
		22.22	− 0.118			8.47	0.381
22.26	0.4401			22.30	0.31		
		22.38	− 0.068			14.71	0.657
22.50	0.4238						

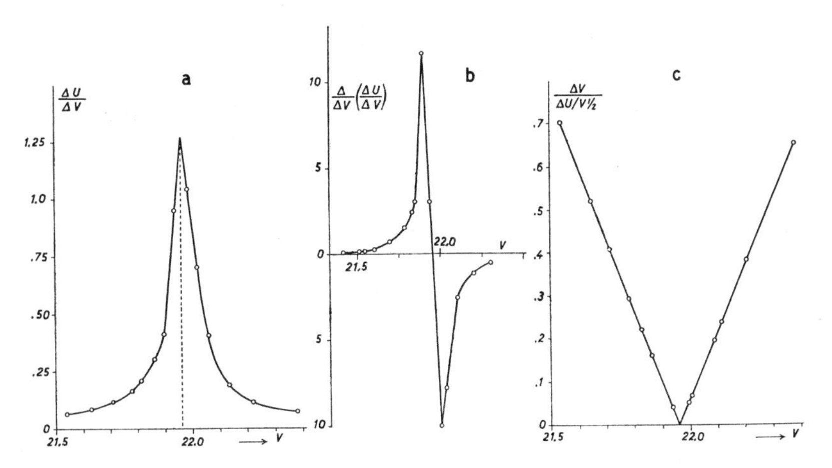

Fig. V,9. Graphical methods for the detection of the end point

form two straight lines whose intersection gives unequivocally the end point. In this way the uncertainties due to extrapolation of curves which are encountered when the changes of tension during the titration are very small, are eliminated.

Analytically, the end point may be interpolated between the points at which the second difference becomes zero ($\Delta^2 U = 0$) and changes sign. From the data given in the above table, the volume corresponding to equivalence is

$$21.96 + 0.04 \cdot \frac{3.05}{3.05 + 10.05} = 23.98$$

Experimental methods

A group of procedures is based on the estimation of the electric tensions of an indicator electrode immersed in a solution of the identical composition to that which will be produced at the end point. Knowing this value it is sufficient to set up a simple experimental arrangement to allow the addition of the reagents until this value is reached. Pinkhof's procedure is shown in Fig. V, 10.

The solution to be titrated is placed in the vessel A and the reference solution in the vessel B. These vessels are connected together by a salt bridge. Identical electrodes are placed in each vessel and connected

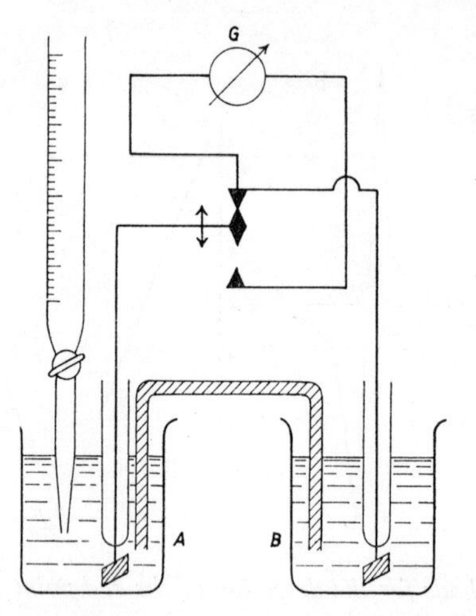

Fig. V, 10. Pinkhof's circuit for potentiometric titration

together through a galvanometer G by means of a key. The deflection of the galvanometer diminishes during the titration and becomes zero at the equivalence point, when both solutions have the same composition. The further addition of reagent causes a deflection in the opposite sense.

It is possible to eliminate the principal defect of this procedure, which is the necessity to prepare a special electrode for each titration, by introducing a modification consisting of the coupling by means of a salt bridge of a comparison electrode with the indicator electrode (Fig. V, 11).

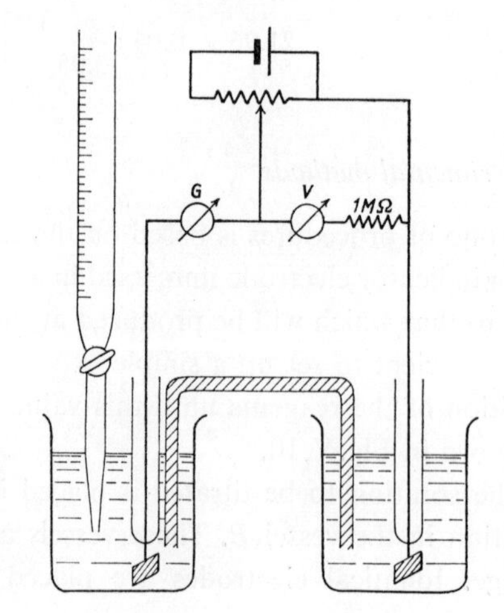

Fig. V, 11. Müller's circuit for potentiometric titration

An electric tension, from a battery, is applied in opposition to that of the galvanic cells thus formed; this tension is equal to that which would be established between the electrodes at the equivalence point. A galvanometer inserted into the circuit indicates the zero. There is no flow of current at the equivalence point and there is an inversion of the deflection on the addition of excess reagent.

TABLE V, 6

STANDARD TENSIONS OF COMPARISON ELECTRODES AT 25° C

Electrode	*U*	*Titration*	
Hg/Hg_2CO_3 in 1 *N* Na_2CO_3	+ 0.048	I^-	with $AgNO_3$
$Hg/Hg_2C_2O_4$ in $Na_2C_2O_4$ soln.	+ 0.182	Br^-	with $AgNO_3$
		SCN^-	with $AgNO_3$
		I^-	with $Hg_2(ClO_4)_2$
		Pb^{2+}	with $K_4[Fe(CN)_6]$ at 75°
		I_2	with $Na_2S_2O_3$
$Hg/Hg_2(CH_3COO)_2$ in 2 *N* CH_3COONa	+ 0.208	I_2	with $Na_2S_2O_3$
$Hg/Hg_2(CH_3COO)_2$ in 1 *N* CH_3COONa	+ 0.245	Cl^-	with $AgNO_3$
		Pb^{2+}	with $K_4[Fe(CN)_6]$ at 18°
Hg/Hg_2SO_4 in 1:2 H_2SO_4	+ 0.313	Zn^{2+}	with $K_4[Fe(CN)_6]$ at 75°
		Cl^-	with $Hg_2(ClO_4)_2$
		Br^-	with $Hg_2(ClO_4)_2$
		Sn^{2+}	with $K_2Cr_2O_7$
Hg/Hg_2SO_4 in sat. K_2SO_4	+ 0.36	Zn^{2+}	with $K_4[Fe(CN)_6]$ at 75°
Hg/Hg_2SO_4 in 1 *N* K_2SO_4	+ 0.365		
Hg/Hg_2SO_4 in sat. Na_2SO_4	+ 0.358		
Pt/I_2 in 2 *N* KI	+ 0.265	Cl^-	with $AgNO_3$
Pt/Br_2 in 0.1 *N* NaBr	+ 0.893	Fe^{2+}	with $KMnO_4$
Pt/Br_2 in 1 *N* NaBr	+ 0.846	$C_2O_4^{2-}$	with $KMnO_4$
Pt/Br_2 in NaBr sat. at 18°	+ 0.798		
Pt/Br_2 in NaBr sat. at 18°	+ 0.798	As^{3+}	with $KBrO_3$
Sat. calomel electrode	+ 0.2444		
0.1 *N* calomel electrode	+ 0.3356		
Silver–silver chloride electrode	− 0.2223		

Another experimental arrangement consists in constructing a cell from an indicator electrode and a comparison electrode having the same electric tension as that assumed by the indicator electrode at the end point; the circuit is completed with a galvanometer as in the preceding system. The titration reaches its end point when no flow of current can be observed in the external circuit. Table V,6 gives comparison electrodes for some of the more common titrations, as described by Lanz[1].

The procedures described are particularly attractive for the simplicity of the apparatus required, which allows a titration to be performed very rapidly. This, however, is achieved at the expense of accuracy both because the electric tension does not immediately assume its equilibrium value in many titrations, particularly near the end point, and because polarization phenomena may interfere since the system is balanced only at the equivalence point. It must also be remembered that the value of the electric tension at the equivalence point may be markedly influenced by the presence of electrolytes and very often by changes in the hydrogen ion concentration, so that the procedures described are particularly useful for routine determinations but are not recommended for very accurate measurements.

The value of the ratio $\Delta U/\Delta v$ (with constant increments in volume) may be directly obtained experimentally with the so-called *differential titration method*, or 'derivative potentiometric titration'. The apparatus required is shown in Fig. V,12. The electrode 1 is immersed in the medium separated from the bulk of the solution and connected with it only through a small pin-hole. The addition of reagent produces the reaction in the bulk of the solution, but it does not reach the solution in which the electrode 1 is immersed in sufficient quantity to provoke a detectable reaction. Thus, an electric tension is established between electrodes 1 and 2 which corresponds to the added volume. By keeping the volumes of reagent added constant, the electric tension measured between the two electrodes is proportional to the ratio $\Delta U/\Delta v$. At the end of each measurement a little gas is passed through a three-way tap to renew the solution about electrode 1 and to bring it into equilibrium with the bulk of the solution. Plotting the values of the added volume and the measured electric tension between the two electrodes gives a curve of the type illustrated in Fig. V,9a.

[1] H. LANZ, *Die Anwendung der Umschlagselektrode bei der Potentiometrischen Massanalyse*, Dissertation, Dresden, 1929.

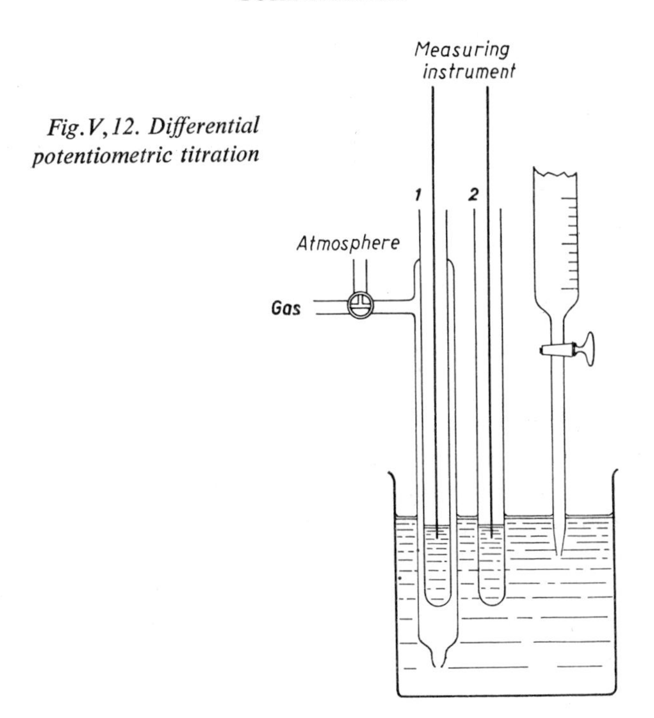

Fig. V, 12. Differential potentiometric titration

Another experimental procedure employs a bimetallic electrode system. If indicator and inert electrodes are inserted into a solution, the electric tension established between the electrodes corresponds to the change in electric tension of the indicator electrode. If, for example, electrodes of platinum and palladium are inserted into an oxidizing or reducing solution, the former rapidly reaches the equilibrium electric tension whilst the latter does so only very slowly and there is practically no change in the electric tension during the course of a titration whilst there is a marked change at the end point. A bimetallic system may be prepared with two equal electrodes, to which a small polarization tension is applied. An application of such a procedure is described on p. 360 (titration with polarized electrodes).

Potentiometric titrations constitute the most versatile method of electrochemical analysis and practically every analytical determination of both inorganic and organic systems carried out with classical methods, could be performed potentiometrically. The technique offers notable advantages in extending the range of usefulness of conventional methods.

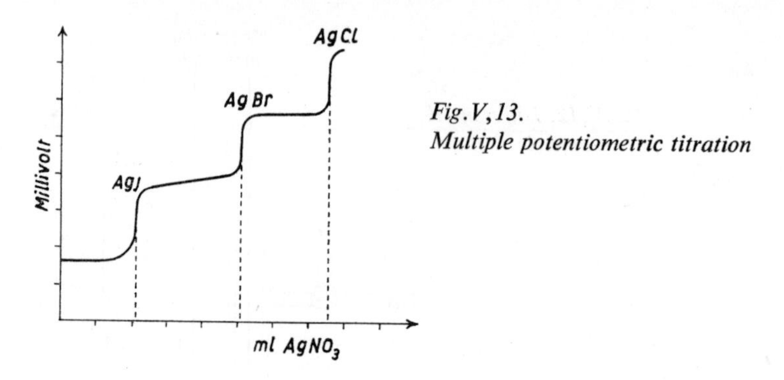

Fig. V, 13.
Multiple potentiometric titration

For example, it is possible to carry out acid–base titrations with weak acids or bases, or in coloured solutions, or in nonaqueous media in which the use of indicators is not possible; differential determinations may be made of substances with different solubility products or with different redox tensions or formation constants. Finally, under suitable conditions, the method also allows the simultaneous determination of several species in a single operation. For example, using a silver electrode as the indicator, it is possible with a single potentiometric titration to determine iodide, bromide and chloride since their different solubility products are reflected in the titration curve as three distinct steps in electric tension, corresponding to the quantitative precipitation of the silver salts in the order AgI, AgBr and AgCl, as shown in Fig. V, 13.

Similarly, the concentrations of stannous and ferrous ions present together in a solution, may be determined by a single titration with permanganate using a platinum wire as the indicator electrode as a consequence of the different redox tensions of the two systems Sn^{2+}/Sn^{4+} and Fe^{2+}/Fe^{3+}. The concentrations of many cations may be determined by titration with ethylenediaminetetracetic acid, using a mercury electrode as an indicator of pM.

The accuracy of potentiometric titrations is essentially a function of the selected analytical procedure. The selectivity of the method is notably better than that of classical ones, in that it is often possible to obtain differentiated titration curves for the various electrode processes.

A special case is that of the determination of the end-point of a redox reaction, which can also be carried out with a suitable indicator, called a redox indicator. A substance can act as a redox indicator if its oxidized form In_{ox} has a colour distinctly different from that of the reduced form

In_{red}, and if the relevant oxidation–reduction reaction of the system ($In_{ox} + ze^- \rightleftharpoons In_{red}$) is sufficiently rapid. The electric tension of the corresponding galvanic half-cell is given by Nernst's equation:

$$U = U_0' + \frac{RT}{z\mathrm{F}} \ln \frac{[In_{ox}]}{[In_{red}]}$$

where U_0' represents the 'formal tension' of the system[1]. If it is assumed that the colour intensities of the reduced and oxidized forms are proportional to their concentrations, and if the colour of one form is apparent rather than that of the other when the two forms occur in the ratio of 10 : 1, then the indicator will undergo its full colour change for a variation in electric tension of $2 \cdot 0.059z$ *i.e.* about 0.12 V when $z = 1$.

The range of electric tensions over which the indicator changes its colour is thus

$$U = U_0' \pm 0.059/z$$

at 25° C.

For a given redox reaction the indicator chosen must have a U'_0 value close to the electric tension corresponding to the end point of the titration or within the region of rapid change of the titration curve. An indicator can be used however only if there is a considerable change in electric tension at the end point so as to give a definite colour change.

Informations about many redox indicators have been collected by Tomiček[2]; many other redox indicators have, however, been studied and used more recently[3].

[1] The term formal tension is used to indicate that electric tension which can be measured for a redox electrode – in which the electrochemically active substances are at unit concentration – in the ionic medium and under the actual conditions of use, rather than under the conditions which would correspond to unit activity of the oxidized and reduced forms (the standard electric tension). The formal tension is more useful in practice for analytical purposes than is the standard tension, in that it indicates the oxidizing or reducing power of a system under the conditions in which it is used. The term U_0' includes variations in the standard electric tension dependent on the inter-ionic forces (activity coefficients) and on the formation of complexes at a particular hydrogen ion concentration, if the H^+ ion takes part in the redox reaction.

[2] O. Tomiček, *Chemical Indicators*, Butterworth, London, 1951.

[3] See for example F. T. Eggersten and F. T. Weiss, *Anal. Chem.*, 28 (1956) 1008.

3. Conductometric Methods at Low and at High Frequency

(I). *Conductometry*

Various problems in scientific investigations may be solved by conductance measurements, *e.g.* the determination of a given ionic species, the basicity of polybasic acids, the degree of dissociation of nonbinary electrolytes, the solubility of electrolytes, the degree of hydrolysis, the velocity of reactions etc. Conductometric titrations for analytical purposes are particularly important and are usually called simply conductometric analyses; here, use is made of conductance measurements to detect the end points of analytical reactions.

The performance of a conductometric analysis is possible only if, at the end of the analytical reaction, there is a marked change in the conductance: a change which actually serves for the detection of the end point of that reaction. If in fact a suitable reagent is added to the solution to be analyzed, the composition of the original solution changes both qualitatively and quantitatively. Correspondingly, it may be possible to observe a change in the conductance if the chemical species added or formed by the reaction have different mobilities. If the reagent used is suitably selected, there is a sudden change in the pattern of the conductance at the end of the reaction; this change will be increased with successive additions of reagent. Plotting graphically the conductances, as ordinates, against the respective volumes of reagent added, the points will be found to fall on to two branches, through which two straight lines may be drawn. The endpoint of the reaction corresponds to the point of intersection of these lines. Titration curves are of different types depending on the particular reaction. In general, however, they may be reduced to three fundamental schemes as far as the behaviour of the conductance up to the end point is concerned. After the end point, the second branch of the graph is always ascending; in fact the reagent used is generally a strong electrolyte whose addition leads to an increase in the conductance of the solution. An example of the first type in which there is a fall in conductance is the neutralization of strong acids with strong bases, since the H^+ and OH^- ions have very much greater mobilities than any other ions. In titrating a strong acid such as hydrochloric acid with sodium hydroxide, the reaction up to the equivalence point may be considered as a replacement of the very mobile H^+ ion by the much less mobile Na^+ ion; the H^+ ion reacts with the OH^- ion forming water which is only very slightly disso-

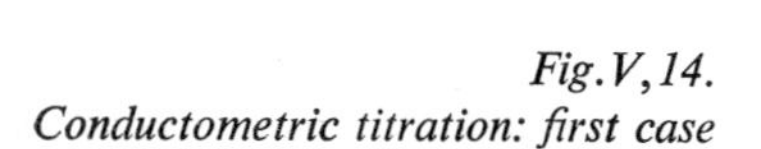
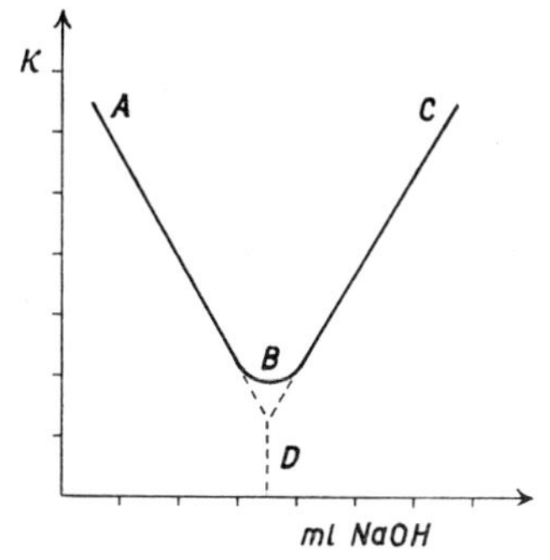

Fig. V, 14.
Conductometric titration: first case

ciated. Thus, whilst at the beginning of the reaction H^+ and Cl^- ions are present in the solution, at the end of the reaction Na^+ and Cl^- ions are present and so the conductance diminishes until this point is reached. The further addition of sodium hydroxide does not produce any further reaction but the ionic concentration of the solution increases and since, moreover, the increase is partly due to the OH^- ion, with its very high mobility, there is an increase in conductance. If the conductance is plotted as a function of the volume of standard sodium hydroxide solution added, a graph of the type shown in Fig. V, 14 is obtained.

Here, the descending branch *AB* corresponds to the change of conductance between the initial value and that at the equivalence point; whilst the ascending branch *BC* corresponds to the change in conductance after this point. Extrapolating the two branches of the graph gives a point of intersection *D*, whose abscissa gives directly the volume of standard sodium hydroxide solution required to neutralize the acid originally present. Exactly the same reasoning applies for the titration of a strong alkali with a strong acid.

The second type is usually encountered when an ionic species is replaced during the titration by another species whose mobility does not greatly differ from that of the first. This happens when the substance being analyzed is a salt which is titrated with a reagent to form an insoluble or poorly dissociated product, as for example in the formation of complexes. Generally, the portion of the curve *AD* from the beginning to the equivalence point is more or less horizontal, whilst the further addition of reagent after the equivalence point *D*, provokes a rapid increase in conductance with a corresponding ascending curve *BC*; the point of inter-

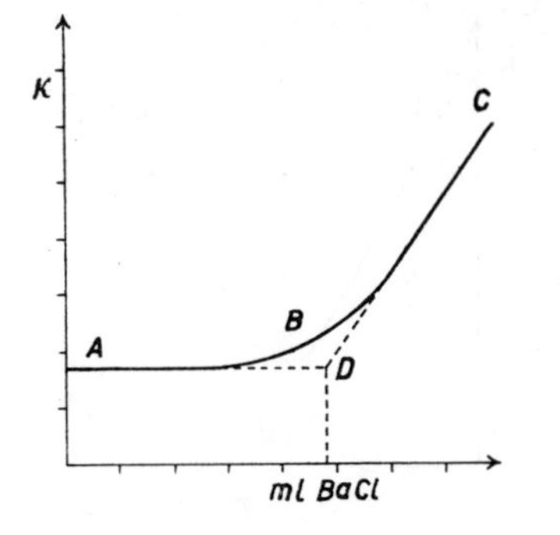

Fig. V, 15.
Conductometric titration: second case

section of the two branches of the graph indicates the end point of the reaction.

On titrating a solution of silver nitrate with a solution of barium chloride, the following reaction occurs.

$$2\,Ag^{+} + 2\,NO_3^{-} + Ba^{2+} + 2\,Cl^{-} \rightarrow 2\,AgCl\ (\text{solid}) + Ba^{2+} + 2\,NO_3^{-}$$

As this reaction proceeds the silver chloride separates and the Ag^{+} ions in the solution are replaced by an equivalent amount of Ba^{2+} ions. The equivalent limiting ionic conductances of the Ag^{+} and Ba^{2+} ions at 18° C are respectively 53.25 and 54.35, *i.e.* almost equal. The Cl^{-} ion added is immediately precipitated by the Ag^{+} whilst the quantity of NO_3^{-} ion remains unchanged, and hence the conductance of the final solution up to the end point of the reaction is virtually unaltered. After this point, the further addition of barium chloride increases the ionic concentration of the solution and thus rapidly augments the conductance. The titration diagram has the appearance shown in Fig. V, 15.

If a solution of sodium chloride were used in place of the barium chloride, since the limiting equivalent conductance of the Na^{+} ion is a little less than that of the Ag^{+} ion (42.6), the portion *AB* would fall slightly. Using potassium chloride, where the limiting equivalent conductance of the K^{+} ion (63.65) is a little greater than that of the Ag^{+} ion, the portion *AB* would rise slightly. The actual solubility of the product of the reaction is a decisive factor in this type of titration and manifests itself as the rounded portion of the curve passing through point *B*. The greater the solubility of the product, the longer is this rounded portion and the more difficult is the extrapolation to the point of intersection *D*. According to Kolthoff, the conductometric titration by a precipitation reaction, of a uni–univalent electrolyte in about 0.1 *N* solution requires a solubility of the reaction product of not more than $5 \cdot 10^{-3}$ moles/litre, and if the original solution were 0.01 *N* the value would fall to $5 \cdot 10^{-4}$ moles/litre.

The following conditions are required for a satisfactory conductometric analysis using a precipitation reaction.

1. The precipitate must rapidly assume its final composition and must not react further with the solution.
2. It must be rapidly deposited.
3. It must not absorb appreciable quantities of ions from solution.

The third type of conductometric analysis, with an ascending initial branch, is generally found when titrating a weak acid or base *i.e.* one which is poorly dissociated and thus has a low initial conductance. As the neutralization proceeds, the corresponding salt is formed which may be considered as virtually completely dissociated; thus, the increase in conductance due to the increase in total ionic concentration is greater than the diminution in conductance due to the disappearance of the H^+ (or OH^-) ions, since these are present in very small concentration due to the slight dissociation of the weak acid or base which is being titrated. After the end point is reached, however, the increase in conductance is more rapid, so that it is still possible, although with diminished accuracy, to locate the end point of the reaction. Naturally, between the two limiting cases there is a whole range of acids and bases of medium strength which can give rise to all sorts of behaviour, including even a slight fall in the

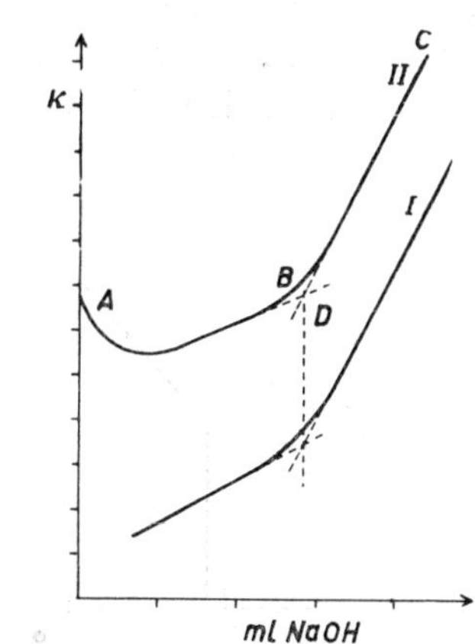

Fig.V,16.
Conductometric titration: third case

first branch of the diagram, depending on the strength of the acid or base and on the concentration (Fig. V, 16). The figure shows diagramatically the progress of conductometric analyses of a weak acid (curve I) and of an acid of medium strength (curve II) titrated with a strong base.

The analysis of a salt of a weak acid and a strong base (or *vice versa*) by titration with a strong acid (or a strong base), *e.g.* the titration of sodium acetate with hydrochloric acid, can in practice only be performed conductometrically. The technique is called 'displacement analysis'. The addition of hydrochloric acid (which is completely dissociated) to sodium acetate (completely dissociated) forms sodium chloride (completely dissociated) and acetic acid (poorly dissociated). In effect, then, the titration replaces the CH_3COO^- ion by the Cl^- ion; or in general terms the anion of the strong acid replaces that of the weak acid (or the cation of the strong base replaces that of the weak base). Depending on the respective mobilities, the conductance may rise or fall slightly, allowing also for the effective degree of dissociation of the weak acid (or weak base) which is formed and its own conductance, which must be added to that of the new salt formed. At the end of the titration the further addition of strong acid sharply increases the conductance. The diagram obtained resembles that in Fig. V, 15.

It is apparent from what has been said, that in particular cases it is possible to titrate mixtures of electrolytes conductometrically. A typical case is a mixture of a strong acid (or base) with a weak acid (or base). The strong component is first neutralized and only when this has been completely achieved does the neutralization of the weak component begin. The analysis is more exact as the difference between the two strengths of the components is greater. The relevant diagram is shown in Fig. V, 17. This is obviously a combination of the diagrams of Fig. V, 14 and Fig. V, 16.

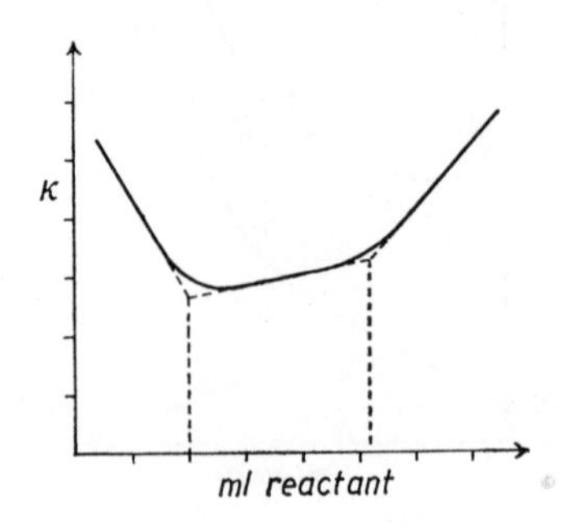

Fig. V, 17
Double conductometric titration

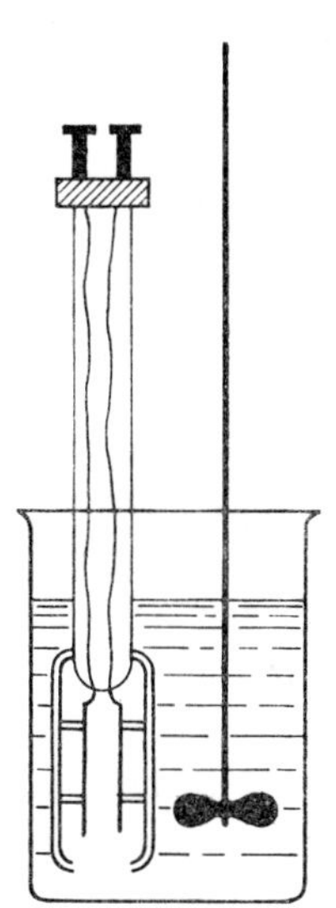

Fig.V,19.
Dipping electrodes for conductometric analysis

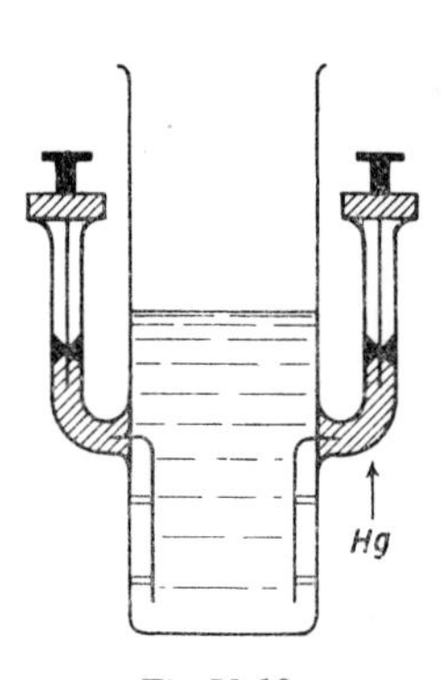

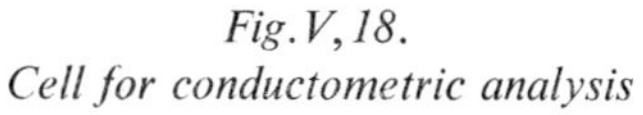

Fig.V,18.
Cell for conductometric analysis

Volume–conductance diagrams are actually of the type described only if the total volume of the reaction remains unchanged. In practice, this condition is sufficiently fulfilled when the reagent is concentrated with respect to the solution to be titrated, so that the change in volume is negligible (2.5 ml titrant for 50–100 ml of solution). To avoid too high percentage errors in the reading of the volume it is preferable to add the reagent from a high precision microburette. The temperature from the beginning to the end of the analysis must also be kept constant. In general, in view of the rapidity of these determinations, the temperature is sufficiently constant provided that the analytical reaction is not strongly endo- or exo-thermic. If this is not so, however, it is necessary to immerse the titration vessel in a thermostatically controlled water bath.

The titration may be carried out in a conductance cell like for example that illustrated in Fig.V,18, or else in an ordinary beaker using a pair of platinized platinum electrodes as shown in Fig.V,19.

The conductance of the solution is measured with one of the apparatus described on page 47 *et seq.* using alternating current of about 1,000 cycles/sec frequency. It is not absolutely necessary to measure the true conductance of the solution at every point provided that the change in conductance is known. However, when an acoustic method is used it is necessary to find the minimum after every addition of reagent; when a visual method of reading is used, it is sufficient to plot on the diagram the change read on the measuring instrument, which is normally proportional to the conductance. The conductometric method has several advantages. It can be used for very dilute solutions (using cells with very small resistances); for coloured or turbid solutions; for solutions with which the analytical reaction does not give a well defined end point, since linear initial and final portions of the curve are available which can be extrapolated to give the end point at their intersection; and it can also be used for titrations in nonaqueous solvents. It does, however, have the inconvenience that it cannot be used in the presence of significant concentrations of electrolytes other than those being determined, since the high conductance of the solution would hinder the detection of changes in conductance due to the addition of the reagent. This is the most serious limitation to the wider use of this method.

In some cases a direct measure of conductance may give some indication of the electrolyte content of the solution. This procedure which is fairly simple and rapid, gives very good results for the determination of the degree of purity of distilled water, for the control of wash liquids from precipitates, for the determination of the salt content of natural waters and for tests of the purity of poorly soluble substances, etc. Further details of the technique of such measurements and various electrical circuits may be obtained from specialized texts[1].

(II). *High frequency titrations* (*oscillometry*)

If a solution is introduced in a suitable way into the electromagnetic field of an inductance or between the plates of a capacitor which form part of a high frequency (several MHz or more) electrical circuit — without however making any contact with the metallic conductors — the electrical size of one of the components of the circuit will be changed, and consequently the electrical characteristics of the oscillating circuit will alter. The change in electrical conductance and dielectric constant of the solution during a chemical reaction leads to a further, reproducible, change in the electrical parameters of the oscillating circuit (frequency, electric tension, current).

[1] H. T. S. Britton, *Conductometric Analysis*, Chapman and Hall, London, 1934; J. J. Lingane, *Electroanalytical Chemistry*, Chapter IX, Interscience Publishers, New York, 1953.

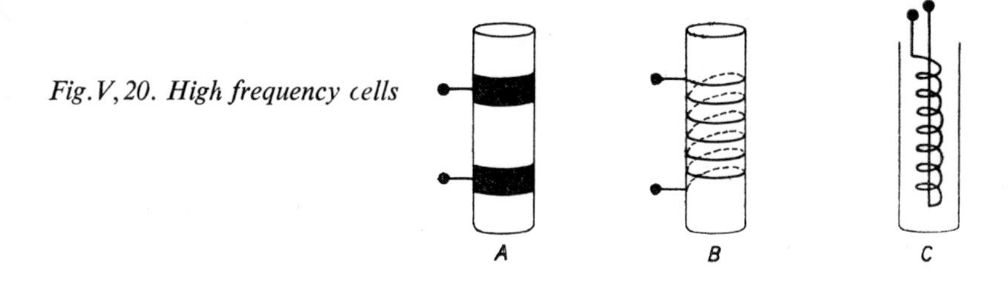

Fig. V, 20. High frequency cells

From such measurements it is possible to obtain the content of a particular ionic species in the solution.

The apparatus required consists of a titration cell (a suitable vessel), in which is placed the solution, connected with an oscillating circuit. An analytical cell for high frequency may be considered as a component of an electrical circuit, with a definite impedence to the passage of an alternating current at radio frequency. This impedence changes with variations in the conductive and capacitive properties of the substance contained in the cell. The various types of cell may be classified into two categories: (a) capacitance cells, where the vessel is surrounded by two metallic bands (Fig. V, 20*A*), to make a condenser which forms part of the oscillating circuit, and (b) inductance cells where the vessel containing the sample is introduced within an inductance or *vice versa* (Fig. V, 20*B* and *C*).

In the types shown in Fig. V, 20*A* and *B*, which are the most common, the material of the vessel (glass, polythene) insulates the solution from the metallic conductors of the electrical circuit. The capacitance type of cell is most generally used and may be conveniently described here; those of the inductance type follow more complex laws which are not yet fully known. The behaviour of a capacitance cell may be considered as equivalent to that of the circuit shown in Fig. V, 21*a*.

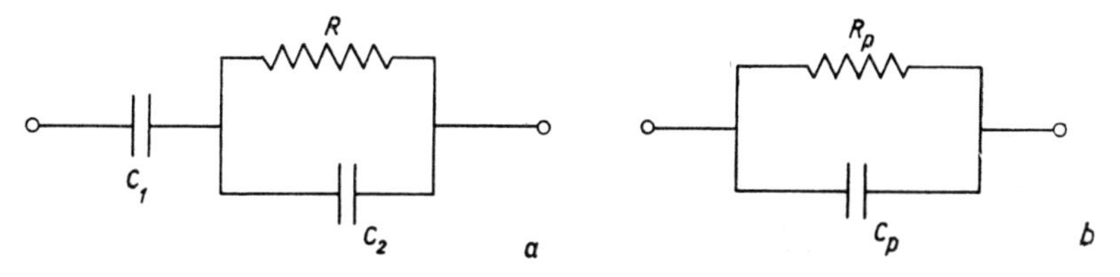

Fig. V, 21. Equivalent circuits of H.F. cells

C_1 represents the capacitance of the dielectric which forms the vessel; C_2 represents the capacitance of the dielectric which consists of the solution and R is the resistance of this solution; the resistance of the walls of the cell is assumed to be infinite. This circuit is equivalent to one in which a resistance R_p is in parallel with a capacitor C_p (Fig. V, 21*b*).

The conductance G_p of a high frequency circuit of this type is given, according to Reilley and McCurdy[1], by

$$G_p = \frac{1}{R_p} = \frac{\varkappa \omega^2 C_1{}^2}{\varkappa^2 + \omega^2 (C_1 + C_2)^2} \tag{1}$$

where $\varkappa$ is the low frequency conductance of the solution itself in a cell of the same shape and dimension as the high frequency cell, ω is equal to $2\,\pi$ times the frequency, and C_1 and C_2 are the values of the capacitance due respectively to the walls of the cell and to the solution within the cell. Variations of G_p with conductance obtained experimentally, may be interpreted by this equation, and are reported in Fig. V, 22.

From the equation it may be deduced that when $\varkappa$ is small, the term $\varkappa^2$ will be negligible and G_p will become directly proportional to $\varkappa$. However,

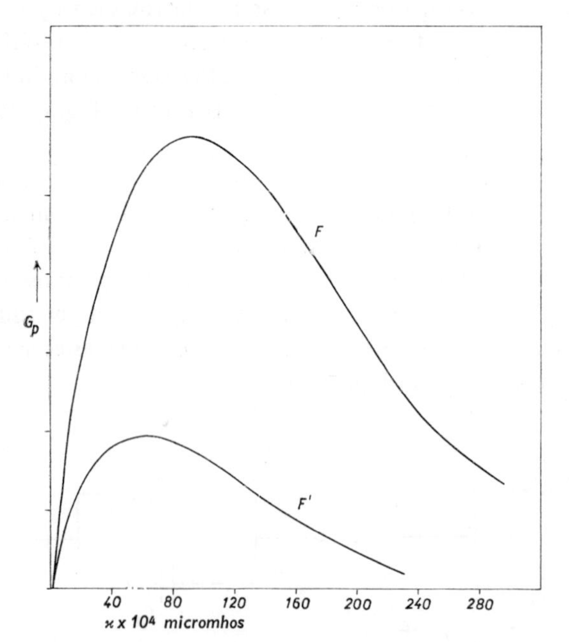

Fig. V, 22
H.F. Conductance

[1] C. N. Reilley and W. H. McCurdy, *Anal. Chem.*, 25 (1953) 86.

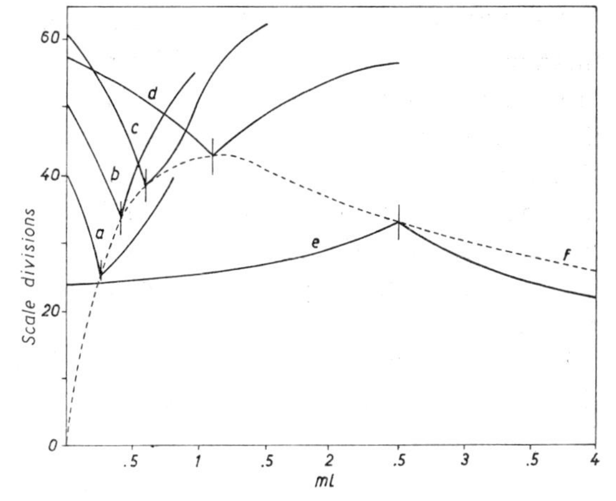

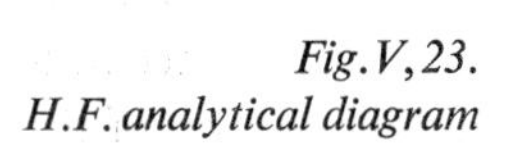
Fig. V, 23.
H.F. analytical diagram

when $\varkappa$ is high the term $\omega^2(C_1 + C_2)^2$ becomes negligible with respect to $\varkappa^2$ and G_p will become inversely proportional to the low frequency conductance. The curve $G_p - \varkappa$ thus passes between these two extreme cases through a maximum. The two curves of the graph, for two experiments at different frequencies ($F > F'$), indicate that the conductance at high frequency increases with the frequency and thus the nature of the titration curve varies as a function of the selected working frequency. The shape of the titration curve depends also on the selected concentration, reaching an 'optimum' value in a concentration range which is characteristic for the particular system. This gives rise to the so-called 'charging curve' or 'concentration curve', which is obtained by introducing into the cell containing the solvent, equal amounts of the solution and of the titrating substance. Curves are obtained which are analogous to those in Fig. V, 22 and to the broken curve *f* in Fig. V, 23.

The behaviour of these curves varies markedly according to the selected working frequency, depending on the type of oscillometric circuit used and especially on the particular type of cell. It also depends on the way in which the cell is inserted into the circuit and on the value of the capacitor in parallel with the cell. The shape of the charging curve also depends on the electrical parameter which is measured (conductance, current, electric tension, capacitance or frequency). Consequently, the titration curve at high frequencies also shows different appearances depending on the concentration of the solution examined and thus on the selected range of the charging curve.

Fig. V,23 reports a series of graphs (continuous lines) taken from the titration of increasing amounts of hydrochloric acid with sodium hydroxide. The titration curves *a*, *b*, *c*, and *d* were obtained with concentrations of hydrochloric acid corresponding to the ascending portion of the charging curve, whilst *e*, which is bell-shaped, corresponds to a concentration falling on the descending part of the charging curve. For an accurate determination of the end point of a titration, it is advisable to select the range of concentration which shows the greatest slope on the charging curve[1].

It has been found that oscillometric methods are less sensitive than the classical conductometric ones. If the most favourable case is considered for equation (1), where $\varkappa$ is very small

$$G_p = \frac{1}{R_p} = \varkappa \frac{C_1^2}{(C_1 + C_2)^2}$$

and since the term $C_1^2/(C_1 + C_2)^2$ is less than 1, the changes in G_p are less than those observed in the conductometric titration. However, oscillometry has the peculiar characteristic that it may be used without inserting any metallic conductor into the solution, and this eliminates difficulties due to the formation of interfacial phenomena, to the deposition of precipitates on electrodes or to the modification of electrode surfaces. Its use for the titration of solvents of low conductance (alcohol, benzine, dioxane, acrylonitrile) is particularly interesting and thus the use of oscillometry is of general advantage in the analysis of binary organic mixtures, in the determination of the humidity of various products and in kinetic studies of various reactions and finally in testing products in sealed vessels. Measurements of conductance at high frequency are also used in chromatographic studies on paper or on columns and for ion exchange resins, since changes in dielectric constant caused by the migrating substance enable the chromatographic zones to be located. The greatest limitation of this experimental technique is the fact that the presence of strong electrolytes or of liquids with high dielectric constants masks small changes in conductance and makes the determination inaccurate. Further information can be obtained from specific texts on this subject[2].

[1] For greater details on this technique see for example K. CRUSE and R. HUBER, *Hochfrequenztitration*, Verlag Chemie, Weinheim, Bergstr., 1957.

[2] P. H. SHERRICK *et al.*, *Manual of Chemical Oscillometry*, Sargent, Chicago, 1954; C. N. REILLEY, in P. Delahay, *New Instrumental Methods in Electrochemistry*, Interscience Publ., New York, 1954.

4. Polarography and Voltammetry

(I). *Theory*

Polarographic, or voltammetric analysis involves the recording of current intensity–electric tension diagrams obtained under suitable conditions, and forms an important analytical technique. The term polarography is properly reserved for the interpretation of *I–U* curves obtained with dropping mercury electrodes, and the more general term of voltammetry should be used for the study of *I–U* curves obtained with any indicator electrode. Such a procedure, in addition to allowing the simultaneous quantitative and qualititative determination of many cations and anions, and of a large number of organic substances which can be electrochemically oxidized or reduced, may also aid the resolution of various chemical problems such as the determination of the structure of substances, the study of reaction kinetics, the evaluation of equilibrium constants, the stability constants of complexes, etc.

The principle of the analytical polarographic technique may be explained by considering a cathodic reduction bearing in mind that the same reasoning applies to an anodic process. The electrolytic cell consists of an anode with a large surface area and a cathode formed of a capillary from which very pure mercury drops regularly. A cathode of this type has the advantage of presenting a continuously renewed surface to the solution, in virtue of the dropping; it is not affected by a preceding polarization and is not contaminated by any product of electrolysis. Moreover, the over-tension of hydrogen on this electrode is fairly high so that many substances can be discharged, including even cations of the alkaline metals, without any evolution of gaseous hydrogen. It should be noted also that the metals discharged at the constantly renewed surface of the mercury form very dilute amalgams. This, and the small area of the cathode, lead to complete cathodic polarization.

The anode used must have a very large surface area with respect to the dropping electrode, so that the current density on the anode is very small and its electric tension may be considered as constant. This means that any change in the current intensity in the electrolytic cell is due solely to a change in the electric tension of the dropping electrode which is induced by the cathodic process. The anode may consist of a well of mercury or of any electrode of the second kind. With the former, the anodic tension depends on the nature of the solution used and its value must be determined with an auxilary comparison electrode. Knowing the value of this allows the electric tension of the anode to be determined, which then gives the electric tension of the dropping electrode by difference. Whenever possible, a saturated calomel electrode is used within the cell as the comparison electrode and hence the electric tensions of the dropping electrode are generally referred to this calomel electrode.

The continuous renewal of the polarizable cathodic surface leads to perfect reproducibility of the results so that — provided that the experimental conditions are kept constant — the current intensity depends exclusively on the electric tension and not on the duration of the electrolysis. It should also be observed that under these conditions, the quantity of electrolyte decomposed is so small as to provoke no detectable change in the composition of the solution even if the measurement of the *I–U* curve is repeated.

The arrangement of the polarograph is illustrated in Fig. V, 24). The electric tension of an accumulator *Acc* is applied to the ends *C*, *D* of a resistance, from which a sliding contact *E* allows a desired electric tension to be taken and varied continuously. A layer of mercury is placed at the bottom of the electrolytic cell *A* and functions as the anode. The solution

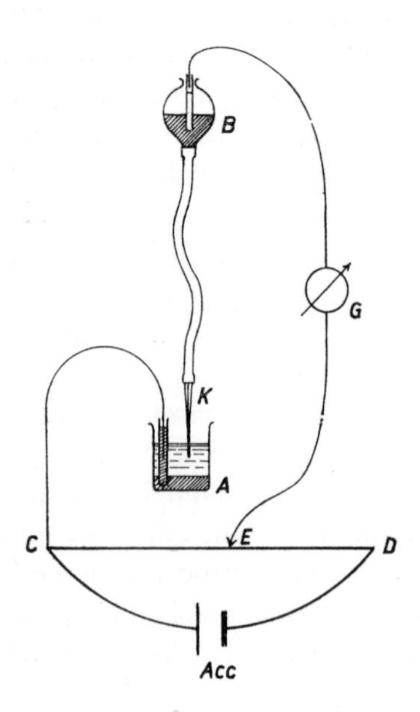

Fig. V, 24.
Schematic circuit for polarography

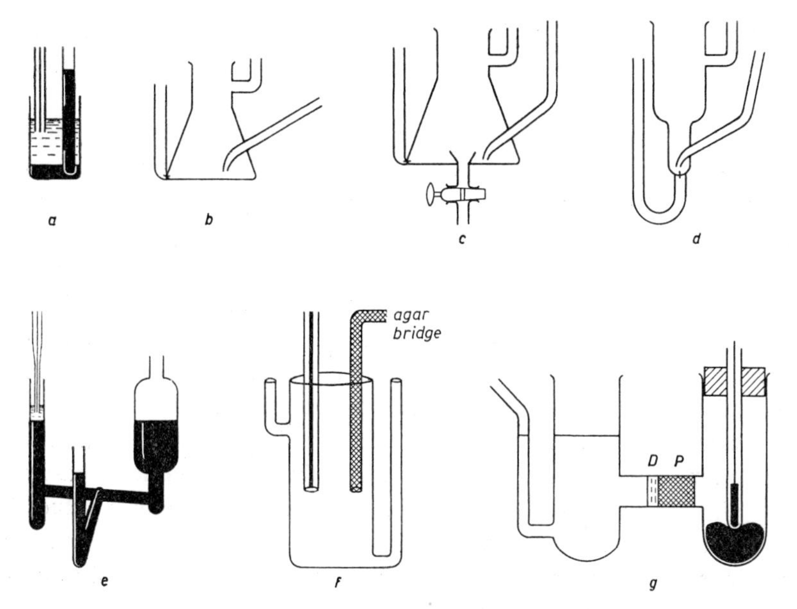

Fig. V, 25. Cells for polarography

is in contact with the anode and through the solution mercury from the reservoir *B* is dropped via the cathodic capillary *K*. The passage of current through the cell is measured with the galvanometer *G*. The circuit shown may be conveniently modified so that the polarity of the dropping electrode, referred to the electric tension of the comparison electrode, may be reversed. In this way, anions and substances which can be electro-oxidized may be also determined.

A wide variety of cells has been used for polarographic analysis (Fig. V, 25). In practice, any vessel is suitable into which the dropping electrode and the comparison electrode can be inserted, and which can be freed from air by bubbling an inert gas through it. If the comparison electrode consists of mercury, the types of cell which may be used, depending on the volume of solution required, are shown in the figure at *a*, *b*, *c*, *d* and *e*. If a calomel or silver chloride electrode is used, cells of type *f* are required and the connection is made with an agar bridge. If the cathodic and anodic compartments must be separated without increasing the resistance of the cell, one of type *g* may be used; this is H-shaped and *D* is a porous septum and P a buffer in agar (Lingane and Laitinens cell). Types d and e are particularly useful when the volume of solution to be analyzed is very small.

The measurement of the current intensity — controlled by the polarization — through the electrolytic cell, may be carried out accurately not only with a galvanometer but also by using an auxiliary potentiometer to measure the electric tension across the ends of a resistance of known value connected in series with the cell; Ohm's Law can then be used to calculate the current which flows through both this resistance and the cell. The advantage of this latter procedure lies in the fact that the galvanometer, used in this way, functions only as a zero indicator so that the determination of the current is not dependent on the linearity of the response of the instrument. Such a circuit is shown in Fig. V,26. U_1 is the polarizing tension measured with the divider P_1, U_2 is the electric tension measured with the divider P_2 across the ends of the resistance of known value R_s and G is the galvanometer, whose sensitivity can be reduced with the resistance R_g. The circuits shown are normally used for manual measurements with determinations made at various points, but they could be modified so that the *I–U* curve could be recorded automatically. The circuit of a polarographic apparatus with photographic recording, suggested by Heyrovsky, is shown in Fig. V,27. The known electric tension of an accumulator *Acc* is applied to the ends of a homogeneous calibrated wire *CD*, wound around a rotatable cylinder. The positive end of the resistance is connected to the anode *A* of the electrolytic cell. The capillary cathode *K* is connected, through the mercury reservoir *B*, the galvanometer *G* and the sensitivity control of the galvanometer *R*, to

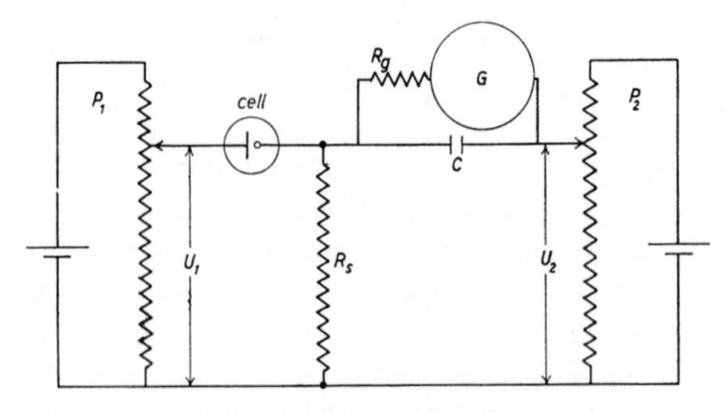

Fig. V,26. Potentiometric polarograph

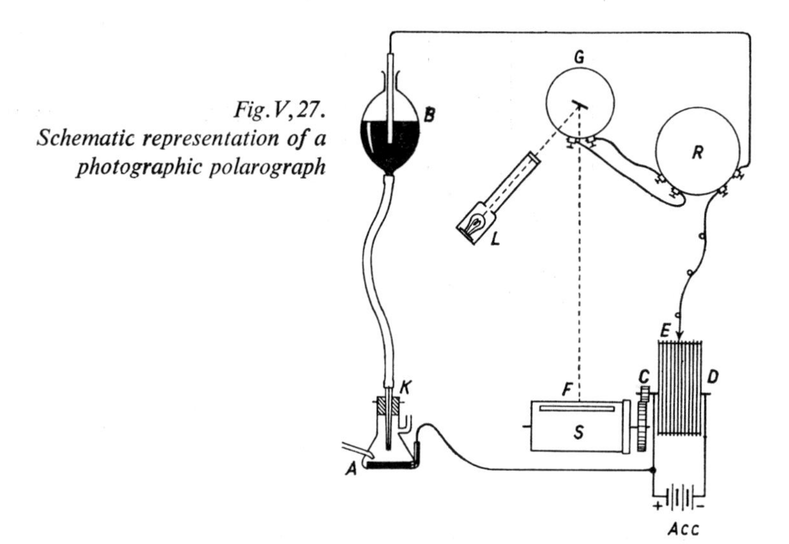

Fig. V, 27. Schematic representation of a photographic polarograph

the sliding contact E. The galvanometer is of the mirror type. A beam of light from the lamp L is deflected by the mirror and falls on a slit F in a box S; inside this box is a sheet of photographic paper coiled around a drum. The drum and the rotating cylinder are connected in such a way that the drum makes one rotation when the cylinder makes as many rotations as there are turns of wire on it. In other words, the drum makes one rotation when the contact E has passed from one end of the calibrated wire CD to the other. An auxiliary lamp, which is not shown, lights up in front of the slit F for a brief instant at every completed revolution of the cylinder and thus leaves a mark on the photographic paper. If the electric tension applied to the ends of the wire is, for example, 2.0 V and if the wire is wound around the cylinder 10 times, it is clear that one revolution corresponds to a movement of the mobile contact by one-tenth of the total length of the wire and that the marks induced by the auxiliary lamp at the completion of each rotation of the cylinder, lie at 0.2 V apart. As a motor rotates the cylinder and the drum, the cathodic polarization increases and the intensity of the current which is characteristic of the value of the polarization is automatically recorded by the deflection of the galvanometer. The photographic diagram obtained (Fig. V, 28) is called a polarogram. Knowing the sensitivity of the galvanometer the current intensity can be read directly from this for any given polariza-

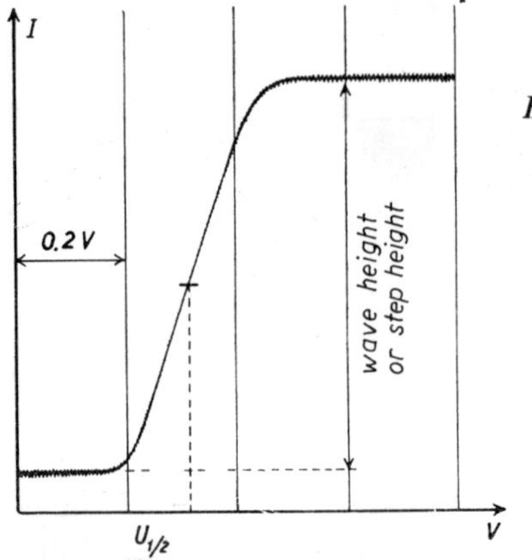

Fig. V, 28. Polarographic wave

tion; the value of the latter is directly recorded by the marks made on the photographic paper at each revolution of the cylinder.

Writing polarographs use a similar mechanical arrangement to move the sliding contact and the paper. The current intensity is measured with a recording millivoltmeter after suitable amplification.

As was previously noted, the current intensity through an electrolytic cell is given by Ohm's Law, if account is taken also of the cathodic and anodic polarization. The equation

$$V = U_a + U_c + RI$$

leads to

$$I = \frac{V - (U_a + U_c)}{R}$$

In polarographic analysis, the anodic tension U_a is constant and hence does not affect the changes in the intensity of the current I. If then the anodic tension is taken as zero, *i.e.* if the cathodic tension is measured with respect to the anodic as the reference point, and if the value of IR is negligibly small (in general I is of the order of a few microamperes and R is about 1,000 ohms)

$$U_c = V$$

Thus, the current intensity–electric tension curve depends in practice solely on the cathodic tension. The electric tension of a reversible amalgam electrode (p. 126 *et seq.*)

$$Me^{z+} + ze^- + Hg \rightleftharpoons Me(Hg)$$

is given by the equation

$$U = U_{0\,amalg.} - \frac{RT}{zF} \ln \frac{f_a[Me]_0}{f_a'[Me^{z+}]_0} \tag{1}$$

The various symbols have their usual significance, so that $[Me^{z+}]_0$ and $[Me]_0$ indicate respectively the concentrations of the metallic ion in the solution and of the metal in the amalgam at the mercury-solution interface, whilst f_a' and f_a are the corresponding activity coefficients. In each drop of mercury, forming on the cathodic capillary, $[Me]_0$ is proportional in accordance with Faraday's First Law, to the intensity of current flowing through the cell, assuming that the time between the falling of the drops is constant and that the size of each drop is also constant, *i.e.*

$$[Me]_0 = K'I \tag{2}$$

The proportionality constant K' allows for diffusion phenomena of the discharged metal at the mercury surface towards the interior of the drops, as well as for the other characteristic constants of the process. Substituting (2) into equation (1) gives

$$U = U_{0\,amalg.} - \frac{RT}{zF} \ln \frac{K'If_a}{f_a'[Me^{z+}]_0} \tag{3}$$

$$\ln \frac{K'If_a}{f_a'[Me^{z+}]_0} = - \frac{(U - U_{0\,amalg.})zF}{RT}$$

$$\frac{K'If_a}{f_a[Me^{z+}]_0} = e^{-\frac{(U - U_{0\,amalg.})zF}{RT}}$$

$$I = \frac{f_a'[Me^{z+}]_0}{K'f_a} e^{-\frac{(U - U_{0\,amalg.})zF}{RT}} \tag{4}$$

This expression indicates that the current intensity–electric tension curve is exponential but that this will be true only up to the value of the limiting current. In this case the ionic concentration $[Me^{z+}]_0$ at the mercury-solution interface tends towards zero and the current intensity thus tends to become constant since it is controlled only by diffusion phenomena. In fact, if the solution to be subjected to polarographic analysis is prepared in the presence of an excess of an inert electrolyte whose ions do not take part in the electrode reactions, at a concentration of at least 50–100 times greater than that of the substance being analysed, the resistance of the solution becomes very small. The electrical field within the bulk of the solution becomes negligible and consequently the electrical transport of the ions involved in the electrochemical process is eliminated. These ions will then move only as if they were subjected to diffusion caused by the difference in concentration between the bulk of the solution $[Me^{z+}]$ and the interface $[Me^{z+}]_0$. The current intensity at every point of the polarographic wave will be determined by the discharge process of the cations Me^{z+}, and will be proportional to the concentration difference.

$$I = K([Me^{z+}] - [Me^{z+}]_0) \tag{5}$$

When $[Me^{z+}]$ tends to zero, the current intensity tends towards a limiting value, since it is due solely to the diffusion of ions from the interior of the solution, so that

$$I_d = K[Me^{z+}] \tag{6}$$

This value is independent of the cathodic tension. The current intensity–electric tension diagram thus assumes the shape of a step[1], as shown in Fig. V, 28. The limiting current intensity obtained under these conditions (*i.e.* with a dropping mercury electrode, in nonagitated solution, in the presence of a high concentration of an inert electrolyte) is the sum of two currents — the residual or charging current I_r (*cf.* p. 192 *et seq.*) and the diffusion current I_d — so that $I_l = I_r + I_d$. The residual current refers to that which can be measured even in the absence of substances which can be electrolytically reduced or oxidized; it is due to the continuous renewal of the mercury drops as a consequence of the applied electric tension. This current is nonfaradic, and is a fraction of a microampere for a tension of 0 volts with respect to the saturated calomel electrode. It is zero at − 0.52 volts, which is called the isoelectric point of the mercury, where the surface tension at the mercury-solution interface reaches its maximum. The current increases almost linearly with the applied electric tension.

Subtracting the value of this current intensity from that of the limiting current gives the intensity of the diffusion current. In practice, this is cal-

[1] This step is usually called the 'polarographic wave'.

culated by measuring, on a polarogram, the distance between the base line and the trace in the diffusion region as shown in Fig. V, 28.

On the basis of the diffusion theory, Ilkovic derived an equation for the calculation of the diffusion current intensity, I_d, obtained with a dropping mercury electrode; the equation gives results in good agreement with the experimentally determined values.

$$I_d = K'z\mathrm{F}cD^{\frac{1}{2}}m^{\frac{2}{3}}t^{\frac{1}{6}}$$

In this equation $z\mathrm{F}$ is the number of faradays of electricity per mole of reaction at the electrode; D is the diffusion coefficient of the substance which is oxidized or reduced; c is the concentration of this substance; m is the weight of mercury which flows through the capillary and t is the dropping time. If I_d is expressed in microamperes; c in millimoles per litre; D in $cm^2 \cdot sec^{-1}$; m in $mg \cdot sec^{-1}$ and t in sec; then the proportionality constant K is 607. The various factors of Ilkovic's equation may be divided into two groups: the terms $n\mathrm{F}cD^{\frac{1}{2}}$, which refer to the properties of the solution, and the terms $m^{\frac{2}{3}}t^{\frac{1}{6}}$, which are specific characteristics of the capillary. Knowledge of m and t allows the correlation of measurements of diffusion currents obtained with different capillaries. Thus, Ilkovic's equation gives

$$I_d = Kc$$

So that the intensity of the diffusion current obtained with a given capillary is proportional to the concentration of the electrolytically active species. This equation constitutes the basis for the use of polarographic analysis in the quantitative determination of a substance.

Numerous modifications of Ilkovic's equation have been proposed to allow for differences between theoretically calculated and experimentally measured values for I_d. Equations proposed by LINGANE and LOVERIDGE (*J. Am. Chem. Soc.*, 72 (1950) 438); STREHLOW and VON STACKELBERG (*Z. Elektrochem.*, 54 (1950) 51) and others are of the type

$$I_d = 607\ zD^{\frac{1}{2}}cm^{\frac{2}{3}}t^{\frac{1}{6}}\ (1 + AD^{\frac{1}{2}}m^{-\frac{1}{3}}t^{\frac{1}{6}})$$

where A is a constant whose values lie between 17 and 39.

The considerations described above allow much information of theoretical and practical interest, to be obtained from the I–U curve. Consider the reduction of a substance at the dropping electrode according to the following equation

$$Ox + ze^- \rightleftharpoons Red$$ [1]

If the electrode reaction is reversible, the electric tension at any point on the polarographic wave is given by

$$U = U_0 - \frac{RT}{\mathrm{F}} \ln \frac{f_{Red}[\mathrm{Red}]_0}{f_{Ox}[\mathrm{Ox}]_0} \tag{7}$$

where $[\mathrm{Red}]_0$ and $[\mathrm{Ox}]_0$ are the concentrations at the mercury-solution interface. As shown by equations (5) and (6)

$$I = K([\mathrm{Ox}] - [\mathrm{Ox}]_0)$$

and

$$I_d = K[\mathrm{Ox}] \tag{8}$$

where [Ox] is the concentration of Ox in the solution. From equation (8)

$$[\mathrm{Ox}]_0 = I_d - I/K \tag{9}$$

If none of the reduced substance were initially present in the solution, its concentration at the interface is always proportional to the current:

$$[\mathrm{Red}]_0 = K'I \tag{10}$$

Substituting equations (9) and (10) into (7) gives

$$U = U_{0\,amalg.} - \frac{RT}{z\mathrm{F}} \ln \left[KK' \frac{f_{Red}\, I}{f_{Ox}\,(I_d - I)} \right]$$

$$= U_{0\,amalg.} - \frac{RT}{z\mathrm{F}} \ln KK' \frac{f_{Red}}{f_{Ox}} - \frac{RT}{z\mathrm{F}} \ln \frac{I}{I_d - I} \tag{11}$$

The I–U curve is thus expressed by equation (11); to determine particular points in this, the points of inflexion may be determined analytically by calculating where the second derivative becomes zero. The first derivative is given by

$$\frac{\mathrm{d}U}{\mathrm{d}I} = \frac{\mathrm{d}\left\{U_0 - \frac{RT}{z\mathrm{F}} [(\ln \frac{KK' f_{Red}}{f_{Ox}} + \ln I - \ln (I_d - I)]\right\}}{\mathrm{d}I}$$

In a solution with a constant supporting electrolyte composition, the ionic strength remains virtually constant and the activity coefficient f_{Ox} will

[1] In this equation the symbols Ox and Red include any ionic charges.

also, to a first approximation, be constant; the value of the activity coefficient f_{Red} may also be taken as constant so that

$$\frac{\mathrm{d}U}{\mathrm{d}I} = -\frac{RT}{z\mathbf{F}}\left(\frac{1}{I} - \frac{-1}{I_d - I}\right)$$

$$= -\frac{RT}{z\mathbf{F}} \frac{I_d}{I_d I - I^2}$$

The second derivative is given by

$$\frac{\mathrm{d}^2U}{\mathrm{d}I^2} = -\frac{RT}{z\mathbf{F}} \frac{I_d(I_d - 2I)}{(I_d I - I^2)^2} \tag{12}$$

Equation (12) becomes zero when

$$I_d - 2I = 0$$

i.e.

$$I = \tfrac{1}{2} I_d$$

This expression indicates that the polarographic wave has a point of inflexion at the electric tension corresponding to that current for which $I = \frac{1}{2} I_d$. The electric tension at this point is given the symbol $U_{\frac{1}{2}}$ and is defined as the half-step or half-wave electric tension[1]. The equation to the polarographic wave for a reversible process may be formulated as

$$U = U_{\frac{1}{2}} - \frac{RT}{z\mathbf{F}} \ln \frac{I}{I_d - I} \tag{13}$$

When $I = \frac{1}{2} I_d$, $U = U_{\frac{1}{2}}$ *i.e.* the half-wave electric tension has a constant value at constant temperature and supporting electrolyte composition. As indicated in equation (11) this tension depends on only two characteristic constants of the electrode process and on the ratio of the activity coefficients, which is also constant. This equation indicates that the value of the half-wave electric tension is independent of the concentration of the species which may be electrolytically reduced and also of any other variable due to the experimental measuring arrangement (dropping rate, type of reading, sensitivity, measuring instruments etc.). Since a polarographic reading, carried out as described, gives a well defined step it is possible to identify

[1] This quantity is usually called 'half-wave potential' by polarographers, but coherently with the IUPAC–CITCE general definitions it will be called throughout 'half-wave electric tension' because it is really an electric tension, as defined in Chapter III. like the electric tension of any other individual electrode.

ionic species unequivocally from the electric tension at the half-height. Since $U_{\frac{1}{2}}$ is a characteristic value for a given species under particular conditions, it is possible to determine simultaneously more than one electrochemically active substance, since each substance will give a step at a constant electric tension which depends on the process itself.

If the reaction at the electrode is reversible, it is possible to derive information to aid the interpretation of the nature of the electrode reaction. From equation (13), if $\log [(I_d - I)/I]$ is plotted on a graph against the electrode tension, a straight line will be obtained with a slope of $0.059/z$. This mathematical analysis of a polarogram allows one to establish, if a given process is reversible, the number of electrons involved and the accurate value for the half-wave electric tension.

(II). *Qualitative and Quantitative Polarographic Analysis*

Polarographic analysis is a very important analytical technique since it can be applied to the determination of a large number of inorganic and organic substances which can be reduced or oxidized at the dropping electrode. It is also possible to determine substances which are not electrolytically active by first converting them into derivatives which are so.

The polarographic analysis of a particular substance can give rise to steps or waves which may be classified as follows.

Cathodic waves

These are given by inorganic cations and anions, by non-ionic substances and by organic compounds.

The majority of inorganic cations can be reduced at the dropping electrode; in view of the high overtension of hydrogen on mercury it is even possible to determine the ions of alkali and alkaline-earth metals, provided that suitable inert electrolytes are used, such as salts or hydroxides of substituted tetra-alkyl ammonium compounds. In most cases reduction leads to the formation of the metals so that a single polarographic step is observed ($Bi^{3+} \rightarrow Bi^0$; $Pb^{2+} \rightarrow Pb^0$ etc.). Sometimes the reduction takes place through more than one step corresponding to the presence of intermediary oxidation states ($Cu^{2+} \rightarrow Cu^+ \rightarrow Cu^0$); in other cases the reaction may stop at a lower oxidation state ($Fe^{3+} \rightarrow Fe^{2+}$). Since every electrode reaction is characterized by a particular value for the half-wave tension,

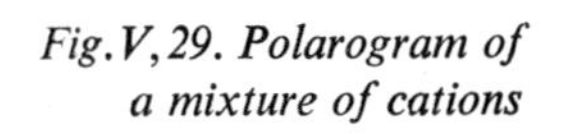
Fig. V, 29. Polarogram of a mixture of cations

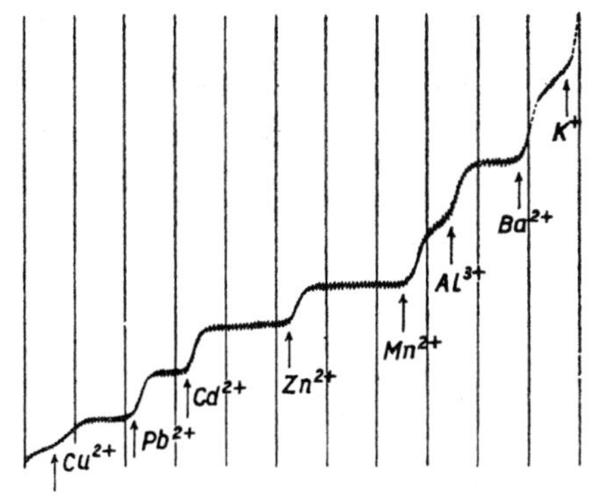

it is apparent that it will sometimes be possible to determine more than one cation in a single operation. Fig. V, 29 shows the polarogram of a solution containing Cu^{2+}, Pb^{2+}, Cd^{2+}, Zn^{2+}, Mn^{2+}, Al^{3+}, Ba^{2+} and K^{+}; eight steps are obtained, each of which represents a particular $U_{\frac{1}{2}}$ value. For this to be achieved, it is necessary that no ionic species is present in a concentration preponderant over that of the others and that the half-wave tensions are sufficiently different to avoid overlapping of two steps. Table V, 7 shows the half-wave tensions of the principal inorganic systems. The values have been selected from those reported in the literature[1] and refer to reversible steps or steps which are sufficiently so to be used for analytical purposes. Where more than one value is reported it has been found that varying the composition of the supporting electrolyte or the pH of the solution, markedly alters the value of $U_{\frac{1}{2}}$. This is because, since the electrolytes used are often binding agents, the ions are complexed and with increasing stability of the complex form, the value of $U_{\frac{1}{2}}$ will be moved towards more negative values. The use of suitably selected supporting electrolytes is the most important criterion for obtaining well-defined polarographic steps. With mixtures of extreme complexity this procedure may not be sufficient and it may then be necessary to carry out a preliminary chemical separation.

Many inorganic anions which contain oxygen, such as iodate, nitrite, tellurite and selenite, may be reduced by the dropping electrode. The

[1] I. M. Kolthoff and J. J. Lingane, *Polarography*, Interscience Publishers, New York, 1952; L. Meites, *Polarographic Techniques*, Interscience Publishers, New York, 1955; A. A. Vlcek, *Tafeln der Halbstufenpotentiale Anorganischer Depolarisatoren.* Akademie Ceskoslovensk, Prague, 1956.

TABLE V,7

HALF-WAVE ELECTRIC TENSIONS OF THE PRINCIPAL INORGANIC SYSTEMS

(The values of $U_{\frac{1}{2}}$ are expressed in volts and referred to the saturated calomel electrode)

Element	*Oxidation state*	*Supporting electrolyte*	$U_{\frac{1}{2}}$
Aluminium	Al^{3+}	0.025 *m* $BaCl_2$ or 0.1 *m* LiCl	−1.75
Antimony	Sb^{3+}	1 *m* HCl 0.01% gelatine	−0.15
		0.1 *m* NaOH 0.003% thymolphthalein	−0.37* - −1.07
	Sb^{5+}	6 *m* HCl	−0.26
Arsenic	As^{3+}	1 *m* HCl 0.0001% methylene blue	−0.43 - −0.67
		2 *m* AcOH 2 *m* $AcONH_4$	−0.92
		0.5 *m* KOH 0.025% gelatine	−0.26*
	As^{5+}	11.5 *m* HCl	−0.52
Barium	Ba^{2+}	0.1 *m* LiCl or $N(C_2H_5)_4I$	−1.92
Beryllium	Be^{2+}	Solutions of $BeCl_2$ or $BeSO_4$	−1.8
Bismuth	Bi^{3+}	1 *m* HCl 0.01% gelatine	−0.09
		0.1 *m* potassium hydrogen phthalate	−0.23
		0.5 *m* Na_2 tartrate 0.01% gelatine pH = 4.5	−0.29
		0.1 *m* EDTA 1 *m* K_2CO_3 pH = 9.5	−0.78
Bromine	Br^-	0.1 *m* KNO_3	+0.12*
	BrO_3^-	0.1 *m* H_2SO_4 0.2 *m* KNO_3	−0.41
		0.1 *m* $BaCl_2$ or $CaCl_2$	−0.53
Cadmium	Cd^{2+}	0.1 *m* KCl or HCl	−0.60
		1 *m* NH_4OH 1 *m* NH_4Cl	−0.81
		1 *m* ethylendiamine 0.1 *m* KNO_3	−0.93
Calcium	Ca^{2+}	0.08 *m* $N(C_2H_5)_4I$ with traces of Ba^{2+} or La^{3+} to suppress the maximum	−2.22
Cerium	Ce^{3+}	0.1 *m* LiCl or $N(CH_3)_4Br$	−2.0
	Ce^{4+}	0.1 *m* strong acids	> 0
Caesium	Cs^+	0.1 *m* $N(CH_3)_4Cl$ or $N(CH_3)_4OH$	−2.09
Chlorine	Cl^-	0.1 *m* KNO_3	+0.25*
	ClO^-	0.5 *m* K_2SO_4	+0.08
	ClO_2^-	1 *m* NaOH	−1.0
Chromium	Cr^{2+}	0.1 *m* NH_4OH 5 *m* NH_4Cl 0.005% gelatine	−1.14*
		1 *m* KCN 0.01% gelatine	−1.38*
		0.1 *m* Na-salicylate 0.1 *m* NaOH 0.005% gelatine	−1.17*
	Cr^{3+}	2 *m* AcOH 2 *m* AcONa 0.01% gelatine	−1.2
		7.3 *m* H_3PO_4	−1.02
		1 *m* KCN	−1.38
	Cr^{6+}	1 *m* NaOH	−0.85
		1 *m* NH_4OH 1 *m* NH_4Cl 0.1 *m* KCl	−0.2 - −1.6
		1 *m* Na_3 citrate 0.1 *m* NaOH	−0.83 - −1.49
Cobalt	Co^{2+}	0.1 *m* AcONa	−1.19
		1 *m* NH_4OH 1 *m* NH_4Cl 0.004% gelatine	−1.29

Element	*Oxidation state*	*Supporting electrolyte*	$U_{\frac{1}{2}}$
Cobalt		1 *m* Na_3 citrate 0.1 *m* NaOH	−1.45
		1 *m* KCNS	−1.04
	Co^{3+}	2 *m* NH_4OH 2 *m* NH_4Cl	−0.28
Copper	Cu^{+}	0.1 *m* KCl 50% pyridine	−0.52
	Cu^{2+}	1 *m* NH_4OH 1 *m* NH_4Cl 0.004% gelatine	−0.24 – −0.51
		1 *m* Na_3 citrate 0.1 *m* NaOH	−0.50
		0.1 *m* potassium hydrogen phthalate	−0.1
		1 *m* KOH	−0.41
Europium	Eu^{3+}	0.1 *m* EDTA pH = 6–8	−1.17
Gadolinium	Gd^{3+}	0.1 *m* KCl or LiCl 0.01% gelatine	−1.75
Gallium	Ga^{3+}	0.1 *m* KCl	−1.1
Germanium	Ge^{2+}	6 *m* HCl	−0.45
	Ge^{4+}	0.5 *m* NH_4OH 1 *m* NH_4Cl	−1.45 – −1.70
		0.1 *m* EDTA pH=6–8 10^{-4} *m* fuchsin	−1.30
Gold	Au^{+}	0.1 *m* KCN	−1.46
	Au^{3+}	2 *m* AcOH 2 *m* $AcONH_4$ 0.01% gelatine	> 0
		0.1 *m* KCN	> 0 – −1.4
Indium	In^{3+}	2.0 *m* AcOH 2 *m* $AcONH_4$ 0.01% gelatine	−0.71
		1 *m* KCl	−0.60
Iodine	I^{-}	0.1 *m* KNO_3	−0.03*
	IO_3^{-}	0.1 *m* acetate buffer pH = 4.9	−0.50
		0.1 *m* NaOH 0.1 *m* KCl	−1.21
	IO_4^{-}	0.2 *m* HBO_2 + KOH pH = 10 0.08% thymol	+0.02
Iridium	Ir^{4+}	1 *m* NaF 0.01% gelatine	−1.4
Iron	Fe^{2+}	0.05 *m* $BaCl_2$ or 0.1 *m* KCl	−1.3
		1 *m* NH_4OH 1 *m* NH_4Cl	−0.34* – −1.49
		0.5 *m* Na_2 tartrate pH = 5–6 0.005% gelatine	−0.17* – −1.50
		1 *m* KCNS	−1.4
	Fe^{3+}	0.15 *m* Na_3 citrate pH = 6	−0.18
		0.2 *m* $K_2C_2O_4$ pH = 4 0.005% gelatine	−0.24
		0.1 *m* EDTA 2 *m* AcONa	−0.12
Lead	Pb^{2+}	2 *m* AcOH 2 *m* $AcONH_4$ 0.01% gelatine	−0.50
		1 *m* Na_3 citrate 0.1 *m* NaOH	−0.78 – −1.5
		1 *m* NaOH 0.005% gelatine	−0.76
Lithium	Li^{+}	0.1 *m* $N(C_4H_9)_4OH$	−2.33
Magnesium	Mg^{2+}	0.1 *m* $N(CH_3)_4Cl$	−2.3
Manganese	Mn^{2+}	1 *m* NH_4OH 1 *m* NH_4Cl 0.004% gelatine	−1.66
		1 *m* KCNS	−1.54
		0.25 *m* Na_2 tartrate 2 *m* NaOH	−0.39*
	Mn^{7+}	Neutral solution of $BaCl_2$	−1.5
Mercury	Hg_2^{2+} Hg^{2+}	0.1 *m* HNO_3	> 0
Molybdenum	Mo^{6+}	12 *m* HCl	−0.63
		0.1 *m* H_2 tartrate pH = 2	−0.22 – −0.52
		0.1 *m* EDTA 0.1 *m* AcOH–$AcONH_4$	−0.63

Element	*Oxidation state*	*Supporting electrolyte*	$U_{\frac{1}{2}}$
Neodymium	Nd^{3+}	0.1 *m* KCl, LiCl or $N(CH_3)_4I$	−1.83
Neptunium	Np^{4+}	1 *m* HCl	−0.10
Nickel	Ni^{2+}	1 *m* NH_4OH 1 *m* NH_4Cl	−1.10
		1 *m* KCNS 0.01% gelatine	−0.68
		Saturated tartaric acid	−1.05
Niobium	Nb^{5+}	0.3 *m* K_3 citrate pH = 6.8	−1.73 - −2.03
Nitrogen	NH_2OH	1 *m* NaOH	−0.43
	N_3^-	0.1 *m* KNO_3	+0.25*
	NO	Dilute HCl	−0.9
	HNO_2	0.1 *m* H_2SO_4 0.2 *m* Na_2SO_4	−0.98
	NO_3^-	0.1 *m* $CeCl_3$ or 0.04 *m* $LaCl_3$	−1.58
	NH_4^+	0.1 *m* $N(CH_3)_4Br$	−2.21
Osmium	Os^{6+}	$Ca(OH)_2$ saturated	−0.40 - −1.16
	Os^{8+}	$Ca(OH)_2$ saturated	> 0 - −0.40 - −1.16
Oxygen	OH^-	0.1 *m* KNO_3	+0.08*
	H_2O_2	0.1 *m* NaOH	−0.18 - −1.0
	O_2	0.1 *m* KNO_3 or any electrolyte	−0.05 - −0.9
Palladium	Pd^{2+}	1 *m* pyridine, 1 *m* KCl	−0.34
Platinum	Pt^{2+}	1 *m* KCl	−0.1 - −0.98 - −1.35
Potassium	K^+	0.1 *m* $N(C_2H_5)_4OH$	−2.14
Praseodymium	Pr^{3+}	0.1 *m* LiCl 0.01% gelatine	−1.80
Radium	Rd^{2+}	Dilute KCl	−1.84
Rhenium	Re^-	2.4 *m* HCl	−0.17 - −0.34
	Re^{3+}	2 *m* $HClO_4$	−0.47 - −0.66
	Re^{4+}	2.4 *m* HCl	−0.53
	Re^{7+}	2.4 *m* HCl	−0.40
Rhodium	Rh^{3+}	1 *m* NH_4Cl	−0.93
Rubidium	Rb^+	0.1 *m* $N(CH_3)_4OH$	−2.03
Ruthenium	Ru^{4+}	1 *m* $HClO_4$	> 0 - +0.20 - −0.34
Samarium	Sm^{3+}	0.1 *m* $N(CH_3)_4I$ 0.0005 *m* H_2SO_4 0.01% gelatine	−1.80 - −1.96
Scandium	Sc^{3+}	0.1 *m* LiCl or HCl	−1.80
Selenium	Se^{2-}	0.05 *m* NH_4OH 1 *m* NH_4Cl	−0.84*
	Se^{4+}	1 *m* NaF 0.01% gelatine	−0.8 - −1.44
Silver	Ag^+	0.1 *m* KNO_3 or any common electrolyte	> 0
Sodium	Na^+	0.1 *m* $N(C_2H_5)_4$ OH	−2.12
Strontium	Sr^{2+}	0.1 *m* $N(C_2H_5)_4I$	−2.11
Sulphur	S^{2-}	0.1 *m* KOH or NaOH	−0.76*
	S^0	1.1 *m* pyridine 0.06 *m* pyridine hydrochloride in methanol	−0.50
	SCN^-	0.1 *m* KNO_3	+0.18*
	$S_2O_3^{2-}$	0.1 *m* KNO_3	−0.14*
	$S_4O_6^{2-}$	1 *m* H_3PO_4 $(NH_4)_2HPO_4$ pH = 1–8 0.001% quinoline	−0.26
	$S_2O_4^{2-}$	1 *m* NH_4OH 0.5 *m* $(NH_4)_2HPO_4$ 0.01% gelatine	−0.43*
	SO_2	0.1 *m* HCl or HNO_3	−0.37

Element	*Oxidation state*	*Supporting electrolyte*	$U_{\frac{1}{2}}$
Tantalum	Ta^{5+}	0.9 *m* HCl	−1.16
Tellurium	Te^{2-}	0.1 *m* NH_4OH 1 *m* NH_4Cl	−1.1*
	Te^{4+}	1 *m* NH_4OH 1 *m* NH_4Cl pH = 9.4	−0.67
	Te^{6+}	Acetate buffer pH = 5.6	−1.18
Thallium	Tl^{+}	0.1 *m* NH_4OH and many common electrolytes	−0.46
	Tl^{3+}	1 *m* $HClO_4$	> 0 − −0.48
Tin	Sn^{2+}	2 *m* AcOH 2 *m* $AcONH_4$ 0.01% gelatine	−0.16* − −0.62
	Sn^{4+}	12 *m* HCl 0.002% triton	−0.50 − −0.83
Titanium	Ti^{3+}	Saturated tartaric acid	−0.44*
	Ti^{4+}	Saturated tartaric acid	−0.44
		EDTA pH = 1–2.5	−0.22
Tungsten	W^{3+}	3 *m* HCl	−0.65*
	W^{5+}	12 *m* HCl	−0.56
	W^{6+}	7.3 *m* H_3PO_4	−0.59
Uranium	U^{3+}	1 *m* HCl	−0.94
	U^{4+}	0.5 *m* H_2SO_4	−1.10
	U^{5+}	0.5 *m* $NaClO_4$ 0.01 *m* $HClO_4$	−0.18 − −1.57
	U^{6+}	2 *m* AcOH 2 *m* $AcONH_4$ 0.01% gelatine	−0.45
Vanadium	V^{2+}	Na_3 citrate pH = 7	−0.87*
	V^{3+}	1 *m* HCl or $HClO_4$ or 0.5 *m* H_2SO_4	−0.51
	V^{4+}	1 *m* AcOH 1 *m* AcONa 0.0003% gelatine	−1.18
	V^{5+}	1 *m* NH_4OH 1 *m* NH_4Cl 0.005% gelatine	−0.96 − −1.26
Zinc	Zn^{2+}	1 *m* KCl 0.0003% gelatine	−1.00
		1 *m* N_2H_4 1 *m* $NaClO_4$ pH = 8–9	−1 13
		1 *m* NaOH 0.01% gelatine	−1.53
Zirconium	Zr^{4+}	0.1 *m* LiCl in absolute methanol	−1.4

* anodic process.

reduction tension is markedly affected by the pH since the hydrogen ion takes part in the reaction. The reduction of these anions is generally irreversible, but in well-buffered solutions the diffusion current is reproducible and proportional to the concentration of the reducible ion.

Many uncharged substances such as oxygen, hydrogen peroxide, cyanogen, etc. may be determined polarographically. This method is particularly important for the determination of oxygen dissolved in a solution. The reduction of oxygen occurs in two stages and so gives rise to two steps. The first of these is due to the reduction of oxygen to hydrogen peroxide ($O_2 + 2\,H^+ + 2\,e^- \rightarrow H_2O_2$); and the second is due to the reduction of this to water ($H_2O_2 + 2\,H^+ + 2\,e^- \rightarrow 2\,H_2O$). The

TABLE V, 8

HALF WAVE ELECTRIC TENSIONS OF ORGANIC SYSTEMS

($U_{\frac{1}{2}}$ with respect to the saturated calomel electrode)

Compound	$U_{\frac{1}{2}}$	pH	*Supporting electrolyte*
Aldehydes			
Formaldehyde	−1.67	12.7	0.05 *m* KOH, 0.1 *m* KCl
Acetaldehyde	−1.89	6.8	LiAc. AcOH
Propionaldehyde	−1.9	—	0.1 *N* LiOH
Glyoxal	−1.41	3.4	Buffer
Acrolein	−0.83	4.8	Buffer
Croton aldehyde	−1.37 − −1.80	—	0.2 *N* $N(CH_3)_4OH$
Citral	−1.93	—	1 *N* NH_4Cl, 75% C_2H_5OH
Benzaldehyde	−1.00 − −1.46	2.2	McIlvaine's buffer
Salicylaldehyde	−1.03 − −1.27	2.2	McIlvaine's buffer
Anisaldehyde	−0.97	2.2	McIlvaine's buffer
Vanillin	−1.05	2.2	McIlvaine's buffer
Ketones and Chalcone			
Acetone	−2.20	—	0.025 *m* $N(CH_3)_4I$
Dihydroxyacetone	−1.64	7	Veronal buffer
Benzoylacetone	−1.10	0.6	Buffer
Methylethylketone	−2.25	—	0.025 *m* $N(CH_3)_4I$
2,3-Butanedione	−0.84 − −1.59	—	0.1 *N* NH_4Cl
Acetophenone	−1.12	1.3	McIlvaine's buffer
o-Hydroxyacetophenone	−1.07 − −1.45	—	0.1 *N* NaOH, 50% isopropanol
Chalcone	−0.57 − −1.00	1.3	McIlvaine's buffer
Ninhydrin	−0.70 − −0.85 − −0.99	2.3	Phosphate buffer
Carbohydrates			
Ribose	−1.81	7	Phosphate buffer
Glucose	−1.58	7	Phosphate buffer
Fructose	−1.80	—	0.1 *N* LiCl
Sorbose	−1.80	—	0.1 *N* LiCl
Ketoacids			
Pyruvic acid	−1.07	4	Britton–Robinson buffer
Phenylglyoxylic acid	−0.52	2.2	Britton–Robinson buffer
Folic acid	−0.41	1.8	Britton–Robinson buffer
Heterocyclic compounds containing oxygen			
α-Pyrone	−1.91	—	0.2 *m* $N(CH_3)_4OH$, 50% ethanol
Flavone	−1.30 − −1.42	6.1	Acetate buffer $N(CH_3)_4OH$, 50% isopropanol
Flavanol	−1.29	5.6	Acetate buffer, 50% isopropanol
Apigenin	−1.61	5.6	Acetate buffer, 50% isopropanol
Quercetin	−1.57	5.6	Acetate buffer, 50% isopropanol

Compound	$U_{\frac{1}{2}}$	pH	*Supporting electrolyte*
Rutin	−1.50	6.3	Acetate buffer, 50% isopropanol
Morin	−1.61	6.3	Acetate buffer, 50% isopropanol
Flavanone	−1.33	6.1	Acetate buffer, $N(CH_3)_4OH$, 50% isopropanol
Coumarin	−1.57	6.8	Phthalate buffer
Unsaturated acids			
Fumaric acid	−0.83	2.56	HCl–KCl buffer
Maleic acid	−0.70	2.0	Britton–Robinson buffer
Citraconic acid	−0.58	–	1 *N* HCl
Mesaconic acid	−0.57	–	1 *N* HCl
Aconitic acid	−0.45	–	1 *N* HCl
Cinnamic acid	−1.99	–	0.1 *N* LiOH
Crotonic acid	−2.9	–	0.1 *N* LiOH
Aliphatic nitrogen compounds			
Nitromethane	−0.72	1.8	Britton–Robinson buffer, 30% ethanol
Nitroethane	−0.73	1.8	Britton–Robinson buffer, 30% ethanol
1-Nitropropane	−0.69	1.8	Britton–Robinson buffer, 30% ethanol
1-Nitro-2-butanol	−0.56	2.0	McIlvaine's buffer
2-Methyl-2-nitropropanol	−0.82	5.0	McIlvaine's buffer
Dimethylglyoxime	−1.63	9.6	2 *N* NH_4OH, 2 *N* NH_4Cl buffer
Chloramin T	−0.13	–	0.5 *N* K_2SO_4
Aromatic nitrogen compounds			
Nitrobenzene	−0.34	2.5	Phthalate buffer
-Dinitrobenzene	−0.16 – −0.36 – −1.3	2.5	Phthalate buffer
,3,5-Trinitrobenzene	−0.12 – −0.19 – −0.28	2.5	Phthalate buffer
Trinitrotoluene	−0.14 – −0.24 – −0.35	2.5	Phthalate buffer
-Nitrophenol	−0.27	–	Britton–Robinson buffer, 8% ethanol
icric acid	−0.11	1.7	McIlvaine's buffer
-Nitropyrocatechin	−0.20	2.1	McIlvaine's buffer
typhnic acid	−0.11 – −0.16 – −0.27	2.0	Britton–Robinson buffer, 8% ethanol
-Nitroaniline	−0.16 – −0.57	1.8	Britton–Robinson buffer
icrolonic acid	−0.34 – −0.75	3.6	Acetate buffer
-Nitroso-β-naphthol	−0.02	4.0	Buffer, 48% ethanol
I-Phenylhydroxylamine	−0.84	3.0	McIlvaine's buffer
enzoin oxime	−0.88	2.0	Buffer
zobenzene	−0.20	4.0	Buffer, 48% ethanol
nsaturated hydrocarbons			
tyrene	−2.39	–	0.17 *m* $N(C_4H_9)_4I$, 75% dioxane
tilbene	−2.18	–	0.17 *m* $N(C_4H_9)_4I$, 75% dioxane

Compound	$U_{\frac{1}{2}}$	pH	*Supporting electrolyte*
Butadiene	−2.59	—	0.05 *m* $N(CH_3)_4Br$, 75% dioxane
Vinylacetylene	−2.40	—	0.05 *m* $N(CH_3)_4Br$, 75% dioxane
Naphthalene	−2.50	—	0.17 *m* $N(C_4H_9)_4I$, 75% dioxane
Anthracene	−1.94	—	0.17 *m* $N(C_4H_9)_4I$, 75% dioxane
Halogenated compounds			
Methylchloride	−2.23	—	0.05 *m* $N(CH_3)_4Br$, 75% dioxane
Chloroform	−1.67	—	0.05 *m* $N(CH_3)_4Br$, 75% dioxane
Bromobenzene	−2.32	—	0.05 *m* $N(CH_3)_4Br$, 75% dioxane
Chloronaphthalene	−2.10 – −2.38	—	0.05 *m* $N(CH_3)_4Br$, 75% dioxane
Hexachloroethane	−0.62 – −1.73 – −1.96	—	0.05 *m* $N(CH_3)_4Br$, 75% dioxane
D.D.T.	−0.90	—	0.1 *m* $N(CH_3)_4Br$, 80% ethanol
α-Hexachlorocyclohexane	−2.02	—	0.1 *m* $N(C_2H_5)_4I$, 80% ethanol
Trichloracetic acid	−0.93	4.0	Phosphate buffer
Monoiodoacetic acid	−0.79	11.3	1% $N(CH_3)_4Br$, 0.5 *N* Na_2CO_3, 20% isopropanol
Monoiodotyrosine	−1.64	11.3	1% $N(CH_3)_4Br$, 0.5 *N* Na_2CO_3, 20% isopropanol
Heterocyclic compounds			
Isatine	−0.26 – −0.46	2.9	McIlvaine's buffer
Nicotinic acid	−0.89	—	0.1 *N* HCl
Quinolinic acid	−0.82	—	0.1 *N* HCl
8-Hydroxyquinoline	−1.03	2	Acetate buffer
Quinaldinic acid	−0.67 – −1.01	2	Acetate buffer
Acridine	−0.34	2	Citrate buffer
Sulphydryl compounds			
Cystein	−0.30	3.8	Acetate buffer
Thioglycolic acid	−0.25	3.8	Acetate buffer
Thiourea	−0.90	—	2 *N* $H_2SO_4 - K_2SO_4$
Glutathione	−0.12	1.8	Britton–Robinson buffer
Mercaptobenzothiazol	+0.03	1.1	Britton–Robinson buffer
Thiamine	−0.42	11.0	Sörensen's buffer
Anodically oxidizable compounds			
Ascorbic acid	−0.06	7.0	Phosphate buffer
Dihydroxymaleic acid	+0.02	1.8	Britton–Robinson buffer
Pyrocatechin	+0.36	3.7	Phosphate buffer
Pyrogallol	+0.31	2.9	Britton–Robinson buffer
Adrenalin	+0.15	1.8	Phosphate buffer
α-Tocopherol	+0.37	1.7	Aniline $HClO_4$, 75% ethanol
p-Azophenol	+0.17	9.2	Borate, 50% ethanol
Alkaloids			
Lobeline	−1.12 – −1.16	1.8	Britton–Robinson buffer
Quinine	−1.60	12	Britton–Robinson buffer
Cinchonine	−1.02 – −1.27	10	Britton–Robinson buffer

Compound	$U_{\frac{1}{2}}$	pH	*Supporting electrolyte*
Hydrastinine	−1.09	3	Acetate buffer
Colchicine	−0.99	2.2	McIlvaine's buffer
Disulphides			
Cystine	−0.72	3.8	Acetate buffer
Oxidized glutathione	−0.26	1.0	Buffer
Tetramethylthiourane disulphide	−0.94	4.8	Buffer, 50% dioxane
Dyes			
Neutral red	−0.25	2	Britton–Robinson buffer
Pyocyanine	−0.19	5	Phosphate buffer
Gallocyanine	−0.26	7	Phosphate buffer
Rosindulin GG	−0.32	2.17	McIlvaine's buffer
Methylene blue	−0.02	2.9	Britton–Robinson buffer
2, 6-Dichlorophenol-indophenol	−0.2	—	0.1 *N* NH_4Cl, 50% ethanol
Indigo-disulphonate	−0.41	7	Buffer

ability of the dropping electrode to reduce oxygen is the essential reason why it is necessary in most cases to deaerate the solution before making measurements.

Many organic compounds may also be determined polarographically. The criterion necessary to establish if a compound can be reduced by the dropping electrode, is the presence in its molecule of reducible organic functions such as $-NO_2$, $-NO$, $-CO$, $-N{=}N-$. Thus, the reducible substances include nitro, azo and diazo compounds, quinones, aldehydes, ketones and unsaturated compounds with double bonds. The reduction of these compounds is always markedly influenced by the hydrogen ion concentration of the solution since the hydrogen ion takes part in the electrode process; it is thus necessary to use highly buffered solutions.

Table V,8 reports $U_{\frac{1}{2}}$ values for the most important organic compounds. In most cases, the pH of the supporting solution is indicated. In general the $U_{\frac{1}{2}}$ values become more negative as the pH of the supporting electrolyte is increased.

Complete collections of $U_{\frac{1}{2}}$ tensions of organic compounds may be found in specialized publications[1].

[1] P. ZUMAN, *Collection Czech. Chem. Commun.*, 15 (1950) 375; G. SEMERANO and L. GRIGGIO-FORRESE, *Selected Values of Polarographic data*, in Contributi Teorici e Sperimentali di Polarografia Vol. III. Rome, Consiglio Nazionale delle Ricerche 1957.

Anodic waves

Anodic waves observed and used for polarographic determination of substances may be classified into two types: those due to actual oxidations of the substance at the dropping electrode and those due to the depolarization of the dropping electrode through the formation of insoluble salts or of complexes with the mercury ions. The first type includes the waves observed during the oxidation of ferrous ions to ferric, of ferrocyanide to ferricyanide, of hydroquinones to quinones etc. In waves of the second type the dropping electrode behaves virtually as a mercury electrode and the concentration of the mercurous ions is determined by the electric tension applied to the electrode, according to Nernst's equation.

$$U = U_{0\,Hg/Hg_2^{2+}} + (RT)/(2\,F) \ln a_{Hg_2^{2+}}$$

These ions — formed by the anodic dissolution of mercury — diffuse into the solution and give rise to a current whose intensity is proportional to their concentration. If ions or substances which react with the mercurous ions are present in the solution, the concentrations of both may diminish at the interface through the formation of insoluble or poorly dissociated complex products. If the electric tension of the dropping electrode is made more positive, the concentration of the depolarizer at the interface tends towards zero and there is then a passage of a current whose intensity is determined by the velocity of diffusion of the depolarizer from the bulk of the solution. Thus, a region is observed where the current intensity is practically constant; a marked increase in current will accompany any further positive polarization due to the anodic dissolution of mercury. These waves may be used for the determination of halogens, sulphides, cyanides, sulphocyanates, mercaptans and sulphydryl compounds.

Compounds which give anodic waves are indicated in Table V,7 with asterisks. In addition to the waves mentioned above — whose formation may be ascribed to diffusion phenomena — the polarization of the dropping electrode may give rise to waves for which these considerations are not valid. These other waves may also be of two types: catalytic or kinetic. Substances which can diminish the overtension of hydrogen on the dropping electrode may be responsible for the formation of catalytic waves. This effect leads to a step at a polarization tension which is less than that which would be observed with hydrogen in the absence of such substances. The current intensity for these steps is greater than that which would be found if the electrodic depolarization were due to the substance which acts as the catalyst. Catalytic waves are found in the *I–U* curves of solutions of metallic salts such as platinum, ruthenium, and palladium and of many organic substances such as proteins, sulphydryl compounds, quinoline derivatives etc.

Kinetic waves refer to those steps due, not to a diffusion process, but to a reaction taking place at the mercury-solution interface. Typical kinetic waves are observed in the polarographic reduction of substances such as pyruvic acid, phenylglyoxylic acid, nitrosophenylhydroxylamine etc. In such systems, in addition to the normal reduction waves, small steps are observed due to the recombination of ions at the interface to form undissociated acid. These small steps differ from those due to diffusion phenomena because their height is independent of the height of the mercury column standing above the dropping electrodes.

A particular phenomenon is sometimes encountered in polarograms, the *I–U* curve being accompanied by a maximum which may alter the shape of the curve itself, invalidating the results of the analysis or even making it completely impossible. The cause of this maximum is rather complex. It may generally be attributed to adsorption of a depolarizer at the mercury surface. Such maxima may be eliminated by adding a small amount of a surface active agent to the solution being analyzed, to lower the surface tension of the mercury without giving rise to electrochemical processes which would be detected polarographically in the range of electric tensions used in the analysis itself. For such purposes gelatine, carboxymethylcellulose and traces of dyes such as methyl red, methyl orange, and fuchsin have been used.

Quantitative analysis

With normal oxidation or reduction waves it is possible to obtain the qualitative composition of the system, responsible for the electrodic depolarization, from the $U_{\frac{1}{2}}$ values; and by using Ilkovic's equation on the proportionality of step height to depolarizer concentration, the quantitative composition may also be derived.

Methods for the determination of the concentration of a species from the wave height may be either absolute or comparative. In the first, the data obtained with an unknown solution and a capillary of known characteristics are compared with those obtained with a solution of known concentration and a different capillary, by using Ilkovic's equation. This is,

$$c = \frac{I_d}{Km^{\frac{2}{3}}t^{\frac{1}{6}}} \tag{14}$$

where the proportionality constant K is $607zD^{\frac{1}{2}}$. Knowing K[1] for a particular standard solution, the concentration of an unknown solution, which gives a current I_d with a capillary of characteristic values for m and t, may be derived from equation (14).

In the comparative methods the data obtained for the test solution with a given capillary are compared with those obtained in solutions of known concentrations of the same substance, using the same capillary under absolutely identical experimental conditions. This comparison may be made in various ways. One procedure consists in preparing a series of solutions identical in composition with the test solution but with different concentrations of the substance to be determined. Measurements are made of the diffusion currents with these solutions and a graph of c against I_d is drawn. The value of c for the test solution is then taken from this graph.

Another method is known as that of standard additions. For this, a certain amount of the component which is to be determined is added to the test solution, after the polarogram has been obtained, and the measurement is repeated. If v_x is the volume of test solution of unknown concentration c_x giving a current of intensity I_1, and if the addition of a volume v_n of a solution of concentration c_n leads to a current intensity I_2 then

$$I_1 = Kc_x \tag{15}$$

and

$$I_2 = \frac{K(v_x c_x + v_n c_n)}{(v_x + v_n)} \tag{16}$$

Substituting the value of K from (16) into (15) gives

$$I_1 = \frac{I_2(v_x + v_n)c_x}{(v_x c_x + v_n c_n)}$$

whence

$$c_x = \frac{I_1 v_n c_n}{v_x(I_2 - I_1) + I_2 v_n}$$

[1] Values for K are reported in L. Meites, *Polarografic Techniques*, Interscience Publ., New York, 1955; I. M. Kolthoff and J. J. Lingane, *Polarography*, Interscience Publ., New York, 1952.

This relationship depends on the validity of the assumptions made about the proportionality of the wave height—and hence the diffusion current—to the concentration.

A further method is that of the so-called pilot ion. This is based on the fact that the ratio of the wave heights obtained with two substances in the same solution is virtually independent of the temperature, of the capillary characteristics and of the composition of the supporting electrolyte. Hence, if the concentration of a substance added to the test solution is known, together with the ratio of the two polarographic waves given by the added substance and the substance to be determined, it is possible to calculate the concentration of the latter in the test solution. When this procedure can be used it gives reasonably accurate results more rapidly than the other methods described.

Quantitative polarographic analysis may be applied within the concentration range of 10^{-5}–10^{-2} *m*, with an optimum of 10^{-4}–10^{-3} *m*. Allowing for the various factors which affect the production and measurements of diffusion current, the degree of precision which can be obtained in quantitative analytical determinations is under the most favourable conditions about 1.5%. Thus the method furnishes excellent results for the determination of elements or substances present as minor constituents. For this degree of accuracy to be achieved, however, it is necessary for the experimental conditions to be critically controlled; the test solution must contain a concentration of inert electrolyte of at least 50–100-fold greater than that of the component to be determined; the wave must not be disturbed by maxima or other adsorption phenomena; the polarogram must be measured on a solution whose temperature is thermostatically controlled and using a capillary which drops regularly (1 drop every 3–5 seconds) and hence the mercury pressure must be kept constant.

(III). *Voltammetry*

Current intensity–electric tension curves may be obtained by polarizing any solid electrode. The useful limits of every electrode are set by the negative polarization for the discharge of hydrogen or of the supporting electrolyte and by the positive polarization at which electric tension the metal is anodically oxidized, or in the case of the noble metals (platinum, gold) at which the water is oxidized to oxygen (about +1.1 volts to the saturated calomel electrode). In view of the low overtension of hydrogen on most metals, practical interest is confined to *I–U* curves obtained with noble metal electrodes. These can be used for strongly oxidizing agents or organic systems which can not be easily reduced and can only be oxidized with difficulty.

The *I–U* curves obtained with stationary platinum or gold electrodes are similar to those obtained with the dropping electrode; the currents produced however are much smaller and the measurements of the curves are of little analytical interest in view of the time required to reach equilibrium. For practical purposes voltammetry with rotating electrodes of platinum or gold or of their amalgams, is of more interest. These electrodes consist of fine wires of the metal (about 4 mm long and 0.5 mm diameter) fused into the end or side of a small glass tube which is rotated at a constant velocity of about 600 r.p.m. by a synchronous motor. Electrical connection is made by putting a little mercury in the tube and dipping into it a nickel or tungsten wire. The reducible or oxidizable substance is brought to the electrode by convection at a constant velocity and gives rise to a curve whose limiting current is proportional to the concentration of the depolarizer. The current also depends upon the surface area of the electrode and on the rotation velocity. All the other factors involved, however, are not completely known so that a simple equation cannot be derived. Equilibrium is reached very rapidly with these electrodes which although limited in use to the indicated range of electric tensions, make possible a simple and rapid determination of various substances such as oxygen, thallium, lead dioxide and iodine. The behaviour of platinum microelectrodes is markedly influenced by the formation of a thin film of oxide. They are particularly important as indicator electrodes in amperometric titrations.

(IV). *Miscellaneous Polarographic and Voltammetric Techniques*

The plotting of *I–U* curves constitutes, as indicated, the classical polarography or direct current polarography. The technique has a series of limitations connected with the experimental arrangement used. The intensity of current which is measured is not constant but oscillates between a maximum and a minimum value in time with the growth and periodic renewal of the drops. The relatively high value of the residual current prevents the use of this technique for the determination of traces of substances. Direct current polarography also suffers from a limited resolution which is particularly inconvenient when it is necessary to determine an ion whose wave is preceded by that of other ions present in high concentrations relative to that of the first. Many polarographic techniques have been proposed to overcome these limitations and some of these will be briefly mentioned, without however entering into details of the principles or their practical use, for which reference should be made to specialized works.

Derivative polarography

Derivative polarography consists in the construction of the derivative dI/dU curves with respect to the applied electric tension instead of the current intensity–electric tension curves. The derivative of the current intensity with respect to the electric tension is practically equal to zero at the points immediately preceding and following the polarographic wave and reaches a maximum value at the half wave tension. If, then, the I–U diagram consists of several steps, its derivative will show maxima at the points of maximum slope and will return to zero at the intervals between the steps.

An apparatus for such measurements may be constructed by placing an electrolytic capacitor of a fairly high capacitance (1,000–4,000 μF) in series with the circuit used to measure the current. If a recording polarograph is used in which the electric tension is varied at a constant rate, a dI/dt curve will be obtained which will be proportional to the dI/dU curve.

Oscillographic polarography

Oscillographic polarography may be defined as polarography with a varying electric tension such that the entire range of electric tensions to be examined is explored during the life of a single drop of mercury, *i.e.* in the time between the fall of one drop and the fall of the next from the capillary. It is however preferable to shorten the period of exploration still further and to confine it to the last part of the drop life, keeping the electric tension constant equal to zero for the rest of the time. In this way it is possible to work with an electrode whose surface area is virtually constant during the scanning of the whole range of electrical tensions and to eliminate the charging current of the drop. This process is automatically repeated for each drop so that the traces of the electric tension–current intensity diagram are superimposed when displayed on an oscillograph.

A similar arrangement is used to that for classical polarography with the difference that the polarization tension increases rapidly and linearly, and explores the whole field during the last phase of the drop life and then returns instantaneously to zero. In this way, curves are obtained similar to polarographic ones, with the advantage of rapidity and greater resolving power. Other types of oscillographic polarography exist: with sinusoidal tensions, with symmetrical sawtooth tensions, with imposed current intensity etc., which are, however, of a special nature and fall outside the limits of this work; specialized textbooks should be consulted.

Polarography with a direct current and a superimposed alternating current

This technique consists in applying a small alternating tension (1–90 mV) to the polarizing electric tension. The alternating current which passes is amplified and measured with an electronic voltmeter. This technique is considerably more rapid than the classical one, the current intensities measured are greater and hence the method is more sensitive. The polarograms obtained are not influenced by the preceding discharge of other ions.

Square-wave polarography

This technique is in principle identical to that described in the preceding paragraph with the sole difference that in place of a sinusoidal alternating tension, a square-wave tension is superimposed on the polarizing tension. It is possible with this method to determine substances in concentrations of the order of 10^{-7} *m*.

(V). *Chronopotentiometry*

In conventional voltammetry the electric tension of the working electrode is controlled and the electrolytic current is measured. However, constant current voltammetry may be used in which a current of a constant preselected intensity is passed through a cell containing an excess of a supporting electrolyte, and the changes in electric tension of the working electrode, (mercury pool) are measured with a recorder or cathode ray oscillograph. This technique is called chronopotentiometry and consists in determining an electric tension–time curve like that shown in Fig. V, 30.

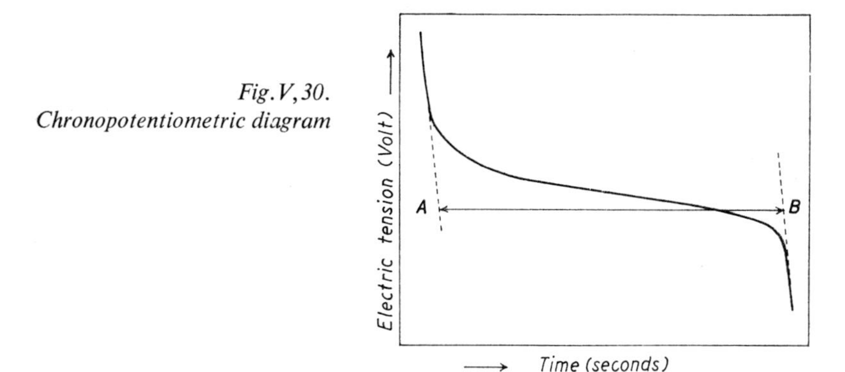

Fig. V, 30.
Chronopotentiometric diagram

These curves may be readily interpreted by considering a reversible process for which the electric tension of the mercury electrode may be expressed by Nernst's equation. In the case of a solution of lead ions, considering the activity of the mercury (dilute amalgam) as unity, gives

$$U = U_{0\,amalg.} + \frac{RT}{2\,\mathrm{F}} \ln \frac{f_{Pb^{2+}} c_{Pb^{2+}}}{f_{Pb} c_{Pb}}$$

The variations of electric tensions during electrolyses are thus determined by the ratio $c_{Pb^{2+}}/c_{Pb}$. During electrolysis at constant current, the concentration of the lead ion (Pb^{2+}) at the electrode surface diminishes as a consequence of the electrodic process whilst the concentration of lead (Pb) within the mercury increases. The electrodic tension becomes increasingly negative passing from a theoretical value of $+\infty$ (when no lead is present in the mercury) to one of $-\infty$ (when the concentration of lead ion at the electrodic surface reaches zero). The time required for the concentration of an electrolizable species to become zero at the electrode surface is defined as the transition time; in Fig. V, 30 it is represented by the segment AB. The analytical importance of this procedure lies in the fact that the square root of the transition time is generally proportional to the concentration of the substance being analysed. Chronopotentiometric analysis may be used even in the presence of two or more electro-reducible or oxidizable ions over a wide range of concentrations (10^{-6}–10^{-2} m) and gives a high degree of accuracy (about 1–2%)[1].

[1] C. N. Reilley, G. W. Everett and R. H. Johns, *Anal. Chem.*, 27 (1955) 483.

(VI). *Amperometric and voltammetric titrations*

The principle underlying polarographic and voltametric analysis, that the diffusion current obtained with a dropping electrode and the limiting current obtained with the rotating electrode are proportional to the concentration of a reducing or oxidizing substance and that changes in concentration during a titration can be readily followed, forms the fundamental theoretical basis of amperometric and voltammetric titrations; originally these were known as polarometric titrations.

Consider a substance A which can be polarographically reduced giving a well-defined polarographic wave (Fig. V, 31 *A*) acting on a reagent B which can also be reduced by the dropping electrode but at a more negative tension (Fig. V, 31 *B*).

If A and B form a precipitate or a poorly dissociated complex which is not reduced by the dropping electrode, it is possible to titrate A with B in a simple manner. The solution of A is placed in a suitable polarographic cell and the dropping and comparison electrodes are inserted into it. An electric tension lying between the values *a* and *b* or *c* and *d* is applied to the electrodes. With successive additions of the reagent B there are corresponding decreases in the diffusion current due to the substance A until the current intensity becomes very small at the equivalence point. The further addition of reagent will not lead to any increase of current if the polarizing tension lies between *a* and *b* whilst the current increases if the tension lies between *c* and *d*, since in this case the reagent will be reduced. Plotting the values of the diffusion currents as ordinates against the

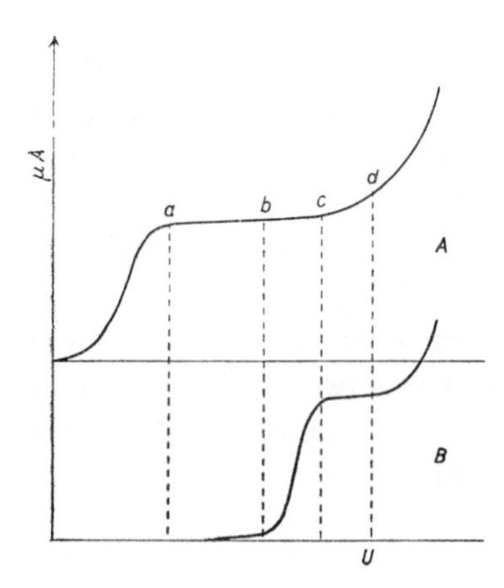

Fig. V, 31. Principle of amperometric titration

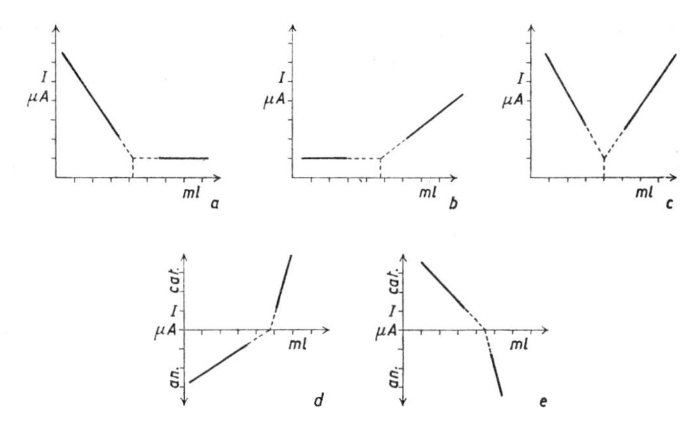

Fig. V, 32. Diagrams of amperometric titrations

volume of reagent solution added, gives two series of points. Straight lines drawn through these points will intersect indicating the end point of the reaction which coincides with the equivalence point if the reaction between the substance and the reagent is stoichiometric. If the electric tension applied lay between *a* and *b*—which is the usual case for an electro-reducible substance titrated with a reagent which is not reducible within that electric tension range — the graph obtained would be similar to that of Fig. V, 32 *a*.

If the applied electric tension lay between *c* and *d*—which is the general case when both the substance and the reagent are reducible—the graph obtained would be similar to that of Fig. V, 32 *c*. A remarkable variety of titration curves can be obtained. If the substance is not electro-reducible but the reagent gives a diffusion current at a particular electric tension, the titration curve resembles that of Fig. V, 32 *b*. If the substance gives an anodic wave and the reagent a cathodic wave at a given electric tension the graph obtained is that of Fig. V, 32 *d* whilst in the opposite case it is that of V, 32 *e*; In these last two cases the titration curve may correspond to a straight line, which intersects the abscissa axis, if the substance and the reagent have the same polarographic characteristics.

If in all these cases the experimental values are plotted without allowing for the effect of the dilution caused by the addition of the reagent, the points will lie on a curve instead of on a straight line. The introduction of the reagent actually dilutes the substance remaining to be titrated and there is thus a gradual fall in the diffusion current. The dilution effect may be corrected for by using the expression

$$I_{\text{corrected}} = I_{\text{measured}} (v_x + v_n)/v_x$$

where v_x is the volume of the solution to be analysed and v_n is the volume of reagent added.

Similarly to conductometric titrations, the end point may be obtained by drawing straight lines through the points preceding and following it. In fact, the points close to the end point do not fall on a straight line because of the slight solubility of the precipitate or dissociation of the complex. Further away from the end point, this is not observed because the presence of a common ion suppresses the solubility or dissociation. Amperometric titrations may be performed using a rotating platinum or gold micro-electrode as indicator electrode. These are of less general use than the dropping electrode but when they can be used they offer the advantage that they can detect smaller concentrations of depolarizer, can give a current free of oscillations and can be used for titrations involving powerful oxidizing agents such as bromine or cerium (IV).

Any suitable circuit may be used for measuring the *I–U* curve in an amperometric titration; a suitable electric tension is applied to a cell containing the solution to be assayed and then changes in the current intensity are measured, after each addition of the titrant from a burette dipping into the cell. Since the diffusion current is often independent of the applied

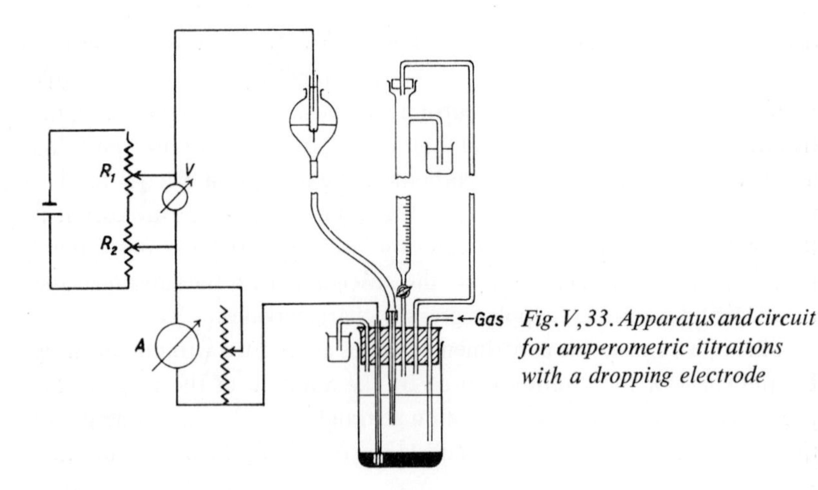

Fig. V, 33. Apparatus and circuit for amperometric titrations with a dropping electrode

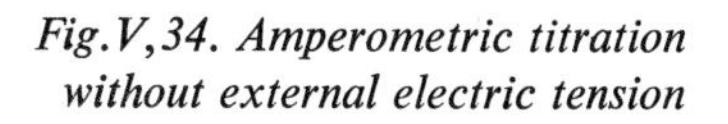
Fig. V, 34. Amperometric titration without external electric tension

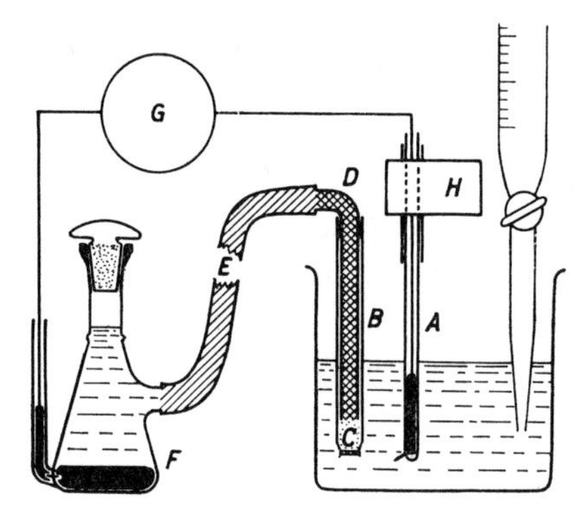

electric tension over a wide range and since it is changed in current intensity, and not absolute values, which are of interest the circuit may be much simplified. The polarizing tension may be taken from a battery through a potential divider and be measured with a voltmeter; the circuit current may be measured with an inexpensive microammeter. Such an arrangement is shown in Fig. V, 33.

In some cases the apparatus may be simplified further if the comparison electrode is selected to give an electric tension at which the reagent ions can depolarize the indicator electrode. Fig. V, 34 shows the apparatus necessary for the amperometric determination of mercaptans using the rotating platinum microelectrode as the indicator. The titration is carried out in an ammoniacal medium using silver ions as the reagent, following the reaction

$$RSH + Ag(NH_3)_2^+ \rightarrow RSAg + NH_4^+ + NH_3$$

and a mercury–mercury iodide electrode as the comparison electrode; this gives an electric tension at which the $Ag(NH_3)_2^+$ ions depolarize the indicator electrode. No current passes on the addition of Ag^+ so long as mercaptans are present in the solution. After the equivalence point is reached, the current rises proportionally to the excess of silver ions added.

Amperometric titration represents an extension of polarography and voltammetry but is more accurate than these and also permits the determination of substances which cannot be electrolytically reduced or oxidized. The errors encountered in such titrations are usually less than 1% and may fall below 0.2%. This is because the graph is drawn through experimental points which represent changes in the current and not absolute measures of it, so that it is not necessary strictly to observe those experimental conditions which are critical in polarographic analysis. Amperometric titrations may also be used even when the limiting current is not perfectly defined because of the dilution; such titrations, unlike conductometric ones, are not affected by high concentrations of electrolytes and may be performed in a short time particularly if the rotating platinum microelectrode is used as the indicator.

Significant concentrations of ions which can be reduced at the same, or at a more negative, electric tension than the substance being titrated form the principal limitation to the use of amperometric titrations; such ions must first be removed. The formation of supersaturated solutions may also be a serious cause of error in precipitation reactions. Amperometric titrations may be used for precipitation reactions, redox and complex formations; they constitute a very simple and accurate method of performing many volumetric titrations. They permit the use of organic reagents such as cupferron, α-nitroso-β-naphthol, benzoin oxime, quinoline and EDTA for the determination of many cations. An up to date review of amperometric titrations has been published by Konopik[1].

Amperometric titrations with two polarized electrodes

Instead of using a polarized and a comparison electrode for amperometric titrations, two identical polarized metal electrodes may be used. This procedure was first suggested by Foulk and Bowden who called it the 'dead stop endpoint' technique; it forms the basis of the polarovoltmetric technique of analysis. If small electric tensions of some 10–100 mV are applied between two identical, usually platinum, electrodes in a solution a small current will pass provided that both electrodes are depolarized. This current may be detected with a galvanometer or a microammeter. If such conditions are obtained in the solution to be titrated and a reagent is then added which removes the species to be assayed, this reagent will polarize one electrode at the end point and the current will no longer pass. Such a system is found in the titration of iodine with thiosulphate. With the inverse titration of thiosulphate by iodine, the inverse system is found with no current passing during the

[1] N. Konopik, *Amperometrische Titrationen*, Österreichische Chemiker-Zeitung, 54 (1953) 289, 325; 55 (1954) 127; 57 (1956) 181.

titration until the equilibrium is reached. In general terms, if the system to be titrated forms a reversible redox couple, the electrodes will be depolarized by a small applied electric tension since the oxidized form will be reduced and the reduced oxidized. If the titrant forms part of a nonreversible couple the electrodes will be polarized as soon as the former system is no longer present, and no current will pass. This behaviour is found in the titration of ferric ions (reversible system) with titanium (III) (irreversible system), whilst the inverse case is found in the titration of thiosulphate (irreversible system) with iodine (reversible system). This method may also be used for the titration of one irreversible system by another if there is a sufficient difference between the oxidation and reduction tensions of the two systems at a particular current density. The titration may also be carried out with two reversible systems: the current corresponding to the titrated substance decreases to zero at the equivalence point, whilst the current due to the titrant and its reaction products increases.

This procedure is not of general use but is particularly helpful in certain cases (iodimetry, cerimetry, bromometry etc.) because of its simplicity, the absence of comparison electrodes and the rapidity with which the end point is shown. The titration of water with the Karl Fischer reagent is of special interest.

CHAPTER VI

THE ELECTROCHEMISTRY OF COLLOIDS AND ELECTROKINETIC PHENOMENA[1]

1. Colloids in general

A century ago in 1861, Graham obtained the first demonstration of those peculiar characteristics which are now referred to as the colloidal state. Some of these characteristics were rapidly related to electrical phenomena. It is now known, in fact, that many colloidal phenomena and all electrokinetic phenomena are principally electrochemical in nature and can be completely described and interpreted by considering colloids as electrolytes and the electric tension existing at the surface of all colloidal particles as due to the presence of substances dissociated electrolytically. It would thus seem opportune to deal with colloidal systems and electrokinetic phenomena from a rather wider viewpoint than is normal in electrochemical works.

According to Ostwald, colloidal science deals with the physical and chemical properties of substances in the form of particles whose dimensions lie between 10^{-2} and 10^{-7} cm. The colloidal state may thus be defined as that in which the number of superficial atoms in each particle is comparable to the number of internal atoms and as that in which the state of disorder of the structure of each particle is medium. In reality, the colloidal state of a substance is characterized by a series of properties which markedly distinguish it both from monophasic solutions on the one hand and from coarse biphasic suspensions on the other. In a certain sense it may be considered as a limiting state possessing some properties of solutions, others of suspensions and yet others of its own. The principal characteristics which differentiate the colloidal state from solutions are its inability to be dialyzed, its very low diffusion velocity, its low tendency to crystallize and its frequent optical heterogeneity. The colloidal state is distinguished from biphasic dispersions by its energy content; the dispersed part cannot be considered as a true phase in the sense defined by

[1] The treatment of colloidal and electrokinetic phenomena in this chapter is limited to aqueous solutions which have long been the most important and the most widely studied.

Gibbs, since the surface energy and entropy of the particles is not negligible with respect to the total entropy and energy. This is a consequence of the fact that, as mentioned above, the number of superficial atoms is not negligible in comparison with the total number of atoms.

In fact, one of the primary characteristics of colloidal systems lies in the particle size. These particles are large compared with ordinary molecules in true solutions, but are small compared to the particles in a biphasic suspension which can be seen under the microscope. However, the transition from biphasic suspensions to true solutions is so gradual and continuous with regard to the size of the dispersed particles, that it is impossible to make any fundamental theoretical separation between biphasic suspensions, colloidal solutions and true solutions. The size of the particles varies in a continuous fashion right down to ionic dimensions so that this criterion may not be used as a theoretical basis to distinguish these various systems.

Another characteristic is the optical heterogeneity, *i.e.* the visibility under the microscope or ultramicroscope. This is a function not only of particle size but also of the wavelength of the light used for the observation and of the refractive indices of the dispersing medium and of the dispersed solution; so that this also cannot be used as a criterion. This is in fact true of almost all the properties: dialyzability, crystallizability, osmotic pressure, insolubility in the molecular state etc. In practice, recourse must be made to somewhat arbitrary criteria and colloidal systems will be considered as those in which the dimensions of the dispersed particles lie within the limits 10^{-7} and 10^{-5} cm, or as those systems composed of dispersed particles each containing between 10^3 and 10^9 atoms, or yet again, as those systems composed of dispersed particles whose volumes lie between 10^{-21} and 10^{-15} cm^3 (*i.e.* between 1 and 10^6 mμ^3). The arbitrary nature of such limits is obvious. Stauff[1] has given a tentative statistical thermodynamic treatment.

The properties of the colloidal state may theoretically be studied in two ways, either starting from the macroscopic state and following the evolution of the properties as the particle size diminishes, or else starting from true solutions and following the evolution of the properties as the particle size increases. Both these methods should converge to give the same result. In practice, one or other route is followed depending on

[1] J. Stauff, *Kolloidchemie*, Springer Verlag, Berlin, 1960, p. 429.

experimental convenience, but it must be emphasized that when both methods are applied, the same result is not always obtained because of the inevitable approximations which differ in the two cases. For electrochemical properties the first approach follows the theory of the electrochemical double layer of Helmholtz–Smoluchowski, and the second approach uses the Debye–Hückel theory of electrolytes. In passing through the continuous series of states from coarse, macroscopically biphasic suspensions to true solutions, the complexity of the properties reaches a maximum at the colloid solutions. This is due to the fact that suspensions have many properties in common which are to some extent independent of the chemical nature of the dispersed substance whilst the behaviour of molecular solutions is relatively simple in consequence of the simplicity of their physico-chemical constitution. In the field of colloidal solutions however there is a superposition of the characteristic physical properties of suspensions on to the characteristic chemical properties of solutions.

The ensemble of particles present in a colloidal solution shows a greatly extended surface and hence possesses a very high surface energy, so that it is at first sight strange that the particles do not reunite to form aggregates of increasing size, with a concomitant diminution in the surface energy content, until finally two macroscopically distinct phases are formed. In other words, it is strange that they are stable. One factor responsible for the stability of the dispersed particles in a colloidal solution is certainly the electrical charge which each particle possesses. The existence of an electrical charge on each particle may be readily demonstrated by electrophoresis (*cf.* p. 373 *et seq.*). All the colloidal particles carry a quantitatively variable electrical charge, but in a given solution the charges on all the particles have the same sign. These electrical charges give rise to repulsive forces between the particles which prevent them from approaching each other sufficiently closely to reunite into a larger aggregate, liberating the surface energy and hence passing into an energetically lower and more stable form[1].

Some colloids however are stable even when virtually uncharged. The stability of such colloids is due to the affinity between the molecules com-

[1] Colloids for which the electrical charge is the principal factor determining their stability belong mainly to the lyophobic group (see below). E. J. W. Verwey and J. Th. G. Overbeek in their monograph *Theory of Stability of Lyophobic Colloids*, Elsevier, Amsterdam, 1948, have given a good theoretical discussion of the lyophobic colloids.

posing their particles and the molecules of the solvent; this affinity must be greater than that existing between the solvent molecules themselves. The affinity in effect, leads in a certain sense to a surface reaction between solvent and colloid, so that the particle becomes covered by one or more layers of solvent molecules which prevent too close an approach between two particles. The colloids for which the solvation is a fundamental factor leading to their stability belong mainly to the lyophilic group. These colloids form spontaneously when the disperse substance and the dispersing medium come into contact. This implies that the formation process is accompanied by a diminution of free enthalpy ($\Delta G < 0$) in the passage from the system of two separate components to the colloidal system. The reverse is true of the lyophobic colloids. The former are thus thermodyna-

TABLE VI,1

THE DISTINCTIVE FEATURES OF LYOPHOBIC AND LYOPHILIC COLLOIDS

*Lyophobic colloids** (*e.g. metallic colloids*)	*Lyophilic colloids** (*e.g. proteins*)
Thermodynamically unstable	Thermodynamically stable
Stable only at low concentrations	Often stable at high concentrations
Unstable on prolonged dialysis	Stable on prolonged dialysis
Precipitate readily on the addition of small quantities of electrolytes	Insensitive to traces of electrolytes, precipitate only with large amounts of electrolytes
Coagulation cannot be reversed**	Coagulation can be reversed**
After drying do not swell on contact with the dispersing medium	After drying swell on contact with the dispersing medium and finally return to the sol state
Do not gel	May gel
Always migrate under the action of an electrical field	May not migrate under the action of an electrical field
Viscosity little greater than that of the dispersing medium	High viscosity; greater than that of the dispersing medium
Surface tension practically equal to that of the dispersing medium	Surface tension often low; less than that of the dispersing medium
Strong Tyndall effect	Weak Tyndall effect

* This sub-division also is not rigorous in that colloids exist with intermediary properties forming transition members between the two classes.

** *Cf.* p. 388 *et seq.*

mically stable, but the latter are thermodynamically unstable and owe their existence to an energy barrier which must be overcome in passing from the unstable thermodynamic system (the solution) to the stable thermodynamic system (the separate phases). This barrier consists of the electrostatic forces of repulsion.

Many other factors, however, influence the stability of colloidal solutions, such as the chemical nature of the particles, the surface tension, the attractive forces between the colloidal particles, the presence of electrolytes in the dispersing medium, the structure of the electrochemical double layer (*cf.* section 2, below) and the hydration energy. These factors are so numerous and so intricately interdependent that it is not as yet possible to give a complete, satisfactory theory of the stability of colloids.

The electrical charge is one of the most important factors involved in stability and on this basis, colloids may be divided into two major groups: lyophobic colloids, for which the electrical charge is essential for stability, and lyophilic colloids, whose stability under certain conditions continues even when they are discharged. Normally, however, even the lyophilic colloids carry an electrical charge. Table VI, 1 shows the characteristics which distinguish these two classes.

2. Theories about the Electrical Charges of Colloids

There are two theories of the origin of the electrical charges of colloids.

According to Helmholtz–Smoluchowski the electrical charge of each colloidal particle is due to the presence of an electrochemical double layer which is almost rigid and is formed by the selective adsorption on each colloidal granule of one ionic species present in the electrolyte, so that the granule assumes a charge of a certain size; the other ionic species remains in the disperse medium and gives to it a charge of the opposite sign. This double layer has been treated by Helmholtz and Smoluchowski with the laws of electrostatics as if it were a capacitor whose plates were separated by a distance of the order of molecular dimensions. Gouy considered the double layer as not rigid but diffuse, with a charge density decreasing towards the interior of the disperse liquid. This theory, which was developed principally on the basis of the phenomena observed at the boundary between two macroscopic phases provided a description, which was even quantitative, of such phenomena and has been extrapolated by diminution of the dimensions involved until it can be applied to particles of colloidal size.

The other theory was first formulated by Malfitano and Duclaux but has been thoroughly elaborated by Pauli and his school. They consider colloids as true electrolytes which dissociate into normal ions and ions of much greater, colloidal, size. In this way, too, an electrical double layer arises analogous to that mentioned above. According to this theory the properties of true electrolyte solutions may be extrapolated by increasing the size of one of the two ionic species to colloidal dimensions, which are very much greater than those of ions in solution.

From a theoretical standpoint, the treatment of colloids by the theory of Helmholtz–Smoluchowski or by that of Pauli gives inexact results since both theories use extrapolations which are not strictly admissible. In practice, however, Lindestrom–Lang showed that for many phenomena, especially of a physical nature, the end result is the same if colloids are treated either by the theory of surface adsorption or by that of dissociation. This is further evidence that colloids form limiting systems sharing the properties both of monophasic solutions and biphasic suspensions. In general, however, and in particular for the description and interpretation of the chemical behaviour of colloidal solutions, the dissociation theory of Pauli has proved to be superior to the adsorption theory of Helmholtz; even though certain thermodynamic concepts about solutions – such as dissociation constants, activities, solubility products – which are perfectly applicable to lyophilic colloids (like soaps and proteins) encounter difficulties in their application to certain typically lyophobic inorganic colloids.

A further argument in favour of the dissociation theory is that the laws of electrochemistry based on thermodynamics are also valid for colloidal systems provided that allowance is made for the specific properties of these systems. In particular, Hartley[1] and more recently Audubert[2] applied the Debye–Hückel theory to colloids; when suitably modified, significant results were obtained, but this work lies outside the scope of this book. The attribution of the electrical charge of a colloidal particle to an ionogenic complex (*cf.* p. 369 *et seq.*) within the particle, provides a reasonable explanation of the differences in behaviour of different col-

[1] G. S. Hartley, *Trans. Faraday Soc.*, 31 (1935) 31.
[2] R. Audubert, *Trans. Faraday Soc.*, 36 (1940) 144.

loidal solutions of the same substance; the differences being due to the existence of different ionogenic complexes.

It is probable, however, that for some typically lyophobic systems (*e.g.* silver iodide under certain conditions, paraffin droplets or air bubbles dispersed in water), phenomena due to the selective adsorption of ions are involved, as in Helmholtz's theory. Such adsorption is caused by the polarization induced by the charge on the ion acting on the colloid particle; a dipole moment is thus induced in this, whose size depends on many factors (susceptibility to polarization, hydration, distance between the ion and the colloidal particle, charge and dimensions of the ion and of the particle etc.). If the force of attraction between the ion in solution and the dipole in the colloidal particle is much greater than the kinetic energy (due to thermal motion) of the ion and of the particle, stable adsorption will occur and the colloidal particle will thus assume a definite charge.

Naturally, in considering colloids as electrolytes it is necessary always to bear in mind their fundamental differences from solutions of true electrolytes. These differences include the much greater size of the particle compared to normal ions and the variation in the size of the charge from particle to particle; under suitable conditions, this charge may even be inverted in sign and it has no stoichiometric relationship with the mass of the particle itself. However, under a particular set of conditions, all the colloidal ions present in a solution will have a charge of the same sign. Another difference is of a thermodynamic nature; the internal energy of a colloidal system cannot be defined simply by using the variables required to define purely monophasic or polyphasic systems in which the area of the boundary between the phases is comparable with their volume. When colloidal systems are considered as biphasic a term must be introduced into the expressions of the thermodynamic functions to take account of the surface energy. This may be expressed as[1]

$$\mathrm{d}E_s = T\mathrm{d}S_s + \sigma \mathrm{d}s + (\Sigma \tilde{\mu}_i \mathrm{d}n_i)_s$$

where E_s, S_s, σ and $\tilde{\mu}_i$ are respectively the surface energy, entropy, surface tension and electrochemical potential (*cf.* Chapter III), s is the surface area and n_i is the number of moles of component i. The use of the electrochemical potential $\tilde{\mu}$ rather than the chemical potential μ allows also for the electrical charge of the particles. (For the definition of a mole of a

[1] Both entropy and surface area appear in this equation and since they normally both have the symbol S, the latter quantity has been given that of s instead.

colloidal component see footnote 1 p. 371). The additional term zF which appears in the definition of the electrochemical potential may be ignored in any process in which equal amounts of positive and negative particles of equal valency, pass from one phase to the other[1] but it cannot be ignored in colloidal systems in which one of the electrically charged species (the colloidal particle) is not diffusable[2].

Still another difference is that the colloidal ion is not generally composed of a single substance; account must also be taken of other differences and of other distinctive characteristics such as the actual polyphasic nature of colloidal systems, the asymmetry of the colloidal electrolyte, the polyvalent nature of the colloidal ion, and the fact that not all of the mass of a colloidal particle but only its surface is active in defining equilibria (see following section).

The electrochemistry of the colloids and of electrokinetic phenomena is in many ways still being developed. In recent years particular attention has been given to the systematic quantitative study of the electrochemical double layer. This layer exists at the boundary surface between two phases of which one at least is liquid and contains ions in solution. The results obtained in the quantitative study of the double layer concerning the structure of the layer itself, are also applicable when Pauli's theory is used for the study of colloids. The existence of the layer cannot be doubted.

3. Colloids as Electrolytes

According to Pauli, a colloidal particle is not perfectly homogenous chemically but is composed of a neutral part (C) and an ionogenic part (I) which is also known as the ionogenic complex. The neutral part represents in general almost all of the colloidal particle and consists of a certain number of atoms or molecules which are usually insoluble in the solvent of the colloid and cannot give rise to ions by electrolytic dissociation. The ionogenic part is normally present in the colloidal particle to a far lesser extent and lies at the surface of the colloidal ion. This part is formed of soluble compounds which dissociate into ions of which one species passes into solution whilst the other remains fixed to the colloidal

[1] Since in this case the two terms zF and $-zF$ cancel out.

[2] For the definition of the positive or negative valency z of a colloidal particle, see p. 371.

particle by chemical forces or by adsorption. This gives to the particle its charge. A schematic formula for the empirical composition of the colloidal particle is $x\mathrm{C}y\mathrm{I}$ where $x \gg y$; part of the ionogenic complex dissociates

$$\mathrm{I} \rightleftharpoons \mathrm{A}^+ + \mathrm{G}^- \text{ (or } \mathrm{A}^- + \mathrm{G}^+\text{)}$$

where G^- represents the compensating ion which passes into the dispersing liquid (see below). The empirical formula for the composition of a colloidal particle thus becomes

$$(x\mathrm{C} + y\mathrm{I} + z\mathrm{A})^{z+} + z\mathrm{G}^-$$

Considering the colloidal electrolyte as completely dissociated as if it were a strong electrolyte gives

$$(x\mathrm{C} + z\mathrm{A})^{z+} + z\mathrm{G}^-$$

The solution remains electroneutral which implies that the charge on the colloidal ion is completely compensated for by the ions in solution; these are called *counter-ions* (Pauli). Such behaviour is perfectly understandable on the theory of dissociation. Colloids may be classified into three main groups on the basis of the chemical nature of the ionogenic complex:

I. Isomolecular colloids. In these the neutral part has the same composition as the ionogenic complex before this is dissociated. This group includes the soaps, certain dyes, silicic acid etc. A particle of colloidal silicic acid, for example, would have the composition

$$(x(\mathrm{SiO_2} \cdot n\mathrm{H_2O})\ z\mathrm{HSiO_3})^{z-} + z\mathrm{H}^+$$

Amongst this group of colloids, a fraction of the ionogenic part may be in true solution in chemical equilibrium with the neutral part and in this case, prolonged dialysis will make all the colloid disappear. This is true for example with soaps.

II. Heteromolecular colloids. In these the ionogenic part does not have the same composition as the neutral part. The colloidal metals, their oxides, hydroxides, halides and sulphides, etc. belong to this group. A particle of colloidal ferric hydroxide would have the composition

$$x(\mathrm{Fe(OH)_3} \cdot n\mathrm{H_2O} \cdot z\mathrm{FeO})^{z+} + z\mathrm{Cl}^-$$

In this group the neutral part and the ionogenic part have different but nevertheless somewhat related compositions.

III. In a third group of colloids the ionogenic part and the neutral part are linked together through primary valencies, so that every particle may be a single molecule containing ionizing groups. This group comprises the so-called macromolecules, *e.g.* proteins, gums, pectin, starch, etc.

The ionogenic complexes cannot always be separated from the neutral parts and isolated in stable forms. Often their stability and ionizability depend upon electrostatic interactions with the neutral parts due to molecular polarizations, or shifts and redistributions of electron clouds. Once the ionogenic complex has been identified by chemical analysis it is possible to derive an empirical formula for the composition and, in particular, for the ratio $x : y : z$, or $x : z$ respectively. Allowance must also be made for the possibility of the coexistence of various ionogenic complexes which influence each other.

The valency of a colloidal ion is also defined on the basis of the dissociation. Its effective valency z is given by the number of electrical charges which remain on it after the dissociation of the ionogenic complex. Values varying between 10 and 100 are very common for the effective valencies of colloidal ions. To obtain the mean valency $\bar{z}$ it is necessary to determine other quantities. If a is the number of colloidal particles per litre, then the molecular concentration c_{mol} of the colloidal solution[1] is given by

$$a/N = c_{mol}$$

where a is the number of individual colloidal particles and N is Avogadro's number. The mean valency $\bar{z}$ is defined by the equation

$$\bar{z} = \Sigma z/a$$

so that it is equal to the total charge on all the particles contained in a litre, divided by the number of particles. The equivalent concentration is given by the equation

$$c_{eq} = c_{mol}\bar{z} = a\bar{z}/N$$

Since the solution is electroneutral, the equivalent concentrations of the colloid and its counter-ions must be the same.

[1] The molar concentration of a colloidal solution must be understood not in the usual chemical sense of moles of substance per litre of solution, but rather by using the definition of a molecule given by Avogadro; molecules must be considered as individual free particles constituting a kinetic entity independent of other particles existing in the solution, without taking account of the number and species of the atoms forming the particle nor of the type of bonds which hold it together.

On this definition, every colloidal particle is a colloidal molecule. Thus, a molar colloidal solution will contain N ($= 6.02.10^{23}$) colloidal particles per litre. The molar concentration in the usual chemical sense will be much greater and will depend upon the state of aggregation of the particles. No stoichiometric relationship exists between the colloidal molar concentration and the molar concentration in the classical chemical sense.

Another characteristic quantity of colloidal solutions is the colloidal equivalent K which is the ratio of the number of molecules in the classical chemical sense, to the number of charges, *i.e.*

$$K = \frac{x + z}{z} = \frac{x}{z} + 1$$

Introduction of Avogadro's number and division by the volume gives

$$K = \frac{c^*}{c_{eq}} + 1$$

where c^* represents the true molar concentration, in the chemical sense, of the neutral part. In this way, the equivalent is referred to the total number of charges. If it is wished to refer it to the number of active free charges, this must be obtained by measurements on the counter-ion. Here, it is necessary to take account also of the activity coefficient f_a, or of the coefficients of conductance f_Λ or of osmosis f_0 depending on whether the number of active free charges is obtained by measurements of activity (when it is possible) of conductance or of the osmotic pressure of the counter-ions. However the values of the three coefficients f_a, f_Λ and f_0 are not normally the same so that the colloidal equivalent depends to some extent upon the method of measurement. Moreover, it is necessary to bear in mind the possibility of an association of ions in the sense of Bjerrum (*cf.* p. 91 *et seq.*) which is not unlikely, given the high value of the effective valency of the colloidal ions. If the three coefficients f_a, f_Λ, f_0 have the same value it would be an indication that the colloid involved should be treated as a weak electrolyte rather than a strong electrolyte and then the normal values of the three coefficients could be considered as numerically equal to the degree of dissociation. Very probably, the chemical nature of the ionogenic part influences the behaviour of colloids as either strong or weak electrolytes.

Multiplying the colloidal equivalent K by the true molecular weight[1]

[1] Strictly, account must be taken of the fact that a colloidal particle is normally formed of x molecules of the neutral part and z molecules of the ionogenic complex so that the molecular weight to be used in calculations should be the mean molecular weight obtained from the expression

$$\overline{M} = (xM_x + zM_z) / (x + z)$$

where M_x is the molecular weight of the substance forming the neutral part and M_z is the molecular weight of the ionogenic complex. Since, however, $x \gg z$ the above definition of the equivalent weight of colloids is sufficiently exact.

of the substance forming the colloidal particle, gives the equivalent weight of the colloidal electrolyte. The two characteristic quantities determining the behaviour of a colloidal ion are, however, always its dimensions and effective valency. Numerous methods have been attempted to determine these two quantities, such as conductance, osmotic pressure, migration velocity, particle counting, sedimentation equilibria, membrane electric tension, diffusion, etc. So far no method has proved not to be open to fundamental criticism and the results obtained with these various methods differ markedly from each other. The main cause of such deviations probably lies in the fact that each method involves the use of simplifications which are not strictly permissible, *e.g.* assuming the validity of Stokes' Law supposing the particle to be spherical, ignoring the amicrons which are invisible even in the ultramicroscope[1], ignoring the polarization of the adherent solvent etc.

4. Electrochemical Quantities of Colloids

The existence of an electrical charge on colloidal particles may be readily demonstrated by electrophoresis. On applying an electric tension to two electrodes immersed in a colloidal solution it is possible to see, through a suitably adapted microscope[1], not only the irregular Brownian movement but also a continuous migration of the colloidal particles in one direction or the other depending upon the sign of the charge. This phenomenon which was discovered by Linder and Picton in 1893 is called electrophoresis and serves primarily to determine the sign of the charge on colloids. The colloids are classified on the basis of the sign as negative, when they migrate towards the anode, and positive when they migrate towards the cathode. The electrophoresis of negative colloids is also called anaphoresis and that of positive colloids cataphoresis. Thus electrophoresis corresponds exactly to the migration of ions and although the phenomenon is still called electrophoresis to distinguish it from ionic migration, it is quantitatively characterized by a quantity which is perfectly analogous to the velocity of ionic migration u (*cf.* p. 29 *et seq.*) and assumes the same name and the symbol u_{coll} for both positive and negative colloids. The experimental measurement of migration velocity is performed with analogous criteria

[1] The ultramicroscope, which uses the Tyndall effect to recognize particles of dimensions less than are necessary for visibility, when suspended in a medium whose refractive index is different from that of the particles themselves.

to those used for true solutions (*cf.* p. 65 *et seq.*) but with apparatus adapted to the particular characteristics of colloids[1]. The possibility that the results are falsified by electroosmotic effects (*cf.* p. 395 *et seq.*) must be borne in mind with colloids and in such cases the necessary corrections must be applied. The migration velocities of colloidal ions are of the same order of magnitude as those of true ions, *i.e.* 10^{-4} cm sec^{-1}. Walden's Rule is also approximately valid at different temperatures, *i.e.* the product of the migration velocity and the viscosity of the solvent at the same temperature is constant.

The material movement of a colloidal ion and of its electrical charge under the action of an electrical field shows that a centain conductance must be attributed to colloidal solutions, which are classified as conductors of the second class because the electrical conductance is accompanied by a migration of material.

The specific conductance of a colloidal solution is measured in a similar manner to that of a solution of an electrolyte and the same equations apply.

$$\varkappa = \lambda c_{eq} \cdot 10^{-3} \quad {}^{2}$$

$$= 10^{-3}\, \mathrm{F}(u_{coll} + u_{ion})c_{mol}\bar{z}f_\lambda \quad {}^{3}$$

$$= 10^{-3}\, \mathrm{F}(u_{coll}f_{\lambda\, coll}\bar{z}a)/(N + u_{ion}c_{eq}f_\lambda)$$

However, this equation is not readily used since the mean valency $\bar{z}$, the values of the migration velocity u_{coll} at infinite dilution and the conductivity coefficient $f_{\lambda\, coll}$ of the colloid are not known. When other electrolytes are present allowance must also be made for their conductance. The best method for measuring the true conductance of a colloidal electrolyte is normally to purify it until the conductance is constant and to consider this final value as the true one for the colloidal electrolyte after deducting any conductance due to the solvent. This method is applicable when it is possible to exclude the presence of true electrolytes other than the compensating ions in equilibrium with the colloid and on which its stability depends (*cf.* p. 388 *et seq.*).

The conductance of a colloid shows certain peculiarities in comparison

[1] See, for example, Wo. Pauli and E. Enghel, *Z. physik. Chem.*, 126 (1927) 247; D. C. Henry and J. Brittain, *Trans. Faraday Soc.*, 29 (1933) 798; A. P. Brady, *J. Am. Chem. Soc.*, 70 (1948) 911.

[2] Taking the litre as the unit of volume rather than the cm^3.

[3] u_{ion} indicates the velocity of migration of the compensating ions, independently of their sign.

with that of a solution of an electrolyte. Firstly, its equivalent or molar conductance tends towards zero as the concentration tends towards zero instead of tending towards a higher finite figure. This can be explained by the peculiar constitution of colloidal ions. In very dilute solutions every colloidal ion is surrounded by its own ionic atmosphere. In the presence of an electrical field, the colloidal ion is induced to migrate in one direction and its atmosphere in the other; but in view of the high charge on each colloidal particle, the attractive force between the colloidal ions and the counter-ions of the atmosphere is very intense so that a strong electrical field is necessary to separate them. Since the colloidal ion is surrounded by this atmosphere, it behaves as a neutral particle at high dilutions. In the presence of an excess of electrolyte every colloidal ion can readily reform its ionic atmosphere. If this excess of electrolyte contains ions differing from the counter-ion itself, changes in the chemical nature of the colloid may follow, in some sense equivalent to an electrodialysis (*cf.* p. 384 *et seq.*). If, however, instead of being in a high dilution the colloid is in such a concentration that the ionic atmosphere of the individual particles overlap with each other, their motion is facilitated in that each particle tends to reform its atmosphere at the expense of neighbouring ones. Its velocity then increases more rapidly than even the increase in concentration. The equivalent conductance of a colloid thus depends to a significant extent on the presence of other electrolytes which can influence its migration velocity.

Another anomaly which is not encountered with true solutions is that the equivalent conductance may markedly increase with increasing concentration over a small concentration range and then fall again. A typical example of colloids showing this is given by certain dyes, where the colloidal particles are in chemical equilibrium with the same dye dispersed in the molecular state. According to Robinson this anomaly may be interpreted by the aggregation of ions dissolved as in a true solution, to form colloidal particles composed of a certain number q of the aggregated ions. Stokes' Law (*cf.* p. 29 *et seq.*) referred to a unit potential gradient, for a colloidal ion of an approximately spherical form may be written as

$$u_{coll} = (zeU)/(6\pi\eta rl) \tag{1}$$

where z is the effective valency of the colloidal ion, e is the elementary electrical charge, η is the viscosity of the solvent and r is the radius of the supposedly spherical colloidal ion and U/l is the strength of the electric field.

In fact, the effective charge as far as the migration of a particle is concerned is not that corresponding to its valency, but is less. Since a colloidal particle is normally polyvalent, a certain number of counter-ions are found in the solvation layer which is firmly attached to each particle, and these ions will also be firmly attached. The ions and particle together will form a single kinetic entity with a total free charge equal to the difference between the charge on the colloidal particle and the sum of the charges of the counter-ions firmly attached in the solvation layer. Equation (1) should thus more strictly be written in the form

$$u_{coll} = (\Delta zeU)/(6\pi\eta r_t l)$$

where Δz is the number obtained by this difference and r_t is the overall radius of the kinetic entity formed by the colloidal particles and the solvation layer attached to it. The number z is not constant since it depends on the concentration of foreign electrolytes. To a first approximation, however, equation (1) may be considered as valid. Various methods exist for the measurement of the free charge of a colloidal particle for which reference should be made to specialized texts.

The valency z of the colloidal ion increases in direct proportion with the number of aggregations q whilst the radius increases in proportion only to the cube root of q; thus the ratio z/r is proportional to $q^{\frac{2}{3}}$. As the number of aggregations q, which is always greater than 1, increases, the migration velocity will also increase in the ratio $q^{\frac{2}{3}}$ until other factors intervene to cause the falling pattern of equivalent conductance. These factors may be, for example, the diminution of the coefficient of conductance, the partial discharge of the colloidal ions through associations with compensating ions, etc. Fig. VI, 1 shows the changes in the equivalent

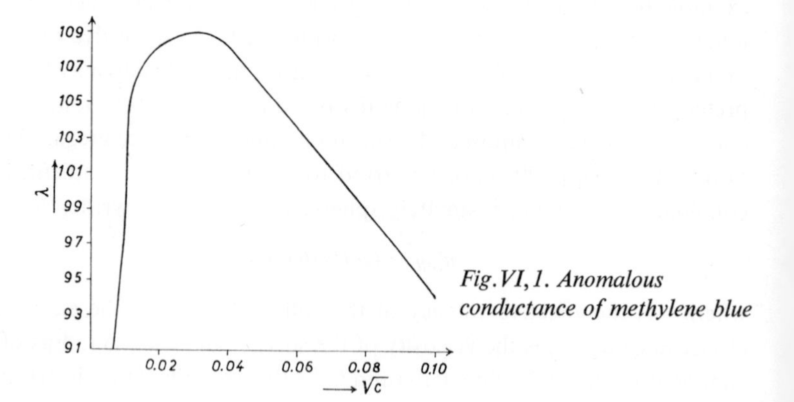

Fig. VI, 1. Anomalous conductance of methylene blue

conductance as a function of the square root of the concentration with methylene blue. A study of the changes in concentration can thus be a valid means of investigating the state of aggregation of some colloids.

With colloids, also, transference numbers represent the fraction of the electrical current carried by the colloid. As defined by Hittorf and using his type of calculation, the transference number of a cation may be expressed as the cationic enrichment of the cathodic region on the passage of 1 **F** of electricity (*cf.* p. 29 *et seq.*). The equation

$$t_i = \frac{u_i c_{eq,\,i}}{\Sigma u_i c_{eq,\,i}}$$

is also valid for mixtures of electrolytes which include colloids. Thus, it is necessary to know the equivalent concentrations of every species which shares in carrying the current. The equivalent concentrations and the equivalent weight of colloids cannot be determined exactly experimentally. The transference number may be calculated, at least approximately, from the increase in concentration on the passage of 1 **F** of electricity, *i.e.* with the equation

$$t_{coll} = \frac{\Delta c_{eq,\,coll}}{n}$$

where t_{coll} is the transference number of the colloid; $\Delta c_{eq,\,coll}$ is the increase in the equivalent concentration of the colloid (*cf.* p. 371) and n is the number of **F** of electricity passed through the cell. But the increase in the equivalent concentration of the colloid is equal to the increase in weight Δm found analytically, divided by the equivalent weight of the colloid. This in turn derives from the true molecular weight M, of the substance forming the colloidal particle, multiplied by the colloidal equivalent, *i.e.*

$$t_{coll} = \frac{\Delta m}{nMK}$$

Uncertainties in the determination of the colloidal equivalent K (*cf.* p. 372 *et seq.*) are naturally reflected in the determination of the transference number of colloids t_{coll}.

It is, however, possible by transference experiments and by measuring the specific conductance to find the equivalent conductance of the colloidal ion in a fairly simple and exact manner, provided that the solution contains only the counter-ions and no foreign electrolyte. From the definition of transference number as equal to that fraction of the current transported by the given ion,

$$\lambda_{coll} = \Lambda t_{coll} = \frac{\varkappa}{c_{eq}} t_{coll} = \frac{\varkappa}{c_{eq}} \frac{\Delta m}{nMK}$$

However the expression MKc_{eq} is only the total concentration c expressed in grams, so that

$$\lambda_{coll} = \frac{\varkappa \Delta m}{nc}$$

In this equation the quantities of the second members are all accessible to direct experiment. The transference numbers also depend on the equilibrium of the aggregation of those colloids which show this phenomenon since, by definition, the transference number of an ion is a function of its migration velocity. In consequence, measurements of transference numbers may be used for the study of aggregation equilibria. It is clear from this, that by combining potentiometric, conductometric and transference measurements, it is possible to determine many physicochemical quantities of colloidal ions with a good degree of accuracy.

The existence of an electrical charge on colloidal particles, as shown by electrophoresis, implies the existence of an electric tension between the colloidal particle and the dispersing medium. By treating the surface of a colloidal particle as a capacitor according to the Helmholtz–Smoluchowski–Gouy theory, the electrophoretic migration velocity under the action of a unitary potential gradient is characteristic for the charged state of the boundary surface and is directly proportional to the electric tension existing between the two surfaces in motion relative to each other, *i.e.* to the sliding surfaces, and not to the galvani tension between the two phases. In fact, every colloidal particle is always enveloped by a more or less thick layer of solvent molecules which adhere to the particle during its motion. The active electric tension as far as the motion of the particle is concerned, exists at the sliding surface and is called the *electrokinetic potential* (although it is a tension); it is represented by the Greek letter ζ and cannot be directly measured. It is defined by the work produced or consumed in carrying a unit electrical charge from any point in the sliding surface (which is an equipotential surface) to infinity. It can be calculated by the theory of the double layer starting from the migration velocity. The calculation for small spherical particles is simple. Assuming the validity of Stokes' Law, the force acting on a colloidal particle in a medium of very low ionic concentration is given by the product of the free charge on the particle Δze and the potential gradient. Thus,

$$\Delta z\mathrm{e}U/l = 6\pi\eta rw$$

Since the effective velocity w is given by uU/l, substituting and rearranging gives

$$u = (\Delta z\mathrm{e})/(6\pi\eta r)$$

This equation thus permits an immediate calculation of the free charge of a small spherical particle in dilute solutions. The potential ζ at the surface of a sphere is connected with the total charge Q and the dielectric constant ε by the equation

$$\zeta = Q/(\varepsilon r)$$

Bearing in mind, that $Q = \Delta ze$ for spherical colloidal particles

$$u = (\varepsilon\zeta)/(6\pi\eta)$$

This final equation allows the calculation of the electrokinetic potential from measurements of the migration velocity of small spherical particles. For large spherical particles or particles of other shapes to which Stokes' Law does not apply, the calculation becomes much more complex and the numerical factor 6π changes in relation to the different values of the frictional constants. On the other hand, by starting from a macroscopic model of a flat condenser and decreasing the dimensions to molecular size a factor of 4π is obtained. Henry[1] tried, and partially succeeded, to eliminate such discrepancies but the description of the procedure he adopted falls outside the limits of this book. The electrokinetic potential depends, for particles of equal size and total free charge, on the structure of the ionic atmosphere surrounding them, *i.e.* on the structure of the electrical double layer. It thus depends also on the presence of foreign electrolytes, on the dielectric constant within the double layer and on the viscosity of the medium. The electrokinetic potential is normally smaller and may even be of opposite sign to the galvani tension. Fig. VI,2 shows the variations in this potential in two possible cases.

In view of the many uncertainties about the diameters of the particles there is a tendency at the present time, not to attribute too great a value

[1] O. C. Henry, *Proc. Roy. Soc. London, Ser. A*, 133 (1931) 106.

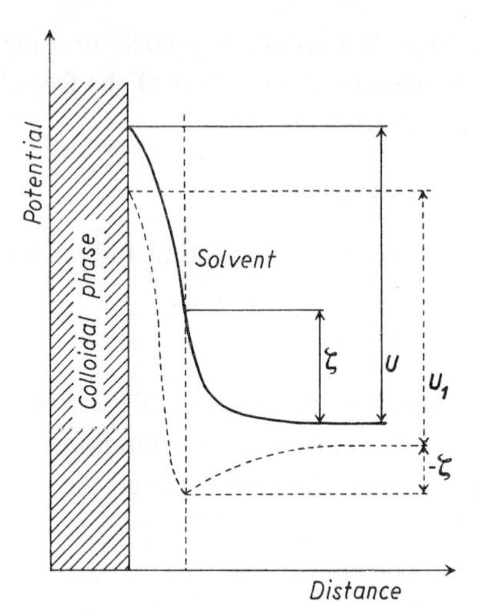

Fig. VI, 2. Electrokinetic potential

to the figures calculated by one or other equation for the electrokinetic potential ζ. An improvement in this situation could probably be brought about by utilizing the methods of the thermodynamics of irreversible processes. Attempts in this direction have been made by Overbeek and his colleagues[1].

5. Membrane Equilibria

The nondialyzability of colloids leads as a consequence to the establishment of particular membrane equilibria which were first studied by Donnan. A solution is in true thermodynamic equilibrium when the concentration of every dissolved species is constant throughout every element of volume. If the solution is divided by a membrane which is permeable to the solvent and to some but not all of the dissolved species, only those for which the membrane is permeable can distribute themselves so that their concentrations are constant throughout every element of volume on both sides of the membrane. If, however, one of the nondialyzable species carries an electrical charge, *i.e.* if it is an ion, colloid or otherwise, the thermodynamic equilibrium distribution of the other ionic species, and of these alone, will be affected; since the solution must remain electro-

[1] J. Overbeek *et al.*, *Rec. trav. chim.*, 70 (1951) 83; *J. Chem. Phys.*, 20 (1952) 1825; *J. Colloid Sci.*, 8 (1953) 420.

neutral throughout its entire volume. This condition implies that the diffusion of a single ionic species is not possible; otherwise there would be an unequal distribution of electrical charges on the two sides of the membrane–destroying the electroneutrality. Ionic species can thus diffuse only as pairs of opposite sign in equivalent amount so that the electroneutrality is maintained.

For simplicity, it may be supposed that a solution contains the colloid NaColl with a compensating ion Na^+ and with sodium chloride as the foreign electrolyte on the other side of the membrane; this is permeable only to the solvent and the Na^+ and Cl^- ions. The distribution at equilibrium will be as follows

I	II
$Coll^{z-}$	
$Na^+{}_I$	$Na^+{}_{II}$
$Cl^-{}_I$	$Cl^-{}_{II}$

Since the solution is electroneutral, the following equations must hold for each side of the membrane

$$[Na^+{}_I] = [Coll^{z-}] + [Cl^-{}_I]$$

$$[Na^+{}_{II}] = [Cl^-{}_{II}]$$

in which the subscripts I and II distinguish the concentrations on the two sides of the membrane. Since at equilibrium the electrochemical potential of the Na^+ ions must be the same on the two sides of the membrane and this is also true of the Cl^- ion

$$\mu_{0Cl^-} + RT \ln a_{Cl^-{}_I} - \varphi_I F = \mu_{0Cl^-} + RT \ln a_{Cl^-{}_{II}} - \varphi_{II} F$$

$$\mu_{0Na^+} + RT \ln a_{Na^+{}_I} + \varphi_I F = \mu_{0Na^+} + RT \ln a_{Na^+{}_{II}} + \varphi_{II} F$$

Summing these two equations and rearranging gives

$$a_{Na^+{}_I} a_{Cl^-{}_I} = a_{Na^+{}_I} a_{Cl^-{}_{II}} \tag{1}$$

Assuming for the sake of simplicity that the volumes on each side of the membrane are both unitary and that the coefficients of activity are also unitary and that the quantity of sodium chloride which has passed from one side of the membrane to the other is x, then the following final concentrations are obtained:

c_1 is the concentration of Na^+ ion which compensates the charge of the colloidal ions;

x is the concentration of the Cl^- ion in zone I which is equal to the concentration of the Na^+ ions in excess of those required to compensate the charge on the colloid in zone I, and corresponding to the sodium chloride which has migrated to zone I from zone II;

$(c_2 - x)$ is the concentration of Cl^- ions in zone II, if c_2 were the initial concentration of sodium chloride in zone II. $(c_2 - x)$ will also be equal to the concentration of Na^+ ion in this zone at equilibrium. Equation (1) now becomes

$$(c_1 + x)x = (c_2 - x)^2$$

and hence

$$x = c_2^2/(c_1 + 2c_2)$$

But since

$$\frac{[Cl^-_{II}]}{[Cl^-_{I}]} = \frac{[NaCl_{II}]}{[NaCl_{I}]}$$

and since

$$Cl^-_{II} = (c_2 - x) \text{ and } Cl^-_{I} = x,$$

$$\frac{[NaCl_{II}]}{[NaCl_{I}]} = \left(c_2 - \frac{c_2^2}{c_1 + 2c_2}\right) \Big/ \left(\frac{c_2^2}{c_1 + 2c_2}\right) = 1 + \frac{c_1}{c_2}$$

i.e.

$$[NaCl_{II}] > [NaCl_{I}]$$

If $c_2 \gg c_1$, the ratio c_1/c_2 tends towards zero and thus the ratio between the sodium chloride concentrations in zones I and II tends towards unity; in other words, the sodium chloride tends to distribute itself equally between the zones I and II. If, however, $c_2 \ll c_1$, the ratio c_1/c_2 tends towards a very high value and thus the sodium chloride concentration assumes very much greater values in zone II than in zone I, in other words it may be almost completely eliminated from the zone containing the nondialyzable ion and in particular, therefore, if this zone contains colloidal ions.

A special case applies when the dialysis is performed against pure solvent. If the dialyzing ion is, for example, a cation but not H^+, it can not diffuse by itself but will diffuse as a pair with the OH^- anion since both the OH^- and H^+ ions are always present in water. Hence, the solution in zone I becomes more rich in H^+ ions, in other words it becomes more acid. Thus, the presence of the membrane can provoke a particular type of hydrolysis called membrane hydrolysis.

By applying the equilibrium equations for electrochemical potentials to the Na^+ ions which have different concentrations in zones I and II, it can readily be seen that the membrane between the zones will at equilibrium be the location of an electric tension π called the *membrane tension*[1].

The equilibrium equations must be satisfied which, referred to one mole, assume the form

$$\mu_{\mathrm{I}} + \mathbf{F}\varphi_{\mathrm{I}} = \mu_{\mathrm{II}} + \mathbf{F}\varphi_{\mathrm{II}}$$

and hence

$$\begin{aligned} \pi &= \varphi_{\mathrm{I}} - \varphi_{\mathrm{II}} \\ &= (1/\mathbf{F})\,(\mu_{\mathrm{II}} - \mu_{\mathrm{I}}) \\ &= (1/\mathbf{F})\big[(\mu_0 + RT\ln a_{\mathrm{II}}) - (\mu_0 + RT\ln a_{\mathrm{I}})\big] \\ &= (RT/\mathbf{F})\ln (a_{\mathrm{II}}/a_{\mathrm{I}}) \end{aligned} \tag{2}$$

For the general case of ions with valencies other than 1 this equation assumes the general form

$$\pi = \frac{RT}{z\mathbf{F}} \ln \frac{a_{\mathrm{II}}}{a_{\mathrm{I}}}$$

An analogous expression is readily derived for negative ions.

This expression signifies that the membrane electric tension is equal, but of opposite sign, to the electric tension of the concentration cell formed as a consequence of the unequal distribution of dialyzable electrolyte on the two sides of the membrane. (*cf.* p. 142 *et seq.*)[2]. This is a reflection of the fact that at thermodynamic equilibrium between the zones the total electric tension measured by two equal electrodes immersed in each zone must be zero. Otherwise, there would be a passage of current and external work would be performed with a change in free enthalpy so that the compartments on each side of the membrane could not be in thermodynamic equilibrium.

The membrane electric tension, which is very important in many biological equilibria, may be determined by measuring separately, as far as possible, the ionic activities (*cf.* p. 148 *et seq.*) on each side of the membrane with a suitable comparison electrode. In the presence of other ionic species of opposite sign similar equations must be satisfied for every

[1] Some texts talk of 'membrane potentials' which is incorrect since they are actually dealing with an electric tension according to the CITCE's definitions and conventions.

[2] Naturally, in the present calculation no allowance is made for an electric diffusion tension, since in this case the membrane prevents diffusion.

combination of each anion with each cation. Such equations must take account of the valencies of the ions according to the general principles already described. The general equilibrium condition may thus be summarized in an equation which is readily derived from (1) and (2)

$$a_{-,\mathrm{I}}/a_{-,\mathrm{II}} = a_{+,\mathrm{II}}/a_{+,\mathrm{I}} = e^{\frac{\pi z \mathrm{F}}{RT}} = \lambda$$

In which $a_{\pm,\mathrm{I}}$ and $a_{\pm,\mathrm{II}}$ represent the activity of particular ionic species in zones I and II respectively, z represents the valency, π represents the membrane electric tension and λ represents the Donnan partition coefficient which generally assumes the form $\lambda = ([a^{z+}{}_{\mathrm{II}}] / [a^{z+}{}_{\mathrm{I}}])^{1/z}$, and the other symbols have their normal significance.

Osmotic measurements, dialysis and work on colloidal phenomena in general must all take account of Donnan equilibria.

6. The Preparation and Purification of Colloids

The preparation of colloidal solutions is in some sense characteristic of the individual colloids in that no general methods are available which can be used indiscriminately for all colloids. On the other hand, the method of preparation may, to a marked extent, influence the properties of colloidal electrolytes depending on the nature of the ionogenic complex which may differ with different methods of preparation. The various techniques may be separated into two general classes – dispersion methods and condensation methods – depending on whether the colloidal particles are formed by the sub-division of a macroscopic phase or by the combination of smaller particles. Examples of the first method are simple contact with the dispersing medium[1] and peptization of precipitates[2] etc. Examples of the second method are reduction or oxidation and subsequent precipitation within colloidal dimensions by other substances, changes in solvents, hydrolysis etc. Bredig's method with a submerged electric arc is probably an example of the coexistence of the two methods: dispersion following the direct pulverization by the arc and condensation to colloidal dimensions of the metallic vapour generated in the arc.

[1] Only for lyophilic colloids.

[2] Peptization is the dispersion of a, normally amorphous, precipitate in a suitable solution of an electrolyte.

The methods of preparation must allow for the electrical charges necessary for the stability of colloids, other than lyophilic colloids under certain circumstances. Suitable conditions must be created for the formation of the necessary ionogenic complex which are often satisfied by the presence of an electrolyte in the medium in which the formation of the colloid takes place. It is, for example, impossible to obtain noble metals in the colloidal state by Bredig's method in media absolutely free of suitable electrolytes such as hydrochloric acid. The peptization of freshly precipitated metallic hydroxides is another typical case; in the absence of a suitable electrolyte it is not possible to disperse the hydroxides in the colloidal state.

For the study of colloidal systems it is essential that they should be as pure as possible. This condition has long been recognized as important in the chemical study of substances but has only in the last twenty or thirty years been recognized as equally essential for the study of colloids. Another necessary condition is that they should have the most homogeneous dimensions possible since many properties are influenced by the particle dimensions. The simplest method of purification is by dialysis through membranes permeable to electrolytes and crystalloids in general but impermeable to colloids; the external liquid should be continuously changed. Purification of colloids by dialysis must allow for phenomena caused by the presence of the membrane. The laws of membrane equilibrium define, however, only the equilibrium state and do not influence the kinetics of the dialysis process; the decisive factors here, other than the concentration differences, are the diffusion and membrane tensions. With some colloids, which are very sensitive, precautions must be taken against possible changes in acidity produced temporarily by membrane hydrolysis (*cf.* p. 380 *et seq.*). Membrane hydrolysis may, for example, denature a protein by excessive acidification of a solution, subjected to prolonged dialysis, during the actual dialysis itself. It may also change the species of the counter-ion and even the nature of the ionogenic complex with consequent changes in the properties of the colloid. It is possible, for example, to obtain acidoid colloids[1] by the hydrolysis of typically negative colloids as with Congo Red whose counter-ion Na^+ may be replaced during the dialysis by H^+. Solutions of several dyes belonging to the Congo Red group change colour after prolonged dialysis; Congo Red itself becomes blue. According to Pauli, the neutral part of the Congo Red

[1] With H^+ as the counter-ions.

after prolonged dialysis of the solution enters the zwitterion quinonoid form[1]

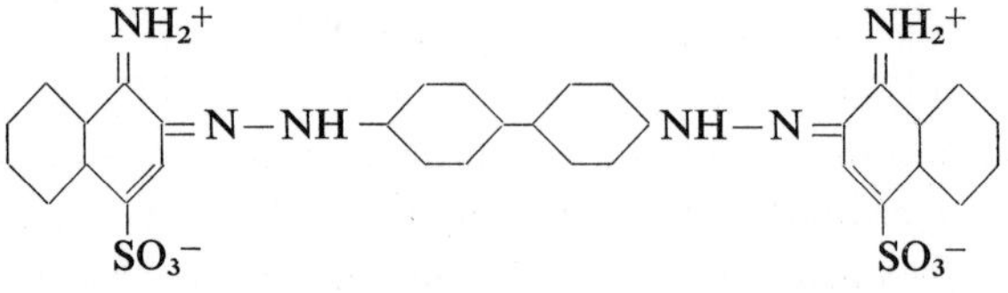

NH_2^+ NH_2^+

=N–NH– –NH–N= I

SO_3^- SO_3^-

whilst the ionogenic complex consists of the same dye in the azoide red form of sulphonic acid

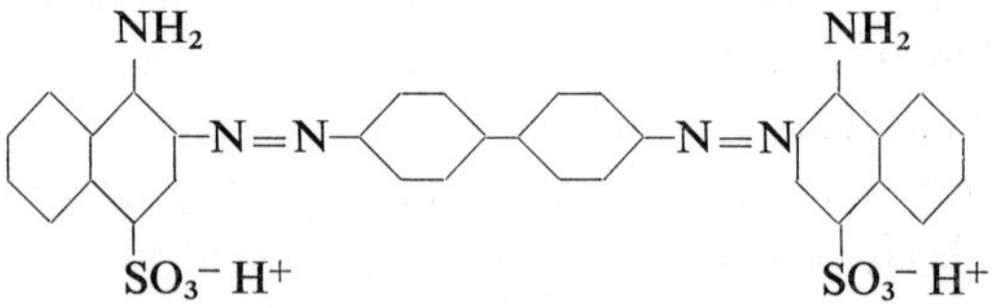

NH_2 NH_2

–N=N– –N=N II

$SO_3^-\ H^+$ $SO_3^-\ H^+$

This dissociates giving H^+ as a counter-ion. It is thus an acidoide corresponding to disulphonic acid whose sodium salt is in fact Congo Red.

The azoide form II may be considered as derived from the quinonoid zwitterion I by dismutation with production of the $-SO_3H$ groups and the formation of poorly dissociated acids. This process may thus be considered as a kind of internal hydrolysis. The quinonoid form I in view of the free localized positive and negative charges present in the same molecule can associate to a far larger extent than the azoide form. The particles of colloidal dimensions thus formed may undergo the hydrolytic process at their surface becoming changed into the azoide form. This represents the ionogenic complex capable of dissociating H^+ ions, which holds the colloidal particles in solution. The transformation of a chloridoide form[2] into a hydroxy-sol in the presence of an excess of base by replacement of

[1] Substances are called *zwitterions* which possess basic and acidic groups *e.g.* amine and sulphonic acid groups, in their molecules. Depending on the pH of the medium and the degree of dissociation of the corresponding base or acid which determines their degree of hydrolysis, these substances may behave either as bases or acids. The main characteristic of these substances is that at particular pH values they may behave simultaneously as acids or bases and may even give rise to internal salts if the acid and basic groups are sufficiently close together and have acid and basic dissociation constants which are not too small. The simultaneous presence of basic and acidic properties differentiates these substances from amphoteric substances like aluminium hydroxide, which may behave either as acids or as bases but not simultaneously as both. Many proteins have a zwitterion nature.

[2] With Cl^- ions as counter-ions.

chloride of the ionogenic complex by hydroxyl is an example of the same type. This transformation of complexes during dialysis confirms the theory of dissociation which considers colloids as electrolytes. Such changes play an important role in conditioning the charge on the colloid surface, and in colloidal reactions.

When the electrolyte concentration of a colloidal solution becomes very small, dialysis becomes very slow. Moreover, the possibility of contamination of the electrolyte by ions such as NH_4^+ or HCO_3^- becomes of some importance. It is thus preferable to use *electrodialysis* which may be considered as dialysis accelerated by the action of an electrical field applied across two electrodes, both placed outside the colloidal solution which is contained within a membrane. The ions of the electrolytes to be removed migrate, one in each direction, under the action of the electrical field and this accelerates the rate of purification. Electrodialysis gives pure colloidal solutions much more rapidly than does simple dialysis and is not subject to the laws of membrane equilibria. It must be noted, however, that electrodialysis may cause coagulation of some colloids whose stability depends upon the presence of small quantities of electrolytes (see below).

Another important process is *electrodecantation*[1] which is a direct development of the electrophoresis of colloidal particles. When colloids are electrodialysed, not only the ions of the electrolyte migrate under the action of the electrical field, but also the colloidal particles. These cannot pass through the membrane and accumulate in the film of solution adjacent to it. This solution becomes heavier and forms a descending liquid film which tends to collect on the bottom of the container. At the other membrane, however, the liquid film adherent to it loses its colloidal ions and becomes lighter tending to collect on the surface of the solution. This is shown diagrammatically in Fig. VI, 3. Some colloids exist, such as latex, with a specific weight less than that of water and in these cases the electrodecantation is inverted. If the process is continued for some time the solution becomes divided into two layers, one of which contains the concentrated colloid and the other lacks the colloid but contains the impurities in about the same concentration as the rest of the solution. By removing this part and replacing it with pure solvent and repeating the electrodecantation a sufficient number of times, the colloidal solution becomes finally free from foreign substances. Electrodecantation also

[1] Wo. Pauli, *Helv. Chim. Acta*, 25 (1942) 137.

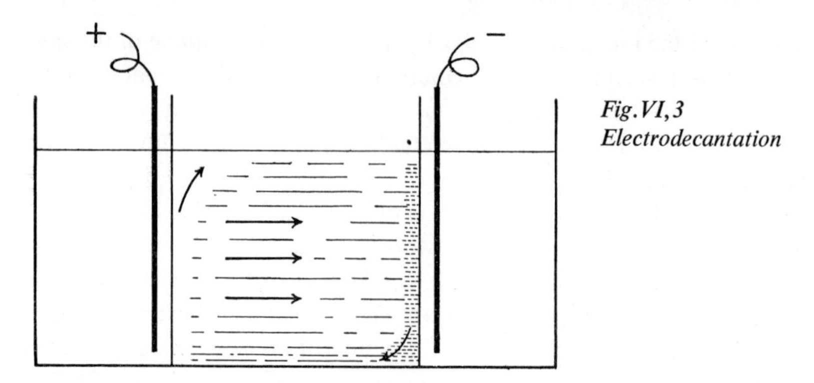

Fig. VI, 3
Electrodecantation

offers the advantage of being able to prepare colloidal solutions which are much more concentrated than those given by other methods.

The purification of some colloids cannot be extended until all the electrolyte is eliminated because they would then coagulate (*cf.* section 7 below).

Another method which has been particularly developed for the study and separation of proteins is the direct application of electrophoresis. Since different colloidal species migrate at different rates it is possible with this method not only to separate the various species for analytical identification by observing the formation and movement of boundaries with high refractive index gradients, but also to use the different migration velocities for preparative purposes to obtain single fractions in high purity from mixtures of proteins[1] (*cf.* p. 399 *et seq.*).

7. The Stability and Reactions of Colloids

It has already been mentioned that the stability of colloidal solutions is linked with the electrical charge of the colloidal particles particularly with lyophobic colloids. For this reason, it is often not possible to obtain colloidal solutions completely free of electrolytes. In fact, in many cases the counter-ions are in equilibrium with an electrolyte present in the solution and it is not possible completely to eliminate this electrolyte without eliminating also the counter-ion and the colloid would then lose its stability and coagulate (see below). In a colloidal solution of chloridoids, for

[1] See particularly the work of A. Tiselius and his colleagues from 1930 onwards.

example, the Cl^- ions are in equilibrium with hydrochloric acid, and it is not possible to lower the acid concentration below 10^{-6} *N*, without the colloid coagulating.

Verwey and Overbeek[1] have developed a quantitative theory for the stability of lyophobic colloids, on the basis that an equilibrium is established between the repulsive forces acting between the electrical charges of the particles, and the attractive forces; these are assumed to be of the Van der Waals–London dispersion forces type. The repulsive forces between the electrical charges of the particles are treated with the Gouy–Chapman equation for the double layer and allowing for Stern's theory of the diffuse double layer. The combination of these forces yields potential curves which explain the stability of lyophobic colloids as well as certain other properties such as the Schulze–Hardy rule (see below). The complete treatment of the Verwey–Overbeek theory, which involves difficult mathematical manipulations, falls completely outside the scope of this book and will not be dealt with further.

In contrast to true solutions, the stability of a colloid is markedly, and in some ways nonspecifically, dependent on the presence of electrolytes. The stability does not show a continuous variation as a function of the electrolyte concentration, but shows a singular point *i.e.* an abrupt change in stability at a certain electrolyte concentration. This particular concentration depends on many factors: the nature of the electrolyte, the valency of the ions, their radius, their hydration, the temperature etc. (see below). By such changes in stability, electrolytes can cause coagulation; *coagulation* is the separation of colloidal solutions into two macroscopically distinct phases (see below).

Changes in the stability of a colloid depending on the reactions which it undergoes, may also be demonstrated by electrophoresis since as the free charge on the particles diminishes so does their migration velocity; when the stability becomes zero, the colloid coagulates and the migration velocity also becomes zero. In other words, stability and migration velocity follow a similar pattern.

According to Weiser and Merrifield[2] the electrical charge does not stabilize lyophobic colloid particles in its own right, but rather because in combination with the other factors discussed (*cf.* p. 373 *et seq.*) gives each

[1] E. J. W. VERWEY and J. TH. G. OVERBEEK, *Theory of the Stability of Lyophobic Colloids*, Elsevier, Amsterdam, 1948.

[2] H. B. WEISER and P. MERRIFIELD, *J. Phys. & Colloid Chem.*, 54 (1950) 990.

particle a certain electrokinetic potential ζ. For a colloid to be stable this electrokinetic potential must have a value greater than a particular minimum, or critical, value below which the particles coagulate. The critical value has been shown to be constant in the presence of different coagulating electrolytes, provided that the colloid is homodisperse. The existence of the critical value explains the existence of abrupt changes in stability.

In some cases, the migration velocity assumes negative values compared to those normally encountered with the same colloid, so that the particles have reversed the direction of their motion and hence must have inverted the sign of their charge. This inversion of charge is easily interpreted on Pauli's theory as reactions of the ionogenic complex with added substances and the formation of new ionogenic complexes at the surface of the colloidal particle. The inverting reagent is always of an ionogenic nature and gives rise to polyvalent ions; for example the inversion of the positive ferric hydroxide colloid by pyrophosphate is due to the reaction of the $Fe(OH)_2^+ \cdot Cl^-$ according to the equation

$$Fe(OH)_2Cl + Na_4P_2O_7 \rightarrow FeP_2O_7^- + Na^+ + 2\ NaOH + NaCl$$

A new ionogenic complex $FeP_2O_7^-$ is thus formed and remains attached to the colloidal particle giving it a negative charge which is neutralized by the Na^+ counter-ions. Inverted colloids of this type may be purified by normal methods.

Inversions of this type cannot be brought about by the gradual addition of polyvalent electrolytes because the colloid would first coagulate and could not then be peptized by further addition of the same electrolyte. It is necessary to add all the electrolyte required to invert the sign by reacting with the ionogenic complex to form a new one, in one single portion. The valency of the ion which brings about the inversion is of primary importance. For example, the inversion of colloidal ferric hydroxide cannot be achieved with $[Fe(CN)_6]^{3-}$ ions; at least tetravalent ions are required, and $[Fe(CN)_6]^{4-}$ or $P_2O_7^{4-}$ ions are effective. Colloidal thorium oxide cannot even be inverted by the $[Fe(CN)_6]^{4-}$ ion but requires at least a pentavalent ion such as the pentavalent hexatungstic ion.

Another type of inversion may be induced with positive colloids by the action of bases, or salts of weak acids (which on hydrolysis act as if they were basic). In this case, new and much less stable ionogenic complexes are formed, *e.g.* the ferrile complex is converted to the less stable ferrite

$$FeO^+ \cdot Cl^- + OH^- \rightarrow (FeO_2)^- + HCl$$

$$FeO^+ + 2\,OH^- \rightarrow (FeO_2)^- + H_2O$$

Inverted colloids of this second type are also very sensitive, will react with atmospheric carbon dioxide and cannot be completely purified.

One of the main characteristics of colloids is their tendency to coagulate, *i.e.* to separate into two macroscopic phases formed of the solvent and the coagulant; this latter contains the substance originally dispersed in the colloidal state. This phenomenon must be interpreted once more as a consequence of the reaction of the ionogenic complex, even when the coagulation takes place as a result of purely physical factors which simply alter its energetic content, such as changes in temperature (coagulation by heating or freezing[1]), violent agitation (coagulation by beating) or the action of ultrasonics. The commonest way of causing the coagulation of a colloidal solution, however, is the addition of an electrolyte. Coagulation consists essentially in an aggregation of the primary particles into increasingly greater sizes until finally two macroscopically distinct phases separate. During this aggregative process the particles at first maintain their individuality but may then undergo a secondary process of sintering and crystallization with similar particles. This process is called *aging* and the particles continue to grow in this way. Coagulation may or may not be inverted depending on whether the elimination of the factor provoking the coagulation will lead to the original colloidal state being regained.

Problems, such as the causes of coagulation, why colloids should coagulate especially on the addition of electrolytes, and the relationship between coagulation and stability are readily answered qualitatively. Quantitatively, however, the treatment becomes rather difficult and requires fairly complex mathematical processes. Up to the present, only Verwey and Overbeek have attempted a rigorous quantitative treatment of lyophobic colloids, and have already obtained valuable results. Such problems in general and particularly the interpretation of coagulation due to physical causes still await a definitive solution. The last cause of coagulation, *i.e.* that which destroys the stability, is the diminution or total loss of the electrical charge as can be shown by electrophoresis. When the

[1] This last could also be interpreted as due to an increase in the concentration brought about by the crystallization of the pure solvent.

electrical repulsive forces between individual particles disappear, they may join together with increasing ease to form ever greater aggregates under the action of the attractive forces, in other words the particles coagulate.

As has been mentioned, it is not the electrical charge, as such, which prevents coagulation so much as the electrokinetic potential which the charge confers, and the size of this potential depends on other factors. When the electrokinetic potential falls below a certain critical value, the attractive forces overcome the repulsive forces and the colloidal particles join together to form the coagulum.

Coagulation on the addition of electrolytes is common to all types of colloids. Such coagulation may be considered as an adsorption of ions with opposite sign to that of the particle. When the concentration of these ions is sufficient they penetrate into the inner part of the electrochemical double layer, *i.e.* into the part forming the charge of the colloidal particle, its number of free charges diminishes and hence so does the charge density at the sliding surface. This diminution in turn leads to a fall in the electrokinetic potential, which may be considered as a more or less accurate measure of the stability of the colloid. In fact, the electrokinetic potential as calculated by the migration velocity is proportional to the electrical charge on the particle provided that other factors, and particularly the dimensions of the particle, are equal. The greater the charge, the greater are the repulsive forces and the greater is the electrokinetic potential. The diminution of the electrokinetic potential thus indicates a diminution of stability. The lyophilic colloids are an exception since their stability does not depend to a significant extent on the electrical charge. The adsorption theory interprets several of the phenomena of coagulation fairly well, but not all of them. The dissociation theory, on the other hand, gives a more complete and detailed explanation of coagulation phenomena although it too is not definitive.

According to the dissociation theory, coagulation is due to an electrostatic interaction between the counter-ions and the colloidal particles. On increasing the concentration of the electrolyte, the counter-ions become progressively deactivated so that their association and the consequent diminution of the free charge of the colloid becomes increasingly probable. This interpretation is supported by the finding that those ions which give insoluble complexes with the ionogenic complex and thus discharge the particles, most readily lead to coagulation. The Schulze–Hardy Rule applies to such coagulations. According to this, the threshold, or minimum value, of electrolyte concentration which can lead to coagulation within a particular time with a given colloid, depends only on the valency of the ions of opposite sign to that of the colloid. This value increases as the valency falls but is independent of the valency of the ions of the same sign as that of the colloid[1]. This rule would also support the adsorption theory but is often not valid; Hofmeister found differences between precipitating ions of the same valency. This implies that the individual properties of particular ionic species may in some cases overcome the influence of the greater charge on polyvalent ions and this indirectly confirms Pauli's theory. In fact, the theory can well interpret the Schulze–Hardy Rule as due to the greater inactivating action of polyvalent ions against the counter-ions.

Finally, there remains the effect of the radius and of the hydration of the precipitating ion and the protective action of ions of the same sign as that of the colloidal ion. Increasing the concentration of these protective ions increases the threshold value for the concentration of coagulating electrolytes. If the threshold value, with some polyvalent electrolytes, is greatly exceeded then coagulation may not occur but rather an inversion of charge. The dependence of the inversion of the charge on a colloid on the chemical nature of the colloid and of the coagulating ion is a further argument in favour of coagulation being a specific chemical reaction as described above.

Coagulation by the addition of another colloid of opposite sign to that originally present may also be considered as an electrolytic coagulation. In this case the coagulation must be due to reactions between the colloids based on the reciprocal interactions of the ionogenic surface groups in one colloid on the other. Complete coagulation occurs when the two colloids are present in equivalent amounts with respect to their total charges. If one of the two colloids is in excess, coagulation does not occur or is incomplete, and the colloid which is not in excess inverts its charge. The ionogenic complex may remain attached to the coagulated particles or be separated from them and appear in the solvent.

The addition of one colloid to another in nonequivalent amounts with

[1] More concordant results would probably be obtained by treating the phenomenon from the standpoint of the kinetics of the process rather than considering simply a threshold value for a single arbitrary time (*cf.* J. STAUFF, *Kolloidchemie*, Springer Verlag, Berlin, 1960).

respect to charge may lead to a sensitizing or a protective action, *i.e.* it may diminish or may increase the threshold limit for the concentration of a coagulating electrolyte. The mechanism of sensitization may be interpreted as a partial reciprocal discharge of the surface charges with the formation of larger, secondary particles which are more sensitive; but the mechanism of the protective action is much more complex. Above all, it must be emphasized that really reliable information on the protective or sensitizing action of one colloid on another is very scarce; many direct experiments in this field have been carried out with colloids which were not sufficiently free of electrolytes. The charges of these often markedly altered the results. The protective action of gelatine and of other proteins on colloidal solutions of gold and of acid Congo Blue is typical; both these colloids are lyophobic and no primary protective action exists. In fact, if the colloidal gold solution prepared by Bredig's method and the colloidal Congo Blue are purified so as to eliminate foreign electrolytes almost completely, they will be coagulated by colloidal solutions of proteins over a wide range of concentrations whereas if the same solutions are not fully purified they will be protected by a gelatine solution. However, definite protective actions between colloids of extreme purity certainly do exist due to interactions between definite surface groups in the two colloids. The protective action of one colloid on another of opposite sign, *e.g.* gum arabic on ferric hydroxide, may be interpreted as an interaction of an ionic nature or of an electrostatic nature induced by polarization. With colloids of the same sign the interpretation is rather more difficult and is based on specific interactions of a more purely chemical nature between surface groups in the two colloids, which may not in fact involve the ionogenic complexes. In such cases, particularly if the protecting colloid is of large size, the protected colloid behaves as if it were enveloped by a layer of the other and the new colloidal particle behaves like a particle of the protecting colloid. The specificity of such protective and sensitizing action confirms the interpretation that their mechanism is a true surface chemical reaction of the colloids.

Coagulation by purely physical factors must be related also to reactions of the ionogenic complex involving changes in its stability such as, for example, hydrolysis with the liberation of electrolytes and the simultaneous discharge of the colloidal ion itself.

Another type of colloidal reaction occurs during electrolysis; the end result of this in a colloidal solution is often coagulation. Sometimes, with certain amphoteric or zwitterion colloids, inversion of the charge may occur. Faraday's Law is probably valid also for colloidal electrolysis reactions leading to primary coagulation although its experimental verification is not possible for several reasons. Firstly, colloids often begin to coagulate before they are fully discharged; secondly, the valency of every ion would have to be known; and finally it is not possible to control secondary peptization reactions of the already coagulated colloid due to the electrolyte of the medium. Coagulation by electrolysis is, however, rarely a primary process, *i.e.* it rarely occurs by the direct discharge of the colloidal ion on the electrode. Normally, coagulation is a consequence of a secondary reaction due to the production of large amounts of H^+ ions at the anode or of OH^- ions at the cathode through the electrolysis of water as the primary reaction. Particularly in the presence of other electrolytes, changes in their concentration due to electrolysis may actually lead to coagulation.

8. Electroosmosis and Other Electrokinetic Phenomena

The colloidal solution constitutes a limiting state between two types of system: ions-solvent, and liquid in contact with another phase. If a colloid is allowed to coagulate or gel it behaves in many respects as a diaphragm and in particular under these new conditions it is possible to observe other electrokinetic phenomena such as *electroosmosis*[1]. Such phenomena may therefore be considered from the standpoint already used for the treatment of other colloidal phenomena.

If an electric tension is applied to the two sides of a clay diaphragm contained within a tube full of water, as shown in Fig. VI, 4, the level of

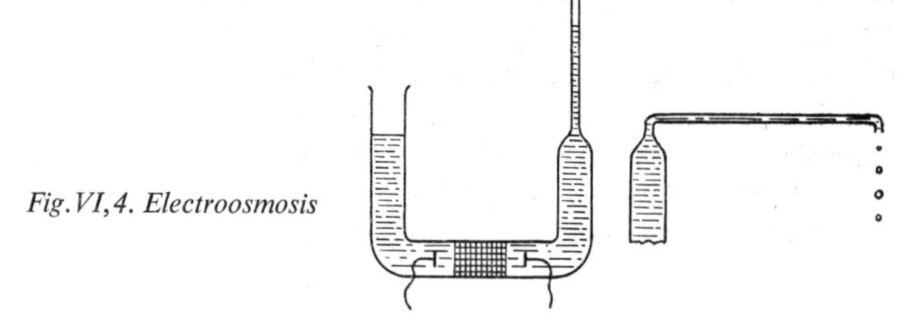

Fig. VI, 4. Electroosmosis

[1] In some texts electroosmosis is called electroendosmosis.

water in the tube will rise until a certain pressure is reached. If the manometric tube is replaced with another bent through two right angles as shown on the right of the figure, water will drip continuously from exit hole. This indicates that the applied electric tension forces water through the diaphragm. On replacing the diaphragm by a glass capillary, a similar phenomenon occurs but to a far lesser extent. In particular, other conditions being equal, the phenomenon is more apparent for a greater ratio between the contact area of the fixed wall surface and the liquid volume. A diaphragm may be considered as a bundle of capillaries in which this ratio assumes high values and so the phenomenon is much more marked than with a single capillary. The direction in which the water[1] moves depends on the chemical nature of the diaphragm and on the nature of any electrolytes present. The quantity of water transported depends on the chemical nature of the diaphragm, its physical characteristics, the applied electric tension and various other factors.

Electrochemical phenomena were first observed by Reuss and were interpreted theoretically by Helmholtz. He assumed that an electric tension existed at the wall–liquid surface which originated in an electrical double layer; this he treated with the electrostatic laws of capacitors combined with the laws of hydrodynamics.

According to Pauli electroosmotic phenomena, like other electrokinetic phenomena, may be interpreted by the same theory of dissociation which has been so fruitful in the study of colloids. It can be assumed that ionogenic complexes exist at the fixed surface similar to those existing on colloidal particles. These complexes dissociate electrolytically and confer an electrical charge on the walls, whilst a charge of opposite sign goes into solution in the form of counter-ions; these are distributed with their greatest density near to the walls. In this way, the electrochemical double layer is formed and the actual electric tension arises. This is well shown by the movement of the liquid with respect to the walls under the action of an electric field. This acts on the counter-ions forcing them to migrate and these, due to their hydration and to the viscosity of the liquid, draw a certain amount of it along with them and thus move it with respect to the walls. Even if Pauli's theory is admitted to be the more probable, the quantitative description of electroosmotic phenomena remains that due originally to Helmholtz and since developed.

[1] With other liquids the direction also depends on the chemical nature of the liquid.

To calculate the quantity of liquid transported electroosmotically, a capillary is considered of cross-sectional area S subjected to an electric tension U applied to the electrodes. The liquid is thus subjected to the force acting on the counter-ions and assumes a velocity which increases until the applied force is balanced by the frictional resistance; the velocity w now becomes constant. The sliding surface is not the wall–liquid interface since a very thin film of liquid remains completely adherent to the wall. The frictional resistance A is proportional to the viscosity η and to the velocity of the liquid flow and is inversely proportional to the thickness of the layer d[1]

$$A = \eta w/d \tag{1}$$

The velocity w can be replaced by v/S which is the ratio of the volume of liquid transported in unit time to the cross sectional area of the capillary

$$A = \eta v/(dS) \tag{2}$$

The motive force F is proportional to the total free charge Q per unit of surface area, *i.e.* to the charge density, and to the intensity of the electrical field U/l, where l is the distance between the electrodes

$$F = QU/l \tag{3}$$

The limiting velocity will be reached when the force F equals the frictional resistance A, *i.e.* when equations (2) and (3) have the same value

$$\eta v/(dS) = QU/l \tag{4}$$

If the double layer is considered as a condenser, its capacitance per unit area is given by the equation

$$Q/\zeta = \varepsilon/(4\pi d) \tag{5}$$

where ζ is the electric tension between the plates and is hence the electric tension at the sliding surface. This, of course, is the electrokinetic potential (*cf.* p. 373 *et seq.*); ε is the dielectric constant of the solution. Substituting the value of Q from equation (5) into equation (4) gives

$$v = (\zeta \varepsilon S U)/(4\pi\eta l) \tag{6}$$

Since the electrical resistance R is given by the equation (p. 47)

$$R = \frac{1}{\varkappa} \cdot \frac{l}{S}$$

[1] Since different symbols are required in the following equations for length and thickness, the latter has been represented by *d*.

and since by Ohms' Law

$$U = IR = (Il)/(\varkappa S) \tag{7}$$

substituting the value of U from equation (7) into equation (6) gives

$$v = (\zeta\varepsilon I)/(4\pi\eta\varkappa) \tag{8}$$

Equations (6) and (8) show the quantity of liquid transported per second as a function of the applied electric tension or of the current intensity through the capillary. The characteristics dependent on the chemical nature of the substances of the capillary walls and of the liquid phase exert their effects through the electrokinetic potential, the dielectric constant and the viscosity.

As a function of the current, however, the quantity of liquid transported is independent of the dimensions of the capillary, and is dependent only on the current intensity. Again, in equation (8) the characteristics of the substances of the capillary wall and the liquid exert their effects through the values of the electrokinetic potential, the dielectric constant, the viscosity and, in addition, the specific conductance.

If, instead of a single capillary, the calculation applies to a diaphragm, the sole approach is to consider it as approximately equivalent to a bundle of parallel capillaries (*cf.* p. 422 *et seq.*).

Poiseuille's equation is used to calculate the hydrostatic pressure Δh at equilibrium by defining the volume of liquid which passes through a capillary; this equation, referred to unit time, becomes

$$v = (\pi r^4 \Delta h)/(8\eta l)$$

$$= (Sr^2 \Delta h)/(8\eta l) \tag{9}$$

where Δh is the difference between the hydrostatic pressures due to gravity at the ends of the capillary. The equilibrium pressure Δh is reached when the quantity of liquid which would pass downwards due to the hydrostatic pressure equals the quantity which would pass upwards due to the electroosmotic force, *i.e.* when equations (6) and (9) are equal

$$(\zeta\varepsilon SU)/(4\pi\eta l) = (Sr^2\Delta h)/(8\eta l)$$

From this, the equilibrium pressure is readily derived as a function of the applied electric tension

$$\Delta h = (2\zeta\varepsilon U)/(\pi r^2)$$

$$= 2\zeta\varepsilon U/S$$

For a diaphragm the considerations discussed on p. 422 *et seq.* are applied and it is considered as equivalent to a bundle of capillaries. The experimental results of Wiedemann and Quincke are in agreement with theory.

Equation (6) also allows the ready calculation of the effective sliding velocity. In fact, replacing S by v/w gives the equation

$$v = (\zeta \varepsilon v U)/(4\pi\eta l w)$$

whence

$$w = (\zeta \varepsilon U)/(4\pi\eta l)$$

Making $U/l = 1$, *i.e.* with a unit potential gradient, gives a sliding velocity which, apart from the numerical factor 4π instead of 6π is equivalent to an ionic or electrophoretic migration velocity and has the same order of magnitude. The sliding velocity is strongly influenced, as is the electrophoretic velocity, by the electrolyte content of the solution and may become zero or even be inverted. All these similarities between electroosmosis and electrophoresis clearly justify the analogous treatment of the two phenomena.

The two electrokinetic phenomena of electrophoresis and electroosmosis both involve a symmetrical phenomenon in the electric tension of falling and the electric tension of filtration. On forcing a liquid through a capillary or diaphragm an electric tension is produced at the free end; similarly, particles falling with some velocity into the bulk of a liquid produce an electric tension. However, the study of these phenomena is very specialized and reference should be made to specialized texts.

9. Analytical Applications of Electrophoresis

(I) *Free Electrophoresis of Colloids*

by

STELLAN HJERTÉN[1]

Electrophoresis as already defined has no limitations as to the size of the particles, but its treatment here will be confined to colloids. The aim of an electrophoresis experiment is to give an analysis of a solution containing a mixture of components and/or to separate these for preparative purposes. This analysis can give information on the number and kind of compo-

[1] Institute of Biochemistry, University ot Uppsala.

nents and the relative amounts of them. Further, by determination of the mobility of a particle some knowledge of its surface structure and charge can be obtained. The mobility values are also useful for characterization of a substance.

Electrophoresis is a very gentle separation method and has therefore been widely used within the field of biochemistry, where the purification of labile substances from biological material is a problem often encountered.

The theory of electrophoresis is rather extensive and it is far beyond the scope of this book to relate it in detail. The treatment will therefore be elementary and the formulas not always strictly derived. Nor will all the assumptions upon which the formulas rest be mentioned. Among the many books on electrophoresis, that by Bier[1] is especially recommended.

To make the derivation of fundamental formulae as simple as possible, some already discussed equations should be again born in mind, a spherical shape of the colloid particle being assumed. In an electrical field of strength F, it is acted upon by the force QF, if Q is the total charge of the particle, already referred to as ze. This force is counteracted by the friction force. Applying Stoke's Law gives

$$QF = 6\pi\eta rw$$

(η = viscosity coefficient, r = radius of the particle, w = velocity). As the mobility u of a particle is defined as its migration velocity in unit field

$$u = \frac{Q}{6\pi\eta r} \tag{1}$$

At rest in an insulating medium of dielectric constant ε, the electric potential ψ of the particle is $Q/\varepsilon r$, and equation (1) takes the form

$$u = \frac{\psi\varepsilon}{6\pi\eta} \tag{2}$$

However, in practice an electrophoresis experiment is never performed in an insulating medium, but in a salt solution. When applied to such a system the equations (1) and (2) are only very rough approximations. Before expanding them to be valid also for charged molecules in a conducting medium (salt solution) a brief account of the theory of ionic distribution in the environment of a colloid must be given.

[1] M. BIER, *Electrophoresis, Theory, Methods and Applications*, Academic Press Inc., New York, 1959.

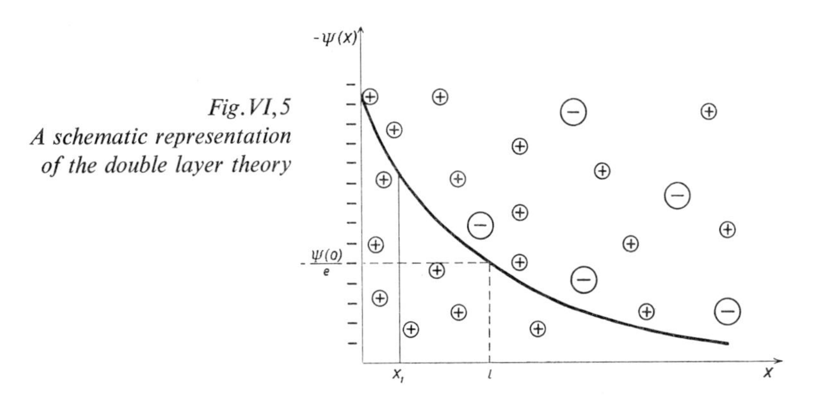

Fig. VI, 5
A schematic representation of the double layer theory

Assume that a particle without any charges on its surface is introduced into a salt solution and that part of the latter's ions are adsorbed to the particle surface. Positive ions are then attracted towards the negative surface, and negative ions are repelled. Thus two oppositely charged layers (the electrochemical double layer) are obtained. The inner layer can be considered as a surface charge, while the outer layer (the diffuse part of the double layer) forms a space charge, which, due to the Brownian motion, can be extended over a thickness of several hundred Ångströms. The double layer theory is schematically presented in Fig. VI, 5, which also shows how the potential $\psi(x)$ changes with the distance x from the surface of the particle. Due to the presence of the double layer the migration entity to be considered in electrokinetics is larger than the particle itself and the existence of a 'slipping plane', localized at a distance x_1 from the surface of the particle and coinciding with the surface of the migrating entity is assumed. The potential of the latter surface $\psi(x_1)$ is called the electrokinetic potential or ζ potential (*cf.* p. 373 *et seq.*). In accordance with this double layer theory we can now modify equation (2) to be valid also for colloids in an ionic medium

$$u = \frac{\zeta \varepsilon}{6\pi\eta} \tag{3}$$

The modification of equation (1) is more complicated and therefore only the result, as obtained by Hückel, will be given

$$u = \frac{Q}{6\pi\eta r} \cdot \frac{1}{1 + \varkappa r} \tag{4}$$

($\varkappa = 1/l$ and l = 'thickness' of the diffuse part of the double layer).

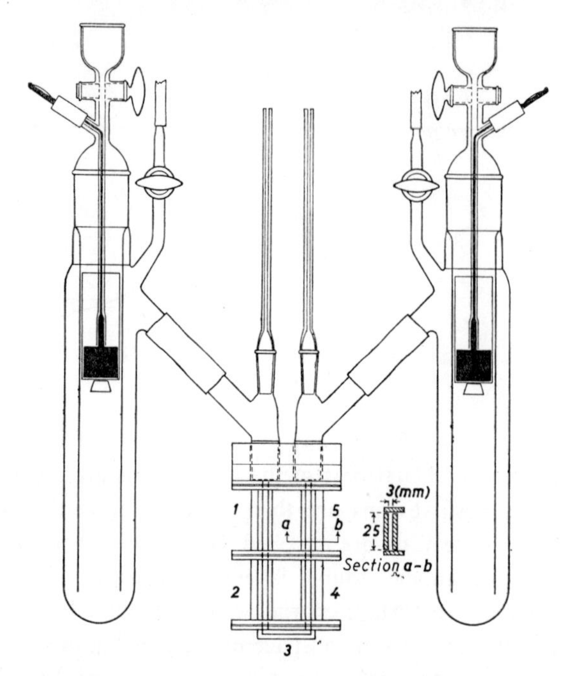

Fig. VI,6. The Tiselius cell and the electrode vessels

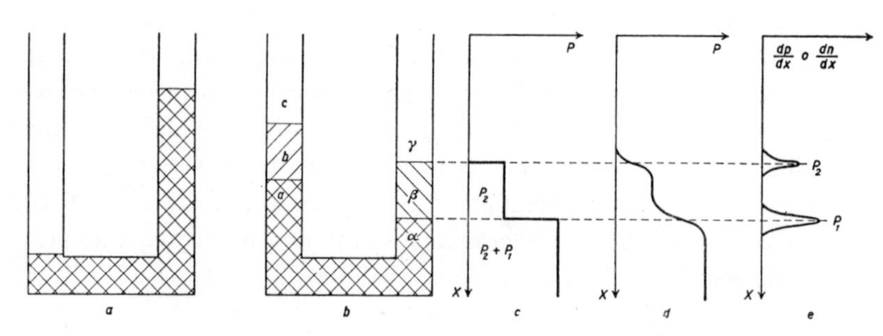

Fig. VI,7. Ideal electrophoresis of a mixture of two proteins
p_1 (////) and p_2 (\\\\\\\\); p = protein concentration, n = refractive index

a. before the start
b. after about one hour's electrophoresis
c. the protein distribution in the right limb in b (without diffusion)
d. the protein distribution in the right limb in b (with diffusion)
e. the derivative of the curve in d. The concentration of a certain protein is proportional to the area of the corresponding peak

The thickness is defined as the distance l from the particle surface, where the potential is $\psi(0)$, to a parallel surface (x) where the potential $\psi(x)$ is $\psi(0)/e$ and $e \approx 2.7$ is the base of the natural logarithmic system. The equations (1) and (4) differ only in a correction factor $1/(1 + \varkappa r)$. The latter equation is therefore transformed to the former when the product $\varkappa r$ approaches zero. There are many other formulae relating charge and mobility, but they will not be treated here. In the double layer theory as described here, the surface charge of the colloid was formed by adsorption of ions. The theory is, of course, applicable also to polyelectrolytes, for instance proteins, which can be charged by dissociation, and in general to colloids whose charge has been considered following Pauli's theory of ionogenic complexes.

One of the most important methods in free electrophoresis is the moving boundary method (*cf.* p. 437 *et seq.*), that will be therefore treated in a more detailed way than other methods. As applied to colloids it is known as the Tiselius method.

The separation chamber has the form of a U-shaped channel, the volume of which is generally 10–20 ml (Fig. VI, 6). The bottom section and one limb are filled with the sample solution, which has been dialysed against a buffer. The rest of the channel and the electrode vessels are filled with the buffer. Fig. VI, 7 shows an electrophoresis experiment with a mixture of two colloids, at the start (Fig. VI, 7*a*) and after some hours' electrophoresis (Fig. VI, 7*b*). As the degree of crosshatching suggests, the solution above any boundary separating two solutions has a density that is lower than that below. Thus there seems to be no tendency to convection. However, when passing an electric current through the channel the heat generated will create a temperature difference between the centre and the wall. This temperature difference is accompanied by a horizontal density gradient, which may cause the solution to rise in the centre and return along the walls. To keep these density differences at a minimum the channel is placed in a water bath at the temperature where the buffer has a maximum density, because at this temperature (about 2° C) the density, of course, is almost independent of temperature. The essential requirement necessary for moving boundary electrophoresis to be an analytical method of high accuracy was created when Tiselius introduced this low temperature bath and a cell of a rectangular cross section (Fig. VI, 6 section *a–b*) to obtain an efficient cooling of the solution in the cell.

Due to diffusion the boundaries are never so sharp as indicated in Fig. VI, 7*b* and *c*. Fig. VI, 7*d* therefore corresponds more closely to reality. Blurring of the boundaries can also be caused by electroendosmosis. The positions of the boundaries *i.e.* the inflexion points of the curve in Fig. VI, 7*d* are easily obtained by taking a derivative of it (Fig. VI, 7*e*). As proteins form the largest group of colloids for which electrophoresis has become a routine method, it is natural that many methods have been worked out for observation of protein boundaries. The astigmatic schlieren camera developed by Philpot and Svensson thus permits the direct recording on a photographic film of the curve in Fig. VI, 7*e*, *i.e.* the refractive index gradient dn/dx (which is proportional to the protein gradient dp/dx) against height in the channel. Such an electropherogram is seen in Fig. VI, 8.

In the moving boundary method some anomalies appear. Thus the protein concentrations calculated from the electropherogram obtained by the schlieren camera (Fig. VI, 8) are not the true ones. Further, the migration distance of the ascending boundary differs from that of the descending boundary. For many purposes these anomalies may be ignored, but for analyses which require high accuracy they must be taken into consideration. The moving boundary equation (7) gives an explanation of the cause of the appearance of the anomalies and also affords possibilities for correction of them. Before deducing the equation it may be recalled that

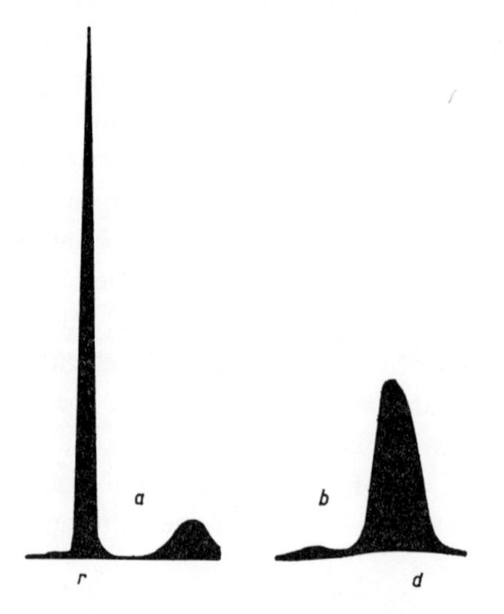

Fig. VI, 8. Electrophoretic patterns of ovalbumin at pH = 3.93: r and d = rising and descending boundaries (after L. G. Longsworth)

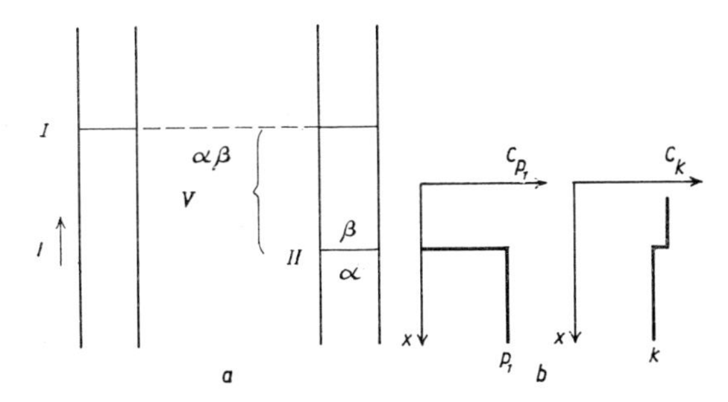

Fig. VI,9. The migration of the boundary $\alpha - \beta$ in Fig. 3b on passage of one Faraday. Displaced volume $= v^{\alpha\beta}$. Curve p_1 for the protein p_1. Curve k for all ions k.

the transference number t_j of an ion[1] j may also be defined as the number of equivalents of the ion that migrates through a plane in a solution on the passage of one **F** (*cf.* p. 29 *et seq.* and p. 373 *et seq.*). This definition is often given the following mathemathical form:

$$t_j = \frac{c_j u_j}{\varkappa} \tag{5}$$

(c_j = concentration of the ion j; u_j = mobility of the ion j; $\varkappa$ = conductance of the solution).

For derivation of the moving boundary equation consider again Fig. VI, 7. The solution α in the right limb is the original dialysed protein mixture and β is the solution that contains the slowest migrating protein.

Consider the boundary between the solutions α and β; on the passage of one **F** the boundary moves from position I to position II (Fig. VI, 9*a*). The direction of the current is indicated by an arrow. The volume between I and II is $v^{\alpha\beta}$. The concentration of the anion j (protein or buffer) in solution α is c_j^{α} and in solution β, c_j^{β}. Before the passage of the current the number of equivalents of the ion j in the volume $v^{\alpha\beta}$ is $c_j^{\alpha} v^{\alpha\beta}$ and after it is $c_j^{\beta} v^{\alpha\beta}$. According to the definition of transference number, t_j^{β} equivalents of the ion j pass the plane at I and t_j^{α} at II. The change in the number of equivalents of the ion j in the volume $v^{\alpha\beta}$ is thus $c_j^{\alpha} v^{\alpha\beta} - c_j^{\beta} v^{\alpha\beta}$ and $t_j^{\alpha} - t_j^{\beta}$ respectively, and the following relationship holds

$$t_j^{\alpha} - t_j^{\beta} = c_j^{\alpha} v^{\alpha\beta} - c_j^{\beta} v^{\alpha\beta} \tag{6}$$

[1] The word ion indicates here both true ions and charged colloidal particles.

By the use of equation (5), and assuming that $u_j^\alpha = u_j^\beta = u_j$, the expression (6) can be transformed into

$$\left(\frac{u_j}{\varkappa^\alpha} - \nu^{\alpha\beta}\right) c_j^\alpha = \left(\frac{u_j}{\varkappa^\beta} - \nu^{\alpha\beta}\right) c_j^\beta \tag{7}$$

This 'moving boundary equation' will form the basis for the following discussion of the anomalies; colloids will be considered in this discussion as strong electrolytes.

Two particular anomalies are important in free electrophoresis. The first one is the *concentration anomaly*. Consider again the boundary between the solutions α and β in Fig. VI, 7. If the ion j in equation (7) corresponds to the protein p_1, *i.e.* $c_j^\alpha = c_{p_1}^\alpha$, and $c_j^\beta = c_{p_1}^\beta = 0$

$$\nu^{\alpha\beta} = \frac{u_{p_1}}{\varkappa^\alpha} \tag{8}$$

For *any other* ion k (protein p_2 or the buffer ions), equation (7) therefore gives

$$\frac{c_k^\alpha}{c_k^\beta} = \frac{\frac{\varkappa^\alpha}{\varkappa^\beta} u_k - u_{p_1}}{u_k - u_{p_1}} \tag{9}$$

As a protein contributes to some extent to the conductance of a solution, $\varkappa^\alpha \neq \varkappa^\beta$ and therefore $c_k^\alpha \neq c_k^\beta$. Thus at a protein boundary *all* ionic species – besides the protein corresponding to the boundary – have different concentrations above and below the boundary. The situation is depicted in Fig. VI, 9*b*. Since the schlieren camera measures all variations in refractive index, concentration gradients originating from all ions k will also be detected and superimposed on that originating from the protein p_1. The magnitude of the error caused by these superimposed gradients amounts in general to 3–10%. From equation (9) it can also be concluded that the most serious deviations appear when there is in the solution an ion having a mobility of the same order of magnitude as that of the protein p_1.

A second type of anomaly is the *mobility anomaly*. The descending boundary between the solutions a and β corresponds to the ascending boundary between the solutions b and c (Fig. VI, 7*b*). Equation (7) applied to these two protein boundaries gives (since $c_{p_1}^\beta = 0$ and $c_{p_1}^c = 0$),

$$\left.\begin{aligned} \left(\frac{u_{p_1}}{\varkappa^\alpha} - \nu^{\alpha\beta}\right) c_{p_1}^\alpha &= 0 \\ \left(\frac{u_{p_1}}{\varkappa^\beta} - \nu^{bc}\right) c_{p_1}^b &= 0 \end{aligned}\right\}$$

Consequently

$$\left.\begin{aligned} v^{a\beta} &= \frac{u_{p_1}}{\varkappa^a} \\ v^{bc} &= \frac{u_{p_1}}{\varkappa^b} \end{aligned}\right\} \qquad (10)$$

and

$$\frac{v^{a\beta}}{v^{bc}} = \frac{\varkappa^b}{\varkappa^a} \qquad (11)$$

Thus $v^{a\beta} \neq v^{bc}$, *i.e.* the migration velocities of the ascending and descending boundaries are different. An idea of the magnitude of their ratio is obtained from Table VI, 2.

TABLE VI, 2

OBSERVED VALUES OF THE RATIO OF THE MIGRATION VELOCITIES OF THE ASCENDING AND DESCENDING BOUNDARIES

The electrophoresis was carried out by L. G. Longsworth in 0.1 *m* sodium acetate buffer, pH = 3.93 and at different concentrations A of the protein (ovalbumin).

A % protein	$\frac{v^{bc}}{v^{a\beta}}$
0·64	1.09
1.36	1.18
2.74	1.33

Of course it is desirable to eliminate these anomalies. When they are completely eliminated the term *ideal electrophoresis* is used. Such an electrophoresis fulfills the conditions $c_k^a = c_k^\beta$ and $v^{a\beta} = v^{bc}$, which according to equations (9) and (11) respectively are equivalent to $\varkappa^a = \varkappa^\beta$ and $\varkappa^a = \varkappa^b$. An ideal electrophoresis could therefore be considered as a limiting case where the conductance of the protein solution in the channel has the same value as that of the buffer. At low protein concentrations this condition is approached. However, at a buffer ionic strength of 0.05, a protein concentration below 0.5% should be avoided, since the instability of a boundary increases when the density difference between the solutions below and above it decreases. Table VI, 2 clearly shows that the deviation from ideal electrophoresis increases with the protein concen-

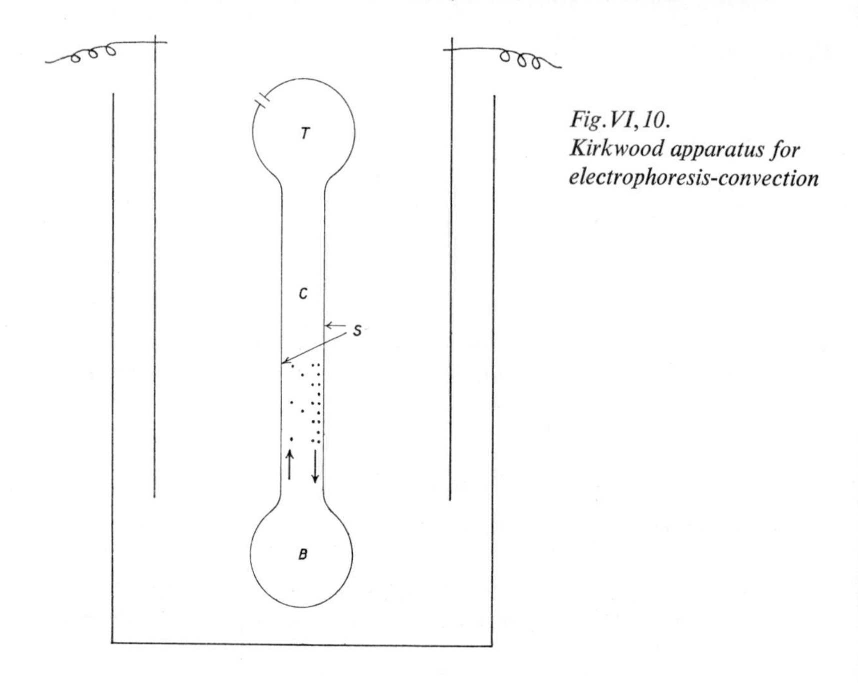

Fig. VI, 10.
Kirkwood apparatus for electrophoresis-convection

tration. Ideally the rising and descending boundaries should be mirror images. In practice this is never attained (Fig. VI, 8).

As Fig. VI, 7*b* suggests, only the fastest and slowest migrating components can be obtained in pure form by moving boundary electrophoresis. Though it was used earlier, it is therefore not very suitable for preparative work. For such purpose it has now been replaced by other methods, especially zone electrophoresis. This latter method has been more widely treated below and therefore only one of these other techniques will be described here: that worked out by Kirkwood and his collaborators. It is schematically presented in Fig. VI, 10. The separation channel *C* and the top and bottom compartments *T* and *B* are filled with the sample solution. To understand how it works assume that one of the colloids in the sample solution is uncharged, *i.e.* the pH of the buffer is equal to the isoelectric point[1] of the colloid. Any mobile colloid migrates under

[1] If the pH of a buffer solution is such that the basic dissociation equals the acid dissociation (see p. 384 *et seq.*) the colloidal particles will possess equal amounts of positive and negative charges and will be effectively uncharged. This pH is called the *isoelectric* point of the colloid.

the influence of the electric field towards one of the semipermeable membranes *S*. The density in the immediate vicinity of these will consequently be comparatively high. This causes a flow of the mobile colloids along the membrane down into the bottom compartment, from which solution will be simultaneously displaced and rise through the channel – where it is affected by the electric field – into the upper compartment. By this circulation the isoelectric colloid will thus be accumulated as the top fraction. This method, called electrophoresis-convection, is very similar to Pauli's electrodecantation and seems to be efficient for the preparation of large quantities of material (on a gram scale). In the explanation of the working procedure the presence of an immobile component was assumed. To have a partial resolution of a mixture this requirement is not necessary.

For determinations of mobilities of colloidal particles, the microscopic method has frequently been used. The migration takes place in a capillary tube or in a flat, thin chamber and the movement of the particles is observed by a microscope.

(II) *Electrophoresis in Porous Media (Zone electrophoresis)*

by

MICHEL LEDERER [1]

The advantages of electromigration methods in porous media over those in free solution have been listed by Tiselius and Flodin:

1. It is possible to obtain complete separation into zones of different migration velocities and thus not only a boundary separation.

2. The so-called boundary anomalies interfere less in zone electrophoresis, and therefore substances of low molecular weight (*e.g.* amino acids, peptides, nucleotides) may also be studied.

3. Zone electrophoresis (particularly in filter paper strips) requires only minute quantities of material, and can be performed with simple and inexpensive equipment.

These advantages are gained, however, at the sacrifice of the greater accuracy of the boundary method (particularly with regard to mobility

[1] Consiglio Nazionale delle Ricerche, Rome.

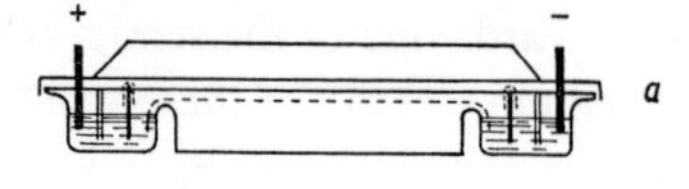

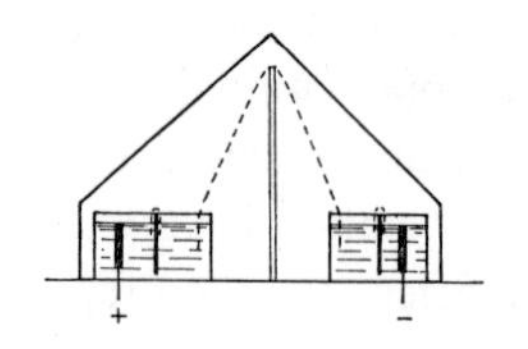

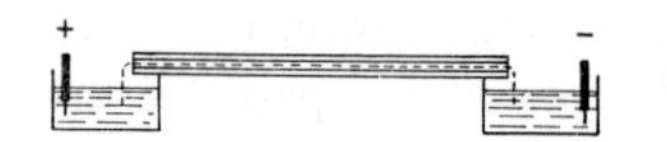

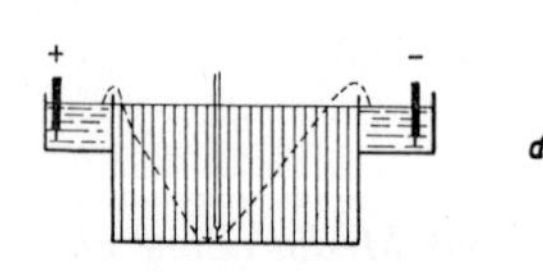

Fig. VI, 11. Various arrangements for paper electrophoresis

and isoelectric point determinations). The supporting medium necessary in zone electrophoresis, introduces new factors which may influence the results in a way which is difficult to control.

A separation by paper electrophoresis is carried out by dipping a paper strip or sheet (usually 30 cm long) into the required electrolyte (for proteins 0.05 *m* veronal buffer at pH 8.6 is mostly used), and drying off the excess of buffer by placing the wet paper on a sheet of blotting paper for a few seconds.

The sample (for example serum) is placed on the centre of the paper strip with a micropipette (0.015 to 0.04 ml is usual). The paper is then placed into the electrophoretic apparatus and the current switched on (for proteins 100 to 200 V for 12 to 20 hours). The electrophoresis apparatus serves to allow contact between electrode vessels and the ends of the paper and to prevent undue evaporation during the migration of the zones. Fig. VI, 11 shows several of the designs of apparatus that are commonly employed.

The paper is either suspended horizontally (Fig. VI, 11*a*) or raised at its centre (Fig. VI, 11*b*). If evaporation is to be diminished the paper may be placed between glass plates (Fig. VI, 11*c*) or into a cooling liquid such as chlorobenzene or carbon tetrachloride (Fig. VI, 11*d*).

With apparatus which cools the paper (Fig. VI, 11 *c* and *d*) currents up to 50 mA may be employed, that is strong electrolytes up to 1 *m* (with 200 V) or weak electrolytes (with correspondingly higher tensions) can be used. When the current is passed for a time which is sufficient for the separation desired, the paper is taken out of the apparatus, dried and suitable reagents are sprayed on the paper or the paper is dipped into the reagent. For detecting proteins a solution of 0.1% bromphenol blue in ethanol saturated with mercuric chloride or a solution of naphthalene black 12B 200 in 10% acetic acid are mainly used. The paper is then washed free of the excess of reagent, and evaluated photometrically if semi-quantitative results are desired.

Paper electrophoretic analysis of serum, haemoglobins and body fluids is now extensively employed in clinical analysis.

Much information also exists on the separation of smaller molecules by paper electrophoresis. Sugars and polyphenols can be separated in borate buffers.

Alkaloids, weak acids, amino acids and peptides separate depending on the pH of the electrolyte and the pK values of the substances to be separated. Inorganic ions may also be separated by employing complexing electrolytes for example 0.1 *m* lactic acid or 1% citric acid although in some cases, such as the alkali metals or some anions the differences in mobilities suffice. Unfortunately the data available in the literature cannot be easily recalculated for equal experimental conditions, so that comparison of data produced by different authors is often very diffucult.

If the mobility in zone electrophoresis is to be related to the ionic mobility the following four factors have to be taken into consideration:

1. Adsorption on the paper (or other support), for large molecules there may even be an ultrafiltration effect (for example in starch gel).

2. Evaporation during electrophoresis with the resultant change in the electrolyte concentration and with the liquid flow from the electrode vessels to compensate the loss.

3. Electroosmotic flow (*cf.* p. 395 *et seq.*). This is of course highest with weak electrolytes and almost negligible with for example 1 *m* KCl.

4. A '*tortuosity factor*' due to the intricate path an ion has to take between the fibres (or particles) of the support.

For an apparatus without evaporation and when adsorption is negligible Kunkel and Tiselius proposed the following equation

$$d = \frac{utI}{q_a\varkappa}\left(\frac{l}{l'}\right)$$

where d = the distance migrated by the substance, u = the mobility, t = the time, l/l' = the correction factor for the 'tortuosity' which may be determined from the resistance of the paper strip, $\varkappa$ = the conductance, q_a = the cross-sectional area of the paper and I = the current intensity. It has later been proposed to change the factor (l/l') to $(l/l')^2$ since the tortuosity not only lengthens the path but equally increases the distance between the electrodes.

The electroosmotic flow is easily corrected for, by running a neutral substance next to the zone and subtracting its movement from that of the moving zone. Dextran is often used for this purpose in protein separations. Some important techniques of zone electrophoresis which differ from the simple migration in paper strips are as follows:

(*a*) *High voltage electrophoresis*

If tensions of 50 to 100 V cm^{-1} are applied the separations are of course much faster and also better, because the increase in zone size due to diffusion is smaller. However highly efficient cooling is necessary. The paper must be only slightly moist and is best placed with uniform pressure on to an insulated metal plate which is cooled with circulating brine at −5 to −10° C. High voltage separations of small molecules such as peptides, amino acids and organic acids are highly successful but proteins separate badly.

(*b*) *Continuous electrophoresis*

An electric tension may be applied to a paper strip in which the electrolyte flows at right angles to the current direction. If a thin stream of the sample is directed into the electric field, separations very much like those in a mass spectrograph can be obtained. The substances will move at angles which depend on the velocity of migration (and thus on the ionic mobility and field strength) as well on the speed of solvent flow.

(c) *Electrophoresis in packed columns or slabs of porous material*

When electrophoresis is carried out on a preparative scale either columns of cellulose powder, starch granules, glass powder or gels as well as horizontal beds or slabs of such materials can be used. The amount of heat generated by the current passed is of course much higher than on paper strips and the columns are usually water-jacketed. In electrophoresis in gels the ultrafiltration effect plays quite a large role with large molecules and improves the separation due to mobility differences.

(d) *Micro-methods*

Separations by electrophoresis on single threads of cellulose or on the moist surface of glass slides has been performed. Both may be observed under the microscope and even quantitative estimates of the constituents in the range of 100 to 1000 picograms ($= 10^{-12}$ g) are possible.

Very many values for electrophoretic mobilities are reported in the literature, but it is difficult to compare these because they have been obtained under too widely differing conditions. Thornburg, Werum and Gordon[1] have recently attempted to reduce these mobilities to homogeneous relative values referred to a standard substance. These authors have published a series of tables, for various groups of substances, containing values of mobilities which are comparable to each other.

10. Industrial Applications of Electrokinetic Phenomena

Attempts have often been made to apply electrokinetic phenomena to industry, but so far have never undergone any major use equivalent to that of electrolysis or of construction of primary cells or accumulators. Such applications of electrokinetic phenomena, however, include the purification of kaolin, the drying of peat, the so-called electrical tanning, the purification of water and the purification and coagulation of latex.

The first of these has been the most widely used. Since these applica-

[1] W. W. Thornburg, L. N. Werum and H. T. Gordon, *J. Chromatog.*, 6 (1961) 131.

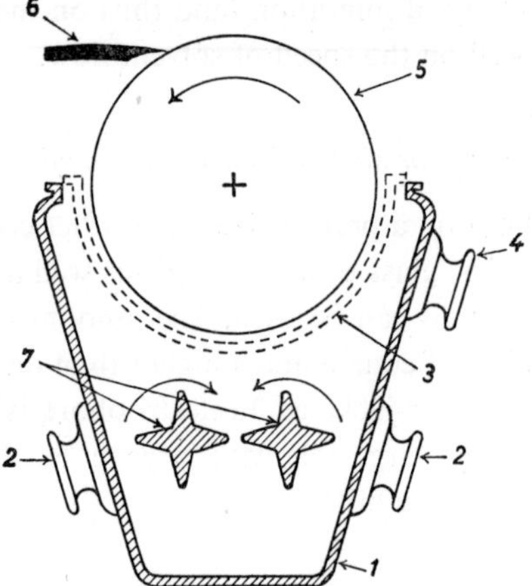

Fig. VI, 12. Purification of kaolin

tions involve all the main electrokinetic phenomena – electrophoresis and electroosmosis together with colloidal electrolysis – it will be worth while examining them further.

Kaolin must be particularly pure for use in the ceramic or pharmaceutical industries. Kaolinite – the main constituent of kaolin – is a hydrated aluminium silicate of the formula $Al_2O_3 \cdot 2\,SiO_2 \cdot 2\,H_2O$ and is a product of the weathering of granite and other feldspar rocks under the action of water and carbon dioxide. The major impurities are quartz and mica and it may contain iron and carbonates. The kaolin in the raw material is peptized in water containing sodium silicate, to form a stable colloidal suspension whose particles are negatively charged. The gross impurities readily precipitate from this but the colloidal kaolin is so finely divided and stable that it cannot be separated by filtration or centrifugation. Coagulation with an electrolyte would be inconvenient as the kaolin would lose the purity and plasticity required for the ceramic fabrication. Use is therefore made of its colloidal properties in the apparatus shown diagrammatically in Fig. VI, 12. The suspension enters at 2, the mills 7 rotate in opposite directions to prevent the stratification of the kaolin particles. A cylinder of hardened lead 5 is immersed in the upper part as an anode and a semi-cylindrical metallic mesh 3 placed coaxially to the anode, and below it, functions as the cathode. Under the action of the electrical field, the colloidal particles migrate to the anode where they are partially or totally discharged and deposited as a fairly compact layer which is further dehydrated by electroosmosis. The layer of kaolin is drawn out of the suspension by a rotating drum and is cut up continuously by the knife 6. The clear liquid is discharged through the tube 4 and recycled to peptize more kaolin. A tension of 70–80 V/cm is used with a current intensity of not more than 1 A/100 cm^2. Other similar substances such as clay, zirconium oxide and amorphous silicic acid may be similarly purified. Kaolin and other clayey substances when thus purified are very pure and have many uses including pharmacy.

CHAPTER VII

GENERAL CONSIDERATIONS ABOUT ELECTROCHEMICAL PLANTS

1. Introduction

Every electrochemical process has its own peculiar characteristics and special requirements, which are naturally reflected in the design of its plants. These must be constructed in accordance not only with the purely technical exigencies of the process, but also with the particular economic conditions under which the plant operates. These latter may vary so much that a particular process may be uneconomic in one place and economic in another. However by suitably modifying the process it may be possible to run it profitably under conditions which made it uneconomic in its original form. It must also be remembered that it is often extremely difficult even on a purely technical basis, and quite apart from economic and environmental reasons, to decide which of a number of plants, designed for the same purpose, will be the best. Thus a very large number of types of plants exist, not only for different electrochemical processes, but even for the same one. Despite all this, electrochemical plants possess certain common features which have maintained their function, even although there has been a continual growth of the technology, and which may therefore conveniently be examined separately, in general terms.

An electrochemical plant for electrolysis in aqueous media has the following components: vessel, electrodes, possibly diaphragms and accessory plant for the circulation of the electrolyte, for the aspiration of any gas produced in the electrolysis, for heating the electrolyte etc. Moreover the outlay must also allow for the plant required for the production of the necessary direct current, for its distribution, for the connections, for the electrical measurements etc. This second part of the plant, although it is no less important than the first, falls more within the field of the electrical engineer than that of the chemist and so will not be discussed in the brief notes which can be given here on this topic.

2. Vessels

The first part of the electrochemical plant is the vessel which must contain the electrolyte. The material which may be used in the construction of the vessel depends both on the particular process and the shape which the vessel is to take. The most commonly used form is a square or rectangularly based tank or cylinder of variable height. Other forms (*e.g.* the filter press type, the vertical frame type with a side opening etc.) are used only when the products of the electrolysis are totally or partially gaseous. A material for the construction of a vessel must resist attack by the electrolyte and the products of the electrolysis; if some corrosion is unavoidable then its products must not contaminate the products of the electrolysis; and it must be as cheap as possible.

Two types of construction may be employed for the vessels. Firstly, the tank may consist of a single material which is resistant to the action of the electrolyte and of the products of the electrolysis; and secondly the tank may be built of two or more materials of which, in general, only one comes into contact with the electrolyte or the products of electrolysis. The materials used in the first case may consist of metals (almost always iron), ceramics (porcelain or earthenware), glass, natural materials (granite, slate, basalt, etc.), synthetic resins or wood. Iron has the advantage that it is easily cleaned, but can only be used for alkaline electrolytes; acid electrolytes can only be used if they have a passivating action. Ceramics, natural materials wood, or synthetic resins are used for acid electrolytes. Ceramics have the disadvantage of fragility and are not suitable for large capacity tanks. Natural stone has the disadvantage that it is practically impossible to make a tank out of one piece and hence cements must be used in the construction. Wood, which is normally taken from resinous timbers, can be used for baths which are not too acid, and its lack of fragility makes it very convenient, but here too some other material must be used between the joints to make them water-tight.

In the second case, a cheap material which is not resistant to the action of the electrolyte is used for the construction of the tank and is lined with a material which will not be attacked. The substances principally used for this purpose are metallic lead, ceramic and painted materials, asphalt and bitumen, ebonite and synthetic resins. Lead gives good results, particularly with electrolytes acidified with sulphuric acid. It has the advantage that the lining may be made perfectly continuous, but has the difficulty of possibly forming a secondary circuit with a consequent current loss. Ceramic materials, particularly in the form of tiles or enamelled metallic vessels, give very good service when the electrolyte is, or becomes, acidified with hydrochloric acid and the process must take place at a relatively high temperature. Tiles have the inconvenience that since they are much smaller than the surface to be covered, some cement must be used in the joints between them which will be resistant to the action of the electrolyte. Asphalt and bitumen are used more rarely because they often release into the solution, substances which unfavorably affect both the electrolytic process and its product.

3. Electrodes and Contacts

This section will not deal with those processes, such as electrolytic refining, electroplating etc., in which the electrodes take part in the electrochemical process as raw materials to be elaborated or as the finished product of the reaction. Only where the electrodes truly form a fixed part of the plant will they be considered, even if the fixed part concerns only one of the two poles. The raw materials for electrodes must have the following characteristics. They must resist the attack of the electrolyte and the products of the electrolysis; if some degree of attack is unavoidable, its products must not contaminate those of the electrolysis; they must have sufficient mechanical strength; must be a good conductor; must have a low overtension for the electrode process for which they are required; and must cost as little as possible. Apart from those for processes which require special electrodes, electrodes may be classified as metallic, oxide, graphite or carbon. The most commonly used metals are iron, nickel, aluminium, lead and platinum, and use is now being made of electrodes of platinum-plated titanium.

Iron can be used only when the electrolyte is alkaline and even then there is a certain unavoidable dissolution of the electrode, particularly with high temperatures and strongly alkaline conditions. When iron would be attacked strongly, it may advantageously be replaced by nickel which is however more expensive. Aluminium is particularly used for certain metallurgical processes where it has been found to be advantageous to strip the product, produced by the cathodic deposition, from the support constituting the true electrode. Lead is an excellent material when the electrolyte is acidified with sulphuric acid and does not contain nitrates;

some chloride can be tolerated but leads to a considerable consumption of electrodes by the anodic formation of lead dioxide through the intermediate production of lead chloride. The maximum amount of Cl^- which can be tolerated depends on the relative costs of preparing new electrodes and of freeing the electrolyte of chloride. Platinum is undoubtedly the most resistent material particularly for anodes, but its extremely high cost greatly limits its use. It is still at present the only anodic material available for use with strongly oxidizing electrolytes and particularly for the preparation of persalts and similar compounds. The use of platinum has greatly increased in the form of plated titanium electrodes.

The oxides used for electrodes and particularly for anodes are essentially lead dioxide and magnetite. The metallic lead electrodes already mentioned are actually lead dioxide electrodes, because when they are used as anodes they become covered with a more or less thick layer of the dioxide. It is however also possible to make compact electrodes of lead dioxide in the form of rods with iron, glass or carbon supports, and to use the rods combined into a comb-shaped electrode, in electrolytes which are particularly rich in Cl^- or NO_3^- ions. Their greatest inconvenience is their fragility. Magnetite electrodes give good service for electrolytes containing chloride in varying degrees of oxidation and also in sulphuric electrolytes when the small quantity of iron which inevitably passes into solution will not upset the reaction. Here too the greatest inconvenience is the fragility as well as the inability to prepare single electrodes of large size.

Another anodic material with excellent properties is carbon in the forms of both retort carbon and graphite. The use of the latter dates from the discovery of a method of preparing it artificially on the industrial scale. Graphite has a low electrical resistance, good mechanical strength and is easily worked. It has a particular resistance to halogen ions but little to oxidants. The average porosity (*cf.* p. 422 *et seq.*) of graphite lies between 19 and 25% and attempts have been made to diminish this by impregnating the electrode with a wide variety of substances: paraffin, bitumen, mineral oil, drying oil, naphthalene, chloronaphthalene, etc. Fairly good results have been obtained with naphthalene and chloronaphthalene, but the problem has still not been fully resolved. The use of definitely porous electrodes has been investigated, whose porosity is utilized to make the electrode function simultaneously as a diaphragm as well[1].

[1] G. W. HEISE *et al.*, *Trans. Electrochem. Soc.*, 75 (1939) 147; 77 (1940) 411; 80 (1941) 121; 88 (1945) 81.

The material of which the electrode is made, particularly influences its form both for mechanical and economic reasons. The most commonly used forms for metallic electrodes consist of plates, which may or may not be pierced, of meshes of more or less heavy wires, of wires wound about a frame of suitable shape, and of brushes etc., and are particularly determined by the cost of the materials. Electrodes made of oxides which cannot be formed into plates are often used as rods either singly or joined into combs. Graphite, because of the ease with which it is worked, can always be made into the most suitable form for the particular electrochemical process for which the electrode is destined.

The electrodes may be inserted into the circuit in two different ways: monopolar or bipolar insertions. The insertion is monopolar when the electrode functions exclusively as either the anode or the cathode; and in this case it is connected with one of the poles of the electric source as

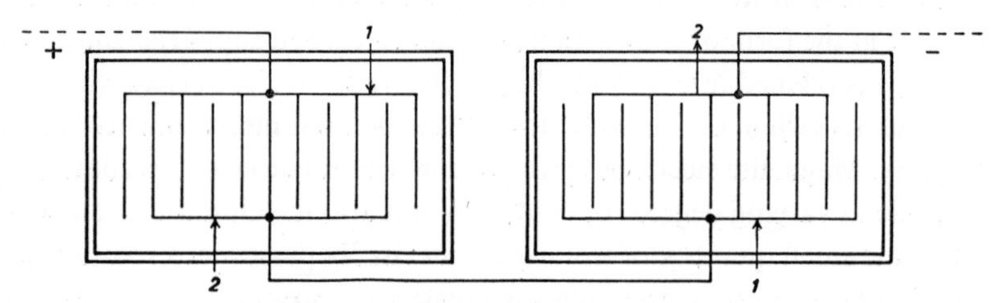

Fig. VII, 1. Unipolar electrode connections

shown in Fig. VII, 1. Because the electric tensions used for electrolytic processes are very small and of the order of a volt, a number of similar cells are usually inserted in series so as to achieve the total electric tension given by the bus-bars for the distribution of the direct current.

With bipolar insertions, however, there are a number of electrodes in each cell which function as anodes on one side and as cathodes on the

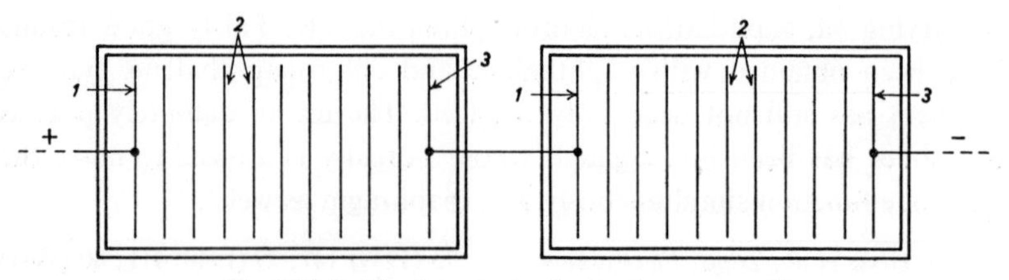

Fig. VII, 2. Bipolar electrodes in the same cell

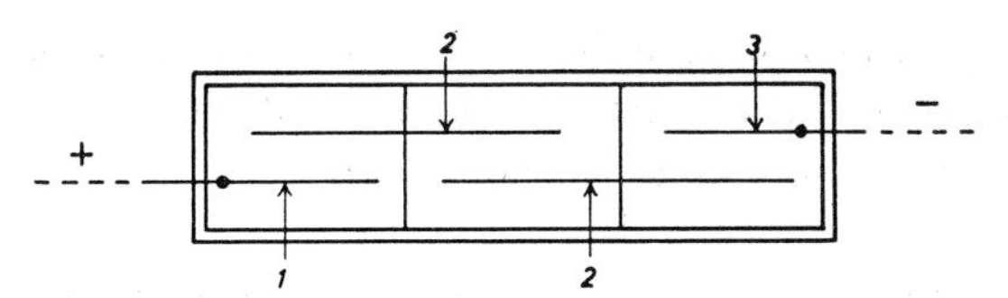

Fig. VII, 3. Bipolar electrodes in different cells

other; apart from the end electrodes, these are not directly connected to the electric source, as can be seen from Fig. VII, 2 and 3. It is as if the cell were divided — which, in some cases it actually is — into a number of unit cells, in which the anode of one cell were short-circuited to the cathode of another. Care must therefore be taken, when a number of electrodes is is immersed in a single tank to function as bipolar electrodes, that they are of sufficient size to occupy the whole cross-section so that the cell is divided into a number of elementary cells; there must be no electrolytic link between different elementary cells, that could give rise to a loss of current.

Since the electric tension required for the electrolytic process must be supplied to every elementary cell, the number of bipolar electrodes placed in each tank is such that they can be directly supplied from the busbars; alternatively, a fraction of this number may be used and then the tanks may be joined in series so as to have the whole electric tension required equal to that available.

Electrodes are usually mounted vertically, and more rarely horizontally; only in special cases are they mounted in an inclined position.

A particularly delicate point in the plant is given by the contacts between the electrodes and the lines. This is because with the high current intensities and low electric tensions used for electrolysis, resistances of the order of $10^{-3} - 10^{-2}$ Ω are sufficient to cause large losses of energy. For example, a current intensity of 1,000 A is not particularly high in ordinary industrial electrolysis plants, but the passage of this current through a resistance of 0.001 Ω will cause a potential drop of 1 V and if the electric tension applied to the cell were of the order of 3 V, 33% of the energy supplied to the cell would be lost as heat dissipated in the resistance. Only rarely is there a single contact between the electrode and the line. More often several contacts are necessary. There are the so-called transition contacts which allow for the operative need for ready replacement of electrodes, there are those due to constructive reasons for the connections between the lines and the true electrodes, consisting of various elements; and there are those due to economic reasons since the true electrodes are of expensive material and hence are connected to the lines by other cheaper materials.

From the point of view of contacts, two fundamental types of electrochemical processes must be distinguished: those in which the electrodes constitute the raw material or the final product of the process and hence must be periodically replaced, and those in which the electrode forms a fixed part of the plant. With processes of the first type, there is a tendency to have very simple contacts, so that the disconnection and replacement of the electrodes can take place as quickly and with as little work as possible. In this case an electrical contact between the electrode and the line which is to some extent imperfect may be tolerated. With processes of the second type, however, the tendency is to make the contacts the best possible, which naturally involves a greater expenditure of time and labour in replacing the electrodes. Although this is necessary only at much greater intervals, their ultimate replacement is unavoidable since they are always consumed to some extent.

4. Diaphragms

The term diaphragm implies a separatory wall which although allowing free passage of the electric current, prevents the products of electrolysis formed at the anode, from coming into contact with those formed at the cathode, so as to avoid, as far as possible, either secondary reactions which would lower the current efficiency, or contaminations of the products which would lessen their value. During electrolysis, in its most general sense, the products formed on the passage of the current may separate in either the gaseous or solid states, or may remain in solution in the electrolyte. Sometimes there are small amounts of solid which are not the principal product of the electrolytic reaction, but rather impurities which could contaminate it. From the point of view of the design of the plant, this case also may be considered in the general case in which it is wished to prevent the solid products formed at one electrode, from coming into contact with the products of the other electrode.

When the products which have to be separated are gaseous or solid, the diaphragm assumes the function of a filter rather than that of a true diaphragm. For gases it is very simple. Often it consists of a metallic division in the form of a plate or mesh, with holes of a slightly smaller diameter than that of the gas bubbles. When such a metallic diaphragm is used, however, it is necessary that the electric tension between the two electrodes is less than twice the decomposition tension of the electrolyte. Otherwise the metallic diaphragm might function as a bipolar electrode. Alternatively, the diaphragm may consist of a fabric which is resistant to the action of the electrolyte. When the products which have to be separated, are solid and it is essential for any reason that they cannot reach the other electrode, the diaphragm may consist of a solid, porous divider; more frequently, however, it consists of a woollen or canvas sack of fairly fine material, enclosing one of the two electrodes. The sack is normally made of an organic material which is as free as possible of ash. In this way, it can eventually be burnt without leaving any foreign residue. Such materials do not have much resistance to the attack of the electrolyte. When it is necessary not to change the diaphragm to frequently, a number of other porous materials are available, which are resistant to either alkaine electrolytes (asbestos) or acid electrolytes (fossil earths, silica powders, glass powder, carborundum, etc.).

The diaphragm actually behaves as such when the products of the electrolysis remain in solution in the electrolyte or are even the electrolyte itself. Such products, by diffusing towards the other electrode, will firstly be lost for the production, and will lower the current efficiency. Secondly, and more importantly they may provoke secondary reactions which will lower still more, and very considerably, the current yield and will also contaminate the products of the electrolysis. Essentially, therefore, the diaphragm has the function of preventing diffusion as far as possible. At the same time, however, it must not have any considerable ohmic resistance and for many processes it must also allow the circulation of the liquid electrolyte. These characteristics, of an ability to prevent diffusion combined with a low ohmic resistance and the capacity to permit circulation of the electrolyte, depend upon two factors: porosity and permeability.

A diaphragm may be considered as a bundle of capillaries extending from one face to the other. The porosity, p is defined as the ratio of the partial volume, v, occupied by the capillary system to the total volume, V, of the diaphragm.

$$p = \frac{v}{V} = \frac{l\Sigma s}{lS}$$

hence

$$\Sigma s = pS \tag{1}$$

Where l is the thickness of the diaphragm, S is its area and Σs is the sum of the right-sectional areas of all the capillaries in the surface S. The term Σs represents the useful area through which diffusion and circulation of electrolyte can take place and through which the passage of electricity also occurs.

The diffusion of a solute through a given plain section of solvent from a zone at greater concentration towards one at lower, is governed by the equation

$$m = \frac{\Delta cst}{l} K_1 \tag{2}$$

where m is the mass of solute which passes in time t from the zone at greater concentration to that at lesser; Δc is the difference in concentrations between the two zones, which are separated by a distance l, where s is the cross-sectional area through which diffusion takes place, and K_1 is a proportionality constant which is a characteristic of the diffusing substance and also takes account of the viscosity of the solvent, and other operational conditions.

For a bundle of capillaries, the value s may be replaced by Σs which in turn is given by equation (1). Thus for unit time and unit total surface area, equation (2) becomes

$$m = \frac{\Delta cp}{l} K_1 \tag{3}$$

For the quantity of substance m which diffuses through a particular diaphragm to be small, the value of the ratio p/l must also be small. This requirement howeveı is in opposition to the other which demands a high conductance from the cell and hence from each of its elements (contacts, electrolyte, diaphragms, etc.). The conductance χ of a diaphragm is, to a first approximation, given by the conductance of the electrolyte within the bundle of capillaries, *i.e.*

$$\chi = \frac{\varkappa \Sigma s}{l} \tag{4}$$

where $\varkappa$ is the specific conductance of the electrolyte and Σs and l have the significance defined above. Replacing the value of Σs with that obtained from equation (1), and referring to unit area of the diaphragm, equation (4) becomes

$$\chi = \frac{\varkappa p}{l}$$

Thus to obtain a high conductance it is necessary for the term p/l to have a high value. The values of the porosity and the thickness must then be so selected that the two requirements are satisfied and a diaphragm is obtained which, whilst having a fairly high conductance, yet presents the greatest possible obstacle to diffusion. The choice of a greater or lesser value for p/l depends largely on the economic conditions governing the local cost of electricity and the value of the electrolysis products.

When a diaphragm must also permit the circulation of the electrolyte it has to be permeable. The permeability of a diaphragm is defined by the quantity of liquid which passes through it under certain conditions, and is expressed by the equation

$$L = \frac{\Delta hSt}{l\eta} K \tag{5}$$

where L is the quantity of liquid which passes through the diaphragm, Δh is the hydrostatic pressure difference across the diaphragm, S is its surface area, t is time, l is thickness, η is the viscosity coefficient of the liquid and K is a constant which is characteristic for the diaphragm. For unit time, area and thickness, equation (5) becomes

$$L = \frac{\Delta h}{\eta} K \tag{6}$$

where K is called the coefficient of permeability. However a diaphragm can also be considered as a bundle of capillaries of mean radius r, through which liquid flows in accordance with Poiseuille's Law. For each individual capillary, the quantity of liquid (L') which flows in unit time is

$$L' = \frac{\pi r^4}{8l} \frac{\Delta h}{\eta} \tag{7}$$

Since the area s of a circle of radius r is πr^2 and assuming, to a first approximation, that the cross-section of each capillary is circular,

$$s = \pi r^2$$

$$r^4 = s^2/\pi^2 \tag{8}$$

Substituting equation (8) into equation (7) gives

$$L' = \frac{s^2}{8\pi l} \frac{\Delta h}{\eta} \tag{9}$$

The quantity of liquid which passes in unit time through 1 cm² of the diaphragm is the sum of the quantities which flow in unit time through all the capillaries found in 1 cm² of the diaphragm:

$$L = \sum_{1\,\mathrm{cm}^2} L' = \sum_{1\,\mathrm{cm}^2} \frac{s^2}{8\pi l}\,\frac{\Delta h}{\eta}$$

$$= \frac{\Delta h}{\eta} \sum_{1\,\mathrm{cm}^2} s^2 \frac{1}{8\pi l} \qquad (10)$$

A comparison of equations (10) and (6) gives

$$K = \sum_{1\,\mathrm{cm}^2} s^2 \frac{1}{8\pi l}$$

i.e. the permeability coefficient of the diaphragm is directly proportional to the sum of the squares of the cross-sectional areas of all the capillaries lying in 1 cm² of the diaphragm's surface, and is inversely proportional to its thickness. The easy passage of liquid through the diaphragm demands a high value for the term $\Sigma s^2/l$, and this in some sense corresponds with the requirement for a high conductance.

It must be emphasized that a term involving the sum of the squares of the cross-sectional areas of each capillary enters into the permeability equation. According to equation (1), for equal porosities this sum term will be less for a greater number of capillaries since each individual value of s will then be smaller. Whereas therefore, the value s of the cross-sectional area of the individual capillaries has no effect on diffusion and conductance — which are controlled by the sum Σs — for permeability, the diminution of the *number* of capillaries, whilst Σs remains constant, will increase the cross-sectional area of the individual capillaries and hence raise the permeability, since $(\Sigma s)^2 > \Sigma s^2$.

The selection of materials for true diaphragms is somewhat difficult, since they act so as to maintain a difference in the chemical composition of the electrolyte on their two faces; thus a material must be chosen which will resist simultaneous chemical attacks of two different kinds. The principal alkali-resistant material used is asbestos either alone or mixed with alumina, barium sulphate or not too acid cement substances. A greater range of acid-resistant materials is available, including especially acidic clays, fossil earths, pure silicic acid, glass powder etc.

The choice of material also depends in part on the type of diaphragm from the mechanical standpoint, depending on whether it is rigid in the form of plates, or flexible or in the form of a paste. In the first case cementing or clayey substances must be used. In the second case, meshes, frames and various fabrics can be used and in the third, a mechanical support must be provided for the diaphragm and this must also be resistant to the action of the electrolyte.

5. Subsidiary Plant

Accessory equipment, which may sometimes be lacking, includes that required for agitation, circulation, heating or cooling of the electrolyte, aspiration of any gases which can not or must not be released into the atmosphere, etc. This means: pumps, piping, gas washers, aspirators, steam generators or water heaters, cooling plant, control instruments etc. As far as materials are concerned, the same considerations apply as in the choice of materials for vessels. To those already mentioned, may be added various types of steels — in general special ones resistant to the corrosive action of various electrolytes — which can be used for parts such as pumps subjected to mechanical strains. However the other details of this part of the plant fall into the fields of technical physics and industrial engineering, rather than into that of electrochemistry, and will not be dealt with further.

CHAPTER VIII

ELECTROMETALLURGY IN AQUEOUS SOLUTIONS

1. Introduction

The origin of the industrial use of wet electrometallurgical processes dates virtually from 1871 with the construction of the first truly large-scale, plant for the electrolytic refining of copper by the Norddeutsche Affinerie at Hamburg. Almost a century has passed since then and the industrial applications of metallurgical electrochemistry using wet processes, have undergone a great development. Not only copper, but antimony, silver, bismuth, cadmium, manganese, nickel, lead, gold and zinc are, or could be, now economically produced and refined by wet electrochemical processes. In view of this wide application the limits of this book will not allow it to deal with all the industrial fields which have so far been achieved; but some examples may be given to illustrate how the industrial processes have derived from theoretical knowledge and laboratory experiments.

For electrometallurgy in particular, as for all industrial electrolytic processes in general, it must be remembered that it is not the cost of the electrochemical treatment alone which determines the economics of the process but the total cost of the whole productive cycle, which in metallurgy is from the mineral or raw material to the commercial metal. Moreover, the electrolysis is not always the most delicate step of this entire cycle. This is obviously true also for other industrial electrochemical processes described in Chapters IX and X.

Electrometallurgy in aqueous solutions has developed along four particular lines which are not all of the same economic importance.

The most important economically, is the electrolytic refining of metals. A metal which contains a certain percentage of impurities is purified from these by anodic dissolution and cathodic deposition. Those impurities which are more noble than the metal, being insoluble, form a deposit which is usually powdery and is called the *anodic sludge*; the less noble impurities remain in solution in the electrolytic bath. The maximum percentage of impurities which can be tolerated in the crude metal used as the anode, depends on the degree of purity required of the cathode metal and on the chemical nature of both the metal to be purified and of the impurities to be eliminated. A refining process is normally required to give a product which is well adherent to the cathode and may even be formed of more or less well defined crystals. These refining processes are also called electrometallurgical processes with soluble anodes.

A second type of procedure which gives an end product comparable to that of refining is the production of metals by the electrolysis of one of their soluble salts obtained more or less directly from minerals. Here, the solution to be electrolyzed must often be drastically purified; particularly it must not contain metals which are more noble than that being produced. Such processes are also called electrometallurgical processes with insoluble anodes.

The third type of procedure aims at the production simply of a fine metallic covering layer on another material to improve the surface characteristics of the latter, from both esthetic and technical viewpoints: by increasing hardness, protecting against corrosion etc. The cathodic deposit in this case must have particularly marked characteristics of adhesion to the support and must show a grain structure which is as micro-crystalline and uniform as possible over all the surface. The electrolytic baths used for these processes vary greatly in composition depending on the metal which must be deposited and on the surface characteristics required. These processes are generally called *electroplating*, and include all those which lead to the direct production of objects, such as seamless tubes, by electrolysis: so-called *electroforming*.

The fourth type differs markedly from the other three in that it aims at producing metallic powders, of a definite granulation and as uniform as possible, for specialized uses. The end product of this fourth type of process must be as little adherent as possible to the cathode, so that it may be easily removed if indeed it does not separate spontaneously on the bottom of the tank.

One of the most important factors determining the characteristics of a process for the electrolytic reduction of metals is the purity desired of the end product. This factor must be considered together with the working cost which includes the interest on the capital required and the depretiation of the plant, particularly if the production or refining process is in competition with other techniques. Similar considerations apply to the preparation of powdered metals. In these three types of process the current and energetic efficiencies must be as high as possible and there is a tendency to operate the cells with the highest possible current density compatible with the characteristics of the process and the end product. The use of a high current density increases the productive capacity of the plant which can be of smaller dimensions and hence reduce the initial outlay and the interest on the capital for the materials immobilized in the work: so markedly diminishing production costs. However, an increase in current density is always accompanied by a fall in energetic efficiency and sometimes by a deterioration in the end product so that it is necessary to strike a balance between these factors.

In galvanotechniques, however, and in particular in electroplating, the cost factor tends to take second place to the quality of the final result and hence lower current and energetic efficiencies are tolerated in order to obtain products of the required characteristics. In order to diminish costs still further, an attempt is made to increase the energetic efficiencies by diminishing to a minimum the fraction of the electric tension absorbed by ohmic resistance and useless electrode polarizations; in industrial practice these latter are themselves often simply considered as ohmic resistances which have to be overcome. Recourse is sometimes made, for this reason, to the addition of inert electrolytes (*i.e.* electrolytes which do not take part in the electrolytic process) to the baths to lower the ohmic resistance of the electrolyte and diminish that fraction of the tension IR, and hence the electric tension applied to the electrodes and finally the energy consumed by the cell. An increase in the conductance can also be achieved by raising the temperature, but in this case a simultaneous discharge of hydrogen becomes possible through the lowering of its overtension and other cations which may be present in the bath may also be deposited and contaminate the final product. It may be noted that in the presence of other cations in the electrolytic liquid, theoretically all the species would participate in the discharge process in proportions determined by their respective standard electric tensions (*cf.* Chapter IV, p. 243 *et seq.* and 265 *et seq.*), concentrations and overtensions. Since H^+ ions are always present in aqueous solutions, their simultaneous discharge is inevitable. During the electrolysis of copper sulphate in 1 N acid solution, for example, it can be calculated that the amount of hydrogen which separates is of the order of 10^{-7} % of the weight of copper separated. An accurate analysis of the electrolytic copper produced under these conditions will disclose the presence of about 10^{-6} % of hydrogen which in view of the difficulties inherent in such analyses is in satisfactory agreement with the theoretical predictions. This observation is very important because the separated hydrogen which remains dissolved in the metal may affect both the structure and properties of the deposit. Electrolytic baths are, therefore, often maintained at constant pH by the addition of buffering substances to avoid changes in the properties of the deposit as a consequence of changes in the acidity of the bath. The dissolved hydrogen can usually be eliminated by heating.

In processes with soluble anodes, it is sometimes possible to lower the electric tension applied to the cell, by diminishing the anodic polarization by adding substances, *e.g.* Cl^+ ions, which facilitate the dissolution of the anode metal and hinder passivation (*cf.* p. 503 *et seq.*).

2. Types of Metallic Deposits[1]

Electrometallurgy in aqueous solutions is generally aimed at the production of metallic deposits which are compact and adherent to the cathode, apart from those cases in which it is wished to prepare the metal in the powdered state. The type of deposit depends on its so-called *structural characteristics*, a term which implies the type of crystallization (number of crystals per unit area, their orientation and mutual connection) and the relationship of the deposit to the metallic base. The final properties of the deposit (porosity, rugosity, hardness, resistance to various stresses such as stretching, torsion etc., response to changes in temperature, purity) depend fundamentally on the structural characteristics. Other factors primarily concerned with the production cycle (current efficiency, ease of collection, etc.) also depend on the structural characteristics. It is thus apparent that a knowledge of these is essential in order to carry out electrolyses in such a way as to obtain the desired type of deposit. Despite this, however, the metallic deposit is not always of the form desired and the various metals present a different behaviour which depends primarily on the characteristic properties of the metal itself (cohesive strength, lattice structure, tendency towards irreversible electrochemical behaviour, overtension of hydrogen on it, etc.) and also on the particular conditions of the electrolysis.

[1] Detailed studies of the cathodic growth of metals may be found in G. I. FINCH, H. WILMAN and L. YANG, *Discussions Faraday Soc.*, No. 1, Electrode Processes, (1947) 144; H. FISCHER, *Z. Elektrochem.*, 59 (1955) 612 and the important monograph of H. FISCHER, *Elektrolytische Abscheidung und Elektrokristallisation von Metallen*, Springer Verlag, Berlin, 1954.

These conditions may be summarized as follows: rate of renewal of metallic ions in the cathode–electrolyte interface, lateral mobility of the discharged metallic atoms on the cathode surface before they are fixed in the crystal lattice, nature of the electrolyte, presence of other components in the solution (of both an electrolytic and a colloidal nature), cathodic polarization, and the nature and crystalline orientations of the support (for very thin deposits)[1].

The properties which affect the type of structure, which are dependent on the characteristics of the metal being deposited, cannot be altered and hence will not be discussed. But it will be interesting to examine the effect of conditions of electrolysis, which within certain limits, may be altered to influence the type of structure.

The first of these conditions is in turn determined in practice by the current density, the concentration of the electrolyte, the agitation and the temperature. The electrolytic separation of a metal occurs when the electrode immersed in the solution is cathodically polarized so that the electrode–electrolyte tension is slightly more negative than that at equilibrium under the experimental conditions, plus any overtension. However, the mechanism by which the discharge of the cations and the separation of metal in the solid state occurs, is not yet known for certain, so that only certain general principles may be discussed. Immediately before the discharge reaction, the cations must lie at the liquid surface in contact with the electrode, *i.e.* in the so-called cathodic film. The cation is normally hydrated (or complexed) so that between the initial state of the cation and the final state of the metallic atoms forming part of a crystal lattice, there must be a certain number of intermediary stages in which various reactions occur: dehydration, discharge and passage from the cathodic surface to the crystal lattice of the metal. When the metal reaches the final state, the cathodic film becomes depleted of cations on the one hand, whereas on the other it is enriched by the arrival of new cations from the bulk of the solution; these move under the influence of both the electrical field and the difference in concentration between the bulk of the solution and the cathodic film. This cathodic film has a thickness of the order of tenths of a micron. In this very thin layer the electrical field is certainly very much more intense than that in the bulk of the solu-

[1] The study of metallic structure dealt with in this section is limited to deposits of a thickness greater than 0.01 mm on which the nature of the support has only a limited effect.

tion, so that the effective velocity of the cations which traverse the cathodic film is also very much greater than that of those in the bulk of the solution. Moreover, the composition of the cathodic film is very probably not constant between one extreme and the other and always in dynamic equilibrium with the bulk of the solution. All these factors, together with the others already mentioned, affect the structure of the metallic deposit and the lack of knowledge about them makes certain aspects of electrolysis obscure.

The deposited metal gives rise to four main types of structure:

1. Single crystals, or crystalline aggregates, which are isolated, well developed, and often orientated along the current-lines.

2. Deposits orientated on the base in continuation of the crystals of the cathodic support, normally with a large-grained compact structure.

3. Deposits orientated in the field in the form of compact bundles of fine fibres parallel to the current-lines with indistinct surfaces between the crystallites.

4. Unorientated deposits of a very fine-grained disordered structure without visible separation of individual crystallites.

Deposition occurs in one or other form, passing from type 1 to 2 to 3 etc. as the deposition process is made more difficult by one or other of the following means:

(a) Increasing hindrance of the desolvation or destruction of complex ions.

(b) Increasing hindrance to traversing of the double layer, *e.g.* by inhibition.

(c) Low surface mobility of the metallic atoms.

Under conditions of a zero net total current, these causes are manifested as a particularly low exchange current density; during cathodic deposition they are manifested as overtensions. The cathodic deposition process of a metal may, in some respects, be assimilated to a process of crystallization and has in fact been called electrocrystallization[1]. It occurs in two distinct steps: the formation of nuclei, and their growth and development into more or less well-formed crystals. These two phases of the crystallization process occur independently of each other and it is the greater or lesser rate of one of the partial processes compared to the other which

[1] A very clear review of electrocrystallization is that of H. FISCHER, *Z. Elektrochem.*, 59 (1955) 612.

forms the ultimate determining factor controlling the type of metallic cathodic deposit[1].

Metals may be classified as *normal*, *intermediate*, or *inert*, depending on their electrochemical characteristics, including their behaviour in cathodic electrocrystallization. Normal metals (*e.g.* Pb, Tl, Cd) rapidly assume the equilibrium electric tension, have a high exchange current density, show very low overtensions and have low heats of fusion and sublimation corresponding to a high surface mobility of the atoms. These characteristics show that the electrocrystallization process is not significantly hindered and the corresponding cathodic deposit is of type 1 which with growing inhibition may pass into type 2. At the other extreme, the inert metals such as those of the Pt group, have the opposite characteristics: slow establishment of the equilibrium electric tension, low exchange current densities, high overtensions, high heats of fusion and sublimation and a strong tendency to form complexes. These characteristics correspond to a strongly hindered electrocrystallization process and the cathodic deposit is of type 3 or 4. The intermediate metals (*e.g.* Zn, Ag, Cu) have intermediary characteristics and crystallize in deposits of type 2 or 3.

Observations of the form of growth lead to the conclusion that the nuclei which initiate electrocrystallization are the growth layer, whose dimensions are of the order of 10^{-5}–10^{-4} cm in thickness and may vary very much (10^{-5}–10^{-2} cm) in extent, from which the discharged atoms are gradually inserted into the crystal lattice of the underlying metal.

Nothing can be said securely on the mechanism of formation of the first nucleus in the growth layer. Probably, in the very earliest step only the top surfaces of those nuclei are active which have a true current density which is very high and much greater than the apparent average density shown by the geometric dimensions of the whole cathode surface. This leads to a rapid exhaustion of the reserve of ions in the cathodic film and hence stops the development of the nuclei in the direction of thickness; this is also due to other reactions of both a primary and secondary nature such as the simultaneous evolution of hydrogen and the formation of hydroxides which block the growth surface. The sides of the growth surface,

[1] The orientation of the crystals in relation to the crystallographic and surface structure of the base metal should be considered as well as the interconnections of the individual crystallites. These two characteristics, although fundamental to the type of deposit, are difficult to influence from the exterior, to change the type of deposit itself and so will not be dealt with further.

however, remain active and it therefore spreads superficially so long as there is no exhaustion or blockage of the type described. As soon as the electrocrystallization process is interrupted during the inactive phase, the ionic exhaustion is compensated for and the electrocrystallization may start again without any new formation of nuclei in the growth layer; this would require activation energy and hence, in electrical terms, a greater polarization than that needed for the simple development of the growth layer.

The factors described previously as influencing the structure of the deposit, act by affecting the rate of formation of nuclei and their growth. In general, therefore, microcrystalline deposits are favoured by conditions which help the formation of new nuclei, whilst deposits of large, more or less well-formed crystals are aided by conditions which favour the expansion of the growth layer. Increasing the current density favours the formation of new nuclei, both directly by increasing the number of discharged ions per unit area, and indirectly because the increase in current density affects various other factors which also determine the nature of the deposit. An increase in the current density leads to a fall in the concentration of cations in the cathodic film so that the individual ions lie at greater distances from the growth surfaces of crystals which have already been formed. The field intensity across the cathodic film also increases so that the cations arriving from the bulk of the solution pass through the film more easily; and the cathodic polarization also increases. All this means that the formation of new nuclei is facilitated and the crystalline grains of the deposit will be finer. Once the value of the limiting current is passed, however, there will be no further decrease in grain size; the simultaneous evolution of hydrogen, which normally occurs in this case, makes the deposit rather porous or spongey and often poorly adherent to the cathode or even powdery.

These considerations apply equally well to the effect of concentration which is normally opposed to that of current density: an increase in the concentration in the bulk of the solution diminishes the thickness of the cathodic film, increases the concentration within it by diffusion, and lowers the cathodic polarization so that the growth of the existing layer is favoured, with the formation of more or less large crystals.

Agitation of the electrolytic bath and increase in the temperature act in the same way, as factors facilitating diffusion and hence opposing the exhaustion of the cathodic film. The increase in the temperature acts, in particular, also by lowering any overtensions and directly affecting both the rate of formation of new nuclei and the rate of growth of the crystals. This latter effect, however, becomes predominant partly because the lateral mobility of the metallic atoms in the growth layer is also increased so that they may the more easily reach the active points of growth of the crystals. Finally, increase of temperature acts, in a way which cannot always be predicted *a priori*, by altering the degree of dissociation or the activity of the electrolyte, by shifting the equilibrium electric tension and by altering the conditions under which any colloids (metallic hydroxides) may be formed or the influence of any colloid already present.

The nature of the anions and the valency of the cations also have an effect which is sometimes obscure. Deposits of lead, silver, cadmium or zinc, for example, from a fluosilicate solution are notably finer grained than those from nitrate solutions whereas deposits of iron from chloride solutions have a coarser grain than those from sulphate solutions. Lead deposited from solutions of Pb^{4+} ions is spongey whereas from solutions of Pb^{2+} ions it consists of relatively large and well-formed crystals.

The electrolyte becomes particularly important when the metal being deposited is present as a complex anion (cyanides, tartrates etc.). Deposits obtained from complex salts are always microcrystalline even when the metal is deposited from a simple salt preferentially as large well developed individual crystals. This behaviour is in agreement with what has been said earlier (*cf.* Chapter IV p. 428 *et seq.*); the more stable the complex, and hence the stronger the bonds between the central ion and the ligand, the more efficient will be the hindrance to electrocrystallization; this will thus be inhibited and give rise to deposits of type 4 which will be more microcrystalline as the inhibition is more efficient. The mechanism of cathodic deposition of metals from complex anions is not yet completely clear. It is improbable that this mechanism involves the direct discharge of metallic cations produced by the dissociation of the complex anions in a similar manner to the electrolysis of simple salts, since this should lead to similar deposits from solutions of the complex ion and from solutions of simple salts of such a dilution as to give the same concentration of the metallic ion as in the complexed ion solution. This is not found in practice. It might be objected, that the conductance of the bath is not the same in the two cases and that the factors which are most important in the cathodic film — ionic concentration, film thickness, electrical field, etc. — are unknown. Another suggestion has been made to interpret the mechanism of the microcrystalline deposition from complex salts. According to this, independently of the particular mechanism of discharge of the metal, the new nuclei which are continuously formed are, at least in part, almost immediately blocked by hydrogen which in view of the high actual current density must certainly take part in the cathodic discharge process. There will thus be a continuous formation of new crystalline nuclei, *i.e.* a microcrystalline deposition. This second interpretation is partly based on observations of the effects, on the electrolytic deposit, of adding certain foreign substances to the electrolyte. Many substances, generally of an organic nature, with a high molecular weight and often truly colloidal, have a marked effect on the physical and chemical characteristics of the deposit and are employed industrially. They always reduce the grain size of the deposit, *i.e.* hinder the growth of the crystals and favour the continuous formation of new crystalline nuclei. The colloids probably act by a mechanism involving a surface adsorption on to the nuclei of the growth layer[1]. Various experimental observations support this. Those colloids which have the most marked action in reducing the grain size of the deposit are actually those which are adsorbed on the surface and so can act as protective colloids. The surface adsorption leads to a partial covering of the cathode with a corresponding increase in the effective current density at the uncovered points and a consequent increase in the polarization. This increase in the cathodic polarization leads, in turn, to more favourable conditions for the continuous formation of new nuclei which requires a more negative electric tension. In effect, an increase in the cathodic polarization is observed experimentally as a consequence of the addition, or formation in the bath, of colloidal substances.

Moreover, it has been observed that the deposited metal often contains a certain, appreciable, amount of the added colloid which is possibly a further confirmation of this hypothesis. In the electrolysis of a solution of copper sulphate containing 25 % $CuSO_4 \cdot 5\,H_2O$ and 0.5 % gelatine at pH 3, with a current density of about 10 mA/cm^2, the deposited copper contains 2.93 % of gelatine. Such foreign substances remaining in the cathodically deposited metal may often alter its properties making it fragile and sometimes causing internal strains etc. In some cases, it is not possible to detect in the cathodically deposited metal, any trace of the colloid added to the bath. It is assumed, here, that the colloid acts as a

[1] F. MÜLLER, *Kolloid Z.*, 100 (1942) 159.

sort of diaphragm, more or less adherent to the cathode, and controlling the discharge phenomenon and the growth of the crystals. Many other theories have been proposed to explain the effect of colloidal substances, so that their mode of action cannot be considered to be completely clear; nevertheless, the interpretation given above seems to be highly probable.

Another type of addition which is sometimes made, particularly to plating baths, is of inert electrolytes, to increase the conductance of the bath. However, the addition of inert electrolytes may also affect the structure of the cathodic deposit. Above all, their presence acts on the activity of the metallic cations to be discharged, and often — if these cations are produced by the dissociation of an electrolyte of medium strength, like most salts of heavy metals — the degree of dissociation, *i.e.* the ionic concentration, may be varied. Moreover, the addition of a strongly dissociated electrolyte acts also on the conductance of the cathodic film and on the density of the bath which will change less rapidly than it would in the absence of an added inert electrolyte. If now the electrolyte has a common ion an increase in the polarization will often occur. Thus, an inert electrolyte acts indirectly on the type of deposit by altering certain conditions which in turn affect it directly.

Sometimes specific actions are found with small additions with respect to some characteristics: hardness, surface brightness etc. A common additive is strong mineral acids which besides their general action as inert electrolytes also have a specific action due to the greater probability of the simultaneous separation of hydrogen which may lead to an undesirable, porous or spongey deposit. It is thus necessary to use optimum conditions of pH since this factor also may affect the type and structure of the deposit.

A final factor which may affect the structure is the electrode polarization. When this increases, a finer grained deposit is obtained since the formation of new nuclei is facilitated; this requires a greater activation energy and hence a more negative polarization. In this respect, it is understandable that metals which have high overtensions, and in particular those of the iron group, are deposited as adherent and compact layers with crystalline grains which are so fine that they may often only be detected by X-ray analysis (type 4 structure); however, cations which do not have significant overtensions give rise to individual, well-formed crystals, *e.g.* Pb, Tl, Cd, etc. The relationship between polarization and

TABLE VIII, 1

THE EFFECTS OF ELECTROLYSIS CONDITIONS ON THE STRUCTURE OF THE CATHODIC DEPOSIT

<table>
<tr><th rowspan="2">Conditions of electrolysis</th><th colspan="4">Structure</th></tr>
<tr><th>1</th><th>2</th><th>3</th><th>4</th></tr>
<tr><td>Increase in current density</td><td></td><td colspan="2">⟶</td><td></td></tr>
<tr><td>Increase in concentration</td><td></td><td colspan="2">⟵</td><td></td></tr>
<tr><td>Increase in agitation</td><td></td><td colspan="2">⟵</td><td></td></tr>
<tr><td>Increase in temperature</td><td></td><td colspan="2">⟵</td><td></td></tr>
<tr><td>Increase in polarization</td><td></td><td colspan="2">⟶</td><td></td></tr>
<tr><td>Addition of colloids</td><td></td><td colspan="2">⟶</td><td></td></tr>
<tr><td>Addition of inert electrolytes</td><td></td><td colspan="2">⟶</td><td></td></tr>
</table>

structure of deposits is not, however, absolutely strict. Various theories have been proposed to explain this which take account of any activation energies but reference should be made to specialized texts on electrometallurgy for these.

Table VIII, 1 summarizes the general effects of the various discharge conditions on the type of structure of the deposits.

3. Electrolytic Refining of Copper; Reactions

The most important application of copper is in electrical engineering. For this purpose it must be extremely pure — better than 99.5 % — which is difficult to achieve by purely thermal metallurgical techniques, whereas electrolytically it is easy to obtain 99.90–99.98 % pure copper.

Copper gives rise to two series of salts in which it is mono- and divalent respectively. In the presence of metallic copper an equilibrium is set up

$$Cu^{2+} + Cu \rightleftharpoons 2\,Cu^{+} \tag{1}$$

which is strongly displaced towards the left (*cf.* p. 260 *et seq.*) at room temperature. With rising temperature, the equilibrium moves increasingly towards the right. A solution of a cupric salt containing a metallic copper electrode always contains Cu^{+} ions. The following anodic processes are thus possible, arranged in order of their standard electric tensions shown at the side.

$$\text{(I)} \quad Cu^+ \rightarrow Cu^{2+} + e^- \qquad + 0.153 \text{ V}$$

$$\text{(II)} \quad Cu \rightarrow Cu^{2+} + 2e^- \qquad + 0.337 \text{ V}$$

$$\text{(III)} \quad Cu \rightarrow Cu^+ + e^- \qquad + 0.52 \text{ V}$$

At the start of electrolysis when the quantity of Cu^+ ions present is very small, process (I) is not possible and in practice process (II) occurs almost exclusively, leading to a marked increase in the concentration of Cu^{2+} ions in the immediate neighbourhood of the anode. The increase in the concentration of Cu^{2+} ions has a double effect. Firstly, because of equilibrium (1) Cu^+ ions may be formed by a secondary reaction between the Cu^{2+} ions and the metallic copper of the electrode; secondly, since the increase in concentration of Cu^{2+} ions shifts the electric tension towards more positive values, process (III) may start forming primarily Cu^+ ions, so that the anodic dissolution of the copper is accompanied by a slight overtension, as shown in Fig. VIII, 1, particularly if the solution contains gelatine. Under practical conditions of electrolysis with a current density which does not tend towards zero, the formation of Cu^{2+} ions at the anode is accompanied by a simultaneous formation of Cu^+ ions in an amount greater than that corresponding to equilibrium (1) in the bulk of the solution (*cf.* p. 260 *et seq.*). Hence, when the Cu^+ ions move away from the anode they pass into solution at a concentration greater than their equilibrium (1) one, so that they react according to reaction (1) from right to left precipitating metallic

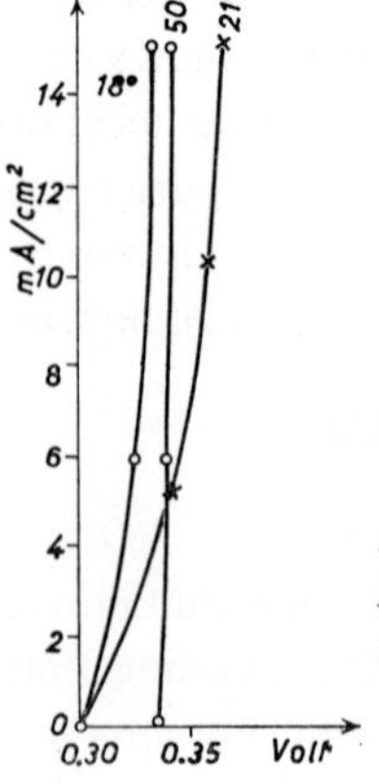

Fig. VIII, 1.
Polarization during copper electrolysis: o — o without gelatine; x — x with 0.002 g/l gelatine

copper in a powdered form. This is deposited on the bottom until equilibrium is obtained again. Part of the Cu^+ ions formed at the anode disappear in this way. Another part is oxidized by atmospheric oxygen when a free acid is present, according to the reaction

$$2\,Cu_2SO_4 + O_2 + 2\,H_2SO_4 \rightarrow 4\,CuSO_4 + 2\,H_2O \tag{2}$$

If the solution does not contain free sulphuric acid, the cuprous sulphate hydrolyzes, according to the reaction

$$Cu_2SO_4 + H_2O \rightarrow H_2SO_4 + Cu_2O \tag{3}$$

precipitating solid cuprous oxide. These two reactions, however, lower the concentration of Cu^+ ions below that of equilibrium. At the cathode, the three effective electric tensions: of oxido-reduction and discharge of the Cu^+ and Cu^{2+} ions – which at equilibrium between the Cu^+ and Cu^{2+} ions are the same – now become different and the electrochemical cathode reaction tends to re-establish the equilibrium by reducing some of the Cu^{2+} ions to Cu^+. The principal cathode reaction is, however, always the discharge of the Cu^{2+} ions, which also occurs with a slight overtension and is favoured by a low ratio of the concentrations $[Cu^+] / [Cu^{2+}]$. If this ratio falls below the equilibrium value, the reduction $Cu^{2+} + e^- \rightarrow Cu^+$ occurs, but at high current density this is overshadowed by the discharge $Cu^{2+} + 2\,e^- \rightarrow Cu$. In fact, only by electrolyzing at high temperature in a neutral medium and with a low current density is it possible to obtain on a platinum cathode, the exclusive reduction of Cu^{2+} ions to Cu^+ with the successive separation of the oxide by hydrolysis according to reaction (3). The *J–U* diagram under suitable conditions, shows two distinct steps, the first of which is probably due to the $Cu^{2+} + e^- \rightarrow Cu^+$ reaction, although the standard electric tension for this is considerably more negative than that for the $Cu^{2+} + 2\,e^- \rightarrow Cu$ reaction. This may be understood by considering that the effective electric tensions for these two processes become the same when there is equilibrium between the Cu^{2+} and Cu^+ ions, but that the electric tension for the $Cu^{2+} + e^- \rightarrow Cu^+$ reaction becomes more positive when the concentration of the Cu^+ ions falls below the equilibrium value, through reactions (2) and (3).

On raising the current density, the process of discharge of the Cu^{2+} ions takes place simultaneously with the process of reduction, $Cu^{2+} + 2\,e^- \rightarrow Cu^+$, and this occurs to a greater extent as the current density increases. The increase in the current density is naturally limited by the start of the simultaneous discharge of the H^+ ion which makes the catholyte more alkaline and facilitates the hydrolysis of the cuprous sulphate. The simultaneous discharge of hydrogen can also make the deposit porous or even spongey.

The theoretically necessary conditions for the best electrolysis may be deduced from this argument so as to obtain the best current yield. These are:

1. a fairly high cathodic current density, which is not so high however as to cause the simultaneous discharge of the H^+ ion;
2. not too high a temperature;
3. not too high a copper sulphate concentration;
4. an acidified electrolyte.

Some of these conditions, however, contradict other requirements which must also be respected to obtain a good energetic efficiency. Thus, a low temperature increases the current efficiency but diminishes the energetic efficiency since the ohmic resistance of the electrolyte is greater at low temperature than at high and the electrode polarizations also diminish with increase in temperature. The same applies to the concentration, too dilute a solution also has a considerable ohmic resistance. However, a high concentration of copper sulphate would again diminish the conductance. Normally, the conductance of the bath is controlled by acidification, *i.e.* by the addition of an electrolyte which although not taking part in the electrolysis process will lower the resistance of the bath. Acidification also serves to bring into solution part of the powdery copper separated in the bulk of the bath by reaction (1). This dissolution takes place with a consumption of atmospheric oxygen, according to reaction (4).

$$Cu + H_2SO_4 + \tfrac{1}{2}\,O_2 \rightarrow CuSO_4 + H_2O \qquad (4)$$

4. Electrolytic Refining of Copper; Anodes and Electrolyte

The raw material for the electrolytic refining of copper is the crude metal from converters or blast furnaces containing from 98–99.5 % copper. The impurities consist of silver, gold, lead, bismuth, arsenic, antimony, iron, nickel, selenium, tellurium, oxygen, sulphur, zinc, cobalt, platinum, and traces of other elements. Of these, the more noble (Ag, Au, Pt) are not anodically attacked and pass undissolved into the anodic sludge. The selenium and tellurium also remain undissolved as Ag_2Se and Ag_2Te; selenium and tellurium in excess of the silver combine with copper to form the equally insoluble CuSe and CuTe. Cuprous sulphide Cu_2S and, to some extent, cuprous oxide Cu_2O also remain undissolved. Other elements (Fe, Ni, Co, Zn) pass into solution but, being much less noble, are not cathodically deposited. Lead, in a sulphuric electrolyte, gives the insoluble sulphate. The arsenic, antimony and bismuth remain and are the most difficult to eliminate and also the most damaging. The standard electric tensions of these three elements are very close to that of copper, as shown in Table VIII,2.

TABLE VIII,2

STANDARD ELECTRIC TENSIONS[1] OF ANTIMONY, BISMUTH, ARSENIC AND COPPER

Element	U_0
Sb/Sb^{3+}	+ 0.1
Bi/Bi^{3+}	+ 0.2
As/As^{3+}	+ 0.3
Cu/Cu^{2+}	+ 0.34

It would thus be extremely difficult to separate copper from antimony, bismuth and arsenic by purely electrolytic means. However, good results can be obtained by selecting a suitable electrolyte so that the greater part of these three elements passes into the anodic sludge as poorly soluble compounds.

The most suitable electrolyte for this is a solution of copper sulphate containing about 40 g/l of copper and 200 g/l of free sulphuric acid, and this also has other fundamental advantages. The antimony and bismuth, in this solution, form poorly soluble compounds which collect in the anodic sludge.

The arsenic also collects here, particularly if some antimony is present. It is notable that a certain significant amount of arsenic oxidizes to arsenic

[1] The values of the standard electric tensions serve only as a guide since the effective ionic activities of the Sb^{3+}, Bi^{3+}, As^{3+} and Cu^{2+} ions in the electrolytic bath are unknown.

acid; arsenic is deposited electrolytically from this acid only with difficulty partly because the arsenic acid forms, with the bases present in the electrolyte, arsenates which tend to collect in the anodic sludge.

The other two impurities which largely collect in the electrolyte are nickel and iron. The former interferes because it markedly lowers the conductance of the bath. Even 10 g/l of nickel will increase the resistance of the electrolyte by about 7.5 %, reducing the energetic efficiency at the same time. The iron, on the other hand, lowers particularly the current efficiency because it consumes electricity in oxidizing from the di- to the trivalent state at the anode and reducing from the tri- to the divalent state at the cathode.

The other advantages which sulphuric acid has as an inert electrolyte are as follows:

1. It increases the conductance of the bath
2. It is inexpensive
3. It strongly inhibits the hydrolysis of the cuprous sulphate
4. It is nonvolatile and may be used at high concentrations and temperatures
5. It does not attack lead, so that it is possible to use this for plant construction

The acidification of the electrolyte with sulphuric acid, *i.e.* with another electrolyte having a common ion, increases the cathodic polarization. This increase is however largely compensated for by the diminution in the resistance of the cell and hence in the ohmic tension given by the product IR; thus the energetic efficiency of the cell is finally improved.

As the electrolysis proceeds, the quantity of free sulphuric acid falls because it is fixed by all the less noble metals which remain in solution as well as by the Cu^{+} ions which are converted into Cu^{2+} ions by atmospheric oxygen. The fall in concentration of the sulphuric acid causes an immediate increase in the resistance and may also lead to the hydrolysis of cuprous sulphate with precipitation of cuprous oxide. To avoid this difficulty an insoluble anode is often inserted; the anodic process on this is the evolution of oxygen with the simultaneous regeneration of free sulphuric acid. Otherwise, the acid must be directly replaced.

The only inconvenience of sulphuric acid is that copper dissolves in this essentially as the divalent ion, which implies a current consumption which is double that which would occur if the electrolysis were conducted in an electrolyte solution containing Cu^{+} ions. However, various attempts in this direction have not led to concrete results, so that at present the use of a sulphuric electrolyte is unchallenged.

The concentration of Cu^{2+} ions must not be too high in order to obtain a good current efficiency. However, since electrolysis in practice is carried out in a strongly acid electrolyte, the concentration cannot fall below a certain limit in order to avoid the simultaneous discharge of H^{+} ions; this would diminish the current efficiency and spoil the mechanical characteristics of the deposited copper. This requirement is particularly felt when the refining is applied to anodes which are relatively rich in arsenic, which remains in solution to a considerable extent. An excessive quantity of arsenic in the bath not only leads to a danger of contaminating the copper by electrolytic deposition and mechanical occlusion of the anodic sludge and the electrolyte, but also lowers the voertension of hydrogen on the copper cathode and thus facilitates its discharge. The maximum limit which can be tolerated without damaging the cathodic deposit of copper is about 17 g/l of arsenic in the bath. The electrolyte must thus be continuously examined for the presence of arsenic.

As the electrolysis proceeds, the electrolyte becomes increasingly impure and also increasingly concentrated in copper sulphate as a consequence of reaction (4). The periodic replacement of part of the electrolyte with fresh solution keeps the concentration of impurities below the maximum limit for a certain period of time; but at more or less regular intervals the electrolyte must be completely renewed. The number of times which a bath may be used before it is necessary to replace part of the electrolyte with fresh solution or after which the exhausted electrolyte must be sent for recovery, depends on the quantity of impurities which collect in it. A small amount of colloidal substances such as gelatine and gums is normally added to the electrolyte to improve the cathodic deposit by making it flat and smooth and avoiding the formation of short-circuits between adjacent electrodes through the irregular growth of the cathode.

Finally, it should be noted that the presence of Cl^{-} ions in the electrolyte in a concentration of about 0.2–0.3 g/l both facilitates the anodic attack and contributes to making the antimony and bismuth insoluble.

The temperature of the electrolyte is maintained at about 50–55° C as a compromise between the various requirements already discussed: conductance, hydrolysis and polarization. The electrolyte is kept in continuous circulation from one tank to another by a system of pumps to avoid the crystallization of copper sulphate at the anode. This circulation must be arranged so that the anodic sludge is not disturbed, to avoid depositing it on the cathodes. The capacity of the circulation will naturally depend on the size of the tanks.

5. Electrolytic Refining of Copper; Cathodes and Processes

The cathodes consist of sheets of electrolytically prepared copper, produced in separate baths from electrolytes and anodes of highly purified copper so as to reduce to a minimum the formation of anodic sludge and the concentration of less noble impurities in the electrolyte. The electrodes in these baths are more widely spaced than in the normal production baths. Under such conditions sheets of uniform thickness are obtained which are suitable for use as cathodes.

The progress of the electrolysis during refining depends on the arrangement of the electrodes. The most widely-used type has a unipolar arrangement, joining a number of tanks in series in such a way that continuous electric tension from the bus-bars may be applied directly to the end tanks. The tanks are normally constructed of wood or concrete and lined inside with lead, lead hardened with 6 % antimony or acid resistant asphalt. The anodes are first suspended in the tanks with their centres about 10 cm apart. The sheets of electrolytic copper used as the cathodes, are suspended exactly mid-way between the anodes. Each tank normally contains n anodes for $n + 1$ cathodes so that all the anodes will be uniformly attacked on both faces. The electric tension applied to each individual tank lies between 0.25 and 0.5 V and the mean current density is about 200 A/m^2. Lowering the current density leads to less loss of precious metals on the cathode, diminishes the cost of the electrolysis by lowering the electric tension applied to the cell and diminishes the probability of the simultaneous separation of impurities or hydrogen on the cathode. However the general costs rise as discussed on p. 427 *et seq*. The choice of the most suitable current density thus also depends on the composition of the anodes and the local cost of electricity. The current efficiency for this type of arrangement is better than 0.90 and may reach 0.98. The energy consumption for each metric ton of copper refined lies at about 250 kWh.

Cells with bipolar electrodes generally contain an insoluble anode, a pure copper cathode and a number of plates of crude copper which are not connected to the electrical supply and function as the bipolar electrodes. They act as cathodes with their face directed towards the anode end, and as anodes with their other face. In these electrodes, it is as if there were a transformation of crude copper into pure copper without any practical change in weight of the individual electrodes; at the end of the electrolysis each plate has maintained unchanged its dimensions, has lost in weight only the impurities originally present and is composed of electrolytic copper. This type of connection leads, as a first consequence, to a diminution in the current efficiency which is markedly less than 1 because, since the electrodes cannot be made of the same size and shape as the tank, part of the current passes directly from the insoluble anode to the terminal cathode through the electrolyte without carrying out the work of dissolution and cathodic deposition for every bipolar electrode. The electrodes cannot be made of the same size and shape as the bath because the anodic sludge collects on the bottom of the tank; to avoid contamination of the cathodes it is necessary to keep their lower edges at a certain distance from the bottom of the tank, so that between the level of the lower edges of the bipolar electrodes and the floor of the tank remains a throughway of electrolyte which forms a short-circuit for the current.

The current efficiency for this type of arrangement is about 0.7–0.75. Another possible cause of the diminution in current efficiency is the difficulty in determining exactly when the purification of the electrode is complete; continuing electrolysis after this point will redissolve pure copper and all the current which flows after the instant in which the last particle of crude copper has passed into solution from the bipolar electrode, is completely lost. On the other hand, if the electrodes are removed too quickly from the bath it is necessary to remove the unchanged portion of the original electrode mechanically. To simplify this procedure, the cathodic face of each plate is sprayed with a resinous soap or a graphitic oil before electrolysis.

The distance between the electrodes is a little less than with the unipolar arrangement – about 33 mm – which allows the electric tension between neighbouring electrodes to be reduced to about 0.11–0.13 V; it also allows the volume of electrolyte to be decreased with a consequent reduction in the area required for the electrolysis tanks and also in the plant for the circulation, re-heating and recovering of the electrolyte. The current density and temperature are the same as with the unipolar system. The energy consumed for each metric ton of copper refined is considerably less: about 160–170 kWh, but falling as low as 130 kWh. To facilitate a regular attack on the electrodes over their whole surface, it is necessary to construct them of sufficiently pure copper (99.2–99.5 %) so that they can be formed into plates which will give electrodes of constant thickness. The number of electrodes immersed in each tank and the number of tanks collected in series, are so arranged that the end tanks may be connected directly to the bus-bars. It is important for this process, that there should be a regular and even growth over the whole surface of the cathodic copper layer so as to avoid short circuits; these easily occur in view of the small distance between the electrodes. Wooden guides are fixed to the walls of the tanks to keep the electrodes at equal distances apart. Moreover, at regular time intervals, a small amount of gelatine is added to the electrolyte to give a more even deposit. Tanks for the bipolar process cannot be lined with lead because this would function as a short circuit of considerably less resistance.

A brief comparison of the advantages and disadvantages of each process will bring out their characteristics. The *unipolar* electrodes have the following advantages:

1. The anodes may be constructed of copper with a higher percentage of impurities since in view of the greater distance between the electrodes, there is no fear of a mechanical deposit of impurities on the cathode; this implies on the one hand, a purer cathodic copper and on the other a smaller loss of precious metals.

2. Working costs are lower as the anodes are prepared by simple casting and are of greater size, and it is possible to use unskilled personnel to tend the tanks and to control the notably simpler process.

3. Lead-lined tanks may be used.

These advantages are opposed by the following disadvantages:

1. A greater current intensity is absorbed by each tank, which involves a greater loss of energy in the various circuit resistances and a greater quantity of copper immobilized in the bars of the distribution network.

2. A large number of cathodes have to be prepared.

3. There are a large number of contacts which involve a loss of energy in the corresponding resistances.

4. The greater distance between the electrodes leads to an increase in the ohmic resistance of the electrolyte, in the applied electryc tension for each cell, in bulk, in the amount of electrolyte in circulation and in plant outlay.

5. There are a greater percentage of anodic residues to recover.

6. The refining cycle lasts longer.

The *bipolar* arrangement has the following advantages:

1. Less amperage is absorbed in each tank for an equal number of electrodes so that there is a smaller loss of energy through the Joule effect and a smaller amount of copper immobilized in the distribution network.
2. Most of the contacts are eliminated.
3. There is a smaller distance between the electrodes and hence a diminution of the ohmic resistance, the tension, the bulk, the volume of electrolyte, and plant outlay.
4. There is a smaller energy consumption per metric ton of copper refined.
5. There is a smaller percentage of anodic residues to recover.
6. The refining cycle is shorter.

These advantages are opposed by the following disadvantages:

1. The total process is more complex and requires more careful control and the use of trained personnel.
2. There is a greater loss of precious metals.
3. A mechanical removal of anodic residues from the plates of pure copper is required.
4. The fabrication of the anodes, by rolling, is more costly.

TABLE VIII,3

TYPICAL ANALYSES OF THE PRODUCTS OF COPPER REFINING

<table>
<tr><th>Element</th><th>Anodic Cu
%</th><th>Cathodic Cu
%</th><th>Sludge
%</th></tr>
<tr><td>Cu</td><td>99.3</td><td>99.98</td><td>40</td></tr>
<tr><td>O</td><td>0.2</td><td></td><td></td></tr>
<tr><td>S</td><td>0.004</td><td>< 0.001</td><td></td></tr>
<tr><td>As</td><td>0.08</td><td>< 0.0005</td><td rowspan="9">9</td></tr>
<tr><td>Sb</td><td>0.01</td><td>< 0.0005</td></tr>
<tr><td>Pb</td><td>0.05</td><td>< 0.0005</td></tr>
<tr><td>Se</td><td>0.1</td><td>< 0.0005</td></tr>
<tr><td>Te</td><td>0.0122</td><td>< 0.0002</td></tr>
<tr><td>Ni</td><td>0.2</td><td>< 0.0005</td></tr>
<tr><td>Bi</td><td>0.01</td><td>< 0.0005</td></tr>
<tr><td>Sn</td><td>0.002</td><td>< 0.0005</td></tr>
<tr><td>Ag</td><td>0.08</td><td>< 0.0005</td><td>10</td></tr>
<tr><td>Au</td><td>0.007</td><td>< 0.0005</td><td>1</td></tr>
</table>

7. The starting copper must already be fairly pure.
8. The electrolyte must be maintained at a high degree of purity.
9. Lead cannot be used to line the tanks.

It may be concluded from this comparison of the two methods that the unipolar system is preferable for those plants which have to work with copper of an inconstant quality, containing considerable amounts of impurities, whereas the bipolar system is more economical for plants refining large quantities of copper which is already fairly pure and of approximately constant composition.

Table VIII, 3 gives some typical analyses.

The electrical values are summarized in Table VIII, 4.

TABLE VIII, 4

ELECTRICAL VALUES FOR COPPER REFINING

	Unipolar system	*Bipolar system*
Electric tension	0.2–0.5 V	0.11–0.13 V
Current density	~ 200 A/m²	~ 200 A/m²
Current efficiency	0.90–0.98	0.70–0.75
Consumption	0.25 kWh/kg	0.16–0.17 kWh/kg

In practice, various other expedients are used both to improve the product and ease work. Reference may be made to specialized texts of electrometallurgy[1] for details of these practical measures.

6. Electrolytic Refining of Copper; Side Products

The side products of the electrolytic refining of copper are the exhausted electrolyte and the anodic sludge.

The impure electrolyte contains significant quantities of valuable products such as copper, free sulphuric acid, nickel and arsenic which can be usefully recovered and employed. Various processes have been designed to recover these from the exhausted electrolyte. The simplest consists

[1] See for example J. BILLITER, *Technische Elektrochemie*, Vol. I, Elektrometallurgie wässeriger Lösungen, W. Kapp. Hall, 1952.

in merely recovering the copper by cementation with scrap iron or by crystallization of copper sulphate after neutralization of the free acid with scrap copper. These, relatively uneconomic, methods may be used in small plants in which the simplicity of the recovery procedure partly compensates for its lowered economy. A more rational method is the electrolysis of the impure electrolyte in three cells successively with insoluble anodes. A large part of the copper separates in the first cell as cathodic copper of a quality practically equivalent to that obtained during refining and with a current efficiency of 0.85. In the second cell this diminishes to 0.50 and the cathodic deposit consists of a not very pure copper which cannot be considered as electrolytic copper but is nevertheless sufficiently good to be used in the melting furnaces to make anodes. In the third cell, the cathodic deposit consists of about 50 % copper and 50 % arsenic and is used for the extraction of arsenic.

The residual electrolyte is concentrated until all the impurities, except the remaining arsenic and any sodium and potassium salts, are precipitated as sulphates. This mass of sulphates which consists principally of nickel sulphate is used for the extraction of this metal and the mother liquor is returned into the cycle, since it consists virtually of sulphuric acid possibly containing some alkaline sulphates.

The anodic sludge which represents about 0.8–1.2 % of the copper used consists mainly of copper and silver. It also contains all the gold originally present in the anodes and the insoluble impurities. The extraction of the precious metals and other valuable components is extremely difficult and complex and for this reason various processes have been proposed.

The sequence of operations in one of the less complex processes is as follows. That part of the copper which is present as fairly large fragments is removed by sieving, and the remainder is washed and filtered. The product which still contains 25–40 % water is heated in reverberatory furnaces at low temperature to oxidize any oxidizable elements as far as possible; these partly volatilize (arsenic, selenium etc.) and may be recovered from the fumes. The roasted product is lixiviated with sulphuric acid which brings almost all the copper and part of some other impurities into solution. The residue is fused with an alkaline and oxidizing flux to give a slag, which is further processed to extract some elements (mainly selenium and tellurium), together with crude silver which is fused into anodes and electrolytically refined, or else treated chemically to separate the silver from the gold and other precious metals.

7. Electrolysis of Cupric Solutions with Insoluble Anodes

The electrolysis of cupric solutions with insoluble anodes is normally used both for ores of such compositions that the extraction of copper by a wet process is more economical than by a thermal process, and for the recovery of copper from effluent waters of various industries and from side products of metallurgical industries. The process consists in bringing the copper into solution, if it is not so already, and then electrolyzing the solution, making the copper deposit cathodically and re-using the acidic exhausted electrolyte to bring fresh copper into solution.

After what has been said about the composition of the electrolyte for the electrolytic refining of copper, it will be clear that the difficulties to be overcome to achieve an economic extraction of copper by a wet method are much more conspicuous than in the case of refining. It may also be noted that whereas for refining the plants employed are already highly standardized, for extraction every plant differs in practice from every other as a consequence of the variable composition of the starting material. This process has undergone a marked development and is especially important for the extraction of copper from poor ores and from ashes of copper pyrites. The thermal metallurgical treatment is more economical for processing copper-rich ores to give black copper which may then be refined. The minerals which may be used in the wet process can be divided into two groups. The first includes cuprite (Cu_2O), malachite ($CuCO_3 \cdot Cu(OH)_2$), azurite ($2\,CuCO_3 \cdot Cu(OH)_2$), atacamite ($CuCl_2 \cdot 3\,Cu(OH)_2$), chrysocolla ($CuSiO_3 \cdot 2\,H_2O$), chalcanthite ($CuSO_4 \cdot 5\,H_2O$), and brochantite ($CuSO_4 \cdot 3\,Cu(OH)_2$). Ashes of copper pyrites which contain copper essentially in the form of the oxide (CuO) may be considered as related to these minerals in that they can be directly lixiviated with sulphuric acid. The second group of minerals includes chalcocite (Cu_2S), covellite (CuS), chalcopyrites ($CuFeS_2$), bornite (Cu_3FeS_3) and bournonite ($CuPbSbS_3$). The minerals of the first group may be directly lixiviated with dilute sulphuric acid to form cupric sulphate and liberate the corresponding free acid (HCl, H_2O–CO_2, H_2SiO_3) whereas those of the second group cannot be lixiviated if they are not first roasted to convert the copper present in the form of sulphides into the oxide. If, however, the lixiviating solution also contains ferric sulphate some of the minerals of the second group, and particularly chalcocite, which is the most important, are brought into solution according to the reaction

$$Cu_2S + 2\,Fe_2(SO_4)_3 \rightarrow 2\,CuSO_4 + 4\,FeSO_4 + S$$

In practice the liquid is circulated in a cycle which brings acid in contact with the mineral, neutralizing the former and extracting the copper. During the electrolysis the copper separates at the cathode and the acid is regenerated at the anode. The acid electrolyte, which has been exhausted of copper, is used to attack more mineral. The scheme is as follows:

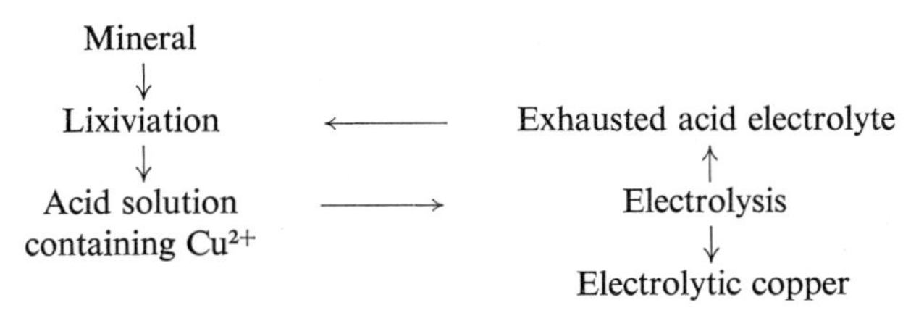

The gangue which forms the greater part of the material extracted from the deposit and treated, must be particularly low in carbonate and in alkaline earth metals to avoid an excessive consumption of sulphuric acid and accumulation of alkaline earth sulphates; these would disturb the progress of the electrolysis and would be difficult to eliminate. Since a certain amount of sulphuric acid is lost each time, both in the attack on impurities and carbonates contained in the mineral and as solution remaining in the mass of treated material, it is necessary to replace this for every cycle. In order to avoid as far as possible bringing impurities into solution, the lixiviating solution is not very acid and contains about 80–100 g of free acid per litre.

After lixiviation, the electrolyte must normally be purified, particularly if it contains an excess of ferric iron or arsenic, and of Cl^- and NO_3^- ions. The presence of these two anions makes the electrolysis particularly difficult because they attack the common anodic materials which can be used in sulphuric electrolytes: lead, alloys of lead with antimony and with silver etc. Much of the Cl^- ion present may be eliminated, and simultaneously the Fe^{3+} ions may be reduced to Fe^{2+} ions, by treating the electrolyte with copper powder obtained by cementation. The metallic copper reacts with the ferric iron to reduce it to ferrous, and passes into solution partly as Cu^+ ions; these precipitate the Cl^- ions present as poorly soluble cuprous chloride. A high concentration of Fe^{3+} ions must in any case be avoided because it would diminish the cathodic current efficiency both through the primary reaction

$$Fe^{3+} + e^- \rightarrow Fe^{2+}$$

and through secondary reactions with copper deposited at the cathode

$$Cu + 2\ Fe^{3+} \rightarrow Cu^{2+} + 2\ Fe^{2+}$$

If the purified solution is fairly low in Cl^- and NO_3^- ions, anodes of lead or its alloys may be used; otherwise, magnetite may be useful because it is fairly resistant, but it has the disadvantage of being fragile and not suitable for anodes of large area; still better is Chilex alloy formed of copper (60 %), iron (8 %), silicium (25 %) and lead (2–3 %), the remainder being tin and manganese. This alloy which mainly consists of the two compounds $CuSi_2$ and FeSi, well resists anodic attack by Cl^- ions.

Electrolysis is carried out in wooden or concrete containers lined with lead or asphalt. The theoretical decomposition electric tension of copper sulphate in slightly acid solution, like that used for electrolysis, is about 1.49 V. In practice, however, to overcome the discharge overtension, the concentration polarization and the electrical resistance of the electrolyte, it is necessary to apply 1.8–2.5 V to the cell.

As far as the applied electric tension is concerned, a certain amount of ferrous iron in the electrolyte is useful because it acts as a depolarizer for the anodic process and thus lowers the electric tension applied to the cell. The regeneration of ferric salt is thus particularly useful for processes destined for the treatment of sulphide minerals with a mixture of sulphuric acid and ferric sulphate. If the regeneration of ferric salt is not useful it can be reduced either by copper during the cementation operation of removing chloride, or with sulphur dioxide.

Electrolysis, begun with an electrolyte containing about 5 % of copper

TABLE VIII, 5

TYPICAL COMPOSITION OF AN ELECTROLYTE (g/l) FOR THE ELECTROLYTIC PRODUCTION OF COPPER

	Cu	H_2SO_4	Cl^-	*Total* Fe	Fe^{3+}
Crude electrolyte	35.39	47.85	0.59	4.60	1.37
Electrolyte purified with Cu	36.19	46.09	0.13	4.60	0.01
Exhausted electrolyte	14.44	78.54	0.21	4.60	1.13

TABLE VIII, 6

ELECTRICAL VALUES FOR THE PRODUCTION OF COPPER

Electric tension	1.8–2.5 V
Current density	~ 130 A/m^2
Current efficiency	0.80–0.90
Consumption	2.0–2.2 kWh/kg

is continued with a current density of about 130 A/m^2 until a final copper content of not less than 1.5 % is reached, with a mean current efficiency of 0.8–0.9. If more copper were deposited, the current efficiency would fall too far. The exhausted electrolyte contains 1.5 % of copper, 80–90 g/l of free acid and any ferric salt regenerated. The energy consumption under these conditions lies around 2.0–2.2 kWh/kg of copper.

Table VIII, 5 summarizes some typical values for the composition of the electrolyte.

The electrical values are shown in Table VIII, 6.

8. Electrolytic Refining of Silver

The electrolytic refining of silver is very simple because silver under normal conditions gives only monovalent ions in solution. At the anode, Ag_2^+ ions with a stoichiometric valency of $\frac{1}{2}$, are also formed but these are not stable and readily decompose according to the equation

$$Ag_2^+ \rightarrow Ag^+ + Ag \qquad (1)$$

This occurs so close to the anode that the metallic silver produced adheres to the anode itself. Only a very small quantity of Ag_2^+ ions pass into the electrolyte and can be detected with a very dilute solution of permanganate. The quantity of Ag_2^+ ions is always so small that they have no significant effect on the progress of the electrolysis or the current efficiency. Thus, this current efficiency is always very close to unity although it never actually reaches it since, as a consequence of reaction (1), a little silver always passes into the anodic sludge. The optimum conditions are therefore dictated by economic considerations rather than by theoretical ones.

The raw material mainly consists of silver–gold alloys obtained by processing of anodic sludges from the refining of various metals (copper, lead, bismuth, etc.), of side-products of metallurgical processing of almost all metallic minerals containing precious metals, of the treatment of silver minerals and of the processing of scrap from gold smiths etc. Before being fabricated into anodes, these materials undergo a preliminary treatment which varies with their composition, in order to raise the silver content as far as possible. Thus, the copper is largely eliminated by lixiviation with sulphuric acid after an oxidizing roast. Part of the lead, the selenium and the tellurium pass into the slag during the refining treatment in a Doré furnace and arsenic and antimony are largely removed in the fumes during the treatment in this furnace. The metal thus obtained is finally cast into anodes. The mean composition of the anodes is given by Table VIII, 7.

TABLE VIII, 7

COMPOSITION OF SILVER ANODES

Element	%
Ag	95–98
Au	0.5–3
Cu, Bi, Pb, Pt, Zn etc.	1.5–3

The silver content must never fall below 70 %.

The simultaneous cathodic separation of impurities along with the silver during the refining is highly unlikely since the standard electric tension of silver is much more noble than that of the other metals present, with the exception of the gold. This in turn, however, has an electric tension which is much more noble than that of the silver. Moreover the cathodic deposition of silver occurs without any significant overtensions. In the anodic dissolution of the crude silver, the gold remains undissolved and passes into the anodic sludge together with metals of the platinum group, tellurium and part of the lead (as its peroxide) whereas the residual copper, the rest of the lead and any other impurities which may be present pass into solution and stay there.

A sulphuric electrolyte cannot be used in the refining of silver because of the low solubility of silver sulphate. Amongst the soluble salts of silver the most economical is the nitrate. The electrolyte thus consists of a solution of silver nitrate acidified with nitric acid and augmented with sodium or potassium nitrate to raise the conductance of the bath and diminish the quantity of silver immobilized as electrolyte. The composition of the resulting electrolyte is shown in Table VIII, 8.

TABLE VIII, 8

COMPOSITION OF THE ELECTROLYTE FOR THE REFINING OF SILVER

Substance	*Concentration* g/l
Ag	15–30
$NaNO_3$	100
Free HNO_3	2–10

The presence of free nitric acid in the bath has the additional advantage that it prevents the separation of copper. The cathodic reduction tension of nitric acid is intermediate between the discharge tensions of silver and copper, so that the reduction of the free acid would occur before the separation of copper at the cathode. Since the reduction tension of nitric acid is fairly close to the deposition tension of silver, a certain amount of nitric acid will always be reduced. The nitric acid concentration in the electrolyte must therefore always be fairly low so as not to consume acid uselessly nor to lower the current efficiency. The acid consumed must be continuously replaced; about 8 g of acid are used for each kg of electrolytic silver.

Under these conditions the concentration of copper which passes into solution from the anode, may reach high values without causing difficulties. To avoid its simultaneous separation through uncontrollable variations in concentration in the immediate vicinity of the cathode, however, the copper content should not exceed 80 g/l. Normally, as soon as it reaches 4 %, part of the electrolyte is replaced with fresh electrolyte which is free from copper.

Silver separates from a nitric solution in an adherent form only at very low current densities; and at current densities of the order of magnitude normally used in electrolytic cells it separates as large crystals which are usually elongated in the cathode–anode direction (type 1 or type 2) and could readily form metallic bridges and hence short-circuit the cell. Because of this, two types of cathodes have been mainly adopted: vertical cathodes of pure rolled silver or steel arranged between vertical anodes (Moebius cell) over which swings a wooden fork which continuously displaces the crystals separated at the cathode and lets them drop on to the floor of the cell, and horizontal graphite cathodes on the bottom of the cell with horizontal anodes suspended in the cell within a wooden cage (Balbach–Thum cell). Since the refined silver is always collected as separate crystals detached from the cathode and deposited on the floor of the cell, it is necessary to avoid any possibility of their contamination by the anodic sludge. For this reason, a diaphragm is always inserted between the anode and the cathode; it consists of very thick material, which is resistant to the electrolyte, in the shape of a sack enclosing the anodes in the Moebius cell or simply lying on the floor of the wooden cage in the Balbach–Thum cell. Despite this, a careful washing of the crystals of cathodic silver is always necessary to eliminate both adherent electrolyte and any particles of anodic sludge which have filtered through the material. This is especially important when the anodic sludge is very rich in gold. The wash water must naturally be free of sulphates so as to avoid the precipitation of lead sulphate.

The cells are tanks of fairly small size formed of porcelain or glazed earthenware or even asphalted concrete. The starting temperature is that of the room, but may rise as the electrolysis proceeds to 40–45° C. The electrical values for the two main types of cell are shown in Table VIII, 9.

The silver obtained is 99.98–99.99 % pure.

If the silver to be refined is less than 70 % pure it is first dissolved anodically in a strongly acid, nitric anolyte. The silver is separated from this solution by cementation with copper and as it is still too impure, is

TABLE VIII, 9

ELECTRICAL VALUES FOR THE REFINING OF SILVER

Quantity	*Moebius Cell*	*Balbach–Thum Cell*
System	Unipolar	Unipolar
Electric tension	2–2.5 V	3.2–3.8 V
Cathodic current density	Up to 500 A/m^2	200 A/m^2
Anodic current density	(Up to 500 A/m^2)	300–400 A/m^2
Current efficiency	~0.95	0.88–0.90
Consumption	0.5–0.6 kWh/kg	0.9–1.1 kWh/kg

further refined electrolytically whilst the nitric acid is recovered from the solution by depositing the copper cathodically in a cell with inert anodes. Electrolytic refining, however, is not economic for alloys which are very poor in silver.

The anodic sludge containing all the gold, platinum and palladium and a small proportion of other impurities, is treated with boiling sulphuric acid to eliminate copper and silver and then fused into anodes for the extraction of the gold (*cf.* the following section).

9. Electrolytic Refining of Gold

The electrolytic refining of gold has presented various practical difficulties due to its tendency to become passive. In practice, it will not pass into solution unless certain ions such as Cl^- or CN^- are present. In a solution of auric chloride or chloroauric acid a gold anode will not be attacked except to a very slight extent because the dissociation of these two complex ions is very limited and hence the concentration of free Cl^- ions which could anodically dissolve the gold, is extremely low. On the addition of a small quantity of HCl or NaCl, however, the anodic dissolution of the gold takes place immediately and easily. During anodic dissolution in the presence of an excess of Cl^- ions, the gold may form complex anions in which it is mono- or trivalent, according to the following reactions whose standard electric tensions are shown at the side:

$$AuCl_2^- + 2\,Cl^- \rightarrow AuCl_4^- + 2e^- \qquad +0.94$$

$$Au + 4\,Cl^- \rightarrow AuCl_4^- + 3e^- \qquad +0.99$$

$$Au + 2\,Cl^- \rightarrow AuCl_2^- + e^- \qquad +1.11$$

There is thus a coexistence of mono- and trivalent, and metallic gold in an equilibrium similar to that described for copper but involving complex ions. The position of this equilibrium depends on the temperature and when this increases, shifts towards a greater concentration of monovalent gold. The behaviour of the gold thus, to some extent, resembles that of copper and in fact the anodic sludge from the refining of gold contains a certain amount of gold powder due to a reaction analogous to reaction (1) cited in the refining of copper. The quantity of gold which passes into the sludge also depends on the concentration of free hydrochloric acid, the temperature, the agitation of the liquid, the composition of the anode, and the current density. Gold, however, differs from copper in its tendency to become passive. In a neutral aqueous solution of auric chloride between gold electrodes there is an exclusive evolution at the anode of gaseous chlorine with possibly some oxygen. If, however, Cl^- ions are present in excess with respect to the Au^{3-} cations, the gold anode will pass into solution without becoming passive and it is possible to use a higher current density as the temperature and the concentration of Cl^- ions are increased. The mechanism of the cathodic reaction is still not clear because both tri- and monovalent gold cations are present near the cathode in only very low concentrations because of the high stability of the gold-containing complex anions. It is possible that the cathodic deposition is not primary but secondary to a reaction between the gold-containing anions and the product of a more likely primary reaction such as the discharge of the H^+ ion. Despite this uncertainty it can be seen from what has been said that the conditions required are firstly, a high temperature and secondly, a high concentration of Cl^- ions in excess.

Further conditions must be added to achieve a cathodic deposit which is well adherent and the maximum rapidity of the process – in view of the high cost of immobilizing the treated material. These additional conditions are thirdly, a high gold concentration and fourthly, a high current density. The maximum current density depends on the temperature, the concentration of Cl^- ions and the composition of the anodes.

The anodes are alloys of gold and silver often containing copper, lead, metals of the platinum group and various other metallic impurities; this material comes mainly either from anodic sludges from the electrolytic refining of silver or from the processing of residues and scrap from mints and gold smiths. The normal refining process without the use of alternating currents (see below) may be applied to anodes with a gold content of not less than 90 %. The normal average composition is given in Table VIII, 10.

TABLE VIII,10

COMPOSITION OF GOLD ANODES

Element	%
Au	~94
Ag	~5
Other metals	~1

The choice of electrolyte is simple in view of the small number of possibilities; a sulphuric electrolyte is useless in view of the low solubility of gold sulphate; the chloride $AuCl_3$ in simple aqueous solution gives rise to a complex anion. The anodic reaction in this case does not lead to the dissolution of the anode but to its passivation with an evolution of gaseous chlorine and oxygen. If Cl^- ions are present in excess the anode dissolves and will support a current density, without becoming passive, which increases with increasing excess of Cl^- ions in solution. Gold shows a similar behaviour in cyanide solution with an excess of CN^- ions. The cathodic deposit from a cyanide solution is also superior so that this is normally used for gold plating but since the discharge tensions of gold, silver and copper from cyanide solutions are too close together, the chloride solution is normally preferred. The electrolyte normally used contains from 50–100 g/l of gold and 100–130 g/l of free hydrochloric acid. To diminish the passivity of the anodes still further, increase the current density and simultaneously improve the cathodic deposit, the temperature of the bath is kept at about 65–70° C. At this relatively high temperature there is a rapid evaporation of water which must be periodically replaced in the bath.

The metallic impurities in this electrolyte, partly pass into solution and partly collect in an anodic sludge; the copper, part of the lead, the platinum, the palladium and a trace of the iridium and the rhodium pass into solution and the other metals of the platinum group (osmium, ruthenium, and the rest of the iridium and rhodium) collect in the anodic sludge with the silver and the rest of the lead; the silver is present as the chloride and the lead as the chloride or sulphate depending on whether SO_4^{2-} ions are present.

Anodes which are rich in copper or lead are not suitable for electrolytic refining because an excess of copper would rapidly lower the concentration of gold in the electrolyte and give a cathodic deposit of gold which was nonadherent; and an excess of lead would provoke the formation of a fine adherent film on the anode which, increasing the effective current density, would facilitate passivation. Moreover, an excess of lead would make the cathodically separated gold rough and uneven with an attendant danger of including both electrolyte and crystals of lead salts which would not only contaminate the purified gold but would make it difficult to work even at a concentration of 0.001 % of lead. A silver content of up to 5 % in the anode does not lead to any difficulties but from 5–10 % it would form a film of chloride which would not adhere well to the anode and could be removed mechanically; above 10 % and up to a maximum of 20 % it is necessary to impose an alternating current on to the direct electrolyzing current. If the raw material consists mainly of anodic sludge from the refining of the silver, it normally contains too much of this to be used directly. It is then subjected to a preliminary treatment with hot sulphuric acid in cast-iron pots to dissolve the silver, lead and any copper. After washing, the residue is melted and cast.

The platinum and the palladium may reach considerable concentrations – 50 g/l and 15 g/l respectively – without causing any difficulties. These metals, together with the residual gold, are recovered chemically from the exhausted electrolyte. The refining of gold alloys containing much copper is also inconvenient in this respect because, since the electrolyte must be frequently replaced, the advantage of the costless and automatic concentration of the platinum and palladium is lost. The electrolyte is prepared very simply by dissolving anodically pure gold in a diaphragm cell (to avoid diffusion and the consequent separation of gold towards the cathodic region) containing initially 25 % hydrochloric acid.

The cathodes consist of laminae of fine gold and the vessels are small tanks of porcelain or glazed earthenware. The electrical conditions are shown in Table VIII, 11.

TABLE VIII, 11

ELECTRICAL VALUES FOR THE REFINING OF GOLD

Arrangement	Unipolar
Electric tension	0.5–3.5 V
Current density	1,000–1,600 A/m^2
Current efficiency	1 or more
Consumption	0.3 kWh/kg

A current efficiency of more than 1 may seem peculiar, but it must be noted that it is calculated with reference to the Au^{3+} ion whereas in reality the closeness of the standard electric tensions of the mono- and trivalent ions mean that a not insignificant amount of the former exists in the solution and takes part in the refining process with a consumption of only one third of the current required to transfer gold from the anode to the cathode in the form of trivalent ions.

The cathodic gold thus obtained is 99.97–100 % pure.

This method of refining may be applied to anodes containing not less than 90 % of gold. Anodes containing 10–20 % of impurities, which must however be mainly silver, can still be refined electrolytically by applying an alternating current to the continuous electrolyzing current. This alternating current must have a low frequency and an intensity about 10 % greater than that of the direct electrolyzing current so as to obtain a periodic, asymmetrical inversion of the current. This imposition of an alternating current has various advantages. Firstly, it makes the anodic film of silver chloride loose and poorly adherent so that it can be easily removed. Secondly, it acts as a depolarizer and lowers the anodic polarization and hence hinders passivation, making possible the use of greater current densities. It also diminishes the formation of Au^+ ions and thus lowers the loss of gold in the form of a powder into the anodic sludge. Finally, it heats the bath.

If the anodes contain less than 80 % of gold the purity may be increased by treatment with nitric acid (quartation) or boiling concentrated sulphuric acid (affination).

In view of the high cost of gold it is necessary to select the materials, shape and dimensions of the vessels so that there shall be no loss. Similarly, the working conditions must be such that the refining process does not require more than 24 hours for each batch of gold, including all the

TABLE VIII,12

COMPOSITION OF THE ANODIC SLUDGE AND EXHAUSTED ELECTROLYTE FROM GOLD REFINING

Anodic sludge		*Exhausted electrolyte*	
Au	1–80 %*	Au	30–60 g/l
AgCl	95–20 %	Pt	40–80 g/l
		Pd	5–20 g/l
Metals of the Pt group, salts of Pb, Sb etc.	~5 %	Other metals of the Pt group, Cu, Pb etc.	small quantities down to traces

* Depending on whether the refining is carried out with an imposed alternating current.

accessory operations, so as not to immobilize for too long, large sums of capital with consequent loss of interest. The vessels are normally porcelain or stone tanks which can be heated to 60–70°.

The side-products consist of the exhausted electrolyte, which is periodically replaced when the quantity of impurities which has accumulated in the solution becomes too high, and of the anodic sludge. The composition of these are shown in Table VIII, 12.

If the sludge contains much gold, it is melted and then after solidifying, the still liquid silver chloride is removed from the crucible and the underlying gold is remelted and formed into anodes. If, however, the silver chloride is the most abundant, it is treated with zinc and hydrochloric acid to reduce the silver chloride and then fused and used for silver refining. Gold is recovered from the exhausted electrolyte by treatment with sulphur dioxide, and the platinum and palladium by the addition of ammonium chloride followed by concentration until the ammonium chloroplatinate and palladium chloride crystallize.

10. Electrolytic Preparation of Zinc; Theory

The electrolytic preparation of zinc with insoluble anodes has been especially developed both because the electrothermal process is very much more difficult with zinc than with other metals, and because the electrolytic process allows the exploitation of ores of complex minerals of zinc whose thermal treatment would be particularly difficult. The electrolytic process was faced with two major problems. The first was to obtain a cathodic separation of the zinc from a solution which was initially slightly acid, in a satisfactory form and with an industrially economic current efficiency. The second was the need to purify the electrolyte very extensively so as to obtain a good cathodic deposit free particularly of metallic impurities.

Essentially, the process used consists in bringing the zinc contained in the minerals into solution as zinc sulphate after converting it into the oxide or directly into the sulphate by roasting or sulphating roasting. The solution thus obtained is electrolyzed with insoluble anodes. During the electrolysis, the acid is anodically regenerated and returned into the cycle to bring more zinc into solution. The process can thus be represented diagrammatically as follows.

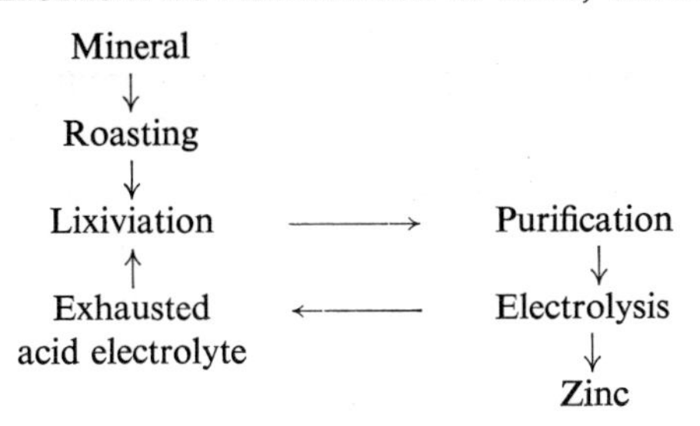

The most important minerals for the extraction of zinc are blend (ZnS), calamine ($Zn_2SiO_4 \cdot H_2O$), smithsonite ($ZnCO_3$), zincite (ZnO) and hydrozincite ($2\ ZnCO_3 \cdot 3\ Zn(OH)_2$). The most common impurities in these are iron as its sulphide, which may also form solid solutions with the blend (marmatite), or as its silicate or hydrated oxide in oxidized minerals; together with lead, copper, cadmium, arsenic, manganese and more rarely, silver, gold, tin, cobalt, nickel, thallium, germanium and traces of other metals. Other impurities from the gangue such as silica, calcium, magnesium and aluminium minerals in a more or less soluble form, may also occur. If these latter impurities are present in considerable amounts, *i.e.* if the mineral is not sufficiently rich, a preliminary concentration by means of flotation is carried out.

The mineral in its original form or enriched by flotation, must first be transformed into the oxide by roasting[1]. The roasting reactions of the most common sulphide minerals are as follows. The zinc sulphide when brought to a fairly high temperature in the presence of air, burns to form zinc oxide and sulphur dioxide; the latter, under the catalytic action of the ferric oxide present, is partially oxidized to sulphur trioxide which reacts with the zinc oxide to form zinc sulphate.

$$2\ ZnS + 3\ O_2 \rightarrow 2\ ZnO + 2\ SO_2$$

$$2\ SO_2 + O_2 \rightarrow 2\ SO_3$$

$$ZnO + SO_3 \rightarrow ZnSO_4$$

[1] Calamine and smithsonite could be lixiviated by direct treatment with sulphuric acid but this route is not usually chosen. It is preferred to convert the zinc of these minerals also into oxide both to avoid an excessive quantity of silicic acid passing into solution – which would be difficult to separate – and because the calamine and smithsonite ores are fairly poor and the treatment with sulphuric acid would consume too much acid. These minerals are therefore normally treated with carbon and an excess of air in rotating furnaces at fairly high temperatures so that the zinc formed by reduction of the mineral by the carbon, will vaporize and reoxidize in the excess of air in the fumes and may then be collected. A product is thus obtained consisting mainly of zinc oxide but still containing other volatile metallic oxides particularly that of cadmium, and readily soluble in sulphuric acid.

The ratio of zinc oxide to sulphate in the final product of the roasting depends on a number of factors (some of which are uncontrollable), such as the roasting temperature, the amount of air available to the mass during roasting, and the presence and activity of catalysts favouring the formation of sulphur trioxide. This partial sulphation of the zinc oxide during roasting is useful to compensate to some extent for the continuous and inevitable loss of sulphuric acid.

During roasting, the iron impurities present – particularly if they form more than 5 % – may cause a further reaction with the formation of zinc ferrite $ZnO \cdot Fe_2O_3$. The rate of formation of this compound increases with temperature and is favoured by an intimate contact of the particles of zinc and ferric oxides. Since it is poorly soluble in dilute sulphuric acid but more soluble in concentrated acid, it is necessary to avoid or at least to limit the formation of ferrite during roasting, particularly if the lixiviation is to be carried out with dilute acid. In this case, the temperature must not rise above 650°. If the lixiviation is carried out with more concentrated acid, a greater roasting temperature may be tolerated. The temperature of roasting is in any case made dependent on the percentage of iron in the mineral.

The mineral is finely powdered and dried before roasting, and during this the mass is continuously agitated to facilitate complete oxidation which takes a fairly long time in the interior of the granules. It will be interesting to briefly describe a particular roasting process designed to use calamine in a calcarious-dolomitic gangue, containing a relatively high proportion of chlorides and fluorides. The presence of high levels of calcium and magnesium carbonates would lead to an excessive consumption of sulphuric acid and the presence of Cl^- and F^- ions would cause marked disturbances during electrolysis. Such minerals, however, can be used after a suitable roasting process which differs from the normal one in that it is designed to give as highly sulphated a product as possible. The calamine is, therefore, not roasted by itself but in admixture with a certain amount of blend (enriched by flotation) and in the presence of an excess of air. The calcium and magnesium carbonates are thus converted into sulphates:

$$CaCO_3 \rightarrow CaO + CO_2$$

$$2\,SO_2 + O_2 \rightarrow 2\,SO_3$$

$$CaO + SO_3 \rightarrow CaSO_4$$

Similar equations apply for the magnesium. At the temperature of the furnace, the zinc sulphate is practically all dissociated into ZnO and SO_3. The anhydrous $CaSO_4$ will not pass into solution during the successive attack with H_2SO_4 and the magnesium sulphate is simply eliminated by washing the product of roasting with water, before the acid treatment. Under these conditions of sulphating roasting, the chlorides and fluorides are also converted into sulphates:

$$4\,NaCl + 2\,SO_2 + O_2 + 2\,H_2O \rightarrow 2\,Na_2SO_4 + 4\,HCl$$

$$2\,CaF_2 + 2\,SO_2 + O_2 + 2\,H_2O \rightarrow 2\,CaSO_4 + 4\,HF$$

$$4\,HF + SiO_2 \rightarrow SiF_4 + 2\,H_2O$$

The chlorine and fluorine are eliminated as HCl and SiF_4 in the fumes. In this way the fluorine content – which is particularly damaging – is reduced to 5 g per metric ton or even less.

The product of the roasting is lixiviated with sulphuric acid. During this, elements other than the zinc which are present as impurities in the mineral also pass into solution; they include iron, copper, cadmium, arsenic, antimony, cobalt, nickel, traces of lead and silver, part of the silica and possibly Cl^- ions if they were present in the starting material or in the diluting and wash waters. The amount of zinc and impurities which pass into solution depends on the composition of the starting mineral, on its granulation, iron content and the temperature and length of roasting; but above all, on the free acid content of the lixiviating solution. The yield of extracted zinc increases with the concentration of free acid in the solution used for treating the roasted mineral, but the quantity of impurities dissolved also increases. It is not possible to extract all the zinc present in the original material both because a certain proportion remains unattacked, particularly if the iron content is relatively high, and because another part remains trapped in the solid residue of the lixiviation; this is gelatinous in nature due to the presence of silicic acid and ferric hydroxide. A thorough washing of the lixiviation residue is inconvenient since it would dilute the solution too much and excessively increase the amount of water in cycle in the plant. Depending on the starting material and the concentration of acid used, the yield of zinc extracted varies between 75 and 90 %. Still higher yields can be obtained in particularly favourable cases.

The impurities dissolved in this operation must be eliminated before electrolysis. This is essential to obtain a compact cathodic deposit of zinc with a good current efficiency. Such a profound purification of the solution to be electrolyzed is necessary because almost all of the impurities, apart from the manganese, are electrochemically more noble than the zinc and would thus be deposited either before or at the same time as this, and would form local galvanic cells; these would favour the re-dissolution of the zinc by the acid electrolyte in contact with it and hence would markedly diminish the current efficiency and would make the zinc obtained poorly resistant to corrosion. The anions present, apart from SO_4^{2-}, which have to be eliminated are Cl^- and F^-; the first would attack the Pb anodes and convert them more or less rapidly into PbO_2 through the intermediate formation of chloride and its successive anodic oxidation; the F^- ions would interfere principally with the cathode by dissolving the protective film of aluminium oxide. Deprived of its protection, the aluminium would alloy with the cathodically deposited zinc making it almost impossible to remove it from the aluminium support. The F^- ions are also damaging to the anodes because they favour the formation of a film of $PbSO_4$ and strongly assist the formation of MnO_2 with a consequent disturbance of the flow of electrolyte through the interelectrode spaces. Moreover, this would raise the anodic polarization and thus lower the energetic efficiency. The maximum concentrations of impurities which can be tolerated in electrolysis tanks are of the order of mg/l. They are shown in

TABLE VIII, 13

LIMITS OF TOLERANCE OF IMPURITIES IN $ZnSO_4$ SOLUTION

Element	*mg/l*
Mn	350
Fe	30
Cd	12
Cu	10
As	1
Sb	1
Co	1
Ni	1
Ge	< 1
Cl^-	50
F^-	30

Table VIII, 13. If more than one impurity is present at the same time the limits shown in Table VIII, 13 are too high. According to Steintveit and Holtan[1] the electrolysis of a solution of pure $ZnSO_4$ to which 0.003 g/l of Co have been added, gives a current efficiency of 0.903; if the impurities consist of 0.05 mg/l of Sb instead of the Co, the current efficiency is 0.917; however, in the presence of both impurities at these concentrations, the current efficiency falls to 0.768. In other words, the combined effect of the various impurities is not simply additive. Table VIII,13 shows that the least damaging impurity is manganese; a certain amount of manganese is in fact useful to cover the anodes with a protective oxide. The rest of the manganese which becomes anodically oxidized to permanganate, remains in solution and is used in the exhausted electrolyte to oxidize ferrous salts to ferric.

The purification of the solution is carried out by two distinct treatments: firstly, a treatment with zinc oxide, *i.e.* an excess of roasted mineral, and after filtration a second treatment with powdered zinc. If any Cl^- ions should be present it would be necessary to remove them by a special treatment to avoid the too rapid consumption of the anodes. The treatment with zinc oxide, possibly accompanied by a treatment with strong oxidants (O_2, MnO_2, PbO_2, $KMnO_4$ etc.) to bring all the iron to the ferric state, leads to a slight alkalinization of the solution[2], as a consequence, the iron precipitates as ferric hydroxide $Fe(OH)_3$, the arsenic and the antimony precipitate, partly as basic salts, but mainly adsorbed on the ferric hydroxide, the silica precipitates as gelatinous silicic acid and the nickel as its hydroxide. Sometimes, this treatment precipitates the silica in a form which may be difficult to filter or in excessive amount depending on the type of silicate present in the mineral. In such cases, it has proved useful to disperse the roasted material in a neutral electrolyte or in a neutral wash solution and then to add the acid solution slowly so as not to allow the pH to fall below 4.5. In this way, up to 80 % of the zinc may be attacked without bringing significant amounts of silica into solution. If the oxidation is sufficiently powerful to bring the cobalt to the trivalent state, it will be precipitated together with the iron as the hydroxide ($Co(OH)_3$). The germanium is also adsorbed on the ferric hydroxide. This first step of the purification takes place at the same time as lixiviation since the exhausted electrolyte which forms the acid attacking solution,

[1] G. Steintveit and H. Holtan, *J. Electrochem. Soc.*, 107 (1960) 247.
[2] The alkalinization is reinforced, if necessary, by a small addition of chalk.

may be treated with a slight excess of the roasted material. If the quantity of iron present in this is not sufficient to adsorb all the arsenic, antimony and germanium, some ferrous salts may be added. After the removal of the excess of roasted mineral and the precipitated impurities, the solution is treated with very pure zinc dust obtained by powdering molten electrolytic zinc with a jet of air. The zinc dust precipitates, by cementation, the impurities of all the more noble metals (Cu, Cd, Co, Ni etc.) which remain. These more noble metals than zinc, may also be eliminated by the formation of complex salts such as cobalt *α-iso*nitroso-*β*-naphtholate; nickel dithiocarbamate or xanthogenate etc. Any chlorine present may be removed by procedures involving the precipitation of silver or mercurous chlorides[1].

The next operation is the electrolysis, which is carried out with a solution which is slightly acid initially and strongly acid at the end; the anodes are of lead and the cathodes initially of aluminium which almost immediately become zinc cathodes. Theoretically the electrolysis of a neutral solution of zinc sulphate with a cathodic separation of the metal should be impossible because the discharge tension of zinc is much more negative than that of hydrogen in the same solution; the former lies at about -0.75 V depending on the effective concentration of Zn^{2+} ions whilst the latter is about -0.42 V. The difference of 0.33 V at the start of the electrolysis is already rather unfavourable for the deposition of zinc and becomes more so during the electrolysis because free sulphuric acid is formed, increasing the concentration of H^+ ions. The discharge tension of the H^+ ions is therefore shifted, although slightly, towards more positive values and hence the difference between the two discharge tensions increases. The cathodic deposition of zinc is, however, made possible by the overtension of hydrogen on the zinc cathode. This overtension depends on a number of factors whose influence must be allowed for in order to obtain the electrolytic separation of the metal.

In general, an increase in temperature lowers the hydrogen overtension and thus favours the simultaneous discharge of H^+ ions and lowers the current efficiency. An increase in current density, however, increases the hydrogen overtension and raises the current efficiency; but, a greater pro-

[1] It may be recalled here that the process for the preparation of the solution to be electrolyzed has been described schematically to illustrate the requirements to be satisfied. The practical industrial procedure undergoes variations adapted to the particular raw materials and the particular economic conditions of the plant.

portion of the energy is lost as heat in the various resistances of the circuit. The presence of traces of impurities, particularly about 0.0005–0.10 % of metals more noble than zinc, not only causes the difficulties mentioned above but markedly diminishes the hydrogen overtension. Copper is typical: at a current density of 1,000 A/m^2 the overtension of hydrogen on a zinc cathode in a 1 N solution of sulphuric acid falls from 0.97 V to 0.74 V for an increase in copper concentration from 0.001 % to 0.01 %.

A high concentration of Zn^{2+} ions about the cathode is necessary to obtain a compact deposit of zinc. Moreover, the cathodic solution must be at least faintly acid right from the beginning because zinc tends to separate in a spongey form from neutral solutions. The separation of spongey zinc is due to the simultaneous deposition on the cathode of zinc oxide, in a more or less hydrated form, and of basic salts, formed by hydrolysis; these interfere with the crystallization of the metal. A vigorous evolution of hydrogen at the cathode not only diminishes the current efficiency but tends to make the cathodic film alkaline and thus to facilitate the formation of spongey zinc. The neutral solution from the purification may be mixed with a suitable quantity of exhausted electrolyte to give it the slight acidity required. Certain colloids, such as gelatine, gum arabic, glue or colloidal silica may advantageously be added. These colloids favour the formation of compact deposits of the metal with smooth surfaces, *i.e.* they oppose the deposition of zinc in a rough or spongey form; these forms, having a large surface area, would diminish the effective current density and lower the hydrogen overtension. The addition of such colloids, moreover, normally raises the hydrogen overtension.

Zinc, with possibly a small amount of hydrogen, separates at the cathode during electrolysis. At the insoluble anode there is a discharge of OH^- ions (which are always present in the water) with an evolution of gaseous oxygen and the regeneration of free sulphuric acid.

From this discussion it is possible to deduce the most favourable electrolysis conditions to give the highest current efficiency. These are:

1. the lowest possible temperature;
2. the highest possible current density;
3. the maximum purity of the solution to be electrolyzed;
4. agitation of the electrolyte;
5. the addition of certain colloids.

The cathodic current efficiency is always significantly less than 1 even when it is possible to carry out the electrolysis without the simultaneous evolution of hydrogen by primary reactions at the cathode. The main reason for this is that zinc is normally attacked to a certain extent, by sulphuric acid, forming zinc sulphate and gaseous hydrogen; this is particularly so when, as in electrolysis, there is an inevitable deposition of traces of elements more noble than zinc. These greatly enhance the phenomena of corrosion and re-solution. In this way, some of the deposited metal returns into solution. The amount of zinc redissolved is a function not only of the actual acidity of the bath, but also of the time and the cathodic surface area. For this reason, the cathodic deposit is removed from the aluminium cathodes at short intervals, depending on the current density used, as soon as it reaches a thickness of 2–3 mm, to avoid the zinc remaining too long in the acid. The current efficiency also depends on the current density, both as shown on p. 243 *et seq.* and because the amount of zinc deposited per unit area in unit time increases whereas the amount dissolved remains virtually constant. It is not possible to electrolyze a solution of zinc sulphate until all its zinc content is quantitatively deposited, both, because the concentration of anodically generated acid would become too high and lead to the discharge of H^+ ions as a primary reaction, and because a certain concentration of Zn^{2+} ions is required below which the compactness of the cathodic deposit would be damaged. For this latter reason also it is necessary to circulate the electrolyte, both to satisfy the fourth condition required to obtain a good current efficiency, and to improve the characteristics of the cathodic deposit. The circulation of the electrolyte keeps the Zn^{2+} concentration fairly high by continuously renewing the cathodic film and also avoids any increase in alkalinity through a simultaneous discharge of H^+ ions. Moreover, by circulating the electrolyte, the process is made continuous.

The energetic efficiency is also considerably less than 1. The decomposition electric tension of zinc sulphate is 2.35 V so that the theoretical energy consumption should be 1.927 kWh/kg of zinc. However, the electric tension which must actually be applied is significantly greater because of the anodic overtension of oxygen and the cathodic overtension of zinc — which is not negligible — and finally, because it is necessary to overcome the resistance of the electrolyte itself. This latter changes during the electrolysis because both the composition of the electrolyte and the distance between the electrodes vary. However, the entire process must be carried out with the electric tension required at the beginning of the electrolysis when the resistance of the electrolyte is highest, because of its low acidity, and when the distance between the electrodes is greatest.

The problem of the anodes is also important. Nowadays, these are composed exclusively of sheets of very pure lead, possibly alloyed with 1 % of silver, on which a layer of dioxide forms in time. Two conditions are necessary for the use of lead anodes; firstly, the solution electrolyzed must not contain more than 50–70 mg/l of Cl^- ions, or the anode would rapidly be destroyed, and secondly, the lead used for the anodes must be extremely pure or its impurities would lead to a very rapid corrosion of the zinc deposited on the cathode and would lower the efficiency.

The vessels used for the electrolysis consist of wooden tanks covered with lead or sometimes of asphalted concrete tanks. The zinc obtained electrolytically has a purity of 99.95–99.99 %.

11. Electrolytic Preparation of Zinc; Processes

Various industrial processes have been proposed and used for the electrolytic preparation of zinc; these follow essentially the same pattern but resolve the difficulties described above by somewhat different means. The process which at the present day is used in the great majority of cases has been developed from the oldest one, the Anaconda process. The zinc is extracted from the roasted mineral with a relatively weak acid solution (100–110 g/l of free sulphuric acid)[1] and the electrolysis is carried out with a low current density ($\sim$ 350 A/m^2). In view of the low acidity of the exhausted electrolyte, which serves as the solution for attacking the roasted mineral, an accurate control of roasting is required so that the temperature does not become too high with a consequent formation of an excess of ferrite which would not be attacked by acid at the concentration used. In order to extract the maximum possible quantity of zinc, the lixiviation is performed in two stages, as shown in Scheme I which also includes all the other phases of the process.

During the neutral lixiviation with an excess of roasted mineral over free acid, the first purification for the elimination of iron, arsenic, antimony, silica etc., takes place, whilst the treatment with zinc dust forms a separate operation. The electrolysis is performed in tanks through which the elec-

[1] The modern tendency is to increase the acidity of the solution used to attack the mineral.

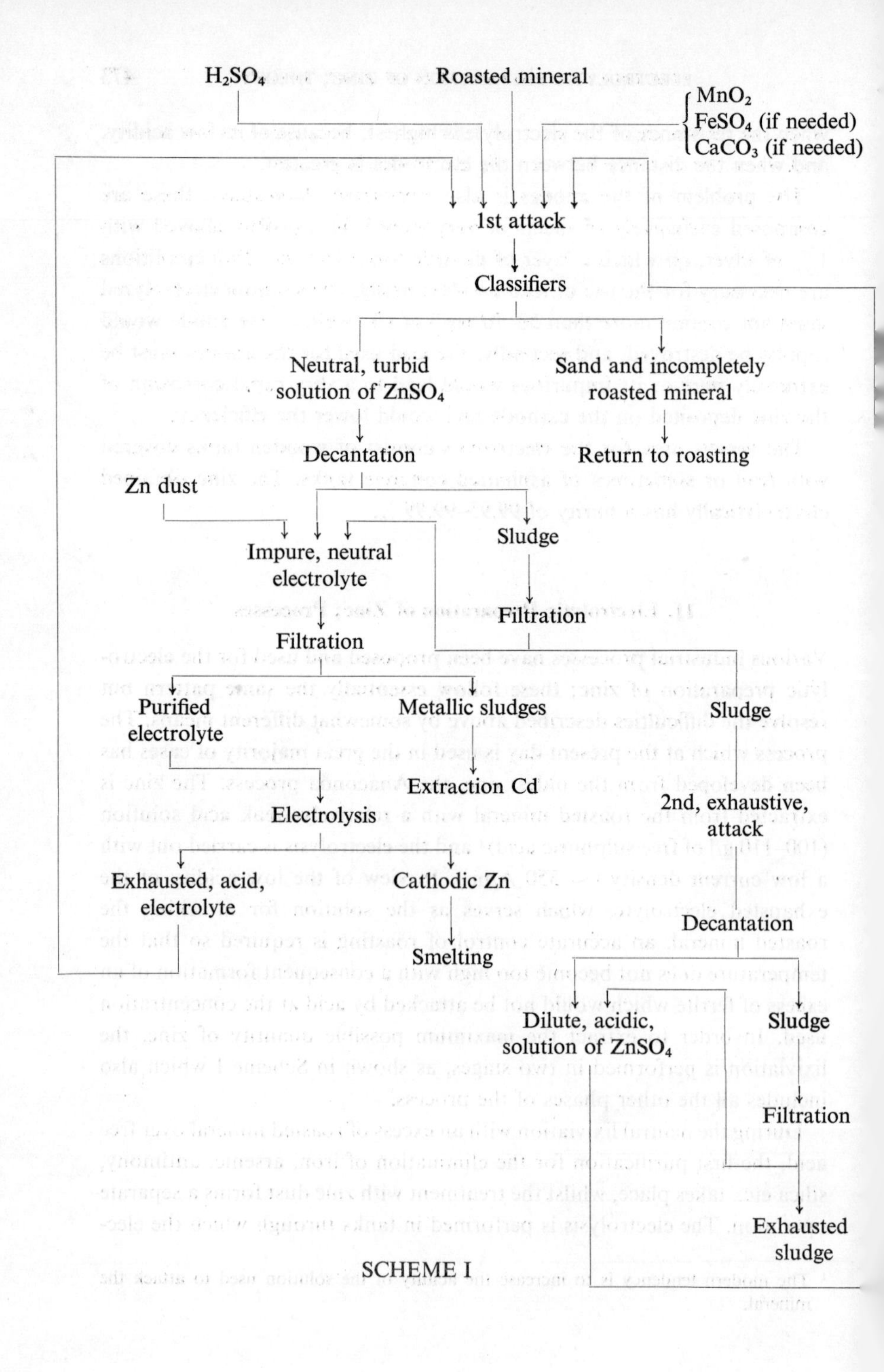

H_2SO_4
Roasted mineral
MnO_2
$FeSO_4$ (if needed)
$CaCO_3$ (if needed)
1st attack
Classifiers
Neutral, turbid solution of $ZnSO_4$
Sand and incompletely roasted mineral
Decantation
Return to roasting
Zn dust
Sludge
Impure, neutral electrolyte
Filtration
Filtration
Purified electrolyte
Metallic sludges
Sludge
Extraction Cd
2nd, exhaustive, attack
Electrolysis
Exhausted, acid, electrolyte
Cathodic Zn
Decantation
Smelting
Dilute, acidic, solution of $ZnSO_4$
Sludge
Filtration
Exhausted sludge

SCHEME I

trolyte circulates in series. This has two purposes: firstly, the electrolyte enters freshly into the first tank and leaves the last one exhausted to return into the lixiviation cycle, and secondly, the average zinc content of the electrolyte in each tank is constant. The rate of circulation is regulated with regard to the current intensity adsorbed by the electrolysis cell and to the initial concentration of zinc, so that the electrolyte leaves the last cell in an exhausted state. Cooling is obtained in the electrolysis tanks themselves by means of a flow of water through lead coils immersed in the electrolyte. Table VIII, 14 shows some typical compositions for the Anaconda process.

TABLE VIII, 14

TYPICAL COMPOSITIONS IN THE ANACONDA PROCESS

Mineral composition			*Electrolyte composition*		
reated	*Component*	*Roasted*	*Fresh*	*Component*	*Exhausted*
55.6 %	Zn	61.5 %	100–120 g/l	Zn	20– 40 g/l
)	Soluble Zn	97.5 %*	5 g/l	Free H_2SO_4	100–110 g/l
.8 %	Pb	3.2 %	500 g/T Zn	Glue	—
).98 %	Cu	0.99 %			
.3 %	FeO	3.2 %			
.17 %	Mn	0.2 %			
9 %	S	1.7 %**			
.9 %	Insoluble	8.4 %			
00 g/T	Ag	743 g/T			
.15 g/T	Au	1.18 g/T			

* Referred to total zinc.
** 0.2 % as sulphide.

The anodic sludge consists almost exclusively of manganese dioxide. The relevant electrical values are shown in Table VIII, 16 with those for a somewhat different process, which was developed more recently than that just described and is called after its inventor: the Tainton process. Although it has never, so far, been very widely used, certain particulars of its operation are interesting and make it worth a short description. It is characterized by a high concentration (280–300 g/l) of free sulphuric acid in the exhausted electrolyte and by a high current density (~ 1,000

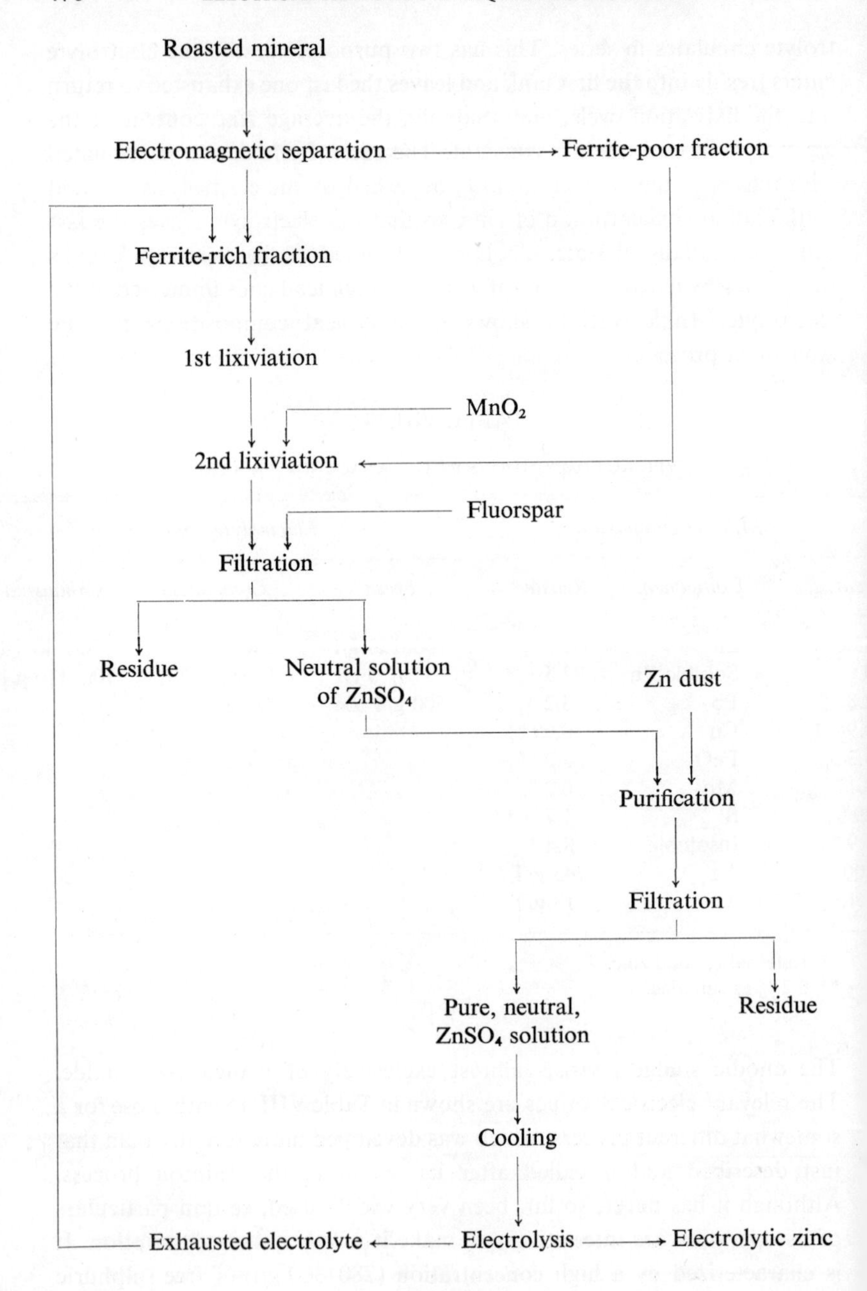
Roasted mineral
Electromagnetic separation
Ferrite-poor fraction
Ferrite-rich fraction
1st lixiviation
MnO_2
2nd lixiviation
Fluorspar
Filtration
Residue
Neutral solution of $ZnSO_4$
Zn dust
Purification
Filtration
Pure, neutral, $ZnSO_4$ solution
Residue
Cooling
Exhausted electrolyte
Electrolysis
Electrolytic zinc

SCHEME II

A/m^2) during electrolysis. The high free acidity of the lixiviating solution, which will also attack zinc ferrite, allows the use of ores which are very rich in iron; the roasting temperature may also be raised and will favour the complete transformation of the original zinc mineral into oxide. Another working characteristic — rather than principle — of this process occurs in the lixiviation, which is preceded by an electromagnetic separation of the ferrite. The ferrite-rich fraction is first treated with hot (60° C), strongly acid, exhausted electrolyte to facilitate solution; only after this is the ferrite-poor, zinc oxide-rich, fraction added together with manganese dioxide to act as an oxidant and favour the precipitation of iron as ferric hydroxide.

The working cycle for the Tainton process is shown in Scheme II.

Because of the high current density and the need to keep the temperature down during electrolysis, the circulation of the electrolyte must be very rapid and the cooling so powerful that a separate refrigerating plant is needed. With such a rapid circulation it is not possible to deposit all the available zinc in a single electrolysis cycle so that the electrolyte must be passed several times through the cells. In the Tainton process therefore, the various cells are connected into the electrolyte circuit in parallel, and not in series as in the Anaconda process. Once a day, part of the electrolyte is diverted to the lixiviation plant and replaced with fresh electrolyte from the purification plant. To improve the deposit, glue or gum arabic is added to the electrolyte at the rate of 1 kg per metric ton of zinc; a

TABLE VIII, 15

TYPICAL COMPOSITIONS IN THE TAINTON PROCESS

Component	*Treated mineral*	*Electrolyte*		*Electrolytic Zinc*
		Fresh	*Exhausted*	
H_2SO_4	—	—	280 g/l	—
Zn	46–48 %	220 g/l	35 g/l	99.99 %
Fe	10–11 %	—	—	0.0085 %
Pb	3– 7 %	—	—	traces
Cu	0.14 %	6 mg/l	—	0.0019 %
Cd	0.14 %	5 mg/l	—	traces
Co	trace	2 mg/l	—	traces
S	28–31 %	—	—	—
SiO_2	3.2 %	—	—	—

suitable foaming agent is also added, to form a dense layer of foam at the surface and prevent the production of sulphuric acid spray by the powerful evolution of anodic oxygen.

The Tainton process should have the following advantages:

1. Minerals more rich in iron could be treated.
2. The roasting furnaces would be more economically used, saving fuel, because roasting takes place more rapidly at a higher temperature.
3. A greater yield of zinc would be extracted.
4. Bringing much of the iron into solution and then reprecipitating it would lead to a better purification from arsenic and antimony.
5. The addition of fluorspar would improve the filtration of the residue, (see Scheme II) and would allow the exploitation of ores rich in attackable silicates.
6. The strongly oxidizing treatment would allow the use of ores rich in cobalt and nickel.
7. The volume of liquid treated would be reduced.
8. The greater acidity and current density would diminish the risk of forming spongey zinc.

Nowadays the advantages of the Tainton over the Anaconda process are, probably however, partially compensated for by improvements in the flotation process, in the preliminary treatment of the mineral, and in the attack on the roasted mineral. Even the advantage that the higher current

TABLE VIII, 16

ELECTROLYTIC VALUES FOR THE PREPARATION OF ZINC

	Anaconda process	*Tainton process*
Cathodes	Aluminium	Aluminium
Anodes	Lead; Pb–Ag alloy	Pb + 1 % Ag alloy
Distance between electrodes	32 mm	20 mm
Electric tension	3.4–3.7 V	3.2–3.6 V
Current density	320–450 A/m^2	~ 1,000 A/m^2
Current efficiency	0.90–0.91	0.88–0.93
Consumption	3.4–4 kWh/kg	3.4–4 kWh/kg
Temperature	4° C	36° C
Cathode life	24 h	~ 10 h
Anodic sludge	MnO_2	MnO_2

density allows the cells to be made smaller is partly offset by the need to provide powerful cooling for the cells; the amount of heat evolved by the Joule effect increases as the square of the current intensity. Some typical analyses are shown in Table VIII, 15, and electrolytic values for the two processes are in Table VIII, 16.

It can be seen from this last mentioned table that the energy consumption is virtually the same for both processes. This is because the greater energy dissipation, through the Joule effect, in the Tainton process is compensated by the lesser electric tension required, in view of the greater specific conductance of the electrolyte and the shorter distance between the electrodes.

Side products of the electrolytic preparation of zinc include the filter residue from the purification of the solution with zinc dust, which in addition to the excess of zinc contains a significant amount of cadmium and is used for its extraction.

12. Electrolytic Preparation of Cadmium

The electrolytic preparation of cadmium has become of increasing importance in the face of a rapid increase in the demand for pure cadmium; this has been due to the recognition of its excellent properties for plating — for which it is superior to zinc — and to the possibility of using its alloys as solders.

The conditions required for the cathodic deposition of cadmium are very similar to those for zinc, with the difference that the more positive discharge tension of cadmium makes it easier. The theoretical conditions for electrolysis, the composition of the electrolyte and the materials used are all analogous to those for zinc and will not be discussed again in detail. Cadmium ores do not exist in nature; in general the cadmium accompanies zinc in whose minerals it is always present in more or less considerable amounts. The raw material for the preparation of cadmium may thus be either (*a*) the powders recovered during the thermal purification, mainly of zinc but also of copper and lead; and (*b*) the residues obtained during the purification with zinc dust of zinc sulphate solutions for electrolysis (*cf.* the two preceding sections).

In general, the same difficulties are encountered during the preparation of cadmium as during that of zinc. The raw materials (*a*) or (*b*) are subjected to various chemical treatments to remove metals more noble than cadmium. The product is a spongey mass of cadmium containing zinc as the main impurity with possible traces of iron and copper. The mass is dissolved in acidic, exhausted electrolyte and the solution is freed of iron and copper, if necessary, and then electrolyzed. The chemical treatments differ somewhat depending on whether the raw material consists of the powders (*a*) or the residue (*b*), because of differences in their composition. The powders recovered from the fumes of zinc furnaces contain zinc, cadmium, copper, iron, manganese, arsenic, selenium, tellurium, cobalt, nickel, silver, gold, bismuth, thallium, etc.[1]

If the arsenic content is very high, the material is treated with concentrated sulphuric acid in low temperature reverberatory furnaces until there is no further evolution of fumes. Much of the arsenic is thus eliminated. The residue is lixiviated with dilute sulphuric acid to dissolve the zinc, cadmium, arsenic, cobalt, nickel and traces of the bismuth, silver, thallium and tellurium. The clear solution is then treated with powdered chalk and air is blown in to oxidize the iron; this precipitates together with part of the copper and the arsenic, which is adsorbed by the ferric hydroxide. After filtering, the solution is treated with zinc dust to precipitate the remainder of the copper and the cadmium; on suspending the precipitate in 25 % sulphuric acid at 60° C, the zinc and cadmium redissolve. The solution is neutralized with calcium hydroxide and carbonate, and the cadmium is cementated with strips of electrolytic zinc. The spongey mass of cadmium formed is washed and dissolved in hot, exhausted electrolyte from the electrolysis of cadmium. The residual iron is removed by blowing air through the solution and treating with calcium hydroxide; any thallium present is removed as the chromate by treatment with sodium dichromate. The final solution is then electrolyzed.

The residue (*b*) from the purification of zinc sulphate solutions with zinc dust, is first roasted at about 700° C to make the iron insoluble and to remove any arsenic or antimony present. It is then lixiviated with exhausted electrolyte from the electrolysis of zinc (10–12 % free H_2SO_4) and decanted. The clear solution contains all the zinc and the cadmium and part of the copper. It is carefully cementated with zinc powder so as to separate only the copper. After filtering this, more zinc is added to separate the cadmium. The spongey cadmium obtained is dissolved in ex-

[1] Many of these elements are present, if at all, only as traces.

hausted electrolyte from the electrolysis of cadmium and the solution is treated with cadmium to separate any remaining traces of copper, filtered and electrolyzed. The solutions containing the zinc used for the separation of the various impurities and then to cementate the cadmium, are normally sent to electrolytic plants for the recovery of the zinc.

The processes described will have made clear the usefulness of a sulphuric electrolyte and the need for high purity of the cadmium sulphate solutions used for electrolysis; impure solutions would lead to difficulties like those discussed for the electrolysis of zinc sulphate solutions.

The electrolyte has a cadmium content of from 90 to 200 g/l and a variable zinc content which may reach 20 % of that of cadmium without causing trouble; since cadmium is more noble than zinc it will be preferentially deposited at the cathode. About 1 kg of glue is dissolved in the electrolyte for each metric ton of cadmium to be separated so as to obtain a compact cathodic deposit. With cadmium again, deposition from an acid solution would be impossible were it not for the high overtension of hydrogen on a cadmium cathode; in neutral solution the discharge tensions of the Cd^{2+} and H^+ ions are almost equal.

It is not easy to obtain a compact, adherent deposit of cadmium on stationary cathodes and cells have been built in which the cathodes form cylinders with horizontal axes, which are rather less than half immersed in the electrolyte and slowly rotate. The deposited cadmium is conti-

TABLE VIII, 17

ELECTROLYTIC VALUES FOR THE PREPARATION OF CADMIUM

Starting electrolyte	90–200 g/l Cd 20– 40 g/l Zn
Exhausted electrolyte	10– 20 g/l Cd 20– 40 g/l Zn 60–140 g/l H_2SO_4
Anodes	Pb; Pb + 1 % Ag
Cathodes	Al
Electric tension	2.5–4 V
Current density	40–250 A/m²
Current efficiency	0.85–0.90
Consumption	1.4–2.25 kWh/kg
Temperature	30–35° C

nuously stripped from these cylinders. Other types of cells use stationary electrodes and obtain fairly compact and adherent deposits by various expedients such as the addition of colloids, agitation of the solution etc. Solutions are usually electrolyzed almost quantitatively. The electrolytic values are shown in Table VIII, 17.

The electrolytic cadmium obtained is melted under caustic soda, to avoid oxidation, and cast into bars, rods or any other desired shape.

Exhausted electrolyte is recycled to lixiviate more spongey cadmium. When the zinc content of the electrolyte becomes too high, a part of it is replaced with fresh exhausted electrolyte from the electrolysis of zinc.

13. Electroplating; Theoretical Principles

One of the objects of electroplating is to cover metallic articles with a more or less fine layer of another metal so as to improve their surface properties either decoratively or protectively by increasing hardness, resistance to corrosion etc. The treated objects need not be actually metallic provided that they are conducting or can be made so. Another object of electroplating is the fabrication, by electrolytic deposition on suitable formers, of articles of particular shape and characteristics such as seamless tubes, parabolic mirrors, clichés for printing, matrices for gramophone records etc. Of these two uses, the former — true electroplating — is much more important than the latter: galvanoplasty or electroforming.

Electroplating aims at producing an electrolytic deposit which is as constant in thickness as possible over the whole of the treated surface, is smooth, regular and very adherent to the support, nonporous and of the most compact structure possible with a very fine crystalline grain; foreign inclusions and surface defects such as cracks and pinholes, must be absent. The type of structure obtained in electroplating obeys the general rules described on page 431 *et seq.*; these rules must be expanded by certain special considerations concerning galvanic deposits in thin layers.

In electrolytic refining and preparation of metals, small differences in the thickness of the cathodic deposit are of no great importance but in electroplating where the thickness of the deposited layer is of the order of 10^{-1}–10^{-2} mm, it is very important to obtain deposits of constant thickness even on articles of varying shape such that the distance from the anode to each part of the article varies. In every case, the differences in thickness must be kept within rather narrow limits to avoid deposits which are too thick at some points and too thin at others. Differences in thickness are due to the fact that the current density tends to increase towards the edges of articles which are being plated and at projecting points which are closer to the anode particularly if their radius of curvature is small, whereas it tends to diminish within cavities. However, it depends on a number of factors which tend to oppose the production of such differences. The main factors acting to maintain the constancy of the thickness of the deposit are as follows.

1. An increase in current density leads to an increase in polarization.
2. An increase in current density can lead to a diminution in current efficiency.
3. Suitably selected electrolytes can cause an increase in polarization.

The mechanism of the first factor is obvious. There will be an increase in the current density at a point on the cathode which is closer to the anode than other points, through the smaller ohmic resistance of the electrolyte which has to be overcome. However, the polarization at this point increases as a consequence. This increase in polarization may be considered, in practice, as a further resistance to be overcome at that point; thus the increase in current density of that point compared with neighbouring points, and hence the increase in the thickness of deposited metal, will be opposed. The second factor acts in a similar way. If, through the third factor, the polarization necessary for the discharge of the cations from a particular electrolyte is high, then the differences existing in ohmic resistances between the anode and the various points in the cathode, become negligible. Thus, the current is distributed more uniformly over the whole cathodic surface. The characteristic of giving more or less regular deposits is designated as the *covering power* or *penetrating power*. The covering power will be increased by an increase in the conductance and by the addition of substances which raise the polarization (colloids etc.); it will be depressed by any factor which tends to lower the polarization. It also depends on the absolute values of the current density etc. So many factors must be taken into consideration as affecting covering power, that, although it is possible to recognize certain general rules it is very difficult to express the covering power as a figure which can be calculated. In general, therefore, covering power is determined by certain empirical tests: tests on plates with similarly shaped depressions which are equidistant but of variable depth; tests on cathodes bent through a right angle and with the vertical limb parallel to the anode; tests of the covering of the inside of tubes of particular dimensions etc. The method proposed by Haring and Blum[1] is still the best; it consists in measuring the deposit obtained on two cathodes placed at different distances from the same anode. Bianchi[2] proposed a new type of cell and a new definition for the penetrating power of plating baths.

In order to make the deposit as regular and smooth as possible, which requires a very fine crystalline grain, other expedients may be used in addition to those discussed on p. 431 *et seq*. Pulsating currents or superimposed alternating currents may be used to obtain a type of electrolytic polishing (see below). Such superimposed alternating currents tend to equalize differences in concentration, any differences in polarization induced at various points on the cathode, and differences in thickness, by partial anodic dissolution at each alternation. The surface structure of the underlying metal also has an effect; to obtain a deposit which is sufficiently uniform and smooth to be used as a mirror surface, it is necessary to thoroughly polish the underlying metal.

Finally, for the deposit to be perfectly adherent to the support, it is necessary to subject this to a surface treatment to 'anchor' the deposit to the support (so-called pickling, see the following section); alternatively the two metals of the support and the surface deposit must be able to form an alloy of the solid solution type. In this case, the first layer deposited will diffuse slightly into the support and so form an intermediate layer of alloy of very fine thickness which is however sufficient to achieve perfect and very tenacious adherence of the deposit to the support. When the supporting metal cannot form an alloy with the deposited one, it is convenient to interpose another metal which can form alloys with each of the others. For this reason, for example, many deposits adhere better if the article is first superficially amalgamated or, as in some cases of nickel plating, is first copper plated. In every case, an essential condition for good adhesion is the perfect cleanliness of the surface to be covered and its suitable preparation (see the following section).

[1] H. E. Haring and W. Blum, *Trans. Am. Electrochem. Soc.*, 44 (1923) 313. For a discussion of the method, see T. P. Hoar and J. N. Agar, *Discussions Faraday Soc.*, No. 1, Electrode Processes (1947) 162.

[2] G. Bianchi, *Chim. e ind. (Milan)*, 35 (1953) 414.

An electroplating bath must contain the following constituents:

1. a compound which, on dissociation, will give the cations whose discharge forms the galvanic deposit;
2. possibly an inert electrolyte to increase the conductance if necessary;
3. a substance to facilitate the dissolution of the anodes, if necessary;
4. possibly correctives, added in small quantities to influence the surface characteristics of the cathodic deposit in some desired manner (colloids, brightening agents, etc.);
5. substances to buffer the solution at a particular pH, if necessary.

Sometimes certain of these components may not be necessary and sometimes a single substance may simultaneously fulfil more than one of the indicated functions. For these reasons, electroplating baths are always much more complex than the electrolytic solutions used for the refining or preparation of metals.

In general, an electroplating process is analogous to a refining process in that the composition of the electrolyte stays constant; whereas cations are discharged on the cathode, the anode of the same metal dissolves in an equivalent amount; a characteristic example is silver plating. The use of insoluble anodes is more rare; an example is chromium plating in which the anode is insoluble and the electrolyte forms both the source and the reserve of cations to be discharged.

14. Electroplating; Practical Considerations[1]

A careful preparation[2] of the article is of fundamental importance for a good result. Before proceeding to the electrolytic deposition, the metallic surface to be covered must be completely exposed, *i.e.* all substances which normally or occasionally adhere to the surface to be plated must be removed, since they could prevent the electrolytic deposition or rather,

[1] This section refers particularly to certain fundamental operations in true electroplating. For electroforming, which is strictly a particular working technique, reference should be made to specialized texts.

[2] The operations described in this section with the general term of preparation are discussed so as to emphasize principles which apply as general rules. The exact conditions must be found for particular cases and may depend on so many other factors (the substance forming the object to be plated is fundamental for example) that it is not possible to indicate a single procedure which will always be suitable nor is it possible in many cases even to predict *a priori* what the best conditions will be.

make the deposit nonadherent to the support at a particular point. Substances adherent to the surface to be plated may be classified into two general groups: firstly, substances deriving from the metal to be plated (oxides, salts generated by corrosion such as carbonates, sulphates etc.); secondly, extraneous substances such as residues from the mould, grease, oil, powder and in general, dirt. All these must be quantitatively eliminated. Normally, a mechanical treatment (grinding, brushing etc.) is first used to remove everything which is strongly adherent particularly if it is resistant to chemical attack, *e.g.* the particles of sand remaining attached to the article from the mould used in casting. The mechanical treatment also serves to prepare the surface and make it as regular and smooth as desired.

After this first operation, or operations, the true cleaning of the article is carried out with suitable solutions. These can act efficiently only if the article is completely wetted so that the chemical treatments may be divided into two types: those aimed at eliminating the last traces of all substances such as fats, oils and minerals which would interfere with wetting and those aimed at removing impurities of the first group (the oxides etc.) Baths used for degreasing may be of two types: organic solvents (benzene, toluene, benzine, trichlorethylene etc.) and more or less strong solutions of caustic soda or sodium or potassium carbonate, with a normal addition of emulsifying agents such as sodium silicate, trisodium phosphate, organic detergents etc.; these latter baths are used hot. Sometimes the alkaline baths also contain solvents for some oxides, *e.g.* cyanides.

The true fats are saponified by the alkaline treatment and pass into solution but the mineral greases and oils are not chemically attacked; however, they are normally removed from the surface by the emulsifying agents. By immersing the article in the alkaline washing solution, and polarizing it cathodically, the evolution of small bubbles of hydrogen aids the emulsification in the bath and acts mechanically to remove the adherent film. This last treatment, which is called electrolytic washing[1], is further assisted by the cathodic film becoming an alkaline caustic solution through the discharge of the H^+ ions. To be effective, the evolution of hydrogen in electrolytic washing must be vigorous and hence the current density

[1] Electrolytic washing is also used for purposes other than preceding a plating operation *e.g.* to rapidly and easily remove paint from tin cans which are to be re-used and to clean the insides of metallic containers which are difficult to reach.

must not fall below 1 A/dm^2. Normally, values of 2 A/dm^2 are used.

Electrolytic washing is difficult for zinc, tin, lead and its alloys which tend to pass into solution in the alkaline liquid and to redeposit cathodically in thin, poorly adherent layers which are detrimental to the following plating operation. This inconvenience may be avoided by polarizing the article anodically for a short time so as to redissolve the layer of cathodically deposited metal.

Apart from a few rare cases, the true washing operation follows the treatment for the elimination of impurities of the first group carried out in acid conditions. This last operation, called pickling, is performed with sulphuric, nitric, hydrochloric or hydrofluoric acids either alone or in mixtures with the possible addition of sodium chloride; the treatment is normally carried out in hot solution, sometimes almost boiling. This acid treatment makes part of the impurities of the first group dissolve and part separate from the article because the acid attacks the metallic surface immediately in contact with the layer of oxide. The support to which the impurity adheres is thus destroyed and it is readily removed by the bubbles of hydrogen evolved in the reaction between the metal and the acid.

The attack on the metallic surface with acid also serves to slightly roughen it, which increases the adherence of the deposit, tending to anchor it to the underlying surface. This treatment is indispensable when the metal of the deposit and that of the article cannot form solid solutions and it is not possible, or desirable, to insert an intermediate metallic layer capable of forming solid solutions with both of the other metals. The pickling can also be carried out electrolytically by polarizing the articles cathodically in the acid bath; the discussion given for electrolytic washing generally applies here also. In particular, electrolytic pickling shows no advantage in acid consumption, it is however preferable because it is more uniform and rapid, can be carried out at lower temperatures and gives a better surface. Immediately after pickling the object is immersed in the electrolytic bath and cathodically polarized for the electroplating.

When the article has to be prepared by passing it through a number of baths of different composition, it is always thoroughly washed with boiling water after each treatment so as not to contaminate the different solutions; if necessary, it is also dried but this must be done with care to avoid reforming a superficial film of oxide.

In practice, it is essential to be able to determine *a priori* various quantities associated with the electrolysis operation such as the time of electrolysis, the thickness of the deposit obtained and its weight. These quantities can be approximately determined by very simple calculations. If

E represents the electrochemical equivalent in g/Ah (*cf.* Table IV, 1)
I the current intensity in A
S the surface area of the article in dm^2
J the cathodic current density I/S in A/dm^2
d the density of the deposit
l its thickness in mm
$R_{curr.}$ the current efficiency under the conditions of electrolysis
t the time in hours
G the total weight of the electrolytic deposit in g
G' the weight of the deposit in g/dm^2

then the following equations apply. The total weight of the electrolytic deposit is given by

$$G = EItR_{curr.} \tag{1}$$

The weight per dm^2 is given by

$$G' = G/S = EItR_{curr.}/S = EJtR_{curr.} \tag{2}$$

The time required to obtain a deposit of a given weight per dm^2 is given by re-arrangement of equation (2)

$$t = G'/(EJR_{curr.}) \tag{3}$$

To calculate the time necessary to obtain a given thickness, the density is introduced by the following equation.

$$G' = G/S = Gl/(Sl) = 10\ ld = EJtR_{curr.}$$

$$\therefore \quad t = 10\ ld/(EJR_{curr.}) \tag{4}$$

Simple rearrangements of equations (1), (2), (3) and (4) give all the desired quantities: current intensity, thickness of the deposit etc. depending on which other quantities are known. These equations apply exactly for stationary electrolytes and articles, when these are is completely immersed in the electrolyte. However, the equations become more complex when the object moves continuously through the electrolyte as in the electroplating of wires. The most commonly used metals for electroplating are Ag, Au, Cd, Cr, Cu, Ni, Pb, Pt, Rh, Sn, and Zn to which has recently been added Al.

A special type of bath for electroplating is based on cyanide. A discussion of the composition and properties of such baths has been published by Thompson[1].

15. Electrolytic Polishing

by

ISRAEL EPELBOIN[2]

Electrolytic polishing is a selective electrolytic dissolution of metals or certain semiconductors which leads to smooth and brilliant surfaces being obtained, since the raised points and projections are attacked more rapidly than the depressions.

The article to be treated is almost always made the anode in an electrolytic cell fed by a continuous or intermittent current, and for this reason the process is also called 'anodic polishing'. It is also sometimes possible to obtain selective dissolution, and even a good polish using a variable, and especially an alternating, current. Finally, a so-called 'chemical polishing' process exists which requires no external source of current and which is allied to electrolytic polishing in that the electric tension which is established spontaneously at the metal–electrolyte interface is itself sufficient to induce selective dissolution.

There is no general agreement on the properties required of a well-polished surface. Some require that a good electrolytic polishing should make the treated surface smooth, uniform and shiny. Some require that it should reveal the graininess and other structural details. And finally, some consider that the aim of the polishing is to give the surface the most brilliant appearance possible, even if a microscopic examination should show pitting or traces of oxides or salts. In this last case, however, it is preferable to call the operation 'brightening'. Mention should also be made of thinning and shaping by electrolytic polishing.

In the discussion of electrolytic polishing it is then necessary, not to lose sight of the characteristics which it is desired to give to the surface.

Once the phenomenon had been recognized, various theories were

[1] M.R. Thompson, *Trans. Electrochem. Soc.*, 79 (1941) 417.

[2] Director of Research, C.N.R.S., Paris.

proposed to explain its mechanism. Theories which related the levelling of the surface to a temperature gradient or to the formation of complexes at the anode–electrolyte interface may be simply mentioned, to be discarded.

At present, most authors agree on the essential role of a viscous layer over the surface which is to be treated; this layer is often visible to the naked eye and disappears instantaneously when the conditions for polishing are lost. The composition of the layer is still under discussion and depends, moreover, on the electric tension, on the metal to be polished and on the solution used. Some authors consider that the layer is formed of oxides but others believe that it is found only in the presence of salts or that it consists of adsorbable ions. It has been shown that the layer has a low ohmic resistance and contains very little water. It is also known that once a certain thickness has been acquired, this will remain constant provided that the conditions of polishing are not altered. Thus, the layer is formed at the common interface with the metal and dissolves on the electrolyte side. It has been shown that the levelling mechanism is linked with a diffusion process which is set up through the layer. Since the current density is proportional to the concentration gradient, the density is lower in the depressions and higher at the crests. A selective dissolution results which leads to the levelling: firstly, on the microscopic scale and then on the macroscopic scale. One hypothesis, which is already obsolescent, presumes that the ions of the dissolved metal diffuse towards the solution. However, recent studies have shown that the anions of the solution lose their water of hydration in the neighbourhood of the anode and as they become adsorbed at the metal–electrolyte interface, the water which is liberated diffuses towards the solution. Sometimes, the anions combine with the metallic ions to form a crystallized salt which can be observed with a microscope in polarized-light. The use of the microscope during electrolysis has also shown that this crystalline layer disappears as soon as the current is interrupted.

The formation of the anodic layer may be followed with the help of the I–U curve which gives the electrolysis current I as a function of the anodic tension U. The current varies during the formation of the layer but it then becomes stabilized and the curve shows a plateau. Polishing could be said to occur for all values of U corresponding to this plateau, but the best results are obtained at the right hand end of the plateau, *i.e.* with the electric tension which gives the maximum apparent resistance U/I. At still higher electric tensions the current increases again, the layer is destroyed and the anode is attacked irregularly over its surface.

When the *I–U* curve is constructed, the independent variable is the anodic tension and it is essential to respect this condition. If not, even a slight variation in the current *I* or the overall tension *V* at the terminals of the cell, may cause marked variations in the anodic tension and so mask the phenomenon. When the solution used allows polishing to be carried out with slight changes in current density and under a high electric tension, it is possible to make the anodic tension *U* proportional to the electric tension at the terminals *V*. For this, it is essential that the anode should be much more conducting than the solution and that its surface area should be much less than that of the cathode. Under these conditions, a rheostat, placed as a potentiometer across the terminals of the cell is sufficient to plot the *I–U* curve, *e.g.* with a perchloric acid bath. On the other hand, if the solution used leads to marked variations in current density or if the anodic tension is very feeble, *e.g.* with chromic–sulphuric baths, the anodic tension will not be proportional to the electric tension at the terminals and it will be necessary to use a potentiostat to follow the development of the diffusion layer.

For an electrolyte to be useful as a bath for electrolytic polishing, it is necessary that at certain electric tensions it can give rise to the formation of a compact diffusion layer; moreover, this diffusion layer must dissolve in the bath at a rate which is proportional to that of its formation, at least after it has attained a certain optimum thickness. Other factors are concerned in the choice of a polishing solution. As explained at the beginning, it is essential first of all, to be clear what surface state is to be obtained; the bath will not be the same for a brightening operation or for polishing for a metallographic examination. Finally, it is necessary to take account of the rate of dissolution, of the ease of control, of the price of the installation and of the bath, of the ease of regenerating the latter, and of safety regulations etc.

Baths can be classified according to the substances used in their composition and this allows a fairly small number of types of bath to be distinguished. Nevertheless, the constituents may have rather different concentrations in the solution depending on the metal or the alloy which is to be polished; thus, the tables found in specialized works show a rather large range of compositions. Even these compositions, however, are not rigorously fixed and may be slightly varied depending on the chemical composition of the substance to be treated; this is not always constant when dealing with an alloy and impurities may be present.

Once the type of bath has been selected, the best composition is often determined empirically. However, a knowledge of the role of water in the levelling process has allowed more rational methods to be established. These are based on studies of the current–tension curve and especially on the determination of the maximum of the apparent resistance U/I. In order to determine the composition of a bath in a precise manner, it is often useful to construct the R–U curve (R = apparent resistance of the electrolytic cell) for baths of very similar composition. That bath whose apparent resistance is greatest will be the most suitable.

An important group of baths for electrolytic polishing consists of solutions based on ClO_4^- ions. The oldest of these was invented by P. Jacquet, 30 years ago, and consists of perchloric acid and acetic anhydride. It is still in use but the group has been extended by perchloric acid–acetic acid, perchlorate–acetic anhydride, and perchlorate–alcohol solutions. All such baths require a high anodic tension (10–50 V). If the surface is properly washed after the electrolysis, no traces of impurities will remain, and moreover, these solutions have the important advantage of not making the treated surface passive. The polishing is generally achieved without any release of gas so that Faraday's Law may be used to calculate the rate of dissolution of the metal; but it should be noted that the calculation is not as simple as it might seem, particularly for certain metals. The ClO_4^- anions effectively lead to the passage of metallic ions into solution at a low valency and when the metals are oxidizable (Al, Be, Mg, Zn, La, U, Ce, Ti, etc.) their dissolution valency will be even lower than the normal one. The metallic ions thus pass into solution with an unstable valency and regain a stable one by reducing the ClO_4^- ions at the anode. In this case, Cl^- ions are found near to the anode. On the other hand, if the anodic layer contains water this will be reduced and hydrogen will be evolved at the anode.

Baths based on ClO_4^- ions are very useful for laboratory work and for micrographic control. They are also advantageous for industrial use since they considerably improve the properties of metallic surfaces such as their reflecting power. Unfortunately, however, their use in industry has a serious drawback. Perchloric acid baths must never, under any account, exceed a concentration of 40 % by weight, or a temperature of 50° C even locally. If these precautions are not scrupulously respected a very grave risk of explosion will be encountered. The perchlorates, however, are much less dangerous. Magnesium perchlorate in particular, can be handled without special precautions but must never be used for polishing bismuth or tin.

Other mixtures are based on phosphoric, sulphuric and chromic acids. These form the industrial baths which have been in principal use for the last twenty years. Compared with earlier ones, they have the advantage of requiring a low anodic tension (a few volts) but they must be used at a temperature of between 60 and 90° C and the current density must be high. The polishing is accompanied by a production of gas which must be allowed for when Faraday's Law is used to determine the rate of dissolution; however, the valency of the metal is always the normal one. Surfaces polished in these solutions are always passive.

Certain noble metals may be polished, at 80° C, using salts which melt in their water of crystallization, *e.g.* ferric chloride for polishing gold. Polishing can also be carried out in molten salts, without any solvent. Certain semi-conductors and platinum can be polished well, in a eutectic of sodium and potassium chlorides at 750° C. Mention should also be made of baths based on potassium cyanide for the polishing of silver and of those based on soda for the polishing of tungsten.

The different compositions used for chemical polishing are shown in the technical literature. Many contain more or less large amounts of nitric acid. The nitric acid concentration is very important since it is this which determines the properties of the diffusion layer which surrounds the treated article. For example, copper is polished chemically at room temperature in a solution containing 160 cm^3 of acetic acid, 40 cm^3 of orthophosphoric acid and 5 cm^3 of nitric acid, and the latter is required for the formation of the diffusion layer. However, a film remains on the treated surface even when the article is taken out of the bath. It is necessary to immerse it in a second solution containing the same acetic–phosphoric acid mixture to which has been added 29 cm^3 of nitric acid. The previously insoluble layer now dissolves and the metal becomes bright. In general terms, baths for chemical polishing contain an oxidant and if nitric acid is removed it is replaced by chromic acid, hydrogen peroxide, etc.

Although a patent was deposited in 1909 for the polishing of precious metals, it is only after the work of P. Jacquet (1929) that electrolytic polishing has developed.

One of its essential applications is the micrographic control of metals and semi-conductors both in the laboratory and in industry, during their fabrication. Its utilization is often recommended for the preparation of surfaces for electroplating, for the finishing of mechanically-worked pieces and even between different mechanical treatments in order to facilitate them. In general, it has become the custom to prepare articles by electrolytic polishing when this will improve their physical properties by eliminating interfering layers, or when the quality of the surface state has a considerable effect on the end result or on the reproducibility of the working conditions. Finally, electrolytic polishing is increasingly frequently used as a very powerful procedure for electrolytic dissolution for edging off, cutting and drilling.

When it is necessary to maintain and reproduce definite conditions of polishing, an ammeter and a voltmeter are often sufficient.

Nevertheless, it is always necessary to carefully determine the electric tension at which the apparent resistance of the electrolytic cell reaches a maximum value. This method can be used to select the composition of a bath; it is also frequently used for the rapid polishing of specimens which require a good surface finish. To determine this maximum, the cell can be inserted in a Wheatstone bridge. A feeble alternating current can also be superimposed on the continuous electrolysis current; studies have shown that the apparent resistance and the impedence of the cell both show a maximum at the same electric tension. Various types of apparatus can now be obtained which allow nonspecialists to use this method. It is particularly fruitful when it is wished to carry out the metallographic control of articles of varying nature and shape.

A number of types of apparatus exist which allow this control to be performed economically. The article is generally placed under electric tension by an electrolytic jet, produced by a pump or by centrifugal force, and the microscopic examination may be made directly. When the articles to be tested are of large size a procedure of polishing under a pad is used, which allows a local treatment. For microscopic studies it is generally necessary to use intermediary replicas. The automatic installations and the necessary accessories for industrial polishing on a large scale are analogous to those used for electroplating. The similarity is particularly marked with industrial polishing in phosphoric–sulphuric–chromic acid baths and this explains their very common utilization throughout the world.

Chemical polishing, particularly of light metals and of certain semiconductors, has expanded greatly in recent years since its application is simpler. It avoids the difficulty of controlling the current but unfortunately the control of the best conditions for polishing is much more delicate than for the electrolytic treatment.

In an electrolytic polishing installation, the good distribution of the current is very important and determines the form of the connections. In the laboratory, advantage is taken of recent progress in the varnish and plastics industry for the convenient protection of surfaces which must not be treated.

However, when electrolytic polishing is used as a forming procedure the apparatus becomes very important. The shape of the electrodes must also be particularly studied. Thus, the cathode employed is filiform when it is wished to cut through a bar; but it is a hollow cylinder of revolution when it is wished to thin down a wire or a ribbon.

For a satisfactory use of electrolytic polishing, it is then necessary to have not only general ideas about the phenomenon but also to be fully informed about apparatus particularly as a function of the end in view.

16. Metallic Powders

The preparation of metallic powders is becoming increasingly important because by their use, it is possible to obtain articles which are difficult to fabricate by other methods such as casting or forging; to prepare porous articles (which have to be impregnated for bearings etc.) which cannot otherwise be obtained; to work very hard metals; and to prepare powders for catalytic reactions, etc.

Various methods exist for the preparation of metallic powders: electrolysis, chemical reduction of powdered oxides, fragmentation of brittle metals and the pulverization of molten metals, etc. The electrolytic procedure is, however, qualitatively superior to the others because it gives particles which are finer and more uniform in size lying between 0.1 and 30 μ in diameter. To prepare metallic powders electrolytically, the operation is carried out under conditions which are as opposite as possible to those used for the metallic deposits of electroplating, always bearing in mind for the choice of conditions, the general rules on the structure of metallic deposits discussed on page 431 *et seq.*

It may be important in preparing powders to obtain a uniform grain size. In general, powdery deposits are produced with high current density, low ionic concentration of the cation to be deposited, low temperature, low conductance and the addition of colloids. It is obviously necessary to carefully select the composition of the electrolyte which must then be accurately controlled during the process, particularly with regard to pH. In every case, it is necessary to exceed the limiting diffusion current density. Metallic powders are also obtained by the electrolysis of molten electrolytes whenever their melting point is below that of the metal.

The powders may either be obtained directly or through a fragile cathodic deposit which can be further processed mechanically. In both cases, an abundant cathodic evolution of hydrogen is very helpful. The size of the particles obtained directly by electrolysis may be varied by altering the factors which facilitate the formation of new crystalline nuclei and hinder their growth. In this respect, in order to keep the current density constant, it is necessary to remove the powdery cathodic deposit continuously from the cathode since its large surface area would tend to lower the effective current density. The addition of reducing substances has sometimes proved to be useful; these will be anodically reoxidized. The addition of colloidal substances makes it possible to produce extremely fine powders by diminishing or completely eliminating the simultaneous evolution of hydrogen; the current efficiency will thus be increased. The main metals obtained industrially in the powdered state are copper, zinc, iron, cadmium, tin, antimony, silver, nickel, and tungsten.

17. Corrosion and Passivity

by

GIUSEPPE BOMBARA[1] and GIULIO MILAZZO

Corrosion and passivity are obviously not phenomena which are used in the electrochemical industry for production purposes and they do not have characteristics which are of particular theoretical interest, but since they are of tremendous economic importance and are underlain by those same electrochemical phenomena which are used in electrometallurgy in aqueous solutions, it seems opportune to deal with them here at the end of this chapter.

(I) *Corrosion*

An essentially electrochemical phenomenon which is of particular importance for metallic materials is that of corrosion. Corrosion leads to direct

[1] Laboratori Riuniti Studi e Ricerche E.N.I., S. Donato Milanese.

expenses, linked with the cost of protective measures and the replacement of corroded material, and also to indirect expenses which can not be estimated but are almost always higher than the direct; these indirect expenses are the result of installations and plants being put out of service, of diminutions in yield, of explosions and contaminations, and sometimes of a loss of life.

The economic importance of this phenomenon, to the use of metallic materials, can be clearly seen from the figures collected by various authors. In the United States alone, H.H. Uhlig calculated by an analysis of the various reports, that the direct loss due to corrosion in 1949 exceeded $5,000 millions. Some $2,000 millions were represented by the cost of protective paints and $600 millions by the cost of maintaining and replacing underground piping.

W.H.J. Vernon estimated that the direct cost of corrosion in the United Kingdom in 1956 was £600 millions and H.K. Worner put the corresponding figure for Australia, for that year, at £120 millions.

Corrosion converts metals into various chemical compounds, depending on the nature of the corroding agent, and these compounds usually have very much less mechanical and chemical resistance than the original metal, which thus disintegrates so that the object is destroyed. Corrosion in the strict sense refers to the attack which metals suffer from substances of an electrolytic nature with which they are in contact, or else to the attack provoked by primary electric currents at the points at which they leave the metal in an environment containing electrolytes, so that the metals function as electrolytic anodes. The attack starts at the surface of the metal and may increase in depth until it is holed.

Corrosion is thus always due to an electrochemical cause: either *galvanic* or *electrolytic*. In the former case the corrosion is a consequence of the formation of local galvanic cells in which the passage of the current is caused by chemical reactions between the metal and the corrosive agents. In the latter case it is the passage of a primary current which provokes a true electrolysis with the anodic dissolution of the metal. The two cases may conveniently be dealt with separately.

In the first case the corrosion is due to one or more chemical reactions occurring between the metal and the substances of an electrolytic nature dissolved in the medium in contact with the metal. The presence of water or another ionizing solvent is essential for corrosion to occur. In many cases, in which the metal does not apparently come into contact with water, it is atmospheric humidity whose condensation produces a liquid film, which is sufficient to act as solvent for the actual corrosive agents when it does not fulfil this role itself.

For metals less noble than hydrogen, in the scale of electric tensions in water, corrosion may be interpreted according to the reaction

$$2\,M + 2z\,H^+ \rightarrow 2\,M^{z+} + z\,H_2$$

This reaction is possible when the equilibrium electric tension of the H^+ ion under the actual conditions of acidity, together with any overtension of hydrogen on the metal, is greater than the electrode tension of the metal itself. The metallic ions thus formed interact with the excess of OH^- ions left in the water by the discharge of the H^+ ions, to form hydroxides. These are usually poorly soluble and precipitate, so that the reaction may continue indefinitely, since the original conditions are, at least partially, restored. If acid gases which are normally present in the atmosphere are dissolved in the water (CO_2, H_2S, SO_2, etc.), the respective salts are formed and are also usually poorly soluble. This type of corrosion is obviously facilitated and enhanced by the presence of dissolved oxygen, since this acts as a depolarizer for the discharge of the H^+ ion.

This theory can obviously not be applied to metals which are more noble than hydrogen. The theory which truly copes with all possible cases of spontaneous corrosion *i.e.* all those not due to an external electric current as a primary cause, is that which considers the formation of a local galvanic cell as an essential condition. According to this interpretation, the fundamental cause of the corrosion is the electric current generated by the local galvanic cell formed. Metals, in practice are never perfectly pure nor perfectly homogeneous, either chemically or physically, particularly if they are alloys rather than pure metals in the chemical sense. A variety of constituents are usually found in alloys: solid solutions of different compositions, eutectics, definite chemical compounds, gas dissolved in varying amounts in different regions, impurities deposited from the surroundings on the metallic surface, etc.; similarly there may be variations in the liquid which wets the metal. Two such different phases – consisting for example of two crystals of an eutectic, or of the metal and an occluded impurity etc., wetted by an electrolyte solution – form an asymmetric electrochemical system in which an electric tension develops. In other words, one of the two phases assumes a positive electrode tension with respect to the other. Since the two metallic phases are in immediate contact, *i.e.* since the external ohmic resistance to the local cell is normally very small, the intensity of the direct current which tends to reestablish electrical equilibrium between the two phases may become very high; a chemical reaction of considerable size must then correspond to this. The chemical reaction on the phase which acts as the negative pole of the galvanic cell, will lead to the formation of cations, *i.e.* metal will pass into solution. It is easy to see from this why the purer a metal is, the higher is its resistance to corrosion.

Other factors may also facilitate corrosion, particularly depolarizing substances; and amongst these the most important is oxygen dissolved in the liquid which wets the metal. A particularly common type of galvanic cell consists of a single metal in contact with two zones of liquid which are differently aerated. The theory of corrosion formulated by U. R. Evans under the name of 'Theory of differential aeration' was developed for this very common type of cell. The theory falls into the general pattern of operation of local galvanic cells dealt with above. The electrolytic solution containing dissolved oxygen, and wetting a metallic phase, leads to the formation of an oxygen electrode whose electric tension is given by the equation

$$U_{rev} = U_0 + \frac{RT}{4\mathbf{F}} \ln \frac{p_{O_2}}{[OH^-]^4}$$

where p_{O_2} is the partial pressure of gaseous oxygen in equilibrium with the oxygen dissolved in the electrolyte. It can be seen from this that an increase in the concentration of dissolved oxygen makes the electric tension of the electrode more positive. If then a metal is wetted, in two different zones, by the same liquid which is however differently aerated in the two zones, an asymmetric electrochemical system will be produced. This will take the form of a concentration cell whose positive pole will be the zone in contact with the electrolyte richer in oxygen *i.e.* more aerated. There is thus the apparent paradox that the metal will be oxidized at the point where the oxygen concentration is lower. This explains why the corrosion of water pipes should be more intense within crevices than at the surface which is directly wetted by the running water.

Different thermal or mechanical treatments at different points in the same metallic object may also lead to differences in electrode tensions between the two points when these come into contact with the same electrolyte. An asymmetric system is thus formed here also, and tends to function as a galvanic cell. This explains why metals which are cold worked are generally more susceptible to corrosion.

It must finally be mentioned that a difference in temperature between two points in a single piece of metal in contact with the same electrolyte may be manifested as a corrosion if the electric tension of the metal–electrolyte half-cell has a considerable temperature coefficient (*cf.* p. 133 *et seq.*).

From what has been said it will be clear that the factors which will diminish the susceptibility of a metal to spontaneous corrosion are: a smooth and homogeneous surface, uniform aeration of the surface exposed to the electrolyte, the absence of galvanic coupling with less noble metals, and uniformity of temperature and thermal or mechanical treatments of the individual object. The most readily corroded alloys include those which are not totally miscible in the solid state.

A considerable amount of corrosion is also due to stray currents dispersed by electric traction systems with ground return, by telephone wires, by radio-communication appliances and in general by all installations which require an earth connection for their use. When the dispersed current encounters an uninsulated metallic conductor (rails, piping etc.) which generally present a much lower ohmic resistance than the electrolytic medium, the current will enter the metallic conductor and leave it at some other point such that the resistance to be overcome to complete the circuit is minimum. The points through which the current leaves the metal to enter the surrounding medium will thus function as the anodes of an electrolytic cell; since the buried or submerged objects are most usually of a ferrous metal, corrosion will occur by anodic dissolution.

With the increasing use of electrical appliances in general, the corrosion due to stray electric currents has assumed alarming proportions. A special case is the framework of reinforced concrete, whose iron in some particular conditions may be completely dissolved in a relatively short period after its construction.

Like all other phenomena, those of corrosion are governed by the laws of thermodynamics. Unfortunately however, chemical thermodynamics which applies the laws of energetics to chemistry on the basis of the concept of chemical equilibrium, is certainly not sufficient for the study of corrosion. Chemical thermodynamics may certainly express the equilibrium conditions for each particular corrosion reaction, but it can give virtually no information on the control of the reaction. The main reason for this is that the corrosion of metallic materials in electrolytic solutions is not strictly a chemical phenomenon but, as has been shown, is more an electrochemical one. Recourse must then be made to electrochemical thermodynamics, which expresses equilibrium conditions in terms not only of fugacities and activities, but also of electrode tensions. It is for this reason that the series of electric tensions (Chapter III, Tables III,3; III,4; III,5; and III,6 on pages 156, 157, 159 and 160 respectively) are of great interest in the study of corrosion.

However this scale gives only an approximate idea about the possibility of corrosion and about relative corrodibilities. Often, in fact, metals which according to the series of electric tensions should be corroded, remain virtually unattacked whilst other metals which should be immune are attacked. The position is complicated by the fact that there are generally multiple mechanisms of attack. It is enough to consider the corrosion of iron in aerated water. Very many substances take part in the reactions; they may be dissolved (H^+, OH^-, Ca^{2+}, Fe^{2+} and Fe^{3+} ions, and H_2O_2 from the reduction of dissolved oxygen), they may be solids ($Fe(OH)_2$, Fe_3O_4 anhydrous and hydrated, $Fe(OH)_3$, Fe_2O_3, $CaCO_3$ etc.) or gaseous (O_2 and CO_2 from the air, and H_2 evolved during the corrosion itself).

These various substances react together chemically and electrochemically to give rise to a system which can not be resolved if each reaction is considered separately. It is thus necessary to use graphical methods which allow the simultaneous consideration of the equilibria for all the reactions, both chemical and electrochemical, which can take place at once.

A satisfactory graphical method consists in the use of an electrochemical equilibrium diagram whose abscissae show pH and ordinates electrode tensions; this technique was invented and systematically applied by M. Pourbaix[1]. With suitable assumptions it is possible to derive practical diagrams of thermodynamic stability from those of electrochemical equilibria, as shown in Fig. VIII, 2 for iron in aqueous solution at 25° C. In such diagrams, the left side naturally refers to acid and the right to alkaline media, whilst the upper part refers to oxidizing and the lower to reducing conditions. The lines *a* and *b* represent respectively the variations of the electric tensions of the reversible hydrogen and oxygen electrodes at 25° C at their respective unitary partial pressures, with respect to pH. These electric tensions are given by the equations

$$U_{rev\,O_2} = 1.228 - 0.0591 \text{ pH V} \qquad (a)$$

$$U_{rev\,H_2} = 0.000 - 0.0591 \text{ pH V} \qquad (b)$$

[1] M. Pourbaix, *Thermodynamique des Solutions Aqueuses Diluées*, Thesis, Delft, ed. Béranger, 1945.

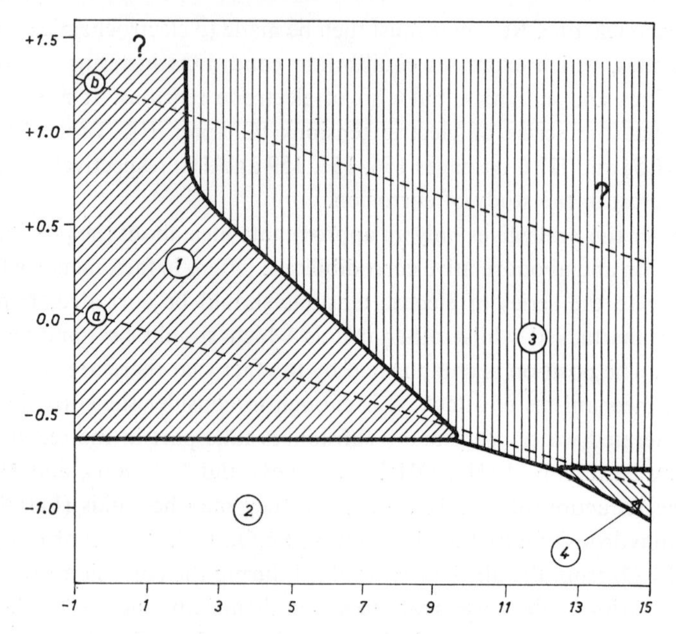

Fig. VIII, 2. pH–electric tension diagram of iron

1 = corrosion; 2 = immunity; 3 = passivity; 4 = corrosion

a = H_2 evolution at p = 1 atm; b = O_2 evolution at p = 1 atm.

Points lying outside the area enclosed by the lines *a* and *b* represent conditions of thermodynamic instability of water. Above *a* water can be oxidized by oxygen at one atmosphere pressure; below *b* it can be reduced by hydrogen at one atmosphere pressure. Between *a* and *b* water is thermodynamically stable under one atmosphere pressure and hence the electrolysis of water is thermodynamically impossible.

If now all the possible electrode reactions between metallic iron, ferrous and ferric ions and the oxidation products mentioned are considered, it is possible to introduce into the *U*–pH diagram a series of lines for each reaction corresponding to changes in the electric tension of the electrode for changes in pH; each line will refer to a particular concentration of dissolved iron.

The following assumptions must now be made. Firstly, the passivation of the iron takes place through the formation of an adherent film of $Fe(OH)_2$ or Fe_2O_3 with a solubility product of 10^{-42} (thus cases in which passivation is due to the formation of Fe_3O_4 are ignored). Secondly, as an approximation, the iron is considered to be corrodible when the solution can dissolve more than the, arbitrarily selected, low value of 10^{-6} g atoms/l. The series of possible lines can now be reduced to those few shown in Fig. VIII, 2. These provide a clear demarcation of the regions in which corrosion is possible (corrosion domains) from those in which it is not.

When corrosion is not possible, two further states can be distinguished; in one of these the stable solid form is the metal itself (domain of immunity, or, in the case of iron, of cathodic protection) and all corrosion reactions are energetically impossible; in the other, (domain of passivation) the solid stable form is not the metal itself but an oxide, hydroxide, hydride or salt. This tends to cover the metal forming either an impermeable film, which prevents all contact between metal and solution (perfect protection), or a porous deposit (partial or imperfect protection). In this sense, passivation does not necessarily imply the absence of corrosion.

Nevertheless, with certain reservations about the effectiveness of passivation protection – both about its mechanism, which is not always simple, and about the general lack of sufficiently precise information on the composition and thermodynamic properties of the protective films – the deductions which can be drawn from the theoretical corrosion, immunity and passivation diagrams constructed by Pourbaix for metals and metalloids, are almost always in reasonable quantitative agreement with practical observations. Thus these diagrams provide information of indisputable use in the study of the electrochemistry of corrosion.

(II) *Passivity*

From the point of view of the study of corrosion, passivity is the phenomenon by which a metal remains unattacked in a medium in which, thermodynamically, a net diminution of free enthalpy is associated with the corrosion reaction or reactions, *i.e.* with the passage from the metallic state to the corresponding product of corrosion. In electrochemical terms, a passive metal when acting as the anode in an electrolysis, does not pass into solution even although the value of the equilibrium electric tension for the dissolution reaction seems to indicate that it should.

The passivity may be merely localized or may be distributed more or less uniformly. It is not an intrinsic and invariable characteristic of the electrode metal: many metals can be made passive by suitable treatment but on other treatment may become active once more. Apart from the nobility of the metal, the passivity is linked not only to the medium and working conditions, but also very closely to the initial surface state and to the physico–chemical properties of any covering layers: chemical nature and structure, type of bonding, degree of cover, porosity, conductance etc. For a preliminary treatment, two substantially different types of passivity may be distinguished: mechanical passivity and chemical passivity.

Mechanical passivity is characterized by the formation on the electrode of a protective film, which is usually insoluble and is relatively thick and apparent. This film increases the reactional resistance for the process of anodic attack, either by a purely ohmic effect or by concentration polarization. If the film is porous it only partially separates the metal from the solution and the protection can only be of the dynamic type *i.e.* by (anodic) polarization of concentration. If the film is impermeable and conductive, the electric tension of the electrode will vary in a manner analogous to that of the naked metal; if the film is not only impermeable but is a perfect insulator and covers the surface completely, then the term electrode tension loses all significance.

On an anode the film in every case greatly diminishes the area of the electrode available for the passage of the current; thus the true current density and hence the concentration polarization, become very high. When the true current density exceeds the value of the limiting current, the overtension will rapidly rise and at the same time the total current intensity will diminish and the anode metal will effectively not pass into solution, provided that other electrode processes are not made possible by the greater positive electric tension reached.

There are two possible mechanisms of film formation. Firstly, it can be readily formed when the solubility product of certain, poorly soluble, substances is exceeded; these substances can originate by reaction of the cations with the anions present in solution. Since the concentration of cations is greater immediately next to the anode than in the bulk of the solution, it is easier to reach the solubility product here, with the consequent deposition of the poorly soluble compound on the anode. Secondly, the film may be produced by electrophoresis (*cf.* p. 373 *et seq.*) of colloidal particles already present, or formed, in the bulk of the solution.

Mechanical passivity is utilized for the manufacture of low power electrolytic rectifiers from aluminium, since the passive layer has the peculiarity of offering a high resistance to current in one direction and a much lower one in the other. Apart from this application to rectifiers, mechanical passivation by anodic oxidation or simple chemical means, is a protective measure of enormous technical importance for aluminium and its alloys. Aluminium is self-protective because a layer of oxide is rapidly formed on its surface. This layer is often invisible to the naked eye but protects the metal from further oxidation provided that it is not dissolved by some outside agent. By making an article, made of aluminium or one of its alloys, the anode in an electrolytic bath of suitable composition and choosing suitable conditions for the electrochemical factors in the treatment, it is possible to obtain a layer of oxide; it is often called an eloxal layer after one of the patented procedures for this.

The formation of such layers can be interpreted by the first of the mechanisms discussed. The aluminium passes into the form of Al^{3+} ions, by anodic dissolution, and these react with the anions present to form the corresponding salts. It is highly probable that the solubility product of the aluminium salt will be exceeded in the immediate neighbourhood of the anode surface, with the consequent separation of the salt on the metallic surface as a layer which is initially porous. This layer greatly diminishes the anodic surface area, increasing the current density and hence the anodic polarization to such a degree that the discharge of anions begins, with the evolution of oxygen. This in turn acts directly on the aluminium to form the oxide. At the same time however the local temperature within the layer increases, because of the high current density, favouring the hydrolytic decomposition of the salt and the partial dehydration of the hydroxide formed. This origin explains both the variable composition of the eloxal layer, from the exterior to the interior and also as a function of the electrolyte composition, as well as its powerful adherence to the metal.

The principal characteristic of *chemical passivity* is that it can be produced on a number of metals (*e.g.* iron, cobalt, nickel, chromium etc.) by treatment with strong oxidants (fuming nitric acid, chromic acid, permanganates etc.) without causing any visible change in the surface characteristics of the treated metal. An electrode of active iron, for example, becomes passive when immersed in fuming nitric acid. The metal becomes insoluble in acids, can not displace the cations of more noble metals from their solutions and does not pass into solution when used as an anode for the electrolysis of dilute oxyacids; the sole electrode process is the discharge of the OH^- ion. It is not however possible to observe the forma-

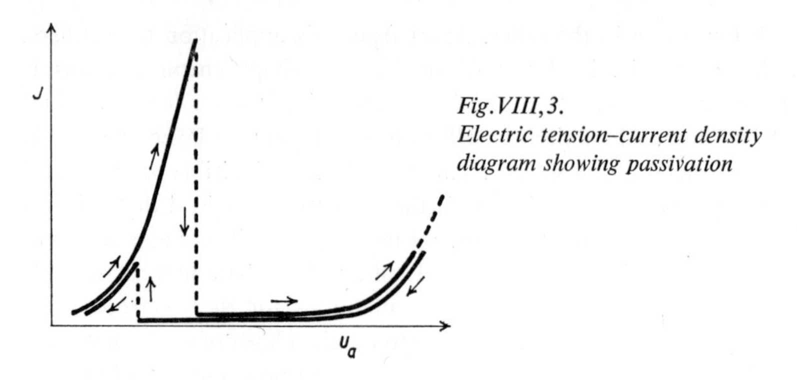

Fig. VIII, 3.
Electric tension–current density diagram showing passivation

tion of any film on the surface, which remains clear and apparently unaltered, even when examined by very sensitive means.

Chemical passivity, which can be produced on some metals (Fe, Co, Ni, Cr, Mo, W, V, Ru) by simple exposure to air, can also be obtained by treatment with a chemical oxidant. Anodic polarization corresponds in fact to a strong oxidation, so that chemical and electrochemical passivities may reasonably be considered as equivalent and differing only in their origin. Passivation by anodic polarization may be readily followed by means of current density–electrode tension diagrams of the type shown in Fig. VIII, 3. At low current densities the anode is active and the anode metal dissolves; as the anodic polarization increases, the current density increases normally until a certain limit is reached. It then falls abruptly to a very low value whilst the anodic polarization reaches a high value. If the applied external electric tension is increased still further, the current density rises only very slowly in a manner similar to that of a residual current, until after a certain value a new anodic process begins, which generally consists of an anionic discharge; the current density now increases normally once more.

As soon as the current density maximum is exceeded, the anodic metal becomes practically insoluble *i.e.* passive, and behaves as an inert electrode. If the anodic tension is now gradually lowered, the diagram assumes a different form as shown by the arrows. It is interesting to note that the active state is regained at values of the anodic tension lower than those corresponding to the maximum of the current density in the increasing phase of the anodic polarization. The shape of the decreasing polarization diagram is naturally the same whether the electrode has been made passive by chemical or electrochemical treatment. Passivity can be easily destroyed by chemical or electrochemical action, as well as by mechanical treatment. An electrode of passive iron, for example, regains its activity on treatment with a solution of caustic soda at a sufficiently high temperature, or on cathodic treatment, or on superficial abrasion. This implies that passivity is a chemical phenomenon involving only the surface of the metal.

The production of passivity on a particular metal depends on various factors. An increase in pH may facilitate passivation, provided that the metal is not amphoteric. Anions of oxidizing oxyacids have an analogous action (NO_3^-, ClO_4^-, CrO_4^{2-} etc.), whereas halogen ions and reducing substances inhibit it. A rise in temperature hinders the appearance of the passive state and a fall in temperature facilitates it.

Many theories have been advanced to explain the nature of passivity, but none of them can yet give a complete and satisfying explanation of all the manifestations of this phenomenon. However all the theories substantially fall into two groups: the generalized *film theory* and the *theory of electronic configuration.*

The film theory owes its origin to the first hypothesis of Faraday on the passivation of iron in nitric acid through the formation of a film of oxide. When suitably generalized, these hypotheses led to the present day theory which has been advanced in various ways by U.R.Evans, E.S.Hedges, S.Glasstone, R.B.Mears and others. According to this, in most if not all cases, passivity can be directly or indirectly attributed to a protective film which is not necessarily an oxide. It should be noted, in fact, that passivity is always connected with oxidizing treatments and that the surface of a passive electrode does not reflect polarized light like the active electrode. This observation makes the existance of an extremely thin film of oxide on the surface of the electrode seem very probable, and it is supported by other observations. An important fact is that extremely fine films of ferric oxide have been demonstrated on passive iron by carefully dissolving the underlying metal. Moreover, an electrode of passive iron can be activated by cathodic treatment in a solution of nitric acid; in fact, with impulses of sufficiently brief duration, about 10^{-4} coulombs/cm^2 are required, and this is the amount which would be needed to reduce a monomolecular layer of oxide. Again, the passivity of iron can be eliminated by treatment with concentrated caustic soda at the same temperature at which ferric hydroxide begins to dissolve. Molybdenum and tungsten, which give oxides of a more strongly acid nature, remain active in alkaline solutions but become readily passive in acid solutions. Finally it should be noted that the presence of adsorbed oxygen hinders the emission of electrons from incandescent tungsten (strongly blocked thermionic effect) and that such films of adsorbed oxygen do not react with hydrogen, even at a temperature of 1200° C, probably because the valencies of the oxygen are fully saturated by the surface atoms of the tungsten. The photoelectric emission from the surface of a passive metal is also much less than that from the active form.

All these observations have led to the so-called catalytic theory, according to which oxygen in the form either of a definite oxide or of an adsorbed film, strongly retards the process of ionization of the metallic atoms. It does this either by acting itself as a negative catalyst for the process, or by eliminating some positive catalyst. This could be hydrogen in the form of more or less labile hydrides, of an adsorbed film, or even dissolved in the mass of the metal. The possibility of regaining the active state by reductive chemical treatments or by cathodic polarization is further support for this theory.

Various attempts to reconcile the numerous theories of passivity with the film theory led to the electronic configuration theory. Based on the first qualitative ideas of A. Russell, R. Swinne and U. Sborgi, on the relationship between the electronic configurations of metallic atoms and passivity, this theory was developed quantitatively for alloys and then generalized by H. H. Uhlig. It is founded on the observation that the electrochemically active metals – those metals, and their alloys, which on passivation acquire properties corresponding to significantly higher positions in the scale of thermodynamic nobilities – are in the large majority, metals in the transition groups of the periodic classification of the elements. Atoms of metals such as Cr, Ni, Co, Fe, Mo and W are characteristically incomplete in the *d* energy levels in the shell below that of the valency electrons and hence they have an incomplete filling of the *d* band in the crystalline structure. According to electronic theory these incomplete energy bands tend to fill with electrons. The passive condition corresponds to that with incomplete *d* bands and the active condition to that when the bands are filled. The adsorption of oxygen or oxidizing substances, on to the metallic surface, induces passivity since they behave as electron adsorbers with no tendency to supply electrons to the surface atoms of the metal. Dissolved hydrogen and certain components of alloys, however, furnish electrons and hence favour the stabilization of the active state.

According to this theory, in stainless steel the chromium is inherently passive and confers passivity on the iron by its strong tendency to adsorb electrons. This view is supported by the observation that whereas active iron undergoes anodic dissolution as the Fe^{2+} ion, stainless steel, in which the iron is passive, gives rise to Fe^{3+} on anodic attack, *i.e.* to an ion with one electron less. Finally, the passivation of iron in nitric acid could indeed be considered as due to the formation of a film of oxide, but only insofar as the tightly adherent film on the surface of the metal can powerfully adsorb electrons.

(III) *Methods of Protection*

From a general viewpoint, protective methods may be classified depending on whether they control the physico–chemical characteristics of the metal, the corrosive agent, or the external source of electric tension.

The first group comprises methods of control of the intrinsic characteristics of the metal (purification, homogenization, finishing etc.) and methods consisting in covering the outside of the material with either other metallic materials (by rolling, electroplating, immersion in molten metals etc.) or with nonmetallic materials (by chemical or electrochemical reaction or oxidation, painting, coating with plastics, enamels, ceramics etc.). As regards the surroundings, the effective corrosive power can be controlled by the addition of inhibiting or passivating substances, or the corrosive power can be modified by controlling the aeration, pH, types of anions (water treatment), the possibilities of galvanic contacts, the formation of foams (which produce phenomena of differential aeration) etc.

The various methods to be considered in the second group include those of protection against the corrosion caused by stray currents, by increasing the ohmic exchange resistance between the material and the corroding medium and by drainage of the stray current itself, and finally those of active protection by applied currents. This last group of methods includes cathodic protection – by simple coupling with less noble anodes and by the imposition of currents from external sources – and anodic protection to favour or maintain the state of passivity.

As far as the very large number of types of nonmetallic coatings are concerned, it is of interest to note only that the protection is due essentially to a barrier which isolates the material from the corroding agent, more or less completely. Naturally, the coating must not only resist attack from the medium with which it is in contact, but must adhere perfectly to the underlying metal, must be perfectly continuous and must have some mechanical resistance.

Essentially the same characteristics are required for metallic coatings. The mechanism of protection may however differ depending on whether the coating is of a metal more, or less, noble than that underlying it. In the former case – a cathodic coating, such as tin on iron – the protection is due exclusively to the resistance to corrosion presented by the coating metal, which must be absolutely continuous. If for any reason the coating should be interrupted, exposing the underlying metal, then a local galvanic cell will be formed at that point and its negative pole will consist of the underlying metal; this will then be corroded much more vigorously than if the coating were absent. If the coating metal is less noble than that underlying it – anodic coating – it will function in the same way as a cathodic coating, so long as it remains intact. If however the continuity of the coating is breached, and the underlying metal is exposed, protection will still be obtained from the coating by an electrochemical effect. This time the coating metal will function as the negative pole of the local cell, and hence will be subjected to anodic corrosion whereas the underlying metal will function as the cathode and thus not be attacked. The most suitable metals for protective coatings of the cathodic type are those which are very hard, and thus resistant to abrasion so that there is little probability of the underlying metals being exposed, and which readily tend to become passive (Cr, Ni, etc.).

Protection based on an anodic, or simply chemically, produced layer of oxidation is particularly used for aluminium and some of its alloys. Table VIII, 18 shows information about the preparation and use of layers of eloxal already mentioned in the preceding section. Other processes for the production of layers of eloxal use oxalic acid, with or without the addition of chromic acid, and direct or alternating currents or a combination of the two. The characteristics of the eloxal layer depend not only on the composition of the electrolyte, but also on the underlying metal, its preliminary treatment, the current density and the temperature. The thickness of the layer normally varies between 0.010 and 0.030 mm, but layers up to 0.6 mm thick have been produced.

Depending on the thickness and on other general conditions, the layers may be hard and fragile (thick layers) or fine and flexible, and able to

TABLE VIII, 18

ELECTROCHEMICAL VALUES FOR ANODIC OXIDATION PROCESSES FOR THE PRODUCTION OF ELOXAL LAYERS

Process	*Electrolyte*	*Current*	*Electric Tension* V	*Current Density* A/dm²	*Uses*
engough Stuart	CrO_3	Direct	0–40–50	0.3–0.5	1, 2
loxal	H_2CrO_4 + oxidants + organic acids	Direct or alternating	20–60	0.7–12	1,2,3,4,5
rotka	HNO_3 + oxidants	Direct	—	—	6, 7
eppard	H_2SO_4 + glycerol + ethanol	Direct	12–16	1.0	1, 3, 5
umilite	H_2SO_4+HCl+glycerol or ethanol	Direct	12	0.7–1.5	1, 5, 6
coa	NH_3 or $(NH_4)_2SO_4$	Direct	150	4–40	1
toh Miyota	H_2CrO_4	Direct with imposed alternating	70–120	1.0–3.0	3
H.R. Kenkyujo	H_2CrO_4	Alternating	60–120	5–15	1, 3

1 = Protection against corrosion.
2 = Impregnation with dyes, oils and other substances.
3 = Electrical insulation.
4 = Surface hardening.
5 = Decorative effects.
6 = Thermal insulation.
7 = Dyeing.

withstand successive working by plastic deformation. These layers are extremely resistant to temperature and can be heated to the melting point of the aluminium (660° C) without suffering any chemical change.

These characteristics not only assure good protection against corrosion, but make it possible to use layers of eloxal as heat and electric insulators, as current rectifiers, anti-abrasive layers, and as layers whose porosity and absorbent capacity make them suitable for impregnation with various substances both to make them more effective for the purposes mentioned, and for decorative effects etc.

A typical method of protection using purely chemical oxidation, is the M.B.V. process, which consists in immersing an article made of aluminium or one cf its alloys, in a bath of sodium chromate made strongly alkaline with sodium hydroxide. The alkali attacks the aluminium, liberating 'nascent' hydrogen which reduces the CrO_4^{2-} and Cr^{3+} ions. The latter of these, precipitates in the presence of an excess of OH^- ions, in the form of chromium hydroxide, adherent to the article; because of the high temperature, this precipitate is at least partially dehydrated. In this case the protective layer consists of the oxide Cr_2O_3. Protective processes utilizing surface oxidation have also been developed for zinc, magnesium, iron and steel.

Amongst the methods based on the control of the environment, particular importance attaches to the use of inhibitors, which is widespread in the chemical, petroleum and petro-chemical industries. Inhibitors include all those substances which on addition to the corrosive medium in very small amounts, chemically or physically alter the electrochemical corrosion process, acting on the anodic or cathodic areas or both, and substantially diminishing the rate of corrosion of the material. Although it is difficult to classify substances with inhibiting action, it is possible to distinguish the chemically active agents from those which facilitate the production of a film (filming agents) and from the oily inhibitors.

The action of the chemically active agents (*e.g.* chromates, nitrites, polyphosphates, molybdates, tungstates etc. for iron) may always be related to one of two properties; either they have sufficient oxidizing power to cause passivation, bringing the electrode tensions of all exposed points in the metallic surface within the zone of passivation of the metal in the relevant electrochemical equilibrium diagram, or they cause the formation of solid products – by reaction or reduction of the inhibitor – which are thermodynamically stable under the conditions in which the material is used.

Filming agents which are particularly used for the control of acid attack are, on the other hand, organic substances containing sulphur or nitrogen. They are adsorbed on the metallic surface, or on its anodic and cathodic areas, forming a monomolecular film which can isolate the metal electrochemically from the corrosive agent, or can produce a marked increase in the overtension of the hydrogen produced in the cathodic reaction of attack.

The oily inhibitors, or soluble oils, are surface active substances which when added in small quantities to aqueous solutions form stable water in oil emulsions and considerably diminish the corrosive action. The emulsified oil acts at least in part, as a filming inhibitor by forming a thin film of oil adsorbed on the metallic surface.

Amongst methods of protection based on the control of external sources of electric tension, that of cathodic protection is today used on a large scale throughout the world for the prevention of corrosion in big, submerged or buried structures. This method was originally used for the protection of long, buried structures against stray currents, and was rapidly extended to the protection of structures of any type even against spontaneous corrosion, *i.e.* that which takes place with a diminution of the free enthalpy of the system. Thermodynamically, the immunization of a metallic material corresponds to the achievement of thermodynamically stable conditions for the material *i.e.* to conditions which are represented in the corresponding electrochemical equilibrium diagram by points falling within the zone of immunity (Fig. VIII, 2). Since such zones are normally situated below the corrosion region, immunization is generally obtained by lowering the electrode tension of the material, sufficiently to protect it, by the imposition of a cathodic current.

The minimum electrode tension required for the cathodic protection of a metal is naturally a function of the nature and composition of the medium, of the pH and of the temperature. With some metals and alloys however, the cathodic protection must be suitably regulated to avoid intensive cathodic corrosion. Thus, for example with lead and tin, too great a cathodic polarization can lead to attack by gasification with the formation of gaseous hydrides; many other metals can give solid hydrides. With some materials the formation of hydrides can also lead to weakening, with consequent disintegration. With stainless steels, and passive alloys in general, an applied cathodic current by leading to reducing conditions, could destroy the passivity or at least hinder its conservation; if the polarization is too weak, the only result is to expose the material to the attack of the corrosive medium.

Apart from such inherent limitations, the method of cathodic protection is certainly the most advantageous of all those available. It is in fact the perfect method in that it can completely annul the effects of corrosion and not simply diminish the rate of the attack. Moreover there is the enormous practical advantage that the efficacy of the protection can be estimated at any point in the protected structure by simple measurements, and possibly recordings, of the electrode tensions; this makes possible the automatic control of the applied electric tension.

Finally mention should be made of the use of corrosive processes for practical purposes, by making use of differing susceptibilities to attack of different points in a metallic surface caused by differences in composition or mechanical treatment.

CHAPTER IX

NONMETALLURGICAL ELECTROLYTIC PROCESSES

by

PATRIZIO GALLONE[1] and GIULIO MILAZZO

A. ELECTROLYSIS OF ALKALI HALIDES[2]

1. Primary Reactions

Industrial electrochemistry took its origin from the technical electrolysis of alkali halides. This process was used even before the invention of the dynamo (Pacinotti, 1860), at a time when means for the production of electricity in substantial and continuous amounts were provided by galvanic cells only.

The first patent relating to the production of caustic soda solutions by electrolysis of sodium chloride on an industrial scale dates back to 1851. More than a century has elapsed since that time during which this industry has acquired a prominent position, by virtue of the variety of its products (hydrogen, alkali solutions, chlorine, hypochlorites, chlorates and perchlorates) which can be directly obtained in high purity with relatively simple methods and equipment, as well as by virtue of the importance of these same products, some of which are basic materials for other chemical industries, so that their annual output is increasing at an ever faster pace.

The economic interest of this particular technology can also be appreciated through the large contribution of scientific research to this field and the exceedingly large number of patents related thereto, so that an outstanding variety of electrolytic cell models have been developed for this purpose and proved to be efficient from the economic stand-point.

In the introduction to this book it was pointed out that it was impos-

[1] Technical Director O. De Nora, Impianti Electrochimici, Milan.

[2] A very good survey of the actual situation and future possibilities of alkali halide electrolysis has been given in several papers devoted to this problem and read at a meeting of the group "Angewandte Elektrochemie" of the German Chemical Society, 5–6 Oct., 1961. All papers are published in full in the May issue (1962) of Chemie Ingenieur Technik.

sible to give a complete picture of the latest technical developments, and this is even more true in this particular context. Indeed, because of the variety of industrial processes in which alkali halide electrolysis is used, it is impossible to give here a detailed description of every individual cell. However, the principles underlying this technology will be particularly stressed. Such an approach is all the more justified as the several models differ from one another, in most cases, only in constructional details, rather than in principle.

Since, starting from the same compound, *i.e.*, an alkali halide, it is possible to obtain such a great number of different products, simply by changing the operating conditions, it is evident that the primary reactions must be accompanied by a multiplicity of side-reactions, any of which may be induced to occur by the particular conditions used.

As the reactions characterizing the electrolysis of any one alkali halide take place in much the same way, irrespective of the alkali metal and the halogen that form the components of the salt, the following discussion will be referred to sodium chloride, which is by far the most important and common in this class of salts.

In an aqueous solution of NaCl the ionic species Na^+, H^+, Cl^-, OH^- are always present. The primary cathodic reaction involves those cations that are characterized by the most positive discharge tension. The standard electric tension of the H_2/H^+ electrode is 0.000 V; in consequence, the reversible hydrogen electrode tension in a neutral solution, on platinized platinum, is −0.41 V.

The standard electric tension of the Na/Na^+ electrode is −2.71 V. Therefore, in order to obtain a condition of reversibility for both H^+ and Na^+ ions at the same time, the concentration of the latter should be raised so as to satisfy the equation,

$$-0.41 = -2.71 + 0.059 \log [Na^+]$$

in which, for sake of simplicity, activities are assumed to be equal to corresponding concentrations, since these considerations are restricted to the purpose of establishing some relevant orders of magnitude. The Na^+ concentration, derived from this equation, comes to about 10^{39} g equivalents/liter, *i.e.* an absurdly high value. It follows that, even taking into account the hydrogen overtension on iron, which is the usual cathode material in this process, the effective electrode tension can never be raised sufficiently to enable the electrodeposition of sodium metal. In consequence, hydrogen evolution will be the only possible outcome, unless, by some special provisions, the effective discharge tension of H^+ ions can be rendered more negative than that of sodium ions. This is indeed the case when the cathode is made of mercury, on which hydrogen shows a strong overtension, whereas the sodium discharge tension becomes more positive by virtue of the depolarizing effect of amalgam formation (see p. 265 *et seq.*).

Similar consideration can be made for the anodic process. Indeed, the standard electric tension of the oxygen electrode referred to the OH^- ion is +0.40 V, which involves, in neutral solution, a reversible electric tension of +0.81 V. On the other hand, the standard electric tension of the chlorine electrode is +1.36 V. Therefore, reversible conditions for the simultaneous discharge of both ionic species at pH 7 could be obtained only if it were possible to raise the Cl^- ion concentration so as to satisfy the equation,

$$0.81 = 1.36 + 0.059 \log \frac{1}{[Cl^-]}$$

which gives about 10^9 g equivalents/liter.

In practice, however, the effective discharge tension of the OH^- ion is much higher than the reversible value considered above. This is mainly due to the fact that the overtension required to discharge the OH^- ion is in general much higher than that for the Cl^- ion, despite any depolarization which might be brought about by such oxidizable materials as carbon or graphite. Moreover, the anodic discharge is always accompanied by a considerable decrease in pH, which further shifts the reaction conditions in favour of chlorine gas evolution. In conclusion, the latter predominates over oxygen gas evolution, even though small quantities of oxygen are to be found as impurities in the chlorine gas product. Besides oxygen, some carbon dioxide formation accompanies the chlorine gas evolution, because of gradual oxidation of the anodic material.

2. Production of Caustic Soda and Chlorine; Theoretical Aspects

The primary reaction, occurring at a cathode material other than mercury, consists of the discharge of H^+ ions originating from the dissociation of water. For each hydrogen equivalent that is thus set free as hydrogen gas, one hydroxyl equivalent is left behind in the solution and combines with the Na^+ ions migrating to the cathode, thus forming the desired caustic product.

The OH^- ions, whose concentration gradually increases in the cathodic region as the primary reaction proceeds, have however a tendency to migrate toward the anodic region, under the influence of the electric field as well as of the concentration gradient that builds up. As soon as they succeed in reaching the anodic region, the hydroxyl ions give rise to several primary and secondary reactions. Such migration must be hindered as far as possible, as it involves a loss in the yield of both desired products, *i.e.*, caustic soda and chlorine.

The chlorine evolved in the primary anodic reaction is partially dissolved in water and reacts, according to the equation,

$$Cl_2 + H_2O \rightleftharpoons HClO + H^+ + Cl^- \tag{1}$$

The equilibrium constant for this side reaction at 25° C, assuming the water concentration to be constant, is,

$$K_1 = \frac{[HClO]\,[H^+]\,[Cl^-]}{[Cl_2]} = 4.84 \cdot 10^{-4} \tag{2}$$

where $[Cl_2]$ indicates the chlorine concentration in the dissolved state.

The hypochlorous acid formed in reaction (1) is a weak acid that dissociates as follows:

$$HClO \rightleftharpoons H^+ + ClO^- \tag{3}$$

for which reaction the dissocation constant at 25° C is,

$$K_2 = \frac{[H^+]\,[ClO^-]}{[HClO]} \cong 4.4 \cdot 10^{-8} \tag{4}$$

The hypochlorous acid and the hypochlorite ions originating from (1) and (3) give rise to a third reaction, which is also purely chemical in nature and produces chlorate, as follows,

$$2\,HClO + ClO^- \rightleftharpoons ClO_3^- + 2\,H^+ + 2\,Cl^- \tag{5}$$

By combining together reactions (1), (3), (5), it is possible to write down the following, significant equation,

$$3\,Cl_2 + 3\,H_2O \rightleftharpoons ClO_3^- + 6\,H^+ + 5\,Cl^- \tag{6}$$

whose equilibrium constant at 25° C is,

$$K_3 = \frac{[H^+]^6 [Cl^-]^5 [ClO_3^-]}{[Cl_2]^3} = 5.75 \cdot 10^{-12} \tag{7}$$

The significance of (7) is that the concentration of dissolved chlorine depends basically upon the partial pressure of chlorine gas, which, under conditions of technical electrolysis, is practically 1 atmosphere, whereas the Cl^- ion concentration is dictated by the sodium chloride content in the feed brine. In consequence, (7) implies that equilibrium conditions, as regards ClO_3^-, vary virtually only with pH and brine strength, so that, the lower the pH and the higher the brine strength, the lesser is the maximum amount of chlorate which, due to side reaction (5), may build up in brine. Chlorate formation is undesirable not only because it means a loss in current efficiency, as regards the anodic process, but also because the formed chlorate may subsequently be reduced at the cathode, with the result of a loss in current efficiency for the cathodic process as well.

Chlorate may be formed also electrochemically at the anode, *i.e.*, according to a primary reaction involving hypochlorite ions according to the reaction:

$$6\,ClO^- + 3\,H_2O \rightarrow 2\,ClO_3^- + 4\,Cl^- + 6\,H^+ + 1.5\,O_2 + 6\,e^- \tag{8}$$

due to the presence of a small quantity of ClO^- ions in the anodic region. This mode of chlorate formation, too, depends mainly on pH in that it is favoured by decreasing the acidity of the anolyte. However, under conditions prevailing in modern cells for the production of chlorine and caustic, primary reaction (8) has a negligible effect so that chlorate formation is mainly controlled by side reaction (5).

A further cause of a fall in current efficiency manifests itself when the OH^- ions formed at the cathode pass into the anodic region. First they may go into a primary discharge reaction, so as to involve a current loss corresponding to evolution of oxygen and possibly also of carbon dioxide, which thus also decreases the chlorine gas purity. Secondly, the presence of OH^- ions in the anolyte gives rise to secondary reactions with the dissolved chlorine, according to the equation

$$Cl_2 + 2\,OH^- \rightleftharpoons ClO^- + Cl^- + H_2O \tag{9}$$

This is formally identical to the combination of (1) and (2) with the dissociation reaction of water, written in the form:

$$2\,H^+ + 2\,OH^- \rightleftharpoons 2\,H_2O$$

The equilibrium constant relating to (9),

$$K_4 = \frac{[ClO^-]\,[Cl^-]}{[Cl_2]\,[OH^-]^2} = 2.1 \cdot 10^{17} \qquad (10)$$

can therefore be easily calculated as the product of the equilibrium constants relating to equations (1), (3) and to the water dissociation reaction respectively[1]. Equation (10) clearly shows that, by operating at low pH in the anodic region, with other factors such as brine strength and chlorine partial pressure equal, not only chlorate (reaction (5)) but also hypochlorite concentration may be kept within acceptable limits. This is obtained by hindering the migration of OH^- ions from the catholyte, where they form, into the anolyte.

It can be deduced from this that, to obtain a high current efficiency, there are a number of operating conditions to be satisfied as far as possible. They can be summarized as follows:

1. Sodium chloride concentration in the feed brine must be kept near to saturation point, so as to have the maximum possible concentration of Cl^- ions; this promotes the discharge of chlorine rather than any other anionic species. Moreover, since the greatest salt concentration corresponds to the smallest chlorine solubility, formation of hypochlorous acid ((1) and (2)), chlorate ((6) and (7)) and hypochlorite ((9) and (10)) is thereby inhibited.

2. The anode current density must be kept as high as possible, in order to increase the overtensions of other competing anionic species in favour of Cl^-.

3. By operating at a high temperature, the solubility of sodium chloride is increased, whereas that of chlorine is diminished, thus fostering the favourable conditions described under 1, beside improving the electrolyte conductance.

4. Migration of OH^- ions, forming at the cathode, toward the anodic region must be hindered by appropriate means.

When the cathode consists of sheet-metal (iron), instead of mercury, the most important ways of achieving condition 4 are provided by the interposition of a diaphragm between the anode and the cathode and by

[1] $$\frac{[HClO]\,[H^+]\,[Cl^-]}{[Cl_2]}\;\frac{[H^+]\,[ClO^-]}{[HClO]}\;\frac{1}{[H^+]^2\,[OH^-]^2} = K_1 K_2 10^{28} = 2.1 \cdot 10^{17}.$$

the application of the countercurrent method, according to which the catholyte is continuously withdrawn from the cell in a direction opposite to the electric field, such flow being sustained by a continuous feed of concentrated solution (brine) into the anodic compartment.

Both methods are always applied together in modern steel-cathode cells, which are therefore more usually termed *diaphragm cells*.

Another way of preventing OH^- migration is, as previously mentioned, the use of a mercury cathode in so called *mercury cells*. In fact, sodium is thus allowed to discharge instead of hydrogen and amalgamate with mercury, so that caustic is prevented from being generated inside the cell itself. Mercury is continuously fed to the cell and withdrawn in the form of sodium amalgam, which is reacted with water to yield the desired caustic product in a separate reaction vessel.

The reason why sodium can be deposited on the mercury cathode, while preventing hydrogen evolution, is as follows. Firstly the formation of sodium amalgam, starting from metallic sodium and mercury, is accompanied by a very considerable decrease of the free enthalpy and it is therefore to be expected that cathodic deposition of sodium on mercury, instead of on an unreactive material, should be depolarized, *i.e.* shifted toward less negative values of the cathodic tension. In fact, the value for the discharge tension of the Na^+ ion, when dealing with saturated brine and 0.2 % Na amalgam, is about −1.83 V. On the other hand, the discharge tension for hydrogen under such conditions rises to as high as −2 V, essentially because of the exceedingly high transfer overtension of hydrogen on mercury, which further rises on increasing the current density, whereas the amalgamated cathode is practically impolarizable in respect of sodium discharge. In consequence, under appropriate operating conditions, the primary reaction may be made to consist exclusively of the deposition of sodium and only to a very minor degree of hydrogen evolution.

The overall cell voltage must be equal to the reversible decomposition electric tension, plus the anodic and the cathodic overtension, plus the resistive electric tension (*IR*) within the electrolyte and the electrodes.

The reversible electric decomposition tension (equal to the electric tension of the corresponding galvanic cell) may be calculated either from the inner energy change, according to Thomson's rule, or, more accurately, from the change in free enthalpy as established by the Gibbs–Helmholtz equation (p. 106 *et seq.*). This value, at room temperature, lies between 2.2 and 2.3 V when the primary cathodic reaction is the discharge of hydrogen, and rises to about 3.4 V when the primary cathodic reaction results in the formation of sodium amalgam. The calculations leading to these results inevitably contain some inaccuracies, for several reasons. In the first place, the activities of the ionic species involved, *i.e.*, Cl^-, Na^+, H^+, cannot be exactly known in concentrated brine; secondly, it is impossible to determine with the required accuracy the temperature coefficient for the electric tension, which is necessary to apply the Gibbs–Helmholtz equation. However, the above values can be taken as sufficiently accurate for practical purposes, all the more so as the other terms contributing to the overvoltage cannot be accurately determined either, being themselves affected by several factors, such as temperature, current density and electrode material. With modern equipment, the overall voltage is on the average about 3.7 V for diaphragm cells and 4.5 V for mercury cells. It is however noteworthy that, in spite of continuous progress in design, the general tendency during recent years has been toward an increase in voltage, on account of the ever higher current density desired for a given cell capacity, in order to reduce capital investment for a planned annual output of products.

The quality of the anode material is of particular importance, as it must be unreactive toward chlorine, oxygen and the acidic conditions in the anolyte. On account of the outstanding reactiveness of atomic chlorine the choice of materials is limited to platinum, magnetite, retort carbon and graphite. On platinum and magnetite the Cl^- and OH^- overvoltage is higher than on graphite, so that, when using the former two materials, hypochlorite formation and consequently the electrochemical reaction leading to chlorate buildup and oxygen evolution are facilitated. On the other hand, retort carbon and graphite show some undesirable qualities too, in that, due to their porosity, ion discharge takes place not only at their surface but also within the bulk of these electrode materials. In consequence, the solution stagnating within the pores is rapidly depleted of its Cl^- ions, which cannot be easily replenished on account of the high resistance to diffusion of the porous structure of these materials. Thus the electrochemical formation of chlorate, with oxygen evolution, is enhanced in the interior of the anode, which thus acquires a tendency to become even more rapidly oxidized and to disintegrate.

Graphite is preferable to retort carbon, for its greater chemical resistance; to make the best use of it and prolong its operating life, it must be shaped into plates of considerable thickness or rods of relatively large diameter. Furthermore, in some applications, especially for diaphragm cells, the graphite anode is impregnated with some sort of polymerizable oil (p. 518 *et seq.*).

This, and other advantages make graphite the only anode material of industrial use in chlorine cells. However, an interest in the particular benefits that might be obtained by using permanent or nearly permanent anodes made of sheet metal, especially as regards cell design and maintenance, has recently been revived by the latest developments in the metallurgy of inoxidizable metals, such as titanium, and in the technology of platinizing them. Titanium could not be used as such to make anodes, because of the nonconductive oxide film that would immediately form on the metal at the very onset of the anodic reaction; this film is responsible for the outstanding passivity of titanium. On the other hand, due to such passivity, which would protect any part of the surface that might still remain exposed by an extremely thin coat, the platinum layer can be made so thin that the cost involved may also be acceptable for industrial plants.

The cathode material, as mentioned before, is always made of iron plate or wiremesh in chlorine cells of the diaphragm type, since this metal, beside being the cheapest, has a relatively low hydrogen overtension and suffers very little attack in the alkaline environment of the catholyte.

3. Diaphragm Cells with Stationary Electrolyte

These cells, which are now completely obsolete, were operated batchwise, *i.e.*, they were periodically filled with brine, and the alkaline liquor forming in the subsequent electrolytic process was withdrawn as soon as salt depletion and caustic buildup had reached a certain stage. Since no countercurrent movement of the solution was counteracting the hydroxyl ion migration, the diaphragm between the anodic and the cathodic region had to hinder diffusion between the catholyte and anolyte as far as possible and was therefore made of a very impervious material. Beside the historical interest of this sort of equipment, since it was the first ever to achieve industrial production of electrolytic chlorine and caustic, it still deserves a particular interest in that, by examining its mode of operation, it is possible to get a deeper insight into the role of ionic migration in affecting the current efficiency.

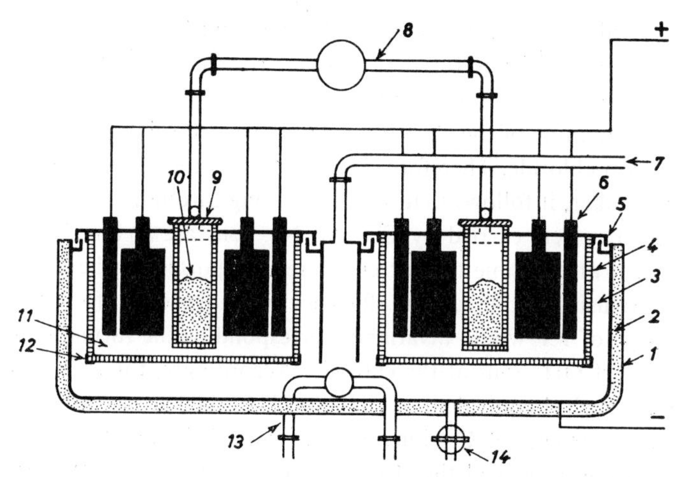

Fig. IX, 1. Griesheim cell for Cl_2–caustic production. 1. thermal insulation; 2. iron container; 3. cathodic region; 4. diaphragm; 5. hydraulic seal; 6. graphite anodes; 7. H_2 outlet; 8. Cl_2 outlet; 9. salt container; 10. solid NaCl; 11. anodic region; 12. supporting frame for diaphragms; 13. heating; 14. NaOH outlet

The prototype of this kind was the Griesheim-Elektron cell (1890). It consisted (Fig. IX, 1) of a rectangular steel tank, which was thermally insulated and also functioned as a cathode. The tank contained twelve steel frames holding the diaphragms. These were made of a mixture of portland cement, sodium chloride and hydrochloric acid, which was poured into a mold and, after setting, was treated with water, to wash out the salt crystals embedded in the mass and thus obtain the desired porous structure. Each frame holding a set of diaphragms formed an anode compartment containing six retort carbon or magnetite anodes facing the diaphragms; at the center of the anode compartment was located a perforated ceramic pot containing solid salt to compensate for the gradual anolyte depletion in the course of the electrolysis.

The free space between the tank and the anode compartments formed the cathodic compartment, in which a number of iron plates were arranged, acting as auxiliary cathodes. The tank was also provided with a heating pipe system, through which steam or hot water could be circulated, and the tank top was fitted with a gas-tight cover, carrying outlets for chlorine and hydrogen gas from the anodic and cathodic compartments respectively.

The overall cell dimensions were 4.8 m : 3.8 m : 1 m depth.

The current efficiency for the cathodic process, under stationary electrolyte conditions, can be assessed as follows.

At the beginning of the electrolysis the fresh electrolyte batch contains only sodium chloride. Therefore, denoting by t_{Cl} the chlorine ion transference number, it follows that, for 1 **F** passing through the solution, the amounts of current carried by Cl^- and Na^+ ions are t_{Cl} and $1 - t_{Cl}$ respectively. If at any time the current were transported by these two ionic species exclusively, the current efficiency would be equal to unity, since the passage of 1 **F** would invariably correspond to the formation of one equivalent of OH^- ions at the cathode. Accordingly, for each **F** passing through the cell, 1 equivalent of OH^- ions is formed at the cathode, while only t_{Cl} equivalents of Cl^- ions migrate toward the anodic region. The alkali concentration will therefore increase at a faster rate than the rate of decrease in chloride concentration, so that the density will rise while the electrolysis proceeds.

If, now, the OH^- ions were assumed to be completely replacing the Cl^- ions in the transport of negative charges, the current fraction carried by the former would be t_{OH}, while the Na^+ ions would carry the fraction $1 - t_{OH}$. Accordingly, for 1 **F** passing through the cell, t_{OH} equivalents of hydroxyl ions would migrate out of the catholyte and be discharged at the anode, or else react with the discharging chloride ions. On this assumption the cathodic current efficiency would be $1 - t_{OH}$.

Actually, only a certain fraction x of the current is transported by the alkali hydroxide and the remainder $1 - x$ by sodium chloride. The ratio $(1 - x)/x$ is equal to the ratio between the conductivities of the two constituents in the solution, so that,

$$\frac{\varkappa_{NaCl}}{\varkappa_{NaOH}} = \frac{(cf_\lambda \lambda_0)_{NaCl}}{(cf_\lambda \lambda_0)_{NaOH}}$$

or,

$$\frac{1 - x}{x} = \frac{(cf_\lambda \lambda_0)_{NaCl}}{(cf_\lambda \lambda_0)_{NaOH}} \quad ; \quad \frac{1}{x} - 1 = \frac{(cf_\lambda \lambda_0)_{NaCl}}{(cf_\lambda \lambda_0)_{NaOH}}$$

$$x = \frac{1}{1 + \dfrac{(cf_\lambda \lambda_0)_{NaCl}}{(cf_\lambda \lambda_0)_{NaOH}}}$$

The cathodic current efficiency R_{curr} can be expressed by the sum of two terms, which are related to the current fractions transported by the two electrolytes, *i.e.*, NaCl and NaOH respectively. Since the alkali chloride transports the fraction $(1 - x)$, which leads completely to the formation of NaOH,

$$R_{NaCl} = 1 - x$$

whereas

$$R_{NaOH} = x(1 - t_{OH})$$

Therefore,

$$R_{curr} = R_{NaCl} + R_{NaOH} = 1 - x + x(1 - t_{OH}) = 1 - t_{OH} \cdot x$$

By replacing x in the last equation with the expression formerly written for it,

$$R_{curr} = 1 - \frac{t_{OH}}{1 + \dfrac{(cf_\lambda \lambda_0)_{NaCl}}{(cf_\lambda \lambda_0)_{NaOH}}}$$

It can now be seen that the ratio between the two activity coefficients may be considered approximately equal to unity, since one is dealing with two strong electrolytes, whose concentrations are of the same order. In consequence, if a denotes the ratio of the conductivities,

$$R_{curr} = 1 - \frac{t_{OH}}{1 + a \dfrac{c_{NaCl}}{c_{NaOH}}}$$

Since, as the electrolysis proceeds, the concentration decreases for the chloride and increases for the alkali hydroxide, the former equation shows that the current efficiency must become progressively smaller.

When numerical values, valid at room temperature, are substituted for the constants appearing in the above expression, this becomes,

$$R_{NaOH} = 1 - \frac{0.799}{1 + 0.50 \dfrac{c_{NaCl}}{c_{NaOH}}}$$

For potassium chloride, the expression for R_{curr} is,

$$R_{KOH} = 1 - \frac{0.726}{1 + 0.55 \dfrac{c_{KCl}}{c_{KOH}}}$$

A comparison between R_{NaOH} and R_{KOH} shows that the cathodic current efficiency is higher for potassium chloride than for sodium chloride. When the temperature increases, the OH^- ion transference number tends to 0.5, and hence the cathodic current efficiency decreases in both cases.

The anodic current efficiency cannot be equal to unity either, due to OH^- ion migration toward the anode. Indeed, it has already been shown that these anions, beside taking part in the primary anodic reaction, which gives rise to oxygen evolution, are also reacting with dissolved chlorine, producing chlorate ions, which in turn will share in the primary reaction too. The foregoing discussion points out the limitations inherent in this electrolytic method. In fact, the highest concentration that is thus obtainable for caustic soda is only 40 to 50 g/l, as, beyond such limit, the current efficiency would become even lower than 0.80, a value that was considered acceptable in the past but would be no longer economic.

Another considerable drawback of the method lies in the need for batchwise operation, which required the emptying of the product liquor from the cell and replenishing with fresh electrolyte every three or four days. Furthermore, since salt replenishment during the process was carried out by adding salt into the cell itself, no preliminary brine purification was possible, so that several impurities, notably sulphates, contributed to impair the current efficiency of the overall process and the purity of the product.

4. Cells without Diaphragm and with Flowing Electrolyte

Like the type discussed in the former section, these cells are also completely obsolete. Instead of counteracting the diffusion of the OH^- ions by the use of a diaphragm, in this sort of equipment the effect of diffusion as well as of the electric field was opposed by moving the electrolyte in the opposite direction.

From the value of the OH^- ion mobility it may be calculated that, in a dilute solution at room temperature, the migration velocity of this ion is 0.0018 mm/sec for a potential gradient of 0.1 V/cm, *i.e.*, a value of the same order as may be found in most industrial cells. As a first approximation, this velocity, which corresponds to 6.48 mm/h, may be assumed to be valid also for a saturated NaCl solution. On the other hand, an electric field of 0.1 V/cm, in a medium of the same conductivity as the solution that is being considered, brings about a current density of 2 A/dm^2. In other words, a quantity of electricity of 2 Ah will be crossing an area of

1 dm^2 normal to the electric field in the bulk of the solution, which involves the decomposition of 4.4 g NaCl and the production of 3 g NaOH. Accordingly, if the solution is moved in the opposite direction to and at the same speed as the migration velocity, which corresponds to a flow-rate of 65 cm^3/h through a cross-section of 100 cm^2, then the NaOH concentration would be about 48 g/l, which is insufficient for a commercial product.

If, on the other hand, one calculates the flow rate which, on the same assumptions as above, would allow all the chloride content to be electrolysed, the resulting velocity is 1.6 mm/h, *i.e.*, only one fourth of the migration speed. It follows that the corresponding current efficiency would be substantially lesser than unity.

From the foregoing argument it can be concluded that a working condition which may be practically acceptable will be one of compromise between two requirements opposing each other, so as to obtain a sufficiently concentrated caustic solution with a tolerable loss in current efficiency. Moreover, the optimum flow-rate will further depend upon the cell shape, in that the liquid stream must be distributed as uniformly as possible over the whole cross-section facing the electrodes.

It is important to observe that a uniform and homogeneous flow of solution can be promoted by the stratification that tends to build up spontaneously. In fact, as already explained, the hydroxide concentration increases rapidly at the cathode, while the chloride content decreases, although not so quickly. It follows that the overall concentration in the catholyte will gradually rise together with the density, so that the catholyte will tend to stratify in the lower layers of the cell liquor, while the lighter anolyte will occupy the upper layers. It is thus possible, by a suitable cell design and by letting the flow take place from the anode toward the cathode, to check the OH^- ion migration efficiently and hence obtain an alkaline solution of sufficient strength with a satisfactory current efficiency. In the cell types that are expressly designed to exploit the advantages afforded by stratification, this is clearly evidenced by the formation of an intermediate neutral layer, which divides the acidic liquor, forming the anolyte, from the alkaline liquor, constituting the catholyte.

The most typical example of the way in which stratification and countercurrent can be applied at the same time is afforded by the Aussig bell-jar cell, of which Fig. IX, 2 gives a sectional and diagrammatic view.

Other well known models in this class are those developed by Billiter–

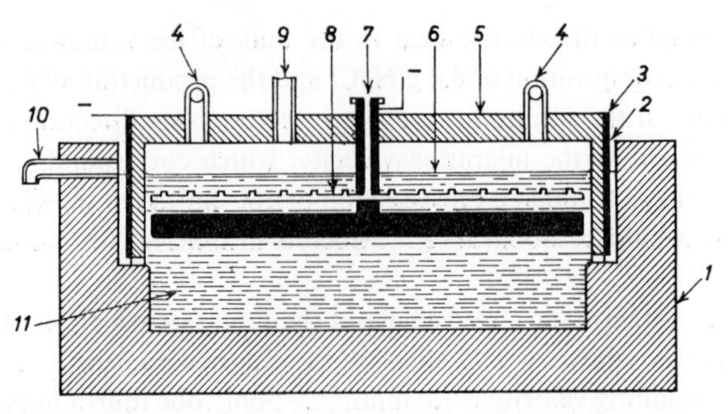

Fig. IX, 2. Aussig cell for Cl_2–caustic production. 1. concrete container; 2. cathodic region; 3. iron cathode; 4. connection tubing of the bells; 5. concrete bell; 6. anolyte; 7. electrolyte feeding inlet; 8. distribution of salt solution; 9. Cl_2 outlet; 10. NaOH outlet; 11. catholyte

Leykam, and by Pestalozza. They were both characterized by an horizontal arrangement of the cathodic surface underneath the anode plate, so as to allow stratification to form, while keeping the smallest possible and most uniform gap between the electrodes, with the further purpose of reducing resistive tension losses. Such an arrangement, however, requires the removal of the hydrogen gas from the cathodes by some suitable means. These consisted of a sort of reversed gutter, made of a porous material pervious to the electric current and sloping over the cathode surface. Owing to this characteristic, this type can be considered as an intermediate stage between the bell-jar model and the modern diaphragm cells, which will be discussed in the next section.

The cells belonging to this class had the advantage over the Griesheim-Elektron type of continuous operation. However, due to the absence of a diaphragm proper, forming a mechanical barrier throughout between the anolyte and the catholyte, these cells were all rather sensitive to any turbulence in the flow of the solution that might upset the stratification and consequently the current efficiency. Also on this account the current density was quite low, thus requiring a comparatively large floor space per unit of product. Furthermore, caustic concentration could not exceed 100 to 120 g/l.

5. Diaphragm Cells with Countercurrent Flow

As previously mentioned, the purpose that fostered new developments in cell design was to exploit at the same time the advantages inherent both in the use of a diaphragm to counteract physical diffusion, and in countercurrent flow of the electrolyte, to oppose the migration of the hydroxyl ions under the effect of the electric field.

With one sole exception that will be discussed at the end, all the models that have been developed so far and are now used in modern plants are of vertical construction, in that the anodic and cathodic surfaces, as well as the diaphragm, extend vertically. A typical arrangement, which was the first ever to be adopted and extensively used in a number of models for many years, is the one illustrated in Fig. IX, 3.

The cell consists of a U-shaped frame of insulating material, both sides of which are covered with perforated steel cathode plates, or steel wire mesh, with the interposition of a diaphragm, which is basically made of asbestos and faces the inner side of the cathodic structure. The enclosed space thus formed within the frame constitutes the anolyte compartment, containing the anode and the anolyte, which is continuously replenished by adding fresh salt solution. Brine is fed through a pipe dipping into the anolyte (not shown in the figure) while the chlorine gas is withdrawn through an outlet on the cell top cover.

The outer side of each cathodic surface is enclosed by a lateral cover-

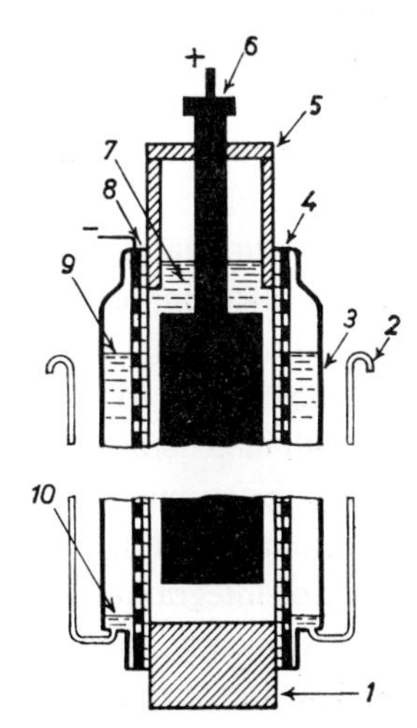

Fig. IX, 3. Vertical diaphragm filter cell for Cl_2–caustic production. 1. concrete frame; 2. syphon for caustic; 3. lateral closing wall of iron; 4. iron cathode; 5. closing bell; 6. graphite anode; 7. anolyte; 8. diaphragm; 9. nonelectrolytic liquid; 10. NaOH solution

plate, so as to form the catholyte compartment, from which the caustic solution is continuously withdrawn through a syphon which is adjustable in level. The hydrogen gas developing at the cathode is prevented by the surface tension forces within the pores of the wet diaphragm from leaking through it and is thus set free, through the perforated metal surface, into the catholyte compartment, from the top of which it is withdrawn.

Diaphragm cells of vertical type were in the past subdivided into two sorts, with or without their diaphragm-cathodes submerged on the catholyte compartment side. The reason for such classification was mainly due to the fact that, during the first decades of this process, mistaken importance was attached to the idea that by keeping the catholyte compartment empty, so as to remove the caustic solution from contact with the cathode as soon as it was formed, hydroxyl ion migration into the anolyte compartment might thus be more effectively hindered. This incorrect opinion, on which the use of unsubmerged cathodes was primarily based, prevailed for more than three decades, after the advent of the prototype developed by Hargreaves and Bird (1890), and was subsequently dispelled by a scientific approach to this matter, as was applied for the first time in the development of the Giordani–Pomilio cell, *i.e.*, the first model working with submerged cathodes.

Indeed, that the idea underlying the use of unsubmerged cathodes was deceptive, as far as quick removal of hydroxyl ions is concerned, can be immediately realized, when it is considered that the cathode will in either case be wet with a layer of catholyte on both its sides, so that the diffusion forces will always be effective in the same way, whether the cathode and the diaphragm be submerged or not; on the other hand, the electric field causing ion migration is always zero beyond the cathode surface next to the diaphragm.

The diaphragm material usually consists of asbestos, in either sheet or woven form, often impregnated or coated with some sort of plaster made of cementing materials, such as barium sulphate admixed with asbestos powder or fiber, so as to obtain the required resistance to diffusion. Such resistance will in general increase during the first weeks of operation, due to the clogging action exerted on the porous structure by the several impurities entrained with the electrolyte, such as graphite powder, coming from anode disintegration, magnesium and calcium hydroxides, and iron oxides. Consequently, the behaviour of the diaphragm will gradually improve, in that the current efficiency will substantially increase, although this is accompanied by a slight increase in voltage drop. However, under the combined action of the acidic attack on the anolyte side and of the alkalinity prevailing on the catholyte side, as well as of the crumbling effect due to the evolving hydrogen, the coating will gradually disintegrate and the antidiffusive properties of the diaphragm will diminish accordingly. When current efficiency decreases beyond an acceptable minimum value, operation must be discontinued and diaphragms must be removed.

The current capacity in these older cell types rarely exceeded 3000 A and the current density was also quite limited. Indeed, a limitation in the latter was mainly dictated by the requirement of not increasing the overall voltage beyond the economic limit, especially on account of the tension loss through the diaphragm and, in spite of the fact that according to theoretical arguments and experimental evidence, the current efficiency is not affected by current density provided that other conditions such as anolyte and catholyte concentration, temperature and diaphragm permeability are equal. In consequence, the current density (calculated on the total surface, *i.e.*, including empty spaces at the cathodic structure) was usually less than 6 A/dm^2 on the cathode and less than 8 A/dm^2 on the anode. Among the best known cells of rectangular and vertical shape, beside those already named, *i.e.* Hargreaves–Bird, Townsend and Giordani–Pomilio, were the models developed by Allen–Moore, Ciba–Monthey, de Nora, Krebs, Nelson.

Other cells, operating according to the same principles described above, are characterized by having a cylindrical shape. Typical of this kind are the Gibbs and the Vorce models, which in some countries became quite popular, owing to their simple construction, and in spite of their very limited current capacity ($\sim$ 1000 A). In the Gibbs cell the auolyte compartment is located at the center of the cylindrical and vertical shell; the catholyte compartment is formed by the annular space between the latter and the diaphragm-cathode, which has the same cylindrical shape but a smaller radius. The anodes consist of graphite rods arranged all around the inner surface of the diaphragm; in consequence, the anodic surface is not completely used to carry the current. Such a shortcoming is partially eliminated in the Vorce cell by arranging a second diaphragm-cathode in the most central part of the cell so that the anolyte compartment acquires an annular shape, since it is delimited between a cylindrical catholyte compartment at the cell center and by another annular catholyte compartment at the cell periphery.

In order to make the best use of floor space and decrease the labour requirements for maintenance and operation, new ingenious models have been further developed, allowing an increase in current capacity. Typical of this kind are the Hooker and the Diamond cells, presenting a cubical shape. The cathode is made out of one single wire screen with a winding configuration, which makes it look like a reversed basket with many rectangular and narrow compartments developing vertically, side by side. These compartments occupy the free spaces between each pair of parallel flat blades of graphite forming the anodic system, so as to build up a catholyte chamber for each anode pair. In order to apply the diaphragm material on this complicated cathodic structure, it is necessary to resort to a special technique, which consists of immersing the structure in an aqueous suspension of asbestos and applying a suction to the inside of the cathode frame, which will thus behave in much the same way as a vacuum filter and retain part of the asbestos fibers passing through it in the slurry, until it is completely coated. With such models it is possible to reach current capacities as high as 30,000 A per unit. The other operating characteristics are about the same as for the rectangular and cylindrical cell shapes. Accordingly, also with these more recent models the average current efficiency is about 0.95 and the alkali concentration in the catholyte effluent is not higher than 140 g/l. The diaphragm cells with vertical electrodes and diaphragms, especially in their modern versions, offer two main advantages in comparison with other designs; such advantages reside first in the relatively small floor space requirement, in that a considerable current capacity can be installed in quite a small area; and secondly, in easiness of operation and maintenance, since inner cleaning and diaphragm replacement can be carried out in the simplest way, with a considerable saving in labour.

On the other hand, such advantages are partly offset because the distance of the anodes from the cathodes cannot be adjusted during operation to compensate for the gradual anode wear; moreover, the diaphragm is exposed on one side to a strongly alkaline liquor, while on the other side it is in contact with a slightly acidic electrolyte containing dissolved chlorine (see Fig. IX, 6). On the other hand, no material, either natural or synthetic, has been found that can withstand the simultaneous attack by the anolyte and the catholyte for a long time so that, after a certain operating period, the diaphragm becomes defective, and must be repaired or replaced.

The filter-press type of assembly, with bipolar electrodes, which has proved so advantageous for water electrolysis (*cf.* p. 561 *et seq.*), has not encountered much favour in the chlorine–alkali industry, in which field the only outstanding example of such a design is afforded by the Dow cell. The difficulty of setting up a bipolar assembly for this particular purpose can indeed be easily appreciated by considering that, beside the other complications involved in this sort of arrangement, each bipolar electrode must be composed of two different materials such as graphite on the anodic side and iron on the cathodic side. Furthermore, the countercurrent method is not easily applicable in a filterpress construction.

Before ending this brief review of the most typical diaphragm cell designs, it is worth while mentioning also the Siemens–Billiter cell, *i.e.*, the only horizontal model that enjoyed a well deserved reputation for many years and was widely used up to the early forties. Even though this cell became obsolete because of its relatively large floor-space requirements, it was built in a particularly ingenious way, with the aim of applying at the same time all the means that may be advantageously brought to bear for the separation of the catholyte from the anolyte, including stratification. Its arrangement is schematically shown in Fig. IX,4. The cell consists of an elongated steel trough, throughout which, at some distance from the bottom, the cathode extends horizontally, in the form

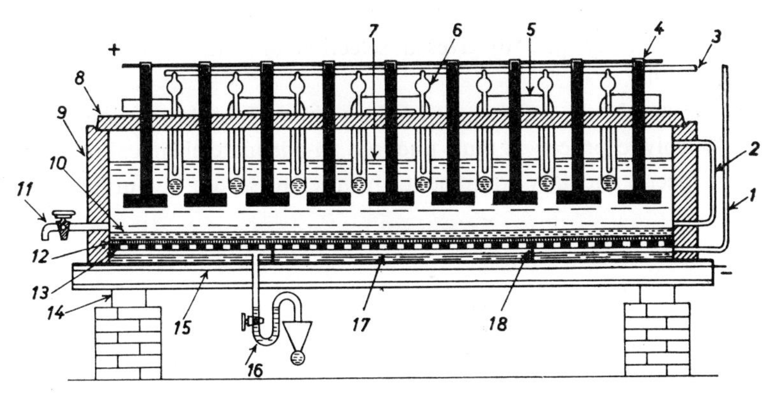

Fig. IX,4. Siemens–Billiter cell for Cl_2–caustic production. 1. H_2 outlet; 2. liquid level; 3. electrolyte feeding inlet; 4. graphite anodes; 5. heating; 6. distributor of salt solution; 7. anolyte; 8. cell cover; 9. concrete container; 10. powder diaphragm; 11. outlet; 12. asbest fabric; 13. iron cathode; 14. insulating supports; 15. iron floor; 16. NaOH outlet; 17. catholyte; 18. cathode supports

of a perforated steel plate or a wire screen. The diaphragm lies on the cathode itself and consists of asbestos fabric, coated with a layer of barium sulphate admixed with asbestos fibers. The thickness of this layer is about 2 cm.

The graphite anode plates hang horizontally, by means of anode stems, from the cell top cover, according to an arrangement similar to that usually employed for mercury cells (see next section). It is thus possible to provide for easy and accurate adjustment of the anode-to-cathode spacing, as the thickness of the graphite plate thins away under the chemical and mechanical action of the anodic gas. This possibility of anode adjustment, provided only by the horizontal arrangement, is another advantage over the vertical one.

The hydrogen gas collects over the caustic effluent in the compartment between the trough bottom and the cathode, and is separately withdrawn from the cell. The catholyte, before percolating through the diaphragm and the cathode, can stratify above the diaphragm itself, so as to build up an alkaline layer between the latter and the acidic anolyte bulk. The variation of pH with distance from diaphragm is diagrammatically illustrated in Figs. IX, 5 and IX, 6, for a vertical and a horizontal arrangement respectively. By thus preserving the diaphragm from any contact with the anolyte, its life can be substantially prolonged before renewal is required. Indeed such a life which is only of a few months in vertical cells, would extend for some years in the horizontal arrangement.

Table IX, 1 (*cf.* p. 550) gives a selection of data for various alkali-chlorine cells.

As a general rule, before passing the sodium chloride solution through the electrolytic process it must undergo an adequate chemical treatment, within special ancillary equipment, in order to secure favourable condi-

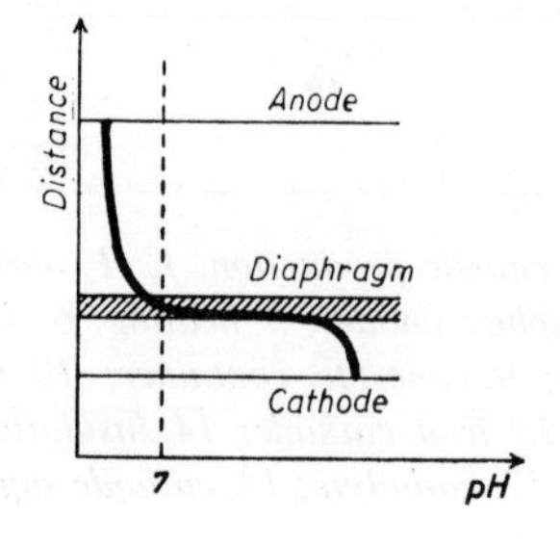

Fig. IX, 5. pH distribution in cells with vertical diaphragm

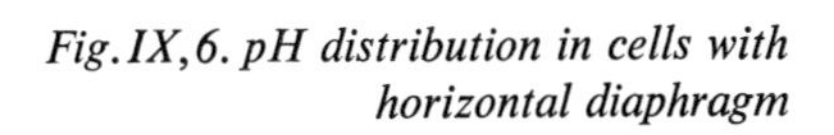

Fig. IX,6. pH distribution in cells with horizontal diaphragm

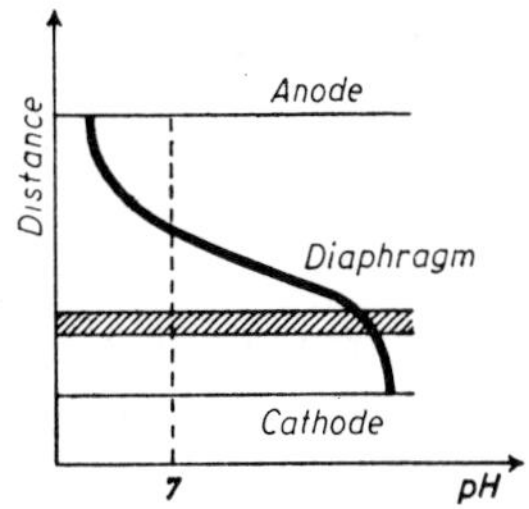

tions for the electrolytic process itself, as well as to obtain a satisfactory quality for the caustic product. Indeed, sodium chloride, whether it be rock-salt or of marine origin, always contains a certain amount of impurities, notably magnesium, calcium and iron, in the form of carbonates, sulphates and hydroxides. Sulphate, if left in the cell feed-brine in amounts greater than 5 g/l, is harmful, in that it will take a considerable part in the anodic process, giving rise to elementary oxygen and thus accelerating the anode consumption. Magnesium and other insoluble hydroxides tend to be retained on the diaphragm; if present in considerable amounts, such impurities would cause rapid diaphragm clogging so that special provisions must be made in order to remove them. Sulphate is precipitated as insoluble barium sulphate, by making it react with barium chloride or barium carbonate. Most of the calcium can be eliminated as carbonate by adding soda ash. Magnesium, iron and heavy metals are allowed to settle and filtered off as flocculating hydroxides, after raising the brine pH to 10 or 11 by the addition of caustic soda.

6. Mercury Cathode Cells

As mentioned above, by using a mercury cathode it is possible to avoid the shortcomings involved in the formation of OH^- ions at the cathode and their migration toward the anode.

The mercury cathode process is rendered possible only by the exceptionally high overtension of hydrogen on mercury and by the depolarizing effect on the discharge of the sodium and potassium ions that is brought about by the formation of sodium or potassium amalgam.

The half cell electric tension of a system consisting of an alkali metal amalgam in contact with a solution containing a salt of this metal is given by the equation,

$$U = U_0 + \frac{RT}{\mathbf{F}} \ln \frac{a_2}{a_1} \tag{1}$$

in which a_1 and a_2 are the activities of the alkali metal in the amalgam and of the alkali ions in the solution respectively and U_0 is the standard electric tension for $a_1 = a_2 = 1$.

In order to enhance the discharge of the alkali metal, the brine to be electrolysed is kept throughout the process nearly saturated with salt at the operating temperature of 70 to 80° C, while the alkali metal concentration in the amalgam is not allowed to exceed an average concentration of 0.1 % by weight. Under such conditions the value for U is about −1.8 V, both for sodium and potassium, and remains practically unchanged on raising the current density. Accordingly, as the alkali metal discharge takes place, hydrogen must be simultaneously evolved at such a rate that its discharge tension, including the overvoltage η, will be equal to U. If hydrogen is assumed to develop in the form of gas bubbles at the mercury surface — which implies a hydrogen pressure of 1 atmosphere — the relationship between the cathodic tension U and the hydrogen overtension η can be expressed as follows.

$$-\frac{RT}{0.4343\ \mathbf{F}}\ \mathrm{pH} - \eta = U \cong -1.8\ \text{(volt)} \tag{2}$$

In this equation the pH value to be considered is obviously that prevailing at the interface between the cathode and the solution. By assuming pH = 14; t = 70° C, the value for η becomes −0.85 V, which, by applying the proper value for the parameters a and b in the Tafel equation, (*cf.* Chapter IV) can be related to a hydrogen discharge current density J_H of about 1 mA/cm^2. In a horizontal mercury cell operated at a cathodic current density of 0.3 A/cm^2, the value for J_H would correspond to a volume content of 0.33 % H_2 in the chlorine gas, which is in fact a usual figure for normal operating conditions.

In the light of the former calculations, even though based on simplified assumptions, the two processes of amalgam formation and hydrogen evolution are clearly controlled by the kinetics on the mercury cathode in such a way that the pH at the interface is substantially greater than in the bulk of the solution, at least whenever the hydrogen content in the cell gas is within acceptable limits. Furthermore, experimental evidence shows that with other factors equal the rate of hydrogen evolution is independent of the overall current density at the cathode: in other words, the absolute amount of hydrogen developed per unit surface and unit time is substantially the same, whether a current is made to pass from the anode to the cathode or not. This seems to indicate that the parameter controlling the hydrogen evolution is closely related to the cathodic tension as determined by the alkali metal concentration in the amalgam.

In fact, the rate of hydrogen evolution independently of current density and for an established set of other operating conditions, such as cell geometry, temperature, brine purity, alkali chloride concentration and pH, is virtually dependent upon the concentration a_1, *i.e.*, the only quantity among those governing the electric tension U in (1) that is still considered as a variable. Therefore, with higher values of a_1 more negative values of U will be associated and thus higher rates of hydrogen evolution. The latter is therefore one of the limiting factors in raising the concentration, which is accordingly hardly ever allowed to exceed 0.2 % by weight at the amalgam outlet — a value definitely below the point at which the amalgam fluidity begins to decrease drastically, which occurs at about 1 % Na. Almost the same operating conditions are valid also for the electrolysis of potassium chloride, although the cathodic current efficiency is, in this case, about 2 % less than for sodium chloride due to the lower stability of the potassium amalgam.

If the electric tension of the cathode, at which electrolysis takes place, is now considered as constant (because it is chiefly determined by the activity of the alkali metal dissolved in the amalgam) each increase of the pH value in the cathodic film, shifting the hydrogen equilibrium electric tension toward more negative values, leaves a smaller margin of electric tension (see (2)) utilizable as overtension for the discharge of the H^+ ions and thus a lesser discharge rate of hydrogen. On the other hand, the agreement between the experimental value of the hydrogen content in chlorine gas and the corresponding value calculated from (2), by assuming in the cathodic film a pH approximately equal to 14, shows that provided the hydrogen content remains within normal limits, the pH in the cathodic film is considerably higher than in the bulk of the solution. This latter value, independently of the initial pH in the feed brine, tends to become slightly acidic, due to the reaction of the dissolved chlorine gas with water. It is therefore of paramount importance to keep this film as undisturbed as possible, by making the mercury flow as evenly as possible over a perfectly smooth surface.

The alkali metal amalgam obtained in the cathodic process is continuously circulated to a denuding cell, or decomposer, from which flows the caustic solution produced in the decomposition reaction. This cell contains a mass of conductive material with a relatively low hydrogen overtension. Graphite is universally adopted for this purpose, because, although it does not present the lowest overtension, compared with other materials such as iron, it has the advantage of not becoming coated with mercury, which would render its surface inactive. The system consisting of the alkali metal amalgam, the caustic solution and the graphite mass, forms a short-circuited galvanic cell, as follows,

$$^{+}C\text{–}H_2 \,/\, NaOH \,/\, Na\text{–}Hg \,/\, C^{-}$$

in which the anodic process takes place at the amalgam, with dissolution of the alkali metal, whereas the graphite performs a cathodic function, since hydrogen is discharged over its surface. Consequently, alkali hydroxide and hydrogen are the final products of the reaction between amalgam and water,

$$Na \cdot x\,Hg + H_2O \rightarrow x\,Hg + NaOH + \tfrac{1}{2}\,H_2$$

These two products, together with the chlorine gas directly delivered by the electrolytic cell, are the same final products as are obtained by electrolysis in diaphragm cells; however, there is an important difference as regards the caustic effluent, since, when it is coming from the mercury process, it is completely free of alkali chloride and can be obtained at much higher concentration.

By considering the three following systems,

$$Na \cdot x\,Hg\,(\text{amalgam}) \,/\, H_2O \,/\, Cl_2 \qquad \text{(I)}$$

$$Na_{aq}^{+} \qquad /\, OH_{aq}^{-} \,/\, H_2 \,/\, x\,Hg \,/\, Cl_2 \qquad \text{(II)}$$

$$Na_{aq}^{+} \qquad /\, Cl_{aq}^{-} \,/\, x\,Hg \qquad \text{(III)}$$

one can easily establish that the transition from any one to the next involves an exothermal reaction: this means that these systems follow each other in the sense of decreasing energy levels. Accordingly, any transition in the reverse direction, *i.e.*, from any one system to the preceding one, requires a contribution of energy from outside. In the diaphragm process such a contribution brings about the passage from system (III) to system (II), in which (on the assumption of thermodynamic reversibility) the unreactive component Hg can obviously be replaced by the unreactive component C, *i.e.*, the carbonaceous (graphite) anode that is used in the process. The energy required is equal to the product of the quantity of electricity involved in the reaction and the difference between the overall cell voltage and a term representing the energy dissipated as heat, which includes ohmic drops and cell overtension.

In the mercury process the first transition to take place is from system (III) to system (I), which is at a higher energy level than system (II). The voltage required in this process is therefore correspondingly higher than for diaphragm cells, since the same amount of electricity is involved in both cases to transform the same amounts of reactants (Na^{+}_{aq} and Cl^{-}_{aq}). On the other hand, the difference in voltage between the two processes must correspond to the difference in energy between system (I) and system (II), *i.e.*, to the transition occurring within the amalgam denuding cell. It ensues that the additional requirement of energy for electrolysis on a mercury cathode could be theoretically recovered in the amalgam decomposition step. This might, for instance, be achieved by arranging the amalgam denuder in such a way that instead of building up a shortcircuited galvanic system, as formerly described, the amalgam is kept separate from the cathodic mass; this might then consist of an iron electrode. Accordingly, direct recovery of the energy released in such a galvanic cell might be obtained by achieving electrolysis and subsequent deamalgamation in two serially connected sections, as diagramatically shown in Fig. IX, 7. Indeed, the denuding element would thus be seriesconnec-

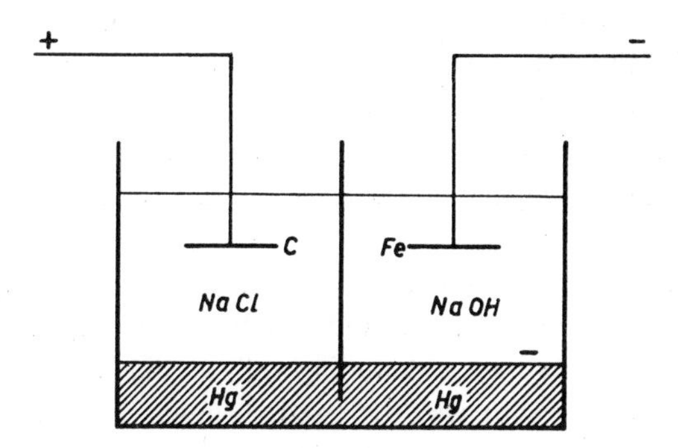

Fig. IX, 7. Electric circuit for principle of electrolysing and denuding cells in series

ted also with the external source supplying the energy for the electrolysis, so that the electric tension to be provided from such a source might be decreased by an amount equal to the electric tension of the denuding element

$$^{+}Fe{-}H_2/NaOH/Na{-}Hg^{-}$$

However, such a simple arrangement as outlined above would in practice be affected by two main shortcomings that render it inapplicable in this form. The first drawback lies in the fact that the cathodic current efficiency of the electrolytic process is hardly as high as 0.95, which means that a considerable fraction (5 % or more) of the current passing through the electrolysis as well as the decomposition cell must carry out some other anodic reaction in the latter, beside the redissolution of sodium from the amalgam whose production rate is only proportional to the current efficiency.

The second drawback is due to the impossibility, with a practically acceptable electrode size in the denuding cell, to keep the hydrogen overvoltage within a sufficiently low limit, so as to prevent polarization from exceeding the equilibrium electric tension of the galvanic element formed. This was not avoided even in the Castner cell, *i.e.*, the first model ever to be put to industrial use for the mercury process and consisting of an arrangement similar to that illustrated in Fig. IX, 7, in which the mercury was moved to and fro between the electrolysis and the decomposition compartments, by applying a rocking motion to the whole system. However, Castner's device was unsuccessful in recovering the amalgam decomposition free enthalpy, due to the difficulties inherent in the overtension, as explained above; on the contrary, the only possibility of achieving decomposition with such a system was to apply an additional voltage from the outside for the deamalgamation, so that the overall energy consumption was actually greater than with the arrangement developed at a later date and now universally adopted, which consists of forming a short-circuited galvanic element, with a large contact area between graphite and amalgam.

In order to overcome the first drawback, of the different current efficiencies of sodium discharge and amalgam decomposition, Castner placed a resistor R across the denuding cell, as illustrated in Fig. IX, 8, so that the excess current was bypassed through this shunt, whose resistance was adjusted accordingly. It is worth noting that such a device was justified only because the voltage applied to the denuding cell

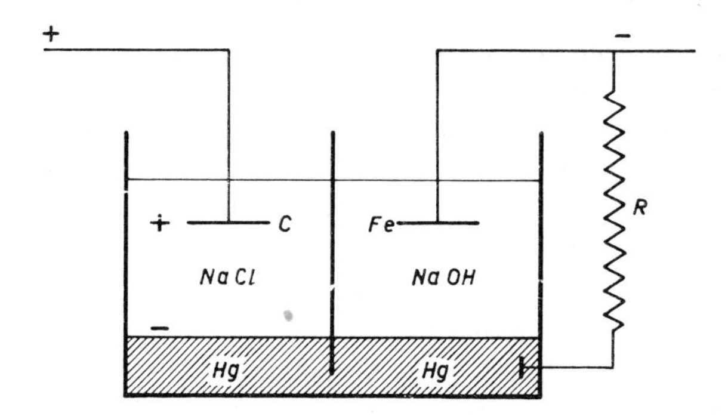

Fig. IX,8. Electric connections of principle of electrolysing and denuding cells in series with branched resistor

was greater than its equilibrinum electric tension, so that the current would pass also through the shunt in the direction from the amalgam to the negative cell terminal. Indeed, on the assumption that the denuding cell could supply energy to the circuit, instead of consuming it, the use of a shunt would be a mistake, since the cell would thereby be partially shortcircuited, so that its current would not be smaller as required, but even greater than in the rest of the circuit.

New prospects, in the matter of recovering a substantial fraction of the free enthalpy of the amalgam decomposition reaction, have been opened by the recent developments of fuel cell technology. However, all the methods proposed so far involve the consumption of the hydrogen by-product, so that they can be economically interesting only when there is no further scope for hydrogen in other chemical processes at the site. The most straightforward system would still consist in using the same basic arrangement as in Fig. IX, 7, with the denuding cell equipped with a porous cathode suitable for the depolarization of hydrogen by means of (atmospheric) oxygen admitted through the porous material. It is expected that it would thus be possible to produce 20 to 25 % of the energy required for the electrolysis in modern cell plants, although the practical applicability of the method still seems somewhat premature both from the economic and the technical standpoints.

Among the many cell models that have been developed, the ones that are most extensively used belong to the horizontal type, which was formerly adopted by Solvay, illustrated diagramatically in Fig. IX, 9. The electrolytic and denuding sections consist of elongated troughs, assembled side by side with a slight slope in opposite directions, so that the continuous layer of mercury flowing over the bottom of the electrolytic (chlorine) cell gradually increases its alkali metal concentration, until it leaves through the lower end in the form of amalgam and passes, by gravity, through an appropriate interconnecting system, into the higher end of the denuding (caustic) cell. The stream of depleted amalgam leaving the other end of the denuder is lifted by mechanical means, such as a scoop-wheel or a pump, up to the higher end of the chlorine cell, so as to be continuously circulated between the two sections. The amalgam flows countercurrent to the alkali chloride solution in the chlorine cell and to the caustic solution in the denuder, which is fed with pure water. Chlorine and hydrogen are released through pipe-lines connected with the tops of these two sections respectively.

The electrolytic cell is equipped with a set of anodes suspended from the top cover and connected with the positive bus-work; the negative line is in electric contact with the mercury cathode through the steel bottom or a special steel structure embodied in the cell. The normal anode material, which was formerly platinum, is now graphite.

In the horizontal denuder the amalgam layer and the alkaline solution

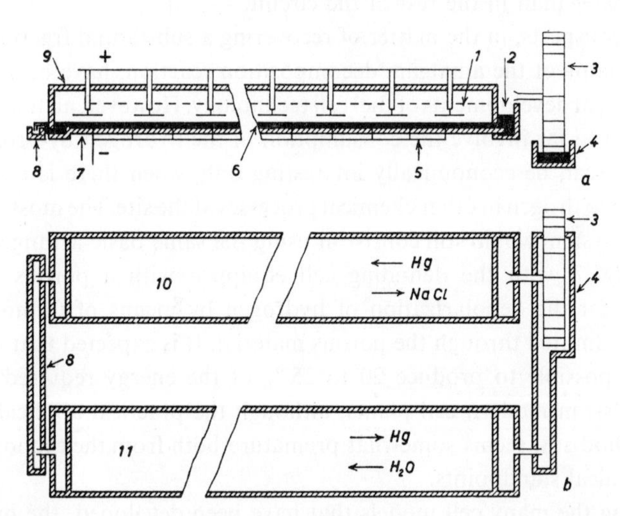

Fig. IX,9. Solvay cell for Cl_2–caustic production. 1. Pt anodes; 2. cathodic Hg; 3. Hg elevator; 4. Hg reservoir; 5. cathodic connection; 6. Hg cathode; 7. concrete container; 8. connection canal between electrolysing and denuding cell

flow countercurrently in mutual contact, as well as in contact with a set of longitudinal blades of graphite, which are metallically connected together through the steel bottom or, possibly, through some other device, in order to facilitate the shortcircuiting of the current. In fact the current after leaving the upper layer of the amalgam and passing through the caustic solution, must enter the cathodic mass of graphite and finally reach again the mercury through the contact surface between the latter and the graphite, as well as between the mercury and the steel bottom if this is unlined and in direct contact with the graphite mass.

Many patented improvements have been made in modern cell construction by several manufacturers, with particular regard to the following points: possibility of continuous adjustment of anode-to-cathode spacing (normally to be kept at about 5 mm) in order to compensate for chemical attack and mechanical wear of the anodes; operation safety; ease of cell maintenance; floor space saving; chlorine- and alkali-resistant cell lining; cell covering and sealing, so as to keep the several liquid and gaseous streams well separated from one another. Accordingly, each particular model has its own valuable characteristics, even though the operating principles are basically the same.

In order to save floor space, there is a modern tendency to place the horizontal denuder under the chlorine cell (Solvay, Uhde). In other constructions (de Nora, Mathieson) the amalgam denuder consists of a vertical tower packed with graphite lumps, down which the amalgam trickles countercurrent to the caustic solution; in the de Nora design the vertical decomposer may embody more than one stage, which gives the possibility of producing caustic solution with up to 73 % NaOH.

As formerly noted, mercury cells are operated at a higher current density than diaphragm cells; indeed, according to modern practice, the former may have an anode current density as high as 60 A/dm^2, with a corresponding cathodic current density of about 55 A/dm^2. Such values are ten times greater than those applicable in the diaphragm process. Unit current capacities in mercury cells of latest design may reach values up to 160,000 A (de Nora).

In spite of such large current densities and unit capacities, the geometry of the horizontal mercury cell involves considerable floor space requirements. In consequence, several efforts have been devoted to the designing of vertical cells, in which a mercury cathode film is supported by a vertical structure of solid metal, such as iron. The Honsberg–Messner cell, developed by Badische Anilin- und Sodafabrik (BASF), is the only model of this sort that has so far found large scale application in industrial plants, even temporarily. The cathodes consist of rotating steel discs, partly submerged in an amalgam pool. A graphite plate is interposed between each pair of discs above the amalgam pool; the intermediate spaces between the graphite plates and rotating discs are fed with the alkali chloride solution.

Although several other vertical cell models were developed in the past fifteen years, none seems to have been completely successful; indeed, beside requiring more careful operation and maintenance, no vertical construction has yet been able to embody satisfactory facilities for anode adjustment in order to keep the cell voltage within acceptable limits in spite of anode wear, and for an economic use of the anode material.

As already noted, the substantial difference in results between the diaphragm and the mercury processes is that the alkali hydroxide is not produced in the salt solution itself during electrolysis. On the other hand, the spent brine that leaves the mercury cell, after having been depleted of about one sixth of its initial salt content, is saturated with chlorine gas. Most of this is recovered by passing the cell effluent through flashing towers under a vacuum corresponding to about 400 mm Hg absolute pressure, after addition of hydrochloric acid, so as to decrease the free and active chlorine content from 0.3–0.4 g/l down to 0.1 g/l. This residual amount is further reduced to 0.01 g/l by blowing air through the brine stream in appropriate equipment. The brine is thus ready to be resaturated, by letting it flow over a bed of solid salt, after adjusting its alkalinity to pH = 10–11. The resaturated brine stream is then submitted to chemical purification and filtration, according to methods that are basically the same as for the diaphragm process.

When the electrolysis is to be carried out by the mercury cathode process, the alkali chloride solution must be purified more carefully than for the diaphragm process, with special regard to magnesium and iron salts, which unfortunately, are the most common impurities to be found in the original salt. If these cations are left in the solution in considerable quantities, the hydrogen evolution over the cathode may become much more active, which is interpreted as being due to the lowering of the hydrogen overtension on mercury caused by a synergistic catalytic action of these metals. Indeed, when present separately, magnesium and iron do not seem to exert any remarkable influence, but the current efficiency is considerably diminished when they are present together. Accordingly, the contents of iron and magnesium in feed brine should never exceed the limits of 100 mg/l and 5 mg/l respectively.

The catalytic action of some other heavy metals, such as vanadium, chromium and molybdenum, is so strong that even when present as a small fraction of a milligram per liter they may bring about such an active hydrogen evolution as to produce an explosive mixture in the cell gas. Vanadium may be present in brine due to leaching from graphite anodes, when the starting material is petroleum coke. Chromium and molybdenum will not be introduced into the brine system, if any sort of alloyed steel is thoroughly avoided among the construction materials of the equipment.

When the cell plant is equipped with graphite anodes, the sulphate content also must not be allowed to exceed 3 to 5 g/l.

Before being recirculated to the electrolysis plant, the saturated brine is preferably slightly reacidified to pH 5. This preacidification prevents the formation of hypochlorite and chlorate in the course of electrolysis, so that, beside increasing the anode current efficiency and diminishing the anode oxidation, the cathodic efficiency is also improved. In fact, the main reactions that are responsible for current losses at the cathode are the following:

$$2e^- \,(Hg) + 2\,H_2O \rightarrow 2\,OH^- + H_2\,(g)$$

$$2e^- \,(Hg) + Cl_2\,(aq) \rightarrow 2\,Cl^-$$

$$2e^- \,(Hg) + ClO^- + H_2O \rightarrow Cl^- + 2\,OH^-$$

$$6e^- \,(Hg) + ClO_3^- + 3\,H_2O \rightarrow Cl^- + 6\,OH^-$$

The first reaction, associated with hydrogen evolution, has already been discussed. The second, involving a reversion of the chlorine gas produced to chlorine ion, can be checked by keeping not too close a spacing between anode and cathode surfaces, so as to prevent chlorine gas bubbles, from reaching the underlying mercury layer as soon as they develop at the anode (the appropriate distance is about 5 mm). The effect of the third and the fourth reactions, corresponding to the reduction of hypochlorite and chlorate ions respectively, can be restrained, as explained above, by feeding the cell with a slightly acidic brine, which limits the formation of such ions (See (1), (2), (8) and (10) on page 516 *et seq.*).

The other advantage of making the feed brine slightly acidic is that it is thereby possible to hinder the discharge of calcium, normally present as an impurity, on the mercury cathode. Indeed, calcium tends to amalgamate and although it does not directly affect the hydrogen discharge, it renders the mercury flow sluggish and irregular, due to the formation of so called 'amalgam butter', which indirectly disturbs the smoothness of the cathodic interface and therefore contributes to the decrease in current efficiency, beside complicating the operation and maintenance. The sodium content in the amalgam is kept quite low, usually at no more than 0.15 %, because the amalgam reacts more easily with water when it is concentrated than when it is diluted. Raising the concentration further would entail the risk that the amalgam might start decomposing while still inside the electrolysis cell, which would involve a loss in current efficiency, with the formation of an explosive mixture of hydrogen in chlorine gas and the production of alkali inside the electrolytic cell itself, with the harmful effects already discussed.

The caustic concentration thus obtained is greater than that obtained by means of cells of other kind; it is 300–400 g/l.

Table IX, 1 collects some significant information concerning the most usual mercury cells.

7. Finishing of Cell Products
Comparison between the Diaphragm and the Mercury Processes

The most usual methods for the production of chlorine and caustic are based on diaphragm cells and mercury cathode cells. Both types yield hydrogen gas, as well as chlorine gas and caustic in aqueous solution. The latter still contains some undecomposed chloride if coming from the diaphragm process, whereas it is practically free of any residual salt if produced in amalgam decomposers.

The hydrogen product will be advantageously sent to adjoining processing plants, if any; otherwise it may be compressed in steel cylinders and sold as such.

Chlorine is usually liquefied before being used for any further manufacturing process, except for the chemical production of hypochlorite. For storage or transportation purposes, it must be liquefied and transferred into storage tanks, tank cars, or cylinders. Liquefaction is achieved after precooling to atmospheric temperature and drying through a packed tower, in countercurrent to a stream of concentrated sulphuric acid. The liquid state may then be reached by two different methods, *i.e.*, either by a mild cooling of the chlorine gas down to about 0° C, followed by a compression sufficient to reach the saturation vapour pressure corresponding to such a temperature range (3 to 5 atm); or else, by deep cooling, which may bring the temperature down to as low as – 50° C, so that the pressure required for liquefaction will be just above the atmospheric value and easily obtainable by rotary compressors.

The caustic solution leaving the diaphragm process and having a strength of about 140 g NaOH/l, with a residual content of 200 g NaCl/l, is first concentrated to 40° Bé by evaporation under reduced pressure to raise the concentration to 500 g NaOH/l. The solubility of sodium chloride is thus drastically decreased and the precipitated salt is separated by centrifuging or any other convenient method. The concentrated caustic solution, with a residual content of 2 % chloride, is then passed to direct-fired pots made of cast iron, where the concentration is raised to 50° Bé, and eventually to final concentrators made of cast-iron, or nickel, or nickel-alloyed steel, where any residual water is eliminated, so as to obtain the final product in the form of fused caustic. This is poured into black steel drums that can be hermetically sealed and shipped as such; or else the molten product is flaked, or allowed to solidify in shallow molds; the solid cakes thus obtained are broken into lumps and these are packed in wooden drums.

If the residual chloride content in the final caustic product has to be diminished to less than 1 %, the 50 % solution is rediluted to 37.5 % and cooled to 5° C; the precipitate is then separated from the solution, which is sent back to the evaporators.

The crystallized salt obtained from the caustic finishing process is of high purity, except for a slight alkalinity (0.2 % NaOH) which can be neutralized with hydrochloric acid. Therefore, rather than recycling it to the electrolytic process, the recovered salt can be profitably sold as a valuable by-product.

The finishing equipment is submitted to a considerable attack by the hot, concentrated alkali. The most efficient and practical system to counteract it is provided by cathodic protection. This is achieved by connecting the negative terminal of a direct current supply with the finishing vessel, and the positive terminal with a platinum anode dipping into the fluid mass; the vessel will thus be cathodically protected and the attack substantially reduced, and electrolysis of the residual water will take place, accelerating the final dehydration of the processed material.

The processing of the caustic product obtained by amalgam decomposition is similar to the procedure described for the diaphragm-cell effluent, except for the preliminary steps of concentration under reduced pressure, crystallization and separation of chloride. Sometimes, the pure and concentrated caustic solution is used at the site (for instance in the rayon industry) or sold as such, without any further operation.

Although the above considerations have particularly concerned the electrolysis of sodium chloride, they can also be wholly applied to potassium chloride.

Whenever a comparison between the diaphragm and the mercury processes is to be made, local conditions and requirements must first be considered; indeed, these will eventually determine the choice, and in particular the cost of electric power compared with the cost of steam needed for caustic evaporators, together with the final concentration and purity wanted for the caustic product.

The main points of difference between the two processes that have to be kept in mind for such a comparison may be summarized.

The outstanding advantages afforded by mercury cells are as follows.

1. The caustic product, as obtained from the electrolysis process, is at a relatively high concentration (50 % or higher) and already separated from the depleted salt solution. Therefore, the evaporation plant, if required at all, is simpler, of reduced size and also cheaper in operation; in several cases it can even be omitted, as for instance in rayon factories, where the caustic solution can be used as such at the site.

2. The purity grade of the caustic product is substantially better, with particular regard to the Cl^- ion content, so that further purification is always unnecessary.

3. Cell maintenance costs are smaller, by virtue of the larger unit capacity which means a lesser number of cells for an established daily production, and on account of the fact that there are no diaphragms to be periodically renewed.

4. Cell shutdowns for maintenance are less frequently needed and maintenance operations are much faster, so that the number of standby units required to allow a constant rate of production is reduced, with a corresponding reduction in additional plant cost.

On the other hand, diaphragm cells are characterized by more favourable operating conditions in the following respects.

1. The electric energy required for electrolysis in diaphragm cells is about 20 % less than in mercury cells.

2. Operation is relatively safer, due to the impossibility of explosive gas mixtures being formed, although the dangers involved in such a possibility may be considered negligible also for some of the most modern mercury cell plants.

3. The diaphragm process is less exacting, as far as brine purification is concerned.

If all considerations in favour of either process are taken into account, the following conclusions can be drawn.

1. The overall plant cost (including caustic fortification) is of the same order for either system.

2. For the production of technical grade caustic soda the diaphragm process may, under some particular local conditions, be cheaper.

3. For the production of high purity (rayon grade) caustic soda the mercury process is in general more economic.

4. When costs are low for electric energy and high for fuel, the mercury process is cheaper.

5. The mercury process can be more easily adapted for the production of other alkali products beside caustic soda, such as caustic potash, or lithium hydroxide.

6. The mercury process can be used also to exploit the outstanding reduction properties of alkali metal amalgams, in order to obtain, beside chlorine and caustic solutions, other valuable products, such as alkali alcoholates.

7. In some cases, as when salt is directly available in the form of natural saturated brine, it is highly advisable to associate the two processes, in such a way that the natural brine is fed to a diaphragm cell plant and the highly pure salt obtained from the evaporation plant handling the alkaline cell effluent is used to feed an adjoining mercury cell plant, without any further necessity for brine treatment of the latter.

Table IX,1 lists a selection of comparative values for some of the best known types of diaphragm and mercury cells.

TABLE IX,1

OPERATING CHARACTERISTICS OF TYPICAL DIAPHRAGM AND MERCURY CELLS

Cell Types	Diaphragm arrangement	Operating period	Electric tension V	Current (rated) A	Current density A/dm^2		Current efficiency %	Energy consumpt. kWh/kg		Lye concentr. g/l		% CO_2 in Cl_2
					Anodic	Cathodic		NaOH	Cl_2	NaOH	NaCl	
Diaphragm cells												
Griesheim*	v.s.**	3– 4 days***	3.6–3.8	2,500	2	3	80	3.1	3.5	40– 60	150–175	10–12
Aussig	—	years	3.5–4	500	4	2	85	3	3.3	120	—	2
Billiter–Leykam	—	years	3.1–4.5	1,000	—	2–4	89–93	2.4	2.8	100–120	—	4–5
Townsend	v.s.	1 month	4.6–5.2	3,000	10–12	10–15	93–97	3.2	3.6	140–200	135	2
Gibbs	v.u.	4 months	3.4–3.6	800	1.9	4.8	92–95	2.7	3	120	140–170	1.5
Vorce	v.u.	6– 9 months	3.7	1,600	5.2	6.6	94–97	2.4	2.7	106	170	1.7
Krebs	v.u.	7–10 months	3.3–3.7	5,000	4.3	4.5	90–94	2.6	2.9	85–135	130–170	1.2
De Nora	v.u.	15 months	3.3–3.6	3,000	7	6.6	90	2.7	3	110	160	1
Giordani–Pomilio	v.s.	7–10 months	3.3–3.8	3,000	7	5	92–95	2.8	3.3	120–180	110–150	1
Ciba	v.s.	12–15 months	3.3–3.5	7,000	13	4.5	90	2.5	2.9	110–130	—	2
Siemens–Billiter	h.e.	years	3.4–3.6	12,000	13	12	94–96	2.5	2.8	120–150	—	1.2
Diamond	v.s.	8 months	3.8	30,000	13	13	96.5	2.7	3	133–140	210	0.9
Hooker S–3 B	v.s.	4– 7 months	3.8	27,000	13.3	13.3	96	2.7	3	135–138	150	—
Mercury cells												
De Nora 18 SGL	—	12 months****	4.4	120,000	41.2	36.9	94–96	3.1	3.5	750	0.02	0.3
Krebs–BASF	—	—	4.25	80,000	36.6	34.8	94–95	3.2	3.6	750	0.02	—
Solvay V–200	—	12 months	4.4	170,000	58.9	55.8	96	3.1	3.5	750	0.02	—
Uhde	—	11–12 months	4.3	80,000	41.5	40	95–97	3	3.4	750	0.02	—

* Diaphragm of cement; anodes of carbon.
For all other cells: diaphragm basically of asbestos; anodes of graphite.

** v.s. = vertical, submerged on both sides
v.u. = vertical, unsubmerged on catholyte side
h.e. = horizontal, with empty space under diaphragm and cathode (catholyte side)
— = without diaphragm.

*** Batch process. In other diaphragm cells (working on continuous process) the operating period mainly depends on diaphragm life.

**** In mercury cells the operating period mainly depends on anode life.

8. Production of Hypochlorites and Chlorates

(I) *Theoretical Aspects*

In section 2 the conditions were discussed that must be maintained during the electrolysis of an alkali chloride in aqueous solution, in order to obtain the highest possible current efficiency, *i.e.*, the maximum yield of chlorine gas and alkali hydroxide. Among such conditions, it was seen that the most important was to keep the OH^- ions, forming in the catholyte, away from the chlorine-saturated anolyte.

If, on the contrary, the anolyte and the catholyte are allowed to mix together, several secondary reactions will take place, as described by equations (1) to (10) in section 2. The final products of such reactions are the hypochlorite and the chlorate of the alkali metal, which is normally sodium or potassium, present in the starting salt.

Since the reaction rates at which hypochlorite and chlorate are formed depend to a large extent upon the operating conditions, it is possible, by a proper control of certain variables, to keep the reaction from reaching thermodynamic equilibrium, corresponding to equations (7) and (10) in section 2 and thus to obtain either one product or the other at will.

Even though some kinetic details of the processes here considered are not yet fully understood, it is possible to form a fairly accurate idea of the several steps according to which the reactions proceed, thanks to the scientific investigation that was carried out, in particular, by E. Müller and his coworkers, as well as to the technical experience that was gathered in the painstaking efforts to bring this branch of electrochemical industry into being, shortly before the turn of the century.

In principle, hypochlorite formation is promoted whenever the operating conditions are kept not too far from alkalinity. This is achieved in practice by allowing the OH^- ions formed at the cathode to react with the elemental chlorine produced at the anode

$$Cl_2 + 2\,OH^- \rightleftharpoons ClO^- + Cl^- + H_2O \qquad (1)$$

The hypochlorite ion tends to further react with water to yield free hypochlorous acid

$$ClO^- + H_2O \rightleftharpoons HClO + OH^- \qquad (2)$$

It will be seen that in both reactions the Na^+ ion takes no part; it will also be noted that the second reaction is shifted toward the right by substracting OH^- ions, or, what is equivalent, by adding H^+ ions, as occurs when under slightly acidic conditions the free hypochlorous acid and the hypochlorite ion react together with formation of chlorate

$$2\,HClO + ClO^- \rightleftharpoons ClO_3^- + 2\,Cl^- + 2\,H^+ \qquad (3)$$

This, as already explained, is the equation representing the secondary reaction controlling the chemical mode of chlorate formation.

The basic experimental observation underlying the assumption that hypochlorite and chlorate are formed according to the mechanism described by equations (1), (2) and (3), is that, whereas a hypochlorite solution is quite stable when slightly alkaline, its concentration will quickly decrease, with a corresponding accumulation of chlorate, as soon as the pH is lowered to weak acidity. In fact, since hypochlorite is the salt of a weak acid and a strong base, it is hydrolized to a large extent according to equation (2). In consequence, the lower the OH^- concentration, the higher the amount of free hypochlorous acid in equilibrium with the ClO^- ion. On the other hand, the rate of reaction (3) can be expressed by the rate of decrease of the ClO^- ion, which obeys the following equation:

$$-\frac{d\,[ClO^-]}{dt} = k\,[HClO]^2\,[ClO^-]$$

Thus that the rate of chlorate formation, beside being proportional to the ClO^- ion concentration, varies as the square of the free hypochlorous acid content, which however increases at the expense of ClO^- ions, as shown by (2). Therefore, an optimum condition will be reached, depending upon pH, when free hypochlorous acid will be present in sufficient amount to maintain chlorate formation at the same rate as the production of hypochlorite, which in turn is controlled by the electrolytic process.

When in a system, beside Cl^- and OH^- ions, there is also an appreciable quantity of ClO^- ions, other reactions beside (1) and (2) will necessarily take place, the first of which is the primary process consisting of the anodic oxidation of the ClO^- ion.

In fact, other conditions being equal, the discharge of ClO^- ions requires a less positive electric tension than that of Cl^- ions, even if all overtensions are taken into consideration, as shown in Fig. IX, 10.

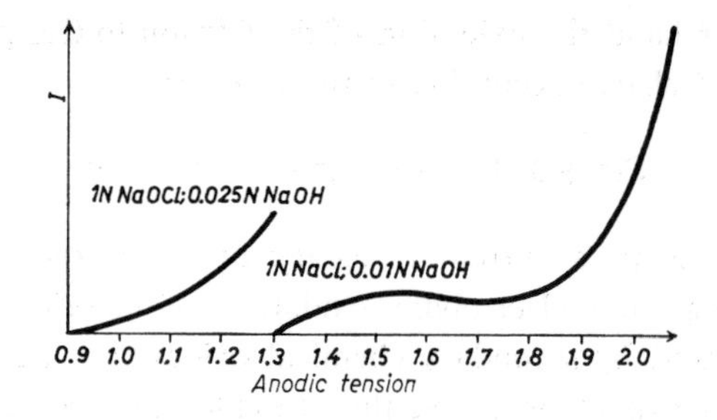

Fig. IX, 10. Discharge diagram for Cl^- and ClO^- ions

This explains the possibility of an electrochemical process taking place, according to the following anodic reaction:

$$6\,ClO^- + 3\,H_2O \rightarrow 2\,ClO_3^- + 4\,Cl^- + 6\,H^+ + \tfrac{3}{2}\,O_2 + 6\,e^- \qquad (4)$$

which is accompanied by the following cathodic reaction:

$$6\,H_2O + 6\,e^- \rightarrow 6\,OH^- + 3\,H_2 \qquad (5)$$

It is worth while pointing out that the formation of chlorates occurs at the expense of further ClO^- ions, which, as in the purely chemical formation, are reduced to Cl^- ions, whereas the electric current evolves oxygen from water.

The experimental basis for such a process is that, by submitting to electrolysis a hypochlorite solution containing no chloride, the oxidation to chlorate is accompanied by the appearance of Cl^- ions, while the quantity of electricity that is passed through the system, exactly corresponds to the quantities of oxygen and hydrogen produced, according to Faraday's Laws. The electrochemical process, as expressed by (4) and (5), requires 6 **F** corresponding to the electrolysis of 3 moles of water, while 2 g ions of chlorate are produced from hypochlorite. On the other hand, these 6 equivalents of reacting hypochlorite must in the first place be made available by the electrolytic production of 12 equivalents of chlorine and of the corresponding quantity of OH^- ions, as shown by reaction (1). Thus the overall amount of electricity required to produce 2 g ions of chlorate is 18 **F** of which 6 **F** are consumed in water electrolysis. This corresponds to a current efficiency (relative to chlorate production) of 0.667.

On the other hand the oxidation of the Cl^- ion to ClO_3^- ion requires 6 equivalents of charge according to the scheme:

$$Cl^- + 3\,H_2O \rightarrow ClO_3^- + 6\,H^+ + 6\,e^- \qquad (6)$$

This corresponds to a current efficiency of 1, for the production of chlorate through the electrolytic oxidation of chloride to elemental chlorine, followed by secondary chemical oxidations, *i.e.* disproportionation, first to hypochlorite and then to chlorate according to the following reactions.

electrochemical process	anodic reaction	$6\,Cl^- \rightarrow 3\,Cl_2 + 6\,e^-$
	cathodic reaction	$6\,H_2O + 6\,e^- \rightarrow 6\,OH + 3\,H_2$
chemical oxidation	1st. secondary oxid.	$3\,Cl_2 + 6\,OH^- \rightleftharpoons 3\,ClO^- + 3\,Cl_2 + 3\,H_2O$
	hydrolysis	$2\,ClO^- + 2\,H_2O \rightleftharpoons 2\,HClO + 2\,OH^-$
	2nd. secondary oxid.	$ClO^- + 2\,HClO \rightleftharpoons ClO_3^- + 2\,Cl^- + 2\,H^+$
overall reaction		$Cl^- + 5\,H_2O + 6\,\mathbf{F} \rightarrow ClO_3^- + 2\,H^+ + 2\,OH^- + 3\,H_2$

The equivalence between (6) and (7) is obvious. It follows that, for technical purposes the establishment of the electrochemical process (4) and (5) represents an unwanted loss of electric energy. However, it is worth while to repeat here that it will mainly depend upon control of pH and ClO^- ion concentration, other conditions being equal, whether chlorate will form from hypochlorite by the chemical or the electrochemical way.

As a rule, for a given current density at the anode, a rise of the ClO^- concentration within the solution will be accompanied by an increase in its discharge rate; this will involve a gradual decline in the Cl^- ion discharge rate and consequently in the further accumulation of ClO^- ions, until a steady state is reached. This is depicted in Fig. IX, 11, the five curves of which were obtained in the course of the electrolysis of a sodium chloride solution having an initial volume of 200 cm^3 and an initial content of 5.1 moles of NaCl, plus 0.44 g of potassium chromate to inhibit the cathodic reduction of the hypochlorite and chlorate ions accumulating in the solution. The temperature was kept at 12.5° C and the current density over a platinum anode was 0.067 A/cm^2. Curves I and II are plotted against the ordinate scale at the right. Curve I represents the change in current efficiency with time and with reference to the active oxygen that is bound either in hypochlorite or in chlorate. Such current efficiency starts with an initial value of about 0.97 or 0.98 and drops to 0.667.

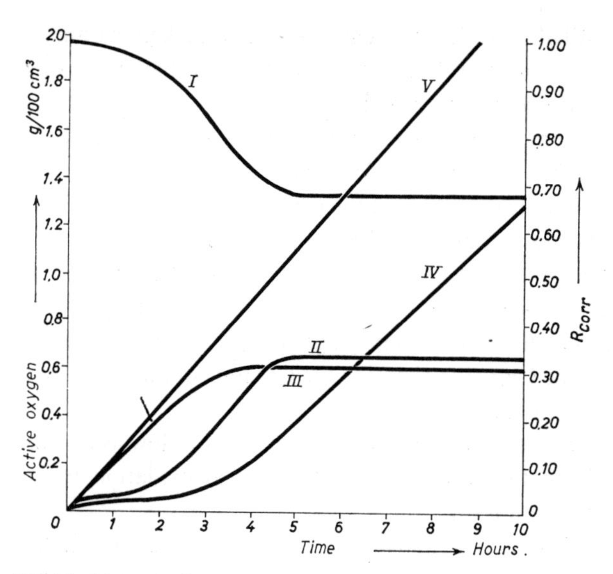

Fig. IX, 11. Diagram for NaClO and $NaClO_3$ electrolytic production

Curve II represents the change in current efficiency for the oxygen liberated in gaseous form; it begins at zero and ends at 0.333.

Curves III,IV,V are plotted against the ordinate scale at the left, which is graduated in grams/100 cm³ solution. Curve III and curve IV respectively represent the gradual increase in concentration of the active oxygen bound in hypochlorite and chlorate; the former rises up to a certain concentration value for hypochlorite, which then remains constant at all times — this ceiling depending upon the several geometrical characteristics and operating conditions; the latter increases steadily up to chlorate saturation point. Curve V gives the sum of active oxygen concentration, according to theoretical calculation.

(II) *Technical Production of Hypochlorite*

The foregoing considerations, supported by experimental evidence as well as by theoretical arguments, show that to obtain hypochlorite by electrolysis of sodium or potassium chloride without allowing the process to continue as far as the oxidation of hypochlorite to chlorate, a set of conditions must be satisfied, which will be outlined as follows.

1. The alkali chloride concentration must be kept as high as possible, so as to promote the discharge of Cl^- ions rather than of ClO^- ions, which would give rise to reaction (4).

2. The geometrical characteristics of the electrolytic cell must as far as possible hinder any turbulent motion in the anolyte, which would favour the anodic discharge of the ClO^- ions forming within the so called diffusion layer, at a distance from the anode.

3. The anodic current density should be kept relatively high. Indeed, when the solution is not stirred and any other cause of turbulence is avoided, the curve of overvoltage *versus* current density rises more steeply for the ClO^- ion than for the Cl^- ion, so that the discharge of the latter is promoted.

4. The operating temperature must be kept as low as possible, in order to keep down the reaction rate of chemical oxidation to chlorate, according to reaction (3).

5. The cell operation must be performed at neutral pH. Alkaline conditions would shift reactions (1) and (2) in the direction of building up such a high ClO^- ion concentration as to promote further oxidation according to the anodic process of (4), whereas acidity would enhance chemical oxidation, according to reaction (3).

6. Since the chlorine that is bound in a higher oxidation state within the hypochlorite ion tends to revert back to the reduced state, *i.e.*, Cl^- ion, upon contacting the cathode, physical diffusion of the hypochlorite to the cathode is usually hindered by means of additives such as potassium chromate, calcium chlorate, Turkey red oil, rosin soap or vanadium salts, which are able to build up an impervious film around the cathode.

It has already been observed that, even though taking place in the electrolytic cell, hypochlorite formation is never electrochemical in nature, since it is a secondary reaction occurring between primary anodic and cathodic reaction products, *i.e.*, chlorine and caustic, according to(1). Therefore, chlorine and caustic can also be withdrawn from the electrolytic chlorine–caustic cell and made to react together within a separate vessel. This is, indeed, practically achieved in most modern plants, specially for large scale production, in view of the fact that the operation of chlorine–caustic cells requires lesser precautions than demanded by hypochlorite cells and may be carried out in equipment of much larger current capacity. In consequence, the advantage of producing hypochlorite directly in specially designed electrolytic cells is by now confined only to plants of very small daily production, since it is thus possible to save the cost of additional equipment.

It may also be noted that the residual content of sodium or potassium chloride in the final bleach liquor, for the same content of active chlorine, must be kept higher in the effluent of hypochlorite cells than in the bleach solution obtained from separate reaction equipment. This means that in the production of sodium hypochlorite the salt consumption for each kg of active chlorine is about 6 kg NaCl in the former case, as against 4 kg NaCl in the latter. On the other hand, specific energy consumptions are not the same in both cases, since 1 kg of chlorine plus the equivalent amount of caustic require 3 to 4 kWh in chlorine–caustic cells, whereas the energy consumption in hypochlorite cells is usually between 5 and 7 kWh per kg of active chlorine.

Hypochlorite cells are always of the bipolar type. In fact, the bipolar arrangement can in this particular case afford all its inherent advantages of simplicity and compactness, due to the lack of diaphragms and of any other requirement arising when anolyte and catholyte must be kept separate. In some cases the bipolar electrodes are made of graphite on the cathodic as well as on the anodic side; in other instances the electrode material is platinum, at least on the anodic side. The distance between each electrode and the next one is of the order of a few millimeters. The cell containers are generally of stoneware. In one type (De Nora) the electrodes, instead of being vertical, are inclined, so as to help the elemental chlorine produced at the anodic side to diffuse into the solution rather than escape to the atmosphere.

The feeding solution consists of 10 to 15 % NaCl, to which reduction inhibitors, are sometimes added such as potassium chromate, in the amount of 0.2 to 0.5 %. The final concentration is kept within the range from 10 to 20 g/l of active chlorine. The single cell voltage can vary considerably for the several models, *i.e.*, between 3.7 and 6.1 V. The number of cells in series within the bipolar unit is usually not more than 40 and the overall voltage not higher than 220 V, while the current capacity does not exceed 50 A. In order to keep the electrolyte as neutral as possible, caustic soda is added periodically, to neutralize the acid forming at the anode due to the reaction of chlorine gas with water.

The Kellner hypochlorite cell is a typical example of this kind. Its top view and vertical section are diagrammatically represented in Fig. IX, 12, *a* and *b* respectively. The bipolar electrodes 7, made of platinum gauze,

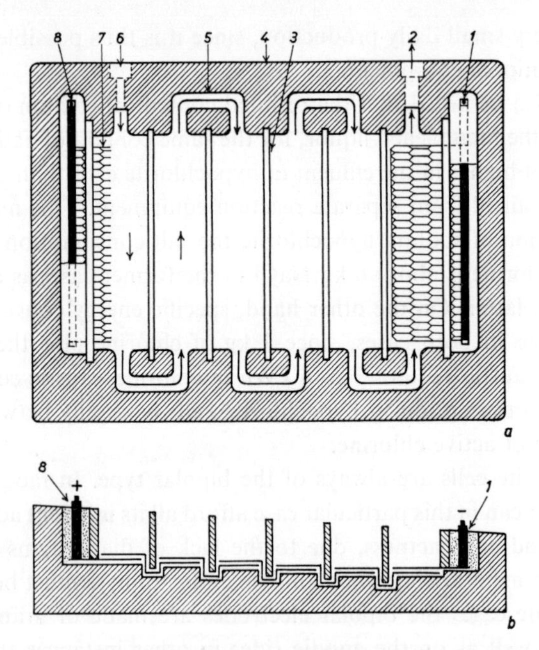

Fig. IX, 12. Kellner cell for hypochlorite production. 1. connection of graphite cathodes; 2. outlet for NaClO; 3. glass walls; 4. concrete container; 5. circulation canals for the electrolyte; 6. NaCl solution feeding; 7. bipolar Pt electrodes; 8. connection of graphite anodes

TABLE IX, 2

OPERATING CHARACTERISTICS OF SOME TYPICAL HYPOCHLORITE CELLS

Model	*Electrodes*	*Initial* NaCl g/l	*Electric Tension* V	*Curr. Eff.*	*Active* Cl g/l	°C	kWh *per* kg *act.* Cl
De Nora	graphite	100–150	3.5	0.78	12–15	18–22	3.4
Kellner	Pt–Ir	100	5	0.55	20	21	6.9
Haas–Oettel	graphite	150	3.7	0.53	14	23	5.5
Schoop	Pt	—	4.5–5	0.75	10	—	5
Schuckert	graphite Pt–Ir	100	5.5–6.1	0.6–0.65	18	20–30	7

are folded over glass partitions 3; the solution is forced to flow from each single cell to the next through side ducts 5. The electrolyte is recirculated from the outlet 2 back to the inlet 5 through an outer cooling coil, until the wanted hypochorite concentration is reached.

Table IX, 2 gives a selection of data for a number of some best known hypochlorite cell models.

(III) *Technical Production of Chlorate*

In order to avoid the anodic oxidation process, which as already explained would entail a loss in current efficiency with a corresponding oxygen evolution, the electrolyte is kept acidic by continuous addition of dilute (0.1 *N*) HCl, so as to maintain the solution within a pH range from 6 to 6.8. A higher acidity would give rise to evolution of chlorine, since in such a case the strong acid, *i.e.*, HCl, would be electrolyzed, while the accumulation of hypochlorite and consequently of chlorate would diminish. In this connection, it is noteworthy that the addition of sulphuric acid beyond the acidic point as indicated above would promote oxygen evolution, due to anodic oxidation of the SO_4^{2-} ion.

Beside a slightly acidic pH, a relatively high temperature is the main factor by which it is possible to control the chemical formation of chlorate at a sufficiently high rate. However, when working with graphite anodes as in most modern plants, the tendency of graphite to become rapidly oxidized at temperatures higher than 45° C makes it advisable never to exceed this limit.

Of the two reactions that are responsible for graphite oxidation, the one involving OH^- ions is anodic in nature, *i.e.*

$$4\,OH^- + C \rightarrow CO_2 + 2\,H_2O + 4\,e^-$$

whereas, according to the second reaction, graphite is directly attacked by hypochlorous acid

$$C + 2\,HClO \rightarrow CO_2 + 2\,H^+ + 2\,Cl^-$$

The latter reaction is independent of current density, so that when this is low, this reaction can account for the major amount of graphite consumption. It is therefore advisable to run the cell at a relatively high anodic current density, such as 7 to 10 A/dm^2, compatible with the limits imposed by temperature and current concentration (see below). Graphite used in chlorate cells is always impregnated with linseed oil or the like, to decrease porosity (p. 418 *et seq.* and p. 521 *et seq.*) which causes rapid deterioration in the bulk of the graphite material, where chloride ion depletion would favour both reactions as written above.

As formerly observed, hypochlorite and chlorate have a tendency to revert back to a reduced state in contact with the cathode. These unwanted reactions are checked to a large extent by keeping 2 to 4 g $K_2Cr_2O_7$ in the solution.

An important role in facilitating chemical oxidation of hypochlorite to chlorate, rather than anodic oxidation, is also played by current concentration. In fact, chlorate cells are normally of such a size that the ratio of their current capacity to their holdup of electrolytic solution does not exceed 2 A/l. By keeping a relatively small current concentration, reactions (1), (2), (3) are given sufficient time to take place in the bulk of the electrolyte, keeping pace with the anodic discharge of the Cl^- ion and the consequent dissolution and diffusion of elemental chlorine away from the anode. On the other hand if the electrolyte is rapidly circulated through the cell the chain of chemical reactions ending with chlorate formation can also be achieved outside the cell, so that the current concentration will acquire a minor significance. For this reason, the tendency in modern plants is to equip the electrolyte system outside the cells and downstream of the electrolytic plant with reaction vessels of considerable retention time.

Chlorate cells, like hypochlorite cells, are without diaphragms. However unlike the latter, they are mostly of unipolar type, even though each cell embodies a number of anodes and cathodes connected in parallel. The cathodes are made of perforated steel plate. The cell trough also is usually of steel and is kept at the same potential as the cathodes, so as to receive cathodic protection. Most cell models embody a cooling coil fed with cold water, which keeps the operating temperature within the desired limits.

The production of potassium chlorate does not differ from that of sodium chlorate, except for the fact that the curve of solubility against temperature is much steeper for the former than for the latter, so that potassium chlorate can be more easily crystallized out of the mother liquor, without any of the previous evaporation or deep cooling usually practiced for sodium chlorate. However in both cases the electrolytic process is carried out continuously in most plants, in that the mother liquor is fed back to the cells on a continuous cycle, after resaturation with sodium or potassium chloride, as the case may be. The resaturated brine, before going back to the cells, is submitted to purification, in much the same way as described for the chlorine–caustic process.

Chlorate cell models differ widely from one another in respect of geometrical and operating characteristics. Therefore, the values contained in Table IX,3 are to be considered as referring only to some typical examples.

TABLE IX,3

CHLORATE PRODUCTION; TYPICAL VALUES

Anode material	graphite; magnetite; platinum
Cathode material	steel
Electric tension	3.5–4.5 V
Anodic current density	2–16 A/dm^2
Cathodic current density	2–10 A/dm^2
Current capacity	5,000–15,000 A
Current efficiency (overall)	0.75–0.90
pH	6.0–6.8
Temperature	35–45° C
Electrolyte composition	
initial	280 g NaCl/l; 140 g $NaClO_3$/l
final	85 g NaCl/l; 500 g $NaClO_3$/l
Energy consumption	6–7 kWh/kg $NaClO_3$
Graphite consumption	10 kg/ton $NaClO_3$

B. OTHER NONMETALLURGICAL PROCESSES

1. Electrolysis of Water

The electrolysis of water is carried out for the production of high purity hydrogen and oxygen. Both gases can also be produced by a number of other methods, such as starting from water gas and liquefied air to obtain hydrogen and oxygen respectively. However, the electrolytic process affords a considerable advantage in that both gases can be immediately obtained in large quantities and at an exceptionally high purity, by using relatively simple equipment.

According to Faraday's Law each amperehour releases 0.037 g H_2

and 0.298 g O_2. Such quantities by weight correspond to volumes of 0.4176 liters and 0.2088 liters respectively, at 0° C and 760 mm Hg. Accordingly, water consumption would amount to roughly 8 liters per m^3 H_2, if evaporation losses could be neglected.

Pure water cannot be used as such for the electrolysis, as its conductivity is too small. It is therefore necessary to resort to a not too dilute solution of an oxyacid, or of a base.

To get a clear concept of the processes that take place at the electrodes, it is important to observe that the decomposition electric tension for most oxyacids and bases, between two smooth platinum electrodes, is about 1.69 V (see Table IV, 2). However, since the decomposition electric tension is the difference between the discharge tensions for the anion and the cation, its constancy means that the overall reaction should be the same whatever the solute, and that the small differences observed are to be attributed to small differences in the overtensions of the two half reactions that can depend on the solute, if it is acidic or alkaline.

In an aqueous solution of an acid or a base there are always at least three ionic species present, since beside the H^+ and the OH^- ion, which are present in both cases, there is either the anion of the acid or the cation of the base. On the other hand, it is difficult to assume that the evolution of hydrogen from an alkaline solution involves a hydrogen ion in the discharge step considering how extremely small is the concentration of this ion under such conditions. Similarly, it does not seem reasonable to suppose that the evolution of oxygen from an acid solution involves the OH^- directly in the discharge step. It is therefore probable that in an acidic solution, where H^+ ions are strongly predominant, the following half-reactions take place:

$$\text{cathodic} \quad 2\,H^+ + 2e^- \rightarrow 2\,H$$

$$\text{anodic} \quad \begin{cases} 2\,H_2O \rightarrow 2\,OH + 2\,H^+ + 2e^- \\ 2\,OH \rightarrow H_2O + O \end{cases}$$

$$\text{overall} \quad H_2O \rightarrow 2\,H + O \rightleftharpoons H_2 + \tfrac{1}{2}\,O_2$$

In alkali solution, where OH^- ions strongly predominate, the following reactions could interpret the process:

$$\text{cathodic} \quad 2\,H_2O + 2e^- \rightarrow 2\,H + 2\,OH^-$$

$$\text{anodic} \quad \begin{cases} 2\,OH^- \rightarrow 2\,OH + 2e^- \\ 2\,OH \rightarrow H_2O + O \end{cases}$$

$$\text{overall} \quad H_2O \rightarrow 2\,H + O \rightarrow H_2 + \tfrac{1}{2}\,O_2$$

so that the overall discharge reaction is in both cases the same, giving H atoms and OH radicals, and producing finally molecular hydrogen and oxygen, whatever the pH of the solution.

The same considerations would still apply if the electrolyte were a salt of an alkali metal and an oxyacid. However, in the technical electrolysis of water, no salt solution is employed in practice, as it is preferable to take advantage of the great mobility of hydrogen and hydroxyl ions, in order to keep the ohmic resistance of the electrolyte as small as possible. On this account the choice of the electrolyte is therefore restricted to an acid or a base.

As mentioned above, the cell tension is practically independent of the nature of the electrolyte. This fact, aside from the irreversible phenomena taking place at the electrodes and causing the overtension, is an obvious consequence of thermodynamic principles, since the final result of the overall process, *i.e.* decomposition of water, is the same in any case. The electric work needed for the electrolysis of 1 mole H_2O is determined by the product $Un\mathbf{F}$, where U is the decomposition electric tension, n the number of faradays involved for each reacting mole (for water $n = 2$) and $\mathbf{F} = 96{,}500$ C.

In practice the voltage at the cell terminals must be greater than the theoretical value, 1.24 V, not only on account of overtensions but also for a number of other reasons. Indeed, the overall voltage consists, in addition to the theoretical value corresponding to the state of thermodynamic reversibility, of several terms: the transfer overtensions for hydrogen and oxygen (each depending upon the current density as well as upon the electrode materials); the internal resistance of the cell, due to the electrolyte; and finally the concentration overtension that builds up as a consequence of the different concentrations that the ionic species carrying the current assume in the two electrolyte regions, facing the anode and cathode respectively, so that in this regard the conditions that arise are equivalent to the establishing of a concentration cell.

The general formulas that have been formerly established for the cathodic and anodic half-reactions clearly indicate why the pH will always increase in the immediate vicinity of the cathode and decrease at the anode — according to a well known phenomenon that is invariably experienced in the electrolysis of aqueous solutions, whatever the electrolyte — because the electrolyte concentration changes more rapidly at one electrode, than at the other.

Other conditions being unchanged, the state described above must however reach a point where no further change in anolyte and catholyte concentration is possible, due to the diffusion forces that are gradually generated and counteract any further development of a concentration cell. The tendency for different concentrations to arise in the anolyte and catholyte can be also artificially counteracted by an appropriate design of the cell, as well as by such devices as introducing the process water at a point where the electrolyte concentration shows a tendency to increase (*i.e.* the anolyte, if the solution is acid, or the catholyte, if the solution is alkaline).

The only electrolytes that are considered in the industrial production of electrolytic hydrogen and oxygen are caustic soda and caustic potash. Indeed, it is easier and cheaper to design the cells out of materials for the various parts, such as containers, electrodes and diaphragms, which can withstand alkali attack far better than acid attack. Furthermore the hydrogen overtension on steel is definitely lower than on any other metal, such as lead, that might be capable of withstanding sulphuric acid, which is the only acid that might be of some technical interest for this particular electrolytic process. For these reasons the choice of an alkaline electrolyte is preferable, although its conductivity is considerably lower than that obtainable with sulphuric acid and in spite of the fact that caustic solutions, when used in open cells, tend to react with atmospheric carbon dioxide, with a corresponding accumulation of sodium carbonate in the electrolyte, which brings about a further decrease in conductivity and the possible development of corrosive conditions for some construction materials.

At normal operating temperatures, 70–80° C, the alkali hydroxide concentration showing a maximum conductance are about 23 % by weight for caustic soda and 27–30 % for caustic potash respectively. In this temperature range and at these concentrations the ohmic drop that is actually measured during operation is about the same for both electrolytes. Corrosion problems are also substantially the same. However, in equipment of modern design, specially of bipolar type, allowing operation at high current density and high temperature (over 70° C) caustic potash is usually to be preferred, even though its cost is considerably higher. The true reason for this choice, which quite often is not sufficiently appreciated, resides in the fact that, under the same temperature conditions, the aqueous vapour pressure over 25–30 % KOH is substantially lower than over 20–23 % NaOH, so that the use of caustic potash reduces the troubles arising when large amounts of condensate have to be recovered from both gaseous streams, in order to avoid the flooding of downstream equipment and pipelines as well as costly losses of distilled or demineralized water which have to be fed to the process.

The evaluation of the ohmic resistance within the cell under actual operating conditions cannot be based only on the resistivity of the electrolyte and the cell geometry. Both gases are developed at the electrodes in the form of small bubbles that are set free from the electrode surface only upon reaching a certain size. A similar phenomenon, known as the anode effect, is experienced also in the electrolysis of fused salts (see p. 592 *et seq.*). The effective area of contact between the metallic electrode and the conductive liquid phase of the electrolyte is thereby decreased, so that the actual current density and the overtension rise correspondingly. Moreover, the bubbles detaching from the electrodes take some time to rise through the electrolyte, until reaching the free space at the top, and the effective cross-sectional area of electrolyte between the two electrodes is diminished. The ohmic resistance therefore becomes higher, the higher is the production rate and the current load. The phenomenon is remarkably dependent upon the cell geometry, in that the cross-sectional area occupied by the gas bubbles is greater the greater the depth-to-width ratio for a given electrode area, as can be easily appreciated by considering that at the higher levels the space between the electrodes is traversed at the same time by the bubbles developing locally, as well as by those that are rising from the lower levels.

The resistance of the conductive medium between the electrodes, beside being affected by gas evolution, is considerably influenced also by the evolution of water vapour at higher temperatures. In fact, since the water vapour pressure over 25 % KOH at 80° C is nearly 250 mm Hg, the volume of aqueous vapour developed at such temperature and at atmospheric pressure is about 50 % of the volume of the useful gases. Moreover, since the vapour pressure increases exponentially with temperature, the ohmic drop and the additional screening effect by the gas bubbles over the electrodes rise so fast that there is virtually no further saving in power consumption on raising the operating temperature above 75 or 80° C. Such counteracting effect by vapour evolution against the overall voltage decrease that otherwise might be expected by raising the operating temperature, becomes even more pronounced, and occurs at even lower temperatures, when the electrolyte consists of caustic soda solution of equal resistivity to that of caustic potash.

The following points summarize the conditions that must be carefully observed, in order to keep down the cell voltage to a value as low as possible, so as to obtain a more favourable energy consumption.

1. A judicious choice of the electrode material is helpful to reduce energy losses due to overtension. The material must of course be selected among those having good resistance to the action of the electrolyte and gas.

2. A higher temperature lowers the overtension and the specific resistance of the electrolyte. However, this practice also has its own limitation, as already noticed.

3. The electrolyte concentration is an important factor also. Indeed, as previously remarked, the specific conductance for a given electrolyte at a given temperature reaches a maximum for a particular concentration, except when by increasing the concentration, the saturation point is reached before attaining the maximum of conductivity. This is not the case with caustic potash and caustic soda between 0° C and 100° C. Within this temperature range the caustic concentration at which maximum conductivity is reached also increases steadily with temperature. Accordingly, by an appropriate choice of temperature and concentration the ohmic drop can be brought down to a minimum.

4. Any sort of means providing fast recirculation of the electrolyte is helpful in preventing the development of a concentration cell. To this purpose the anolyte and catholyte are also allowed to remix together in some cell designs. However, such practice may be objectionable, since the purity of the gas can be impaired, due to the dissolved hydrogen and oxygen that saturate the catholyte and anolyte respectively.

5. Any device providing an aid to the prompt removal of the gas and vapour bubbles from between the electrodes is highly beneficial. The most common method consists of having the electrodes made of perforated plates, supported so as to leave some room behind, through which the gas bubbles, as soon as they are formed can freely escape away from the gap between the electrodes.

6. The inner cell resistance can also be favourably affected by keeping a suitable distance between the electrodes. The best spacing is neither too large nor too small, for the developing gases must have the possibility to

escaping very rapidly from between the electrodes (see also the preceding point 5).

7. The inner cell resistance can be efficiently decreased also by keeping the electrolyte in fast motion and by any constructive device helping a fast removal of the developing gas from the electrolyte.

The electrodes can be arranged according to the unipolar or the bipolar type of assembly (see p. 418 *et seq.*). Because of the particular structure required for either arrangement a unipolar unit is often referred to as belonging to the 'tank type'. Conversely, most of the bipolar models consist of a plurality of frames, so that there are as many cells in series as there are frames, these being assembled together, with the interposition of insulating gaskets, in much the same way as the elements of a filter-press. For this reason these bipolar assemblies are said to belong to the 'filter-press' type. Indeed the first models of this sort, as conceived by Schmidt and Shriver at the turn of the century were adaptations of filter-press apparatus.

Unipolar cell models may show some slightly different features, specially depending upon the way in which the hydrogen and oxygen gases are kept separated from each other. Indeed, means of gas separation may exclusively consist of a bell system and in the following we shall refer to this as Type 1, which is represented in Fig. IX, 13[1]. In other unipolar models gas separation is accomplished only by diaphragms (Type 2, Fig. IX, 14), while in some others bell and diaphragm systems are combined together (Type 3, Fig. IX, 15).

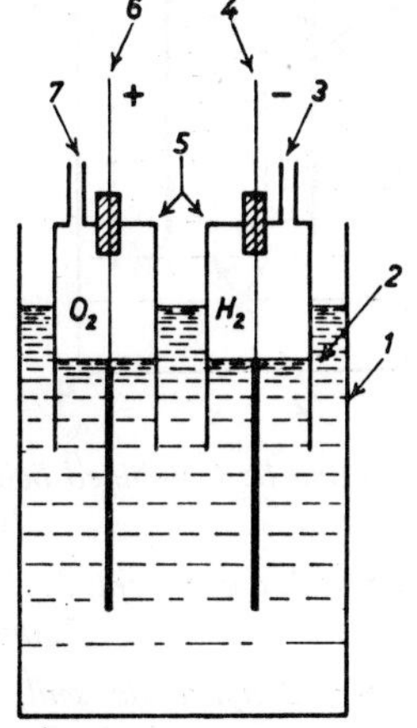

Fig. IX, 13. Bell-type unipolar cell for electrolysis of water. 1. container; 2. electrolyte; 3. H_2 outlet; 4. cathode; 5. gas bells; 6. anode; 7. O_2 outlet

[1] The Figs. IX, 13 ... IX, 17 show only diagrammatically the different cell types, but do not represent constructional drawings.

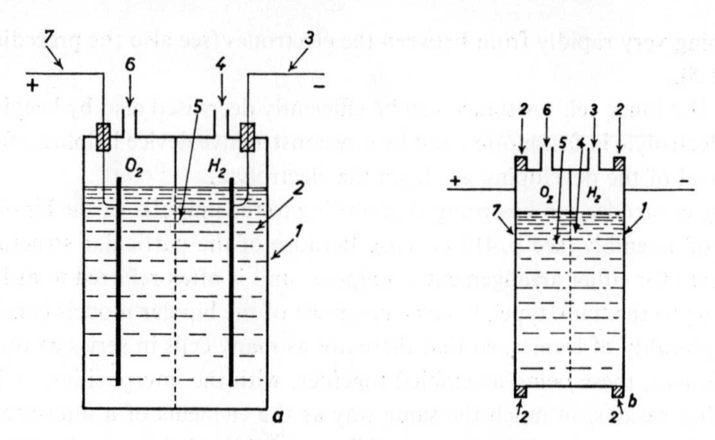

Fig. IX,14. a and b: Diaphragm-type unipolar cell for electrolysis of water.

a. 1. container; 2. electrolyte; 3. cathodic connection; 4. H_2 outlet; 5. diaphragm; 6. O_2 outlet; 7. anodic connection

b. 1. side-wall acting as a cathode; 2. electrical insulation; 3. H_2 outlet; 4. diaphragm; 5. electrolyte; 6. O_2 outlet; 7. side-wall acting as an anode

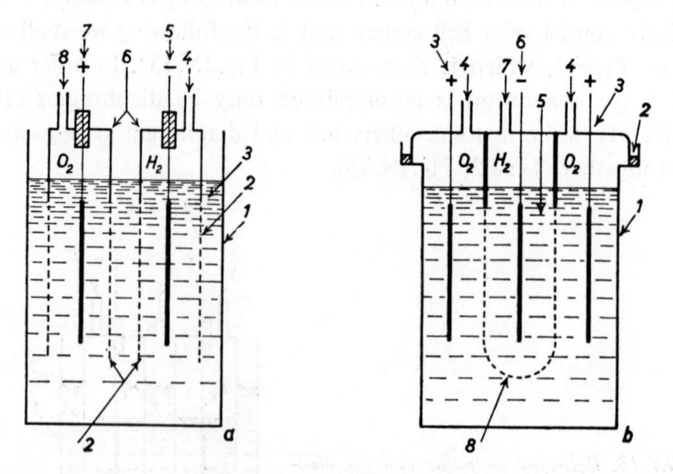

Fig. IX,15. a and b: Combined bell-and-diaphragm type unipolar cell for electrolysis of water.

a. 1. container; 2. diaphragm; 3. electrolyte; 4. H_2 outlet; 5. cathode; 6. gas-bells; 7. anode; 8. O_2 outlet

b. 1. container; 2. hydraulic seal; 3. anodes; 4. O_2 outlet; 5. electrolyte; 6. cathode; 7. H_2 outlet; 8. diaphragm

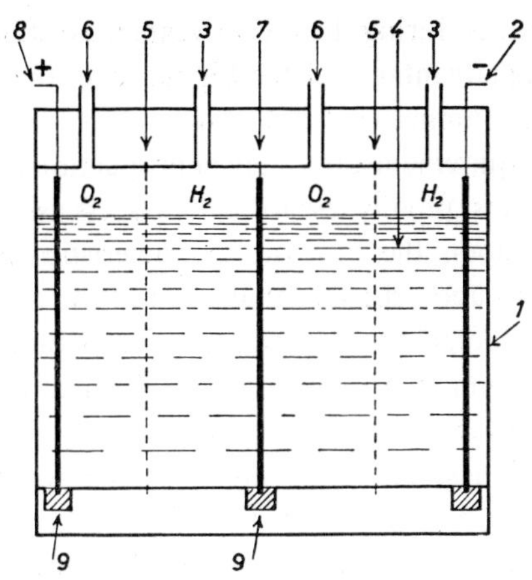

Fig.IX,16. Bipolar tank-type cell for electrolysis of water. 1. container; 2. end-cathode; 3. H_2 outlet; 4. electrolyte; 5. diaphragms; 6. O_2 outlet; 7. bipolar electrode; 8. end-anode; 9. insulators

Bipolar assemblies may also be arranged in a vat (Type 4, Fig. IX, 16), but they are more frequently shaped as a filter-press (Type 5, Fig. IX, 17). Even though the problems to be faced in the design and construction of bipolar electrolyzers are more difficult than for unipolar models, the

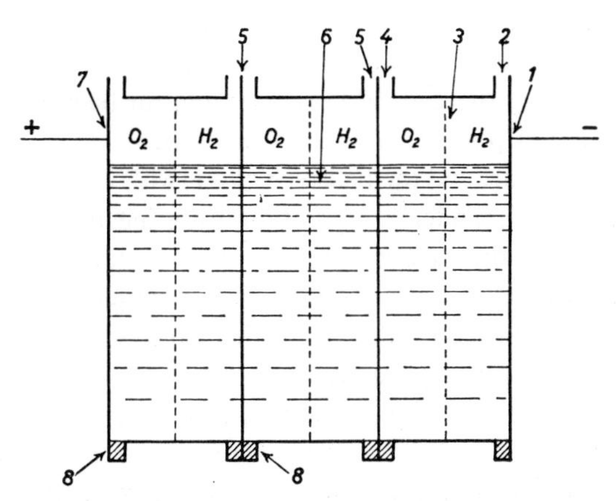

Fig.IX,17. Bipolar filter-press-type cell for electrolysis of water. 1. end-plate acting as a cathode; 2. H_2 outlet; 3. diaphragm; 4. O_2 outlet; 5. intermediate plate acting as a bipolar electrode; 6. electrolyte; 7. end-plate acting as an anode; 8. insulators

gradual improvements during the recent years have still further increased certain outstanding advantages of the bipolar over the unipolar arrangement, which can be summarized as follows:

1. Floor space requirements, for a given production capacity and power consumption, are substantially less for a filter-press assembly. This feature not only depends upon the intrinsic characteristics of the electrolytic cell proper, but also upon the fact that most of the ancillary equipment, such as electrolyte coolers, lye deentrainment apparatus, pressure equalizers, gas washing equipment, gas cooler and water vapour condensers, can be assembled over the cell-block in one compact system extending vertically. Some of these modern units are composed of more than 100 cells in series, having a current capacity of more than 10,000 A, so that the production rate per unit is over 500 m^3 H_2/hour. However, the floor space requirement is not more than 35 m^2.

2. An active electrolyte circulation can be much more easily obtained in a bipolar assembly, so that the current density can be considerably higher, which gives the possibility of an additional saving in floor space.

3. The higher current density at which a filter-press electrolyser can be operated gives a further advantage when electrolytic hydrogen production is associated with production of heavy water, since the separation factor of deuterium (^2H) to protium (^1H) is greater with a greater current density.

4. Filter-press construction allows operation at pressures above atmospheric, so that no boosters are in general required to compensate for pressure drop along the gas delivery lines and for back-pressure at gasholders. This is not possible with unipolar cells.

5. The water seal around the bell-type cover, with which many unipolar models are provided, renders evaporation losses inevitable and promotes the degradation of the caustic solution, due to attack by atmospheric carbon dioxide. Since this drawback is entirely absent in the totally enclosed bipolar construction, this is another point in favour of the latter for the production of heavy water.

6. In a unipolar construction there must be as many branch connections between the bus-bars and the cells as there are electrodes, whereas in a bipolar assembly the electric continuity is accomplished directly and exclusively by connecting the bus-work with the positive and the negative ends of the electrolyzer. This not only allows a substantial saving in copper, but also diminishes the voltage drop along the line and through the

contacts at the joints (which thus become very few) to a negligible fraction of the overall voltage, which is usually several hundred volts.

As mentioned above, the usual fabrication material is steel, which has a satisfactory durability in contact with caustic solutions. The anodes are generally nickel-plated, in order to keep down the oxygen overtension. The steel cathodes are sometimes plated with cobalt, even though the higher cost involved by such practice is usually not justified by the slight advantage that is thus obtained.

Since the diaphragm is an essential part of the system, due consideration must be given to its material. Beside the requirement of having a hardly detectable effect on the ohmic resistance, it must provide a thorough separation of hydrogen from oxygen, so as to ensure the highest purity for both gases, as required by most of the processes in which they are to be further employed. Moreover, the diaphragm must be sturdy enough to exclude any possibility of sudden breakdowns, which would entail the formation of an explosive gas mixture.

The commonly used material for the diaphragm is asbestos fabric, sometimes reinforced with an interwoven mesh of thin nickel wire. Diaphragms consisting of plain wire mesh or perforated metal plates have also been successfully used. In such cases, the metal structure playing the part of the diaphragm does not behave as a bipolar electrode for the obvious reason that the ohmic voltage drop through its empty spaces (holes or meshes) is less than the decomposition electric tension for water.

The specific energy consumption can change considerably from one cell type to another, since, as formerly noticed, the operating voltage depends on many factors. The amount of energy required to produce 1 m^3 of hydrogen and 0.5 m^3 of oxygen under standard conditions, if based upon a minimum discharge voltage of 1.69 V (see Table IV, 2) and 100 % current efficiency, would be 4.02 kWh. However, as already explained, practical conditions are always far from the ideal ones, so that the actual specific consumption of energy is scarcely less than 5 kWh and more usually slightly more.

The gases leaving the electrolysis apparatus contain water vapour in amounts corresponding to its partial pressure over the electrolyte solution at the operating temperature. Additionally, both gases entrain some electrolyte as a mist. The major part of the water vapour is recovered as condensate by appropriate gas cooling. Should either gas be required in a very dry state, further desiccation has to be carried out by some suitable means, such as packed towers sprinkled with concentrated sulfuric acid, followed, if necessary, by treatment over silica-gel or activated alumina.

The electrolyte mist is removed by suitable means, such as passing the gas through a washing system which is fed with the process water to be used for electrolysis. Sometimes the residual particles of caustic are removed by electrostatic precipitation.

Process water is obtained by distillation or from boiler condensate, or even more frequently, it is cheaply provided in the form of demineralized water by ion exchange units. Its residual salt content must of course be very low (no more than 5 to 10 mg/l) otherwise salt accumulation in the electrolyte would be excessively fast. A too high salt content, beside being the possible cause of undesirable side-reactions, might impair the integrity of the electrodes, if oxidizing anions such as chloride were present in considerable quantities.

It can be seen from this that the operation of a water electrolysis plant requires ancillary equipment, such as mist separators, condensers, means for continuous or periodic filtration of the electrolyte, circulation and transfer pumps for the electrolyte and process water, water treatment units, gas boosters and compressors.

As mentioned before, some of these parts of equipment can be embodied in the electrolysis apparatus itself, specially if it is of bipolar type. However, the general arrangement must be carefully designed, for any stray current following some other path rather than passing through the cells proper, beside entailing a loss in current efficiency, might give rise to electrolytic corrosion and foster gas evolution at other points, thus possibly decreasing the purity of the product gas.

Methods have also been devised to allow water electrolysis to be carried out at pressures considerably higher than atmospheric. Indeed, by such a procedure it is possible to reduce considerably the voltage drop through the cell, by virtue of the smaller size thus assumed by the gas bubbles rising in the electrolyte. Such decrease in energy consumption, by producing the gases at a higher pressure, might seem paradoxical, if considered according to the laws of thermodynamic equilibrium alone. In fact, from the equations for reversible electric tensions of the H_2/O_2 cell (see p. 106 *et seq.*) it can be calculated that at 25° C there should be an increase of about 15 mV for each tenfold increase in gas pressure. Yet, under practical conditions, governed by irreversibility, the gain obtainable by the decrease in ohmic drop at higher pressure is considerably greater than the additional work required for compression, so that the overall cell voltage in a pressurized cell is less than under atmospheric conditions, the other factors remaining equal. The most important difficulties that have prevented an industrial development of the water electrolysis at higher pressures are a greater complexity of the cell, the increase of solubility of hydrogen and oxygen in the electrolytic solution, unavoidable losses of current and particularly the difficulty of controlling the pressure, that must be held equal and constant on both sides of the diaphragm. In fact a pressure difference of 1 % between the anodic and cathodic space of a cell working at 200 atm (the usual final pressure of compressed hydrogen and oxygen) is equal to an actual pressure of 2 atm on one side of the diaphragm, *i.e.* too high pressure for this very delicate part of the cell.

Some operating figures for the electrolysis of water are summarized in Table IX,4.

Commercial units of the pressurized filter-press type, delivering gas at about 450 psi, *i.e.* about 30 atm, have been working for a number of years.

A further application of water electrolysis consists of associating this

TABLE IX,4

OPERATING CHARACTERISTICS OF SOME TYPICAL CELL MODELS FOR ELECTROLYSIS OF WATER

odel	*Type*	*Electric Tension* V	*% Gas Purity* H_2	O_2	*Temp.* (°C)	*Nominal Current Density* (A/m^2)	*Current Efficiency*	*Energy Consumption* (kWh/m^3H_2)
wles	1	2.12–2.25	99.95	99.5	60–75	600–700	—	5.5–6.3
trolab	2	—	99.8	99.8	—	250–300	—	—
er	3	2.0	99.9	—	60	400	—	—
nboe	3	1.99–2.09	99.9	99.7	—	—	—	5.02–5.06
ens	4	1.9 –2.3	99.9	98	65–75	800–1,500	0.96–0.98	4.5–5.6
ag	5	2.0 –2.2	99.9	99.8	75–85	2,500	0.99	5.35
Nora	5	2.0 –2.05	99.9	99.8	70–80	1,400	0.99	5.01
kranz	5	2.0 –2.5	99.5–99.9	98.5	80	1,500–2,000	—	5.74
sky	5*	1.87	99.9	99.5	110	1,000	—	4.3

rating pressure 450 psi.

process with the extraction of heavy water, or deuterium oxide D_2O. In natural water the mole fraction of deuterium (a hydrogen isotope with atomic weight equal to 2) is about 1.5 per 10,000 moles of total hydrogen present. During electrolysis this ratio does not remain constant, since the deuterium content in the gas phase is less than in the water undergoing electrolysis. Thus in the latter a gradual enrichment in deuterium takes place, until steady state conditions are attained, when the amount of deuterium leaving with the hydrogen gas is the same as the amount entering with the water to be processed. The suitability of the process to carry out isotopic separation is measured by the *separation factor S*, *i.e.* the ratio

$$S = \frac{\ln \mathrm{H} - \ln \mathrm{H_w}}{\ln \mathrm{D} - \ln \mathrm{D_w}} = \frac{\Delta \ln \mathrm{H}}{\Delta \ln \mathrm{D}}$$

where H and D denote absolute numbers of moles of hydrogen and deuterium evolved in the gas phase, and H_w and D_w are the numbers of moles remaining in the liquid phase, for an infinitesimal degree of advancement of the separation process.

The separation factor as defined above, although independent of deuterium concentration, is affected to some extent by other conditions, such as cathode material, current density and temperature. Indeed, whereas the separation factor is improved by a higher current density, it is diminished by operating at a higher temperature, so that, under normal operating conditions, it may be assumed to vary inversely as the absolute temperature.

The above formula may be in some cases simplified, so as to give the separation factor a more practical significance, and the value thus obtained is sufficiently accurate for technical purposes. The new definition thus obtained is as follows,

$$S^* = (\mathrm{H/D})_{\mathrm{gas}} : (\mathrm{H/D})_{\mathrm{liquid}}$$

where (H/D) indicates the ratio between the molar concentrations of the two isotopic species in either phase. Under the operating conditions used in modern electrolysis equipment the value for the separation factor is usually about 6.

Other chemical elements show a similar tendency for their isotopes to be electrolytically separated. Table IX, 5 lists some values for the

TABLE IX,5

ELECTROLYTIC SEPARATION FACTOR FOR SOME ISOTOPES

H/D	(S*)	2.8–7.6
^{7}Li/^{6}Li	(S)	1.020–1.079
^{16}O/^{18}O	(S)	1.008
^{35}Cl/^{37}Cl	(S)	1.061
^{39}K/^{41}K	(S)	1.054

separation factors determined for the most abundant isotopes of some elements.

A satisfactory theory accounting for all the details of isotopic separation by electrolysis has not yet been completely developed. However, one present view is that one controlling factor in the separation of deuterium from protium is the difference in the activation energies accompanying the transfer of H^+ and D^+ ions to a state of adsorption on the metal surface.

By a suitable cascade arrangement of a number of electrolytic cells, in such a way that the deuterium-enriched condensate withdrawn from any one stage in the cascade feeds the next stage, it is possible to produce heavy water up to an equivalent D_2O content of 99.8 %.

Beside the electrolytic process, which was the first one applied to this purpose, a number of other competitive methods have recently been developed, such as fractional distillation of hydrogen or water, and the dual temperature exchange process. Electrolytic production of heavy water would never be economic, even at the lowest costs for electricity and in plants of largest capacity, unless electrolytic hydrogen had to be produced for some other primary purpose.

2. Anodic Oxidations and Cathodic Reductions

The industrial production of substances by electrochemical oxidation or reduction can afford several advantages over the corresponding purely chemical processes. Firstly the product purity is usually greater when obtained electrochemically, since the substance to be oxidized or reduced can be prevented from reacting with other substances that might lead to contamination of the product and have to be separated by some further operation. A second advantageous and important feature of the electrochemical method is the great ease with which it can be controlled. This makes it possible to obtain different products at will, starting from one substance, when the oxidation or reduction process passes through several steps. In fact, it is generally possible to make the overall reaction reach the desired stage simply by adjusting such electric variables as the electrode tension and the current density. In other words, the final product can thus be obtained in a more dependable way and with a better quality than by chemical methods.

It may also be noted that, beside process dependability and product purity, which generally characterize electrochemical oxidations and reductions, some of the equivalent chemical processes may present several technological difficulties, apart from cost; indeed, the use of strong oxidizing or reducing agents is in general much more expensive than the simple consumption of electric energy.

The choice of the anodic material for anodic oxidation is in general somewhat limited, for the anode must be inert or easily passivated; this is particularly difficult to obtain when the wanted oxidation reaction is characterized by a relatively high electric tension, so that the electrolyte must be kept acidic to raise the discharge electric tension for oxygen evolution above the anodic tension required for the desired reaction.

When the solution may be kept alkaline, the anodic materials usually employed are platinum, iridium, carbon (most usually in the form of retort carbon or graphite), iron (as such or alloyed with nickel), or pure nickel. In acidic solutions the choice is confined to platinum, iridium, carbon and lead, the latter being applicable only when the electrolyte consists of sulphuric acid.

As to the electrode material for cathodic reactions, a greater variety of pure or alloyed metals can be used than listed above; indeed, although cathodic polarization suppresses the effects due to passivation, it exerts a very efficient protective action. Obviously the choice of the metallic material must be made according to each individual case, so that the hydrogen overtension is sufficiently high to bring the actual electric discharge tension of the H^+ ions to a more negative value than that corresponding to the desired cathodic process.

(I) *Perchlorate*

Perchlorate is obtained by further oxidation of chlorate. However, it is impossible to perform this reaction in the same electrolytic cell in which chlorate is produced starting from chloride, because the actual discharge tension of the Cl^- ion is too much lower than the oxidation tension of the ClO_3^- ion; therefore, it is necessary to start again from a chlorate solution that must be as free of chloride as possible.

The mechanism of primary and secondary reactions by which chlorate is oxidized to perchlorate is not yet fully understood. Among the several schemes that have been proposed, the following seems most likely. The primary anodic reaction would be the electrolysis of water, with formation of atomic oxygen adsorbed at the anode,

$$H_2O \rightarrow 2\,H^+ + O + 2e^-$$

followed by a secondary reaction consisting of the direct oxidation of the ClO_3^- ion by atomic oxygen,

$$ClO_3^- + O \rightarrow ClO_4^-$$

On this view, there would be a close similarity between this electro-oxidation and those instances of electro-reduction in which the substance to be reduced reacts with atomic hydrogen adsorbed at the cathode.

The accompanying cathodic reaction consists of the discharge of H^+ ions,

$$2\,H^+ + 2e^- \rightarrow H_2$$

The ClO_3^- ions can be oxidized by the atomic oxygen adsorbed on the anode only if their concentration in the aqueous solution, and more particularly within the anodic film, is sufficiently high (more than 5 %, if expressed as $NaClO_3$) to act as a depolarizer of the anodic process (see p. 265 *et seq.*). Under such conditions a competitive reaction is set up in favour of the secondary reactions, giving rise to the final evolution of oxygen gas. All these reactions require a higher anodic tension. It is therefore clear that, under favourable conditions, the evolution of gaseous oxygen can be completely suppressed. For this reason the electrolysis, which is carried out batchwise, is discontinued as soon as the chlorate concentration, that is initially 700 g $NaClO_3$/l, drops to about 20 g/l, with a corresponding accumulation of approximately 780 g $NaClO_4$/l.

In order to raise as far as possible the anodic tension corresponding to oxygen evolution, it is advisable to work with a slightly acidic solution, *i.e.*, within a pH range from 6.5 to 7. A higher acidity would not be advantageous for it would favour the cathodic reduction of the perchlorate formed. Such a reduction is hindered by the addition of bichromate at the rate of about 5 g $Na_2Cr_2O_7$/l, which builds up an impervious film of chromium hydroxide around the cathode surface, as already explained. Sometimes, instead of using bichromate, a diaphragm is placed between the anode and the cathode, in order to prevent the ClO_4^- ions from diffusing into the catholyte; however, the electric resistance of a diaphragm is considerably greater than that of the chromium hydroxide film.

The optimum conditions for perchlorate production are

1. a high anodic electric tension;
2. a high ClO_3^- ion concentration;
3. absence of Cl^- ions.

The first requirement is satisfied by using smooth platinum anodes, on which the oxygen overtension is high, and by working at high current density and sufficiently low temperature, so as to further increase the oxygen overtension. With regard to temperature in particular, it is therefore necessary to accept some compromise, since not only the current efficiency, but also the electric tension would diminish on raising the temperature, so that there is some sort of mutual compensation between these two factors of energy consumption.

TABLE IX, 6

PERCHLORATE PRODUCTION; TYPICAL VALUES

Anode material	platinum
Cathode material	steel
Electric tension	5–6.5 V
Anodic current density	20–50 A/dm^2
Cathodic current density	7–10 A/dm^2
Current capacity	6,000–12,000 A
Current efficiency (overall)	0.70–0.80
pH	6.5
Temperature	50–60° C
Electrolyte composition	
initial	700 g $NaClO_3$/l; 20 g $NaClO_4$/l
final	20 g $NaClO_3$/l; 800 g $NaClO_4$/l
Energy consumption	3–3.5 kWh/kg $NaClO_4$

The usual starting material for perchlorate production is sodium chlorate, which is more soluble than potassium chlorate and ammonium chlorate. Potassium and ammonium perchlorates are therefore produced by metathetic reaction of sodium perchlorate with potassium chloride and ammonium chloride respectively.

(II) *Permanganate*

The electrochemical production of permanganate has at present almost completely replaced the chemical methods, principally because of the better economy afforded by the former in respect of caustic potash consumption.

In both processes the starting material is potassium manganate, which is first prepared by smelting the MnO_2 ore (pyrolusite) in the presence of caustic potash and oxygen, according to the reaction:

$$MnO_2 + 2\,KOH + \tfrac{1}{2}\,O_2 \rightarrow K_2MnO_4 + 2\,H_2O \qquad (1)$$

In the chemical production of permanganate the manganate produced by reaction (1) is oxidized by means of chlorine or by disproportionation in the presence of carbon dioxide, according to either one of the following reactions:

$$2\,K_2MnO_4 + Cl_2 \rightarrow 2\,KMnO_4 + 2\,KCl \qquad (2)$$

$$3\,K_2MnO_4 + 2\,CO_2 \rightarrow 2\,KMnO_4 + MnO_2 + 2\,K_2CO_3 \qquad (3)$$

By applying reaction (2), 50 % of the alkali is lost in the formation of chloride, whereas, by using reaction (3), the carbonate production involves a loss of alkali as high as 66.7 %.

However no caustic losses are involved in the anodic oxidation process by which permanganate can be obtained from manganate, according to the primary reaction:

$$MnO_4^{2-} \rightarrow MnO_4^- + e^-$$

On the contrary, the accompanying cathodic half-reaction, involving one molecule of water, regenerates caustic, with the evolution of hydrogen:

$$H_2O + e^- \rightarrow OH^- + \tfrac{1}{2}\,H_2$$

The OH^- ions thus accumulating in the catholyte will meet the K^+ ions that have been left in excess after the loss of one negative charge by the manganic ions, on their oxidation to permanganic ions. Accordingly, while the oxidation process is taking place, there is a gradual accumulation of potassium hydroxide, which is withdrawn with the cell effluent and eventually sent back again to the smelting process for reaction (1).

Since the standard electric tension corresponding to the anodic reaction $MnO_4^{2-} \rightarrow MnO_4^- + e^-$ is only about $+0.56$ V, it is not necessary to use any special anode material giving a high oxygen overtension; plain sheet iron may be sufficient, as owing to the alkalinity of the solution, iron can be rather easily passivated, so that its consumption is not very great. However, nickel or nickel plated steel are usually preferred.

The anodic current density must be suitably controlled, in order to keep the anodic tension at the value that is required not only for manganate oxidation, but also for the anode to become passivated and maintain its passivity. Nevertheless, the anodic current density must not be so high as to raise the anodic tension to the point of oxygen evolution.

The cathodic current density can be much higher; in fact with high cathodic current density in cells without diaphragm only a small cathodic reduction of the MnO_4^- ions can be observed, because in such conditions their diffusion from the anolyte is hindered by the strong electric migration of the anions toward the anode. Since preservation of the anode depends upon its ability to acquire the passive state, it is necessary that the starting material used for manganate preparation be as free of chlorides and nitrates as possible: indeed, the presence of such impurities in the solution to be electrolyzed would destroy the passivity (see p. 503 *et seq.*). For the same reason the operating temperature is generally kept not higher than 60° C.

It ensues from the foregoing that the best operating conditions can be summarized in the following points:

1. the anodic tension must have an appropriate value;
2. the anodic current density must not be excessively high;
3. the cathodic current density must be relatively high, so as to decrease the tendency of the MnO_4^- ions diffusing toward the catholyte being cathodically reduced;
4. the operating temperature must be kept below a certain limit;
5. Cl^- and NO_3^- ions must be absent from the solution to be electrolyzed.

In one modern industrial plant the electrolysis is carried out in batches in rectangular cells, which, for a current capacity of 10,000 A, are about

3 m long, 1.25 m wide and 2 m deep. The anodes consist of nickel plated sheet-iron arranged in several parallel rows. Between each pair of anode rows are placed the cathodes, consisting of iron rods; each cathode rod is wrapped in a sheath of asbestos or synthetic fabric, which prevents the diffusion of permanganate ions to the cathodic surface.

The starting solution has a manganate content of 200–225 g K_2MnO_4/l plus 5% KOH; part of the manganate is in the form of an undissolved slurry, so that the electrolyte must be actively circulated, in order to dissolve as much manganate as possible during the electrolysis; this is discontinued when the manganate content has decreased to 20–25 g K_2MnO_4/l, with a corresponding increase in the KOH concentration.

The spent electrolyte is drained out of the cell and its permanganate content is allowed to crystallize and settle in batch crystallizers, after being cooled to 10–15° C. The mother liquor, having a content of 15–20 g K_2MnO_4/l and 190–210 g KOH/l, is sent to evaporators where the alkali concentration is increased to 750 g KOH/l, and the precipitated manganate is added to the solution feeding the electrolytic process. The concentrated caustic solution is then sent back to repeat the next cycle, beginning with reaction (1).

The electrical values for the electrochemical process are summarized in Table IX, 7.

TABLE IX, 7

ELECTRICAL VALUES FOR PERMANGANATE PRODUCTION

Electrodes	Fe; Ni
Electric tension	2.5–3 V
Anodic current density	1.5 A/dm²
Cathodic current density	15 A/dm²
Current efficiency	0.55–0.7
Energy consumption	0.77–0.85 kWh/kg

(III) *Persulphuric Acid and Persulphates*

Persulphuric acid and persulphates are produced exclusively by anodic oxidation. The primary anodic reactions involved have not yet been completely clarified.

In a concentrated sulphuric acid solution there are H^+ and HSO_4^- ions present, so that it might be logical to assume that the primary anodic reaction occurs in the simplest possible way, such as

$$2\,HSO_4^- \rightarrow H_2S_2O_8 + 2\,e^-$$

However, a closer consideration of the several reaction steps shows that the process cannot be as simple as that. In the first place, the following reactions must also be assumed to be possible:

$$2\,OH^- \rightarrow O + H_2O + 2\,e^-$$
$$2\,HSO_4^- + H_2O + O \rightarrow H_2S_2O_8 + 2\,OH^-$$

Furthermore, in neutral sulphate solutions there is also a possibility that the SO_4^{2-} ions will be partially discharged, with a subsequent dimerization to persulphate ions:

$$2\,SO_4^{2-} \rightarrow 2\,SO_4^- + 2\,e^-$$
$$2\,SO_4^- \rightarrow S_2O_8^{2-}$$

The accompanying cathodic primary reaction consists in any case of the discharge of H^+ ions. Beside these primary oxidation reactions, a high current efficiency of persulphate from concentrated sulphuric acid solutions is hindered by the occurrence of secondary decomposition reactions which tend to destroy the persulphuric acid, as soon as it forms, with a production of permonosulphuric acid and the consequent anodic decomposition of the latter also:

$$H_2S_2O_8 + H_2O \rightarrow H_2SO_5 + H_2SO_4$$
$$H_2SO_5 + 2\,OH^- \rightarrow H_2SO_4 + H_2O + O_2 + 2\,e^-$$

The last reactions mean that, through the intermediate degrading to permonosulphuric acid, the persulphuric acid produced will revert to the starting sulphuric acid, which involves the same amount of electric charges as required for persulphuric acid production. Under certain conditions, the current efficiency can become so low as to assume a negative value, which means that, by starting with a solution already containing persulphuric acid, its concentration during the electrolysis may even diminish, instead of increasing. Accordingly, in order to obtain a satisfactory current efficiency of persulphate, the current concentration must be high enough to allow the persulphate concentration to rise rapidly, while using appropriate means, such as the addition of Cl^- ions, to decompose the permonosulphuric acid as soon as it forms.

The oxidation processes leading to the production of persulphuric acid or persulphates, as well as their cathodic reduction, are markedly irreversible, so that the relevant standard electric tensions are not known. However, it is an experimental fact that the anodic tension must be as high as possible in order to obtain a satisfactory current efficiency. On this ground it is advisable to operate at a high current density and to use anodes of smooth platinum, which is characterized by a high oxygen overtension.

Because of the several reactions controlling the formation and the decomposition of persulphuric acid, as because of the other factors affecting the efficiency of the process, its satisfactory performance is subject to a number of conditions, the most important of which are the following:

1. an appropriate sulphuric acid concentration, which must be neither too low nor too high, in order to check the formation of permonosulphuric acid;
2. a high anodic tension;
3. a relatively high anodic current density and current concentration, so that the rate of formation of persulphuric acid is substantially higher than its rate of decomposition;
4. a low operating temperature, since a high temperature would favour the decomposition of the percompound;
5. the elimination of permonosulphuric acid by chemical means;
6. the removal from the electrolyte of any substances, such as arsenic, iron, manganese, finely divided platinum, which might dangerously act as catalysts in the decomposition of persulphate.

When the anolyte is separated from the catholyte by a diaphragm, a very convenient method of satisfying the last condition is to circulate the feeding electrolyte first into the catholyte and then into the anolyte compartment, so as to plate out most of the harmful contents on the cathode. At the same time the catholyte becomes depleted of sulphate ions, which tend to migrate across the diaphragm and can be replenished with fresh sulphuric acid (which serves also to compensate the several losses in the process) and is thus also purified before being admitted into the anolyte stream.

When the incoming electrolyte basically consists of sulphuric acid, the catholyte is invariably kept separated from the anolyte by a diaphragm of porous, unglazed porcelain, which prevents the diffusion of persulphuric acid toward the cathode, where it would be reduced. Under such conditions the maximum concentration of active oxygen is obtained by feeding the electrolytic process with sulphuric acid at a concentration of 500–600 g/l (d = 1.30–1.35); on the other hand, an initial sulphuric acid content of 730 g/l (d = 1.415) corresponds to conditions allowing maximum current efficiency to be reached.

The electrolysis of ammonium sulphate or potassium sulphate solutions allows a substantially higher current efficiency than the electrolysis of sulphuric acid. Accordingly, the use of porcelain diaphragms is not considered so essential, even though it is advisable to wrap the cathodes with asbestos cord or add about 0.2 % of potassium bichromate, so as to build up an impervious film of chromium hydroxide over the cathodic surface. However, the use of potassium bichromate is possible only when the sulphate solution fed to the process is neutral, since otherwise the chromium hydroxide film inhibiting the persulphate reduction would be destroyed.

Due to its much higher solubility, ammonium sulphate is more convenient to electrolyse than potassium sulphate. In a typical plant the feed electrolyte is composed of 230–250 g/l of ammonium sulphate and 250–270 g/l of sulphuric acid corresponding roughly to the composition $(NH_4)HSO_4$. The electrolyzed solution contains 240–250 g/l of ammonium persulphate, about 100 g/l of ammonium sulphate and 200 g/l of free acid. The electrolysis is carried out at a temperature of 30° C. The current density on the platinum anodes is 125 A/dm^2 and the current concentration is 13 A/l. The cathodes may consist of graphite rods or of lead.

The persulphate concentration could be gradually increased by recirculating the outflowing electrolyte into the cell. However, it is found more expedient to pass the electrolyte through a series of cells arranged in cascade, in such a way that the solution first flows through all the catholyte compartments and then is passed back into the anolyte compartments. The cells are provided with cooling coils near the anodes and the cathodes. When the cathodes are of lead, they can perform the function of cooling coils at the same time. The final product is recovered by crystallization. The cell tanks are lined with acid proof bricks or with polyvinyl chloride sheet.

The electrolytic production of percompounds has acquired a great industrial importance in providing the starting materials for the production of hydrogen peroxide, by hydrolysis of the percompound:

$$(NH_4)_2S_2O_8 + 2\,H_2O \rightarrow 2(NH_4)HSO_4 + H_2O_2$$

Hydrolysis is achieved by steam at 110° C and hydrogen peroxide is distilled in vacuo at an absolute pressure of 20 mm Hg. The ammonium hydrogen sulphate is sent back to the electrolysis process, after purification, filtration and readjustment to the required concentration of ammonium salt and sulphuric acid.

TABLE IX, 8

ELECTRICAL VALUES RELATING TO AMMONIUM PERSULPHATE PRODUCTION

Electrolyte: $(NH_4)_2SO_4 + H_2SO_4$

Anodes	Pt
Cathodes	Pb; graphite
Electric tension	5.5–6.5 V
Current density	30–150 A/dm^2
Current capacity	4,500 A
Current efficiency	0.7–0.8
Energy consumption	1.8–2.2 kWh/kg $(NH_4)_2S_2O_8$

(IV) *Organic Oxidations and Reductions*

Many organic compounds can be produced by electrolytic methods, even when the starting materials and products are not themselves electrolytes; the electrolysis of water taking place in all such cases is always accompanied by secondary reactions of the organic materials, acting as depolarizers, with the oxygen or the hydrogen discharged by the anodic or the cathodic primary reaction respectively. However, organic electrochemistry has not found such a wide industrial application as the inorganic. In fact, the products that can thus be obtained are so limited in number and their quantities are so small that it is not possible to speak of any electro-organic industry proper. Among the useful processes, which will not be considered further for these reasons are the oxidation of anthracene to anthraquinone, the reduction of glucose to mannitol and sorbitol, and the reduction of nitrobenzene to *p*-aminophenol.

There are several reasons why organic electrochemistry has not been able to develop on a larger scale. In the first place, as already mentioned, almost all organic processes undergo a number of more or less complex side-reactions, characterized by different reaction rates. This often involves the presence of delayed reactions also, which are the cause of the development of chemical polarizations and overtensions which may direct the process in an unwanted direction. Furthermore, the organic compounds are scarcely soluble and have a poor conductivity. Finally, because of the frequent development of delayed reactions, it is very difficult to adjust the anodic and the cathodic electric tensions in such an accurate way that the process will take place with values for current and energy efficiency which are still economically acceptable.

The electroorganic reactions do not have in themselves any particular feature, beside those already discussed in Chapters IV and IX about electrochemical oxidation and reduction in general. In fact, the possibility of a given reaction taking place, as well as its efficiency, depend, for organic compounds also, upon such factors as: electrode tension, current density, concentration of the depolarizer and its speed of diffusion, temperature, bath agitation, electrode material, catalytic influence of the electrode material as well as of any possible additives in the bath, rate of depolarization, and so on.

3. Electrolysis of Hydrochloric Acid

This process, which, even though technically feasible, might seem at first sight paradoxical from the economic standpoint, has recently come to the fore, because of the large amounts of hydrochloric acid made available in modern chemical industry, in particular as a byproduct of chlorination in organic chemistry. In fact, under many circumstances such large amounts of hydrochloric acid would be difficult to be disposed of in a more economic and practical way than by the electrolytic recovery of chlorine that can be immediately reutilized at the site. The only other alternative would be to run the hydrochloric acid to waste, and this would cause major difficulties in avoiding pollution of rivers.

The electrolysis process, as applied to hydrochloric acid, yields chlorine with its equivalent amount of hydrogen but without the equivalent amount of caustic soda which would be otherwise obtained from sodium chloride. However, the electrolysis of hydrogen chloride is justified under many circumstances, beside the foregoing by the fact that the energy requirement per unit of chlorine produced is about one half of that for the electrolysis of sodium chloride.

Since the two products obtained from the electrolysis of an aqueous solution of hydrogen chloride are both in gaseous form, the technical characteristics of this process have many similarities with those of water electrolysis. The industrial equipment that has been developed for the electrolysis of hydrochloric acid is based upon the bipolar type or filter press assembly for the same reasons that make this type so advantageous in the latter process. However, for this particular application, on account of the corrosive action of the electrolyte and of the chlorine gas, a number of design and constructional problems are to be faced that are not encountered in water electrolysis. In this respect, a great help has been given to the designer by the acid-proof materials made available by the modern synthetic resins industry. Accordingly, the cell frames may be integrally made of such molding compounds as phenol–formaldehyde or urea–formaldehyde, while the diaphragms are generally of polyvinyl chloride fabric.

Each diaphragm is stretched between one pair of frames. Each frame contains a bipolar electrode cemented to it. Each bipolar electrode is made of a graphite plate provided with vertical grooves on the cathode side, so as to facilitate hydrogen evolution; the anodic side of the plate is protected from oxidation, which would accompany any possible discharge of oxygen together with chlorine, by a layer of graphite lumps; these are pressed in contact with the plate by the diaphragm, which in turn is kept against this anodic layer by the grooved cathodic side of the next plate. The graphite lumps are easily replaced, as soon as they are consumed, by introducing a new filling through a stoppered hole in the frame top.

The hydrochloric acid solution is added at about 33 % concentration and withdrawn at 18 %; this depleted solution is refortified by flowing through an absorption column countercurrent to the stream of hydrochloric acid coming from the chlorination process.

At the normal operating temperature of 80° C, both chlorine and hydrogen gas streams leave the electrolyser with a content of 2 to 3 % HCl. This is removed by flowing each gas stream through a separate washing and cooling tower, the acidic effluent of which is sent to the hydrochloric acid absorption column.

Although, at this operating temperature, the average conductivity of the hydrochloric acid solution is 3 to 4 times greater than that of a sodium chloride brine, the current density is usually kept lower than 10 A/dm^2. This is dictated by the considerable resistance offered by the graphite material used for the electrodes, especially if the graphite plates are protected on the anodic side by a layer of graphite lumps.

Since oxygen discharge in an acidic solution can hardly take place, the product chlorine is characterized by a high purity, in that it may contain only traces of carbon dioxide. However, the current efficiency is only exceptionally higher than 90 %, mainly due to the diffusion of hydrogen from the catholyte into the anolyte and recombination of the two gases.

Table IX, 9 summarizes the electrical characteristics for the hydrochloric acid electrolysis process.

TABLE IX, 9

ELECTRICAL VALUES FOR THE ELECTROLYSIS OF HYDROCHLORIC ACID

Electrodes	graphite
Electric tension	2.3 V
Current density	9–10 A/dm^2
Current efficiency	0.90
Energy consumption	1.95 kWh/kg Cl_2

4. Poorly Soluble Metallic Compounds

When a metal is immersed as an anode in an electrolytic solution containing anions that form poorly soluble compounds with the metal considered, this will be scarcely attacked, for the first cations that enter the solution, as a result of the anodic reaction, will form the poorly soluble compounds with the anions present in the solution. Such a compound tends to form on the anode itself an adherent precipitate which confers on the anode a passivity state (see p. 503 *et seq.*). However, if the electrolyte contains also some other anionic species, which is capable of forming an easily soluble compound with the cations released by the anode, then the operating conditions under which the electrolysis is performed can be established in such a way that the poorly soluble compound will not precipitate immediately on the anodic surface but at some distance from it. Under such conditions, the anodic product will not stick to the anode surface and the anodic reaction can proceed with a current efficiency as high as unity, while the product reprecipitates after entering the solution.

On the foregoing principle is based the electrochemical production of such compounds as white lead (basic lead carbonate) and chrome yellow (lead chromate), both of which are produced on an industrial scale[1].

The first process was developed by von Luckow. The electrolyte is composed of 80 % sodium chlorate and 20 % sodium carbonate or sodium chromate, in aqueous solution at a concentration of 1.5 %. The anodes consist of pure lead and the cathodes of hard lead. By carrying out the electrolysis with a current density as low as 0.5 A/dm^2, the basic lead carbonate or the lead chromate thus formed will easily separate from the anode.

As the formation of either compound substracts anions from the solution, so that this becomes alkaline, a part of it is continuously withdrawn from the cell, resaturated with carbon dioxide or chromic acid and then recycled. A later method for the production of white lead was developed with the Sperry process, employing a bifluid cell, in which the diaphragm separates two electrolytes of different composition. The anolyte consists of an aqueous solution containing 4 % sodium acetate, 0.06 to 0.2 % sodium carbonate and 0.05 % sodium bicarbonate. The catholyte also has a 4 % sodium acetate content, but the carbonate concentration is considerably greater than in the anolyte, in that it may be as high as 5 %. The lead anode dissolves according to the reaction

$$Pb \rightarrow Pb^{2+} + 2\,e^-$$

and the lead cations thus formed approach to the diaphragm, where they meet the carbonate ions coming through the diaphragm from the cathodic region. Precipitation takes place according to the reaction

$$3\,Pb^{2+} + 4\,CO_3^{2-} + 2\,H_2O \rightarrow (2\,PbCO_3) \cdot Pb(OH)_2 + 2\,HCO_3^-$$

The anolyte, containing the white lead thus formed in suspension, is passed through a system of concentrators and filters and is then returned clear to the anodic compartment of the electrolytic cell; the anolyte thus acts as a carrier for the continuous extraction of the product white lead.

[1] Process patents have also been granted for the production of basic chromates and sulphochromates (which are likely to be mixtures of sulphates and chromates) of heavy metals, useful in the pigment industry, such processes being based upon the use of cells containing two electrolytes.

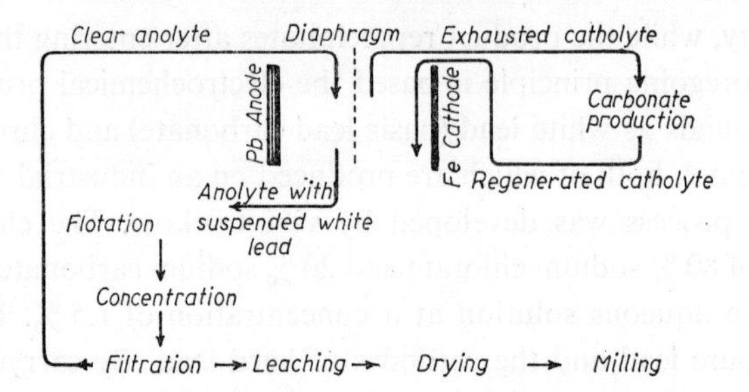

Fig. IX, 18. Flow diagram for white lead production

The catholyte, which becomes alkaline, is cycled through an absorption tower, in countercurrent to a saturating stream of carbon dioxide and is then returned to the cathodic zone. In order to obtain a good product of constant composition it is necessary also that the electrolyte composition be kept constant, which is obtained by properly adjusting the circulation of both anolyte and catholyte in accordance with the intensity of the total electric current.

The process is represented diagrammatically in Fig. IX, 18.

The cell tank is usually made of reinforced concrete, which is lined with bitumen to make the cell walls leak-proof. The anode lead is of a fairly good purity grade. However, high purity is not a strict necessity in this process, as the impurities contained in lead are in general more noble than lead itself; therefore, instead of passing into solution, such impurities remain over the anode surface forming a 'slime blanket' which is periodically removed by brushing. The cathodes are of sheet iron or steel. The diaphragm material is of linen with a sufficiently thick texture to prevent the white lead particles from entering the anodic region. The diaphragm

TABLE IX, 10

ELECTRICAL VALUES FOR WHITE LEAD PRODUCTION
(Sperry process)

Anodes	Pb
Cathodes	Fe
Electric tension	3.5 V
Current density	2.7–2.8 A/dm^2
Current efficiency	0.97
Energy consumption	0.45–0.5 kWh/kg

is stretched between the anode and the cathode without being in contact with either.

The white lead thus obtained is of high purity grade, constant composition, very fine and uniform grain size and glossy whiteness. The anolyte carries it as a 0.5 % suspension, which is removed by passage through a concentrator, a thickener and finally through a filter. The raw product is washed with hot water, dried and pulverized, this last operation only serving to break down the lumps, since the fine grain size is already determined by the electrolytic process. Table IX, 10 summarizes the electrical characteristics relating to the Sperry process.

CHAPTER X

ELECTROLYSIS IN MOLTEN ELECTROLYTES[1]

1. Special Considerations on Electrolysis in the Molten State

Electrolysis of molten electrolytes has been used in technology to obtain those metals whose reduction with carbon, even when it is possible, is difficult and costly and which for various reasons cannot be obtained by the electrolysis of aqueous solutions. The metals which are prepared by the electrolysis of their molten compounds are thus principally the alkali metals, the alkaline-earth metals and some metals of sub-groups IV*a*, V*a*, and VI*a* of the periodic system, like palladium, uranium, titanium, thallium, etc.[2]. Electrolysis in the molten state has many points in common with that of electrolytes in solution and reference should be made to Chapters IV and VIII for all general considerations which apply also to electrolytes in solution. Certain particular points, however, must be considered for the electrolysis in the molten state.

A first, fundamental, difference between electrolytes in solution and in the molten state is that the latter in general do not need solvents to dissociate. There are certain gaps in the present-day knowledge of the state of molten electrolytes (*cf.* p. 79 *et seq.*). All that can be said in this field is that strong, medium, weak and nonelectrolytes also exist in the molten state, *i.e.* electrolytes which are strongly dissociated and electrolytes with intermediate and slight dissociation, together with substances which do not dissociate into ions. However, it is not possible to have an exact quantitative idea of the degree of dissociation since not only does dissociation of molecules into ions occur but there is also probably an association of molecules into di- or pluri-molecular complexes; and the identification and quantitative study of these is very difficult. When an electrolyte exists in the molten state as the only component present *i.e.* not as a solution of one electrolyte in another molten electrolyte, all phenomena due to ionic concentration during electrolysis, such as the concentration polarization, disappear.

[1] *Cf.* general considerations p. 428 *et seq.*

[2] Electrolytic processes have also been designed for the production of the rare earth metals, tantalum and boron.

A method which might give useful indications could be to measure the decomposition tensions of molten salts using current intensity–applied electric tension diagrams (*cf.* p. 192 *et seq.*). Such measurements are, however, very difficult because of depolarization phenomena which make the inflexions of the curves indistinct and hence must be eliminated as far as possible. Depolarization phenomena are due partly to the chemical nature of the system electrolyzed and partly to any impurities present. Phenomena due to the chemical nature of the system are mainly caused by the solubility of the cathodically deposited metal in the electrolyte. This metal diffuses towards the anode – as a true solution, as addition compounds, as ionic complexes or as a metallic cloud – and there reacts with the products of the primary anodic reaction. If now the electrolyte is not chemically pure – or at least of known and defined composition – other depolarization effects may arise; for example, the presence of water absorbed from the atmosphere is one of the principal and most common causes of error because it decomposes at a very low electric tension. In the current intensity–applied electric tension diagram this will be shown as a single point which must not be confused with the decomposition tension of the electrolyte being examined. Water also leads to hydrolysis phenomena which are particularly marked in view of the high temperature; new substances are formed having decomposition electric tensions different from that of the electrolyte and may thus disturb or completely falsify the measurement. All these depolarization phenomena give rise to a residual current whose intensity is not negligible. It is thus necessary to know the internal resistance of the cell, so as to be able to subtract from the applied electric tension the value of the product IR which represents the electric tension lost through the residual current. Thus, to have reliable results, it would be necessary to know not only the applied voltage and the current intensity but also the resistance of the cell, the anodic and cathodic current efficiencies and the products of electrolysis so as to understand if, and to what extent, depolarization phenomena could be falsifying the results.

All these determinations are very difficult and always affected by errors which are not negligible and are, moreover, markedly dependent on the temperature. However, in every case, it is impossible to obtain the electric tensions of individual metals from measurements of decomposition tensions because the electric tensions of the anions of the salts used as electrolytes are unknown at the experimental temperatures; it is, therefore,

TABLE X,1

DECOMPOSITION ELECTRIC TENSIONS U_s OF MOLTEN ELECTROLYTES

Electrolyte	°C	U_s (V)	*Electrolyte*	°C	U_s (V)
LiCl	800	3.17	KBr	800	2.88
NaCl	820	3.15	KI	800	2.40
NaBr	800	2.75	KOH	200	2.4
NaI	800	2.22	KOH	300	2.35
NaOH	200	2.32	$MgCl_2$	800	~2.5
NaOH	300	2.25	$CaCl_2$	800	3.21
$Na_4P_2O_7$	1010	0.71	$BaCl_2$	1005	3.14
Na_2SO_4	890	2.5	$ZnCl_2$	400	1.96
KCl	800	3.10	$PbCl_2$	600	1.28

almost impossible to construct a series of electric tensions for molten electrolytes. Table X, 1 shows values for the decomposition electric tensions in the molten state of some of the more common electrolytes.

Another difference between the electrolysis of molten electrolytes and electrolysis in aqueous solutions, is that it is theoretically impossible to decompose, by electrolysis in solution, a compound whose decomposition electric tension is greater than that of the solvent; or in practice, when it is greater than the decomposition electric tension of the solvent plus the overtensions characteristic for electrolysis under those particular conditions (nature of electrodes, temperature, current density, etc.). With molten electrolytes, since solvents are generally absent except in special cases such as the electrolysis of aluminium oxide, electrolysis is theoretically always possible since the necessary decomposition electric tension can always be attained.

Electrolysis in molten salts obeys Faraday's Laws perfectly although the demonstration of their validity is sometimes very difficult. In fact, often during the electrolysis of molten electrolytes there are considerable and not readily avoidable losses in the current efficiency. These losses have various causes such as the evaporation or distillation of metal separated in the molten state, or secondary reactions between it and the materials with which it comes into contact. However the main cause of a loss in current efficiency lies in the solubility of the metal in the electrolyte and in the formation of metallic clouds in the bulk of the molten electrolyte. The metal which separates at the cathode may redissolve until it reaches saturation in the molten salt and may then diffuse towards the surface where it will be rapidly oxidized by atmospheric oxygen, or towards the anode where it may react with the products of the discharge of the anion – often halogens – reforming the original salt; in this way, the current efficiency falls. Under other experimental conditions, the metal may occur in the electrolyte as a so-called metallic cloud which may also diffuse towards the surface or the anode and be subjected to the same secondary reactions. However even today various workers are not in agreement on the interpretation of the state of the metal in a metallic cloud, *i.e.* whether it is a true solution or rather a colloidal state. Metallic clouds were first observed by Lorenz in 1895 during the electrolysis of molten $CdCl_2$ and studies on the problem have been very numerous since then[1].

Firstly, it must be noted that metallic clouds are often formed when a metal is immersed in one of its molten salts, independently of any electrolytic phenomenon; in this case the cloud diffuses from the surface of the metal into the interior of the molten mass. Clouds may also be formed when a small quantity of a reducing agent is added to the molten salt. Ultramicroscopic studies at room temperature on the mass of an electrolyte in which a metallic cloud has been formed, clearly show a colloidal type of dispersion of the metal which disappears on suitable chemical treatment *e.g.* with chlorine and hydrogen chloride. This observation, however, is not sufficient to draw the conclusion that the metal is actually dispersed in the colloidal state at the temperature of the molten salt also. Often, chemical analysis shows the existence of a compound between the metal and the salt, which is dissolved in the molecular state. Lead, for example, dissolves in molten lead chloride to form a metallic cloud. The free dissolved lead can be titrated with lead dioxide according to the reaction

$$PbO_2 + Pb \rightarrow 2\,PbO$$

By weighing the block of metallic lead before and after its immersion in the molten lead chloride, it is possible to determine the total amount of metal dissolved, from the loss in weight. This does not coincide with the amount of free lead determined by titration with dioxide. It is thus necessary to assume that a molecular compound is formed according to the equation

$$n\,Pb + PbCl_2 \rightleftharpoons Pb_n \cdot PbCl_2$$

[1] A summary of this field has been published by B. Bergland, *Svensk Kem. Tidskr.*, 70 (1958) 3.

With increase in temperature, the equilibrium shifts towards the left.

During solidification, the component $PbCl_2$ crystallizes and is removed from the equilibrium so that the molecular compound disappears. The excess of lead produced by the decomposition of the molecular compound forms a supersaturated solution and crystallizes in a very finely divided state, because the diffusibility of lead dissolved in a medium of high viscosity is very small and there is no possibility of forming large crystal aggregates. Metallic particles are thus formed, dispersed in an already solidified electrolyte, so that a colloidal dispersion is obtained. The molecular compound is effectively formed under the action of the secondary valencies of the molten salt.

If another compound is added to the molten salt which can react with the secondary valencies of the electrolyte to give an addition compound, they may become saturated. The formation of the molecular compound between the metal and the molten salt will then be blocked and the dissolution of the metal in the salt will be prevented or at least diminished. The metallic cloud, if it is still formed, will be produced in much smaller amounts. Addition compounds of this type are known. The present tendency is to consider the metallic cloud as a true solution.

Metallic clouds may sometimes arise through a defective reunion of the individual, submicroscopic, metallic particles formed during the electrolysis.

TABLE X,2

CURRENT EFFICIENCY IN THE ELECTROLYSIS OF MOLTEN LEAD CHLORIDE IN THE PRESENCE OF POTASSIUM CHLORIDE

KCl %	$R_{curr.}$	KCl %	$R_{curr.}$
0	0.921	11.9	0.976
2.4	0.935	16.1	0.983
5.1	0.957	27.3	0.984
8.2	0.969	44.6	0.987

If it is wished to obtain the highest possible current efficiency, the electrolysis must be carried out at the lowest possible temperature, *i.e.* only slightly above the melting point of the salt, so as to minimize the solubility of the metal in the electrolyte. The addition of another salt has two practical advantages:

1. The presence of addition compounds hinders the formation of metallic clouds. Table X, 2 shows the effect on current efficiency of adding potassium chloride during the electrolysis of lead chloride at 600° C between electrodes 35 mm apart.

2. The cryoscopic effect may lower the melting point of the electrolyte and hence the temperature of electrolysis; this effectively, greatly lowers the solubility of the molten metal and the formation of clouds. The effect of temperature on current efficiency during the electrolysis of lead chloride is shown in Table X, 3.

TABLE X, 3

CURRENT EFFICIENCY IN THE ELECTROLYSIS OF MOLTEN LEAD CHLORIDE AT VARIOUS TEMPERATURES

°C	$R_{curr.}$
540	0.963
600	0.926
700	0.876
800	0.659
900	0.380
956*	0

* boiling point.

To increase the current efficiency still further, it is convenient to use the highest possible current density and to move the electrodes far apart. The effect of current density may be explained on the basis of the fact that the efficiency in electrolysis is less than 1 because part of the separated metal dissolves. Since the quantity of metal which dissolves is practically constant and depends only on the temperature, increasing the amount of metal separated electrolytically also increases the current efficiency. However, the current density cannot rise above the limit at which the anodic effect begins (see below).

The effect of the distance between the electrodes is readily explained since, as mentioned already, the metal diffuses towards the anode where it may react once more with the product of anodic discharge to form the starting compound. In this way, one of the components of the equilibrium between dissolved metal and molten electrolyte is removed, *i.e.* the

TABLE X,4

CURRENT EFFICIENCY IN THE ELECTROLYSIS OF MOLTEN LEAD CHLORIDE AT VARIOUS ELECTROLYTE DISTANCES

Distance between electrodes (mm)	$R_{curr.}$
2.5	0.775
5.0	0.792
10	0.813
25	0.854
35	0.876
60	0.875

TABLE X,5

CURRENT EFFICIENCY IN THE ELECTROLYSIS OF MOLTEN LEAD CHLORIDE AT VARIOUS CURRENT INTENSITIES

Current intensity A	$R_{curr.}$
2.0	0.953
1.0	0.926
0.5	0.897
0.3	0.841
0.1	0.728
0.05	0.441
0.03	0.197
0.01	0.1

TABLE X,6

CURRENT EFFICIENCY IN THE ELECTROLYSIS OF MOLTEN LEAD CHLORIDE WITH SEPARATED ELECTRODES

Enclosed electrode	$R_{curr.}$
Anode	0.9795
Cathode	0.9946
Both	0.9998

dissolved metal. To re-establish equilibrium, more metal must dissolve. If the electrodes are placed far apart or separated with a diaphragm, this secondary reaction will be obstructed and the current efficiency will consequently increase. Tables X,4, X,5 and X,6 show the effect of the current density, of the distance between the electrodes, and of the separation of the electrodes on current efficiency during the electrolysis of molten lead chloride.

A last, special phenomenon which sometimes occurs during the electrolysis of molten electrolytes is the so-called anodic effect. In this, the regular evolution of gas at the anode stops, and the molten mass becomes separated from the anode which is no longer wetted. A number of small voltaic arcs are now formed between the molten mass and the anode, the ohmic resistance of the cell increases markedly and in consequence the applied electric tension rises and the current intensity falls. In other words, the anodic effect is characterized by the formation of a gaseous sheet about the electrode which prevents the electrolyte from wetting it.

The anodic effect appears particularly at high current densities; it is also connected, for every electrolyte, with a characteristic current density. This depends not only on the nature of the molten electrolyte, but also on the type of electrode – carbon or graphite – and in particular on the purity of the electrolyte itself. The greater the purity of the electrolyte, the lower is the critical value of the current density above which the anodic effect appears.

The impurities which are thus able to raise the value of the critical current density include the particularly important oxides which can be formed even in the purest of electrolytes by reaction with water (which is always present in traces) or with atmospheric oxygen. If dissolved oxide is completely eliminated, a critical current density is obtained which is so low that pure electrolytes are virtually not susceptible to electrolysis. The anodic effect has not yet been clearly interpreted. Systematic studies aimed at clarifying its causes have shown that one of these is probably a disturbance of an electrostatic nature. It seems very probable that a double electrical layer is set up on the bubbles of gas passing through the electrolyte; this makes the bubbles behave as if they carried an electrical charge whose sign depended on the nature of the electrolyte. Bubbles passing through an electrolyte which was free of oxide would have a negative charge whereas even a small percentage of dissolved oxide would invert the sign of the charge on the bubbles and make them positive. The anodic effect in this case would thus be provoked by an electrostatic attraction between the charge of the electrode and the charges of the bubbles. When these are of opposite signs the electrostatic attraction set up would have the effect of surrounding the anode with a gaseous sheath which would separate the molten mass of electrolyte from the anode; the electrical arcs would thus be caused, with the increase in resistance and all the other phenomena described.

A second effect of the presence of oxides, or other impurities, in the molten electrolyte is given by the corresponding changes in the anode–electrolyte interfacial tension, *i.e.* the susceptibility of the anode to wetting. This depends also on the nature of the anodic material and the greater the 'wettability' of the anode the greater will be the critical current density above which the anodic effect occurs. As soon as the anodic effect begins, the current, passing through a large number of arcs between the anode and the molten mass, strongly heats the gas film adhering to the electrode; as the gas expands the electrolyte is forced further away from the electrode.

Another cause of the anodic effect lies in the local overheating of the electrode which can lead to evaporation and thermal decomposition of the electrolyte with the formation, in both cases, of gaseous products separating the electrode from the molten electrolyte. All conditions which favour local overheating, and in particular the presence of a film of solid material of high electrical resistance, such as solidified electrolyte or insoluble materials present as impurities in the anode or present in the electrolyte and migrating towards the anode, may lead to the appearance of the anodic effect.

The materials most commonly electrolyzed in the molten state for technical purposes are chlorides, oxides and hydroxides. If the electrolyte is not pure but consists of a mixture of electrolytes, electrolysis will first deposit the cations of the most noble metal present bearing in mind their various electric tensions and concentrations at the temperature at which the operation is carried out. The temperature for industrial electrolyses in molten electrolytes is normally above the melting point of the metal deposited at the cathode so that it will separate and remain in the liquid state during the electrolysis. In this way, the electrolysis, and the cathodic separation of the metal, take place very regularly without any of the difficulties discussed on p. 431 *et seq.*, concerning the various types of structure of metallic deposits.

2. Electrolytic Preparation of Aluminium; Electrolytes and Reactions

The electrical preparation of aluminium from an aqueous solution is not possible because of the strongly negative deposition tension of this metal. The preparation from a pure aluminium salt is also not possible for a number of reasons. Firstly, the chloride has a very low conductance and must be considered virtually as a nonelectrolyte, in which the majority of the molecules are undissociated. Moreover, at ordinary pressures aluminium chloride sublimes, *i.e.* it passes into the vapour state before melting. Secondly, other aluminium salts present various difficulties so that recourse has to be made to aluminium oxide which is practically the only aluminium compound stable at high temperature. Its melting point is 2,050° C so that it cannot be electrolyzed in the pure molten state because the technical difficulties of constructing a furnace for such a high temperature, and sufficiently large for an industrial process, are very great. There would also be large losses of energy by radiation. The aluminium oxide must therefore be dissolved in a suitable solvent which satisfies the following requirements:

1. It must be a good solvent of alumina.
2. It must have a greater decomposition electric tension than alumina.
3. It must be a good conductor.
4. It must be free of impurities, particularly metallic ones, which could be deposited at a cathodic polarization less than that required to deposit the aluminium.
5. It must have as low a melting point as possible both to diminish the loss of heat by radiation and to avoid the formation of aluminium carbide through reaction of the metallic aluminium with the carbon of the cathode.
6. It must have sufficient fluidity.
7. It must have a density less than that of aluminium at the working temperature.
8. It must have a low vapour pressure.
9. It must not react with the electrodes or the products of electrolysis.

The electrolyte whose properties best approximate to these requirements is cryolite: the double fluoride of sodium and aluminium, of the composition $3\ NaF \cdot AlF_3$, which should more strictly be considered as a complex fluoride Na_3AlF_6. Alumina is reasonably soluble in molten cryolite and the two components, alumina and cryolite, give rise to a

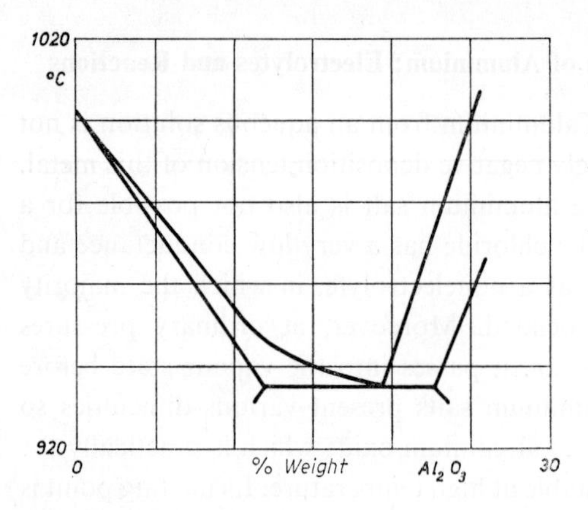

Fig. X, 1.
Phase diagram of cryolite–Al_2O_3

system whose phase diagram (Fig. X, 1) shows complete miscibility in the molten state and partial miscibility in the solid state. The eutectic is composed of 81.5 % cryolite and 18.5 % alumina and melts at 935° C. The eutectic horizontal extends between 12 and 22.5 % of alumina.

According to more recent figures[1] the melting point of the eutectic is actually 962° C and the eutectic composition corresponds to 10 % alumina. This discrepancy in itself will serve to indicate the difficulties of making measurements at such temperatures and the consequent uncertainties. Other authors have given a value for the solubility of alumina in cryolite of not more than 12 %.

Other compounds such as calcium fluoride, sodium fluoride, aluminium fluoride and recently magnesium fluoride, may be added to lower the temperature of the electrolyte. In the preparation of such mixtures with melting points lower than that of the eutectic between cryolite and alumina, allowance must be made for the density of the mixture obtained in the molten state at the working temperature. It is essential that the density of the molten mass should always be less than that of molten metallic aluminium which is 2.29 g/cm^3 at a temperature of 1,000° C; this is necessary to prevent the aluminium floating on the surface of the bath and reoxidizing with the oxygen evolved at the anode or with atmospheric oxygen.

The addition of calcium fluoride is advantageous for lowering the melting point; the ternary eutectic between cryolite, calcium fluoride and

[1] N. W. F. Phillips, R. H. Singleton and E. A. Hollingshead, *J. Electrochem. Soc.*, 102 (1955) 648, 690.

alumina has a remarkably low melting point of 868°. However, the addition of considerable quantities of calcium fluoride causes difficulties because it increases the density of the molten electrolyte. The addition of sodium fluoride would increase the solubility of alumina until the eutectic composition of the system $NaF-Na_3AlF_6$ was reached but a further increase in fluoride concentration would make the solubility of the alumina decrease again. Here too, however, considerable difficulties can arise through the increased density of the resulting electrolyte. The addition of aluminium fluoride not only lowers the melting point but also lowers the solubility of the alumina. However, the density of the electrolyte is decreased. The most common additions are, therefore, calcium fluoride or aluminium fluoride. As already mentioned, the present-day tendency is increasingly to add magnesium fluoride[1] which acts on the melting point, the electrical resistance, the density of the bath and its surface tension to give highly satisfactory practical results. By its use it is possible to lower the working temperature, diminish the simultaneous separation of sodium (probably by diminishing the cathodic concentration of the Na^+ ion) and hence increase the current efficiency, and raise the conductance of the electrolytic system so lowering the specific consumption of energy. Sometimes chyolite[2] is used instead of aluminium chloride, because it is cheaper. The addition of aluminium chloride or chyolite is also helpful in correcting the alkalinity of the electrolyte; this tends to increase as the electrolysis proceeds (*cf.* p. 609 *et seq.*). In fact, all the various additions generally tend to diminish the conductance of the resulting mixture. The most suitable cryolite is an artificial one prepared by the reaction,

$$6\,NaOH + Al_2O_3 + 12\,HF \rightarrow 2\,Na_3AlF_6 + 9\,H_2O$$

since it is not always possible to remove occluded quartz from natural cryolite and this would be inconveniently reduced to silicon which in turn would form an alloy with the aluminium. Moreover, natural cryolite is found in sufficient purity in only one industrially useful deposit in the world: in Greenland.

The aluminium compound which is electrolyzed is the oxide Al_2O_3. A hydrated oxide – bauxite – exists in nature but it cannot be electrolyzed directly since it contains too many impurities. The main impurity is iron,

[1] See, for example, B. PANEBIANCO, *Metallurg. Ital.*, 52 (1960) 531.

[2] Chyolite is another complex fluoride of sodium and aluminium with a composition $5\,NaF \cdot 3\,AlF_3$; it thus contains more aluminium fluoride than does cryolite.

present in considerable amounts, which would be cathodically separated before the aluminium and would form an undesirable Al–Fe alloy. The bauxite must therefore be first converted into very pure anhydrous alumina (*cf.* p. 613 *et seq.*).

The concentration of alumina which is considered as most suitable for electrolysis has varied greatly throughout the years. In the first attempts at the industrial electrolytic production of aluminium, Hall – one of the pioneers of the industry – advised the use of a saturated solution. Later, however, it was realized that it was more suitable to use solutions of lower concentration so as to increase the conductance of the electrolyte. Concentrations in use today vary between about 5 and 7 %. It is inconvenient to exceed these limits because the periodically added alumina would not dissolve sufficiently rapidly and would form a deposit below the layer of molten aluminium, because of its density which is about 4 g/cm^3, *i.e.* greater than that of molten aluminium. In this way, the electrical contact between the cell lining and the molten aluminium which acts as the cathode, would be interfered with. The increase in resistance would then lead to local overheating with a danger of forming aluminium carbide.

The anodic and cathodic processes are still not known for certain. It is not even known what ionic species are definitely present in the electrolyte. The most probable one ones seem to be Na^+ and AlO_2^- and possibly AlF_4^-. But various authors are deeply divided on the presence of the ionic species F^-, AlO^+, AlO_2^-, O^{2-}, AlF_6^{3-}, AlF_4^-, Al^{3+} and $AlOF_2^-$. This will indicate the marked complexity of the system. Under such conditions it is evidently very difficult to identify for sure which are the primary cathodic and anodic processes and to determine if the overall process is simple or consist of a number of consecutive reactions.

According to Piontelli[1] the Al^{3+} ion is formed from one or more chemical species containing aluminium, including undissociated species such as AlF_3 and Al_2O_3, by a direct or indirect electrode reaction or by homogeneous dissociation. After neutralization of the charge on the Al^{3+} with three electrons, it passes into the metallic phase. A possible cathodic mechanism could be as follows. In the cathodic film the elementary capacitor (*cf.* p. 186 *et seq.*) would be formed of electrons – constituting the charge of one plate – and a high concentration of Na^+ ions, constituting the charge of the other plate. In view of the action of the electrical field,

[1] R. Piontelli, *Chim. e Ind.* (*Milan*), 22 (1940) 501; *J. chim. phys.*, 49C (1952) 29; *Alluminio*, 22 (1953) 672.

and the powerful deforming action of the Na^+ ions, even on undissociated molecules, and bearing in mind the decomposition electric tensions of the molecular species present, the primary cathodic process could be described by one of the following reactions.

$$3\,Na^+ + AlF_3 + 3\,e^- \rightarrow Al + 3\,NaF$$

$$3\,Na^+ + Al_2O_3 + 3\,e^- \rightarrow Al + Na_3AlO_3$$

The simultaneous separation of sodium, which occurs to a greater or lesser extent depending on the experimental conditions, would be a concomitant primary reaction made possible by the preponderant concentration of the Na^+ ions and by the slight difference between the deposition tensions of sodium and aluminium. Under the conditions of industrial electrolysis, this difference would be of the order of 0.1–0.2 V[1]. This contemporaneous separation of sodium – which under some conditions may become predominant – may be considered as a concomitant parasitic reaction.

According to Pearson and Waddington[2] the primary cathodic process is the discharge of the Al^{3+} ion. This interpretation is founded on a consideration of the decomposition electric tensions of various fluoaluminates in the presence and absence of alumina, as shown in Table X,7.

TABLE X,7

DECOMPOSITION ELECTRIC TENSIONS OF VARIOUS MOLTEN FLUOALUMINATES

Fluoaluminate	U_s (V) *at* 950°C	U_s (V) *at* 1,080°C
Na_3AlF_6	—	2.07
Na_3AlF_6 + 15% Al_2O_3	2.22	2.01
K_3AlF_6	—	2.13
K_3AlF_6 + 15% Al_2O_3	2.20	2.01
Li_3AlF_6	—	2.20
Li_3AlF_6 + 7% Al_2O_3	2.20	2.03
Al_2O_3 (theoretical)	2.15	—

Since the decomposition electric tensions of different fluoaluminates (with various alkali cations) are constant in the presence of alumina but differ

[1] M. Feinlieb and B. Porter, *J. Electrochem. Soc.*, 103 (1956) 231.

[2] T. G. Pearson and J. Waddington, *Discussions Faraday Soc. No. 1*, Electrode Processes, (1947) 307.

in its absence these authors conclude that the Al^{3+} cations come from the dissociation of alumina

$$Al_2O_3 \rightarrow Al^{3+} + AlO_3^{3-} \quad (1)$$

and not from the dissociation of aluminium fluoride produced by the thermal decomposition of cryolite

$$2\,AlF_3 \rightarrow Al^{3+} + AlF_6^{3-}$$

However, Pearson and Waddington's explanation of the cathodic reaction is in contradiction of the fact that the presence of Al^{3+} ions is problematical and the experimental values in Table X,7 are not opposed to Piontelli's hypothesis but rather support it.

The primary cathodic current efficiency should be 1 but the actual efficiency value is less because of the formation of metallic cloud and the simultaneous deposition of sodium.

The mechanism of the primary anodic process is also not clear, principally for the same reasons as with the cathodic process. Under normal conditions, the primary discharge of the F^- ion to form elementary fluorine can be excluded since it should then be detectable in the anodic gas, either as the element F_2 or the compound CF_4; both of which are absent from the anodic gas. Similarly, the presence of the simple anion O^{2-} can equally be excluded. According to Piontelli[1] the principal anionic carriers of electrical charge are F^- and AlF_6^{3-}; but the AlO_2^- and AlO_3^{3-} anions also occur in the molten mass. The primary anodic process must then be the discharge of these anions evolving oxygen, which may be adsorbed on the anodic surface, by reactions of the type

$$2\,AlO_2^- \rightarrow Al_2O_3 + O + 2\,e^-$$

or

$$6\,F^- + Al_2O_3 \rightarrow 2\,AlF_3 + 3\,O + 6\,e^-$$

without primary discharge of F^- or AlF_6^{3-} ions. Such primary discharge is highly improbable for kinetic and energetic reasons. It must also be borne in mind that the discharge reactions of anions containing oxygen are depolarized on graphite electrodes through the reaction between carbon and oxygen; such reactions are therefore more probable.

[1] R. PIONTELLI, *Chim. e Ind. (Milan)*, 22 (1940) 501.

The anodic gas under normal conditions consists of 70–90 % carbon dioxide and 10–30 % carbon monoxide. Energetic and kinetic studies of the equilibrium reaction

$$CO_2 + C \rightarrow 2\,CO$$

together with a comparison of the analytical composition of the anodic gas with the cathodic current efficiency lead to the conclusion that the primary anodic process must follow a primary formation of carbon dioxide with an anodic current efficiency of 1. The presence of carbon monoxide may be interpreted by reaction between the metallic cloud and carbon dioxide; the metallic cloud consists principally of particles of aluminium and gaseous bubbles of elementary sodium, which has a vapour pressure greater than 1,000 mm Hg at the working temperature of the cell. The carbon dioxide would be almost instantaneously reduced by the metals of the cloud giving carbon monoxide and Al_2O_3 or Na_2O respectively. The sodium oxide could further react with aluminium fluoride according to the reaction

$$3\,Na_2O + 2\,AlF_3 \rightarrow 6\,NaF + Al_2O_3$$

In sum, therefore, alumina would be obtained again from the primary anodic and cathodic products (CO_2; Al or Na) by various secondary reactions. This explains both the presence of carbon monoxide in the anodic gas and the current efficiency of less than 1. This interpretation together with the stoichiometric relationships would lead to the equation[1]

$$\%\,CO_2 = 2\,R_{curr.} - 100 \qquad (2)$$

where the current efficiency is expressed as a percentage. The Table X, 8, taken from industrial figures, will serve to confirm this.

[1] In fact, the theoretical ratio between cathodic aluminium and anodic carbon dioxide with a current efficiency $R_{curr.}$ of 1 in both cases is 4 Al : 3 CO_2. Assuming that the effective current efficiency is 0.75, *i.e.* that 1 part of aluminium out of every 4 parts produced, passes into the metallic cloud and then reacts quantitatively with carbon dioxide according to the reaction

$$Al + \tfrac{3}{2}\,CO_2 \rightarrow \tfrac{1}{2}\,Al_2O_3 + \tfrac{3}{2}\,CO$$

the final composition of the anodic gas should be 50 % CO_2 and 50 % CO, since $\tfrac{3}{2}\,CO_2$, *i.e.* half, is reduced to CO, so that for $R_{curr.} = 75\,\%$, $CO_2 = 50$ as required by equation (2).

TABLE X,8

COMPARISON BETWEEN $R_{curr.}$ AND % CO_2 IN THE ANODIC GAS

$R_{curr.}$	% CO_2 (*theoretical*)	% CO_2 (*experimental*)
0.85 = 85 %	70	71
0.83 = 83 %	66	66
0.72 = 72 %	44	45

The small discrepancy between the theoretical and experimental values may be explained by the fact that equation (2) ignores two phenomena:

1. The percentage of carbon dioxide is less than 100 % even for a current efficiency of 1 because a small amount of carbon monoxide is produced by the reaction between carbon dioxide and the carbon of the electrode.

2. A small loss of current efficiency may be caused by oxidation of the cathodic product without involving carbon dioxide.

According to Beck[1] these two factors can be allowed for by introducing two correction factors, *g* and *z*, to give a new equation for the current efficiency from the percentage of carbon dioxide

$$R_{curr.}\,\% = g\left[50 + \tfrac{1}{2}(\%\,CO_2)\right] - z$$

Obviously the two correction terms *g* and *z* depend on the actual working arrangement and must be determined anew for each set of conditions.

The overall process results in the decomposition of aluminium oxide. This is confirmed by studies of the decomposition electric tensions of the compounds present. In the molten state, cryolite is partially split into its components NaF and AlF_3. It is thus equivalent to simultaneously electrolyzing aluminium oxide, aluminium fluoride and sodium fluoride, together with other possible components. Figures for the decomposition electric tensions of these three compounds are rather uncertain; values for sodium fluoride in the literature vary between 2.80 and 5.92 V. It is, however, certain that the decomposition electric tension of aluminium oxide is lower than that of the other components and bearing in mind the depolarization of the anodic process by the reaction between oxygen and the electrode carbon, it becomes evident that the overall process must be the decomposition of the alumina.

[1] T.R. BECK, *J. Electrochem. Soc.*, 106 (1959) 710.

3. Electrolytic Preparation of Aluminium; Electrodes and Processes

Since oxygen is evolved at the anode at a temperature of the order of 1,000° C, the only material which can be used industrially for the preparation of anodes is carbon. This reacts with the oxygen to give carbon dioxide, and secondarily carbon monoxide, and so does not contaminate the electrolyte or the product of the electrolysis. The carbon used for the preparation of the anodes must be very pure and in particular must have the lowest ash-content possible. Silica and ferric oxide are particularly damaging because their reduction products (Si and Fe) readily alloy with the aluminium separated at the cathode. Physically, one of the most important characteristics of the anodes is their resistance to crumbling during use, because any detached particles would float on the molten electrolyte since they have a lower density; they could then produce short circuits between the anodes and the cell lining and could contaminate the electrolyte.

The cathode is replaced by the actual lining of the cell which is usually a mass of carbon similar to that of the anode, but which may contain a higher percentage of impurities, particularly ash, since it does not take part chemically in the electrolytic process. In actual fact, it is the layer of liquid aluminium collected on the floor of the cell by electrolysis, which acts as the cathode. Even when a cell is put into operation a small amount of pure molten aluminium is added so as to cover the bottom entirely. The lining of the cell must be continuous without any cavities into which the pure molten aluminium could infiltrate because should it come into contact with the iron forming the outer wall of the cell, it could dissolve part of it to make an Fe–Al alloy and contaminate the cathodic product. The lining must also have some mechanical resistance and must not crumble for the same reasons as the anode.

Processes in use at present for the electrolytic preparation of aluminium do not show any practical variations apart from the composition of the electrolyte; this is always based on cryolite but is corrected with various additions as mentioned above. For this reason the various cells used are all very similar. Fig. X, 2 shows a section through a modern American cell which is widely used.

Cells operate at a tension of 4.5–5.5 V which is considerably greater than the decomposition electric tension of alumina. The excess of energy supplied, transformed into heat by the resistance of the electrolyte, serves

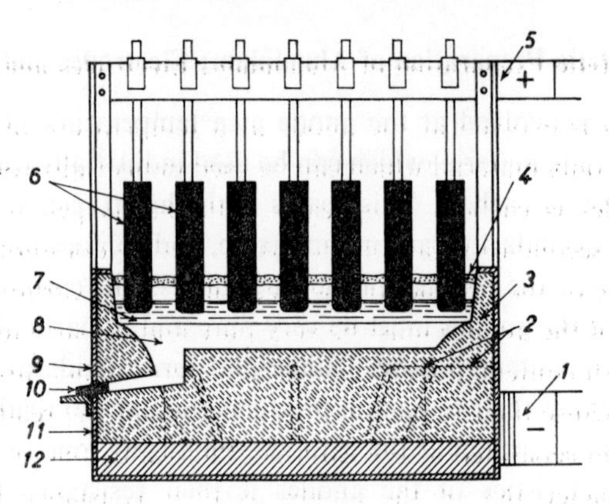

Fig. X, 2. Aluminium production cell. 1. cathodic connection; 2. framework connecting with the carbon lining; 3. carbon lining; 4. solidified electrolyte; 5. anodic connection; 6. anodes; 7. molten electrolyte; 8. molten Al; 9. iron container; 10. graphite spout (Al outlet); 11. iron container; 12. insulating layer

to keep the cell at the working temperature and to compensate for the loss of heat due to radiation and the continuous addition of cold, raw material (Al_2O_3), and to the loss of heat carried away by the products of electrolysis which leave the cell at the working temperature. The cell is kept at this temperature not only by the heat supplied as electrical energy but also, in part, by heat evolved in the anodic reaction of the combustion of the carbon. However, not all the energy of the secondary anodic reaction

$$C + 2\,O \rightarrow CO_2$$

is converted into heat. Part of the free energy released serves to depolarize the anodic process.

The cathodic current density lies at about 2.5 A/cm^2 and the anodic current density is lower, at about 1 A/cm^2, to lessen the consumption of the anode and avoid local overheating by excessive current density. The intensity of the current absorbed by each cell tends to increase. Recently, cells have been constructed with consumptions of up to 100,000 A which leads to new difficulties. The high continuous current intensity generates very strong electromagnetic fields. By interaction with the electrical current itself, these produce considerable forces acting on the boundary between the two liquids (molten electrolyte and cathodic aluminium). This boundary may then undergo distortions which make it impossible to control the distance between the anode and cathode and cause a mixing of the anolyte and catholyte. The cell must therefore be designed to avoid these difficulties. The electrodes are placed about 10 cm apart and it is important that the anodes should be adjustable in height so that the distance between the electrodes can be kept constant to compensate for the consumption of the electrodes. It is convenient if the anodes can be adjusted individually so that they can all be kept at the same distance from the cathode so as to obtain as even a distribution of the current as possible; this will lead to an even temperature and consumption of the anodes. The total current intensity through the cell can then be regulated so as to keep the mass of the electrolyte molten and to form only a thin layer of solidified electrolyte at the surface.

The working temperature varies between 900° and 1,000°C but does not exceed this, so as to avoid too great a loss of electrolyte (particularly aluminium fluoride) through vaporization, and to hinder the formation of clouds. In this case, the formation of clouds would have a different origin from that discussed on p. 592 *et seq.* They would actually be formed from particles of metallic aluminium in a very finely divided state, brought into the bulk of the molten mass by the extremely powerful electromagnetic field existing in the cell. This fine sub-division of the metal becomes easier as the mass of molten aluminium becomes more fluid at higher temperatures. The formation of metallic clouds naturally leads to a lowering of the current efficiency. Under normal conditions, this varies between 0.75 and 0.90. The main, and inevitable, cause is the simultaneous discharge of Na^+ ions together with the aluminium. The energy efficiency is not known and cannot be calculated since the theoretical deposition electric tension of aluminium at the working temperature of the electrolysis is not known exactly.

The product of electrolysis at the cathode is 99.7 % pure aluminium which still contains about 0.14 % silicon and about 0.16 % iron. The purity of the aluminium is tested continuously, particularly for its iron content. An abnormal rise in the iron content indicates that the cell lining has become defective and the cathodic aluminium has come into contact with the iron or steel of the outer wall and has alloyed with it. The cell must then be taken out of service for repair.

The product of electrolysis at the anode consists of a mixture of carbon dioxide and carbon monoxide. On the basis of the amount of oxygen liberated, the consumption of the anode should be about 550 g of anode carbon for each kg of aluminium separated; but in practice, the consumption rises to about 600 g since some slight crumbling of the anode is inevitable. As the electrolysis proceeds, the alumina concentration falls and when it reaches a certain limit, the anodic effect begins. This limit depends not only on the chemical nature of the electrode–electrolyte system but also on the current density and the temperature. As soon as the anodic effect starts, the applied electric tension rapidly rises so that an ordinary electric light bulb placed in parallel with the electrodes will become luminous and will indicate the need to add more alumina.

The alumina added must be sufficiently finely ground so that it will remain suspended in the electrolyte for long enough to dissolve. Otherwise, it would be deposited on the layer of molten aluminium and because of its greater density would sink through this and give rise to the difficulties discussed above: increase in resistance, increase in temperature and a danger of forming carbides, etc. The addition of the alumina is carried out fairly rapidly by sprinkling it on the crust of solidified electrolyte and then breaking it. It is necessary to do this rapidly, because so long as the anodic effect continues, the gas evolved at the anode consists almost exclusively of carbon monoxide so that the consumption of the anode is much greater than normal. Besides additions of aluminium oxide, it is normal to add aluminium fluoride at intervals, partly to compensate for that lost by vaporization and partly to avoid the electrolyte becoming alkaline due to the fact that the alumina always contains some sodium carbonate.

In recent years, the energy consumption has been markedly reduced; in some modern plants, it has fallen to 16 kWh/kg. Table X,9 shows electrical values for the electrolytic production of aluminium.

TABLE X,9

ELECTRICAL VALUES FOR THE PRODUCTION OF ALUMINIUM

Electric tension	4.5–5.5 V
Cathodic current density	250 A/dm^2
Anodic current density	100 A/dm^2
Current efficiency	0.80–0.90
Consumption	16–18 kWh/kg

4. Electrolytic Preparation of Aluminium; Auxiliary Operations

It has proved to be considerably more economic to produce pure aluminium directly by the electrolysis of pure aluminium oxide dissolved in pure electrolyte, using pure electrodes, than to prepare impure aluminium from unpurified, natural raw materials, and then to refine it. This makes it necessary, during the working cycle, for all the substances used in the electrolytic preparation to be extremely highly purified. For this reason, an electrolytic aluminium factory normally contains plant for auxiliary operations which to some extent form an integral part of the productive cycle of the aluminium, in that the products of these auxiliary operations are destined exclusively for the aluminium industry and are not otherwise used.

(I) Alumina

A hydrated oxide of aluminium — bauxite — exists in nature with the average empirical composition $Al_2O_3 \cdot 2\,H_2O$, containing a considerable proportion of ferric oxide, silica and other impurities. Typical analyses of useful bauxites are shown in Table X, 10.

TABLE X, 10

TYPICAL COMPOSITIONS OF BAUXITES

Components	*Bauxite*		
	American %	*French* %	*Italian* %
Al_2O_3	~ 50	50–57	53–54
Fe_2O_3	~ 8	21–24	~ 24
SiO_2	~ 10	3–7	~ 4
TiO_2	~ 3	2–3	3–4
Loss on calcining	~ 29	21–24	10–20

Bauxite consists essentially of a mixture of aluminium mono- and trihydrates (diaspore and boehmite; hydrargillite), which have been detected mineralogically, together with variable amounts of impurities. In view of the high level of impurities, particularly iron and silicon, it is not convenient to use bauxite[1] after a simple dehydration as the raw material for electrolysis; the aluminium obtained by this would be an Al–Fe–Si alloy. The bauxite must first be converted into pure aluminium oxide. Many processes have been developed for this but only a few have been employed on the industrial scale. The principal processes almost all have the common characteristic of utilizing the amphoteric properties of aluminium hydroxide to prepare an aluminate which is readily separated from the other components. The successive hydrolysis of the aluminate gives aluminium hydroxide which on calcination is converted into oxide of a high degree of purity, better than 99.5 %.

Dry process

The crude bauxite is roasted to dehydrate it and destroy any organic substances present, at temperatures not over 600° C so as not to lessen the reactivity with alkalies. After grinding sufficiently finely, it is mixed with anhydrous sodium carbonate — equally finely powdered — containing about 10–20 % of chalk; the mixture is moistened with a solution of sodium carbonate and heated in rotatory furnaces to 1,000–1,100°C. The aluminium and the iron of the bauxite are converted into sodium and calcium aluminates and ferrites respectively and the titanium and silicon mainly stay in the slag as calcium silicate and titanate. The product of this first treatment is broken up and then digested with a weakly alkaline solution at 80° C. The sodium aluminate passes unchanged into solution and the ferrite hydrolyzes to give sodium hydroxide and insoluble ferric hydroxide; that part of the silica which was not removed as calcium silicate in the slag, precipitates as sodium or aluminium silicate. After filtration, the clear solution of aluminate is saturated with carbon dioxide at 80° C or is diluted with continuous stirring and then saturated with carbon dioxide. The aluminium hydroxide is thus precipitated and, after washing, is calcined at 1,100° C. The alumina obtained is 99.5 % pure. The mother liquor is concentrated and the soda is allowed to crystallize; it can be sold as such or returned to the cycle after calcination.

Bayer process

The conversion into aluminate is a wet process in this method. The dried and ground bauxite is treated with a solution of caustic soda of 46–47° Bé

[1] There has been a recent study by G. Wilde (*Chem. Ing. Tech.*, 28 (1956) 371), on the Johnson process for the direct preparation of aluminium from bauxite.

(42–45 %) in autoclaves at 160–170° C. The quantity of soda is such that the ratio $Na_2O : Al_2O_3$ is about 1 : 1.18. By this treatment a little silica is attacked as well as the alumina, so that the concentration of sodium hydroxide, the temperature and the duration of the attack must be regulated depending on the silica content of the original bauxite. The attacked silica, however, does not go into solution but forms insoluble sodium aluminium silicate. Increasing the temperature, the duration of the attack and the concentration of caustic soda increases the yield of soluble aluminium; however, the amount of caustic soda lost as sodium aluminium silicate also increases and filtration becomes more difficult too.

After decanting, the clear solution of sodium aluminate is diluted to 20° Bé and seeded with aluminium hydroxide. The sodium aluminate hydrolyzes and about 70–80 % of the aluminium is precipitated as hydroxide. This is filtered off, washed and calcined at 1,100–1,200° C. Here, too, the alumina has a purity of about 99.5 %. The clear residual solution, after concentrating and enriching with caustic soda to compensate for the loss due to silica, returns to the cycle to attack fresh bauxite.

Pedersen process

In this process, the bauxite is treated in an electrical furnace with iron minerals, carbon and chalk. The iron and silica are both reduced by the carbon and separate as an iron–silica alloy. The aluminium passes into the slag which is composed mainly of calcium aluminate with 30–50 % of aluminium oxide and 5–10 % of silicon and iron. After fragmentation, the slag is lixiviated with a hot 5 % solution of sodium carbonate containing 0.5 % sodium hydroxide. The sodium aluminate passes into solution whilst the calcium carbonate remains insoluble. The filtered solution of sodium aluminate is saturated with carbon dioxide to give aluminium hydroxide which is treated as in the preceding processes.

Haglund process

This process makes use of the solubility of aluminium oxide in aluminium sulphide. The bauxite is treated with ferrous sulphide (pyrites) and carbon in an electric furnace at a temperature of 1,200–1,400° C; the following reaction occurs

$$Al_2O_3 + 3\,C + 3\,FeS \rightarrow Al_2S_3 + 3\,CO + 3\,Fe$$

The molten aluminium sulphide dissolves considerable quantities of aluminium oxide which crystallizes as corundum as the mass cools. The sulphide slag, after thorough cooling, is ground and treated with water and steam to decompose the aluminium sulphide according to the reaction

$$Al_2S_3 + 6\,H_2O \rightarrow 2\,Al(OH)_3 + 3\,H_2S$$

The corundum is first separated from the hydroxide which is then filtered and returned to the cycle. The hydrogen sulphide is recovered by absorption on to fresh bauxite whose iron is converted into sulphide, hence reducing the consumption of ferrous sulphide.

(*II*) *Cryolite*

The synthetic cryolite is normally prepared in independent factories by various processes both by using the reaction cited on p. 603 and by treating aluminium fluoride with sodium chloride in the presence of hydrofluoric acid and by treating the sodium fluoride with aluminium sulphate.

(*III*) *Anodes and Linings*

Anodes are prepared from coke from petroleum or special anthracites, having a very low ash content (0.1–0.4 %). After grinding, the coke is made into a paste with tar pitch (containing a maximum of 1–2 % ash) which acts as a binder. After forming by pressure or extrusion, the anodes are cured at a temperature of 1,000° C or more.

The Södeberg electrode deserves special mention. It can be considered as a continuous electrode; it consists of a kind of tube into which the fresh electrode paste is continually passed and compressed. The lower part of this tube dips into the electrolysis cell so that the curing of the paste is performed by the heat of the electrolysis cell itself, immediately before use. As the electrode is consumed within the cell, it is pushed downwards by the addition and compression of the new paste.

The cell linings are prepared from the same raw materials as are the anodes with the difference that it is not necessary to use such a pure coke and that curing at 700° C is sufficient. The lining may be cured in blocks which are then arranged within the cell and cemented together with a mixture of tar, pitch and powdered coke, or the raw paste may be first arranged within the cell and then cured there so as to give a continuous lining.

5. Electrolytic Refining of Aluminium

It is possible to prepare aluminium which is purer than that given by direct electrolysis, by using a process of electrolytic refining; the starting material for this is not aluminium obtained as above but an alloy of copper–aluminium–silicon prepared electrothermically. For the reasons explained on p. 601 *et seq.*, an aqueous solution cannot be used as an electrolyte and an organic solvent would be too costly. A mixture of aluminium, sodium, barium, calcium and magnesium fluorides with aluminium oxide, similar to that used in cells for the electrolytic preparation of aluminium, is therefore employed.

The refining of aluminium by electrolysis in such a molten electrolyte is an electrometallurgical process with soluble anodes, perfectly analogous to the electrolytic refining of metals in aqueous electrolytes. In this case, also, the aluminium passes into solution from the anodic alloy into a suitable solvent from which it is deposited on the cathode as metallic aluminium. The only electrolytic refining process for aluminium which is in successful use was suggested by Hooper but was first introduced in 1905 by A. G. Betts; it is also called the 'three layer process'. It is based on the use of an electrolyte whose density is intermediate between those of the alloy used as the anode and the pure aluminium obtained at the cathode. In this way, the electrolyte automatically separates the molten anode which lies at the bottom of the cell from the molten cathode of pure aluminium, which floats at the top of the cell. The most commonly used electrolyte, which has given good results, is a ternary mixture of aluminium fluoride, sodium fluoride and barium fluoride which also contains aluminium oxide and calcium and magnesium fluorides. The composition of the electrolyte is shown in Table X, 11.

TABLE X, 11

COMPOSITION OF THE ELECTROLYTE FOR THE REFINING OF ALUMINIUM

Component	%
AlF_3	26–38
NaF	26–31
BaF_2	31–41
Al_2O_3	0.5–3
CaF_2, MgF_2	~2

With this electrolyte, the cell can be operated at a temperature of 900–1,100° C. The densities of pure aluminium, the electrolytes and the anodic alloy are about 2.3, 2.6 and 2.8 respectively at 950° C. Somewhat different compositions have been suggested to enable the temperature to be lowered to 750° C.

The anodic alloy is prepared electrothermally and has the composition shown in Table X, 12.

TABLE X, 12

COMPOSITION OF THE ANODIC ALLOY FOR THE REFINING OF ALUMINIUM

Component	%
Al	30–40
Cu	45–55
Si	5–10
Fe	~ 5
Ti	< 1

It is not possible to obtain an alloy with a greater aluminium content than that shown in Table X, 12 electrothermally, because there would be a danger of forming the carbide which would make the alloy useless. However, the aluminium content can be raised to 60 % by mixing the alloy with further aluminium. It is best if the anodic alloy is prepared so that impurities do not collect in the electrolyte in contrast to refining in aqueous electrolytes. This is particularly true in the refining of scrap which may contain magnesium; in this case, the magnesium must first be removed by reaction in a furnace with aluminium fluoride to form magnesium fluoride and aluminium.

Since the least noble metal in the anode is aluminium, the only possible anodic process is the passage of aluminium from the alloy into the electrolyte and in fact — as is normally the case — if all the impurities consist of Mn, Si, Zn, Fe, Sn, Cu and Ag, the aluminium may be extracted from the anodic alloy and reprecipitated practically free from impurities. The high silicon content of the anodic alloy is required by the need to have a fairly fluid anodic layer even when the aluminium content has fallen considerably during electrolysis. This is necessary to avoid the excessive loss of aluminium from the boundary region of the alloy. If the aluminium which passes into solution is not replenished from the interior of the alloy at the same rate at which it is lost, the surface region will become impoverished and the other components of the alloy might pass into solution and ultimately contaminate the end product. The iron and titanium contents, however, must be as low as possible since these elements tend to raise the melting point of the anodic alloy.

The cathodic process is identical with that in the cell for the electrolytic preparation of aluminium. The discharge of a certain amount of Na^+ ion is unavoidable. The metallic sodium formed diffuses towards the surface where it is immediately oxidized with a consequent drop in current efficiency. For this reason – after a certain period of operation depending upon the amperage absorbed by the cell, the current density, the quantity of electrolyte initially present and the temperature of electrolysis – it is necessary to refurnish the electrolyte with the sodium which has been lost. The cell used is shown diagrammatically in Fig. X, 3.

The operation of this cell is inverted with respect to that used for the preparation of aluminium, in the sense that the lining of the cell is connected to the positive pole and the upper electrode is connected to the

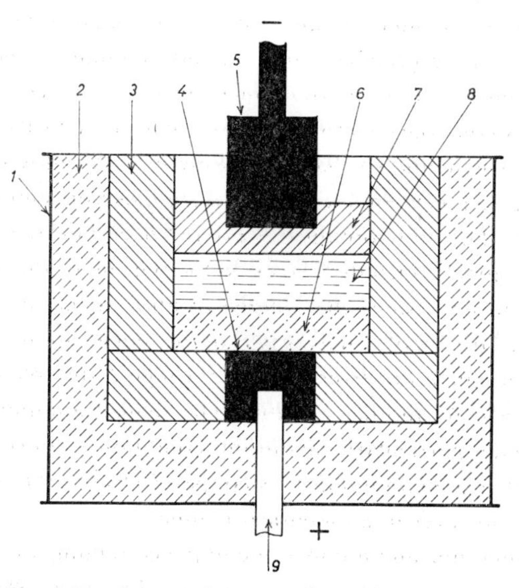

Fig. X, 3. Pechinay Al refining cell. 1. iron cell; 2. thermal insulation; 3. carbon lining; 4. graphite anode; 5. graphite cathode; 6. anodic alloy; 7. refined Al; 8. molten electrolyte

negative pole. Since electrolysis does not lead to an evolution of gas in this cell, and since the upper electrodes are connected to the negative pole, it is not necessary to prepare them from petroleum coke; normal graphite electrodes can be used instead which have the advantage, amongst others, of possessing a lower electrical resistance.

The cell is put into operation by first adding to it sufficient molten electrolyte to form a layer at least 10 cm in thickness. Such a thickness is necessary to avoid remixing of the cathodic product with the anodic alloy through the rather violent movements to which the two metallic layers are subjected; these movements are a consequence of the high temperature, high current density and powerful electromagnetic field. The molten anodic alloy is then gently poured into the cell. Finally, a layer of pure molten aluminium is poured on to the surface. The cathodes are then adjusted in height so that they are immersed only in the layer of pure aluminium. A crust of oxide forms immediately on top of the aluminium layer and protects it from further atmospheric oxidation. The cell is now ready for use.

The anodic alloy may be electrolyzed until its aluminium content falls to 20 %. At this point it must be enriched with fresh aluminium. For this, the current is disconnected and part of the copper-rich anodic layer is poured into a large crucible which already contains a new, molten, charge of aluminium or its alloy, which is to be refined. A sufficient amount is used so that the resulting mixture will have a density greater than that of the electrolyte and so that, on adding it to the alloy remaining in the cell, it will give a new anodic layer of a composition similar to that of the original one. The alloy obtained is poured back into the cell through a carbon-lined funnel which dips down to the bottom of the cell. The rise in level so produced – since the volume of alloy replaced in the cell is greater than that drawn off – makes part of the refined aluminium floating on the electrolyte, flow out from the cell through a suitable spout where it can be collected in another crucible. The refined aluminium is 99.8–99.9 % pure; under particularly favourable conditions, a purity of 99.98 % can be obtained. The main impurity is copper. Cells of other designs exist which function however in an identical manner.

With increasing time and number of completed refining cycles, the iron, titanium and other impurities in the anodic alloy increase in amount; this must therefore be periodically renewed when the concentrations of iron and titanium rise above the maxima indicated in Table X, 12. The

TABLE X, 13

ELECTRICAL VALUES FOR THE REFINING OF ALUMINIUM

Electric tension	6–7 V
Current density	140–150 A/dm^2
Current efficiency	*
Consumption	22–24 kWh/kg

* This value is not reported in the literature but is certainly very high.

exhausted anodic alloy is used for the recovery of the copper. Table X, 13 shows the relevant electrical values.

Other electrolyte systems have been proposed based on aluminium chloride and alkali chlorides, which would have the advantage of melting at a very much lower temperature (120–150° C). At this temperature, both the anodes to be refined and the cathodes of refined aluminium are solid; the aluminium does not give a smooth deposit on this cathode. To obtain smooth deposits it is necessary to add heavy metal salts to the electrolyte. The aluminium obtained, however, is considerably less pure than that given by the Hooper process; its purity varies between 99.5 and 99.9 %. These processes have not yet, however, found any large-scale industrial application.

6. Electrolytic Preparation of Magnesium

Magnesium also must be prepared electrolytically from one of its molten salts since its discharge electric tension is very much more negative than that of the H^+ ion so that it could not be separated from an aqueous solution unless it were present at saturation concentration and a very high current density were used. Even then, the magnesium would be obtained at a very low current efficiency and in a technologically useless state.

The most economical electrolytes would be the chloride and the sulphate but of these only the former can be used; sulphate in contact with metallic magnesium reacts according to the equation

$$Mg + MgSO_4 \rightarrow 2\,MgO + SO_2 \qquad (1)$$

The insoluble oxide would be formed and would cause further difficulties (see below). A process similar to that described for the preparation of aluminium was also attempted. As electrolyte, this used a mixture of magnesium and barium fluorides in equal parts with a small addition of sodium fluoride. Magnesium oxide was suspended in this electrolyte and about 0.1 % of the oxide dissolved. This process has, however, been virtually abandoned for various reasons such as the lower economy of the starting material and the greater working temperature. At present, almost all electrolytically prepared magnesium comes from cells in which the electrolyte is molten anhydrous magnesium chloride.

The electrolysis of anhydrous magnesium chloride does not show any particularly noteworthy aspects from the point of view of theoretical electrochemistry. The Mg^{2+} ions are discharged on the cathode to form metallic magnesium and the Cl^- ions are discharged on the anode, evolving gaseous chlorine. The primary anodic and cathodic reactions are not accompanied by secondary reactions to any appreciable extent.

The electrolyte consists of anhydrous magnesium chloride with the addition of sodium chloride, or other alkali or alkaline-earth chlorides, to lower the melting point, raise the conductance, increase the resistance to hydrolysis (reactions 3 and 4 below) and to make it possible to adjust the density. The electrolyte must be very pure to avoid lowering the efficieny. The most common impurities present in anhydrous magnesium chloride are water, sulphates, traces of iron and calcium. Table X, 14 shows the composition normally used for the electrolyte.

Anhydrous magnesium chloride, prepared for electrolysis, rarely contains less than 1–2 % water; even if this were not already present in the product of the dehydration process it would be readily absorbed from the atmosphere – even if in lesser amount – during the necessary operations of discharging the dehydration apparatus, transport, charging the electrolysis cell etc.

TABLE X, 14

COMPOSITION OF THE ELECTROLYTE FOR THE PREPARATION OF MAGNESIUM

Component	%
$MgCl_2$	20–30
NaCl	50–60
$CaCl_2$	~15

The water present in the magnesium chloride is partially removed from the cell by evaporation and partially decomposed by the current flowing through the cell before the electrolysis of the magnesium chloride begins. The decomposition electric tension of this is certainly greater than that of water which consumes a considerable amount of energy. For example, a cell charged with 1,500 kg of magnesium chloride containing 2 % water (the water lost almost instantaneously by evaporation, when the cell is charged, is not considered) would operate for at least 18 hours at 5,000 A before the magnesium starts to deposit; and this would apply for every successive charge. Moreover, water added with successive charges could react with metallic magnesium which had already been separated, according to the reaction

$$Mg + H_2O \rightarrow MgO + H_2 \tag{2}$$

As far as current efficiency is concerned, this reaction is perfectly analogous to the direct electrolytic decomposition of the water in that the two equivalents of current consumed in the separation of 1 g atom of magnesium will be lost if the magnesium is converted back into oxide. The effect is just as if 1 mole of water were electrolyzed and consumed the two equivalents of electricity itself. Reaction (2) is, however, even more damaging because the magnesium oxide formed cannot be used for electrolysis and there is thus a loss of electrolyte. The water may also react with the magnesium chloride by the two reactions

$$MgCl_2 + H_2O \rightleftharpoons Mg(OH)Cl + HCl \tag{3}$$

$$MgCl_2 + H_2O \rightleftharpoons MgO + 2\,HCl \tag{4}$$

forming the oxychloride and the oxide respectively, both of which cannot be used for electrolysis. However, depending on the characteristics of the magnesium chloride used initially (fused lumps or powdered) a more or less considerable proportion of the water will evaporate before it has time to pass truly into solution in the molten electrolyte and give rise to reactions (2), (3) and (4). This is particularly so if the charge is added to a suitably constructed anodic compartment so that the current of hot gaseous chlorine facilitates the evaporation of water and carries off the steam so as to hinder hydrolysis.

Sulphate ions are also damaging for the same reason that it is not possible to carry out electrolysis in a molten magnesium sulphate electrolyte because of reaction (1). The sulphate content must not exceed 0.05 %.

The formation of magnesium oxide in the bath by reactions (3) and (4) will further lower the current efficiency by hindering the aggregation of the droplets of metallic magnesium by covering them with a very fine layer of powder. Part of the separated metallic magnesium will then be trapped in the mass of magnesium oxide whose density carries it to the bottom of the cell. In fact, if magnesium oxide from the electrolytic cell is analyzed, the magnesium content will usually be greater than the theoretical one through the presence of metallic magnesium. A typical analysis of slag from the bottom of the cell is shown in Table X, 15. Small additions of sodium or calcium fluoride diminish the formation of dispersed metallic magnesium by acting as solvents for dispersed magnesium oxide.

TABLE X, 15

ANALYSIS OF A SLAG FROM THE BOTTOM OF A MAGNESIUM ELECTROLYTIC CELL

Component	%
MgO	18
Mg	5
Heavy metals	1
Flux	76

Iron, and any other heavy metal impurities – which are however normally absent – are damaging in that they would contaminate the final product of the electrolysis. Amongst other impurities, boron is particularly deleterious because by separating at the cathode, it would give metallic borides. These would not only contaminate the product but would have a dispersing action on the cathodic magnesium and would increase its density: both difficulties to be avoided (see below). The presence of other alkali, or alkaline-earth, ions is not disadvantageous since they are not deposited on the cathode until their concentrations reach very high values; such concentrations are never attained in electrolytes for the electrolysis of magnesium, and lie at about 20 % for calcium chloride, 70 % for sodium chloride and 85 % for potassium chloride.

If very pure and truly anhydrous magnesium chloride is used, the current efficiency is very close to the theoretical value and the metal obtained by the electrolysis is very pure containing 99.9 % Mg, or even more.

Although by suitable manipulation of the components of the electrolyte, it is possible to give it a density less than that of molten metallic magnesium, so that the magnesium would be deposited on the bottom of the cell, it is preferable to allow the molten metal to float on the surface so as to avoid mixing it with the oxide which accumulates at the bottom. This oxide would prevent the drops of magnesium, falling from the cathode, from uniting into a mass. Moreover, when the magnesium floats on the surface, it is easier to discharge the cell whose construction is thereby simplified. By using fairly pure raw materials, and by taking advantage of the automatic elimination of the magnesium oxide which sinks to the bottom and of the dissolving, and hence in a certain sense washing action, of the electrolyte and of the absence of any cathodic deposition of other impurities, it is possible to obtain an electrolytic product which is extremely pure containing 99.7–99.9 % magnesium or even better. The principal impurities are aluminium, silicon and iron. Sometimes it is possible to find occluded electrolyte as impurity. Such an impurity, which is often not visible to the naked eye, is very damaging; expressed in terms of the chloride present, it must not exceed 0.005 %.

If the cathodic product is not sufficiently pure it is further refined by sublimation under reduced pressure (0.15–0.5 mm Hg) at a temperature just below the melting point of magnesium (651° C). Under these conditions, it is possible to obtain 99.99 % magnesium from a metal containing about 90 % magnesium.

The decomposition electric tension of magnesium chloride is not known for certain. It is about 2.5 V and is thus less than that of other possible components of the electrolyte. The electric tension actually applied to the cell is, however, considerably greater: 6–7 V. This is because the cell utilizes partially, or totally, electrical energy to keep the electrolyte at the working temperature. The cells normally employ anodic current densities of about 500–1,000 A/dm^2 but the cathodic current density is considerably less: 100 A/dm^2; but these values are not of decisive importance for the efficiency of the electrolytic process. The current efficiency cannot be determined exactly. It depends in the main, as mentioned before, on the state of hydration of the magnesium chloride at the moment of charge. The working temperature must be rather greater than the melting point of magnesium so that the cells operate at about 700–750° C. At this temperature, molten magnesium does not show a strong tendency to react with the anodic chlorine so that the loss from this is quite negli-

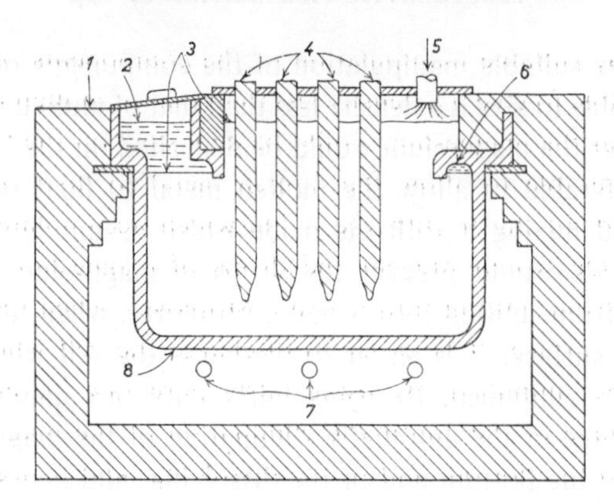

Fig. X,4. Dow Mg cell. 1. refractory casing; 2. molten Mg; 3. fused $MgCl_2$ level; 4. graphite anodes; Cl_2 outlet; 6. molten Mg trap; 7. gas burner openings; 8. cast steel container

gible. On the other hand, the molten magnesium is protected from atmospheric oxidation by a thin layer of molten electrolyte which always rests on the surface.

A typical cell is shown in Fig. X,4, in which the heating is partially supplied externally. The cell is constructed in refractory material which is resistant to molten magnesium chloride and is cased with iron to increase the mechanical strength. The cover is of iron which is lined with a refractory material resistant to the action of gaseous chlorine. The cathodes are made of iron and if the temperature is kept sufficiently low they do not dissolve to any appreciable extent. The anode is made of carbon or graphite. The consumption lies around 17–20 kWh/kg depending upon the effective applied tension, *i.e.* depending on whether the heating is principally external or principally or even totally electrical. The electrical values are shown in Table X,16.

TABLE X,16

ELECTRICAL VALUES FOR THE PREPARATION OF MAGNESIUM

Electric tension	6–7 V
Cathodic current density	~ 100 A/dm²
Anodic current density	500–1,000 A/dm²
Current efficiency	0.75–0.85 *
Consumption	17–20 kWh/kg

* This value will depend on the water content of the electrolyte.

7. Preparation of Anhydrous Magnesium Chloride

The raw materials for the preparation of the electrolyte are obtained from the following sources:

1. Soluble magnesium minerals: bischofite ($MgCl_2 \cdot 6\,H_2O$), carnallite ($MgCl_2 \cdot KCl \cdot 6\,H_2O$) and tachidrite ($MgCl_2 \cdot CaCl_2 \cdot 12\,H_2O$).
2. Easily converted minerals: magnesite ($MgCO_3$), dolomite ($MgCO_3 \cdot CaCO_3$) and brucite ($Mg(OH)_2$).
3. Magnesium-rich saline springs.
4. Seawater.

The preparation of anhydrous magnesium chloride from these raw materials may be carried out by either dry or wet processes.

The first step in the dry process is the preparation of the oxide MgO. If the raw material consists of sufficiently pure magnesite or brucite, it is calcined. If calcium or potassium salts are present these are eliminated by various procedures. The oxide is then treated with carbon and chlorine in furnaces at temperatures above the melting point of the anhydrous chloride $MgCl_2$ (708°C). The reaction

$$MgO + C + Cl_2 \rightarrow CO + MgCl_2$$

takes place and gives the anhydrous chloride directly.

In the wet process, the pure magnesium chloride hexahydrate is obtained by a variety of procedures which generally consist of fractional solution and crystallization together with a possible treatment with magnesium hydroxide to eliminate impurities of silica, iron, aluminium and other metals and with a treatment to recover bromine if it is present. The hexahydrate must then be dehydrated and this is the most delicate phase of the process of preparing anhydrous magnesium chloride. Dehydration to a water content corresponding to the formula $MgCl_2 \cdot 2\,H_2O$ is easy; however, it is not so simple to convert the dihydrate into the anhydrous salt because of the hydrolytic reactions (3) and (4), which are particularly liable to occur at the higher temperatures required for complete dehydration, so that drying is usually stopped at a water content of 1–1.5 %.

To diminish the effects of the hydrolytic reactions, the magnesium chloride is treated with sodium chloride and possibly small quantities of ammonium chloride, or else the final dehydration is carried out in an atmosphere of hydrogen chloride. The anhydrous magnesium chloride, as prepared for electrolysis, normally contains 10 % oxide which however does not cause difficulties during the electrolysis.

It will be interesting to briefly mention the process for the extraction of magnesium chloride from seawater, which has recently been applied in America; this will show how, under suitable conditions, even very dilute solutions may be used. The seawater is passed through screens to remove suspended solids and is then treated with chalk at a rigorously controlled pH to obtain a good precipitation of magnesium hydroxide. After a preliminary decantation, the magnesia milk is filtered to give a cake of the hydroxide. This is dissolved in 10 % hydrochloric acid, obtained by burning natural hydrogen in an atmosphere of chlorine which is in turn produced by the electrolysis of magnesium chloride or alkali chlorides. The dilute solution of magnesium chloride is first concentrated and then dehydrated in desiccators of various types which finally give pure, almost anhydrous, magnesium chloride (with 1.5 % H_2O). It is necessary to process 800 tons of seawater to obtain 1 ton of magnesium.

The process for the extraction of magnesium from seawater can be economical only if the auxiliary raw materials can be obtained locally at low prices. The Dow Chemical Company's plant at Freeport in Texas uses deposits of oyster shells, which are found locally, for the preparation of the calcium oxide. The hydrochloric acid is produced from natural gas and chlorine obtained from the $MgCl_2$ electrolysis and from other industrial electrolyses.

8. Electrolytic Preparation of Sodium

The electrolytic preparation of sodium represents one of the most delicate electrometallurgical processes because of the difficulties which must be overcome in view of the high reactivity of the product at the temperature at which it is formed.

It is not possible to separate metallic sodium directly from aqueous solutions of its salts on an ordinary metallic electrode (*cf.* p. 514 *et seq.*) because its discharge tension is more negative than that of hydrogen. Even if a mercury cathode is used, the overtension of hydrogen on it would not alone be sufficient to compensate for the difference between the discharge tensions of hydrogen and sodium. The discharge of the Na^+ ion on the mercury cathode is made possible by the fact that sodium and mercury form intermetallic compounds (*cf.* p. 535 *et seq.*); these are formed exothermically with a relatively high heat of formation. Moreover, these compounds are soluble in excess of the cathodic mercury so that the effective discharge tension of the Na^+ ion is sufficiently depolarized for the discharge to occur. To prepare metallic sodium it is then necessary to separate it from the amalgam; this is a very expensive procedure so that the overall cost of preparing metallic sodium by this route is prohibitive. It is, therefore, preferable to prepare sodium by the electrolysis of a molten electrolyte and all the metallic sodium now produced is made in this way.

The first difficulty encountered concerns the temperature of the electrolysis in relation to the electrolyte used. Sodium is a metal with relatively low melting and boiling points (97.5° C and 880° C respectively). This makes it necessary to work at the lowest possible temperature to avoid loss of sodium by evaporation and by reaction with anodic products and atmospheric gases, since its reactivity increases with the temperature. The most suitable electrolyte in this sense would be anhydrous sodium hydroxide which melts, when pure, at 318° C. For this reason it has been, and still is, used as an electrolyte for the preparation of metallic sodium. It is possible to prepare metallic sodium by the electrolysis of its less costly salts: chloride, carbonate, nitrate, borate etc. However, electrolysis with a saline electrolyte is carried out only with the chloride since its cost as a raw material is much lower than that of the hydrate. At the present time, the electrolysis of molten sodium chloride is considerably more important than that of the hydroxide.

(I) *Electrolysis of the Hydroxide*

Since this process has now been virtually abandoned, only a brief discussion will be given of the primary and secondary reactions, and of the conditions required; and these will clearly show the difficulties involved which have led to the preferential use of the chloride process.

The primary reactions during the electrolysis of sodium hydroxide are

$$Na^+ + e^- \rightarrow Na \tag{1}$$

at the cathode, and

$$4\,OH^- \rightarrow 2\,H_2O + O_2 + 4\,e^- \tag{2}$$

at the anode.

The water formed at the anode makes another simultaneous primary cathodic reaction possible

$$2\,H^+ + 2\,e^- \rightarrow H_2 \tag{3}$$

The sodium formed at the cathode, however, is considerably soluble in the electrolyte. This solubility diminishes as the temperature rises but since the coefficient of diffusion also increases with the temperature, the overall diffusion is greater at higher temperatures. The metallic sodium which diffuses into the electrolyte may react both with the water and with the oxygen formed at the anode, according to the following secondary reactions.

$$2\,Na + 2\,H_2O \rightarrow 2\,NaOH + H_2 \quad (4)$$

$$2\,Na + O_2 \rightarrow Na_2O_2 \quad (5)$$

$$2\,Na + 2\,Na_2O_2 \rightarrow 2\,Na_2O \quad (6)$$

$$2\,Na_2O + 2\,H_2O \rightarrow 4\,NaOH \quad (7)$$

The group of secondary reactions (4), (5), (6) and (7) consume sodium which has already been separated and so lower still further the current efficiency, which is already low because of concomitant primary cathodic reaction (3). Reactions (4) and (5) are also damaging for another reason. Reaction (4) leads to the evolution of hydrogen in the anodic region. This hydrogen on mixing with atmospheric or anodic oxygen forms a combustible mixture which detonates continuously. Sodium reacts according to equation (5) with atmospheric or anodic oxygen and also causes repeated detonations so that a continuous and careful supervision of the operation of the cell is necessary.

The decomposition electric tension of sodium hydroxide at the working temperature is about 2.2 V but the decomposition electric tension of water at the same temperature is about 1.4 V. Thus, the decomposition of water formed at the anode, and of water which is always present in the electrolyte initially through its hygroscopic nature, is inevitable. Thus, during electrolysis, the sodium hydroxide is completely decomposed into its elements: sodium, hydrogen and oxygen. Hence the decomposition of one mole of hydroxide requires 2 **F** of electricity. The current efficiency cannot theoretically exceed 0.5 unless some water evaporates from the molten mass. In practice, it is difficult to exceed a current efficiency of 0.4 because of secondary reactions which consume sodium, unless the cell is constructed in such a way that the greater part of the anodic water is removed by evaporation (see below).

The amount of metallic sodium consumed in secondary reactions increases rapidly as the temperature rises above the melting point of the electrolyte. At a temperature 25° C above the melting point of the electrolyte, secondary reactions consume virtually all the separated sodium and the current efficiency tends rapidly towards zero. For this reason, the working temperature range for the electrolysis is very restricted – 15–20° C – and must be kept fairly rigorously within these limits; if the temperature were to fall too far, the electrolyte would become too viscous and a fairly rigid foam would be formed around the cathode. This would interfere with the aggregation of the sodium droplets. If the temperature were to rise too high, the surface tension of the liquid sodium would fall too much and would again interfere with the aggregation of the droplets and, moreover, the current efficiency would fall excessively. Within this temperature range, the practical current efficiency varies between 0.45 and 0.36. Changes in temperature can be followed even with a voltmeter placed across the ends of the cell. If the temperature falls the applied electric tension will rise and it will fall again if the temperature rises.

The melting point of pure, anhydrous, sodium hydroxide is 318° C, but since it is always somewhat impure and never perfectly dry, its melting point may fall as low as 300° C. In practice, therefore, the working temperature varies between 310° and 320° C.

The electrolyte used must be very pure and as dry as possible so as not to waste electricity by decomposing water before the electrolysis of the sodium hydroxide begins, and to avoid too great a formation of foam. In particular, the electrolyte must be free of chlorides because otherwise the anodically produced chlorine would attack the metallic parts of the cell, which are normally iron, in contact with the electrolyte to form ferric chloride. In contact with the water this would hydrolyze to ferric oxide which would contaminate the electrolyte, might contaminate the cathodic product and would obstruct the diaphragm. The sodium hydroxide used for the electrolyte is normally obtained therefore, by the electrolysis of the chloride with a mercury cathode (*cf.* p. 535 *et seq.*). The sodium produced has a lower density than the electrolyte and floats on its surface.

(II) *Electrolysis of the Chloride*

The electrolysis of the molten chloride has certain peculiarities which are noteworthy from the standpoint of theoretical electrochemistry. Peroxides, sodium chlorate and sodium chloride may be formed by secondary reactions between metallic sodium dissolved in the electrolyte and dif-

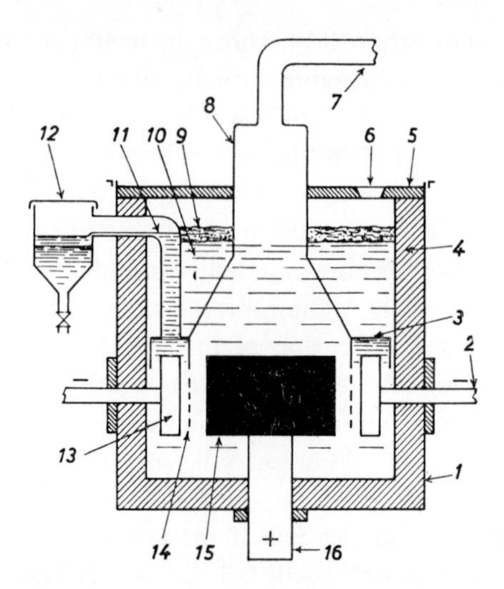

Fig. X,5. Dow Na cell. 1. iron container; 2. cathodic connection; 3. Na collecting space; 4. refractory; 5. cover; 6. opening for NaCl feeding; 7. Cl_2 outlet; 8. Cl_2 collecting space; 9. solidified electrolyte; 10. molten electrolyte; 11. molten Na; 12. Na collecting pot; 13. cathode; 14. anodic connection

fusing to the anode, and anodic chlorine or atmospheric oxygen. Such reactions lower the current efficiency. Carbon monoxide may also be formed at the anode by reaction between the anodic carbon and atmospheric oxygen; similarly, phosgene may be produced by reactions between the carbon monoxide and the anodic chlorine. In view of its toxicity this latter gas is particularly dangerous.

The decomposition electric tension of molten sodium chloride at the melting point is about 3.24 V but the process occurs with considerable overtensions. In practice, the greatest difficulty encountered in this process is caused by the high melting point of sodium chloride (800° C). This means that the cell must operate at a temperature of 850–900° C to keep the molten electrolyte sufficiently fluid. But the boiling point of metallic sodium at atmospheric pressure is 880° C so that in practice it would virtually all vaporize.

Another difficulty arises from the high fluidity of metallic sodium at such a high temperature. The solubility of metallic sodium in the electrolyte must also be allowed for; this rises rapidly at 500–600° C and even at 700° C the formation of clouds is so great as to make the deposition of metallic sodium almost impossible. It is however possible to lower the working temperature of the electrolyte by diminishing its melting point. By suitable additions, such as sodium carbonate, it is possible to lower the working temperature of the cell to 600° C; but this temperature is still too high because the sodium would have a considerable vapour pressure and would burn immediately on contact with atmospheric oxygen. It is therefore necessary for the separated sodium to be cooled within the cell to a fairly low temperature before it comes into contact with the atmosphere.

In cells for the electrolysis of chloride, the cathodes may be of iron or copper and the anodes must be of graphite. Moreover, the cell must be constructed of materials which will resist molten sodium chloride.

A typical cell – the Dow cell – is shown in Fig. X, 5. In this, the heating is electrical so that with the decomposition electric tension of the chloride, which is considerably greater than that of the hydroxide; with the greater overtensions in the electrolysis of the chloride than in that of the hydroxide; and with the greater consumption of energy to keep the cell at the much higher temperature; the applied electric tension must be very much higher than that for cells electrolyzing the hydroxide. The greater current efficiency in the electrolysis of the chloride compensates for the higher electric tension, however, so that the energy consumed in producing a kg of sodium is about the same for both cells. The chloride process becomes advantageous through the lower price of the raw material and the smaller number of secondary difficulties. Table X, 17 shows electrical values for these two processes.

TABLE X, 17

ELECTRICAL VALUES FOR THE PRODUCTION OF SODIUM

	Electrolyte	
	NaOH*	NaCl**
Electric tension	4–5 V	~ 10 V
Cathodic current density	200 A/dm²	—
Anodic current density	150 A/dm²	—
Current efficiency	0.36–0.45	0.75–0.8
Consumption	14–15 kWh/kg	14–15 kWh/kg

* Externally heated Castner cell.
** Dow cell.

CHAPTER XI

PRACTICAL PRIMARY CELLS AND STORAGE BATTERIES (ACCUMULATORS)

by

GIUSEPPE BOMBARA[1] and GIULIO MILAZZO

1. Introduction

A practical primary cell is a galvanic element designed to produce electrical energy from a chemical reaction. Their origin dates from 1800 with the work of Alessandro Volta, and for a long time cells were the only available source of significant amounts of electricity until the invention of the dynamo by Antonio Pacinotti (1860), since frictional machines although providing high electric tensions could not give any notable amounts of electricity. The fundamental electrochemical studies of Faraday, Davy and others were carried out using galvanic cells. Volta's first 'pile' or cell, consisted of discs of copper and zinc separated by a layer of a damp absorbing material. Under the action of the oxygen and carbon dioxide of the atmosphere, small quantities of bicarbonate were formed in solution so that the electric tension of the cell on open circuit was that of the element $Cu/Cu^{2+}/Zn^{2+}/Zn$. However, as soon as the circuit was closed, the traces of copper in solution were deposited and the electrochemical process at the positive pole became

$$2\,H^+ + 2\,e^- \rightarrow H_2$$

This cell proved to be of no practical use. The first cell which was able to function regularly was that of Daniell and this was followed by other combinations.

Galvanic elements which have been suggested and also used for practical purposes and called primary cells are not many in number. The most important are shown in Table XI, 1. All these cells, with the possible exception of the Meidinger cell, have now disappeared from practical use. Their main defect lies in the presence of two electrolytes, which involves a diaphragm or a special construction to prevent the diffusion of one into

[1] Laboratori Riuniti Studi e Ricerche E.N.I., S. Donato Milanese.

TABLE XI, 1

TENSIONS OF CELLS WITH DOUBLE ELECTROLYTES

Cell		*Composition*			*Electric tension* (V)
Bunsen	C	HNO_3 *fuming*	$H_2SO_4 + 12\ H_2O$	Zn *amalg.*	1.94
Bunsen	C	HNO_3 $d = 1.38$ g/cm³	$H_2SO_4 + 12\ H_2O$	Zn *amalg.*	1.86
Poggendorf	C	$12\ K_2Cr_2O_7 + 25\ H_2SO_4 + 100\ H_2O$	$H_2SO_4 + 12\ H_2O$	Zn *amalg.*	2.00
Poggendorf	C	$12\ K_2Cr_2O_7 + 100\ H_2O$	$H_2SO_4 + 12\ H_2O$	Zn *amalg.*	2.03
Daniell	C	$CuSO_4 \cdot 5\ H_2O$ *sat. soln*	$H_2SO_4 + 4\ H_2O$	Zn *amalg.*	1.06
Daniell	C	$CuSO_4 \cdot 5\ H_2O$ *sat. soln*	NaCl *aq.*	Zn *amalg.*	1.05
Daniell	C	$CuSO_4 \cdot 5\ H_2O$ *sat. soln*	$ZnSO_4 \cdot 7\ H_2O$ 5 %	Zn *amalg.*	1.08
Grove	Pt	HNO_3 *fuming*	$H_2SO_4 + 12\ H_2O$	Zn *amalg.*	1.93
Grove	Pt	HNO_3 $d = 1.33$ g/ml	$ZnSO_4$ *aq.*	Zn *amalg.*	1.66
Grove	Pt	HNO_3 $d = 1.33$ g/ml	H_2SO_4 $d = 1.136$ g/ml	Zn *amalg.*	1.79
Grove	Pt	HNO_3 $d = 1.33$ g/ml	NaCl *aq.*	Zn *amalg.*	1.88
Meidinger	Cu	$CuSO_4$ *sat. soln*	$ZnSO_4$ *sat. soln*	Zn *amalg.*	1.18

the other with the consequent exhaustion of the cell due to the chemical reaction occurring without producing electrical energy.

The use of cells as sources of electricity is today limited to batteries for portable lamps, radio receivers and transmitters and for rocket apparatus. Almost all other applications in which they were once important (telecommunications, signalling devices, electrical bells, lighting for vehicles etc.) have gradually adopted the use of continuous current obtained by rectifying the alternating current of the mains distribution, or accumulators.

A good cell for practical uses must have as high and constant a terminal electric tension as possible throughout its use. It must also convert the largest possible fraction of the reactional free enthalpy into externally useful energy. In other words, the efficiency of a cell for practical use is determined by the electric tension of the galvanic element used, the internal resistance and its capacity for continuous and intermittent discharge. With cells in practical use it is not possible to convert all the reactional free enthalpy into external electrical work both because the internal resistance of the cell is not zero, so that a certain fraction of the electric tension is lost by the Joule effect within the cell, and because the chemical reactions are never perfectly reversible particularly when, as in many cases, one of the primary processes is the discharge of the H^+ ion. Failure of reversibility immediately leads to the appearance of overtensions. Two conditions are necessary for a high terminal tension: a high electric tension and a low internal resistance (*cf.* p.98 *et seq.*). For a constant electric tension the cell must have only small overtensions. Only when other substantial advantages are present it is possible to partially sacrifice the constancy of the electric tension, as for example with Leclanché or dry cells. In this respect cells may be classified as constant or inconstant.

Cells show two types of overtensions: polarization of concentration and transfer overtensions. During the operation of a Daniell cell, for example, the solution of zinc sulphate becomes more concentrated. Since the electric tension of a metallic electrode of the first kind depends on the concentrations of the electrochemically active metallic ions, there will be an increase in the electric tension of the zinc electrode during the operation. Since the copper is the positive pole of the Daniell cell and the zinc is the negative, the electric tension of the cell will fall. This phenomenon constitutes the polarization of concentration. It has a small effect since it will alter the electric tension of any electrode by $0.059/z$ V at room temperature for a change in concentration of 1 : 10.

The transfer overtension is much more important; it is caused by the fact that the chemical reactions occurring in the two half cells have a limited velocity particularly for certain processes such as the evolution of hydrogen, the dissolution or deposition of iron etc. Thus, even with low current intensities the reaction velocity may rapidly become so small as to markedly lower the electric tension. It is thus opportune to eliminate or diminish the effects of transfer overtensions using suitable substances which will eliminate or accelerate the retarding processes. Substances added for this purpose are called 'depolarizers'.

The fundamental advantage of cells with only one electrolytic solution lies in the fact that since diffusion phenomena between different solutions are not possible, the chemical reactions which lead to a self consumption of the active substance of the cells and to its exhaustion without being used, are also not possible or are reduced to a minimum. Such phenomena of exhaustion are also called spontaneous discharges. The absence of such phenomena in cells of this type give them a relatively long shelf life. Cells which are today in practical use all have only one solution and their characteristics are shown in Table XI,2. Once the active material of a cell has been exhausted it will no longer work and it is not possible to reverse the electrochemical process, which gave rise to the electrical energy, to restore the cell to its initial state. Accumulators, however,

TABLE XI,2

COMMERCIAL CELLS WITH ONLY ONE ELECTROLYTIC SOLUTION

	Leclanché	*Dry Leclanché*	*Lalande*	*Le Carbone*	*National Carbon Co.*	*Ruben–Mallory*
ositive pole	Zn *amalg.*	Zn *amalg.*	Zn *amalg.*	Zn *amalg.*	Zn *amalg.*	Zn *amalg.*
egative pole	C	C	C	C	C	Hg
epolarizer	MnO_2	MnO_2	CuO	Air	Air	HgO
lectrolyte	NH_4Cl*	NH_4Cl	NaOH	NH_4Cl	NaOH	KOH
	10 — 20 %	10 — 20 %	20 — 25 %		20 %	40 %
ean electric tension V. abs. value)	1.1	1.1	0.70	1.0	~ 1.2	~ 1.3
apacity Ah	~ 30	1 — 2	300 — 1,000	~ 125	~ 500	0.4 — 3.2
apacity Wh	~ 33	1 — 2	210 — 700	~ 125	~ 600	0.5 — 4.1
verage life (years)	2	1.5	indeterm.	indeterm.	indeterm.	> 3

* With the possible addition of other electrolytes, such as magnesium chloride

may be reversed in their operation by passing an electrical current through them so as to flow inside the accumulator from the positive to the negative pole. The accumulator behaves as an electrolytic cell in which the electrochemical process, which occurs during discharge, is reversed; it is thus restored to its initial state and can then give up again part of the energy which it has absorbed. This is the reason for the name 'accumulator', since it stores energy in a chemical form during the period of charge whilst during the discharge it yields electrical energy at the expense of this stored chemical energy.

Theoretically any reversible reaction could be used to construct an

TABLE XI,3

THE COMPOSITION OF ACCUMULATORS

Type	*Composition*	*Mean electric tension* (V)
Lead	$Pb/PbO_2/H_2SO_4$ ($d = 1.20$)/Pb	~ 2.00
Edison	Steel/$Ni_2O_3 \cdot xH_2O$/KOH (~ 20 %) + LiOH/Fe/Steel	~ 1.25
Jüngner	Steel/$Ni_2O_3 \cdot xH_2O$/KOH (20–25 %) + LiOH/Cd/Steel	~ 1.2
Ag–Zn	Ag/Ag_2O_2/KOH (~ 40 %) + K_2ZnO_2/Zn/Ag	~ 1.5

accumulator, but in practice, the necessity of having only one electrolytic solution severely limits the choice of possible chemical reactions to be used. Elements which are so used are shown in Table XI, 3.

2. Leclanché Cells

The most widely used cell at the present day is without doubt the Leclanché or one of its modifications; other types have been developed from this. The original Leclanché cell had the composition

$$^{+}\text{Carbon}/MnO_2/NH_4Cl, \sim 10\,\%/Zn^{-}$$

The reaction at the negative pole (in this case the anode) is the formation of Zn^{2+} ions

$$Zn - 2\,e^- \rightarrow Zn^{2+}$$

The reactions at the positive pole (in this case the cathode) are diverse and complex, but the primary electrochemical reaction is certainly

$$MnO_2 + 4\,H^+ + 2\,e^- \rightarrow Mn^{2+} + 2\,H_2O$$

with the manganese dioxide which acts as the depolarizer.

The electric tension of the manganese dioxide electrode is not perfectly defined, particularly at the beginning of its operation, and depends to a large extent on the mode of preparation of the manganese dioxide and the treatment which it has undergone, as well as on variations in the relative concentrations of active materials, which are inevitable in large scale commercial production.

This dependence must be at least partially related to the capacity to adsorb atmospheric oxygen as a second depolarizer. Atmospheric oxygen, moreover, has been shown[1] to be the main cause of spontaneous discharge, particularly in the period of initial aging. The oxygen is reduced to H_2O_2 at the metallic electrode and the H_2O_2 which reaches the cathode in the presence of NH_4Cl reduces the MnO_2.

Another factor which leads to changes in the electric tension of the manganese dioxide electrode is the possible variation in the physicochemical constitution of the oxides MnO_2, Mn_2O_3 and Mn_3O_4; the latter in particular may be formed at the end of the discharge. An accurate analysis of the overall process of the Leclanché cell was first attempted by Scarpa[2], who particularly distinguished cells with artificial and with activated manganese

[1] W. C. Vosburgh, D. R. Allenson and S. Hills, *J. Electrochem. Soc.*, 103 (1956) 91.
[2] O. Scarpa, *Ricerca sci.*, 12 (1941) 5.

dioxide. The former gave on the average, initial electric tensions of 1.53 V whereas the latter gave initially 1.67 V; with highly activated MnO_2 values up to about 1.9 V were obtained. The isothermal temperature coefficient for cells with natural MnO_2 is on the average $+ 0.00085 \pm 0.00005$ V/° C. Using this figure with the Gibbs–Helmholtz equation, it is possible to calculate that at a room temperature of 20° C, $\Delta E/n$ will have a value of 30,030 cal. (where n is the number of F involved and is assumed to be 2).

According to Scarpa, the possible overall reversible reactions are as follows:

$$Zn + 2\,NH_4Cl + 2\,MnO_2 \rightarrow ZnCl_2 \cdot 2\,NH_3 + Mn_2O_3 + H_2O; \quad n=2 \qquad (1)$$

$$2\,Zn + 4\,NH_4Cl + 3\,MnO_2 \rightarrow 2\,ZnCl_2 \cdot 2\,NH_3 + Mn_3O_4 + 2\,H_2O; \quad n=4 \qquad (2)$$

$$Zn + 2\,NH_4Cl + MnO_2 \rightarrow ZnCl_2 \cdot 2\,NH_3 + MnO + H_2O; \quad n=2 \qquad (3)$$

Of these, only reaction (1) gives values in agreement with the changes in internal energy calculated from the thermochemical and electrochemical data, which give respectively 59,500 and 60,060 cal.; so that reaction (1) can be considered as satisfactory for the interpretation of electrochemical processes for cells with natural nonactivated MnO_2. However, since reaction (1) indicates a reduction product in which the manganese is trivalent, the formulation of this as Mn_2O_3 is not nowadays considered as correct. Numerous experimental studies for the identification of the reaction products at the positive pole have shown by X-ray diffraction, that (MnOOH) and ($ZnO \cdot Mn_2O_3$) are both present as the solid products of the reaction. Soluble manganese is also found in the catholyte.

On the basis of such observations, the process at the positive pole of the Leclanché cell may be represented as follows[1]. Two types of reaction occur. One primary reaction, already mentioned, is truly electrochemical

$$MnO_2 + 4\,H^+ + 2\,e^- \rightarrow Mn^{2+} + 2\,H_2O \qquad (4)$$

and thus occurs only during the actual withdrawal of current. A further two chemical reactions occur: one between Mn^{2+}, which is formed as a soluble manganous salt in the electrolyte, and the residual MnO_2

$$Mn^{2+} + MnO_2 + 2\,OH^- \rightarrow 2\,MnOOH \qquad (5)$$

to form manganite, and the other parallel reaction involving the Zn^{2+} ions from the anode also

$$Mn^{2+} + MnO_2 + 4\,OH^- + Zn^{2+} \rightarrow ZnO \cdot Mn_2O_3 + 2\,H_2O \qquad (6)$$

[1] N. C. Cahoon, R. S. Johnson and M. P. Korver, *J. Electrochem. Soc.*, 105 (1958) 296.

forming heterolite. Both reactions (5) and (6) occur independently of the discharge.

Other products which have been identified such as $ZnCl_2 \cdot 2\,NH_3$ are essentially formed during operation at intense discharge. Under these conditions free ammonia is liberated, probably by the reaction of ammonium ions with the excess of cathodic hydroxyl ions

$$NH_4^+ + OH^- \rightarrow NH_3 + H_2O \tag{7}$$

The successive reaction of the ammonia with the zinc chloride forms the poorly soluble zinco-amine

$$ZnCl_2 + 2\,NH_3 \rightarrow ZnCl_2 \cdot 2\,NH_3 \tag{8}$$

Under normal conditions, however, the zinco-amine is not formed[1] partly because its solubility is greater than that of the product $ZnO \cdot Mn_2O_3$ and because this latter is formed at a lower pH than the amine.

Systematic experiments[2] have demonstrated that equations (4), (5) and (6) represent the exact reaction mechanism at the positive pole. The calculation of the chemical energy from the quantity of chemical products found in the discharged cell is in very good agreement with the value obtained by electrical measurements. Scarpa also calculated for the cell with activated manganese dioxide, using the initial electric tension, a value for the change in internal energy which was too high for that derived from thermochemical data, so that it is not possible to describe the operation of the cell even approximately by reaction (1).

Considering the atmospheric oxygen as a depolarizer, the reaction

$$Zn + 2\,NH_4Cl + \tfrac{1}{2}\,O_2 \rightarrow ZnCl_2 \cdot 2\,NH_3 + H_2O \tag{9}$$

should give an electric tension of 1.99 V. This value is not too far removed from the experimental one when it is considered that an electrode dependent upon reaction (9) would be markedly irreversible and that, in particular, oxygen cannot be considered as simply absorbed but rather as *adsorbed* with an adsorption energy which cannot be neglected. This energy will lower the resultant electric tension particularly if it is measured with a current flow which is not infinitely small. Thus, the initial reaction would be (9) but as this proceeds the electric tension would fall and reaction (1) would become increasingly important. During the operation, the solution around the positive pole becomes alkaline because of the

[1] L. C. Copeland and F. S. Griffith, *Trans. Electrochem. Soc.*, 89 (1946) 495.
[2] N. C. Cahoon, R. S. Johnson and M. P. Korver, *J. Electrochem. Soc.*, 105 (1958) 296.

excess of OH^- ions produced by the primary reaction (4). This leads to a polarization of the electrode with a diminution of the electric tension.

The working C/MnO_2 electrode cannot be considered in any way as in thermodynamic equilibrium and for most of the working periods not even as in a steady state; hence, any application of Nernst's equation or similar thermodynamic relationship is completely invalid because, amongst other reasons, the Mn^{2+} and Zn^{2+} ions are not actually free but are present as hydrated amine complexes of varying composition. However, from a purely qualitative standpoint, the dependence of the electric tension of the manganese dioxide electrode on pH may be demonstrated by assuming that it functions according to the overall reaction (1). The electric tension at 25° C will then be given by

$$U_{rev} = U_0 + \frac{0.059}{2} \log \frac{[MnO_2]^2 [H^+]^2}{[Mn_2O_3] [H_2O]}$$

Collecting in the constant U_0 the values of the concentrations $[MnO_2]^2$ and $[Mn_2O_3]$ – the concentrations replacing the activities in view of the qualitative nature of the equation – since they are in the solid phase, and $[H_2O]$ since it is present in large excess, gives

$$U_{rev} = U_0' + 0.059 \log [H^+]$$

It can be seen from this that diminishing the concentration of hydrogen ions will make the electric tension of the positive pole more negative.

When very low current intensities, tending towards zero, are withdrawn from the cell, the pH of the medium around the MnO_2 electrode, and hence its electric tension, should remain constant since the two OH^- gram-ions formed for every two **F** of discharge will be consumed by the secondary reactions (5) and (6), and (7) and (8).

All of the ammonia formed by reaction (7) will not be fixed by zinc chloride: a part of it will always escape as a gas contributing to the irreversibility of the cell and favouring other reactions which complicate the interpretation of the operation of the Leclanché cell. For example, basic zinc chloride is formed according to the reactions

$$Zn^{2+} + 2\,OH^- \rightarrow Zn(OH)_2$$

$$4\,Zn(OH)_2 + ZnCl_2 \rightarrow 4\,Zn(OH)_2 \cdot ZnCl_2$$

The Leclanché cell is a characteristic representative of cells of the inconstant type. During its operation with finite current intensities it becomes strongly polarized since the OH^- ions formed at the positive pole can only

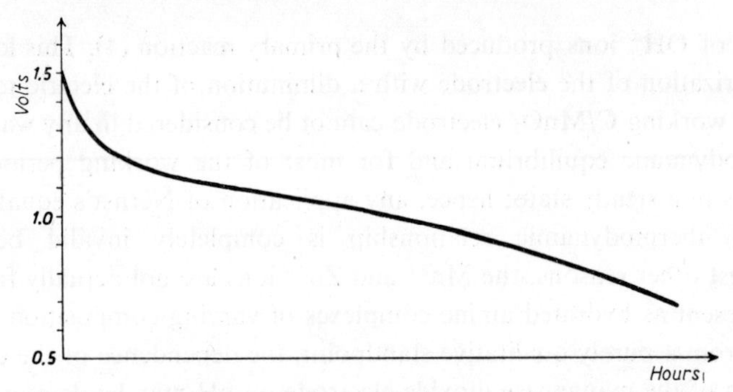

Fig. XI, 1. Uninterrupted discharge diagram of a Leclanché cell

diffuse with difficulty in the solid depolarizer and thus make the cathodic environment strongly alkaline and lower the electric tension. Fig. XI, 1 shows a typical discharge curve for a Leclanché cell. The slope of the curve, because it is caused by overtension phenomena, is a function of the discharge current intensity when other conditions are constant. The greatei is the intensity, the steeper is the discharge curve. If after a certain period of intense current withdrawal which has not exhausted the cell, it is allowed to rest, the diffusion of the OH^- ions from the manganese dioxide into the electrolyte diminishes the alkalinity of the environment of the dioxide and the electric tension rises again. Fig. XI, 2 shows this for three cells in series with a diagram for intermittent discharges which clearly shows the increased electric tension after every resting period.

A variant of the Leclanché cell is the dry cell whose electrochemical functioning is the same as that of the Leclanché cell. It differs from this

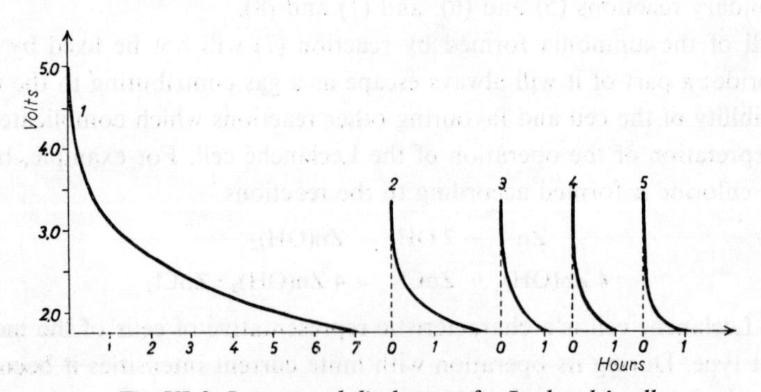

Fig. XI, 2. Interrupted discharge of a Leclanché cell

in that the electrolyte is partially inmobilized by the addition of gelling substances such as flour, agar, tylose etc. The capacity of this type of cell is somewhat empirical since it depends not only on the active mass but also to a very large extent, on the conditions of discharge such as the intensity of the discharge current, intermittency or continuity of the discharge, the value of the electric tension at which in practice the cell is considered to be exhausted, etc.

If it is assumed, for the time being, that the true overall reaction is

$$\underset{65}{Zn} + \underset{174}{2\,MnO_2} + \underset{107}{2\,NH_4Cl} \rightarrow \underset{170}{ZnCl_2 \cdot 2\,NH_3} + \underset{158}{Mn_2O_3} + \underset{18}{H_2O}$$

then the weights of substances taking part in the reaction will be those indicated beneath each component for every 2 F withdrawn. Thus, the various components should be present in the cell in these proportions. In practice, however, particularly when working with a low and intermittent discharge current, the number of Ah obtained is more than that corresponding to the weights of magnanese dioxide and ammonium chloride originally present in the cell. This confirms what has been said about the complexity of the electrochemical reactions in the Leclanché cell and in particular about the participation of atmospheric oxygen in the reaction, the reduction of manganese dioxide beyond the Mn_2O_3 stage to MnO and even Mn_3O_4, and the formation of other basic salts of zinc. Zinc however is utilized to a maximum of 25–30 % since, particularly in dry cells, it is also used as the container and so cannot be consumed until it perforates.

Spontaneous discharge of Leclanché cells is almost negligible in that the reduction of the atmospheric oxygen to H_2O_2 is very slow on metallic zinc, particularly if fairly pure raw materials are used free from more noble metals such as copper, lead and silver. These metals would be deposited galvanically on the zinc and would form local cells of the Daniell type in short circuit — analogous to those which cause corrosion (*cf.* p. 496 *et seq.*) — with a consequent rapid consumption of zinc and the eventual perforation of the container. The essential advantages of the Leclanché cell lie in its relatively long shelf-life and the possibility of mechanically immobilizing its electrolyte, which makes it suitable for use as a portable cell. Its disadvantages are its relatively high internal resistance of the order of tenths of an ohm, the increase in this resistance as discharge continues and the inconstancy of the electric tension at the terminals during its operation. For intermittent use, this last is less important.

Recently, a promising variant of the Leclanché dry cell has come into prominence; it consists essentially in the replacement of the zinc by magnesium. The diagram of the cell is

$$^{+}C/MnO_2 + BaCrO_4\ 3\ \%/MgBr_2\ 19\text{–}25\ \%/Mg\ \textit{amalg.}^{-}$$

with barium chromate added to improve the depolarization of the positive electrode. The net reaction is essentially

$$Mg + 2\ MnO_2 + 2\ H_2O \rightarrow Mg(OH)_2 + 2\ MnOOH$$

The advantages of the use of magnesium as the negative electrode are connected with the more negative electric tension which it has in the series of standard electric tensions ($U_{0,\,Zn} = -0.763$ V; $U_{0,\,Mg} = -2.37$ V), the significantly lower equivalent weight (theoretically a third of that of zinc), and the very low specific weight (Mg = 1.7, Zn = 7.1). Thus in practice the dry magnesium cell has a greater electric tension at the terminals, a significantly lower (by weight) anodic consumption for an equivalent withdrawal of Ah, and a lower net weight of the commercial cell. The electric tension on open circuit is about 1.9 V compared with the 1.6 V of the normal Leclanché. The increase in electric tension is thus much lower than would be predicted from the difference between the standard electric tensions of magnesium and zinc. This is largely due to the high electronegativity of magnesium which means that the measured electric tensions are always tensions during reactions, thus not corresponding to the equilibrium Mg/Mg^{2+} but to a more or less controlled corrosion of the magnesium with the evolution of hydrogen:

$$Mg \rightarrow Mg^{2+} + 2\ e^-$$

$$2\ H^+ + 2\ e^- \rightarrow H_2$$

This gives electric tensions which are much less negative than can be calculated from the equilibrium reaction

$$Mg \rightleftharpoons Mg^{2+} + 2\ e^-$$

For normal use either continuously or intermittently, the constancy of the electric tension is significantly greater than with zinc and the useful capacity is more than doubled for the same dimensions.

3. Ruben–Mallory Cell

A cell which is much used in appliances requiring a high discharge intensity for a considerable time is the Ruben cell, produced by the Ruben laboratories and by P. R. Mallory and Co. under the trade name 'RM' cell.

It consists of

$$^{+}Hg/HgO + Zn(OH)_2 \textit{solid}/KOH\ 40\,\%/Zn\ \textit{amalg.}^{-}$$

The primary process at the negative pole is the same as in the Leclanché cell

$$Zn - 2\,e^- \rightarrow Zn^{2+}$$

The Zn^{2+} ions produced react in the alkaline solution according to the equation

$$Zn^{2+} + 2\,OH^- \rightarrow Zn(OH)_2 \rightleftharpoons ZnO + H_2O$$

so that the overall anodic process is

$$Zn + 2\,OH^- - 2\,e^- \rightarrow ZnO + H_2O$$

Since the electrolyte is substantially saturated with ZnO the quantity of zincate ion is sufficient to ensure that the products of anodic oxidation are virtually only ZnO and $Zn(OH)_2$. The ionic product $[Zn^{2+}] \cdot [OH^-]^2$ is $4.5 \cdot 10^{-17}$ and the molarity of the OH^- ion is 7.7, with an activity coefficient of about 2, so that for the Zn/Zn^{2+} an electric tension of -1.317 V may be calculated with respect to the standard hydrogen electrode. At the positive pole, where the true electrode initially consists only of graphite mixed with the HgO catalyst, the following reactions occur:

$$HgO + H_2O \rightarrow Hg(OH)_2 \rightleftharpoons Hg^{2+} + 2\,OH^-$$

$$Hg^{2+} + 2\,e^- \rightarrow Hg$$

giving

$$HgO + H_2O + 2\,e^- \rightarrow Hg + 2\,OH^-$$

Since the standard electric tension of the $Hg/HgO/OH^-$ electrode is $+0.854$ V and the ionic product $[Hg^{2+}] \cdot [OH^-]^2$ is $1.7 \cdot 10^{-26}$, the half cell will have an electric tension of $+0.021$ V with respect to the standard hydrogen electrode. The total electric tension of the cell should thus be 1.338 V which is in very good agreement with the values of about 1.34 V normally given by the commercial cells.

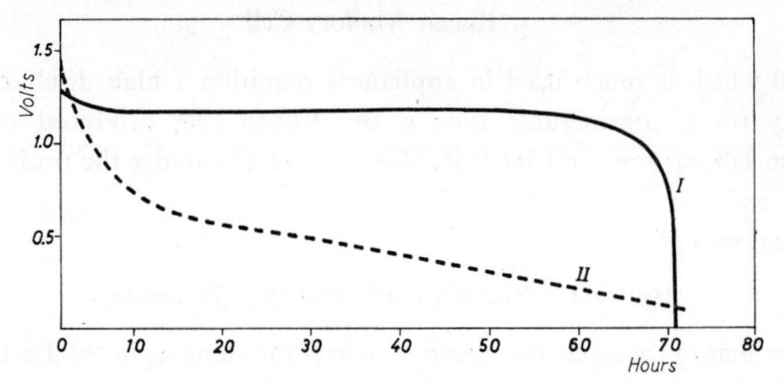

Fig. XI, 3. Comparison of the discharge diagrams of a R.M. cell (I) and a Leclanché cell (II)

Fig. XI, 3 shows a typical curve (I) for the discharge of a Ruben cell through a 5 Ω resistance with a fairly high current intensity, compared with a similar curve (II) for a dry Leclanché cell of the same size and shape.

The superiority of the Ruben cell during intense discharge is apparent both as regards the electric tension at the terminals and its constancy during the discharge. The theoretical coulometric capacity of the cell is 0.223 Ah/g of depolarizer. In practice the capacity efficiency for currents of up to 1.5 A/dm^2 of depolarizer surface is of the order of 90 % of this figure and is only slightly diminished for very much higher currents. The practical capacity of these cells with high current withdrawals is some 5–8 times greater than that of a similar Leclanché cell. As far as shelf life is concerned, storage for three years induces not the slightest change in the electric tension of the cell and diminishes its capacity only slightly. The sole disadvantage of the Ruben cell is its cost, which is much greater than that of conventional cells, but this is offset by a number of advantages: a large capacity, or small specific volume; the long life of the steel container; the long shelf life even at high temperatures and humidities; the low impedance during discharge; and the possibility of withdrawing instantaneous currents of very high intensities. These advantages are such that in terms of the ratio cost: service the Ruben cell is highly competitive with the conventional Leclanché.

The most important applications are of course in the military field, but it is now being widely used in hearing aids and photographic microflash equipment.

Ruben Laboratories have introduced a considerably modified version of this cell which combines some of the advantages of the normal alkaline type described with the constructive economy of the Leclanché cell; this is the mercuric dioxysulphate cell:

$$^{+}C/HgSO_4 \cdot 2\,HgO/ZnSO_4\ 11\,\%/Zn\ \textit{amalg.}^{-}$$

Its overall reaction is

$$3\,Zn + HgSO_4 \cdot 2\,HgO \rightarrow ZnSO_4 + 2\,ZnO + 3\,Hg$$

giving an electric tension of about 1.36 V.

A buffering agent added to the electrolyte raises the pH and maintains it within the optimum range of 5–6 at which the solubility of the depolarizer is minimal. This cell is suitable for all applications requiring a low current intensity for long periods of service, with a flat discharge curve.

4. Other Types of Cell

The Lalande cell is another type which has considerable use and offers the advantage of being partially regenerable; it has the form

$$^{+}Cu/CuO/NaOH\ 20\,\%/Zn\ \textit{amalg.}^{-}$$

The primary process at the negative pole is the same as in the Leclanché cell:

$$Zn - 2\,e^{-} \rightarrow Zn^{2+}$$

The Zn^{2+} ions are unstable in the alkaline medium and react with the OH^{-}

$$Zn^{2+} + 4\,OH^{-} \rightarrow ZnO_2^{2-} + 2\,H_2O \qquad (1)$$

to give poorly soluble sodium zincate; the overall anodic reaction thus becomes

$$Zn + 4\,OH^{-} - 2\,e^{-} \rightarrow ZnO_2^{2-} + 2\,H_2O \qquad (2)$$

The concentration of Zn^{2+} ions is thus very low and the electric tension of the zinc electrode very negative.

The copper oxide acts as depolarizer at the positive electrode, according to the reaction

$$CuO + H_2O + 2\,e^{-} \rightarrow Cu + 2\,OH^{-}$$

For thermodynamic reasons this reaction occurs through different steps

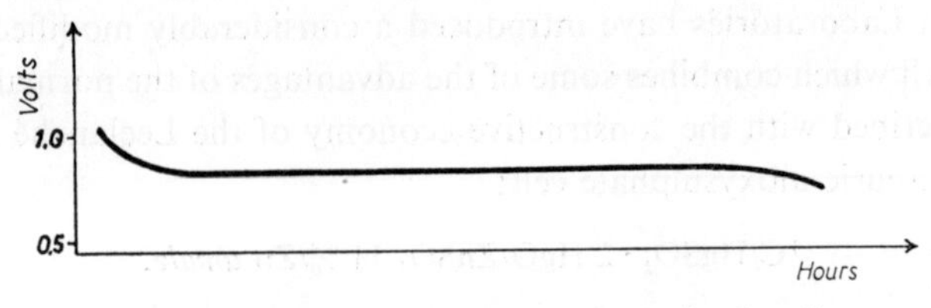

Fig. XI, 4. Discharge diagram of a Lalande cell

with the formation of cuprous oxide as an intermediate. However the net reaction of the Lalande cell is

$$Zn + 2\,OH^- + CuO \rightarrow ZnO_2^{2-} + H_2O + Cu$$

In the Lalande cell, neither the copper oxide nor the sodium zincate are at all soluble in the electrolyte, which may therefore be considered as always saturated with both the initial and final components of the electrode reaction. As a consequence the cell has a practically constant electric tension during its operation. The discharge curve for the Lalande cell, shown in Fig. XI, 4, clearly demonstrates the horizontal portion corresponding to the constant electric tension during the discharge. The electric tension of the cell is lower than that of the Leclanché, but its internal resistance is also much lower ($\sim 10^{-2}\ \Omega$) and these two factors to some extent compensate for each other.

The copper oxide is only slightly soluble in the sodium hydroxide solution, but not so slightly as to prevent the spontaneous discharge of the cell, through the deposition of metallic copper on the zinc, forming local cells in short circuit. This makes it difficult to store the cell. The addition of substances which can dissociate to give S^{2-} or $S_2O_3^{2-}$ ions will reduce the concentration of Cu^{2+} ions still further, and prolong the life of the cell.

The quantity of electrolyte is chosen so that it will be almost all consumed when the copper oxide has been reduced to metallic copper. The discharged cell may still be used however, by allowing the copper to undergo aerial oxidation, preferably at a temperature somewhat above normal, and replacing the exhausted electrolyte. The zinc electrode can also be replaced if it should become excessively consumed. The Lalande cell has the advantages of constancy of tension, possible partial regeneration and simple construction, which make it suitable as a source of power for railway signals and lighting in places lacking mains electricity.

Another type of cell which has considerable use employs atmospheric oxygen as a depolarizer. The prototype was the Féry cell, which can to some extent be derived from a Leclanché cell, in which the secondary depolarization phenomenon due to atmospheric oxygen adsorbed on the mass of manganese dioxide and carbon, becomes completely predominant and the manganese dioxide is eliminated; the cell thus becomes

$$^{+}\text{C/electrolyte/Zn}\ \textit{amalg.}^{-}$$

The primary process at the negative pole is the same as in the Lalande cell whereas at the positive, with atmospheric oxygen as the depolarizer, it is

$$\tfrac{1}{2}\,O_2\,(\textit{ads.}) + H_2O + 2\,e^- \rightarrow 2\,OH^-$$

In the absence of a suitable catalyst, the reaction

$$O_2\,(\textit{ads.}) + H_2O + 2\,e^- \rightarrow HO_2^- + OH^-$$

will also occur and give rise to a lower electric tension.

If the electrolyte is ammonium chloride, the secondary reactions (7) and (8) described on p. 640 *et seq.*, will occur, whereas if it is sodium hydroxide the above secondary reaction (1) will occur.

The carbon electrode of the Féry cell is porous and spongy, and protrudes some way from the electrolyte, to aid the adsorption of atmospheric oxygen on the carbon to act as the depolarizer.

The Féry cell has an initial electric tension of about 1.25 V which however rapidly falls during use to about 0.9 V. Its main drawback lies in the slowness of the depolarizing action of the oxygen which prevents the withdrawal of substantial currents.

A considerable improvement in this cell has been achieved by the Société Le Carbone and the National Carbon Company who use both ammonium chloride and sodium hydroxide as electrolyte. The carbon granules of the positive electrode are pretreated with a hydrophobic substance; this enables the porosity of the electrode to be considerably increased without fear that the electrolyte entering the pores will wet the carbon granules; only the external part of the electrode actually comes into contact with the electrolyte. The effect is as if the surface of the carbon electrode, across which the equilibrium between atmospheric and adsorbed oxygen is established, were greatly enlarged, so aiding the depolarizing action of the oxygen. Because of its greater effective surface, the electric tension of the Le Carbone electrode, with ammonium chloride,

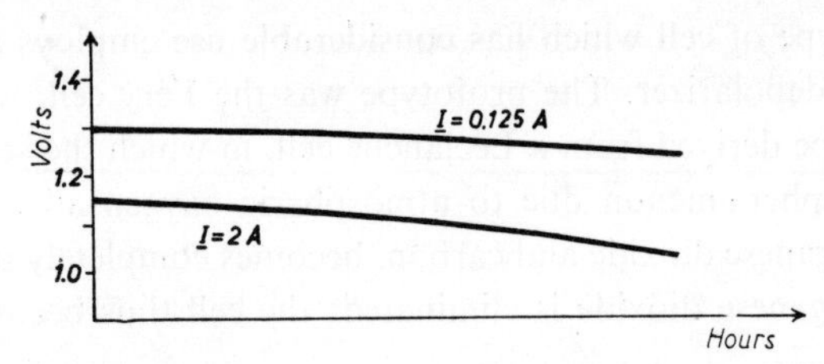

Fig. XI, 5. Discharge diagrams of a National Carbon cell under different conditions

is about 1.0 V *i.e.* 0.1 V greater than that of the Féry cell. Replacing the ammonium chloride by sodium hydroxide leads to a further increase of the electric tension since that of the zinc electrode becomes some 0.2 V more negative, giving an average total electric tension of 1.2 V. With both types of cell, the electric tension remains constant even with discharges of considerable current intensity as shown in Fig. XI, 5.

The National Carbon Company produces a further improvement by adding chalk, which reacts with the sodium zincate forming calcium zincate and regenerating the sodium hydroxide:

$$Ca(OH)_2 + 2\,NaHZnO_2 \rightarrow CaZn_2O_3 + 2\,NaOH + H_2O$$

Of the various types of cells proposed in recent years, two kinds which use chlorides as depolarizers, have had a considerable industrial production. They consist of

$$^{+}Ag/AgCl/MgCl_2/Mg\ \textit{amalg.}^{-}$$

and

$$^{+}Cu/Cu_2Cl_2/MgCl_2/Mg\ \textit{amalg.}^{-}$$

The reaction at the negative pole is the ionization of the magnesium to divalent ions, giving finally, hydrated magnesium chloride.

The reaction at the positive pole is the reduction of the silver or cuprous chlorides by the adsorbed atomic hydrogen produced by the discharge. These cells can supply fairly high currents. The silver chloride cell has an electric tension of about 1.6 V for a current of 0.2 mA/cm^2 of electrode surface, and of about 1.3 V at 70 mA/cm^2; the cuprous chloride cell has a somewhat lower electric tension. The electric tensions are very stable during discharge and the behaviour is excellent at temperatures down to − 50° C, so that these cells are well adapted to use in arctic regions. The possibility of using simple sea water as the electrolyte also makes them suitable as emergency supplies for marine plants.

5. Fuel Cells

The energy of oxidation of conventional fuels, which is normally employed as heat, may be directly transformed into electrical energy in very good yield, in a 'fuel cell'. Since all oxidation reactions involve a transfer of electrons between fuel and oxidant, it is clear that the chemical energy of oxidation may be directly converted into electrical energy. An oxido-reduction is involved in which the oxidation of the fuel may be considered as a loss of electrons and the reduction of the oxidant as a gain. Every galvanic cell has an oxidation at the negative pole and a reduction at the positive. And as in any other galvanic cell, fuel cells tend to separate the two half-reactions, making the exchanged electrons pass through an external working circuit before reacting with the oxidant.

To achieve this separation it is necessary to construct a cell consisting of an anode, a cathode and an electrolyte, which can be fed directly with the fuel, *e.g.* carbon, and with air (Fig. XI, 6). The oxygen required for the combustion of the fuel, is ionized at the cathode and migrates through the electrolyte to the anode, where the fuel present is oxidized. The electrons liberated pass through the external working circuit to the cathode of the cell where they are used for the reduction of the oxygen; other O^{2-} ions are produced and the electrical circuit completed. This outline il-

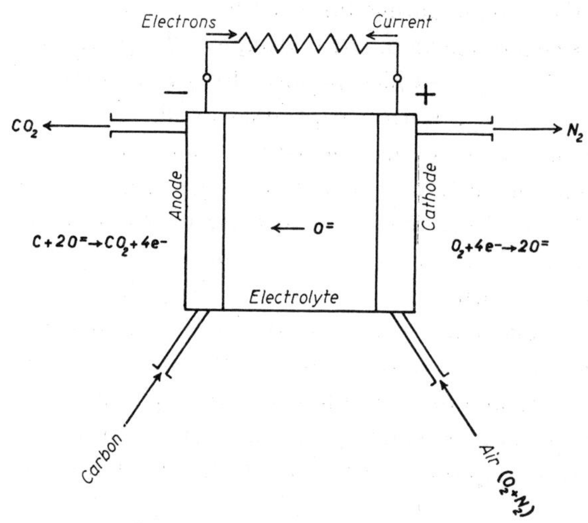

Fig. XI, 6. Diagram of a fuel cell

lustrates the theoretical simplicity of the fuel cell as a generator of electricity.

The most important aspect of this electrochemical exploitation of a fuel, is that the thermodynamic limitations imposed by the Carnot cycle no longer apply, because the energy released is only degraded to a minimal extent in the form of heat. As de Béthune[1] showed very clearly, the thermodynamic difference between the utilization of a fuel in a reversible heat engine and in a fuel cell working under reversible conditions, may be completely expressed in terms of useful woιk. In a heat engine the maximum useful work is the fraction of the enthalpy change ($\Delta H(T_2 - T_1)/T_2$) of the corresponding Carnot cycle; however in the fuel cell under reversible conditions, the useful work is equal to the reactional free enthalpy ΔG corresponding to the isothermal combination of the fuel with oxygen. Thus the interest of the fuel cell as a source of electrical energy essentially lies in its high net efficiency, and applies wherever the cost of the fuel is a determining factor.

From a general standpoint, moreover, it may be mentioned that fuel cells can be used not only to generate energy but also as readily controllable chemical reactors for dehydrogenations, or partial or total oxidations, governed by purely electrical characteristics of the circuit and producing energy as a side product.

In a cell functioning at constant pressure, if ΔE is the fall in internal energy of the system during the overall reaction, then the reactional enthalpy, or heat of reaction, ΔH is given by $\Delta E + p\Delta v$. A certain amount of heat $q = T\Delta S + q'$, which is normally greater than the work $p\Delta v$, will be liberated or absorbed depending on the initial and final components of the reaction ($T\Delta S$), and on the route followed during the reaction (q'). The quantity of heat q' can not be converted into electrical energy and makes the cell either increase or diminish in temperature during its operation. The values of the enthalpy of reaction ΔH and of the term $T\Delta S$, depend solely on the amount of substance reacting, but the practical value of q' also depends on the mode of reaction; it contains, for example, the heat evolved in the passage of the current through the electrolyte (RI^2), and this is lost heat which, increasing more rapidly with increasing current than with the internal resistance of the cell, depends on the rate at which electrical work is supplied. Other terms are subsumed in

[1] A. J. DE BÉTHUNE, *J. Electrochem. Soc.*, 107 (1960) 937.

q' which are best considered kinetically: they comprise secondary reactions; physico–chemical resistances (overtensions) which diminish the rate of electron transfer in the electrode reactions; and mass-transport polarizations, derived from concentration gradients within the electrolyte.

Although it is very difficult to calculate the overall value of q in practical cases, it is possible to calculate the maximum electrical work obtainable under reversible working conditions *i.e.* when the current intensity tends towards zero and the electric tensions towards their maximum theoretical values. Under these conditions, the absolute value of the electric tension $U_{rev.}$ is given by

$$\Delta G = -zFU_{rev}$$

where ΔG is the reactional free enthalpy of the overall reaction; z is the number of electrons involved; and **F** is a Faraday.

The ideal efficiency, R_i for reversible conditions is given by

$$R_i = \Delta G/\Delta H$$
$$= 1 - (T\Delta S/\Delta H)$$

Under practical and hence nonreversible conditions however, the yield, R_p, is expressed by

$$R_p = 1 - (q/\Delta H)$$

where $q > T\Delta S$.

Theoretical values for the electric tensions of various fuels as a function of temperature have been calculated[1] using the most recent data available for enthalpies and specific heats; the results obtained are plotted in Fig. XI,7.

Theoretically it is possible to obtain, with any fuel, electric tensions of the order of a volt, but they strongly depend on the temperature and on the particular fuel.

It can be seen, for example, that with the sole exception of the C/CO cell it would be advantageous to work at the lowest temperature possible to obtain the maximum electric tension but, since the electrode reactions become very slow at low temperatures, a compromise is necessary. Thermodynamically, this is the most critical factor affecting the electric tension of fuel cells.

Other factors, which also involve the temperature, are connected with

[1] V. S. Daniel–Bek and M. Z. Mints, *Zhur. Priklad. Khim.*, 32 (1959) 649.

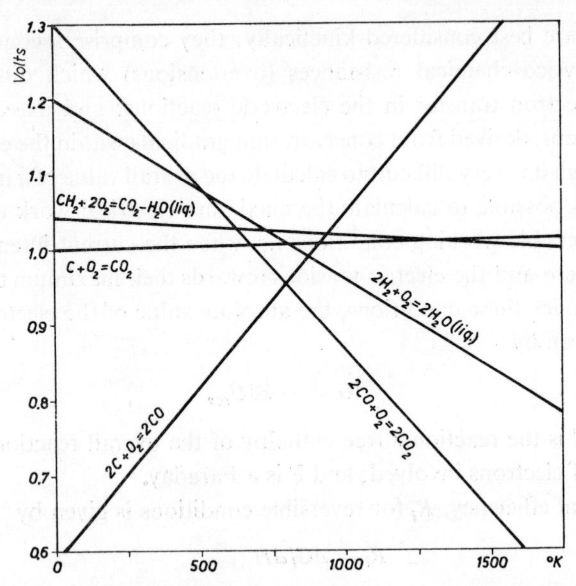

Fig. XI,7. Temperature–electric tension diagrams of different fuel cells

the choice of practical working conditions which although inevitably nonreversible should nevertheless give electric tensions as close as possible to the theoretical value. In practice the internal fall with a useful current density should not exceed 20–30 % of the electric tension on open circuit. Only thus is it possible to obtain a good net efficiency.

An exhaustive analysis of the three main terms of the internal fall – activation overtension, polarization of concentration and ohmic resistance – was carried out by Austin[1].

1. The activation overtension is a direct result of the operation of the cell under conditions far removed from thermodynamic equilibrium with a consequent conversion of part of the reactional free enthalpy into heat. Consider, for example, the case of a hydrogen half cell with a porous carbon electrode, impregnated with catalyst and immersed in an alkaline solution. Catalysts are essential, and have a double function. The catalyst must above all produce a rapid chemical adsorption of the gas so as to make it reactive for the ionization and subsequent reaction with the active species of the electrolyte. At the same time it must reduce to a minimum

L. G. Austin, *Ind. Eng. Chem.*, 52 (1960) 300

the loss of the reactional free enthalpy due to this adsorption. The chemical adsorption of the reacting gas must be preferential over that of the reaction products, to avoid self-poisoning of the catalyst.

With the hydrogen electrode, when the current intensity is zero the rate of the reaction is also zero and a dynamic equilibrium is set up for the passage of hydrogen from the surface layer of chemical adsorption to the electrolyte. There is no change in free energy involved in these interfacial exchanges. When the electrode is working, the reaction rate becomes finite and a certain amount of substance reacts with a corresponding change in free enthalpy and the flow of a current. In consequence the equilibrium conditions no longer exist and part of the reactional free enthalpy is expended in overcoming the activation overtensions of the various partial reactions: chemical adsorption, electrochemical oxidations, reactions with the active species of the electrolyte, and desorption.

Expressions for the reactional free enthalpy involved in the two reactions:

$$H_2 + \text{active centre} \rightleftharpoons 2\,[H]_{adsorbed}$$

$$[H]_{adsorbed} + OH^- \rightleftharpoons H_2O + e^- + \text{active centre}$$

give an equation connecting current density and activation overtension.

$$J = (I_0/S)\,(e^{\alpha z F \eta_A/RT} - e^{-\beta z F \eta_A/RT}) \qquad (1)$$

where

J = current/unit geometrical area of the electrode *i.e.* current density
S = geometrical area of the electrode
η_A = activation overtension
α, β = proportionality constants (*cf.* Chapter IV) lying between 0 and 1, which are generally constant over a wide range of current densities
I_0 = exchange current, $KN_cS_e(a_p)^\alpha\,(a_r)^\beta e^{-\beta\Delta G^*/RT} e^{\alpha\Delta G/RT}$
N_c = number of active centres/internal unit area of electrode
S_e = effective area/unit geometrical area of electrode
ΔG_0 = standard free enthalpy of the reaction
ΔG^* = free enthalpy of activation
a_p = activity of the end product of the reaction
a_r = activity of the reactants

The most important factor is the exchange current which must be as high as possible. The temperature is very important here. The porosity of the electrodes also directly affects the value of I_0 through its effect on the area S_e and the catalyst diminishes the value of ΔG^* to a minimum. Finally the pressure advantageously increases the activity.

When the overtension η_A is high, equation (1) becomes the well known Tafel's equation

$$\eta_A = A + B \log J$$

in which

$$A = (2.3\ RT/\alpha\ \mathrm{F}z) \log (S/I_0)$$

$$B = 2.3\ RT/\alpha\ \mathrm{F}z$$

If the exchange current is very small even the slightest current loss on open circuit is sufficient to induce a significant polarization giving an electric tension on open circuit which is already less than the reversible electric tension.

2. The concentration polarization is connected with the current density by the equation

$$\eta_c = \frac{2.3\ RT}{\mathrm{F}z} \log \frac{J_l}{J_l - J} \tag{2}$$

where J_l is the limiting current density depending on the transport properties of the system. For a gas electrode, for example, the diffusion of the gas through the porous mass towards the face where the reaction with the electrolyte occurs, can only take place under the influence of a pressure gradient. The pressure at the electrode is limited to that at which gas is supplied and the limiting current corresponds to this condition. It will be seen from equation (2) that the closer the current density of the working cell approaches to J_l, the more rapidly does the polarization increase. For a given internal area the electrodes must therefore have pores which are not so small as to provoke an excessive concentration polarization.

3. The ohmic fall, $\eta_0 = IR$, depends not only on the conductance of the electrolyte, but also on the formation on the electrodes of layers which are poorly conducting or even insulating, as happens, for example, in cells at high temperatures.

Other forms of overtension which are still effectively activation overtensions, derive from the occurrence of other electrode reactions having changes in free enthalpy which are lower than that of the theoretical reaction and hence induce electric tensions lower than that predicted. This happens in the ionization of oxygen to peroxide ions rather than to hydroxyl ions in certain electrodes, which therefore require the presence of a catalyst of the reaction $HO_2^- \rightarrow OH^- + \frac{1}{2} O_2$.

The history of fuel cells really originated in 1839 with Groves' observation on the production of currents of short duration at the moment of interruption of the operation of cells electrolyzing water.

However, only in recent years, have fuel cells ceased to be purely objects of scientific curiosity and to be subjected to systematic studies of their applications which have led to various industrial developments. Cells which have now been constructed may be classified on various criteria as follows:

1. The type of electrolyte: aqueous, non aqueous, molten or solid.
2. The oxidation mechanism: direct or indirect (*i.e.* redox cells in which the fuel and the oxidant are not directly consumed in the cell but are used for the chemical regeneration of the reagents which are in turn electrochemically consumed by the two electrode reactions of the cell itself).
3. The type of fuel: gas, liquid or solid; carbonaceous or pure hydrogen.
4. The type of oxidant: air or pure oxygen.
5. The working conditions: temperature and pressure.
6. The nature and form of the electrode materials and catalyst.

A short outline of the situation at present may be given by mentioning the expedients which have been tried for each condition.

1. Aqueous electrolytes have consisted essentially of concentrated (30–50 %) solutions of potassium hydroxide which provide a high conductance without excessive problems of corrosion and whose alkalinity advantageously diminishes the life of peroxide ions at the oxygen electrode. Cells with molten electrolytes generally use an eutectic mixture of sodium and lithium carbonates. Solid electrolytes are used in the Bishoff and Davtyan cells (based on sodium carbonate and silicate, cerium oxide and tungsten trioxide) and, in the form of ion exchange membranes in the experimental cell of Grubb (General Electric Co.).

2. In addition to many types with direct oxidation, various types of regenerative cells have been designed in which the actual cell consists of redox systems such as Fe^{3+}/Fe^{2+}, Sn^{4+}/Sn^{2+} and Cu^{2+}/Cu^{+}.

3. The most satisfactory cells use pure hydrogen as fuel but various designs and studies have been made with propane, ethane, methane, coal gas, carbon, carbon monoxide, methanol, acetylene, ethylene and finally liquid hydrocarbons such as benzene in the Daniel–Bek cell and kerosene in the Chambers cell.

4. The various working conditions of present day cells may be summarized as follows:

Low temperature, low pressure (max. 3–4 atm.): H_2/O_2 (Standard Oil, National Carbon, General Electric, Allis–Chalmers, etc.); hydrocarbon/O_2 (air) cells (Allis–Chalmers, National Carbon, Standard Oil, Leesona etc.).

Moderate temperatures (200–300°C), *high pressures* (30–70 atm.): mainly H_2/O_2 cells (Bacon, United Aircraft, Leesona etc.).

High temperature (300–800° C): hydrocarbon/air cells with molten electrolytes (General Electric, Leesona, United Aircraft etc.).

5. The electrodes vary from the porous carbon of the National Carbon cell to the sintered nickel of the Bacon cell, to silver mesh (for the oxygen electrode only), to zinc oxide- and silver oxide- based ceramics in certain other types of cells.

To clarify what has been said it will be convenient to describe a fuel cell in more detail; this will be the hydrogen/oxygen, moderate temperature, high pressure cell developed by F.T. Bacon at the Cambridge Flying School. This cell, at electric tensions not too inferior to that on open circuit, successfully provides current densities which are significantly higher than in any other cell thanks to the effects of temperature and pressure. At tensions of about 0.7 V, currents of 300–400 mA/cm² are obtained in comparison with the 50–100 mA/cm² of other cells. Fig. XI,8 gives a diagram of the cell.

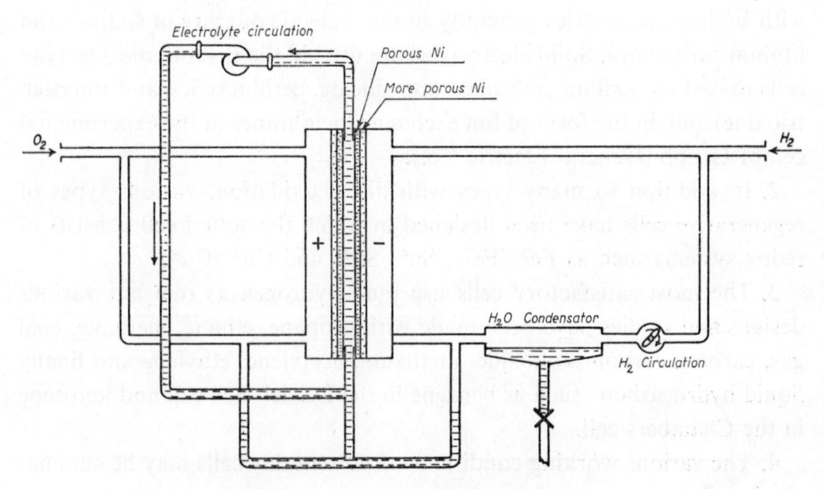

Fig. XI,8. Bacon fuel cell (schematic)

The working temperature is about 200° C and the pressure may vary between 20 and 40 atm. The electrolyte is a solution of potassium hydroxide at 37–50 % which in the cell itself has a thickness of only 3 mm to diminish the internal ohmic resistance to a minimum. The electrodes consist of discs of porous nickel obtained by sintering nickel powder prepared by the thermal decomposition of nickel carbonyl.

The thickness of every disc, about 1.5 mm, is divided into two layers of different porosity: the layer in contact with the electrolyte has pores of 3–5 microns diameter and the layer in contact with the gas has pores of 30 microns diameter. This arrangement satisfies the dual purpose of hindering the flow of liquid, but facilitating the mass transport and diffusion of the gas, into the interior of the electrode. A slight pressure difference is sufficient to expel the liquid from the larger pores without making the gas bubble into the solution. The oxygen electrode is preoxidized at high temperature in air, to diminish the corrosion by the oxygen at the high pressure and by the electrolyte solution. Lithium atoms, previously introduced into the crystal lattice of the nickel oxide (by impregnation of the porous mass with lithium hydroxide) then produce a black double oxide of nickel and lithium, which is a very good semiconductor, instead of the normal green nickel oxide which is an insulator. For the hydrogen electrode, the nickel is activated by impregnating it with a concentrated solution of nickel nitrate, heating it to 400° C in air and finally reducing it with hydrogen at this temperature. The insulating gaskets are composed of asbestos impregnated with neoprene or other halogenated high polymers which are more resistant to alkali at high tempera-

TABLE XI,4

WORKING CHARACTERISTICS OF THE BACON CELL

Current density mA/cm²	*Electric tension* V	*Efficiency referred to reactional free enthalpy* (%)	*Efficiency referred to heat of reaction* (%)
10	1.02	85	76
100	0.905	75	67
250	0.805	66	59
500	0.677	57	51
675	0.585	48	43

ture. The maximum effective electrode diameters obtained are 25 cm.

The behaviour of the electric tension and the efficiency as a function of the current density is reported in Table XI, 4 for working conditions of 200° C and 40 atm.

Increasing the pressure has little effect on the electric tension but markedly increases the power given (current density) as a consequence of the increased reaction velocity. Increased temperature also augments the power given which is increased by a factor of 10 on passing from 100 to 200° C. Increasing the concentration of the electrolyte will again increase the power up to a maximum at a concentration of 35 % at which the thickness of electrolyte in the cell gives a resistance of about 0.25 Ω/cm^2.

Table XI, 5 compares the various types of generators of electrical energy.

TABLE XI, 5

CHARACTERISTICS OF GENERATORS OF ELECTRICAL ENERGY

	Specific power kW/m³	*Efficiency* %
Bacon cell H_2/O_2	150–300	70–55
Lead accumulators	10	75–85
Diesel alternator	5	38
Turbo alternator	35.5	40

The characteristics of the Bacon cell make it clear that fuel cells must be considered as one of the most efficient means of energy production. Besides their interest for fixed plants of medium and large size and for railway traction, they have very many possible military applications for a number of advantageous reasons such as their high power density, small bulk, high energy density and silent operation.

6. Lead Accumulators

Accumulators are a type of galvanic cell in which the electrode processes are almost perfectly reversible. At the end of discharge, the electrochemical process may be inverted by making the cell function as an electrolytic cell with current supplied from outside. When working as an electrolytic cell the electrical energy is, for the greater part, converted into internal energy of the system which is brought back to the initial state and can be used for a new discharge process. The types of cell described so far cannot work as accumulators because the electrode processes are not reversible; at the positive pole of a Leclanché cell, for example, there would be an evolution of chlorine on any attempt to recharge it.

Theoretically, any reversible electrochemical process could be used to construct an accumulator. In practice, however, the tendency to have only one electrolytic solution greatly limits the choice of possible electrochemical processes. Up to the present day only lead accumulators and ferro–nickel alkaline accumulators (whose characteristics are shown in Table XI,6) have proved to be useful on the large scale. Recently, however, zinc–silver, alkaline accumulators have entered into commercial use; they have many interesting characteristics and for many applications are undoubtedly competitive with the other types.

TABLE XI,6

CHARACTERISTICS OF ACCUMULATORS

	Accumulator			
	Pb	Ni–Fe	Ni–Cd	Ag–Zn
sitive pole	PbO_2 on Pb	Ni_2O_3	Ni_2O_3	Ag_2O_2 on Ag
egative pole	Pb	Fe	Cd	Zn
ectrolyte	H_2SO_4 15–40 %	KOH 21 % + LiOH 50 g/l	KOH 21 % + LiOH 50 g/l	KOH ~40%+ K_2ZnO_2 sat.
ean electric tension V	1.95	1.18–1.20	1.20	1.50
apacity Ah/kg	~15	~22	~21	~95
apacity Wh/kg	~29	~26	~25	~140
verage life	15 years	10 years	10 years	100 discharges

The lead accumulator, invented by Planté in 1859, consists of the following galvanic cell:

$$^{+}Pb/PbO_2;\ PbSO_4/H_2SO_4/PbSO_4/Pb^{-}$$

The electrode reaction during discharge, at the positive pole is:

$$PbO_2 + 4\,H^+ + 2\,e^- \rightarrow Pb^{2+} + 2\,H_2O$$
$$Pb^{2+} + SO_4^{2-} \rightarrow PbSO_4$$

giving

$$PbO_2 + 4\,H^+ + SO_4^{2-} + 2\,e^- \rightarrow PbSO_4 + 2\,H_2O$$

There is thus a production of water and a deposition of lead sulphate with a consumption of sulphuric acid. The reaction at the negative pole is

$$Pb \rightarrow Pb^{2+} + 2\,e^-$$
$$Pb^{2+} + SO_4^{2-} \rightarrow PbSO_4$$

giving

$$Pb + SO_4^{2-} \rightarrow PbSO_4 + 2\,e^-$$

also giving a formation of lead sulphate with a consumption of sulphuric acid. The net electrochemical reaction is the sum of the individual electrode processes:

$$PbO_2 + 2\,H_2SO_4 + Pb \underset{\text{charge}}{\overset{\text{discharge}}{\rightleftarrows}} 2\,PbSO_4 + 2\,H_2O \qquad (1)$$

That this double sulphation is the actual overall reaction has been fully confirmed by experimental observations and the thermodynamic calculations of Craig and Vinal[1] and of Beck and Wynne Jones[2].

Craig and Vinal derived the heat of reaction for the double sulphation from equation (1) both from the most reliable experimental data on the electric tension of lead accumulators and from thermochemical data and found a very good agreement particularly using electric tensions obtained, under conditions very close to thermodynamic reversibility, with the electrodes (Pt)$PbO_2/PbSO_4$, H_2SO_4 and $Pb/PbSO_4$, H_2SO_4 at various concentrations of acid.

Beck and Wynne Jones analyzed thermodynamically the effects of changes in temperature, pressure and electrolyte concentration on the electric tension of lead accumulators and obtained results in perfect accord with the theory of double sulphation. They also showed that the double sulphation reaction was not only the reversible reaction which determined the electric tension of the cell but was also the reaction occurring in the discharge (irreversible) process at finite current density. Under normal discharge conditions the dependence of the electric tension on time and on the current density showed no change which could be associated with

[1] D. N. Craig and G. W. Vinal, *J. Research Nat. Bur. Standards*, 24 (1940) 482.
[2] W. H. Beck and W. F. K. Wynne Jones, *Trans. Faraday Soc.*, 50 (1954) 136.

changes in the process or with the intervention of an irreversible stage in the normal process (1).

The electric tensions of each electrode, and hence of the accumulator, depend mainly on the concentration of sulphuric acid. For the positive electrode, the equilibrium Pb^{2+}/Pb^{4+} may be considered through the reaction

$$\begin{array}{lll} & PbO_2 + 4\,H^+ & \rightarrow Pb^{4+} + 2\,H_2O \\ & Pb^{2+} + 2\,H_2O & \rightarrow PbO_2 + 4\,H^+ + 2\,e^- \\ \hline \text{giving} & Pb^{2+} & \rightarrow Pb^{4+} + 2\,e^- \end{array}$$

Hence, at 25° C, the electric tension of the positive electrode is given by the equation

$$U_1 = U_{0Pb^{2+}/Pb^{4+}} + (0.059/2) \log ([Pb^{4+}]/[Pb^{2+}]) \tag{2}$$

in which the activity coefficients are taken as 1 for simplicity. The standard electric tension of the Pb^{2+}/Pb^{4+} electrode is about 1.7 V (*cf.* Table III, 5). In a solution of sulphuric acid of density 1.15 g/cm^3 at 18° C, with PbO_2 and $PbSO_4$ present as solid phases, the concentration of the Pb^{4+} ions is $0.91 \cdot 10^{-4}$ and that of Pb^{2+} ions is $5 \cdot 10^{-6}$. Introducing these values into equation (2) gives, for a temperature of 25° C,

$$\begin{aligned} U_1 &= 1.70 + 0.0295 \log (0.91 \cdot 10^{-4}/5 \cdot 10^{-6}) \\ &\simeq +\,1.74 \text{ V} \end{aligned}$$

For the negative pole it is sufficient to consider the Pb/Pb^{2+} electrode, whose standard electric tension is $-$ 0.126 V. Using the Pb^{2+} ion concentration given, the electric tension of the electrode becomes

$$\begin{aligned} U_2 &= U_{0Pb/Pb^{2+}} + (0.059/2) \log [Pb^{2+}] \\ &= -\,0.126 + 0.0295 \log 5 \cdot 10^{-6} \\ &= -\,0.27 \text{ V} \end{aligned}$$

The total electric tension of an accumulator with sulphuric acid of density 1.15 g/cm^3 at 18° C, becomes at room temperature

$$U = U_1 - U_2 = 1.74 - (-\,0.27) = 2.01 \text{ V}$$

A value of 1.98 V can be obtained by measurement at 20° C. There is thus extremely good agreement with the calculated value in view of the uncertainties surrounding the determination of the concentrations of Pb^{4+} and Pb^{2+} ions.

To calculate the dependence of the electric tensions on the hydrogen ion concentration, for the electrode Pb^{4+}/Pb^{2+}, it is necessary to consider that the lead peroxide forms the solid phase.

Thus, the equilibrium constant for the reaction

$$PbO_2 + 4\,H^+ \rightleftharpoons Pb^{4+} + 2\,H_2O$$

becomes simply

$$K = [Pb^{4+}]/[H^+]^4 \qquad (3)$$

In other words, the concentration of Pb^{4+} ions increases with the fourth power of the H^+ ion concentration. Replacing the concentration of Pb^{4+} ions in equation (2) by the value derived from equation (3) gives

$$U_1 = U_{0Pb^{2+}/Pb^{4+}} + (0.059/2)\log(K[H^+]^4/[Pb^{2+}]) \qquad (4)$$

On the other hand, increasing the concentration of H^+ ions also increases that of SO_4^{2-} ions, and since the solubility product of lead sulphate is given by the equation

$$[Pb^{2+}]\,[SO_4^{2-}] = L = 1.3 \cdot 10^{-8} \text{ at } 25°\text{ C}$$

an increase in the concentration of the SO_4^{2-} ions must diminish that of the Pb^{2+} ions and hence the electric tension of the dioxide electrode will be further shifted towards more positive values. However, a diminution of the Pb^{2+} ion concentration shifts the electric tension of the negative electrode towards more negative values so that the electric tension of the accumulator increases when the hydrogen ion concentration rises.

An expression for the electric tension of the cell may readily be derived from equation (4) for U_1 and from the Nernst formula given for U_2:

$$U = U_0' + 0.059\log([H^+]^2/[Pb^{2+}]) \qquad (5)$$

This clearly shows the dependence on the pH which not only affects equation (5) directly but, as shown above, indirectly determines the concentration of the Pb^{2+} ions.

Sulphuric acid is consumed during the discharge, diminishing the concentration of H^+ ions and hence lowering the electric tension. During the charging process the reverse reaction occurs, forming sulphuric acid, increasing the concentration of H^+ ions and consequently raising the electric tension. The dependence of the electric tension of an accumulator on the density d of the acid used is given by the empirical equation

$$U = 1.85 + 0.917\,(d - 1)$$

During discharge lead sulphate is formed, at both the negative and positive poles, in an extremely divided form so that the reverse charging reaction occurs without difficulty.

The capacity of a lead accumulator depends upon the active mass, the state in which this is present, the type of electrodes, their percentage utilization, the concentration of acid, the intensity of the discharge current, the final electric tension at which the accumulator is considered to be exhausted, and the temperature. Fine, porous electrodes have a greater capacity than thick compact electrodes of the same active mass, but they can never be completely utilized because of variations in volume to which they are subject. The change $PbO_2 \rightarrow PbSO_4$ is accompanied by a change in volume of 164 % and the change $Pb \rightarrow PbSO_4$ by one of 82 %. If the electrode reaction were continued to complete utilization of the active mass, the large variations in volume induced would rapidly disintegrate the electrodes and make the accumulator useless. The utilization of the active mass is not normally continued beyond 25–35 %.

The concentration of the sulphuric acid only indirectly affects the capacity of an accumulator in that the electric tension depends upon it. It should be noted that the concentration which determines the electric tension is not the average concentration of the bulk of the acid between the electrodes but that of the acid saturating the electrodes themselves. The higher is the discharge current intensity, the more rapidly will fall the concentration of the acid saturating the electrode. Since diffusion is a slow process, equilibrium of concentration between the acid within the electrode and the bulk of the acid in which the electrode is immersed, will only be achieved with some difficulty; thus the greater the intensity of current withdrawn, the greater will be the fall in electric tension. In other words, the greater the discharge current, the more rapidly will the final electric tension be reached and the smaller will be the capacity.

The temperature acts by facilitating diffusion phenomena and an increase in temperature leads to an increase in capacity. Assuming, for example, a capacity of 100 at 27° C, this would become 50 at 12° C. The temperature markedly influences the electric tension according to thermodynamic laws and in particular the Gibbs–Helmholtz equation for temperature coefficients. However, the following empirical equation is convenient; it gives the electric tension of an accumulator as a function of the temperature in degrees C,

$$U = U' + at + bt^2$$

where the values of the constants U' (tension at 0°), a and b are somewhat dependent on the concentration of the acid. These values are shown in Table XI,7.

TABLE XI,7

THE DEPENDENCE OF THE CONSTANTS U', a AND b ON THE CONCENTRATION OF SULPHURIC ACID

c (mol/1)	U' (V)	$a \cdot 10^6$	$b \cdot 10^8$
2	1.9666	159	103
3	2.0087	178	97
4	2.0479	177	91
5	2.0850	167	87
6	2.1191	162	85
7	2.1507	153	80

The final value of the electric tension below which it is normal to consider a lead accumulator as exhausted, is 1.8 V. Below this electric tension the granules of lead sulphate begin to grow markedly and make the successive recharge very difficult because, not only do they have a marked ohmic resistance, but they are also slow to react to give the initial substances again (PbO_2 and Pb). Once the electric tension reaches 1.8 V the discharge curve (*cf.* Fig. XI,9) shows a sharply descending branch. This does not imply, however, that the active mass is exhausted but is rather an indication that every granule of it is covered by a layer of lead sulphate of such a thickness that the reaction velocity becomes low and the electric tension of the accumulator diminishes.

Besides never being discharged below 1.8 V, a lead accumulator should never be left for long unused, particularly when discharged or nearly so, to avoid the growth of the granules of sulphate: a phenomenon which is known as sulphation.

Using the equivalent weights of substances consumed for every Ah withdrawn (3.86 g Pb, 4.46 g PbO_2, 9.2 g H_2SO_4, $d = 1.30$), the coefficient of utilization of the active mass ($\sim 35\,\%$) and allowing ~ 18 g/Ah for the supporting plates, the container and the necessary excess of sulphuric acid[1], gives a capacity for a normal accumulator of about 15 Ah/kg total

[1] The sulphuric acid must be in excess with respect to the active mass so that even at the end of the discharge when the active acid has been fixed as sulphate, the electrolyte will still have a sufficiently high concentration of free acid to ensure a good conductance.

weight. Taking its average discharge electric tension as 1.95 V, gives an energetic capacity of about 29 Wh/kg.

The spontaneous discharge of lead accumulators is due to a number of reactions which were theoretically and experimentally analyzed by Ruetschi and Angstadt[1]. At least six reactions are possible. For the positive plate, the spontaneous discharge is essentially due to the reaction between the lead dioxide and the alloy of the supporting armature, according to the equations

$$Pb + PbO_2 + 2\,H_2SO_4 \rightarrow 2\,PbSO_4 + 2\,H_2O$$

$$2\,Sb + 5\,PbO_2 + 6\,H_2SO_4 \rightarrow (SbO_2)_2SO_4 + 5\,PbSO_4 + 6\,H_2O$$

This spontaneous discharge diminishes with increasing acidity through the formation of passive layers of $PbSO_4$ in solution in concentrated sulphuric acid. Nevertheless, the rate of discharge is a function of the type of alloy and is significantly greater for alloys containing antimony which has an activating effect through the formation of pores in the protective layer of $PbSO_4$. Other spontaneous discharge reactions at the positive plate are the decomposition of water by the lead dioxide

$$PbO_2 + 2\,H^+ + SO_4^{2-} \rightarrow PbSO_4 + H_2O + \tfrac{1}{2}\,O_2$$

and the oxidation of the materials of the separators in contact with the dioxide

$$PbO_2 + (\text{oxidizable separator material}) + H_2SO_4 \rightarrow PbSO_4 + (\text{oxidized material})$$

From this point of view, the best material for the separators seems to be micro-porous rubber. Yet another spontaneous discharge reaction of the positive electrode is the oxidation of hydrogen evolved on open circuit by the negative plate

$$PbO_2 + H_2 + H_2SO_4 \rightarrow PbSO_4 + 2\,H_2O$$

However, this reaction makes only a very small contribution to the discharge in view of the very low solubility of hydrogen in concentrated solutions of sulphuric acid.

The principal spontaneous discharge reaction at the negative plate is the attack of the lead by the acid, evolving hydrogen

$$Pb + H_2SO_4 \rightarrow PbSO_4 + H_2$$

[1] P. RUETSCHI and R. T. ANGSTADT, *J. Electrochem. Soc.*, 105 (1958) 555.

In the absence of foreign elements, this reaction is very slow because of the high overtension of hydrogen on lead. In the presence of antimony, however, the overtension diminishes considerably and the spontaneous discharge is significantly accelerated. The presence of traces of metals which are more noble than lead, and of sulphur dioxide, hydrochloric acid, nitric acid and chromic acid has the same effect. A second spontaneous discharge reaction at the negative plate is provoked by oxygen dissolved in the electrolyte.

$$Pb + \frac{1}{2} O_2 + H_2SO_4 \rightarrow PbSO_4 + H_2O$$

This reaction is, in fact, so rapid that the discharge is controlled solely by the rate of diffusion of the oxygen. In practice an accumulator in good condition should not lose more than 30 % of its charge in a month.

The outstanding characteristic of the lead accumulator is the constancy of its electric tension during discharge, as shown in Fig. XI, 9. The slight fall in electric tension during discharge is due to dilution of the acid.

An accumulator has the function of storing electrical energy during the charging process by transforming it into internal energy of the system, to give it up again during the discharge. An ideal accumulator would restore all the energy absorbed during the charging. Naturally, in reality, accumulators restore only a part of the absorbed energy; and the *current efficiency* of an accumulator is the ratio of the quantity of electricity obtained during the discharge to the quantity supplied during charge. The *energy efficiency* is defined as the ratio between the quantities of energy obtained during discharge and furnished during charge.

It must be noted that the reversible discharge electric tension of the OH^- ions is more negative than that of the lead dioxide electrode so that in the absence of an oxygen overtension on this electrode, the charging

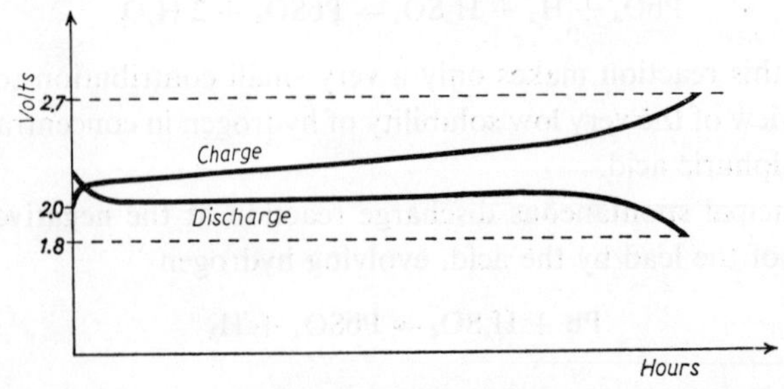

Fig. XI, 9. Charge and discharge diagrams of a lead accumulator

process would be impossible. Similarly, the discharge electric tension of the H^+ ion is more positive than that of the lead electrode. The charging process is made possible by the overtensions of oxygen and hydrogen on the respective electrodes.

As charging proceeds, sulphuric acid is formed and increases the concentration of the H^+ and SO_4^{2-} ions and, as mentioned above, thus increases the effective electric tensions of each of the electrodes until a tension is reached at which the H^+ and OH^- ions are discharged on the respective electrodes. The charging stops at this point, because the accumulator starts to function as an electrolytic cell in which water is decomposed rather than the electrochemical charging process continuing. The end of the charging process corresponds to the step of the charging curve which approaches more or less asymptotically to 2.7 V; this point is indicated by an abundant evolution of gas which irretrievably consumes current. This is the principal cause of the lowering of the current efficiency. Another loss follows from the spontaneous discharge of the electrodes discussed above. The current efficiency from lead accumulators in a good state of repair and properly treated lies between 0.94 and 0.98.

The energy efficiency is the product of the current efficiency and the ratio of the electric tensions at the terminals during discharge and during charge; this ratio is less than 1. During charge the electric tension at the terminals is given by the equation:

$$U_{char.} = U + IR_i$$

and during discharge it is:

$$U_{dis.} = U - IR_i$$

so that the ratio $U_{dis.}/U_{char.}$ is always less than one. The difference between the electric tensions at the terminals during discharge and charge, caused by the internal resistance of the accumulator, is not however sufficient to explain completely the diminution in the electric tension at the terminals during discharge and its increase during charge, as shown in Fig. XI,9, since the internal resistance of the accumulators is very small and of the order of 10^{-3} Ω.

The fundamental cause of the observed difference between the electric tensions at the terminals is rather to be found in the already-mentioned slowness of the diffusion phenomena; these ensure that the concentration of acid within the electrode remains lower during the discharge than that of the bulk of the acid, whilst the reverse is true during the charge. The ratio $U_{dis.}/U_{char.}$ thus becomes considerably less than 1. The energy efficiency of lead accumulators varies between 0.75 and 0.85.

The electrode plates of lead accumulators are of two types depending on the fabrication process. Planté electrodes have the active mass formed by electrolysis of the initial lead mass and Faure electrodes (or paste electrodes) have the active mass formed by electrolysis of a paste of lead oxide and sulphuric acid.

The starting material for Planté electrodes must be very pure lead. If a plate of very pure lead is made the anode for an electrolysis in a bath of pure sulphuric acid, it will rapidly become covered with a layer of dioxide and further electrochemical action will evolve oxygen from the water of the solution. If the sulphuric acid solution also contains acids whose anions can give soluble lead salts (NO_3^-, ClO_4^-, CH_3COO^- etc.), then under suitable conditions of temperature and ratios of the sulphuric and the other acid, the transformation of the lead into dioxide does not stop at the surface but continues into the bulk of the material. This probably occurs through a mechanism involving the formation of soluble salts of lead, the precipitation of lead sulphate on the electrode in a not very compact or adherent form (through the reaction of the soluble salt with sulphuric acid) and the successive conversion of the sulphate to dioxide. On inverting the direction of the current flow, the layer of dioxide is reduced to porous metallic lead; in this way, the negative plates can be prepared. To obtain a rapid and fairly profound conversion, the effective surface area of the plates may be increased by ribbing or similar devices. Both the lead dioxide and the spongy lead formed by this method are fairly porous so that the electrolyte can penetrate into the interior of the active mass, and they are at the same time fairly compact so that the electrode maintains its shape and size. They have the inconvenience that the electrochemical action of formation may continue even when the plates have been mounted to form the accumulator and may gradually transform the metallic frames of the plates and hence diminish their mechanical strength.

Faure plates, however, consist of a frame of a hardened lead grill – normally containing 8 % antimony – filled with a paste of lead oxides (PbO, Pb_3O_4) in sulphuric acid to which may be added inert substances, such as barium sulphate, graphite, sawdust etc. for the paste to be used in the negative plate to maintain the porosity and avoid excessive contraction, or else magnesium sulphate, sugar, lignosulphuric acid etc. for the paste to be used in the positive plate to increase porosity, since these latter substances, being soluble will be eliminated at the end of the process. After drying, the plates consist principally of lead sulphate produced by reaction with the oxides. These plates when immersed in sulphuric acid ($d = 1.1$–1.2) and subjected to electrolysis are converted into spongy lead dioxide. In this type of fabrication it is not necessary for the sulphuric acid of the electrolysis bath to contain solubilizing agents. The electrolysis is not continued until the end of the quantitative transformation of oxides and sulphate into dioxide or lead; normally the finished plates have the composition shown in Table XI, 8.

TABLE XI, 8

THE COMPOSITION OF CHARGED FAURE PLATES

Component	*Positive plate*	*Negative plate*
PbO_2	90 %	—
Pb	—	95 %
PbO	7 %	3 %
$PbSO_4$	3 %	2 %

The small quantity of oxide and sulphate is left because it contributes to the strength and mechanical consistency of the plate.

For Faure plates the fineness of the original particles of oxide, used in the paste, is important because the life of the accumulator also depends on this factor.

Lead accumulators require constant maintenance. The water which evaporates and that which is decomposed towards the end of the charging must be replaced with distilled water (and not tap water to avoid the accumulation of impurities). The density of the acid must be tested repeatedly if the accumulator is not used for some time; and the accumulator cannot be left inactive but must be regularly charged and discharged. Charge and discharge must be carried out according to instructions, both in regard to current intensity and to the quantity of energy supplied or withdrawn. An excessive charge or discharge intensity could lead to too abrupt changes in volume, limited to the surface layers, which would alter the shape of the electrodes and rapidly disrupt the active mass. A lead accumulator can support, for a short time, even very strong discharge currents up to a density of 1 A/cm^2 of electrode surface. When the discharge is continuous, however, it must not exceed 0.1 A/cm^2. An excessive discharge beyond the limit of 1.8 V would lead to the growth of the granules of lead sulphate, mentioned above, whereas an excessive charge would disrupt the electrode through the gas evolved within it.

7. Alkaline Accumulators

The search for an accumulator lighter than the lead one gave rise to the alkaline accumulator, which is also called the Edison or ferro–nickel accumulator. One variety of this is the Jungner, cadmium–nickel accumulator. This consists of one electrode of hydrated nickelic oxide, made conducting by admixture with metallic nickel, and one electrode of finely divided iron, immersed in an electrolyte composed of a 20 % solution of potassium hydroxide containing 50 g/l of lithium hydroxide:

$$^{+}\mathrm{Steel/NiOOH;\ Ni(OH)_2/KOHaq.\ 20\%/Fe(OH)_2;\ Fe/Steel}^{-}$$

This accumulator may to some extent be considered as analogous to, and derived from, the Lalande cell (*cf.* p. 647 *et seq.*) in which the zinc has been replaced by iron and the copper oxide by nickelic oxide. It differs from the Lalande, however, in that both the electrode processes are reversible in the alkaline accumulator.

The reactions which produce the current in these accumulators are as follows. At the positive pole, according to the most recent views[1], the primary product of the oxidation of the $Ni(OH)_2$ during charge, and even strong overcharge, is NiOOH. The reactions would thus be

$$2\,\mathrm{NiOOH} + 2\,\mathrm{H_2O} \rightleftharpoons 2\,\mathrm{Ni^{3+}} + 6\,\mathrm{OH^-}$$

$$* \quad 2\,\mathrm{Ni^{3+}} + 2\,e^- \rightleftharpoons 2\,\mathrm{Ni^{2+}}$$

$$2\,\mathrm{Ni^{2+}} + 4\,\mathrm{OH^-} \rightleftharpoons 2\,\mathrm{Ni(OH)_2}$$

giving

$$2\,\mathrm{NiOOH} + 2\,\mathrm{H_2O} + 2\,e^- \underset{\text{charge}}{\overset{\text{discharge}}{\rightleftharpoons}} 2\,\mathrm{Ni(OH)_2} + 2\,\mathrm{OH^-}$$

[1] S. Uno Falk, *J. Electrochem. Soc.*, 107 (1960) 661.

The reactions at the negative pole are

$$* \quad Fe \rightleftharpoons Fe^{2+} + 2\,e^-$$

$$Fe^{2+} + 2\,OH^- \rightleftharpoons 2\,Fe(OH)_2$$

giving

$$Fe + 2\,OH^- \underset{\text{charge}}{\overset{\text{discharge}}{\rightleftharpoons}} Fe(OH)_2 + 2\,e^-$$

The starred reactions are those which electrochemically determine the electric tensions. The net process is the sum of the anodic and cathodic reactions

$$2\,NiOOH + Fe + 2\,H_2O \underset{\text{charge}}{\overset{\text{discharge}}{\rightleftharpoons}} 2\,Ni(OH)_2 + Fe(OH)_2$$

The positive electrode is thus a redox element whose electric tension is given by the equation

$$U_1 = U_{0\,Ni^{2+}/Ni^{3+}} + (RT/2\,\mathbf{F}) \ln ([Ni^{3+}]^2/[Ni^{2+}]^2)$$

The electric tension of the negative electrode is simply given by the equation

$$U_2 = U_{0\,Fe/Fe^{2+}} + (RT/2\,\mathbf{F}) \ln [Fe^{2+}]$$

The net electric tension is the difference between these two

$$\begin{aligned} U &= U_1 - U_2 \\ &= U_{0\,Ni^{2+}/Ni^{3+}} + (RT/2\mathbf{F}) \ln ([Ni^{3+}]^2/[Ni^{2+}]^2) - U_{0\,Fe/Fe^{2+}} - (RT/2\mathbf{F}) \ln [Fe^{2+}] \\ &= U_0 + (RT/2\,\mathbf{F}) \ln ([Ni^{3+}]^2/[Ni^{2+}]^2\,[Fe^{2+}]) \\ &= U_0 + (0.059/2) \log ([Ni^{3+}]^2/[Ni^{2+}]^2\,[Fe^{2+}]) \end{aligned}$$

The electrolyte in this type of accumulator should function solely as a conductor since it does not take part in the electrochemical processes as in the lead accumulator, and hence the electric tension should be independent of its concentration. In fact, all the hydrated components are always present as solid phases, so that the following three equations[1] hold:

$$[Ni^{3+}]\,[OH^-]^3 = L_1; \quad [Ni^{3+}]^2 = L_1^2/[OH^-]^6$$

$$[Ni^{2+}]\,[OH^-]^2 = L_2; \quad [Ni^{2+}]^2 = L_2^2/[OH^-]^4$$

$$[Fe^{2+}]\,[OH^-]^2 = L_3; \quad [Fe^{2+}] = L_3/[OH^-]^2$$

which gives

$$\begin{aligned} [Ni^{3+}]^2/([Ni^{2+}]^2\,[Fe^{2+}]) &= (L_1^2/[OH^-]^6)/\{(L_2^2/[OH^-]^4)\,(L_3/[OH^-]^2)\} \\ &= L_1^2/L_2^2 L_3 \qquad 1 \end{aligned}$$

[1] For simplicity, activity coefficients are usually taken as 1.

Thus, the electric tension of the accumulator should be independent of the concentration of electrolyte and should also be constant. In practice, however, a slight variation of the electric tension and also of the density of the electrolyte are noted during the operation of the accumulator; after charging the electrolyte is more dilute through the production of water at the positive electrode. Thus, the relationship between the concentrations $[Ni^{3+}]^2/([Ni^{2+}]\,[Fe^{2+}])$ is no longer constant and independent of the concentration of OH^- ions, and this explains the variations in electric tension of the accumulator and of the concentration of its electrolyte during operation.

The initial tension is 1.4 V which rapidly falls to 1.3 V as soon as discharge begins and continues to fall gradually down to 1 V, at which tension the accumulator is considered to be exhausted. The charge and discharge curves are shown in Fig. XI, 10. It appears from these that certain secondary reactions occur together with the principal reaction; these secondary reactions affect the shape of the curves, and hence the energy efficiency, because of the electric tensions to which they give rise.

Firstly, the discharge curve for the iron electrode considered by itself, Fig. XI, 11, clearly shows two steps: the first (I) corresponds to the initial charge of hydrogen, which disappears immediately after the discharge circuit is closed; the second (II) represents the reaction

$$FeO \rightarrow Fe_2O_3$$

which must be avoided as far as possible, principally because the reverse

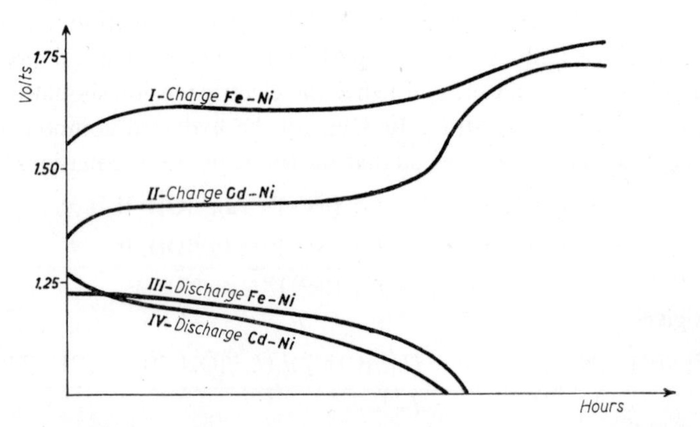

Fig. XI, 10. Charge and discharge diagrams of alkaline accumulators

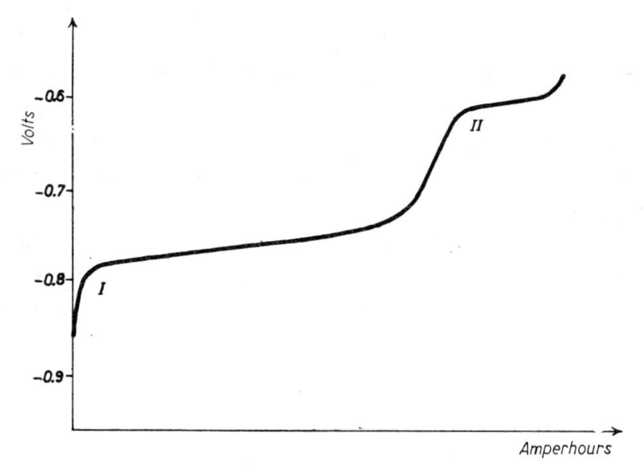

Fig. XI, 11. Discharge diagram of an iron electrode

charging reaction is slow and difficult. The oxidation of iron to the trivalent state would be analogous, in its effects, to the sulphation of lead accumulators. Secondly, the charge curve of the whole accumulator lies at significantly higher electric tensions than the discharge curve. Since the internal resistance of a normal alkaline accumulator is about 5-fold greater than that of a normal lead accumulator, this difference is not sufficient to explain the difference in electric tensions at the terminals during charge and during discharge. This difference also cannot be attributed to changes in the concentration of electrolyte within the electrodes because their effect is minimal. Evidently, these differences must be attributed to an irreversible process, which for the positive electrode is probably an adsorption of oxygen by the primary product of electrolysis, according to the reaction

$$2\,NiOOH + 2\,OH^- \rightarrow (NiOOH)_2 \cdot O_{ads} + H_2O + 2\,e^-$$

The oxygen would then be liberated spontaneously to give primary oxygen once more. At the negative electrode it is probable that the reaction $FeO \rightarrow Fe$ is to some extent retarded due to the tendency of iron to become passive, so that an overtension appears.

From an electrochemical viewpoint sodium or potassium hydroxides could equally well be used but potassium hydroxide is preferred for its greater conductance.

Since the concentration of electrolyte does not affect the electric tension of the accumulator in any way, it would be preferable to choose the concentration corresponding to the maximum specific conductance (*cf.* p. 52 *et seq.*); this is a concentration of 6.58 N (28 %; $d = 1.271$ g/cm^3). However, at this concentration there is a considerable attack on the iron electrode by the electrolyte which damages the capacity of the accumulator since its facilitates the spontaneous discharge reactions. A lower concentration is thus preferred – 20–21 % potassium hydroxide – at which the solubility of the iron electrode becomes negligible.

A certain amount of lithium hydroxide is normally dissolved in the electrolyte, and considerably increases the capacity of the accumulator. The addition of 50 g/l lithium hydroxide – which is virtually a saturated solution – increases the capacity by almost 22 %. This addition of lithium hydroxide increases both the density and the resistance of the electrolyte solution. The resistance is about 20 % greater than that of a solution not containing lithium hydroxide. The mode in which lithium hydroxide acts on the capacity of the alkaline accumulator is still not clear.

The principal disadvantage of the alkaline electrolyte is its tendency to absorb carbon dioxide from the air to form carbonate. This does not affect the electrode reactions but very markedly increases the resistance of the electrolyte. The maximum concentration which can be tolerated is 0.4 N in terms of CO_2 which represents a conversion of 10% of the original potassium hydroxide to carbonate. Once this limit has been reached, the electrolyte must be renewed. Under normal conditions, this is required about once a year. The lithium hydroxide also has a very favourable influence on the carbonation of the electrolyte. The addition of 8 g/l of $LiOH \cdot H_2O$ to the solution of KOH halves the amount of CO_2 absorbed from the air in a given time. For stationary plants it is also useful to place a layer of liquid paraffin over the electrolyte to avoid contact with the atmosphere and hinder carbonation.

The capacity of the iron–nickel cell does not depend to a significant extent on so many factors as that of the lead accumulator; the only physical factor which significantly influences its capacity is the temperature, and an increase in temperature produces a slight increase in capacity.

In view of the equivalent weights of the active mass consumed per Ah withdrawn (1.042 g Fe; 4.094 g $Ni_2O_3 \cdot 3\,H_2O$), the coefficients of utilization (17 % and 45 % respectively), and allowing 40–50 g/Ah for the weight of the supporting frames, the container and the electrolyte etc., the capacity of a normal iron–nickel accumulator is about 20 Ah/kg, *i.e.* greater than that of a lead accumulator. Since, however, the internal resistance of the alkaline accumulator is greater than that of the lead accumulator, it will be apparent that the capacity, expressed in Wh/kg depends to a large extent on the intensity of the discharge current. Assuming an average electric tension of 1.1 V with a normal discharge intensity, the capacity of the alkaline accumulator is about 25 Wh/kg, and is thus of the same order as that of a lead accumulator.

The spontaneous discharge phenomena are of smaller entity than those of the lead accumulator, because, since the electrochemically active components are virtually insoluble in the electrolyte, which does not itself take part in the electrode reactions, the possibility of spontaneous discharge is limited. Moreover, the secondary reactions

$$Fe + 2\,H^+ \rightarrow Fe^{2+} + H_2$$

$$2\,NiOOH + Ni \rightarrow 3\,NiO + H_2O$$

are very slow, so that the alkaline accumulator is obviously more stable than the lead accumulator. One of the principal advantages of the alkaline accumulator, however, is its stability in the discharged state; it may remain inactive for long periods without causing trouble. It is in fact, preferable to discharge an alkaline accumulator completely when it has to be left unused for any time.

Compared with the lead accumulator, the alkaline one has the disadvantage of a lower current and energy efficiency. The current efficiency is about 0.80; and the energy efficiency depends very largely on the intensities of the charging and discharging currents, since the high internal resistance implies that an increase in the current intensity will increase the proportion of energy dissipated in the form of heat by the Joule effect in the resistance of the accumulator. On the other hand, the intensity of the charging current cannot be diminished below a certain limit because of the polarization required to overcome the overtension of the $FeO \rightarrow Fe$ reaction. If the charging current density is insufficient to produce this required polarization, the electrode process will stop at the evolution of gaseous hydrogen without charging the accumulator. Under normal working conditions, the energy efficiency lies between 50 and 65 %.

The active masses of the alkaline accumulator have a gelatinous nature and lack any mechanical coherence; moreover, apart from the metallic iron, they have a very low conductance. It is thus necessary to support them in suitable containers and to add components to augment their conductance. The containers have a tubular or other suitable form, and are made of nickel-steel; they are perforated and reinforced with rings or other armatures to resist the changes in pressure due to variations in volume during operation. They contain compressed nickelic oxide alternated with layers of metallic nickel flakes for the positive electrodes, and powdered iron mixed with mercurous oxide for the negative electrode. The mercurous oxide in contact with metallic iron is reduced to metallic mercury which, being unable to form droplets, constitutes a conductive network of mercury throughout the active mass of the negative electrode. The individual elements are soldered between plates of nickel-steel which are arranged alternately in the electrolyte and joined together in parallel in positive and negative series to form the accumulator. These characteristics form a cell of a more robust and resistant (but also costly) type than the lead accumulator; the alkaline accumulator is thus absolutely immune to shock, excessively prolonged charging, high current densities etc.

The active mass of the positive electrode is prepared initially from pure nickel dissolved in sulphuric acid. This solution is sprayed into a hot solution of sodium hydroxide, precipitating nickelous hydroxide which is filtered off, washed, dried, lixiviated with hot water, redried, ground and sieved. Nickel foil is obtained by electrolysis as films about 1 μ thick which are cut into rectangles about 1.5 mm square. The active mass of the positive electrode contains 14 % of these nickel flakes arranged in alternating layers with the nickel oxide. The tubular containers are filled with this mixture under pressure. After filling, the electrodes are anodized to oxidize the nickelous oxide to nickelic.

The active mass of the negative electrode is produced from ferrous sulphate which is purified by recrystallization and then roasted in an oxidizing atmosphere to convert it to ferric oxide. Any remaining metallic impurities are eliminated by lixiviation and after washing and drying, the oxide is reduced with hydrogen at 480° C, cooled in an atmosphere of hydrogen and finally partially reoxidized to Fe + FeO. Before filling, this is dried, ground and mixed with 3 % of mercuric oxide and compressed into the container.

A variant of the iron–nickel accumulator is the cadmium–nickel or Jungner accumulator which differs from the Edison only in having the negative electrode formed of cadmium rather than iron. Cadmium has certain advantages over iron. Firstly, the oxide formed during the discharge has itself some conductance so that the addition of mercuric oxide is not necessary. Secondly, cadmium does not give rise to any secondary reaction because no trivalent cadmium compounds exist and because the spontaneous discharge reaction

$$Cd + 2\,H^+ \rightarrow Cd^{2+} + H_2$$

is of much smaller entity than the analogous reaction in the discharge of the iron electrode. Moreover, unlike iron, it does not become passive. These differences give rise to a charging curve for the cadmium–nickel accumulator which lies at a significantly lower electric tension than that for the iron–nickel accumulator with a correspondingly rather superior energy efficiency. Finally, the cadmium–nickel accumulator undergoes a considerably lower diminution of its capacity at low temperatures than does the iron–nickel, because of the marked passivity of iron at temperatures below zero.

The main disadvantage of the cadmium cell compared with the iron cell is that on aging, spongy cadmium undergoes a marked contraction. This may be partially avoided by using a mixed iron and cadmium electrode. The discharge curve for such an electrode, shown in Fig. XI, 12, has no steps which implies that the iron oxide does not actually take part in the electrochemical processes and behaves simply as an inert mass.

The alkaline accumulator does not completely achieve the purpose for which it was designed, *i.e.* greater lightness, but has a number of advantages over the lead accumulator which make it particularly suitable for use by persons with little or no practical training.

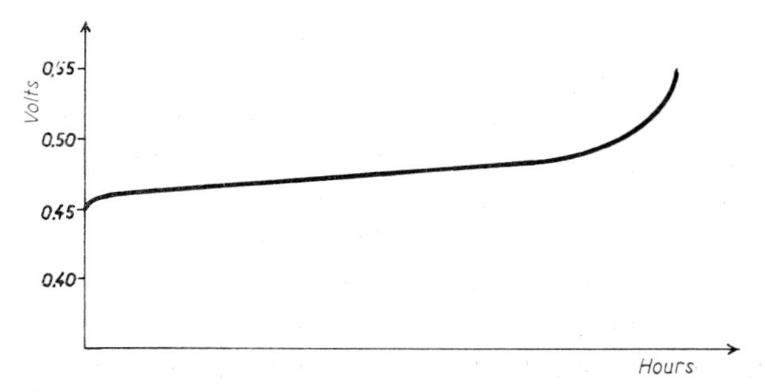

Fig. XI, 12. Discharge diagram of an iron–cadmium electrode

8. Silver–Zinc Alkaline Accumulators

Despite the fact that it was originally considered that only metals insoluble in the caustic electrolyte could be used for alkaline accumulators, many studies have been carried out on the use of zinc in combination with oxides of nickel and silver. The earliest worker was Drumm who considered their use only in cells. The major part of the work on silver–zinc accumulators is due to André. These batteries have already a practical use in a commercial form because, despite their relatively high cost due to the use of silver, their eceptional compactness, rapidity of charge and recharge and constancy of electric tension during discharge make them at least competitive with more classical types of accumulator in various applications.

The element consists of

$$^{+}Ag/Ag_2O_2/KOH \sim 40\,\%/K_2ZnO_2/Zn^{-}$$

The porous zinc anodes are separated from the oxidized silver plates which form the cathodes, by an ion exchange membrane which is usually of cellulosic material; and the volume of electrolyte solution is just sufficient to saturate these membranes.

In the discharge process at the positive pole, the silver peroxide is reduced first to oxide and then to silver. There are thus two, fairly distinct, steps:

$$Ag_2O_2 + H_2O + 2\,e^- \rightarrow Ag_2O + 2\,OH^-$$

$$Ag_2O + H_2O + 2\,e^- \rightarrow 2\,Ag + 2\,OH^-$$

giving

$$Ag_2O_2 + 2\,H_2O + 4\,e^- \rightarrow 2\,Ag + 4\,OH^-$$

During charging, the reoxidation of the silver also occurs in these two steps, the first at a tension of 50–100 mV positive with respect to the electric tension of the Ag/Ag_2O electrode in alkaline solution (considered as the comparison electrode) and the second at a tension of 150–300 mV positive with respect to the same electrode. The conversion of silver to the peroxide is always less than 60 % although the faradic efficiency approaches 100 %. This is because the formation of the layer of intermediary oxide Ag_2O increases the resistance of the electrode and introduces a sufficient ohmic polarization for the second step of oxidation to begin before the first is fully completed. This same phenomenon, during discharge, leads to the persistance of a plateau in the electric tension corresponding to the first reduction Ag_2O_2/Ag_2O, and limits the overall conversion.

The discharge reaction at the negative pole consists in the dissolution of the zinc according to the reactions

$$Zn \rightarrow Zn^{2+} + 2\,e^-$$

$$Zn^{2+} + 2\,OH^- \rightarrow Zn(OH)_2$$

$$Zn(OH)_2 + 2\,OH^- \rightleftharpoons ZnO_2^{2-} + 2\,H_2O$$

Zinc hydroxide and potassium zincate are formed and, since the solution is saturated with these, are precipitated; the lower part of the negative electrode becomes gradually richer in zinc as the cycles of charge and discharge are carried out. This accumulation of zincate at the bottom of the accumulator can cause short-circuiting of the electrode. Moreover, the zincate ions diffuse to the cathode and react with its active mass diminishing proportionately the capacity of the cell[1]. The inter-electrode membrane was therefore introduced to diminish this diffusion of zincate ions towards the positive pole. With the introduction of this membrane, also, the anodic and cathodic spaces become very small and make the precipitation of zincate difficult.

Another indispensible reason for the ionic membranes is the limitation of the corrosion of the zinc electrode[2]. The silver oxide dissolves in the alkaline solution and in the absence of the diaphragm would migrate towards the anode, reduce the zinc and form a solid solution with this in a spongy form or even deposit metallic silver which would lead to the gradual short-circuiting of the plates.

Typical charging and discharge curves for silver–zinc accumulators are shown in Fig. XI, 13. The two levels corresponding to the two stages of

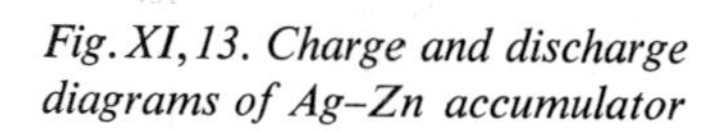

Fig. XI, 13. Charge and discharge diagrams of Ag–Zn accumulator

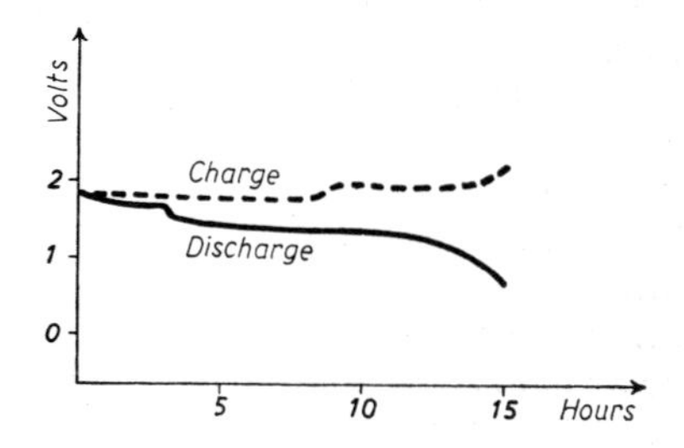

[1] H. WINKLER, *Electrotech.*, 9 (1955) 300.
[2] T. P. DIRKSE and F. DE HAAN, *J. Electrochem. Soc.*, 105 (1958) 311.

oxidation of the silver can be clearly seen. Whether the plateau corresponding to the reduction of Ag_2O_2 to Ag_2O during discharge will be seen or not, will depend on the dimensions of the cell and on the density of the discharge current. The sharp decrease in electric tension will be greater in cells of small size which are slowly discharged. The final rise in the charging curve corresponds to the evolution of oxygen and hence to overcharge.

The maximum permissible density of the discharge current for silver–zinc accumulators is very high – about 80 A/dm² – but is normally limited to about 2 A/dm² with a remarkably constant electric tension of about 1.5 V. On discharging with a current density ten-fold greater, the capacity is diminished by 10 % and the average electric tension by 0.1 V. Spontaneous discharge at room temperature largely depends on the dimensions of the accumulator. A loss of capacity of 25 % will be produced by 1 month of storage for a small cell whereas a larger cell (of about 10 Ah) will require 6 months to suffer a similar loss. At higher temperatures the rate of spontaneous discharge is greater. Low temperatures naturally lead to a diminution of capacity and impose a practical limit of use at − 20° C.

With normal rates of charge and discharge, silver–zinc accumulators give a current efficiency of 90–95 % and the ratio of discharge to charge electric tensions is similarly elevated at about 85 % so that the energy efficiency lies around 80 %.

TABLE XI,9

PRACTICAL CHARACTERISTICS OF ACCUMULATORS

	Pb	Fe–Ni	Ag–Zn
Energy efficiency	0.75–0.85	0.55–0.65	~ 0.80
Fall in electric tension during discharge, %	15	33	15
Capacity, Wh/kg	29	25	135
Monthly spontaneous discharge, %	10–30	2–30	4–25
Average life, years	1–15*	15	indeterm.
Maintenance	frequent	none	—
Sensitivity to overcharge	yes	no	no
Sensitivity to overdischarge	yes	no	no
Sensitivity to inactivity	yes	no	—
Sensitivity in the discharged state	yes	no	—
Sensitivity to shocks	yes	no	no

* The lower figure refers to mobile plant, the upper to stationary.

The essential advantages of this type of accumulator are its compactness (up to 190 Ah/dm^3) and lightness (up to 95 Ah/kg) which are respectively four-fold and five-fold better than those for lead accumulators; the velocity of charge and discharge; and the constancy of the electric tension during discharge. The most important applications thus concern those in which economy of space and weight is important, in particular for aeronautics and portable apparatus.

The disadvantages are principally linked with the cost and the useful life which is at present limited to about 100 discharges during which the capacity diminishes by an average of 50 %.

Table XI,9 illustrates some of the particular characteristics of the various types of accumulator.

CHAPTER XII

THE ELECTROCHEMISTRY OF GASES

1. Ionization and Conductance in Gases

For an electric tension applied across a layer of a gas to give rise to an electric current, the gas must contain ions or free electrons to carry the electricity. Such carriers are always produced by the action of an external agent independently of the nature of the gas, so that a gaseous system which is completely isolated from external influences is a perfect insulator, at least for field intensities which are not too high. This is a first fundamental difference from electrolytic systems where the presence of particular ionic species is a characteristic of the system independently of the more or less fortuitous action of external agents. A second difference is that the ionic concentration of electrolytic systems is also a characteristic of the system determined both by the chemical nature and by the physico-chemical conditions in which it exists; whereas with gaseous systems the ionic concentration is variable and, although depending on the physico-chemical conditions of the system, cannot be quantitatively related to them.

The agents which lead to the formation of ions or free electrons within a gaseous mass vary in nature, but all have a certain characteristic in common. They all carry a certain minimum quantity of energy which is sufficient to provoke the separation of an electron from an atom or molecule of a gas (true ionization) or the emission of a free electron from another substance often in the solid state (emission of electrons by charged metallic points; the thermoionic effect; and the photoelectric effect).

True ionization may be provoked by, firstly, collisions between charged particles whose kinetic energy is sufficiently high, *e.g.* α particles, free electrons and atoms or molecules of the system itself, at a sufficiently high temperature; and secondly, by the absorption of radiation of sufficiently high frequency (extreme ultraviolet; Schumann region; X-rays and cosmic rays). In all these cases a quantity of energy equal to, or greater than, the ionization energy must be given to the gas, *i.e.* greater than the minimum energy required to separate an electron from a gaseous

TABLE XII, 1

IONIZATION ENERGIES OF SUBSTANCES IN ELECTRONVOLTS * **

Atoms	*I*	*Inorganic molecules*	*I*	*Organic molecules*	*I*
A	15.76	Br_2	13	CH_4	13.0
Ba	5.21	Cl_2	13.2	CH_3Cl	11.2
Ca	6.11	CO	14.1	$CHCl_3$	11.5
Cs	3.89	CO_2	13.7	CH_3OH	10.8
He	24.56	I_2	9	HCHO	10.9
Hg	10.44	H_2	15.8	C_2H_4	10.5
K	4.34	HBr	12.0	C_2H_6	11.8
Kr	14.0	HCl	12.8	C_2H_5Cl	10.8
Li	5.40	HCN	15	C_2H_5OH	10.7
Na	5.14	HI	10.7	CH_3CHO	10.2
Ne	21.56	H_2O	12.6	$(CH_3)_2CO$	10.1
Rb	4.17	H_2S	10.4	$(C_2H_5)_2O$	10.2
Sr	5.69	N_2	16.5	C_6H_6	9.2
Xe	12.13	NH_3	11.5		
Zn	6.39	NO	9.5		
		NO_2	11		
		O_2	12.2		
		S_2	10.7		
		SO_2	12.1		

* Other values of ionization energies are given in LANDOLT–BORNSTEIN, *Zahlenwerte und Funktionen*, Springer Verlag, Berlin.

** The electronvolt (eV) is a unit of energy corresponding to the kinetic energy acquired by an electron which is accelerated between two points across which there is an applied tension of 1 V. For corpuscles 1 eV is 23.04 kcal/mol.

atom or molecule leaving it as a charged ion. The ionization energies I of some gaseous substances, or substances which readily become gases, are shown in Table XII, 1.

It is thus possible for a single substance to give rise to positive ions by losing an electron or to negative ions by gaining an additional foreign electron. This property is the third difference between ionizable gaseous substances and electrolytic systems. Hydrogen, for example, gives rise in solution only to positive ions and chlorine only to negative, whereas in the gaseous state positive and negative ions of both hydrogen and chlorine

TABLE XII, 2

MIGRATION VELOCITIES
of monovalent positive and negative ions in the gas itself at 0° C in $(cm\ sec^{-1})/(V\ cm^{-1})$

Gas	*Positive ions*		*Negative ions*	
	$p = 1$ mm Hg	$p = 760$ mm Hg	$p = 1$ mm Hg	$p = 760$ mm Hg
A	$1.2 \cdot 10^3$	1.3		1.7
C_2H_2		0.78		0.84
C_2H_5OH	$0.27 \cdot 10^3$	0.34	$0.27 \cdot 10^3$	0.37
CCl_4		0.30		0.32
Cl_2	$0.56 \cdot 10^3$	0.65	$0.56 \cdot 10^3$	0.51
CO (in CO_2)	$0.84 \cdot 10^3$	1.10	$0.87 \cdot 10^3$	1.14
CO_2	$0.73 \cdot 10^3$	0.76	$0.73 \cdot 10^3$	1.01
H_2	$10 \cdot 10^3$	5.9		10.7
HCl	$0.40 \cdot 10^3$	0.65	$0.47 \cdot 10^3$	0.56
He	$15.4 \cdot 10^3$	17		
H_2O*	$0.47 \cdot 10^3$	0.62	$0.43 \cdot 10^3$	0.56
H_2S		0.71	$0.54 \cdot 10^3$	0.69
Kr	$0.69 \cdot 10^3$	0.94		
N_2	$2 \cdot 10^3$	1.28		1.8
N_2O		0.82		0.90
Ne	$3.3 \cdot 10^3$	6.23		
NH_3	$0.43 \cdot 10^3$	0.74	$0.50 \cdot 10^3$	0.8
O_2	$1.0 \cdot 10^3$	2.18	$1.4 \cdot 10^3$	1.58
SO_2		0.48		0.44
Xe	$0.44 \cdot 10^3$	0.65		

* t = 100° C.

exist (H_2^+ and H_2^-; and Cl_2^+ and Cl_2^-). The mobilities of some gaseous ions are shown in Table XII, 2[1].

It is the presence of gaseous ions, produced as described, which makes gases conducting. Since in practice the effect of cosmic rays cannot be eliminated – unless very special experimental conditions are adopted – every gaseous system shows some conductance even in the absence of other ionizing agents or substances which emit free electrons; this con-

[1] The dependence of the mobility of gaseous ions on pressure and temperature falls outside the limits of this book and reference should be made to specialized texts dealing with discharge in gases.

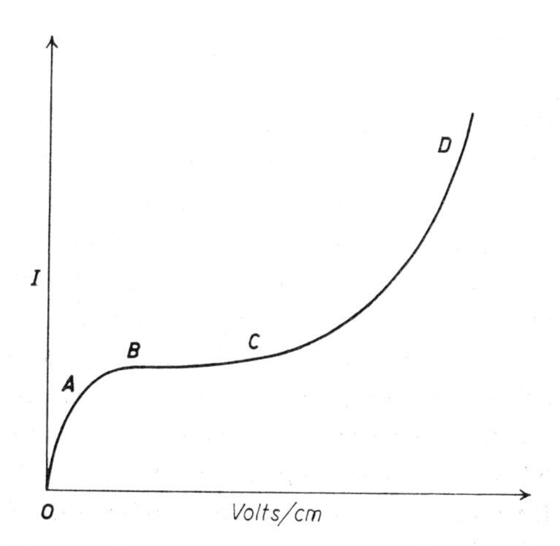

Fig. XII,1
Electric field–current intensity diagram for a discharge in a gas

ductance is due to the motion of charged particles (positive or negative ions or electrons) within the gaseous mass.

The conductance of a gaseous system is influenced by the strength of the applied electric field which is a further substantial difference from electrolytic systems[1]. If the current intensity between two electrodes in a gaseous system is plotted as a graph against the strength of the applied electric field, a curve is obtained like that shown in Fig. XII,1. With low field strengths, the current rises proportionately to the increase in the strength, following Ohms' Law (segment OA), as in an electrolytic system governed by Stokes' Law; this is due to the increasing migration velocity of the particles carrying the electrical charges. With a higher field strength, however, the increase in the current is no longer proportional to the increase in the field strength (segment AB) culminating in the segment BC which is parallel to the abscissa and represents a saturation current. This behaviour is due to the fact that the same number of charged carrier particles is present throughout the whole range from the origin to point C, because of the effects of discharge at the electrodes and of the reciprocal neutralizations by particles of opposite signs. But after point A is reached, the velocity of the particles tends towards a value which is finally independent of changes in the field strength (comparable with the limiting current in electrolytic systems). Thus, the graph shows a concave

[1] Other features distinguishing the electrochemistry of gases from that of electrolytes are discussed in the following section.

segment tending towards the abscissa followed by a segment parallel to it. However, when the field strength is still further increased, the current rises once more – slowly at first and then increasingly rapidly (segment *CD*) – without showing any tendency to reach a maximum. This last segment of the graph represents a completely new phenomenon; the charged carrier particles are accelerated by the field in the time interval between two successive collisions, sufficiently to acquire a kinetic energy which is greater than the ionization energy of the neutral particle which is struck, or greater than the electron extraction energy of the electrode substance. Thus, the collision of the sufficiently accelerated particle with a neutral particle splits the latter, following statistical probability, to give a free electron and a positive ion. The number of charged particles present thus increases and raises the conductance. Collision of a charged particle with the electrode may give either a simple emission of a new free electron or the simultaneous neutralization of the colliding particle. The collision ionization of new atoms or molecules and the emission of new free electrons becomes increasingly probable with increasing applied electric field strength and hence the conductance tends to increase indefinitely. With the increasing current intensity, the current density at the electrode surface also increases proportionately so that the temperature of the electrodes gradually rises through the Joule effect. The increase in electrode temperature leads in turn to a further increase in conductance of the gaseous system, since by the thermoionic effect the electrodes begin to emit electrons when they surmount a certain temperature. This effect depends upon the chemical nature of the substance of the electrodes, and becomes greater as the temperature rises. The number of charged particles thus increases further and the conductance undergoes another increase with a simultaneous fall in the potential gradient[1].

The general behaviour described is especially characteristic of systems at low pressures of the order of a few mm Hg. In such a system the discharge assumes various forms and is given various names: *dark discharge*, *luminous discharge*, *luminous arc*, etc.[2]. At high pressures of the order of an atmosphere other types of discharge may occur.

[1] The segment of the graph corresponding to this phenomenon is not shown in Fig. XII, 1. It is called the descending characteristic.

[2] The detailed description of the various types of discharge and of their characteristics falls outside the limits of electrochemistry and reference may be made to texts of physics or electrical technology.

1. An intermittent current of very high intensity may pass, particularly if a capacitor of large capacitance is connected in parallel with the electrodes. This type of discharge, which occurs at high field strength, is called a *spark*. It is due essentially to collision ionization and also depends on the electrode material, the electrode shape, the nature of the gas, its pressure, temperature and the presence of other ionizing agents.

2. A special type of arc may form between carbon electrodes, when the anodic current density reaches values of the order of 300 A/cm^2. The high anodic current density, and hence the high electron bombardment of the anodic material, increases its temperature very greatly and leads to rapid evaporation of material from the anode itself. This evaporated material becomes visible as an *anodic flame*. Whilst the temperature of the gas in a luminous arc is about 7,000–8,000° the temperature of the anodic flame near to the anode is significantly higher. This type of discharge which shows a further ascending characteristic is called a *high intensity carbon arc*[1].

3. When the current intensity increases further, its own magnetic field leads to a contraction of the diameter of the column of luminous gas through which the discharge passes between the electrodes. The density of the current in the column of gas assumes values of the order of 10^3 A/cm^2 and the temperature rises to about 15,000°. This type of discharge is independent of the electrode material and is called a *negative point flame*.

Discharge can occur in gaseous systems not only as a consequence of the application of a direct electric tension to a pair of electrodes in contact with a gas, but also through applying an alternating electric tension and if the frequency is sufficiently high it will occur even if the electrodes are separated from the system by a dielectric. A particular type of discharge between electrodes separated by a dielectric other than the gas occurs in ozonizers and will be described in the next section.

2. Chemical Reactions in Gaseous Discharges

The chemical reactions which occur during an electrical discharge in a gaseous medium may be divided into two main groups. Firstly, there are those due to the high temperature reached in the reaction medium; their

[1] See footnote 2, page 688.

course is independent of any electrical factors which do not directly affect the temperature to any degree, *e.g.* frequency of the current supply to an arc or a spark. Secondly, there are those whose nature is strongly dependent upon the electrical factors affecting the discharge and upon other factors which may to a greater or lesser extent affect reactions between electrically charged particles.

The former group consists of reactions which are not truly electrochemical, but rather electrothermal in that they occur largely due to the high temperature reached, particularly if the starting material and the final product are in the condensed phase (*e.g.* calcium carbide production and the manufacture of various types of steel, etc.). This group of reactions is of no particular electrochemical interest.

The second type of reaction, however, includes the gaseous reactions which are directly influenced by the electrical factors of the discharge, and it must be assumed that the primary products are gaseous ions, atoms or free radicals which are intermediates in the production of the final chemical products. These ions are produced by collisions of free electrons accelerated by the electrical field. Atoms and free radicals are also the products of a complex series of reactions in the bulk of the gas through which an electric discharge passes and owe their origin in turn to the action of accelerated free electrons. This type of reaction is truly electrochemical and analogous to the forced reactions of electrolytic systems, since these reactions lead from an initial system to a final system which is energetically richer, at the expense of electrical energy provided by an external source. Examples are the formation of ozone, nitric oxide, atoms of hydrogen, oxygen and nitrogen, from free radicals such as OH and NH.

The reaction schemes and the theory of these reactions are still little known. It is unlikely that the equilibria which are established within the mass of the gas through which the discharge occurs are controlled solely by the temperature and the quantity of energy supplied according to the laws of thermodynamics. It is true, for example, that a gaseous mass traversed by an arc discharge must be considered as a high temperature region. But the conditions of the components of the mass and in particular of the charged ions and free electrons, are different from those existing in a simple zone of high temperature in the absence of an electric field, as for example in a flame. In fact, the kinetic energy of the charged particles – which is a function of their velocity and hence, according to thermodynamic concepts, of the temperature – is very much greater in an arc discharge than it is in a simple zone of equal temperature[1], in the absence of an electric field. In the presence of the electric field, the particles are accelerated and hence have a greater mean velocity than that which corresponds to the thermodynamic temperature of the system. This implies that it is extremely difficult, if not impossible, to separate the purely thermal from the electrical effects[2] which are concomitant with strongly endothermic reactions, such as occur within the mass of a gas traversed by an electrical discharge. This also makes it impossible to define such systems thermodynamically.

Discharges can generate directly only ions, which undergo further acceleration, but they do not give rise to any primary reactions in the classical electrochemical sense. However, various short-lived species of particles are generated by a series of secondary reactions depending upon the particular conditions of the experiments. Such particles may be mono- or poly-atomic gaseous ions with positive or negative charges and may be either in the fundamental state or in an electronically excited state; they may also be free atoms in the fundamental or in an excited state; free radicals in the fundamental or excited state; or finally, may consist of a gaseous ion surrounded by a number of neutral molecules. All these unstable, or metastable, particles have a high reactivity which, in distinction to electrolytic systems, is not developed at the electrode interface but rather within the bulk of the gas. This reactivity may, moreover, be directed in one sense rather than another by the catalytic action of minute traces of other substances, which do not themselves take part in the reaction, or even by the chemical nature of the walls of the vessel (wall catalysis). In other words, many reactions coexist and are superimposed one on the other, and moreover because of the instability of one or more of their components, never reach a true state of thermodynamic equilibrium. The difficulty of determining experimentally the actual concentrations of each of the chemical species present and taking part in all of the reactions, makes the solution of the electrochemical problems of gases still more difficult.

[1] That is, in a zone in which the uncharged particles have the same distribution of kinetic energy, *i.e.* have the same mean kinetic energy.

[2] In some cases, allowance must also be made for the superposition of photochemical reactions due to the radiation generated by the electrical discharge; these complicate the picture still further.

Another difficulty in the study of these reactions is that Faraday's Law is not valid in these systems; this is yet another fundamental difference from true electrolytic systems. The quantity of substance transformed in an electrical discharge in a gas has no quantitative connection with the amount of electricity passing through the system and varies greatly with changes in its physicochemical condition (pressure and temperature) and in the electrical factors of the discharge; in fact, it is possible to obtain a current efficiency much greater than 1. In the formation of ozone, the number of coulombs of electricity required to obtain one equivalent of the gas varies according to different authors from 84 to 1,400 and all these numbers are much less than the theoretical figure of 96,500 which would be required if Faraday's Law were valid for the gaseous system. At the same time, it must be emphasized that whilst in an electrolytic system the application of a pure alternating current, lacking any continuous component, does not lead to any permanent electrolytic phenomenon since the process which occurs in the first half-phase is normally inverted in the second, yet with gaseous systems this is not true; in fact the contrary is often found and the application of an alternating current may often increase the efficiency significantly. A final difference between gaseous and electrolytic systems in condensed phases is that in the former the passage of electricity and the production of chemical reactions requires electric tensions which are usually very much greater than those required for electrolytic systems. Thus, in spite of current efficiency which can also be often greater than 1, in gaseous systems the situation energetically is much less favourable than it is in electrolytic systems in the condensed phase.

For all these reasons, the electrochemistry of gases has not undergone a development similar to that of electrolytic systems in the condensed phase and it does not seem unjustified to claim that such processes are still in a primitive state in which the results obtained are almost exclusively descriptive, and a complete theory to interpret and predict the results of such experiments is lacking.

To illustrate this, three gaseous electrochemical reactions will be briefly discussed. These are the formation of ozone, the synthesis of nitric oxide and the formation of hydrazine in a glow discharge at low pressure; of these processes the first two have found industrial application.

(I) *Ozone*

Ozone is formed by a particular type of silent discharge in appliances which have been gradually improved for industrial applications and other types of reactions. By applying a gradually increasing alternating tension to two electrodes separated from each other not only by the gas but also by two layers of a dielectric of high resistance, there is initially a behaviour similar to that of an ideal conductor and the effective current intensity I_{eff} will be given by the equation

$$I_{eff} = U_{eff}\,\omega\, C \tag{1}$$

where ω is the frequency and C is the capacitance. The power consumed W will then be

$$W = U_{eff} I_{eff} \cos \varphi \tag{2}$$

where φ is the phase angle; and since φ is virtually $\pi/2$, the power absorbed will be zero. When the electric tension rises above a certain mimimum critical value, however, a low intensity discharge of very low luminosity, occurs accompanied by a low noise and this is called a *dark* or *silent discharge*. This discharge consists of an enormous number of minute sparks which, starting from the electrodes, disperse within the gaseous mass. It would not be stable if the dielectric were not present because the increase in ionization due to the sparks would markedly increase the conductance of the gas and hence the intensity of the current would more or less rapidly pass into that of a self sustained spark or arc. The presence of the dielectric, with its high resistance, prevents an excessive

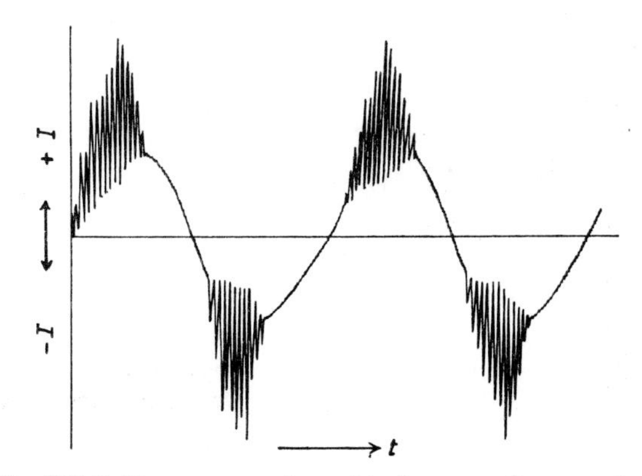

Fig. XII, 2. Time–current intensity diagram of an ozonizer

rise in the current intensity as a consequence of the ionization, and thus keeps the silent discharge stabilized. Under these conditions, the gas no longer behaves as a dielectric since it is traversed by the minute sparks, and the equations (1) and (2) lose all their significance since the concept of capacitance does not mean anything for a gas which is not a dielectric. If an oscillograph is used to follow the discharges with each alternation of the electric tension, diagrams like those shown in Fig. XII, 2 are obtained, in which the more or less irregularly distributed fringes correspond to the time intervals during which the silent discharge takes place. Under such conditions, in which equations (1) and (2) lose their validity the power absorbed is no longer zero and a chemical reaction may occur. A fairly large number of types of such apparatus (called *ozonizers*) have been designed but all have the same underlying principle which is shown in Fig. XII, 3. This illustrates diagrammatically a section through a plate ozonizer. This consists of a frame (5) of an insulating material carrying tubes for the entry (7) and exit (6) of the gas. Two glass plates (4) are applied to the sides of this frame and function as the dielectric of high resistance. Outside these two glass plates are the electrodes (3) one of which is connected to earth and the other to a high tension source which may be the secondary coil of a transformer (1) (2), or a high tension electrical machine, or a high frequency electronic oscillator. A tension of the order of thousands of volts is applied to the electrodes, up to a maximum of 25,000 V. The current density is of the order of 0.1–0.2

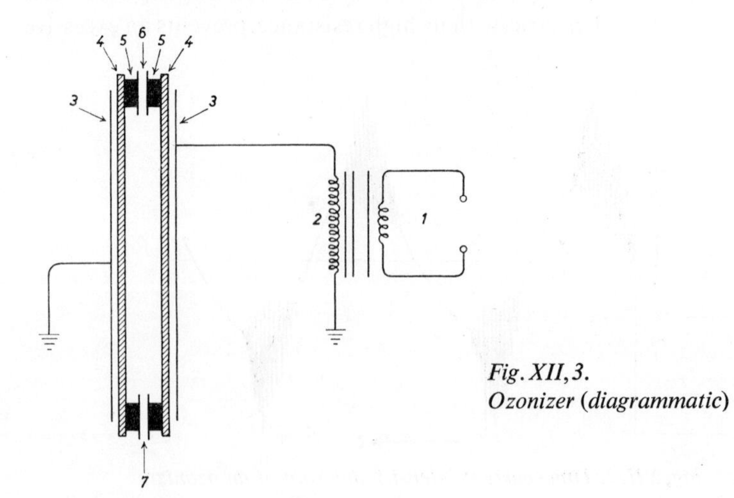

Fig. XII, 3.
Ozonizer (diagrammatic)

A/cm². The frequency used in industrial ozonizers varies between 50 and 1,000 cycles/sec whereas laboratory ozonizers have used frequencies up to 10,000 cycles/sec. Oxygen or air flows between the plates and the silent discharge occurs through it inducing the overall reaction

$$3\,O_2 \rightarrow 2\,O_3$$

This is strongly endothermic since the heat of formation of ozone is + 34.5 kcal and so the reaction should be favoured by high temperatures. However, the formation of ozone by the effect of temperature alone would require such a high temperature that in fact even at that of the normal electric arc no significant yield is obtained. Ozonizers give a good yield at low temperatures, little above ambient and in fact the yield can be increased by working at the temperature of liquid air[1]. The very low yield with the electric arc can be increased by various electrical (*e.g.* increasing the frequency) or physicochemical (*e.g.* lowering the pressure) expedients.

All this indicates that the reaction which forms the ozone is not the simple one shown above but is much more complex and must be affected or even induced by the primary charged or uncharged particles formed in the discharge. An electrical discharge in oxygen can, by splitting the gaseous molecule, by ionization and by the addition of free electrons, lead to the formation of various particles which may be involved in the formation of ozone. These include O, O_2^+, O^+, O_2^-, O^-. Ozone could theoretically be formed from such particles by any one of the following reactions.

$$O_2 + O \rightarrow O_3$$

$$O_2^+ + O^- \rightarrow O_3$$

$$O_2^- + O^+ \rightarrow O_3$$

$$3\,O \rightarrow O_3$$

The actual mechanism of the formation of ozone in a dark discharge is not known for certain but at room temperature probably follows one of these reactions involving free atoms or gaseous ions. Ozone, it will be noticed, could be formed by a triple collision between three oxygen atoms and in this case the reaction would be strongly exothermic. The superposition of a strongly exothermic reaction on to a strongly endothermic

[1] E. Briner, *Arch. sci. phys. et nat.*, 23 (1941) 25, 71.

one, favoured respectively by a low and a high temperature, in addition to the possible gaseous ionic reactions described is typical of the complications which make the theoretical study of electrochemical reactions in the gaseous phase so difficult.

Silent discharge is still the only industrial method for the production of ozone. The thermal data for the reaction indicate that one kWh of electricity should give about 1,200 g of O_3 but the best conditions so far obtained give only 300 g/kWh, *i.e.* an electrical efficiency of about 0.25[1]. Ozone is used for various purposes based on its strong oxidizing and sterilizing activity, but principally for the treatment of air and water both to sterilize and to purify by oxidizing the substances which would give them unpleasant tastes and smells.

(II) *Nitric Oxide*

The fixation of atmospheric nitrogen was for some time carried out with an electrical arc of such characteristics as to favour the direct endothermic reaction

$$N_2 + O_2 + 43.2 \text{ k cal} \rightarrow 2\,NO$$

The nitric oxide thus formed reacted further with oxygen to give nitrogen dioxide NO_2 which in turn, gave with water, nitric acid and other more or less stable nitrogen derivatives. However, at the present time nitrogen fixation on the industrial scale is performed solely by the synthesis of ammonia from its elements and hence the numerous methods proposed for modifying the arc to make it more suitable for the synthesis of nitric oxide will not be discussed. It may be deduced from the reaction given and from the heat of reaction involved that it would be favoured by a high temperature. However, nitric oxide can also be formed directly from the atoms

$$N + O \rightarrow NO + 121 \text{ k cal}$$

by a strongly exothermic reaction which would be favoured by a low temperature. The nitrogen and oxygen atoms produced by the dissociation of the respective molecules can also be formed directly by thermal reac-

[1] The formation of ozone and the effect of various physicochemical and electrical factors on this have been studied in detail by E. BRINER and his colleagues. A review article of this work has been published in *Helv. Chim. Acta*, 38 (1955) 340.

tions and can be produced by fission due to collisions between charged particles which have been sufficiently accelerated by the electrical field. Both these conditions are present in the electrical arc so that the concentration of nitrogen and oxygen atoms is not negligible and increases with increasing temperature and electric tension applied to the electrodes. The formation of nitric oxide is, however, the result not only of these two reactions but also of others involving particles formed by electrochemical action in the gaseous phase (N^{2+} and N^+ ions, and N and O atoms). These have been demonstrated in three ways; firstly, by studies of electron collisions which have shown that the formation of NO begins at a minimal energy for the accelerated electrons of 17 eV corresponding to the ionization energy of the nitrogen molecule, and is accelerated by an energy for the incident electrons of 22 eV corresponding to the fission energy of the nitrogen molecule into a neutral atom N, a positive ion N^+ and a free electron; secondly, by the formation of nitric oxide at low temperature in a silent discharge, which excludes the presence of thermally split atoms and thirdly, by the effect of the discharge frequency on the yield, which shows an increase in the intensity of the emission band for the N_2^+ ions and hence for their concentration. Thus, in the formation of nitric oxide also, various reactions coexist, any of which may become dominant depending upon the experimental conditions.

(III) *Hydrazine*

Recently, Schüler[1] and his colleagues observed another synthetic reaction due to an electrical discharge in the gaseous phase. They used a glow discharge in gas at low pressure and when NH_3 flowed through the positive column of the discharge, appreciable amounts of hydrazine were formed; this was probably due to the dissociation of the NH_3 molecule,

$$NH_3 \rightarrow NH_2 + H$$

by accelerated electrons, followed by the interaction of two NH_2 radicals to give hydrazine N_2H_4. Naturally, at the same time, there would be recombinations of such NH_2 radicals with hydrogen atoms, the formation of H_2 from hydrogen atoms and of nitrogen by a more vigorous rupture of the ammonia and the NH_2 radicals; thus, there would again be a super-

[1] H. SCHÜLER and V. DEGENHART, *Z. Naturforsch.*, 8A (1953) 251.

position of a number of reactions. In this formation of hydrazine in the positive column of a glow discharge, Schüler obtained up to 13 g of hydrazine per kWh which is a markedly better yield than the maximum of 3–4 g/kWh obtained in apparatus of the ozonizer type. This suggests that the process is probably of a different type from that leading to the formation of ozone.

It may finally be mentioned that electrical discharge in the gaseous phase has been used to prepare various gases in the atomic state – H, N, O, Cl – to study their reactivity; to prepare free radicals such as OH and NH; and has been tried in many other reactions of various types with results, however, which are difficult to treat theoretically. Undoubtedly, the electrochemistry of gases offers possibilities for research and unforeseeable applications which will be achieved only when the theory and experimental methods have been developed further than they are at present.

INDEX

PRINTED IN THE NETHERLANDS
BY J. B. WOLTERS, GRONINGEN